Enhancement *in* Drug Delivery

Enhancement *in* Drug Delivery

Edited by

Elka Touitou

Brian W. Barry

CRC Press
Taylor & Francis Group
Boca Raton London New York

CRC Press is an imprint of the
Taylor & Francis Group, an **informa** business

CRC Press
Taylor & Francis Group
6000 Broken Sound Parkway NW, Suite 300
Boca Raton, FL 33487-2742

First issued in paperback 2019

ISBN-13: 978-0-8493-3203-6 (hbk)
ISBN-13: 978-0-367-38982-6 (pbk)

Library of Congress Cataloging-in-Publication Data

Enhancement in drug delivery / edited by Elka Touitou, Brian W. Barry.
p. ; cm.
Includes bibliographical references and index.
ISBN 0-8493-3203-6
1. Drug delivery systems. 2. Drugs--Dosage forms. 3. Drugs--Physiological transport. 4. Absorption (Physiology) I. Touitou, Elka, 1942- II. Barry, Brian W., 1939-
[DNLM: 1. Drug Administration Routes. 2. Adjuvants, Pharmaceutic. 3. Drug Delivery Systems. WB 340 E58 2006]

RS199.5.E54 2006
615'.1--dc22 2006045582

Visit the Taylor & Francis Web site at
http://www.taylorandfrancis.com

and the CRC Press Web site at
http://www.crcpress.com

Preface

Enhancement in drug delivery and absorption via a specific route of administration often becomes essential in the design of novel pharmaceutical products and new therapies. On the one hand, biotechnological therapeutic molecules usually require enhancers for their successful administration by noninjectable modes; alternatively, for the more efficient delivery of conventional drugs, other routes that overcome the disadvantages of traditional administration may also require additional augmentation.

The literature currently available on drug delivery enhancement, including the most recent developments, though extensive, is still fragmented into sources that concentrate on specific administration routes, making it difficult to gain an integrated knowledge in this field.

The concept that guided us while editing this book was that by assembling knowledge on various modes of enhancement for problematic administration routes, we could deliver a multidisciplinary comprehensive review to the readers. Thus, we believe that this volume can be used as a reference book for the research community and pharmaceutical industries as well as an educational tool for senior students and practitioners in the field of pharmaceutics, medicine, and health-related disciplines. More specifically, this book is targeted toward scientists in academia and industry and graduate students in various research-intensive programs in pharmaceutical sciences, biotechnology, and medicine who are dealing with many aspects of drug design, development, and testing.

Within the chapters, the reader may find the same enhancer tested for various administration routes and in diverse experimental models. By understanding the properties and behavior of the accelerants operating within such systems, the scientist may be inspired with new ideas for choosing the ideal promotor for a new application. The contributing authors, in their reviews, discuss not only the achievements, but also the failures and drawbacks of various enhancers that are often mainly related to their toxicities.

Thus, the scheme of this book is to present a comprehensive review of the theory and methods for enhancing drug absorption through various routes of the human body. It is hoped that, by a process of cross-fertilization, investigators primarily involved with one specific route of delivery will find additional stimulation and helpful concepts applicable to their own area of expertise from a reading of the approaches other workers have used within their spheres of activity. Not every enhancement method ever tried, however speculative, for each route of delivery finds a place in the relevant section. We believe that the book presents the most interesting approaches operating at the time the authors prepared their contributions.

The opening chapters deal with the major route of drug delivery, and reflect on gastrointestinal anatomy, physiology, and permeation pathways, together with such considerations as the role of surfactants in accelerating the input of macromolecules, targeted gastrointestinal delivery, inhibition of enzymes and secretory transport, problems with lipophilic drugs, and the use of chitosan and its derivatives. Treatments on permeation pathways in rectal absorption and relevant enhancers provide information for a route whose use varies significantly in different countries. A deliberation on the basic biopharmaceutics of buccal and sublingual absorption precedes a contemplation of the role of chemical enhancers. The problems associated with circumventing the barrier of the skin during transdermal drug delivery bring into sharp focus many problems associated with breaching biological barriers; the stratum corneum has developed an elegant structure with which to limit very significantly the access of most chemicals, while permitting the controlled loss of water. It is therefore in this area of drug delivery that we find use of the widest range of enhancement methods. After a chapter orientating the reader to the structure and barrier function of the stratum corneum, contributors consider the most widely investigated enhancement approaches. These include the very extensive employment of chemical promotors, at least experimentally, electrically assisted methods such as iontophoresis, electroporation, and ultrasound, the promise offered by vesicular carriers, as well as methods that combine liposomes with electrical potentiation. The final transdermal chapter considers the approach of by-passing or removing the major source of our difficulties, the horny layer of the skin.

In recent years, a considerable interest in nasal delivery has developed and the text on the physiological parameters that affect this process leads once again to the use of chemical enhancers. The problem of peptide delivery, which is a concern with respect to all routes of delivery in light of the as yet unfulfilled promises of the biotechnology revolution, completes this section. Consideration of the nature of the vagina and uterus as absorbing organs sets the scene for a contemplation of strategies for improving the bioavailability of a drug when administered via the vaginal route. Details on the structure and function of the eye help us to understand delivery systems for this route and the relevance of chemical permeation enhancers, together with the promise of iontophoresis. The final section of the book deals with structure and function of the blood–brain barrier and strategies for overcoming it.

We express our appreciation to all the authors for contributing outstanding chapters. We would also like to thank Mrs. Yvonne Western and Mrs. Madelyn Segev for their excellent assistance with respect to the manifold duties involved in the preparation of the manuscript.

Elka Touitou
Brian Barry

Editors

Elka Touitou is Professor of Pharmaceutical Sciences, Head of the Dermal/Transdermal Drug Delivery Group, and Head of the Teaching Committee of the School of Pharmacy, The Hebrew University of Jerusalem, Israel. She is the president of the Israeli chapter of the Controlled Release Society (ICRS) and serves as a member of the Scientific Advisory Board of the CRS. She has been a professor at a number of pharmaceutical companies and universities in Europe and the United States, including Hofmann La Roche, American Cyanamid, and the University of Rome. She has a varied and broad experience in collaborating with the pharmaceutical industry in the design of new formulations. Professor Touitou is an internationally recognized authority in the field of drug delivery. She obtained her PhD in 1980 from The Hebrew University of Jerusalem. Her primary research interest is in the field of enhanced drug absorption from various administration routes (oral, transdermal, nasal) and design of novel carriers for enhanced drug delivery. She is the inventor of "Ethosome" and holds 14 patents, has published over 200 scientific works, including original research papers, reviews, and book chapters. She has also coedited *Novel Cosmetic Delivery Systems* (Marcel Dekker, 1999). Professor Touitou is the recipient of a number of awards, including the Jorge Heller Outstanding Paper Award and Kaye Innovation Award.

Brian W. Barry is Professor of Pharmaceutical Technology and Head of the Drug Delivery Group of the School of Pharmacy, University of Bradford, UK. His education includes a BSc (pharmacy) and a DSc (both obtained at the University of Manchester) together with a PhD from the Faculty of Medicine of the University of London. He is a Fellow of the Royal Pharmaceutical Society of Great Britain, a Chartered Chemist, a Fellow of the Royal Society of Chemistry, and a Fellow of the American Association of Pharmaceutical Scientists. Professor Barry is an international authority on drug delivery systems, especially via the topical and transdermal routes. He has over 400 publications, including *Dermatological Formulations; Percutaneous Absorption* (Marcel Dekker, Inc., 1983) to his credit. He served as a member of the UK Chemistry, Pharmacy and Standards Sub-Committee of the Committee on Safety of Medicines and acted as adviser on topical and transdermal delivery of drugs to the Medicines Control Agency of the UK and the Food and Drug Administration in the United States. He has a wide and varied experience of collaborating with and acting as a consultant for some 40 firms in the pharmaceutical industry in the UK, Europe, the United States, and Australia.

Contributors

Muhammad Abdulrazik
Department of Pharmaceutics
School of Pharmacy
The Hebrew University of Jerusalem
Jerusalem, Israel

Hidetoshi Arima
Graduate School of Pharmaceutical Sciences
Kumamoto University
Kumamoto, Japan

John J. Arnold
Department of Ophthalmology
School of Medicine
Duke University
Durham, North Carolina

Brian W. Barry
Drug Delivery Group, School of Pharmacy
University of Bradford
Bradford, UK

Priya Batheja
Ernest Mario School of Pharmacy
Rutgers—The State University of New Jersey
Piscataway, New Jersey

Anupam Batra
University Institute of Pharmaceutical Sciences
Punjab University
Chandigarh, India

Elena V. Batrakova
Department of Pharmaceutical Sciences
University of Nebraska Medical Center
Omaha, Nebraska

David J. Begley
Blood–Brain Barrier group
Center for Neuroscience Research
King's College London
London, UK

Francine Behar-Cohen
Physiopathology of Ocular Diseases: Therapeutic Innovations Unit INSERM U 598
Paris, France

Simon Benita
Department of Pharmaceutics
School of Pharmacy
The Hebrew University of Jerusalem
Jerusalem, Israel

Andreas Bernkop-Schnürch
Department of Pharmaceutical Technology
Leopold-Franzens-University
Innsbruck, Austria

James C. Birchall
Welsh School of Pharmacy
Cardiff University
Cardiff, UK

Maria Cristina Bonferoni
Department of Pharmaceutical Chemistry
School of Pharmacy
University of Pavia
Pavia, Italy

Michael C. Bonner
Drug Delivery Group, School of Pharmacy
University of Bradford
Bradford, UK

Joke A. Bouwstra
Department of Drug Delivery Technology
Leiden/Amsterdam Center for Drug Research
Leiden University
Leiden, The Netherlands

Marc B. Brown
MedPharm, King's College London
London, UK

Carla Caramella
Department of Pharmaceutical Chemistry
School of Pharmacy
University of Pavia
Pavia, Italy

Arik Dahan
Department of Pharmaceutics
School of Pharmacy
The Hebrew University of Jerusalem
Jerusalem, Israel

Miranda W. de Jager
Department of Drug Delivery Technology
Leiden/Amsterdam Center for Drug Research
Leiden University
Leiden, The Netherlands

M. Begoña Delgado-Charro
Department of Pharmacy and Pharmacology
University of Bath
Claverton Down
Bath, UK

Abraham J. Domb
Department of Medicinal Chemistry and Natural Products
School of Pharmacy
Faculty of Medicine
The Hebrew University of Jerusalem
Jerusalem, Israel

Esther Eljarrat-Binstock
Department of Medicinal Chemistry and Natural Products
School of Pharmacy
Faculty of Medicine
The Hebrew University of Jerusalem
Jerusalem, Israel

Franca Ferrari
Department of Pharmaceutical Chemistry
School of Pharmacy
University of Pavia
Pavia, Italy

Ben Forbes
Pharmaceutical Sciences Research Division
King's College London
London, UK

David R. Friend
Vyteris, Inc.
Fair Lawn, New Jersey

Joseph Frucht-Pery
Department of Ophthalmology
Hadassah University Hospital
Jerusalem, Israel

Biana Godin
Department of Pharmaceutics
School of Pharmacy
The Hebrew University of Jerusalem
Jerusalem, Israel

Richard H. Guy
Department of Pharmacy and Pharmacology
University of Bath
Claverton Down
Bath, UK

Amnon Hoffman
Department of Pharmaceutics
School of Pharmacy
The Hebrew University of Jerusalem
Jerusalem, Israel

Patrick M. Hughes
Allergan, Inc.
Irvine, California

Alexander V. Kabanov
Department of Pharmaceutical Sciences
University of Nebraska Medical Center
Omaha, Nebraska

Indu Pal Kaur
University Institute of Pharmaceutical Sciences
Punjab University
Chandigarh, India

Joseph Kost
Department of Chemical Engineering
Ben-Gurion University of Negev
Beer Sheva, Israel

Sian Tiong Lim
MedPharm
King's College London
London, UK

Yoshiharu Machida
Department of Drug Delivery Research
Hoshi University
Tokyo, Japan

R. Karl Malcolm
School of Pharmacy
Queen's University of Belfast
Belfast, UK

Gary P. Martin
Pharmaceutical Sciences Research Division
King's College London,
London, UK

Stephen D. McCullagh
School of Pharmacy
Queen's University of Belfast
Belfast, UK

Elias Meezan
Department of Pharmacology and Toxicology
School of Medicine
University of Alabama at Birmingham
Birmingham, Alabama

Bozena Michniak
Ernest Mario School of Pharmacy
Rutgers—The State University of New Jersey
Piscataway, New Jersey

Ryan J. Morrow
School of Pharmacy
Queen's University of Belfast
Belfast, UK

Blaise Mudry
School of Pharmaceutical Sciences
University of Geneva
Geneva, Switzerland

Orest Olejnik
Allergan, Inc.
Irvine, California

Hiraku Onishi
Department of Drug Delivery Research
Hoshi University
Tokyo, Japan

Dennis J. Pillion
Department of Pharmacology and Toxicology
School of Medicine
University of Alabama at Birmingham
Birmingham, Alabama

Maja Ponec
Department of Drug Delivery Technology
Leiden/Amsterdam Center for Drug Research
Leiden University
Leiden, The Netherlands

Silvia Rossi
Department of Pharmaceutical Chemistry
School of Pharmacy
University of Pavia
Pavia, Italy

Abraham Rubinstein
Department of Pharmaceutics
School of Pharmacy
The Hebrew University of Jerusalem
Jerusalem, Israel

Giuseppina Sandri
Department of Pharmaceutical Chemistry
School of Pharmacy
University of Pavia
Pavia, Italy

Ekaterina M. Semenova
Jules Stein Eye Institute
UCLA School of Medicine
Los Angeles, California

John D. Smart
School of Pharmacy and Biomolecular Sciences
University of Brighton
Brighton, UK

Rashmi Thakur
Ernest Mario School of Pharmacy
Rutgers—The State University of New Jersey
Piscataway, New Jersey

Elka Touitou
Department of Pharmaceutics
School of Pharmacy
The Hebrew University of Jerusalem
Jerusalem, Israel

Kaneto Uekama
Graduate School of Pharmaceutical Sciences
Kumamoto University
Kumamoto, Japan

Yoshiteru Watanabe
Department of Pharmaceutics and Biopharmaceutics
Showa Pharmaceutical University
Tokyo, Japan

Martin Werle
ThioMatrix GmbH, Research Center
Innsbruck, Austria

Adrian C. Williams
School of Pharmacy
University of Reading
Reading, UK

Clive G. Wilson
Department of Pharmaceutical Sciences
Royal College, University of Strathclyde
Glasgow, UK

Lior Wolloch
Department of Biomedical Engineering
Ben-Gurion University of Negev
Beer Sheva, Israel

A. David Woolfson
School of Pharmacy
Queen's University of Belfast
Belfast, UK

Table of Contents

Part I

Promoted Gastrointestinal Drug Absorption

1 Gastrointestinal Anatomy, Physiology and Permeation Pathways

Abraham Rubinstein

CONTENTS

1.1 INTRODUCTION

The oral route is by far the most common means for ingesting drugs into the body. It is also the favored route due to the low cost of drug treatment management and patient compliance resulting from the convenience of oral drug administration. Because the alimentary canal is the functional organ, constructed naturally to absorb nutrition of diverse chemical complexity, oral administration of xenobiotics appears to be simple. However, as pharmacokinetics (PK) developed since 1953 when Gold and coworkers measured digoxin bioavailability after oral administration [1,2] and as the disciplines of drug discovery and drug delivery expanded enormously, it became evident that a profound understanding of the biology and physiology of the gastrointestinal (GI) tract is crucial for optimizing the bioavailability of orally administered drugs.

Two major developments contributed to this realization. The first is associated with the expansion of bioavailability studies as a major tool in drug development. These studies, which employ a limited number of subjects, assume that under certain statistical limitations the future performance of a new drug or drug carrier can be predicted from the data obtained in clinical trials [3]. However, individual variability cannot be underestimated. It is now well established that genetic polymorphisms of cytochrome P450 (CYP450) enzymes are one of the factors that contribute to the PK variability of drugs observed in the bimodal distribution between extensive metabolizers and poor metabolizers. PK variability may also exist between individuals genotyped as homozygous extensive metabolizers and heterozygous extensive metabolizers, implying clinical relevance to drug dosing and different drug responses [4]. Genetic polymorphisms of CYP450 also implicate gut wall metabolism [5] and raise the question regarding other GI physiological constraints that lead to disparities in drug absorption and PK variability. This chapter will extensively review all physiological aspects that govern drug absorption and their involvement in PK variability.

The second evolution occurred with the improvement in new drug synthesis techniques. *In silico* design, robotic combinatorial chemistry techniques and high-throughput screening methods increased dramatically the number of new compounds tested each year for potential pharmacological activity. Very rapidly it was found that whereas the receptor-mediated activity of the new molecules could be predicted, their ability to cross-epithelial barriers in the GI tract was hard to envisage [6,7]. It is believed that approximately 40% of the new compounds synthesized in pharmaceutical laboratories are practically water insoluble and therefore cannot be absorbed when administered orally.

Drug absorption from the digestive tube is subjected to complicated events that, in many cases, are not understood completely [8]. Many absorption mechanisms are based on animal studies and are restricted to those GI regions in which the studies were performed. Drug absorption also changes longitudinally. Usually it decreases in the caudal direction. It should be realized, therefore, that drug blood level after oral administration is, in most cases, a vector of variables that cannot always be predicted. Although huge steps have been made in the past two decades in understanding GI variables affecting drug absorption and their modeling, much is left to be divulged, especially in combining multidisciplinary know-how.

1.2 BRIEF LONGITUDINAL OVERVIEW

1.2.1 The Oral Cavity

In terms of drug delivery and drug absorption, the oral cavity is unique due to its comfort accessibility and thus the ability to control both spatial and temporal drug release and absorption from drug platforms (the only other organ that shares similar advantages in the alimentary canal is the rectal cavity; see below). Residence time and localization of drug carriers in the oral cavity organelles is controllable and depends solely on their specific design. Transverse anatomical aspects of the mouth in the context of drug absorption are discussed later in this chapter.

1.2.2 The Stomach and Secretions Relevant to Drug Absorption

Anatomically, the stomach (which, in upright posture, has a J-shape) is divided into three major regions relevant to food handling and gastric emptying: fundus—the superior part of the stomach, body (corpus)—the larger portion of the stomach, and antrum (also known as the pyloric part of the stomach) (Figure 1.1). The main function of the fundus and the body of the stomach is storage, whereas the function of antrum is propulsion and retropulsion (grinding and sieving of food). The stomach mucosa is devoid of villi but does contain gastric pits. Under fasting condition, the stomach is in the form of a collapsed bag with a small amount of air and gastric juice in the fundus [9].

The average stomach secretes about 3 L of fluid every 24 h. This fluid consists primarily of mucus, the enzymes pepsinogen and lipase, hydrochloric acid, and intrinsic factor [10]. Mucus is secreted by the goblet cells (located at the surface gastric epithelium) and mucus neck cells (located primarily in the antrum). Gastric mucus is saturated with bicarbonate and serves as a thick barrier against hydrogen ions and pepsin (mostly active at low pH) [11]. Gastric mucus also serves as a barrier to drug absorption. This together with the small surface area of the gastric epithelium is responsible for the inability of the stomach to act as an absorptive organ except in few cases.

Hydrochloric acid is secreted by parietal cells through gastric pits (superficial openings of a deep cluster of glands), causing the pH of the stomach to drop to 2–3 right after feeding. The major function of gastric HCl is to activate pepsin and to stimulate pancreatic fluids and bile secretion. Pepsin, a 37-kDa C-terminus protease, is produced in gastric chief cells and secreted

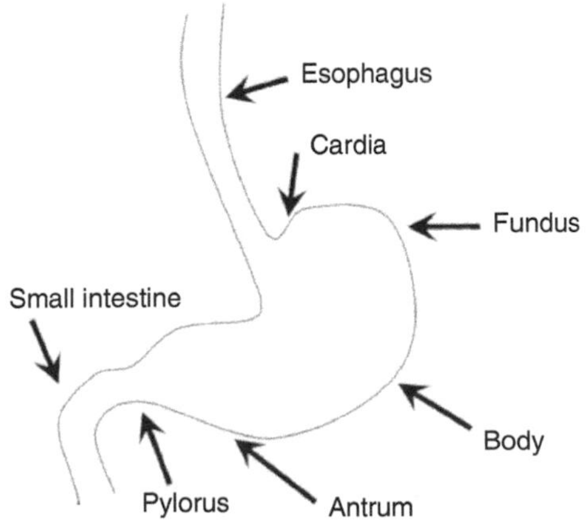

FIGURE 1.1 A scheme of the three major regions of the human stomach.

through the gastric pits as a 42-kDa zymogen. An acidic chopping of nine amino acids is required for its activation. Gastric lipase hydrolyzes (together with pancreatic lipase) triglycerides to fatty acids and glycerol. The parietal cells of the stomach mucosa also produce the intrinsic factor, which is essential for the absorption of vitamin B_{12} in the distal portions of the small intestine (ileum). For a profound review on the hormonal and chemical control of gastric secretion, the reader is referred to Ref. [13].

1.2.3 The Small Intestine and Secretions Relevant to Drug Absorption

The entry of materials into the small intestine is controlled by coordination of the antrum, gastroduodenal junction, and the proximal duodenum. At the other end, traffic between the terminal ileum and colon is regulated by tonic pressure of the ileocecal valve [14,15]. The gastroduodenal junction and the ileocecal valve serve as gates for the entry and exit to and from the small intestine. It is postulated that feedback mechanisms, existing in the terminal ileum, cause a slowing down of both gastric emptying and intestinal transit of nutrients [16,17].

The dimensions of the three major organs of the small intestine are listed in Table 1.1. Because of its large surface area (Table 1.1), the small intestine is the primary organ in which drug absorption occurs after oral ingestion, although drug absorption occurs along the entire GI tract and in terms of extended drug products the colon becomes the major player [18]. The large surface is a result of the unique anatomy of the small bowel. It is covered with millions of small projections (approximately 1 mm long) called villi. The villi house a dynamic, self-renewing population of epithelial cells that include absorptive cells (enterocytes), secretory cells, and endocrine cells (Figure 1.2). The duodenum is considered to be the best absorptive organ in the small intestine and its epithelium is of the leaky type [19].

Pancreatic fluid, secreted in the duodenum is composed of digestive enzymes and bicarbonate. The two major pancreatic proteases are the serine proteases trypsin and chymotrypsin.

TABLE 1.1
Gross Characteristics of the Human Gastrointestinal Tract

Characteristics	Organ	Dimensions
Length (cm)	Entire GI tract	500–700
	Duodenum	20–30
	Jejunum	150–250
	Ileum	200–350
	Colon	90–150
Resting volume (mL)	Stomach	25–50
pH	Fasted stomach	1.5–2
	Fed stomach	2–6
	Duodenum	6–6.5
	Ileum	7
	Colon	5.5–7
	Rectum	7
Bacterial count (CFU/mL)	Stomach	10^2–10^4
	Small intestine	10^4–10^8
	Colon	10^9–10^{11}
Average mucus thickness (μm)	Stomach	600
	Small intestine	200
	Colon	100–200

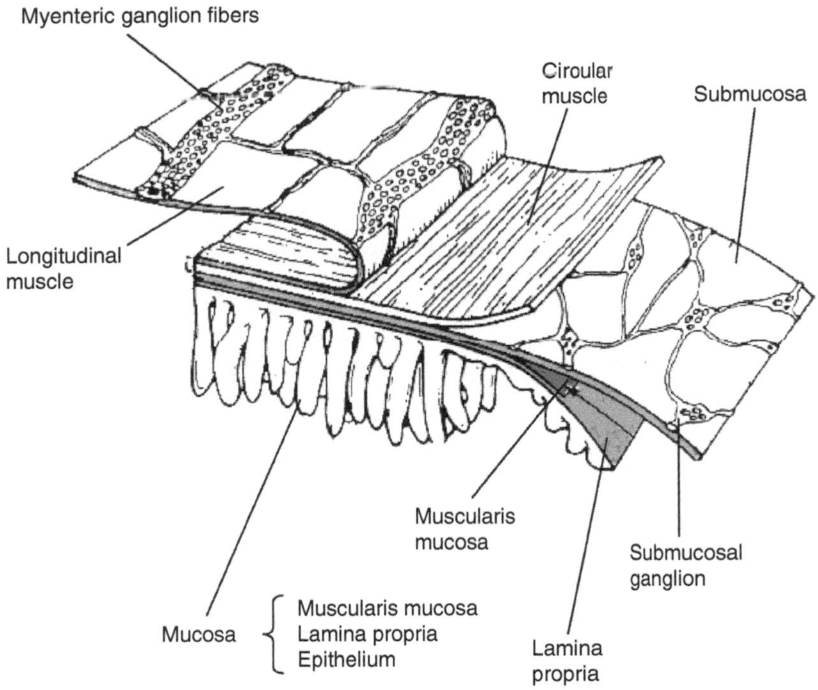

FIGURE 1.2 Schematic cross section of the small intestine wall.

Their mode of appearance in the lumen of the intestine is rather complicated and involves activation of trypsinogen secretion by enterokinase. Once trypsin is formed it activates chymotrypsinogen. Pancreatic lipase is also secreted into the lumen with the pancreatic fluid. The digestion process of fatty acids by their lipase-mediated hydrolysis is completed by bile salts, which are also secreted in the duodenum and are crucial for micellization of lipophilic compounds. The micelles formed in the duodenum enable the absorption of hydrophobic drugs such as steroids. They pose, however, a serious constraint for the stability of drug delivery carriers such as liposomes and emulsions.

The major saccharidase of the small intestine is amylase that digests starch to the disaccharide maltose and the trisaccharide maltotriose. Intestinal mucus is secreted by goblet cells, which either ooze (constitutive basal secretion) or burst as a result of stimuli. In the last mode of secretion condensed mucus gel granules can expand 500-fold within 20 ms [20].

1.2.4 The Colon

The large intestine, 150 cm long, arches around the small intestine. Its primary function is water reabsorption in the ascending part and storage of indigestible food residues in its transverse and descending sections [15,21]. Unlike the small intestine epithelium, the colonic columnar epithelium does not have villi projections. Its surface area is lined with numerous depressions (crypts) and its absorbing surface area per unit length is limited. However, the long residence time of colonic contents makes it possible for drugs to be absorbed, sometimes in significant amounts. Another constraint, which minimizes drug absorption in the colon, is the extensive luminal metabolism caused by its active bacterial flora [18,22,23].

Pertinent to absorption enhancement of drugs it is noteworthy that because of the long residence time of the colon contents, it can be considered as a closed biological compartment;

therefore absorption promoters achieve better results in this organ (and also in the rectum) in terms of relative fraction absorbed, as compared with the small bowel, using the rat as an animal model [24–27].

1.2.5 The Rectum

The rectum, with its small surface area (200–400 cm^2) and columnar epithelial cells without villi, is an interesting absorption cavity because of its venous and lymphatic drainage. The rectal upper hemorrhoidal vein is drained to the portal system, and the middle and lower hemorrhoidal veins are drained to the inferior vena cava (IVC). Therefore drugs that are absorbed in the proximal rectum are more susceptible to first-pass metabolism than drugs absorbed in the lower part of the rectum [28,29]. Because of the extensive lymphatic drainage in the organ, first pass of rectally administered drugs may also be avoided if lymphatic absorption is achieved [30]. As mentioned above, the rectum is a favorable target for drug absorption enhancement [29,31–33].

1.3 RELEVANT LONGITUDINAL ASPECTS TO ORAL DELIVERY OF DRUGS AND DRUG DELIVERY SYSTEMS

1.3.1 Gastrointestinal Motility

A major problem associated with classic oral sustained release drug delivery is the uncertainty of location, and hence environment, of the drug–dosage form at any point in time following ingestion. Therefore, it is important to understand the possible implications of GI motility on the location of drug delivery systems in the GI tract under various physiological conditions.

There are two modes of motility patterns in the stomach and consequently in the small intestine [13,34–36]. The digestive (fed) pattern consists of continuous motor activity, characterized by a constant emptying of chyme from the stomach into the duodenum. The interdigestive (fasted) pattern (commonly called the migrating motor complex, MMC) is organized into alternating cycles of activity and quiescence and can be roughly subdivided into three phases: I (basal), II (preburst), and III (burst) [37] (Figure 1.3).

Typically, the MMC sequence begins in the stomach or esophagus and migrates to the distal ileum. Some MMC, however, originates in the duodenum or jejunum and not all MMC

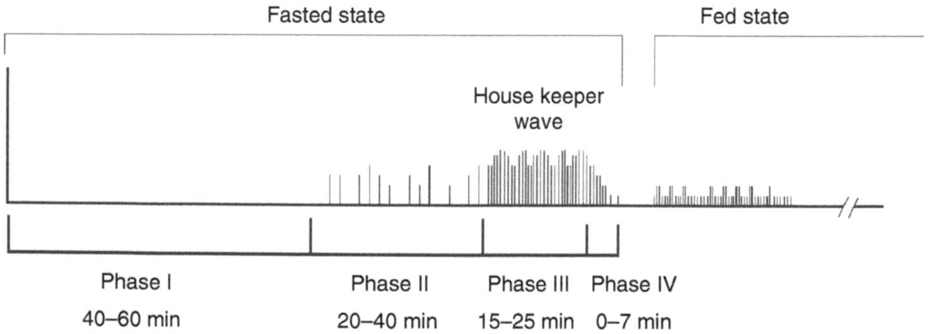

FIGURE 1.3 Pictorial presentation of the typical motility patterns in the fasted (interdigestive) and fed stomach. (From Rubinstein, A., Kin Li, H.K.V., Gruber, P., and Robinson, J.R., in *Oral Sustained Release Formulations, Design and Evaluation* (A. Yacobi and E. Halperin-Walega, Eds.), Pergamon Press, New York, 1988, p. 125.)

traverses the entire small bowel [38]. Both the frequency and amplitude of contraction reach their maximum during phase III. No contraction or secretion occurs during phase I. Irregular contraction and bile secretion begin in phase II whereas mucus discharge occurs any time during phase III. In general, it takes about 2 h for the activity wave to migrate from the stomach to the ileocecal junction [39]. The MMC pattern of the stomach can be disrupted by either the presence of food or infusion of nutrient into the small intestine. The duration of this disruption depends on meal composition. It is also longer for solid meals than for liquids [40].

1.3.2 Gastric Emptying of Liquids and Solids in the Fed State

Incoming food layers upon itself in the fundus and the body of the stomach, which then evacuates it into the antrum where grinding occurs. Liquid passes through the food layer and drains into the duodenum before any solid emptying occurs. Emptying of the stomach usually follows first-order kinetics [41,42]. Food particles are handled differently according to their dimensions. The antral activity in the fed state results in the grinding of large food fragments into smaller pieces. The flow of digested chow into the duodenum depends on (a) the magnitude of antral constriction, (b) the degree of pyloric relaxation, (c) the receptive relaxation of the duodenum, and (d) the type of duodenal contractions [43]. It is generally agreed that the rate of gastric emptying depends primarily on the caloric content of the ingested meal. More specifically, it has been suggested that stimulation of three types of duodenal receptors regulates the rate of gastric emptying [44]: hydrogen ions receptor, carbohydrates and proteins osmotic pressure receptor, and fat products (e.g., triglyceride emulsion) receptor. The extent of delay in gastric emptying of protein, fat or carbohydrate meals is governed by their caloric content or energy density. The higher the energy density of the gastric content, the lower is the transfer rate of digesta into the duodenum. However, on a molar basis, fat causes the greatest delay, followed by protein. The delay by fatty acid appears to increase with chain length [45].

Based on dog studies [46] it was once believed that solids could empty the fed stomach only if they are about 2 mm in diameter. However, using gamma scintigraphy technique, it was found in human studies that nondisintegrating tablets ranging from 3 to 11 mm do not differ significantly in their gastric emptying times when administered after test meals. Profound delay was observed only for 13 mm tablets discharged from the fed stomach [47,48]. There is no exact cutoff size for smaller particles and their emptying from a fed stomach may be uncertain and highly variable [49]. Particles smaller than 1 mm disperse in the liquid content of the stomach and empty as a fluid (square root of the retained volume kinetics) [50].

1.3.3 Nonnutritional Factors Affecting Emptying of the Fed Stomach

Emotional state and exercise influence gastric emptying by altering the neural and hormonal control of the stomach [51]. Stress [52] and aggression increase gastric motility and cause acceleration of the emptying rate. Vigorous exercise reduces gastric emptying [51]. The effect of body posture is known to influence gastric emptying and may alter significantly the oral bioavailability of drugs. This was demonstrated in drugs such as nifedipine and paracetamol [53,54]. An interesting speculation was raised by Backon and Hoffman [55], who suggested that pressure-vegetative reflex is responsible for that phenomenon. Pressure on the right side increases vagal tone, and pressure on the left side increases sympathetic tone.

1.3.4 Emptying of Liquids and Solids from the Fasted Stomach

Much of the experimental work with empty stomach handling of small objects and small volumes of liquid was conducted in the 1980s using dog models because of the similarity in the fasted stomach patterns between human and canine models. It should be realized, however, that the differences between the two species have been identified in terms of empty stomach pH (see below) and discharge patterns, as the research of gastric retention dosage forms progressed during the next 20 years.

Gastric emptying patterns of liquids depend on their volume and viscosity. For small volumes of 150 mL or less the emptying pattern is controlled by the MMC cycle. Thus, little or no liquid empties in phase I. The onset of discharge occurs in phase II and usually finishes before phase III. The discharge kinetics cannot be accurately described by a specific kinetic order. The onset of the emptying of larger volumes of liquids, 200 mL or more, occurs almost immediately following administration regardless of phasic activities. The kinetics of the emptying of large volumes of liquid appears to follow either a square root of retained volume versus time or an exponential relationship [41,56]. It is difficult (but not impossible) to keep solid objects in the stomach for more than one MMC cycle in that animal model. In humans, the length of the MMC cycle is more varied [57], and it may take up to 3 h for complete emptying of large solid objects [58].

Following oral administration with a small volume of water, a dosage form such as a gelatin capsule containing pellets will likely reside first in the fundus. The small volume of coadministered water will cause the gelatin capsule to disintegrate and release its content. Nevertheless, little dispersion of the capsule content will occur due to the quiescent nature of the stomach in phase I. The small volume of coadministered water will empty during phase II without any particles [59]. Even with a large volume of liquid, less than 5% of the ingested particles, mainly the lighter ones will be emptied from the stomach together with the liquid. Graham and coworkers found that in the empty stomach of human volunteers tablets and particulate matter do not differ in their dispersion and concluded that formulation of a drug in a microencapsulated multiple-unit dosage form does not guarantee wide dispersion or absence of high local concentration of drug [60]. A possible explanation for this observation is the entrapment of particles in mucus clots whether in the fasted stomach [59] or the fasted small intestine (Figure 1.4) and the ability of these particle-loaded clots to travel over relatively large intestinal distances [61].

1.3.5 Intestinal Transit of Liquids and Solids in Fed and Fasted States

Transit time of dosage forms in the small intestine determine the extent of drug absorption and the derived bioavailability of orally administered drug molecules. Transit time is even more important for drugs with absorption windows such as amoxicillin [62] because their absorption is limited to their residence time at the intestinal site of absorption. In a number of human studies the mean small bowel transit time for both liquids and solids regardless of particle size is similar, namely around 3–4 h in both the fasted and fed state [63,64]. Deviations from the mean transit time may affect dosage form performance. It has also been shown that common pharmaceutical excipients could alter small intestinal transit. The transit time of iso-osmotic solutions of sodium acid pyrophosphate and mannitol was reduced by 39% and 34%, respectively, in human volunteers compared with aqueous control (iso-osmotic sucrose did not alter the transit time and in all cases coadministered tablets traveled with no change in transit time). It was also shown that the intestinal transit of mannitol solutions at increasing concentrations had a concentration-dependent effect on small intestinal transit. Therefore, small concentrations of mannitol included in a pharmaceutical formulation could lead to a reduced absorption of any drug exclusively absorbed from the small intestine [65–67].

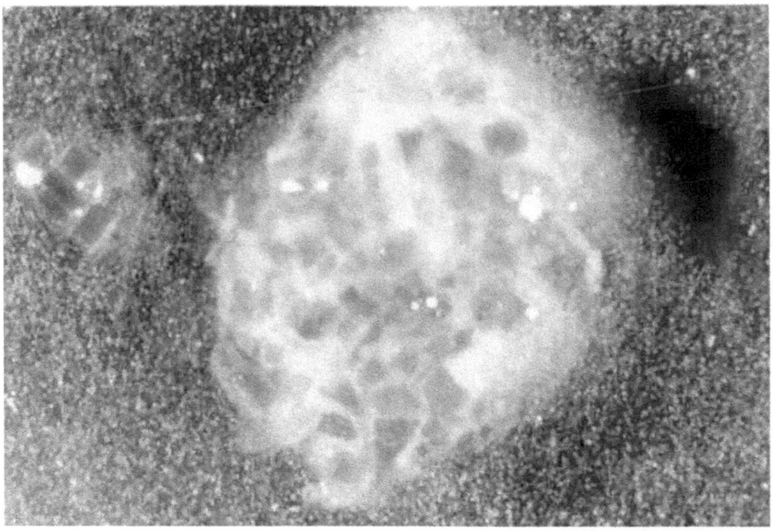

FIGURE 1.4 Agglomerated nonadhesive particles (Eudragit-RL labeled with aminoazotolune) coated with a mucus plug collected immediately following discharge from the proximal jejunum of the perfused rat. Magnification: ×60. (From Tirosh, B. and Rubinstein, A., *J. Pharm. Sci.*, 87, 453, 1998. With permission.)

1.3.6 Intestinal Spreading of Solid Dosage Forms

Spreading of particulate dosage forms such as pellets in the small intestine is highly affected by the rate of gastric emptying in the fasted state. Once the particles have left the stomach there is little, if any, additional spreading [59]. Hardy and coworkers found that following oral administration of nondisintegrating capsules and multipellet dosage forms, no distribution of the latter occurred in the stomach or along the length of the small bowel in fasted volunteers. Both preparations moved together through the stomach and small intestine and reached the colon approximately 4 h after administration (see also Refs. [63,64]). It was only after they reached the colon that they dispersed and moved at a slower rate [68].

1.3.7 Motility of the Colon

All three functions of the colon, water reabsorption [69], residual carbohydrate fermentation, and storage and propulsion of fecal material in its distal portion [15,38,70], are time consuming and require a slow exposure of the colonic mucosa to the luminal content. Therefore, the motility of the colon is different from that of the stomach and small intestine. Although three major contractile activities can be identified in the large intestine (individual phasic contractions, organized groups of contractions, and special propulsive contractions), the resulting contractions are less organized than in the small intestine. Thus, it is believed that the typical colon motility is less propulsive, and results in to-and-fro movements that mix, knead, and churn the luminal content [38].

The motor response of the human colon depends on the caloric density of the ingested meal [15]. A 1000-kcal meal stimulates colonic motor activity but a 350-kcal meal does not. Fat stimulates human colonic motor activity when administered orally, but proteins and amino acids do not [71,72]. Gastric or duodenal mucosal contact with the meal is necessary for colonic motor response because the intravenous administration of lipids, like sham feeding, does not evoke colonic motility [73]. Carbohydrates may stimulate colonic motor activity only when ingested in large quantities [70].

1.3.8 Transit of Liquids and Solids in the Colon

The mean colonic transit time measured by radiopaque markers and confirmed by gamma scintigraphy is 33 h in men and 47 h in women [74]. However, different materials behave differently in the colon. Although it was found that the ileocecal junction does not discriminate among tablets on the basis of their size (diameter) [75], it was observed that the transit time of tablets in the colon depends on their size. The larger the tablet the less time it resided in the ascending and transverse colon. For tablets, the ascending and transverse parts of the colon were found to be the main regions of tablets' stasis [75]. When 0.5–1.8 mm pellets were compared to tablets, the latter moved significantly faster through the ascending colon than the small pellets [68]. Under normal conditions, transit through the ascending colon is similar for solids and small volumes (10 ml) of liquids [76].

1.3.9 Intestinal Microflora

The human GI tract consists of a highly complex ecosystem of aerobic and anaerobic microorganisms [77]. Although the gastric microflora is predominantly aerobic and its bacterial concentration is around 103 colony-forming units (CFU)/mL, the large bowel environment is anaerobic in nature with a typical bacterial concentration of 10^{11} CFU/mL [78] (Table 1.1). Out of the 400 distinct bacterial species in the colon, 30% of fecal isolates are of the genus *Bacteroides*. This phylogenetic group is characterized by a relatively high growth yield even at low rates, which enables it to compete successfully in this region of the intestine [79]. To obtain energy for cellular functions, the colonic bacterial flora ferments various substrates not susceptible to digestion in the small intestine. These substrates may be unabsorbed low-molecular-weight saccharides and polysaccharides from endogenous sources such as mucopolysaccharides from sloughed epithelium. In addition, non-α-glucan polymers that originate in the plant cell wall, such as cellulose and hemicellulose, and pectic substances are susceptible to fermentation in the large bowel [80–82]. Under normal conditions the principal end products of the colonic fermentation processes are short-chain fatty acids such as acetic, propionic, and butyric acids and the gases methane, carbon dioxide, and hydrogen [80].

1.3.10 Luminal pH

The pH of the resting stomach is acidic, ranging between 1.7 in young people and 1.3 in the elderly. Immediately after meal ingestion the pH in the stomach increases sharply (to values of 5–6) and decreases again slowly when gastric acid secretion overrides the buffering capacity of the food [83,84]. This decline lasts between 2 (young people) and 4 (elderly people) h, after which the pH values return to the basal levels of the empty stomach. In fasted humans the duodenal pH is around 6.4 [83,84]. After a meal, the pH of the proximal duodenum drops because of the influx of acidic chyme. The bicarbonate secretion then rapidly buffers the chyme as it moves distally. The intestinal postprandial pH can be as high as 7.8 [85,86].

The pH of the colon varies depending on the type of food ingested. In general, a drop in pH to 5.5–6.5 is monitored when a radiotelemetric capsule enters the colon [85,86]. This pH drop is caused by acidification of the colonic contents by bacterial fermentation (Table 1.1).

1.3.11 Intestinal Mucus and Turnover Rates

Similar to gastric mucus the function of the mucus gel layering the intestinal epithelium is to protect the epithelium underneath against mechanical, chemical, and enzymatic erosion and to serve as a barrier against diffusion of macromolecules (toxins), particles and bacteria [87,88]. It also interacts with luminal reactive oxygen species to function as an efficient scavenger [89]. The mucus gel covering the entire surface of the intestine is made up of highly hydrated mucin (5% w/v), containing sialic or sulfonic acid residues that cause its dense negative charge. The average molecular weight of gastric and intestinal mucin ranges from 250 to 2000 kDa. Mucin monomers (see Figure 1.5) form highly cross-linked gels that can impede the diffusion of noxious agents such a H^+ or bile salts. Mucus is secreted primarily by specialized goblet cells that are located throughout the intestine by means of exocytosis, epical expulsion, and cell exfoliation. The number of goblet cells increases progressively further down the GI tract. Once mucus has been released into the lumen, it anneals to form a continuous layer adhering to the surface epithelium [13]. The primary function of the mucus is to protect the surface of the mucosal cells. Because of its ability to immobilize water and ions, the mucus layer maintains the underlying tissue in a proper state of hydration. The average thickness of the mucus gel that covers the gastric and intestinal epithelium is a consequence of the steady state between the mucus secretion and erosion rates and varies in the various organs of the gut (Table 1.1).

Because of typical gastric motility and proteolytic activity, mucus turnover is most intensive in the stomach [13]. The high gastric mucin turnover rate makes it difficult to use bioadhesion in gastric drug delivery. Indeed, it should be recognized that in the stomach and intestine the attachment of mucoadhesives to the mucus layer is stronger than the attachment among the mucin layers. Therefore, drug delivery residence time at the mucosa of the GI tract is rate limited by mucus-turnover rather than by tightness of polymer adherence to the mucus lining. Lehr and coworkers have calculated the turnover time of

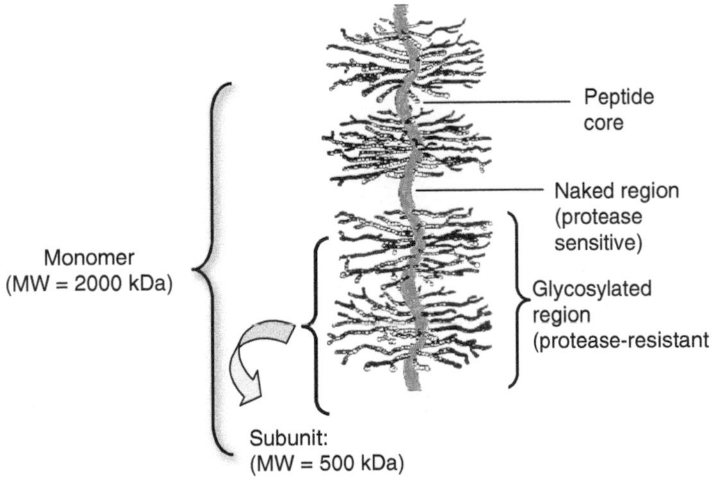

FIGURE 1.5 An illustration of an intestinal mucin molecule. The typical structure comprises a linear protein core with radial arranged oligosaccharides side chains. The major amino acids of the protein core are serine, threonine, proline, glycine, and alanine. The major saccharide components are D-galactose, *N*-acetyl-D-galactoseamine, *N*-acetyl-D-glucosamine, L-fucose, mannose, and sialic acid. The nonglycosylated (naked) regions are susceptible to protease activity, whereas the glycosylated segments are protease resistant.

mucus in the intestine of the perfused rat preparation (a common animal model in the study of bioadhesive drug platforms suggested to enhance drug absorption in the gut) and found it to range between 47 and 270 min [90]. This demonstrates the variability of mucus turnover as another problem in the design of mucoadhesives.

1.4 RELEVANT TRANSVERSE ASPECTS TO DRUG ABSORPTION

1.4.1 Boundary Layers

The boundary layer or unstirred layer refers to the fluid layer adjacent to the surface of the membrane during fluid flow [91,92]. Two major physical obstacles create epithelial boundary layers in the intestine: (a) the hydrodynamic gradient during fluid flow, which causes transverse changes from the center of the lumen to the gut wall (chyme flow is rapid in the center and slower at the periphery, assuming a laminar, nonturbulent flow [93,94]), (b) the physical nature of the fluid layers (e.g., increase in the viscosity of mucus adjacent to the enterocyte membrane due to the cell membrane glycocalix [95], Figure 1.6, or transverse changes in mucin content) and their interactions with epithelial components [96]. It has been shown that the hydrodynamic flow around a dosage form in the human GI tract could be extremely low and differs from values found in animal models [97]. When analyzing a transverse migration

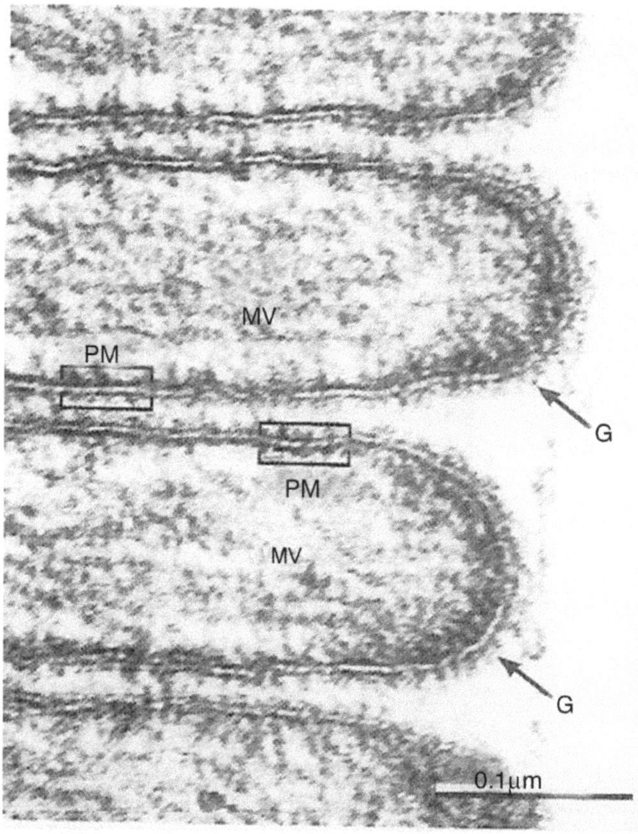

FIGURE 1.6 The glycocalix, G, of the intestinal brush border. MV, microvillous; PM, plasma membrane. (From Wheater, P.R. et al., *Wheater's Functional Histology, A Text and Colour Atlas*, 3rd ed., Churchill Livingstone, Edinburgh, 1993. With permission.)

of a drug molecule toward the epithelium, a two-compartment model should be considered: diffusion in the aqueous part and diffusion in the mucus part, which can be looked upon as the unstirred layer resistant to diffusion [98].

Boundary layers also contribute to the effect of intestinal fluid hydrodynamics on drug absorption by both diffusional- and carrier-mediated processes. In a well-defined isolated *in situ* model such as perfused intestine of the rat, a good estimate of the gut wall permeability, which is the vector of convective diffusive mass transfer, passive diffusion and carrier-mediated transport, can be accomplished [99,100].

1.4.2 Diffusional Path in Mucus Layers and Possible Drug Interactions

The thickness of the mucus layer in the human stomach is about 570 μm [101], whereas in the human colon the thickness is about 100–150 μm [102]. Ryu and Grim [103] suggested that there is little mixing between the villi and the unstirred layer over the tip of the villi of the canine jejunum, which is about 500–1000 μm thick. Therefore, for a solute to reach the lateral surfaces of the villi, an additional barrier of as much as 800 μm needs to be traversed [103]. It was shown that native mucus gel from rat small intestine reduced the diffusion coefficients of ^{3}H-water, urea, benzoic acid, antipyrine, aminopyrine, alpha-methyl-glucoside, L-phenylalanine by 37%–53% compared with buffer solution [104]. For high-molecular-weight substances, such as proteins or microparticulates, mucin can offer considerable resistance to diffusion [105].

Apart from being a diffusional barrier, mucin can also interact with drugs to decrease their bioavailability, as has been shown with tetracycline [106], phenylbutazone, and warfarin [107]. On the other hand, studies in rats showed that binding of some water-soluble drugs to intestinal mucus was essential for their absorption and that damage to the mucus significantly reduced absorption [108]. The acidic mucus is essential for lipid absorption and could be important for the diffusion of lipophilic drugs (see below).

1.4.3 Transverse pH Gradient

In addition to the longitudinal pH gradient, there is a pH gradient starting with the lumen to the absorbing surface [109,110]. Because of this gradient, the pH at the mucosal surface of the small intestine is different from that of the luminal content (a decrease of at least 1 pH unit, e.g., from 7.1 to 6.1) [111,112]. It was shown that the acidic microclimate is an essential determinant in fatty acid uptake after micellization [113]. The presence of a low-pH compartment facilitates the dissociation of mixed micelles made up of taurocholate and oleic acid. The rate of fatty acid diffusion in the mucin layer was estimated to be 400% of that in a buffer solution [114].

1.4.4 Splanchnic Blood Flow and Drug Absorption

The splanchnic circulation (the vasculature which brings blood to and from the major abdominal organs) receives approximately 28% of cardiac output. Fluid exchange in the splanchnic system is very high. The secretion by the digestive tract is 7–9 L/d with 1–2 L/d of ingested water [13,115]. This volume is approximately 2–3 times the body's plasma volume. It is therefore obvious that the absorption of compounds with high intestinal permeability will be affected by changes in blood flow, which occurs daily as a result of normal GI activity. For example, splanchnic blood flow is increased after a meal, both longitudinally and transversally, by approximately 40%, causing an increase in the extent of absorption of drugs with high intestinal permeability [116]. On the other hand, the circulation of the villous is arranged such that the supply of blood runs parallel to the vessels that drain them. Specifically, the

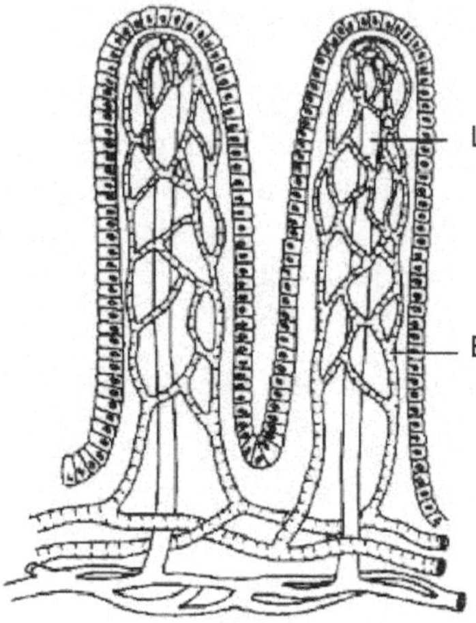

FIGURE 1.7 The blood drainage of the villous and the countercurrent absorption resistance.

arteriole and venule are located in the center of the villi, whereas the capillaries form a net under the submucosa (Figure 1.7). This arrangement creates a countercurrent exchange system between the base and tip of the villi, representing an absorption resistance layer [117].

1.5 MUCOSAL MEMBRANES AND DRUG ABSORPTION

The analysis of transfer mechanisms of drugs across the intestinal epithelial layer has passed a long way since the theory of lipid pore membrane [118] in which the total pore area of the intestinal membranes was calculated (and found to be low compared with the total surface of the mucosal aspect of the gut), through the Fickian diffusion calculations of the transport of unionized moieties of drug molecules (the Henderson–Hasselbach equation), which led to the conclusion that acidic drugs are absorbed in the stomach [119,120].

It is well established today that drug absorption through the alimentary canal walls is a complex event, which involves, in many cases, parallel or sequent microprocesses at the apical membrane of the absorptive cell (enterocyte) or between them (paracellular absorption). In addition to the various types of diffusion processes across the enterocyte membrane, numerous specific proteins—transporters and efflux pumps—are involved in the intricate drug absorption process. In the following sections the various epithelial tissues of the different organs of the GI tract will be looked at briefly. A review of major drug absorption mechanisms across epithelial cells, as they are customary today will follow.

1.5.1 Oral Cavity Mucosa

The pertinent epithelia tissues of the oral cavity are the buccal and the sublingual ones. They are nonkeratinized tissues, both covered with a thin layer of mucus (Figure 1.8). The first is 500–800 μm thick, whereas the latter is 100–200 μm thick, and more vascularized than the former [121,122]. This difference in thickness may be a possible cause for the diversity in their permeabilities: the sublingual is more permeable than the buccal mucosa, and the palatal mucosa is the least permeable. Permeability values of water and horseradish peroxidase (a 40-kDa heme protein commonly used as a permeation label) in the pig oral cavity are shown in Table 1.2. Blood vasculature in both epithelia of the oral cavity does not drain directly to

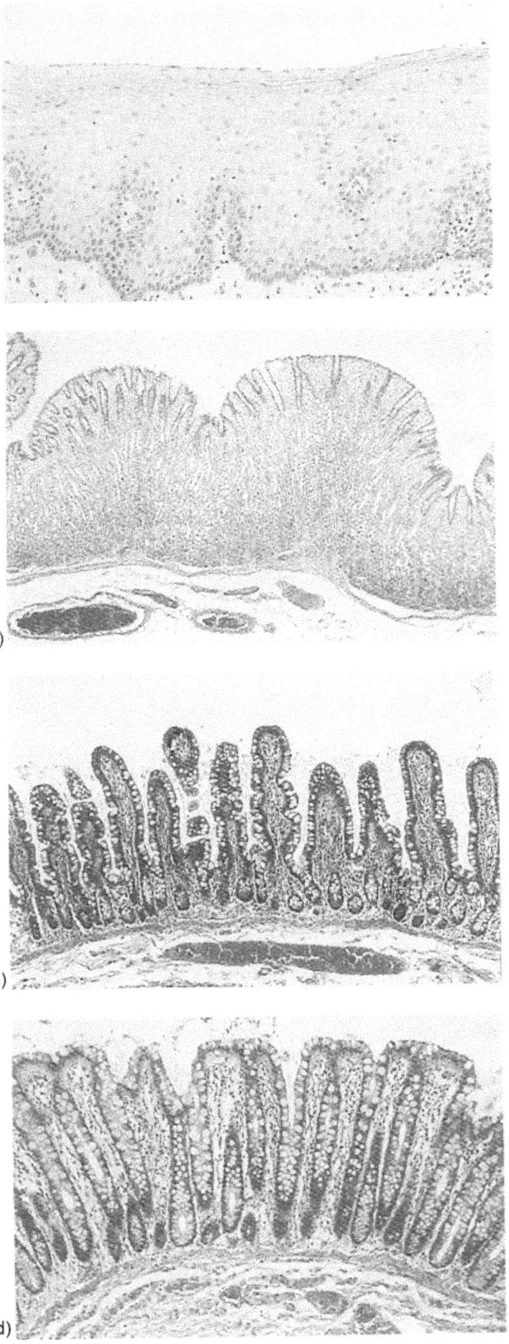

FIGURE 1.8 Basic mucosal forms in the GI tract. (a) Protective: found in the oral cavity, pharynx, esophagus, and anus. The surface epithelium is stratified squamous. (b) Secretory: typical to the stomach. The openings at the surface belong to simple or branched tubular glands. (c) Absorptive: typical to the small bowel. The surface area is increased by the villi projections, separated by crypts. (d) Absorptive/protective: typical to the large bowel: the crypts are dominating the epithelial structure. (From Wheater, P.R. et al., *Wheater's Functional Histology, A Text and Colour Atlas*, 3rd ed., Churchill Livingstone, Edinburgh, 1993. With permission.)

TABLE 1.2
Epithelia Thickness in the Oral Cavity of the Pig (Compared with its Skin, Lower Row) and Derived Permeability Constants (K_p)

	Thickness (μm)		K_p ($\times 10^{-7}$ cm/s)	
Mucosa/Tissue	**Entire Epithelium**	**Permeability Barrier**	**Water**	**Horseradish Peroxidase**
Gingiva	208±9	35±4	280±36	222±30
Buccal	772±20	282±17	451±72	118±18
Floor of the mouth	197±7	23±1	753±107	430±46
Skin	69±4	16±1	132±27	61±12

Source: From Kurosaki, Y. and Kimura, T., *Crit. Rev. Ther. Drug Carrier Syst.*, 17, 467, 2000. With permission.

the liver (as the portal vein does). This makes them attractive targets for drug delivery aimed at bypassing first-pass metabolism [123,124]. Unlike the small intestine mucosa, the buccal epithelium does not contain surface-bound peptidases [125,126]. However, it was shown that buccal homogenates possess greater peptidase activity than homogenates of the small intestine [127].

1.5.2 Gastric Mucosa

The epithelial cells of the stomach are characterized by abrupt transition from stratified squamous epithelium extended from the esophagus to a columnar epithelium dedicated to secretion (Figure 1.8). Unlike the surface of the small intestine, the surface of the gastric epithelium is flat with numerous cavities, being the openings of the gastric pits, which extend into the mucosa to form the gastric secretion glands [13]. The unique structure of the gastric mucosa together with the thick mucus lining practically prevents significant drug absorption from that organ.

1.5.3 Intestinal Mucosa

As mentioned above, the villi of the small intestine (Figure 1.2) house a dynamic, self-renewing population of the epithelial cells that includes absorptive cells (enterocytes), secretory cells, and endocrine cells. The thin lining (height: 25 μM; height of the microvilli is 1.5 μM) of the columnar enterocytes is the only barrier between the intestinal lumen and the muscularis mucosa, which represents, in this context, the entire body interior. The entire epithelial lining of the intestine replaces itself every 3–5 d [128]. It is the enterocyte and its neighboring cells where absorption processes occur and it will therefore be the focus of the mechanistic discussions below.

1.5.4 Transcellular Drug Absorption—Simple and Facilitated Diffusion

Typically to the plasma membrane of the cell, the lipophilic nature of the phospholipid bilayer determines the nature of drugs that would be allowed to diffuse through the apical membrane of the enterocyte. It appears that the more lipophilic the drug, the higher its permeability through the membrane's bilayer. This is demonstrated by the classic absorption study performed in the rat jejunum in which the extent of absorption of morphine, codeine, and thebaine correlated well with the reduction of the hydroxyl groups in the tested molecules (Figure 1.9) [19]. At the same time, the absorption path includes passage through aqueous

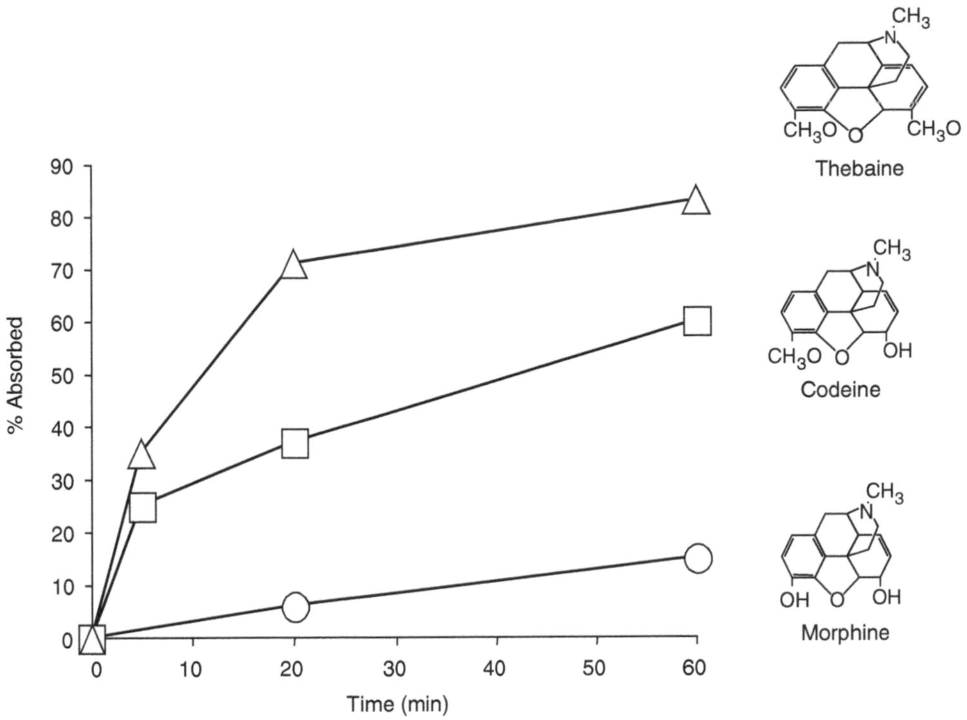

FIGURE 1.9 The absorption of morphine (2-OH groups), codeine (a single –OH group), and thebaine (no hydroxyl) from the jejunum of the rat. (From Kurz, H., in *Pharmacology of Intestinal Absorption: Gastrointestinal Absorption of Drugs* (W. Forth and W. Rummel, Eds.), Pergamon Press, Oxford, 1975, p. 245.)

boundary layers adjacent to the cell membrane. Therefore, it is not sufficient for a drug molecule to be highly water soluble at physiological pH to be absorbed. It should possess an optimal ratio between its hydrophilic and hydrophobic properties, which in turn depends on the pK_a of the drug and the local pH at the site of absorption (see below). It should also reside at the site of absorption for a sufficient period of time, or alternatively, exploit as much tissue surface area as possible. This is the reason that despite the low pH of the stomach contents, acidic drugs, like aspirin, almost do not absorb in this organ. The structure of the stomach epithelium and the thick adjacent carbonated mucus layer, together with the limited residence time of suspended aspirin particles (after tablet disintegration) in its antrum cause the drug to absorb primarily in the duodenum and jejunum.

Upon passage, a poorly water-soluble drug also accumulates in the membrane. Doluisio and coworkers have shown that a biexponential plot can describe the absorption kinetics of lipid-soluble drugs, such as prochlorperazine, whereas a single exponent slope characterizes the absorption of hydrophilic drugs, such as trimeprazine, which does not accumulate in the membrane [129].

It would have been expected that partitioning studies (e.g., octanol–water partitioning test [130]) could predict the exact capability of drugs to pass the intestinal membrane. However, because the enterocyte membrane is a highly complex, multicompartmental organ and because it contains transport proteins as well as efflux pumps, these studies can provide only a general delineation of the potential of a drug molecule to cross the phospholipid barrier of the absorptive cell. Still, with the absence of a better predictor, log *D* studies (plotting the log of the octanol–water partitioning characteristics of a new drug molecule against its *in situ* permeability) became a crucial tool in quantitative structure–activity relationship (QSAR) and high-throughput screening studies in the area of new drug design and synthesis [131]. Thus, in the past 20 years synthetic membrane models (e.g., phospholipid impregnated membranes with controlled pore size), parallel artificial-membrane permeability assay (PAMPA), and predictive systems based on partitioning into liposomal bilayers, as well as complicated, multicompartmental sink models with sophisticated, tailor-made distribution capacities have been developed.

Partitioning techniques have also improved and elaborated. It is common today to add octanol to a series of organic solvents such as cyclohexane, dodecane, propylene glycol dipelargonate (considered as proton acceptor similar to phospholipid membranes), and chloroform, on a routine basis to cover as many biological compartments (e.g., portioning in between serum and plasma membranes) as possible (see Ref. [132] for comprehensive textbook). Hence, the optimal log D range for oral absorption is 1–3. A drug molecule with a negative log D value (highly hydrophilic) is considered to face permeability problems across the enterocyte membrane. Lipophilic drugs with log D values of 3 and above are considered as potential candidates for metabolic processes at the intestinal brush border [131].

An important step in the understanding of partitioning phenomena is the ability to link between a drug molecule's pK_a and its pH-dependent solubility. The major tool for this estimation has been, for the past 80 years, the Henderson–Hasselbalch equation [133]. The ionized fraction of a drug can be calculated for any pH value if its pK_a is known [134], as follows:

Assuming a dissociation of a weak acid drug according to the general reaction:

$$\mathrm{HA} \leftrightarrow \mathrm{H}^+ + \mathrm{A}^-$$

$$\mathrm{HA} \leftrightarrow \mathrm{H}^+ + \mathrm{A}^-$$

(A^- can be looked at as the conjugated base of the acidic drug molecule). The dissociation constant is then

$$K = \frac{[\mathrm{H}^+][\mathrm{A}^-]}{[\mathrm{HA}]}$$

Taking the negative logarithm on both sides of the equations to enable the use of the terms pH and pK and rearranging, the Henderson–Hasselbalch equation becomes

$$\mathrm{pH} = \mathrm{p}K + \log\frac{[\mathrm{A}^-]}{[\mathrm{HA}]} = \mathrm{p}K + \log\frac{[\text{conjugate base}]}{[\text{conjugate acid}]}$$

At pH = pK_a the ratio $[A^-]/[HA]$ becomes a unit, which means that exactly half of the drug is in its ionized form if it is dissolved in a medium with a pH equal to its pK_a (Figure 1.10).

The Henderson–Hasselbalch equation enables predicting of how small changes in the pH values of the intestinal chyme could affect the solubility of poorly water-soluble drugs and hence their partitioning into the enterocyte membrane and their intestinal absorption.

In contrast to Fickian diffusion, which is driven by a concentration gradient across the enterocyte membrane, is nonspecific, and takes place through the entire surface of the epithelium according to the partitioning principles described above, there are molecules that cross the membrane at a rate higher than what would have been expected from concentration flux considerations. A typical example is the passage of glucose through the basolateral membrane

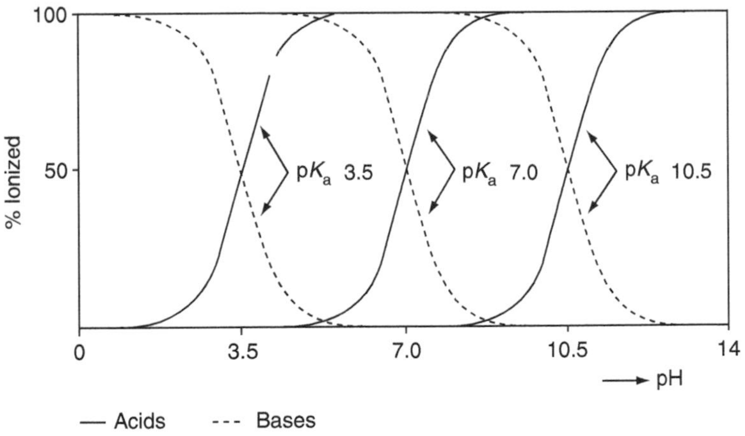

FIGURE 1.10 The pH-dependent extent of ionization of hypothetic acidic and basic molecules, having theoretical pK_a values of 3.5, 7.0, and 10.5. (From Kurz, H., in *Pharmacology of Intestinal Absorption: Gastrointestinal Absorption of Drugs* (W. Forth and W. Rummel, Eds.), Pergamon Press, Oxford, 1975, p. 245.)

of the enterocyte into the interstitial fluid and on to the bloodstream. This speedy passage (facilitated diffusion) involves a uniport protein (GLUT2 in the case of glucose export from the cell), capable of ferrying a number of substrates in a specific manner, but still driven by a concentration gradient. The transport, diffusional in nature (that is, does not require energy), occurs through a limited number of membrane proteins and therefore its kinetics is saturable and can be characterized by Michaelis–Menten constants such as K_m and V_{max}. Another type of protein, which mediates facilitated diffusion, is a channel protein, which simply creates an open pore in the bilayer membrane, allowing molecules to freely diffuse only through these proteins [135,136]. Uptake and metabolism of L-carnitine, D-carnitine, and acetyl-L-carnitine were studied in isolated guinea-pig enterocytes. It was suggested that the uptake mechanism was a facilitated diffusion rather than active transport because L-carnitine did not develop a significant concentration gradient, and was unaffected by ouabain or actinomycin A [137]. The transport of the peptide glutathione (GSH) has been studied with the rat small intestine *in vitro* and the human buccal cavity *in vivo*. Uptake was found to be sodium-independent in both systems. Saturation kinetics was demonstrated and uptake did not require energy in either system. However, it was inhibited by other small peptides, suggesting that a carrier-mediated facilitated diffusion was the cause of the transport [138]. Finally, the mechanism of intestinal absorption of arachidonic acid was studied in everted intestinal sacs of the rat. Metabolic inhibitors and uncouplers did not change its absorption and its absorption increased with thinning of the unstirred water layer. It was concluded that arachidonic acid was absorbed by a concentration-dependent dual mechanism of transport, which was not energy-dependent. At the low micron range of concentrations, facilitated diffusion was predominant, whereas at millimolar concentrations, simple diffusion was the dominant mechanism of absorption [139].

1.5.5 Transcellular Drug Absorption—Nondiffusional Processes

1.5.5.1 Carrier-Mediated Approaches to Drug Absorption

Trans- and paracellular passage routes cannot explain in full the phenomena of drug absorption in the GI tract, as they cannot provide a simple explanation for the absorption of

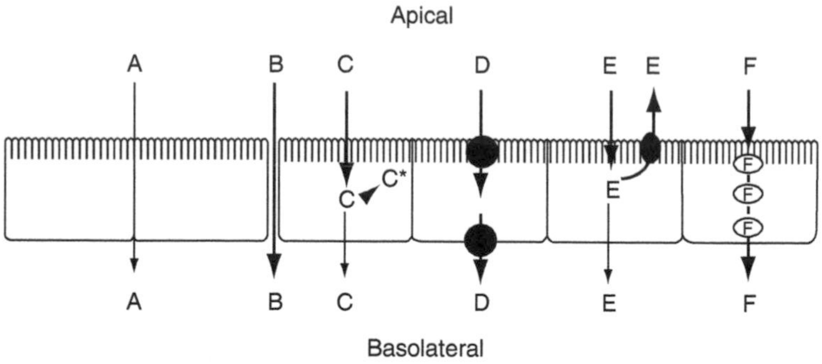

FIGURE 1.11 A scheme of the various absorption routes across the intestinal epithelium and cellular barriers to xenobiotics absorption. A, Transcellular absorption (plain diffusion); B, paracellular absorption; C, carrier-mediated transcellular absorption; D, facilitated diffusion; E, the MDR and P-gp absorption barrier; and F, endocytosis. (From Hunter, J. and Hirst, B.H., *Adv. Drug Deliv. Rev.*, 25, 129, 1997. With permission.)

monosaccharides, inorganic phosphate, monocarboxylic acids (especially in the context of short-chain fatty acids produced during colonic fermentation of polysaccharides), some water-soluble vitamins (such as ascorbic acid in both its reduced and oxidized forms), and di- or tripeptides (Figure 1.11). Indeed, in the past 40 years, it was found that these moieties are absorbed across the intestinal epithelium by carrier-mediated processes (transporters) [140–146]. The most relevant intestinal brush-border transporters for drug absorption are listed in Table 1.3 [147]. This led to the recognition that brush-border transporters could be exploited for the absorption enhancement of drug molecules that share conformational similarity with the substrates of these natural carriers.

The use of xenobiotics that were designed to penetrate through specific interaction with the corresponding transporters was demonstrated in a number of studies. The intestinal absorption of benzyl β-glucoside (commonly used as a neutraceutical agent in Japan) was tested in the rat context of the apical glucose transporters SGLT1 [148]. Insulin, which was conjugated

TABLE 1.3
Some Brush-Border Transporters Involved in Intestinal Absorption Processes, Pertinent to GI Drug Absorption

Transporter	Drug Tested
Hexoses	Benzyl β-glucoside, insulin conjugate with *p*-(succinylamido)-phenyl-α-D-glucopyranoside
Monocarboxylic acids	Short-chain fatty acids, salicylic acid, tolbutamide, atorvastatin, valproic acid
Phosphate	Foscarnet
Vitamins: folic acid, ascorbic acid (reduced and oxidized species), biotin, cholin, pantothenic acid	
Amino acids	α-L-Methyldopa
Di- and tripeptides	ACE inhibitors, β-lactam antibiotics
Bile acids	Somatostatin analogs, cyclosporine
Receptor-mediated endocytosis	Vitamin B_{12}, lectins, drug polymer conjugates

Source: From Tamai, I. and Tsuji, A., *Adv. Drug Del. Rev.*, 20, 5, 1996.

to *p*-(succinylamido)-phenyl-α-D-glucopyranoside, was administered intraintestinally to the rat and was found to be more effective than bovine insulin. In a series of inhibition studies, it was found that this enhanced activity could be related to the active transporting system of the rat brush border [149].

The monocarboxylic acid transporter family MCT (mostly MCT1) was investigated in the past decade for its involvement in intestinal, nondiffusional drug absorption. Typical studies have indicated that salicylic acid [150], tolbutamide [151], atorvastatin (a water-soluble derivative of coenzyme A) [152], and valproic acid [153] are probably uptaken, at least in part, by MCT into the bloodstream.

Phosphonoformic acid (Foscarnet), an antiviral drug, which consists of carboxylic acid and phosphate linked together, was shown to be absorbed in the small intestine by the phosphate carrier-mediated mechanism [154]. The transport of folic acid was investigated in biopsy specimens of intestinal mucosa from healthy volunteers. A pH-dependent, active transport of folic acid was found in the proximal small intestine, whereas in the mucosa of the colon, folic acid uptake was driven by a facilitated diffusion, mediated by a low-affinity carrier [155]. The existence of folate receptors in the human colonic mucosa is logical due to the large amounts of folic acid produced by the colonic flora [145].

The major route for the absorption of the end products of protein digestion in the intestine is the oligopeptide transporter PepT1 (the homologous transporter PepT2 is responsible for the reabsorption of di- and tripeptides in the kidney) [156]. PepT1 is highly relevant for the intestinal absorption of peptides and peptidomimetic drugs such as β-lactam antibiotics (only those which contain free α-amino group, e.g., ampicillin, cephalexin) and the angiotensin-converting enzyme (ACE) inhibitors captopril and enalapril in the intestine [157,158]. This transporter, a H^+, sodium-independent dipeptide symporting protein of 127 kDa, has been frequently mentioned in the last decade as a possible shuttle for the administration of peptidomimetic drugs that share similar conformation with dipeptides such as the L-alanyl-L-alanine, which is used in the transport of the β-lactam antibiotics. Moreover, prodrugs containing peptidomimetic moieties with the ability to recognize the PepT1 transporter have been tested successfully. These include prodrugs of acyclovir, L-DOPA, and some bisphosphonates [158].

1.5.5.2 Receptor-Mediated Endocytosis and Drug Absorption

The last nondiffusional route for drugs and drug carriers is the endocytotic pathway. In endocytosis, the plasma membrane of the enterocyte invaginates to form a cytoplasmic vesicle. The process is specific, saturable, and highly regulated. The binding of a ligand to a membrane receptor triggers an internalization process at the site of contact that occurs within minutes. The ligand–receptor complex then becomes a transport vesicle: the endosome. The receptors are recycled into the plasma membrane, whereas the ligands are further fused with the lysosomes to be degraded by their acidic pH and lysosomal enzymes. Thus the low-density lipoprotein (LDL) receptor binds and internalizes cholesterol-containing particles [159], transferrin receptor internalizes iron [160,161] and insulin, and other protein hormones are internalized by their specific receptors [162]. Glycoproteins whose oligosaccharide side chains contain terminal glucose, mannose, or galactose residues rather than the normal sialic acid also invaginate into enterocytes or colonocytes in the same manner. In the context of drug absorption enhancement, the absorption of vitamin B_{12} along with the intrinsic factor, which escorted it from the stomach, is worth mentioning because its receptor, located at the distal portion of the ileum, has been a target for a number of interesting studies in which poorly absorbed drugs were conjugated to the vitamin to allow their endocytotic uptake after oral administration [163].

The endocytic processes in the epithelial cells are counted on when polymeric drug conjugates are designed for both systemic and oral administrations. Typical examples are the use of plant lectins as drug carriers, aiming at the specific recognition of membrane galactose or galactoseamine residues [164] and synthetic polymers containing either antibodies or sugars, as sophisticated targeting vehicles for poorly absorbed drugs or toxic drugs whose systemic exposure should be minimized [165,166].

1.5.6 Paracellular Drug Absorption

Diffusion through the enterocyte membrane is not the only portal for drugs to the systemic compartment from the intestinal lumen. The junction zones between the membrane cells can also serve as portals for that purpose. In the mid of the last century this option was not realized because of the lipid pore theory, which dominated mechanistic studies in drug transport. According to that premise, the intestinal epithelium was looked upon as a single bilayer membrane, covered with pores that, despite the relative surface area they employed, had to be considered when diffusion constants were calculated. According to Fordtran, the size of the pores decreases in the caudal direction: 0.67–0.88 in the jejunum and 0.3–0.38 nm in the ileum [167]. Once the physiology of junctional complexes among the enterocytes attracted sufficient research, the pore theory was replaced with the paracellular absorption paradigm.

Careful observation at the borders of the enterocytes reveals a close juxtaposition of the membranes of adjacent epithelial cells, identified as zonulae occludens, located at the apical neck of the epithelial cells [168]. The zonulae occludens is composed of apical complex: the tight junctions and more basal structure underneath, the adherens junctions (also referred to as intermediate junctions) [169,170]. The adherens junctions are linked to the actomyosin ring at the cell circumference. This actin cytoskeleton, which forms a perijunctional area at the apical neck of the epithelial cells, is the cause for their polarized nature (Figure 1.12). The classical functions of tight junctions are the regulation of paracellular permeability and the restriction of apical–basolateral intramembrane diffusion of lipids (cell polarity housekeeping).

The tight junctions are constructed of highly complex sets of proteins (many of which are signaling proteins) connected to cytoplasmic components and the actin cytoskeleton. Grossly, they are divided into (a) integral membrane proteins, the actual proteins that form the extracellular tight junctions and (b) scaffolding proteins, the peripheral membrane proteins that interact with both tight junctions and elements of the actomyosin cytoskeleton [170]. The integral membrane proteins are the occludin, and claudin protein family and junctional adhesion molecules that interact with junction-associated proteins. The occludins and claudins include 24 proteins of 17–25 kDa. Their extracellular interactions are strong enough to produce intramembranous particles [171]. The integral membrane protein group (the ZO family) contains primarily the ZO-1, ZO-2, and ZO-3 proteins [172]. Another protein frequently associated with tight junction complex proteins is cingulin, which probably functions as a signaling protein rather than a structural one [169,170].

Accumulating reports suggest that tight junctions' signaling proteins are involved in the regulation of differentiation, polarity, and proliferation of epithelial cells [173]. An interesting observation concerning tight junctions functioning is that bacterial infection can cause a neutrophil to move across epithelia through the paracellular route following a gradient of the bioactive lipid hepoxilin A_3 [174]. Stringently, the entire cell apparently moves between adjacent epithelial cells with only a slight loss of solute barrier integrity [170].

It is often mentioned that when drugs are absorbed paracellularly, they are dragged by a local water flux through the tight junctions. In this context, it is interesting to mention the Pappenheimer hypothesis [169,175,176], which suggests that the absorption of solutes across

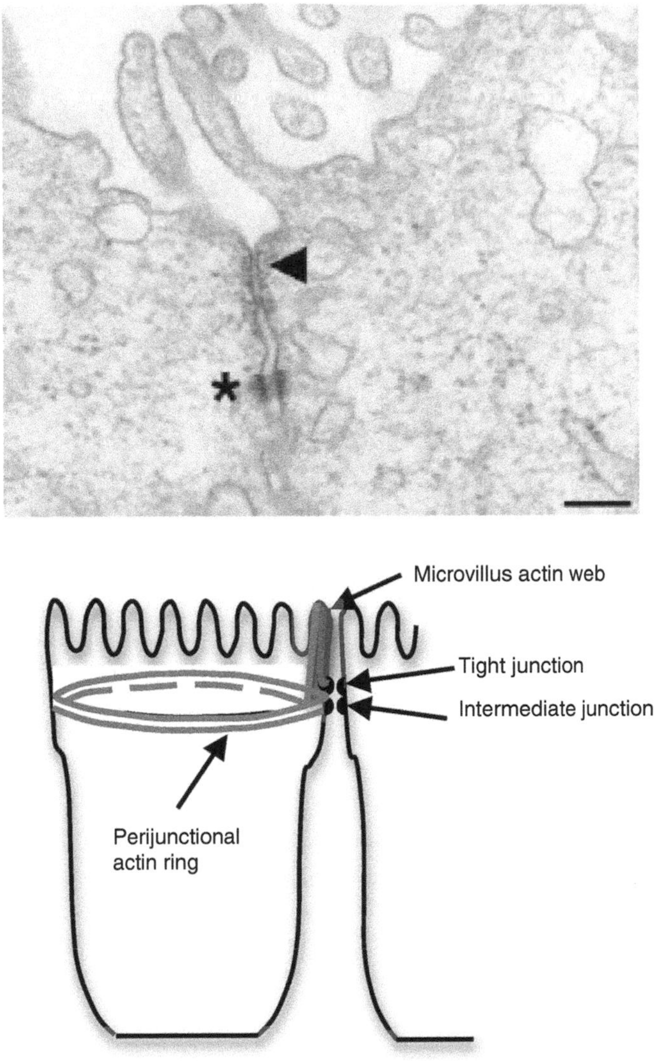

FIGURE 1.12 *Top*: an electron microscopy image of an epithelial tight junction. The position of the tight junction at the apical end of the lateral membrane is marked with an arrowhead, whereas the position of a desmosome is indicated with an asterisk. Bar = 200 nm. (From Matter, K. and Balda, M.S., *Methods*, 30, 228, 2003. With permission.) *Bottom*: a schematic location of the tight junctions, intermediate junctions, and actin cytoskeleton in the epithelial cell. (From Ballard, S.T., Hunter, J.H., and Taylor, A.E., *Ann. Rev. Nutr.*, 15, 35, 1995. With permission.)

the small intestine epithelium is regulated by glucose uptake into the enterocytes and a concomitant widening of the tight junctions (Figure 1.13).

Tight junctions are mentioned frequently in association with facilitated intestinal drug absorption. Their loosening is linked to one of the following: (a) contractions of the acto-myosin ring, which circumscribes the cell, (b) the integrity of cytoskeletal actin filaments, (c) phosphorylation of the tight junction components (protein kinase C activation increases paracellular flux), and (d) calcium ion depletion from the vicinity of the epithelium (thought to be mediated by protein kinase) [177,178]. Table 1.4 presents some modulating agents, relevant for assessment as drug absorption enhancers, reported to open tight junctions as

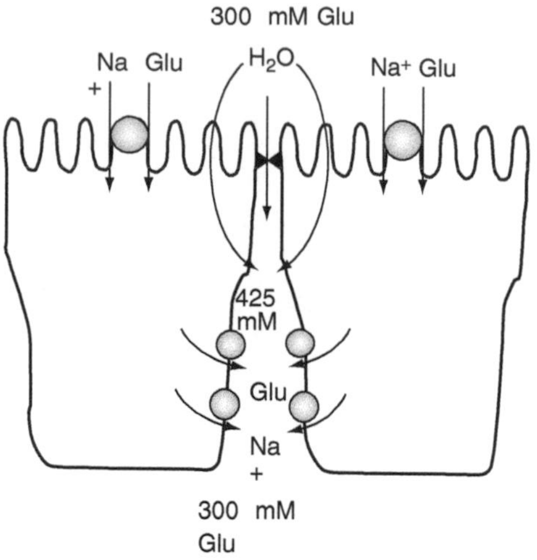

FIGURE 1.13 The osmotic gradient theory for paracellular solute passage. Sodium and glucose are absorbed actively by the apical membrane and secreted into the lateral intracellular space by active transport (sodium) and facilitated diffusion (glucose). The developed osmotic gradient between the apical and basolateral aspects of the enterocyte causes water to flow both through the cell and through the tight junction. Hydrophilic molecules that would have otherwise rejected from the lipophilic cell membrane find an open portal to pass to the lamina propria. (From Ballard, S.T., Hunter, J.H., and Taylor, A.E., *Ann. Rev. Nutr.*, 15, 35, 1995. With permission.)

assessed by the transepithelial resistance (TER) measurements of epithelial cells and the onset times for their activity. Looking at this table, it is tempting to try agents like protein kinase inhibitors as specific paracellular absorption enhancers. However, these agents might interfere with whole-body phosphorylation processes. Mild surface-active agents, such as sodium caprate, could be better candidates because of their multifunctional properties: at the membrane level, tight junction level, and P-glycoprotein (P-gp) inhibition level [179–182].

TABLE 1.4
Loosening Agents of Tight Junctions and the Time Required for Each of Them to Deplete Transmural TER as Analyzed in Cell Cultures and Isolated Epithelial Tissues

Tight Junction Modulator	Approximate Time Required to Deplete TER
Interferon-γ (cytokine)	2–3 d
Cytochalasin B or D (retards actin assembly)	10–30 min
Sodium-dependent glucose transport	10 min
ATP depletion	10–120 min
Protein kinase A inhibitor	1 h
Protein kinase C activator	1–3 h
ZO toxin	30 min
Calcium chelators	5–15 min
Calcium ionophores	30 min

Source: From Hochman, J. and Artursson, P., *J. Contr. Rel.*, 29, 253, 1994.

1.5.7 Intestinal Efflux Processing of Absorbed Drugs and Brush-Border Metabolism

When discussing drug absorption processes in the GI tract and implications on drug bioavailability, efflux pumping of the lipophilic enterocyte membrane should be taken into consideration. It has long been recognized that secretory efflux systems, such as multidrug-resistance associated protein (MRP) [183], flurochrome efflux systems bis-carboxyethyl-carboxyfluorescein (BCECF) [184], and the methotrexate efflux system [185] mediate intestinal secretion of drugs at the enterocyte level. The archetypal membrane protein of this phenomenon is P-gp. It is a member of the ATP-binding cassette (ABC) homolog group of membrane transport proteins, an ATP-dependent secretory transporter, encoded by the MDR1 gene in humans. P-gp acts to clear the membrane lipid bilayer of lipophilic drugs in the same way membrane flippase translocates membrane lipids in erythrocytes [186]. It was first identified in tumor cells but very rapidly discovered in normal adrenal glands, lung, liver, brain, jejunum, and colon. Today, P-gp is considered as an important brick in blood–brain barrier functioning and a limiting factor in intestinal absorption as well as the cause for intersubject PK variability for many drugs such as digoxin, ranitidine, fluoroquinolones (e.g., ciprofloxacin), β-adrenergic receptor antagonists (e.g., celiprolol), and amoxicillin. It should be realized, however, that most of the mechanistic information regarding P-gp activity comes from *in vitro* studies of Caco-2 cell lines. In humans the picture could be more complex as intestinal drug absorption is an interplay of multifactorial parameters. In addition to transcellular absorption of high doses of drugs that could bypass the saturable efflux pumping of intestinal P-gp, gut wall metabolism by the CYP450 family should also be taken into consideration. Their concomitant, regional activity, in concert with the P-gp and other efflux eliminating proteins should not be underestimated [187,188].

Out of the entire family of CYP450, the most frequent enzyme mentioned in the context of drug metabolism is the cytochrome P-450 3A (CYP3A). Although this enzyme constitutes 30% of the total hepatic activity of CYP450, in the human enterocytes it accounts for about 70% of the activity of this group of enzymes and it exceeds the activity of these enzymes in the liver. Intensive research in the area of gut wall metabolism and bioavailability of orally administered drugs has indicated the important role of Phase-I metabolism (oxidation) processes occurring at the villi tips [5]. Much of the information about this unique, presystemic drug metabolism was obtained from studies conducted on cyclosporine by Benet and coworkers [189]. These studies led them to infer that nutrition has an impact on drug bioavailability after concomitant administration with food. The most famous example is the influence of grapefruit juice and related flavonoids (reduce brush-border CYP3A activity) on the modulation of oral drug bioavailability. In most cases drug absolute bioavailability was increased. These studies triggered the assumption that brush-border metabolism could be associated with ethnic differences in drug absorption of drugs like nifedipine [5].

1.6 PHYSIOLOGICAL INSIGHT ON THE USE OF ABSORPTION ENHANCERS IN THE GI TRACT

Drug absorption enhancers have been intensively studied over the past three decades [190–192] in order to increase the oral availability of poorly absorbed drugs (BSC classes II–IV) [193,194]. Much attention has been paid to the synthesis and evaluation of new absorption enhancer molecules. Excluding the preparation of polymer drug conjugates; less awareness was focused on the delivery rate of the molecules tested; that is, the concomitant input of the poorly soluble drug together with its enhancer, so that maximal area of the small intestine would be exploited for absorption. The differences between the dimensions of the GI tract in small animal models (closed compartment conditions in rodents) and large animal models (dog) and even humans (open compartment) could cause problems when scaling up the biological evaluation analysis [195,196]. A careful design, physiologically oriented, should be employed when efficient absorption enhancement is desired. It was demonstrated that, in rat, the constant input rate of the absorption enhancer sodium decanoate (SD) was more important for sodium cephazoline absorption than SD concentration in the perfused solution [197]. It was concluded that a rational oral dosage form of a poorly absorbed drug would be the one that is designed to release both drug and its absorption enhancer at similar rates (synchronized release). Assuming intestinal absorption only, this concept was further tested with a variety of drugs like sulpiride (30% bioavailability) [181] and calcitonin. In all studies, an emphasis was put on the concomitant release of both drugs and absorption enhancers over defined intestinal transit and time slots. It is suggested that an appropriate oral formulation, with the ability to synchronize the release rates of two or more molecules is equally important in identifying a novel absorption promoter.

REFERENCES

1. Gold, H., et al. 1953. Clinical pharmacology of digoxin. *J Pharmacol Exp Ther* 109:45.
2. Wagner, J.G. 1981. History of pharmacokinetics. *Pharmacol Ther* 12:537.
3. Lee, P.I. 2001. Design and power of a population pharmacokinetic study. *Pharm Res* 18:75.
4. Ma, J.D., A.N. Nafziger, and J.S. Bertino, Jr. 2004. Genetic polymorphisms of cytochrome P450 enzymes and the effect on interindividual, pharmacokinetic variability in extensive metabolizers. *J Clin Pharmacol* 44:447.
5. Wacher, V.J., L. Salphati, and L.Z. Benet. 1996. Active secretion and enterocytic drug metabolism barriers to drug absorption. *Adv Drug Deliv Rev* 20:99.
6. Ekins, S., et al. 2000. Progress in predicting human ADME parameters in silico. *J Pharmacol Toxicol Methods* 44:251.
7. Balimane, P.V., and S. Chong. 2005. Cell culture-based models for intestinal permeability: A critique. *Drug Discov Today* 10:335.
8. Stainier, D.Y.R. 2005. No organ left behind: Tales of gut development and evolution. *Science* 307:1902.
9. Khanvilkar, K., M.D. Donovan, and D.R. Flanagan. 2001. Drug transfer through mucus. *Adv Drug Deliv Rev* 48:173.
10. Schubert, M.L. 2004. Gastric secretion. *Curr Opin Gastroenterol* 20:519.
11. Engel, E., P.H. Guth, Y. Nishizaki, and J.D. Kaunitz. 1995. Barrier function of the gastric mucus gel. *Am J Physiol* 269:G994.
12. Nimmerfall, F., and J. Rosenthaler. 1980. Significance of the goblet-cell mucin layer, the outermost luminal barrier to passage through the gut wall. *Biochem Biophys Res Commun* 94:960.
13. Johnson, L.R. 2001. *Gastrointestinal physiology*, 6th ed. St. Louis: Mosby.
14. Rendleman, D.F., et al. 1958. Reflux pressure studies on the ileocecal valve of dogs and humans. *Surgery* 44:640.
15. Phillips, S.F., J.H. Pemberton, and R.G. Shorter. 1991. *The large intestine: Physiology, pathophysiology and disease*. New York: Raven Press.
16. Cherbut, C., A.C. Aube, H.M. Blottiere, and J.P. Galmiche. 1997. Effects of short-chain fatty acids on gastrointestinal motility. *Scand J Gastroenterol Suppl* 222:58.
17. Read, N.W., A. McFarlane, R.I. Kinsman, and S.R. Bloom. 1984. The ileal brake: A potent mechanism for feedback control of gastric emptying and small bowel transit. In *Gastrointestinal motility*, ed. C. Roman, 335. Lancaster: MTP Press.
18. Bieck, P.R. 1993. *Colonic drug absorption and metabolism*. New York: Marcel Dekker.
19. Kurz, H. 1975. Principles of drug absorption. In *Pharmacology of intestinal absorption: Gastrointestinal absorption of drugs*, eds. W. Forth and W. Rummel, 245. Oxford: Pergamon Press.

20. Neutra, M.R., and J.F. Forstner. 1987. Gastrointestinal mucus: Synthesis, secretion, and function. In *Physiology of the gastrointestinal tract*, 2nd ed., ed. R.L. Johnson, 975. New York: Raven Press.
21. Sarna, S.K. 1991. Physiology and pathophysiology of colonic motor activity (1). *Dig Dis Sci* 36:827.
22. Scheline, R.R. 1973. Metabolism of foreign compounds by gastrointestinal microorganisms. *Pharmacol Rev* 25:451.
23. Chadwick, R.W., S.E. George, and L.D. Claxton. 1992. The role of the gastrointestinal mucosa and microflora in the bioactivation of dietary and environmental mutagens or carcinogens. *Drug Metab Rev* 24:425.
24. Sintov, A., M. Simberg, and A. Rubinstein. 1996. Absorption enhancement of captopril in the rat colon as a putative method for captopril delivery by extended release formulations. *Int J Pharm* 143:101.
25. Kim, J.S., et al. 1994. Absorption of ACE inhibitors from small intestine and colon. *J Pharm Sci* 83:1350.
26. Fetih, G., et al. 2005. Improvement of absorption enhancing effects of *n*-dodecyl-beta-D-maltopyranoside by its colon-specific delivery using chitosan capsules. *Int J Pharm* 293:127.
27. Sutton, S.C., E.L. LeCluyse, L. Cammack, and J.A. Fix. 1992. Enhanced bioavailability of cefoxitin using palmitoyl L-carnitine. I. Enhancer activity in different intestinal regions. *Pharm Res* 9:191.
28. de Boer, A.G., E.J. van Hoogdalem, and D.D. Breimer. 1992. Rate-controlled rectal peptide drug absorption enhancement. *Adv Drug Deliv Rev* 8:237.
29. Yoshikawa, H., S. Muranishi, and H. Sezakai. 1983. Mechanisms of selective transfer of bleomycin into lymphatics by a bifunctional delivery system via the lumen of the large intestine. *Int J Pharm* 13:321.
30. Hauss, D.J., H.Y. Ando, D.M. Grasela, and E.T. Sugita. 1991. The relationship of salicylate lipophilicity to rectal insulin absorption enhancement and relative lymphatic uptake. *J Pharmacobiodyn* 14:139.
31. Touitou, E., and M. Donbrow. 1983. Promoted rectal absorption of insulin: Formulative parameters involved in the absorption from hydrophilic bases. *Int J Pharm* 15:13.
32. Lindmark, T., et al. 1997. Mechanism of absorption enhancement in humans after rectal administration of ampicillin in suppositories containing sodium caprate. *Pharm Res* 14:930.
33. Noach, A.B., M.C. Blom-Roosemalen, A.G. de Boer, and D.D. Breimer. 1995. Absorption enhancement of a hydrophilic model compound by verapamil after rectal administration to rats. *J Pharm Pharmacol* 47:466.
34. Mathias, J.R., and C.A. Sninsky. 1985. Motility of the small intestine: A look ahead. *Am J Physiol* 248:G495.
35. Macheras, P., C. Reppas, and J.B. Dressman. 1995. Physiological factors related to drug absorption. In *Biopharmaceutics of orally administered drugs*. London: Ellis Horwood.
36. Kumar, D., and D. Wingate. 1993. *An illustrated guide to gastrointestinal motility*, 2nd ed. Edinburgh: Churchill Livingstone.
37. Code, C.F., and J.A. Marlett. 1975. The interdigestive myo-electric complex of the stomach and small bowel of dogs. *J Physiol* 246:289.
38. Sarna, S.K. 1985. Cyclic motor activity: Migrating motor complex. *Gastroenterology* 89:894.
39. Schemann, M., and H.J. Ehrlein. 1986. Mechanical characteristics of phase II and phase III of the interdigestive migrating motor complex in dogs. *Gastroenterology* 91:117.
40. Rubinstein, A., H.K.V. Kin Li, P. Gruber, and J.R. Robinson. 1988. Gastrointestinal physiological variables affecting the performance of oral sustained release dosage forms. In *Oral sustained release formulations, design and evaluation*, eds. A. Yacobi and E. Halperin-Walega, 125. New York: Pergamon Press.
41. Hunt, J.N., and I. MacDonald. 1954. The influence of volume on gastric emptying. *J Physiol (Lond)* 126:459.
42. Hunt, J.N. 1959. Gastric emptying and secretion in man. *Physiol Rev* 39:491.
43. Keinke, O., M. Schemann, and H.J. Ehrlein. 1984. Mechanical factors regulating gastric emptying of viscous nutrient meals in dogs. *J Exp Physiol* 69:781.

44. Hunt, J.N. 1984. Regulation of gastric emptying by neurohumoral factors and by gastric and duodenal receptors. In *Esophageal and gastric emptying*, eds. A. Dubois and D.O. Castell, 65. Boca Raton: CRC Press.
45. Hunt, J.N. 1980. A possible relation between the regulation of gastric emptying and food intake. *Am J Physiol* 239:G1.
46. Meyer, J.H., J. Dressman, A. Fink, and G. Amidon. 1985. Effect of size and density on canine gastric emptying of nondigestible solids. *Gastroenterology* 89:805.
47. Khosla, R., and S.S. Davis. 1990. The effect of tablet size on the gastric emptying of non-disintegrating tablets. *Int J Pharm* 62:R9.
48. Coupe, A.J., S.S. Davis, D.F. Evans, and I.R. Wilsing. 1993. Do pellets empty from the stomach with food? *Int J Pharm* 92:167.
49. Timmermans, J., and A.J. Moes. 1993. The cutoff size for gastric emptying of dosage forms. *J Pharm Sci* 82:854.
50. Smith, J.L., C.L. Jiang, and J.N. Hunt. 1984. Intrinsic emptying pattern of the human stomach. *Am J Physiol* 246:R959.
51. Minami, H., and R.W. McCallum. 1984. The physiology and pathophysiology of gastric emptying in humans. *Gastroenterology* 86:1592.
52. Kraus, L.C., and J.N. Fell. 1984. Effect of stress on the gastric emptying of capsules. *J Clin Hosp Pharm* 9:249.
53. Nimmo, W.S., and F. Prescott. 1978. The influence of posture on paracetamol absorption. *Br J Clin Pharmacol* 5:348.
54. Renwick, A.G., et al. 1992. The influence of posture on the pharmacokinetics of orally administered nifedipine. *Br J Clin Pharmacol* 34:332.
55. Backon, J., and A. Hoffman. 1991. The lateral decubitus position may affect gastric emptying through an autonomic mechanism: The skin pressure-vegetative reflex. *Br J Clin Pharmacol* 32:138.
56. Gupta, P., and J.R. Robinson. 1988. Gastric emptying of liquids in the fasted dog. *Int J Pharm* 43:45.
57. Thompson, D.G., et al. 1980. Normal patterns of human upper small bowel motor activity recorded by prolonged radiotelemetry. *Gut* 21:500.
58. Park, H.M., et al. 1984. Gastric emptying of enteric-coated tablets. *Dig Dis Sci* 29:207.
59. Gruber, P., et al. 1987. Gastric emptying of nondigestible solids in the fasted dog. *J Pharm Sci* 76:117.
60. Graham, D.Y., J.L. Smith, and A.A. Bouvet. 1990. What happens to tablets and capsules in the stomach: Endoscopic comparison of disintegration and dispersion characteristics of two microencapsulated potassium formulations. *J Pharm Sci* 79:420.
61. Tirosh, B., and A. Rubinstein. 1998. Migration of adhesive and non-adhesive particles in the rat intestine under altered mucus secretion conditions. *J Pharm Sci* 87:453.
62. Barr, W.H., et al. 1994. Different absorption of amoxicillin from the human small and large intestine. *Clin Pharmacol Ther* 56:279.
63. Davis, S.S., J.G. Hardy, and J.W. Fara. 1986. Transit of pharmaceutical dosage forms through the small intestine. *Gut* 27:886.
64. Davis, S.S., et al. 1986. Gastrointestinal transit of controlled release naproxen tablet formulation. *Int J Pharm* 32:85.
65. Adkin, D.A., et al. 1995. The effect of pharmaceutical excipients on small intestinal transit. *Br J Clin Pharmacol* 39:381.
66. Adkin, D.A., et al. 1995. The effect of different concentrations of mannitol in solution on small intestinal transit: Implications for drug absorption. *Pharm Res* 12:393.
67. Adkin, D.A., et al. 1995. The effect of mannitol on the oral bioavailability of cimetidine. *J Pharm Sci* 84:1405.
68. Hardy, J.G., C.G. Wilson, and E. Wood. 1985. Drug delivery to the proximal colon. *J Pharm Pharmacol* 37:874.
69. Debongnie, J.C., and S.F. Phillips. 1978. Capacity of the human colon to absorb fluid. *Gastroenterology* 74:698.

70. Levinson, S., et al. 1985. Comparison of intraluminal and intravenous mediators of colonic response to eating. *Dig Dis Sci* 30:33.
71. Snap, W.J.J., S.A. Matarazzo, and S. Cohen. 1978. Effect of eating and gastrointestinal hormones on human colonic myoelectrical and motor activity. *Gastroenterology* 75:373.
72. Wiley, J.N., D. Tatum, R. Keinath, and C. Owyang. 1988. Participation of gastric mechanoreceptors and intestinal chemoreceptors in the gastrocolonic response. *Gastroenterology* 94:1144.
73. Wright, S.H., et al. 1980. Effect of dietary components on gastrocolonic response. *Am J Physiol* 238:G228.
74. Sarna, S.K. 1991. Physiology and pathophysiology of colonic motor activity (2). *Dig Dis Sci* 36:998.
75. Adkin, D.A., S.S. Davis, R.A. Sparrow, and I.R. Wilding. 1993. Colonic transit of different sized tablets in healthy subjects. *J Contr Rel* 23:147.
76. Proano, M., et al. 1991. Unprepared human colon does not discriminate between solids and liquids. *Am J Physiol* 260:G13.
77. Eckburg, P.B., et al. 2005. Diversity of the human intestinal microbial flora. *Science* 308:1635.
78. Simon, G.L., and S.L. Gorbach. 1984. Intestinal flora in health and disease. *Gastroenterology* 86:174.
79. Salyers, A.A. 1984. Bacteroides of the human lower intestinal tract. *Ann Rev Microbiol* 38:293.
80. Macfarlane, G.T., and J.H. Cummings. 1991. The colonic flora, fermentation, and large bowel digestive function. In *The large intestine: Physiology, pathophysiology, and disease*, eds. S.F. Phillips, J.H. Pemberton, and R.G. Shorter, 51. New York: Raven Press.
81. Cummings, J.H. 1984. Microbial digestion of complex carbohydrates in man. *Proc Nutr Soc* 43:35.
82. Werch, S.C., and A.C. Ivy. 1941. On the fate of ingested pectin. *Am J Digest Dis* 8:101.
83. Dressman, J.B., et al. 1990. Upper gastrointestinal (GI) pH in young, healthy men and women. *Pharm Res* 7:756.
84. Russell, T.L., et al. 1993. Upper gastrointestinal pH in seventy-nine healthy, elderly, North American men and women. *Pharm Res* 10:187.
85. Evans, D.F., et al. 1988. Measurement of gastrointestinal pH profiles in normal ambulant human subjects. *Gut* 29:1035.
86. Evans, D.F., et al. 1986. Gastrointestinal pH profiles in man. *Gastroenterology* 90:1410.
87. Allen, A., and G. Flemstrom. 2005. Gastroduodenal mucus bicarbonate barrier: Protection against acid and pepsin. *Am J Physiol Cell Physiol* 288:C1.
88. Sherman, P., et al. 1987. Bacteria and the mucus blanket in experimental small bowel bacterial overgrowth. *Am J Pathol* 126:527.
89. Lamont, J.T. 1992. Mucus: The front line of intestinal mucosal defense. In *Neuro-immuno physiology of the gastrointestinal mucosa, implications for inflammatory diseases*, eds. R.H. Stead, M.H. Perdue, H. Cook, D.W. Powel, and K.E. Barrett, 190. New York: New York Academy of Sciences.
90. Lehr, C.M., F.G.P. Poelma, H.E. Junginger, and J.J. Tukker. 1991. An estimate of turnover time of intestinal mucus gel layer in the rat *in situ* loop. *Int J Pharm* 70:235.
91. Wilson, F.A., V.L. Sallee, and J.M. Dietschy. 1971. Unstirred water layers in intestine: Rate determination of fatty acid absorption from micellar solutions. *Science* 174:1031.
92. Li, C.Y., C.L. Zimmerman, and T.S. Wiedmann. 1996. Diffusivity of bile salt/phospholipid aggregates in mucin. *Pharm Res* 13:535.
93. Amidon, G.L., J. Kou, R.L. Elliot, and E.N. Lightfoot. 1980. Analysis of models for determining intestinal wall permeabilities. *J Pharm Sci* 69:1369.
94. Levitt, M.D., et al. 1987. Quantitative assessment of luminal stirring in the perfused small intestine of the rat. *Am J Physiol* 252:G325.
95. Ito, S. 1969. Structure and function of the glycocalyx. *Fed Proc* 28:12.
96. Smithson, K.W., and D.B. Millar. 1981. Intestinal diffusion barrier: Unstirred water layer or membrane surface mucous coat? *Science* 214:1241.
97. Katori, N., N. Aoyagi, and T. Terao. 1995. Estimation of agitation intensity in the GI tract in humans and dogs based on *in vitro/in vivo* correlation. *Pharm Res* 12:237.
98. Winne, D., H. Goig, and U. Muller. 1987. Closed rat jejunum segment *in situ*: Role of pre-epithelial diffusion resistance (unstirred layer) in absorption process and model analysis. *Naunyn Schmiedebergs Arch Pharmacol* 335:204.

99. Kou, J.H., D. Fleisher, and G.L. Amidon. 1991. Calculation of the aqueous diffusion layer resistance for absorption in a tube: Application to intestinal membrane permeability determination. *Pharm Res* 8:298.
100. Sinko, P.J., et al. 1996. Analysis of intestinal perfusion data for highly permeable drugs using a numerical aqueous resistance—Nonlinear regression method. *Pharm Res* 13:570.
101. Bickel, M., and G.L. Kauffman, Jr. 1981. Gastric gel mucus thickness: Effect of distention, 16,16-dimethyl prostaglandin e2, and carbenoxolone. *Gastroenterology* 80:770.
102. Pullan, R.D., et al. 1994. Thickness of adherent mucus gel on colonic mucosa in humans and its relevance to colitis. *Gut* 35:353.
103. Ryu, K.H., and E. Grim. 1982. Unstirred water layer in canine jejunum. *Am J Physiol* 242:G364.
104. Winne, D., and W. Verheyen. 1990. Diffusion coefficient in native mucus gel of rat small intestine. *J Pharm Pharmacol* 42:517.
105. Flemstrom, G., et al. 1999. Adherent surface mucus gel restricts diffusion of macromolecules in rat duodenum *in vivo*. *Am J Physiol* 277:G375.
106. Braybrooks, M.P., B.W. Barry, and E.T. Abbs. 1975. The effect of mucin on the bioavailability of tetracycline from the gastrointestinal tract; *in vivo*, *in vitro* correlations. *J Pharm Pharmacol* 27:508.
107. Barry, B.W., and M.P. Braybrooks. 1975. Proceedings: Influence of a mucin model system upon the bioavailability of phenylbutazone and warfarin sodium from the small intestine. *J Pharm Pharmacol* 27(Suppl. 2):74P.
108. Nakamura, J., et al. 1978. Role of intestinal mucus in the absorption of quinine and water soluble dyes from rat small intestine. *Chem Pharm Bull (Tokyo)* 26:857.
109. Flemstrom, G., and E. Kivilaakso. 1983. Demonstration of a pH gradient at the luminal surface of rat duodenum *in vivo* and its dependence on mucosal alkaline secretion. *Gastroenterology* 84:787.
110. Shorrock, C.J., and W.D. Rees. 1988. Overview of gastroduodenal mucosal protection. *Am J Med* 84:25.
111. McEwan, G.T., B. Schousboe, and E. Skadhauge. 1990. Direct measurement of mucosal surface pH of pig jejunum *in vivo*. *Zentralbl Veterinarmed A* 37:439.
112. Shiau, Y.F., P. Fernandez, M.J. Jackson, and S. McMonagle. 1985. Mechanisms maintaining a low-pH microclimate in the intestine. *Am J Physiol* 248:G608.
113. Shiau, Y.F. 1990. Mechanism of intestinal fatty acid uptake in the rat: The role of an acidic microclimate. *J Physiol* 421:463.
114. Shiau, Y.F., R.J. Kelemen, and M.A. Reed. 1990. Acidic mucin layer facilitates micelle dissociation and fatty acid diffusion. *Am J Physiol* 259:G671.
115. Phillips, S.F. 1973. Integration of secretory and absorptive function. *Mayo Clin Proc* 48:630.
116. Crouthamel, W.G., L. Diamond, L.W. Dittert, and J.T. Doluisio. 1975. Drug absorption VII: Influence of mesenteric blood flow on intestinal drug absorption in dogs. *J Pharm Sci* 64:664.
117. Winne, D. 1980. Influence of blood flow on intestinal absorption of xenobiotics. *Pharmacology* 21:1.
118. Hober, R., and J. Hober. 1937. Experiments on the absorption of organic solutes in the small intestine of rats. *J Cell Comp Physiol* 10:401.
119. Schanker, L.S., P.A. Shore, B.B. Brodie, and C.A.M. Hogben. 1957. Absorption of drugs from the stomach I. The rat. *J Pharmacol Exp Ther* 120:528.
120. Hogben, C.A.M., L.S. Schanker, D.J. Tocco, and B.B. Brodie. 1957. Absorption of drugs from the stomach II. The human. *J Pharmacol Exp Ther* 120:540.
121. Landay, M.A., and H.E. Schroeder. 1979. Differentiation in normal human buccal mucosa epithelium. *J Anat* 128:31.
122. Meyer, J., J. Squier, and S.J. Gerson. 1984. *The structure and function of oral mucosa*. Oxford: Pergamon Press.
123. Kurosaki, Y., and T. Kimura. 2000. Regional variation in oral mucosal drug permeability. *Crit Rev Ther Drug Carrier Syst* 17:467.
124. Birudaraj, R., R. Mahalingam, X. Li, and B.R. Jasti. 2005. Advances in buccal drug delivery. *Crit Rev Ther Drug Carrier Syst* 22:295.
125. Ho, N.F.H., C.L. Barsuhn, P.S. Burton, and H.P. Merkle. 1992. Mechanistic insights to buccal delivery of proteinaceous substances. *Adv Drug Deliv Rev* 8:197.

126. Giannitsis, D.J., G.J. Giakoumakis, K.J. Louisos, and A. Antonopoulos. 1972. Aminopeptidasische eigenschaften der mundschleimhaut beim menschen (Properties of aminopeptidase of human buccal mucosa). *Enzymologia* 43:143.
127. Harris, D., and J.R. Robinson. 1992. Drug delivery via the mucous membranes of the oral cavity. *J Pharm Sci* 81:1.
128. Wheater, P.R., et al. 1993. *Wheater's functional histology, a text and colour atlas*, 3rd ed. Edinburgh: Churchill Livingstone.
129. Doluisio, J.T., et al. 1970. Drug absorption III: Effect of membrane storage on the kinetics of drug absorption. *J Pharm Sci* 59:72.
130. Martin, A., P. Bustamante, and A.H.C. Chun. 1993. *Physical pharmacy, physical chemical principles in the pharmaceutical sciences*, 4th ed. Philadelphia: Lea & Febiger.
131. Comer, J.E.A. 2004. High-throughput measurements of log *D* and pK_a. In *Drug bioavailability: Estimation of solubility, permeability, absorption and bioavailability*, eds. H. van de Waterbeemd, H. Lennernas, P. Artursson, R. Mannhold, H. Kubinyi, and G. Folkers, 21. Weinheim: Wiley-Verlag.
132. Avdeef, A. 2003. *Absorption and drug development, solubility, permeability and charge state.* Hoboken, NJ: John Willey.
133. Po, H.N., and N.M. Senozan. 2003. The Henderson–Hasselbalch equation: Its history and limitations. *J Chem Ed* 78:1499.
134. Perrin, D.D., B. Dempsey, and E.P. Serjeant. 1981. pK_a *Prediction for organic acid and bases.* London: Chapman & Hall.
135. Lodish, H., et al. 2000. Transport across cell membrane. In *Molecular cell biology*, 4th ed. New York: Freeman & Co.
136. Kellett, G.L. and P.A. Helliwell. 2000. The diffusive component of intestinal glucose absorption is mediated by the glucose-induced recruitment of GLUT2 to the brush-border membrane. *Biochem J* 350(Pt 1):155.
137. Gross, C.J., L.M. Henderson, and D.A. Savaiano. 1986. Uptake of L-carnitine, D-carnitine and acetyl-L-carnitine by isolated guinea-pig enterocytes. *Biochim Biophys Acta* 886:425.
138. Hunjan, M.K. and D.F. Evered. 1985. Absorption of glutathione from the gastro-intestinal tract. *Biochim Biophys Acta* 815:184.
139. Chow, S.L., and D. Hollander. 1978. Arachidonic acid intestinal absorption: Mechanism of transport and influence of luminal factors of absorption *in vitro*. *Lipids* 13:768.
140. Crane, R.K. 1965. Na-dependent transport in the intestine and other animal tissues. *Fed Proc* 24:1000.
141. Thorens, B. 1993. Facilitated glucose transporters in epithelial cells. *Ann Rev Physiol* 55:591.
142. Tamai, I., et al. 1995. Participation of a proton-cotransporter, MCT1, in the intestinal transport of monocarboxylic acids. *Biochem Biophys Res Commun* 214:482.
143. Gill, R.K., et al. 2005. Expression and membrane localization of MCT isoforms along the length of the human intestine. *Am J Physiol Cell Physiol* 289:C846.
144. Walton, J., and T.K. Gray. 1979. Absorption of inorganic phosphate in the human small intestine. *Clin Sci (Lond)* 56:407.
145. Dudeja, P.K., S.A. Torania, and H.M. Said. 1997. Evidence for the existence of a carrier-mediated folate uptake mechanism in human colonic luminal membranes. *Am J Physiol* 272:G1408.
146. Rose, R.C., et al. 1986. Transport and metabolism of vitamins. *Fed Proc* 45:30.
147. Tamai, I., and A. Tsuji. 1996. Carrier-mediated approaches for oral drug delivery. *Adv Drug Deliv Rev* 20:5.
148. Mizuma, T., et al. 2005. Intestinal SGLT1-mediated absorption and metabolism of benzyl beta-glucoside contained in *Prunus mume*: Carrier-mediated transport increases intestinal availability. *Biochim Biophys Acta* 1722:218.
149. Hashimoto, T., et al. Improvement of intestinal absorption of peptides: Adsorption of B1-Phe monoglucosylated insulin to rat intestinal brush-border membrane vesicles. *Eur J Pharm Biopharm* 50:197.
150. Fisher, R.B. 1981. Active transport of salicylate by rat jejunum. *Q J Exp Physiol* 66:91.
151. Nishimura, N., et al. 2004. Transepithelial permeation of tolbutamide across the human intestinal cell line, Caco-2. *Drug Metab Pharmacokinet* 19:48.

152. Wu, X., L.R. Whitfield, and B.H. Stewart. 2000. Atorvastatin transport in the Caco-2 cell model: Contributions of P-glycoprotein and the proton-monocarboxylic acid co-transporter. *Pharm Res* 17:209.
153. Utoguchi, N., and K.L. Audus. 2000. Carrier-mediated transport of valproic acid in BeWo cells, a human trophoblast cell line. *Int J Pharm* 195:115.
154. Tsuji, A., and I. Tamai. 1989. Na^+ and pH dependent transport of foscarnet via the phosphate carrier system across intestinal brush-border membrane. *Biochem Pharmacol* 38:1019.
155. Zimmerman, J. 1990. Folic acid transport in organ-cultured mucosa of human intestine. Evidence for distinct carriers. *Gastroenterology* 99:964.
156. Nussberger, S., and M.A. Hediger. 1995. How peptides cross biological membranes. *Exp Nephrol* 3:211.
157. Quay, J.F., and S. Foster. 1970. Cephalexin permeation of the surviving rat intestine. *J Physiol (Lond)* 269:241.
158. Brodin, B., C.U. Nielsen, B. Steffansen, and S. Frokjaer. 2002. Transport of peptidomimetic drugs by the intestinal di/tri-peptide transporter, PepT1. *Pharmacol Toxicol* 90:285.
159. Goldstein, J.L., R.G. Anderson, and M.S. Brown. 1982. Receptor-mediated endocytosis and the cellular uptake of low density lipoprotein. *Ciba Found Symp* 77.
160. Larrick, J.W., C. Enns, A. Raubitschek, and H. Weintraub. 1985. Receptor-mediated endocytosis of human transferrin and its cell surface receptor. *J Cell Physiol* 124:283.
161. Qian, Z.M., H. Li, H. Sun, and K. Ho. 2002. Targeted drug delivery via the transferrin receptor-mediated endocytosis pathway. *Pharmacol Rev* 54:561.
162. Fan, J.Y., et al. 1982. Receptor-mediated endocytosis of insulin: Role of microvilli, coated pits, and coated vesicles. *Proc Natl Acad Sci* 79:7788.
163. Carmel, R., et al. 1969. Vitamin B12 uptake by human small bowel homogenate and its enhancement by intrinsic factor. *Gastroenterology* 56:548.
164. Clark, M.A., B.H. Hirst, and M.A. Jepson. 2000. Lectin-mediated mucosal delivery of drugs and microparticles. *Adv Drug Deliv Rev* 43:207.
165. Duncan, R., M.J. Vicent, F. Greco, and R.I. Nicholson. 2005. Polymer–drug conjugates: Towards a novel approach for the treatment of endocrine-related cancer. *Endocr Relat Cancer* 12(Suppl. 1):S189.
166. Nori, A., and J. Kopecek. 2005. Intracellular targeting of polymer-bound drugs for cancer chemotherapy. *Adv Drug Deliv Rev* 57:609.
167. Fordtran, J.S., et al. 1965. Permeability characteristics of the human small intestine. *J Clin Invest* 44:1935.
168. Farquhar, M.G., and G.E. Palade. 1963. Junctional complexes in various epithelia. *J Cell Biol* 17:375.
169. Ballard, S.T., J.H. Hunter, and A.E. Taylor. 1995. Regulation of tight-junction permeability during nutrient absorption across the intestinal epithelium. *Ann Rev Nutr* 15:35.
170. Mrsny, R.J. 2005. Modification of epithelial tight junction integrity to enhance transmucosal absorption. *Crit Rev Ther Drug Carrier Syst* 22:331.
171. Madara, J.L. 1988. Tight junction dynamics: Is paracellular transport regulated? *Cell* 53:497.
172. Itoh, M., et al. 1999. Direct binding of three tight junction associated MAGUKs, ZO-1, ZO-2 and ZO-3 with the COOH termini of claudins. *J Cell Biol* 147:1351.
173. Matter, K., and M.S. Balda. 2003. Signalling to and from tight junctions. *Nat Rev Mol Cell Biol* 4:225.
174. Mrsny, R.J., et al. 2004. Identification of hepoxilin A3 in inflammatory events: A required role in neutrophil migration across intestinal epithelia. *Proc Natl Acad Sci* 101:7421.
175. Pappenheimer, J.R. 1987. Physiological regulation of transepithelial impedance in the intestinal mucosa of rats and hamsters. *J Membr Biol* 100:137.
176. Pappenheimer, J.R., and K.Z. Reiss. 1987. Contribution of solvent drag through intercellular junctions to absorption of nutrients by the small intestine of the rat. *J Membr Biol* 100:123.
177. Hochman, J., and P. Artursson. 1994. Mechanism of absorption enhancement and tight junction regulation. *J Contr Rel* 29:253.
178. Matter, K., and M.S. Balda. 2003. Functional analysis of tight junctions. *Methods* 30:228.

179. Anderberg, E.K., T. Lindmark, and P. Artursson. 1993. Sodium caprate elicits dilatation in human intestinal tight junctions and enhances drug absorption by the paracellular route. *Pharm Res* 10:857.
180. Tomita, M., M. Hayashi, and S. Awazu. 1995. Absorption-enhancing mechanism of sodium caprate and decanoylcarnitine in Caco-2 cells. *J Pharmacol Exp Ther* 272:739.
181. Baluom, M., M. Friedman, and A. Rubinstein. 2001. Improved intestinal absorption of sulpiride in rats with synchronized oral delivery systems. *J Contr Rel* 70:139.
182. Watanabe, K., T. Sawano, T. Jinriki, and J. Sato. 2004. Studies on intestinal absorption of sulpiride (3): Intestinal absorption of sulpiride in rats. *Biol Pharm Bull* 27:77.
183. Zaman, G.J., et al. 1994. The human multidrug resistance-associated protein MRP is a plasma membrane drug-efflux pump. *Proc Natl Acad Sci* 91:8822.
184. Allen, C.N., et al. 1990. Efflux of bis-carboxyethyl-carboxyfluorescein (BCECF) by a novel ATP-dependent transport mechanism in epithelial cells. *Biochem Biophys Res Commun* 172:262.
185. Zimmerman, J. 1992. Methotrexate transport in the human intestine. Evidence for heterogeneity. *Biochem Pharmacol* 43:2377.
186. Hunter, J., and B.H. Hirst. 1997. Intestinal secretion of drugs. The role of P-glycoprotein and related drug efflux systems in limiting oral drug absorption. *Adv Drug Deliv Rev* 25:129.
187. Cummins, C.L., W. Jacobsen, and L.Z. Benet. 2002. Unmasking the dynamic interplay between intestinal P-glycoprotein and CYP3A4. *J Pharmacol Exp Ther* 300:1036.
188. Kivisto, K.T., M. Niemi, and M.F. Fromm. 2004. Functional interaction of intestinal CYP3A4 and P-glycoprotein. *Fundam Clin Pharmacol* 18:621.
189. Benet, L.Z., et al. 1999. Intestinal MDR transport proteins and P-450 enzymes as barriers to oral drug delivery. *J Contr Rel* 62:25.
190. Aungst, B.J., et al. 1996. Enhancement of the intestinal absorption of peptides and non-peptides. *J Contr Rel* 41:91.
191. Swenson, E.S., and W.J. Curatolo. 1992. Intestinal permeability enhancement for proteins, peptides and other polar drugs: Mechanisms and potential toxicity. *Adv Drug Deliv Rev* 8:39.
192. Swenson, E.S., W.B. Milisen, and W. Curatolo. 1994. Intestinal permeability enhancement: Efficacy, acute local toxicity, and reversibility. *Pharm Res* 11:1132.
193. Amidon, G.L., H. Lennernas, V.P. Shah, and J.R. Crison. 1995. A theoretical basis for a biopharmaceutic drug classification: The correlation of *in vitro* drug product dissolution and *in vivo* bioavailability. *Pharm Res* 12:413.
194. Varma, M.V., et al. 2004. Biopharmaceutic classification system: A scientific framework for pharmacokinetic optimization in drug research. *Curr Drug Metab* 5:375.
195. Sutton, S.C., et al. 1993. Enhanced bioavailability of cefoxitin using palmitoylcarnitine. II. Use of directly compressed tablet formulations in the rat and dog. *Pharm Res* 10:1516.
196. Baluom, M., et al. 2000. Synchronous release of sulpiride and sodium decanoate from HPMC matrices: A rationale approach to enhance sulpiride absorption in the rat intestine over time. *Pharm Res* 17:1071.
197. Baluom, M., M. Friedman, and A. Rubinstein. 1998. The importance of intestinal residence time of absorption enhancer on drug absorption and implication on formulative considerations. *Int J Pharm* 176:21.

2 Enhancers for Enteral Delivery of Macromolecules with Emphasis on Surfactants

Biana Godin and Elka Touitou

CONTENTS

2.1 VARIOUS APPROACHES FOR GASTROINTESTINAL-PROMOTED ABSORPTION OF MACROMOLECULES

The oral absorption of biological macromolecules is one of the most challenging and extensively studied topics in drug delivery research. The main obstacles associated with oral administration and absorption of therapeutic macromolecules are the gastrointestinal (GI) membrane impermeability to the large hydrophilic molecules and their susceptibility to inactivation by GI enzymes. Various formulation approaches have been investigated to enhance GI delivery of biological macromolecules including permeability of GI membrane by use of surfactants, inhibition of enzymatic cleavage (reviewed in Chapter 5), enhancement of tight junction permeability, and use of carrier molecules and systems. This chapter will review tight junction modulators and various carriers used in oral drug delivery with a particular focus on the enhancing effects of surfactants on GI delivery.

The role of tight junctions in the paracellular permeability of protein and polypeptide drugs is the subject of intense ongoing research that was launched two decades ago. Generally, the absorption of small hydrophilic molecules is mainly believed to occur through a paracellular pathway, with the exception of those molecules that are transported by active or facilitated mechanisms. The leakiness of cell-to-cell junctions, tight junctions, is mainly open to ions and small molecules with molecular radii <11 Å. A number of excellent reviews on the topic of intestinal absorption of solutes through the paracellular route have been published [1–3]. The macromolecules investigated for their paracellular transport include insulin [4], polyethylene glycol 4000, inulin [5], dextran 4000 [5,6], opioid peptide prodrugs [7], poly-D-glutamate

(PDGlu) [8], and proteolytic enzymes [9]. It is noteworthy that many studies have been performed *in vitro* in Caco-2 monolayers, an acceptable model for measuring paracellular transport of solutes.

Tight junctions are dynamic structures that can adapt to a variety of physiological and biochemical conditions and thus render themselves modifiable by absorption enhancers. One method to modify paracellular transport is by modulating the permeability of tight junctions to reversibly expand their pore diameter. Among the modifiers of the tight junction structure are calcium-chelating agents, phosphate esters, surfactants (medium-chain fatty acids, bile salts), and some cationic polymers (e.g., cationic chitosan derivatives) [10–14]. Calcium-chelating agents, ethylenediaminetetraacetate (EDTA) and egtazic acid (EGTA), perturbate tight junctions by depleting extracellular calcium [12]. These absorption promoters are not specific and were found to be toxic to the structure and functions of the GI membrane [13]. Other agents that are believed to act through the mechanism of calcium depletion are anionic polyacrylate derivatives, which form a complex between endogenous calcium from the intestinal cells and poly(acrylic acid) [3]. More recently, alternative tight junction modifiers have been proposed. These compounds enhance macromolecules delivery through different mechanisms. The most promising candidate among these is zonula occludens toxin (ZOT), which was proven not to be cytotoxic to GI cells [4,15–20]. ZOT interacts with specific receptors mainly in the small intestine and is thought to modify tight junctions through protein kinase C (PKC)-induced effects on the cell cytoskeleton. More specifically, ZOT induces a dose- and time-dependent PKC-α-related polymerization of actin filaments [16]. This process occurs at a toxin concentration of 10^{-13} M and is a prerequisite to the tight junctions opening. The toxin exerts its effect by interacting with a specific surface receptor that is present in the small intestine, but not in the colon [4].

Cationic chitosan derivatives, such as *N*-trimethyl chitosan chloride, increase the paracellular delivery of macromolecules due to the interaction between the positive charges of chitosan and anionic glycoproteins on the surface of the enterocytes. Other chitosans, e.g., chitosan hydrochloride, were reported to shift specific cations required for the efficient closing of tight junctions [21] or reorganize the protein structure of tight junctions [22]. This topic is reviewed in a separate chapter of this book (Chapter 3).

Other agents used to modify the permeability of tight junctions are Pz-pentapeptide (4-phenylazobenzyloxycarbonyl-L-Pro-L-Leu-Gly-L-Pro-D-Arg) [23], thiolated polymers [24], glucose solution [2], and synthetic peptides derived from E-cadherin or occludin [25]. It is noteworthy that the main concern in expanding the tight junction is the potential for systemic absorption of unwanted microbiological and immunological agents. More *in vivo* studies are required to confirm the effects of tight junction modifiers and to evaluate the suitability of introducing these compounds as enhancers in oral drug formulation.

Various carrier molecules and drug delivery systems were also tested to increase oral bioavailability of macromolecules. Among the drug delivery systems investigated to enable GI absorption of these challenging molecules are polymeric micro- and nanoparticles [26–35], polymeric hydrogels [36], modified liposomes [37–39], and cyclodextrins [40]. In one of the very recent works by the group of Peppas [35], smaller-sized insulin-loaded polymer (SS-ILP) microparticles with diameters of <53 μm composed of cross-linked poly(methacrylic acid) and poly(ethylene glycol) were found to be efficient carriers for the intestinal delivery of the protein. The mechanism of their action is based on a rapid pH-dependent insulin release in the intestine, enzyme-inhibiting effects, and mucoadhesiveness. In *in vivo* studies of diabetic rats, SS-ILPs were shown to significantly suppress the postprandial rise in blood glucose with continuous hypoglycemic effects following 3 times per day oral administration (Figure 2.1).

Interesting approaches to affect intestinal absorption include drug physicochemical modification by either chemical derivatization or the use of delivery agents, low molecular weight peptide-like compounds that form a complex with the permeant. Generally, the transport of

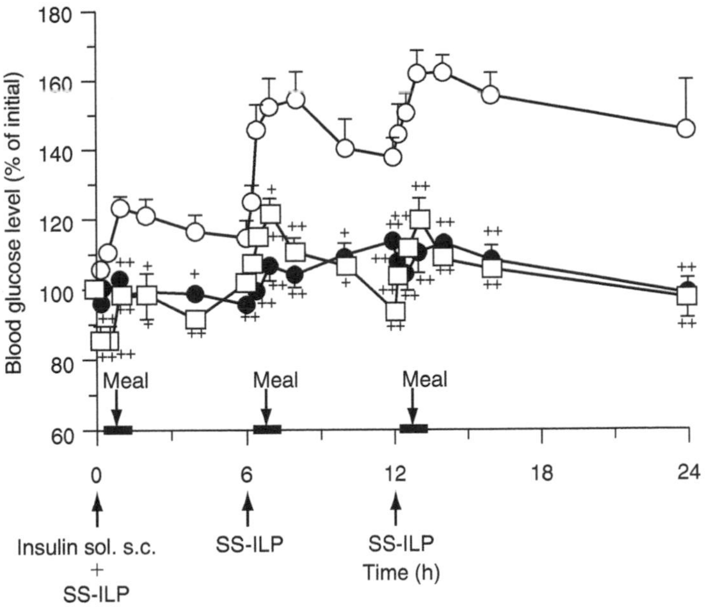

FIGURE 2.1 Changes in blood glucose level versus time profiles in type 1 diabetic rats following multiple oral administration of SS-ILP and subcutaneous insulin (only at the first dosing) (open square), insulin solution (open circles), and subcutaneous administration of insulin solution (closed circles). Insulin solution was used as control (open circles). The dose of insulin was 25 IU/kg (oral) and 0.1 IU/kg (subcutaneous) body weight. Each value represents mean ± SE (n = 5–10). Statistically significant difference from control: *$p < 0.05$; **$p < 0.01$. (From Morishita, M. et al., *J. Control. Release*, 110, 587, 2006. With permission from Elsevier.)

drugs across the GI membrane is highly affected by the intrinsic properties of the molecule such as ionic charge, secondary structure, 3D conformation, molecular weight (MW), and lipophilicity [41–43]. For instance, for small peptides it was reported that increased lipophilicity could shift the permeation pathway of the molecule from paracellular to transcellular [41]. Examples of molecular delivery agents that form a complex with the permeant drug increasing its lipophilicity include acylated non-α-amino acids such as *N*-[8-(2-hydroxybenzoyl) amino]caprylate (SNAC), *N*-α-deoxycholyl-L-lysyl-methyl-ester (DCK), and *N*-[8-(2-hydroxy-4-methoxy)bensoyl]amino caprylic acid (4-MOAC) [44–47]. It was suggested that noncovalent interactions between specific delivery agents and a protein cause conformational changes that are the basis for enhanced delivery. Although the exact mechanism and nature of the conformational change are not known, it was shown that the enhanced oral absorption using these delivery agent complexes is not related to the inhibition of intestinal enzymes by the complex or the damage to the GI membrane [45]. SNAC and MOAC developed by Emisphere Technologies were found to efficiently promote oral absorption of heparin, salmon calcitonin (sCT), insulin, parathyroid hormone (Figure 2.2), recombinant human growth hormone (rhGH), and interferon-α in rodents and primates. Emisphere plans to initiate the Phase III clinical study in the first half of 2006. This trial is designed to determine the safety and efficacy of oral heparin versus titrated sodium warfarin for the prevention of venous thromboembolism following elective total hip replacement [48].

The use of surfactants as intestinal absorption promoters is the most investigated approach. Furthermore, this chapter will review the enhancing effect of surfactants with various structures on the permeability of GI membrane to macromolecules.

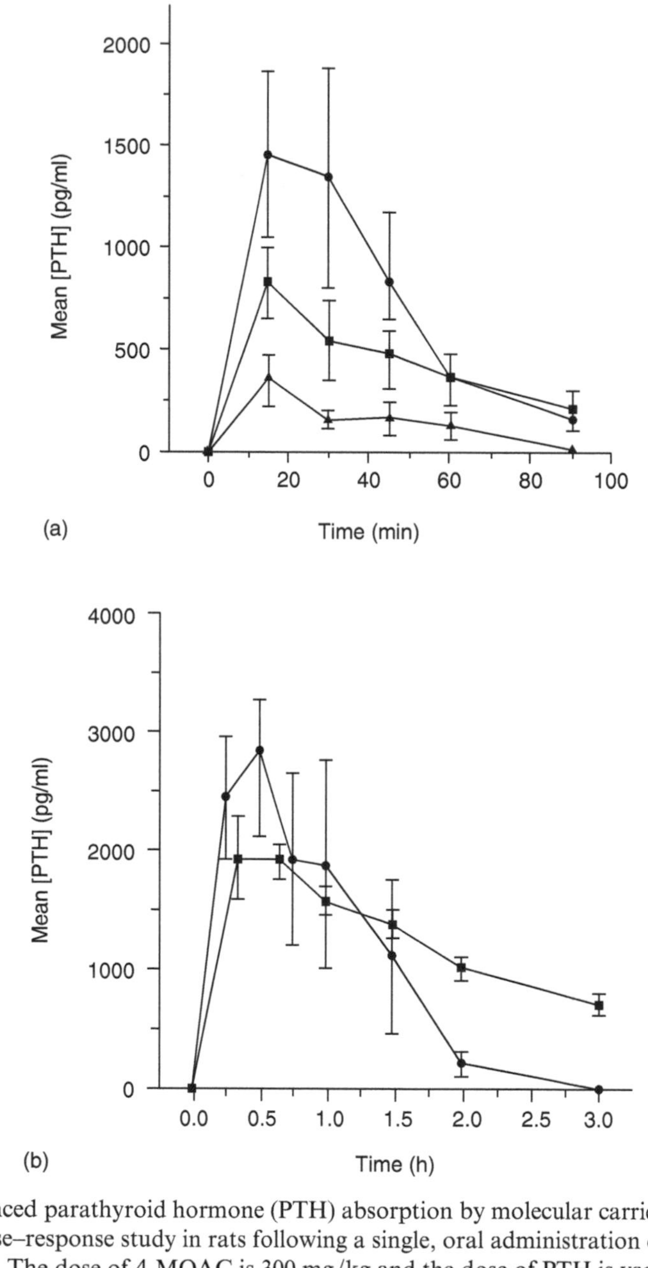

FIGURE 2.2 Enhanced parathyroid hormone (PTH) absorption by molecular carriers: (a) Oral delivery of PTH in rats. Dose–response study in rats following a single, oral administration of the 4-MOAC and PTH combinations. The dose of 4-MOAC is 300 mg/kg and the dose of PTH is varied. The PTH doses are 400 mcg/kg (circles), 200 mcg/kg (squares), and 100 mcg/kg (triangles). (b) Serum PTH concentration time profile in rhesus monkeys. The circles represent the response following a single, oral dose of 4-MOAC (200 mg/kg) in combination with PTH (400 mcg/kg) in water. The squares represent the response following a single, subcutaneous injection of PTH (10 mcg/kg). The 4-MOAC-facilitated oral bioavailability of PTH relative to subcutaneous administration is 2.1%. (From Leone-Bay, A. et al., *Pharm. Res.*, 18, 964, 2001. With permission from Springer Science and Business Media.)

2.2 SURFACTANTS FOR ENHANCED GASTROINTESTINAL ABSORPTION OF MACROMOLECULES

Surfactants have been the most investigated chemicals to promote drug absorption from all body tracts. In this section, we will focus on work carried out from the early stages on the enhancing effects of surfactants on drug GI absorption as well as on their interactions with the GI membrane and their toxicity. Systems with multifactorial effects such as emulsions and microemulsions are not the focus of this review.

Numerous surface-active molecules have been studied as GI absorption promoters in a wide variety of testing conditions, including model membranes, everted intestinal sacs, tissue cultures, intestinal epithelia in diffusion chambers, intact animals, and humans. The physical properties of a chemical enhancer may be strongly dependent on the interactions with the endogenous GI components such as bile salts, pH, and bacteria. Thus the *in vitro* experiments on enhancing GI absorption are not necessarily predictive of the behavior of the promoter in animals or humans, and we will mainly focus on summarizing results from *in vivo* studies.

2.2.1 Early Studies

An extensive body of literature has been published on the effects of surface-active agents on drug absorption from the various sites of the GI tract. Early research mainly focused on the use of these agents with various chemical structures to promote the absorption of small molecules such as vitamin A, spironolactone, griseofulvin, estrogens, and benzoic acid. In these studies, excellently reviewed by Gibaldi and Feldman [49], the main factors that were considered to be responsible for improved GI absorption were the surfactant effects on drug solubility, dissolution, emulsification, micellar solubilization, and gastric emptying. Other reports suggested that surfactants promoted drug absorption following their direct effect on the GI membrane, resulting in a temporary change in its permeability [50–54]. In contradiction, Levy and Perala [55] reported that nonionic surfactants, polysorbate 80 and oleic acid, had no apparent effect on the rat's small intestinal absorption of salicylate, salicylamide, and 4-aminoantipyrine.

In the late 1960s, Engel and coworkers [56–58] proposed sulfated and sulfonated surfactants as absorption promoters for two biological macromolecules, heparin and insulin. In these studies, sodium lauryl sulfate (SLS) promoted the absorption of heparin but not of insulin intraduodenally administered in rats and dogs.

In 1978, two papers, one from Israel and one from Japan, published in the same issue of the *Journal of Pharmacy and Pharmacology* reported successful rectal insulin absorption by use of polyoxyethylene (POE) ether nonionic surfactants [59,60]. These reports led to further investigations on the use of surfactants for enhancing the absorption of macromolecules. Later in this chapter we will discuss the work done on the enteral delivery of large molecules with nonionic and anionic promoters, which have been the most investigated ones to date.

2.2.2 Gastrointestinal Absorption Enhancement by Nonionic Surfactants

Nonionic surfactants generally possess a polar head group and one or more nonpolar hydrocarbon (C_{10}–C_{18}) chains. The hydrophilic and hydrophobic moieties are generally linked by ether, ester, or amide bonds. Among the nonionic surfactants often investigated for enteral drug delivery are POE alkyl derivatives (esters, ethers), POE sorbitan fatty acid esters, caprylocaproyl macrogolglycerides, saturated polyglycolysed C_8–C_{18} glycerides, and others. The chemical structure and electrical neutrality of nonionic molecules impart a lower sensitivity to the presence of electrolytes and these surfactants were also reported as less toxic than the anionic ones [61,62].

As previously mentioned, POE ethers were the first nonionic surfactants investigated for enhanced GI delivery of proteins. In the late 1970s, Touitou et al. [59] reported that effective systemic absorption of insulin is promoted from a microenema containing POE 20–24 monocetyl ether (cetomacrogol 1000). In a further study, these authors showed that intrajejunal administration to diabetic rats of a solution containing insulin and cetomacrogol 1000 resulted in a significant reduction in blood glucose levels (BGL) [63]. The maximum hypoglycemic effect observed in this study, nearly 80% reduction in BGL, took place 2 h after administration. Additional POE nonionic surfactants were tested as enhancers for macromolecules rectal absorption [64]. The agents investigated in this work were POE sorbitan fatty acid esters (polysorbates 20, 21, 40, 60, 65, 80, 81), POE stearates (POE 8, 40, and 50 stearates), and POE alkyl ethers (POE (4) lauryl ether, POE (23) lauryl ether, POE (6) stearyl ether, POE (45) stearyl ether, and cetomacrogol 1000). Most of the tested surfactants coadministered with insulin were effective in promoting insulin absorption (Figure 2.3). In contrast, POE stearates administered with the hormone did not cause any effect. In this study, the time to peak effect and the persistence of effect for at least 4 h were common to all the active promoters. In the same work, the authors demonstrated a strong dependence of absorption extent on pH with the minimum effect near the isoelectric point of insulin (pH 5.2). The effect of surfactant was also strongly dependent on the microenema carrier used. For Carbopol bases, a systematic decrease in the ability to reduce BGL was observed with the increase in chain length of polysorbates from 12 to 18. The hypoglycemic effect was more prominent but not chain-length-dependent when insulin was introduced in a polyethylene glycol base [64].

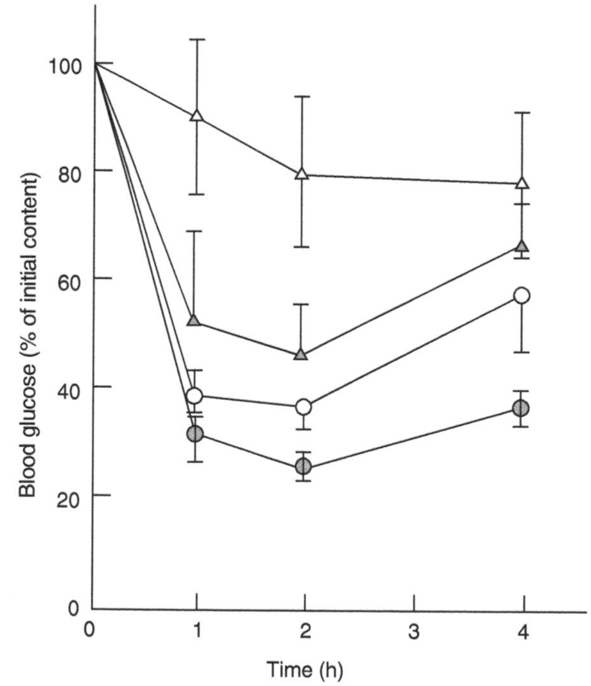

FIGURE 2.3 Hypoglycemic effects of insulin administered rectally to rats in PEG base containing POE monostearate (Myrj 45, open triangles), POE (45) stearyl ether (Texofor A6, close triangles), POE (20) sorbitol monolaurate (Polysorbate 20, open circles), and POE 20–24 monocetyl ether (cetomacrogol 1000, closed circles). (From Touitou, E., *J. Control. Release*, 21, 139, 1992. With permission from Elsevier.)

The results of this study emphasized that the pharmacodynamic activity of the rectally administered insulin is affected by the formulation factors such as the nature and concentration of surfactant used, the type of the base, pH, and the insulin dose. The hypoglycemic effect following enteral absorption of insulin coadministered with cetomacrogol 1000 was evidenced in other studies [63,65]. An interesting approach for promoting insulin absorption via targeted delivery took into consideration the simultaneous and competitive processes of insulin inactivation by GI enzymes and its absorption. In this work, insulin was combined with cetomacrogol 1000 and introduced intrajejunally to rats. The BGL reduction was maintained for 4 h with a peak at 2 h after intrajejunal administration. The GI administration of surfactants prior to insulin administration also resulted in a significant hypoglycemic effect [63]. Another work on cetomacrogol 1000 indicated that this surfactant significantly increased the absorption of gentamicin (GM), an aminoglycoside antibiotic poorly absorbed from the GI tract, after oral and rectal administrations [66]. The oral administration was less effective in promoting the absorption of GM than the rectal suppository, whereas a surfactant dose-dependent effect on drug absorption was observed. The above work evidenced the GI absorption promoting effect of nonionic surfactants [67].

Guzman and Garcia [68] evaluated the effect of POE (20) oleyl ether (Brij 99), POE (23) lauryl ether (Brij 35), and POE (20) cetyl ether (Brij 58) on insulin oral absorption in rabbits. In this work, both glucose and insulin levels in blood were measured at different time intervals following the *per os* administration of granules containing the surfactant along with stearic acid. Although a significant decrease in BGL was obtained after oral administration of each kind of granule containing the protein, only the composition containing Brij 58 (similar in its structure to cetomacrogol 1000) showed an increase of blood insulin level (equivalent to one tenth of the change produced after hypodermal injection).

Extensive work on another hydrophilic nonionic surfactant, caprylocaproyl macrogolglyceride (Labrasol), was conducted by Takada's group [69–78]. Using this surfactant, the authors showed systemic absorption of various orally administrated drugs such as glycyrrhizin, insulin, gentamicin, low molecular weight heparin (LMWH), erythropoietin (EPO), and vancomycin hydrochloride. In one of these studies, the coadministration of 20 mg/kg LMWH and 30 mg/kg Labrasol resulted in an increase in plasma anti-Xa activity to a level above 0.2 IU/mL, which is the critical level for assessment of anti-Xa anticoagulant activity [74]. The authors examined the effect of saturated fatty acids on drug absorption and found that the enhanced LMWH absorption effect was acyl chain-length-dependent: $C_{10} = C_{12} > C_{14} > C_{16} > C_8 >$ or $= C_6$. In further work, formulations containing 200 IU/kg LMWH and 50–200 mg/kg Labrasol were administered to duodenum, jejunum, and ileum of fasted rats [71] (Figure 2.4). Administration to jejunum resulted in 2–3 times higher plasma anti-Xa activity compared to duodenum and ileum. However, the absorption of LMWH was negligible when the drug was administered at 0.5 or 1 h post-Labrasol administration. These data suggest that intestinal membrane permeability changes induced by Labrasol were transient and reversible. These authors also reported the results on the effect of Labrasol on GM absorption, evaluating various aspects of GI delivery [70,72,78]. Colon administration of GM with Labrasol or polysorbate 80 resulted in an absolute bioavailability of 54.2% and 8.4%, respectively [78]. Interestingly, the *in vitro* findings of this study show that Labrasol also inhibited the intestinal secretory transport, thus inhibiting the efflux of GM from the enterocytes to the GI lumen.

A number of recent studies focused on GI delivery of challenging molecules using nano- and microparticulate solid adsorbents in the presence of nonionic surfactants [75,76]. Since some nonionic surfactants studied to improve intestinal absorption are liquids or administered as aqueous systems, the presence of an adsorbent permits formulating them within a common solid dosage form, a tablet, whereas reducing the risk of drug degradation

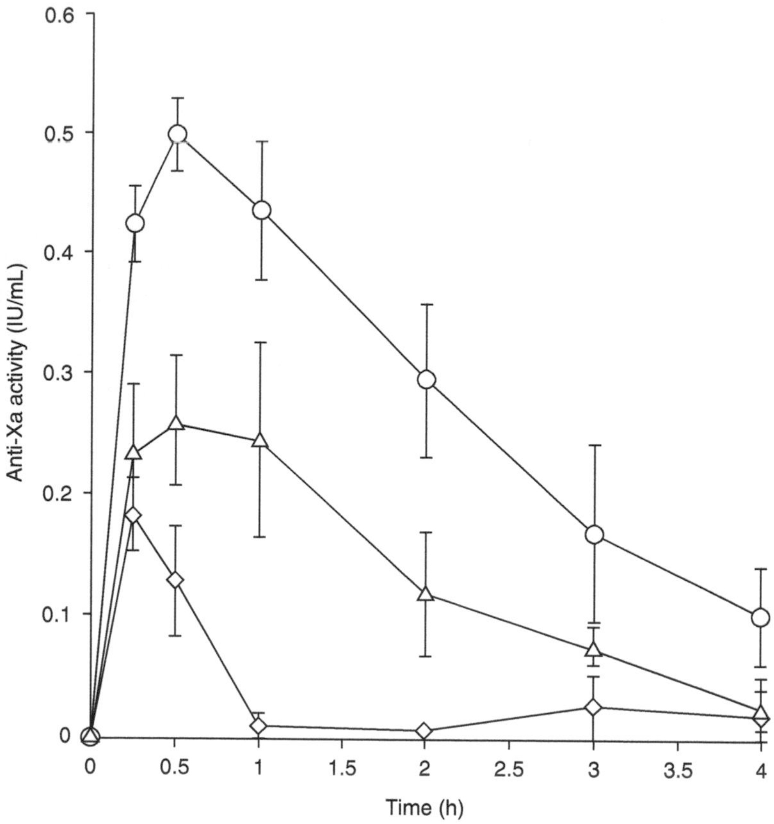

FIGURE 2.4 Plasma anti-Xa activity versus time profiles following administration of LMWH formulation to rat duodenum (rhombus), jejunum (circles), and ileum (triangles). LMWH was dissolved in water and Labrasol was added to it. The ratio of water and Labrasol was 1:1. The dose of LMWH was 200 IU/kg and that of Labrasol was 50 mg/kg. Values are the mean $\pm$ SE of three animals. (From Rama Prasad, Y.V. et al., *Int. J. Pharm.*, 271, 225, 2004. With permission from Elsevier.)

and inactivation. Porous silicon dioxide, carbon nanotubes, carbon nanohorns, fullerene, charcoal, microporous calcium silicate, magnesium aluminometasilicate, and bamboo charcoal were used as solid adsorbents for surfactants. Results obtained in a recent work on nanoparticulate adsorbents as drug delivery carriers for the administration of EPO have shown that the most prominent absorption was obtained with carbon nanotubes. Among the tested surfactants, Labrasol at a dose of 50 mg/kg showed the highest absorption-enhancing effect with jejunum (site with relatively low enzymatic activity) as the best EPO absorption site. Addition of an enzyme inhibitor, casein, further improved the drug plasma levels [77].

A surprising finding was that the enhanced effect of surfactants on the GI absorption of macromolecules could be extended to virus particles, which have a size in the order of magnitude of 100 Å. It was shown that macrogol ether surfactants enhance the absorption of viruses from the rectum. In this work, the *Salmonella* 0–1 of Felix virus incorporated into polyethylene glycol base containing cetomacrogol 1000 was administered rectally to healthy rabbits. Virus particles were detected at high concentrations in the animal blood (2.8×10^6/mL blood), 10 min after the rectal administration of the cetomacrogol 1000 suspension containing virus particles (10^9/mL) [79].

2.2.3 Anionic Surfactants as Promoters of Gastrointestinal Absorption

Bile salts, naturally occurring anionic surfactants in the GI tract, are involved in the process of fat digestion and absorption in mammals. These surface-active molecules are synthesized in the animal liver from cholesterol, stored in the gall bladder and, when required for digestion, delivered to the duodenum in the form of mixed micelles, also containing cholesterol and phospholipids [80]. The physiological concentrations of conjugated bile salts in the upper jejunum of the fasting human range between 5 and 10 mM (values above the critical micelle concentration, cmc), and increase to about 40 mM in the process of digestion.

The common bile salts in humans are glycine and taurine conjugates of sodium deoxycholate (NaDOC) and sodium chenodeoxycholate (CDOC), dihydroxy bile salts, and sodium cholate (NaC, the trihydroxy bile salt). The ability of bile salts to increase transmucosal transport of solutes has been frequently stated [62,81,82]. Generally, the more hydrophobic dihydroxy bile salts act as more effective absorption enhancers in comparison to trihydroxy bile salts. For example, Gullikson et al. [83] have reported that the absorption of inulin, dextran, and albumin in the perfused rat jejunum was enhanced with dihydroxy but not with trihydroxy bile salts.

During the past few decades, a number of important studies were published on the effect of cholic acid derivatives on the absorption of macromolecules [81]. Guarini and Ferrari [84–86] compared simultaneous oral dosing of NaDOC and heparin to pretreatment with NaDOC by oral gavage in dogs followed by oral heparin administration at a 0.5–24-h interval. In all pretreatment regimens, NaDOC enhanced heparin absorption, with the maximum effect observed when heparin was administered 1 h after NaDOC.

Hosny et al. [87–90] systematically studied the effect of bile salts on the GI absorption of insulin. In one of these works, insulin (10 IU/kg) suppositories with deoxycholic acid (DCA), NaC, NaDOC, sodium taurocholate (NaTC), and sodium taurodeoxycholate (NaTDOC) were formulated to investigate the effect of the surfactants on the plasma glucose concentration of diabetic beagle dogs as compared to insulin subcutaneous injections [87]. Of the bile salts or acids studied alone, suppositories containing 100 mg of either NaTC or NaTDOC produced the highest area under the curve (AUC) and relative hypoglycemia (RH) of about 50%. This effect was equivalent to that caused by a mixture of NaDOC (100 mg) and NaC (50 mg), and was dependent on the dose of insulin. However, the addition of NaC (50 mg) did not further improve the hypoglycemic effect of NaTC. Other studies by the same research group revealed that oral administration of insulin from capsules coated with Eudragit S100, a pH-dependent soluble polymer, and containing 20 or 50 mg NaC, produced a surfactant dose-related decrease in the BGL in alloxan-hyperglycemic rabbits [88,89]. At the same time, the capsules containing 100 mg surfactant did not significantly ($p > 0.05$) improve the hypoglycemic effect of insulin more than the smaller dose (50 mg/capsule). In a further study, capsules containing 20 IU insulin and 50 mg sodium salicylate with or without NaC produced a significant reduction in plasma glucose level to 82% and 73% of initial values at 2 and 3 h after administration, respectively. The BGL slowly returned to normal values after 5 h. The $AUC_{0-5\ h}$ was 74 ± 44 mg h/dL compared to 242 ± 71 mg h/dL for subcutaneous insulin administration (20 IU) with a relative hypoglycemia of 30%. A higher dose of oral insulin (40 IU) and sodium salicylate (50 mg) was more effective in reducing plasma glucose level, which steadily decreased and reached 56% of the initial value by 5 h ($AUC_{0-5\ h}$ 132 ± 42 mg h/dL and RH 27%). NaC (50 mg), however, only slightly improved sodium salicylate effect, producing an $AUC_{0-5\ h}$ of 139 ± 37 mg h/dL with relative hypoglycemia of 29% [89]. The results of a recent pilot study on seven fasted insulin-dependent diabetic patients showed that the suppositories containing 100 mg of NaC and 200 IU of insulin produced a maximum reduction in plasma glucose levels of 48% at t_{max} of 1.5 h as compared to C_{max} of 51% at

t_{max} of 3 h after subcutaneous injection of 20 IU insulin. The coadministration of insulin with the surfactant was able to abolish the significant rise in plasma glucose levels after meal, suggesting the possibility to use the insulin–NaC suppository as an effective buffering agent against meal-related hyperglycemia [90].

The efforts of a number of studies concentrated on elucidating the mechanism of permeation enhancement by bile salts. For this purpose, work was carried out *in vitro* and *in vivo* on model animals [91–93]. Bile salts are considered to act by increasing the permeability of the lipid bilayer membranes of the epithelial cells and paracellular transport in the alimentary tract. These mechanisms, in general, involve (a) increased permeability of GI membrane [91,92], (b) inhibition of intestinal proteolytic enzymes [94], and (c) opening transcellular tight junctions by forming complexes with calcium ions [1,93]. In *in vitro* studies it was shown that at low concentrations (below cmc), bile salts are intercalated between phospholipids present in membranes of intestinal cells, probably as dimers or tetramers [91], resulting in increased membrane permeability. At concentrations above cmc, processes such as membrane disruption and solubilization generally occur as will be discussed later [92]. Other possible contributions to the permeation enhancement process by bile salts could be related to the micellization of intestinal contents and enhanced paracellular transport due to increased tight junction permeability [1,49,93].

Beside the bile salts, the anionic surfactants investigated for enhanced intestinal delivery were mainly sodium salts of fatty acids and their derivatives. These include sodium salts of saturated and unsaturated fatty acids (C_8–C_{18}), SLS, dioctyl sodium sulfossuccinate (DOSS), and others.

The enhancement of insulin absorption by sodium caprate (C_{10}) in a closed loop model in rats showed the following rank of efficiency: colon > ileum > jejunum > duodenum. On the other hand, sodium glycocholate affected the intestinal permeability in a different way. With this surfactant, the order of permeation enhancement was colon > jejunum > duodenum > ileum [95]. These results show that the absorption site in the GI tract may strongly affect the effectiveness of permeation enhancer.

It is known that the proteolytic degradation takes place mainly in small intestine and the proteolysis in the colon is relatively low. Touitou and Rubinstein [96], in their study on GI absorption of insulin, combined two approaches: lowering enzymatic degradation by targeting the dosage form to the colon and enhanced delivery by using a combination of sodium laurate with cetyl alcohol (1:4). In this study, small gelatin capsules containing 8 IU porcine insulin with or without the surfactant combination were coated with a mixture of various ratios of Eudragits RS, S, and L. Apparently this was the first dosage form, containing insulin and enhancers, formulated as Eudragit-coated capsules to release the protein as a function of pH. The results of this study demonstrated that significant reduction in BGL (up to 45%) was measured with the insulin capsule containing the surfactant (Figure 2.5). This approach to combine permeation enhancement with prevention of enzymatic cleavage was further explored. Other authors used bile salts and proteinase inhibitor to promote intestinal absorption of two proteins, biologically active insulin, and pancreatic RNase [97,98].

Among the challenging active agents tested with anionic surfactants have been oligonucleotides, which are reported to possess poor intestinal permeability due to unfavorable physicochemical properties (strong charge, high MW). The effect of sodium caprate on the oral absorption of two chemically modified antisense oligonucleotides ISIS 2503 (phosphorothioate) and ISIS 104838 (methoxyethyl-modified phosphorothioate) was studied in an intraintestinal-catheterized pig model [99,100]. At all tested enhancer doses (25, 50, 100 mg/kg), both nucleotides were rapidly absorbed showing short t_{max} and plasma life of 10 min and up to 60 min, respectively, whereas the permeation enhancer dose had no effect on the pharmacokinetic parameters (Figure 2.6). To obtain insights on the relationship between the absorption

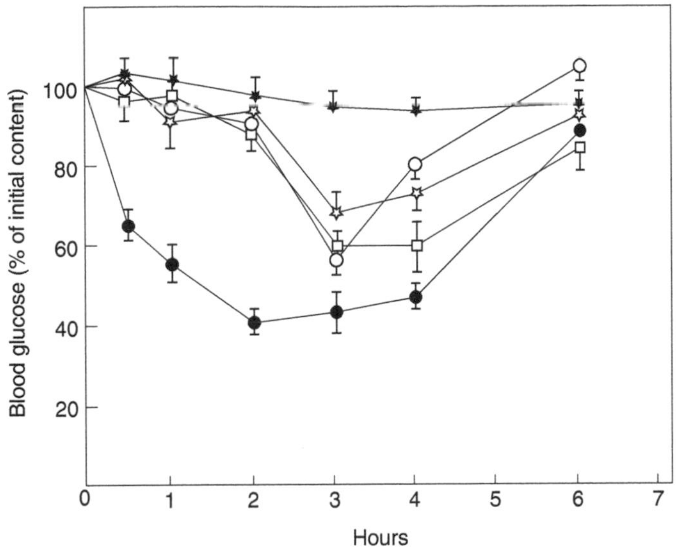

FIGURE 2.5 Hypoglycemic effect of insulin administered orally to normal rats by means of coated soft capsules containing an absorption-enhancing formulation (8 IU porcine insulin, 4 mg sodium laurate (C_{12}) and 16 mg cetyl alcohol): two capsules RSl (open stars); 2 capsules RS2 (open circles); 2 capsules RS2 + 1 capsule surfactant post-insulin administration (open square); insulin i.p. 4 IU (close circles); 2 placebo capsules (no insulin) (close stars). Each point is the mean ± SD of five animals for insulin administration and of four animals for controls. RS1 and RS2 are capsules coated with various mixtures of Eudragits RS, S, and L. (From Touitou, E. and Rubinstein, A., *Int. J. Pharm.*, 30, 95, 1986. With permission from Elsevier.)

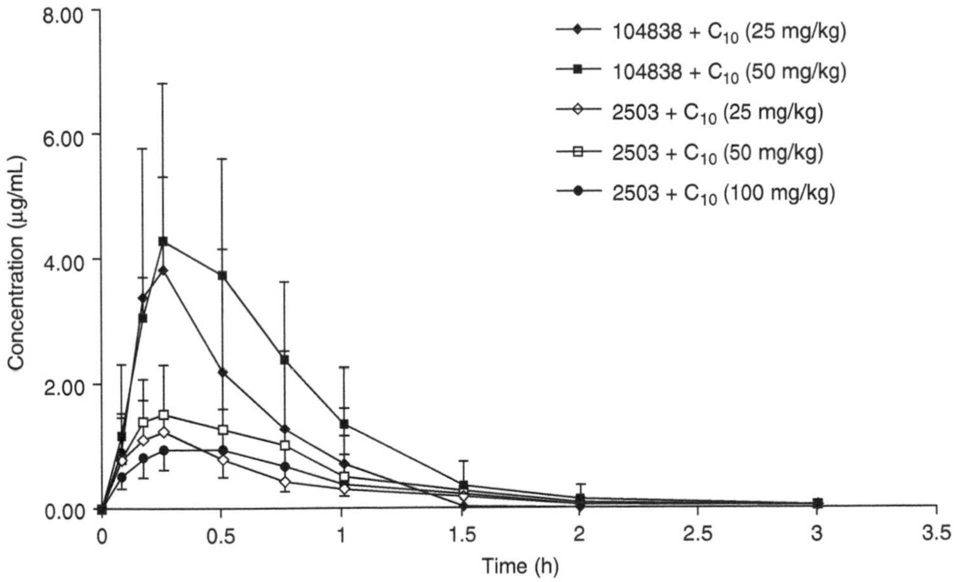

FIGURE 2.6 Mean plasma concentration–time profiles following intrajejunal administration of 10 mg/kg of nucleotides ISIS 2503 and ISIS 104838 with 25, 50, and 100 mg/kg sodium caprate (C_{10}) in pigs ($n = 8$). (From Raoof, A.A. et al., *Eur. J. Pharm. Sci.*, 17, 131, 2002. With permission from Elsevier.)

of the surfactant and its action as a membrane permeation enhancer, the authors investigated the plasma kinetics of sodium caprate. It was found that the surfactant was rapidly absorbed from jejunum ($t_{max} \sim 7$ min) with dose-dependent kinetics [99]. The rapid intrajejunal absorption of the enhancer observed in this work is in agreement with the results obtained in previously published studies in rats (colon) and in humans (rectum) [101], and may explain its fast mode of action.

For enhanced GI delivery, surfactants were also used in combination with other agents. For example, it was suggested that a synergistic permeation enhancement effect could be obtained by combining a surfactant with a mucolytic agent, which can reduce the mucus viscosity and facilitate the drug and enhancer diffusion close to GI membrane. A recent work examined the effect of combination of *N*-acetylcysteine, a mucolytic agent, combined with anionic or nonionic surfactants on the absorption of fluorescein isothiocyanate (FITC)-dextran (MW 4.4 kDa) in the rat jejunum [102]. Surfactants examined in this work were sodium caprate (C_{10}), SLS, NaTDOC, tartaric acid, and *p*-*t*-octyl phenol polyoxyethylene-9.5 (Triton X-100). The results of this study indicate that the bioavailability of this poorly absorbed high-MW hydrophilic compound could be enhanced up to 22.5-fold by using the surfactant–mucolytic agent combination. Although the detailed mechanism for such a drastic improvement is unclear, one possible reason proposed by the authors is the mucus disruption, which could improve the diffusivity of drug–surfactant micelles in the mucus layer.

2.2.4 Toxicity of Surfactants for Gastrointestinal Mucosa

The toxicity of permeation enhancers in the GI tract is one of the main concerns regarding their use in pharmaceutical products [81,103,104]. Numerous works on the effect of surfactants on the morphology and biochemical markers of intestinal wall have been published. It is generally agreed that ionic surfactants possess higher toxic potential than nonionic surfactants [61,62,105,106]. One of the first studies published in 1960 by Nissim [105] in *Nature* explored the effect of cationic, anionic, and nonionic surfactants on the histology of the mouse GI tract. In this study, significant pathological changes were observed following GI exposure to ionic surfactants but not with the nonionic surfactants. Feldman and Reinhard [106] studied the interactions of sodium alkyl sulfates (C_6–C_{14}) with the everted rat small intestinal membrane. Maximum effect on the loss of protein from the tissue was found with sodium decyl sulfate (10 mM), and salicylate permeability across the intestinal membrane was promoted by anionic and cationic surfactants but not by nonionics. These findings were further confirmed by Whitmore et al. [107], who reported that proteins and phospholipids were released from rat small intestine following its treatment with various surfactants. This last group concluded that surfactants of all classes are unlikely to significantly enhance membrane permeability without causing membrane damage. Ionic detergents in concentrations below cmc are generally well known for their ability to disrupt lipid membranes and to denature proteins. In concentrations above cmc, these surfactants dissolve lipid bilayers. Due to the toxic effect of anionic surfactants on GI mucosa, increasingly fewer reports on these initially widely investigated excipients for peroral administration have been published in recent years.

Although bile salts are endogenous substances in the GI tract, high concentrations of these compounds could cause prominent histological and functional damage to intestinal membranes. In fact, bile acids have long been associated with mucosal irritation and damage [82,108]. Bile salt malabsorption due to congenital disruption of Na^+–bile acid symport is associated with gross disturbance in colonic mucosal function and persistent diarrhea [108]. In such cases, two distinct processes may occur: a generalized increase in transintestinal permeability followed by a stimulation of salt and fluid secretion. Unconjugated primary and secondary bile acids may be particularly irritating [109]. Furthermore, dihydroxy bile acids such as CDOC and deoxycholic acid may cause mast cell degranulation after penetration through damaged mucosa [110]. Certain bile salts have been reported to cause acute local damage to intestinal wall. In Caco-2 and T84 cell monolayers, it was shown that cellular exposure and accumulation of cholate are deleterious to mucosal function and caused a dose- and time-dependent disruption of barrier function [92]. Shah et al. [111] evaluated the effect of time of exposure of NaC (3%), glycocholate (3%), glycosursodeoxycholate (3%), and polysorbate 80 (0.1%) on markers of Caco-2 cell viability (mannitol permeability, transepithelial electrical resistance measurements, DNA–propidium iodide staining assay, and tetrazolium salt-based assay). In this work, the toxicity increased directly as a result of an increase in the time of incubation. In other works, local toxicity of sodium glycocholate, NaTC, or NaDOC at concentration of 20 mM *in situ* in the rat small intestine [112] and colon [113] was estimated by assessing protein and phospholipid release. Among the tested surfactants NaDOC, the most powerful enhancer for phenol red absorption, caused the most significant release of protein and phospholipids at both administration sites. On the other hand, in this work NaTC enhanced phenol red absorption from the small intestine but caused little or no protein and phospholipids release [112]. Swenson et al. [104,114] investigated the absorption enhancement of phenol red versus mucosal damage caused by five surfactants: sodium dodecyl sulfate (SDS), NaTC, NaTDOC, polysorbate-80, and nonylphenoxypolyoxyethylene (NP-POE) with an average polar group size of 10.5 POE units. These authors suggested that transient local wall damage may be involved in the mechanism of permeability enhancement [104] (Figure 2.7). The toxic effects and the intestinal wall permeability measured in this study were reversible within 1–2 h after cessation of enhancer treatment. In a study evaluating the release of lactate dehydrogenase (LDH) and mucus as well as GI wall histological changes as potential markers of intestinal damage in a single pass *in situ* rat perfusion model, the release rate of LDH increased in the order of saline < Tween 80 < Triton X-100 in both jejunum and colon. In overall, tissue damage was approximately 3 times lower in the colon than in the jejunum. The rate of LDH release in the jejunum was concentration dependent and in a reverse relationship to the perfusion rate. The histological damage conformed to the results obtained for the release of mucus and the biochemical marker [115].

The intestinal membrane damage induced by absorption promoters is dependent on the structure and intrinsic properties of the surfactant, the concentration, and exposure time. The intrinsic physicochemical properties of surfactants, such as hydrophile–lipophile balance (HLB), surface tension reduction, and cmc can highly influence their permeability, enhancing their toxic characteristics during the contact with GI mucosa [114,116]. In a homologous series of nonionic NP-POE surfactants, only the chemicals possessing HLB < 17 (NP-POE 10.5 and 20) increased the permeability of the intestinal wall to phenol red, which was correlated with acute histological damage and release of biochemical markers (Figure 2.8). None of the above effects were reported for nonionic surfactants with HLB > 17 (NP-POE 30, 50, 100) [114].

2.3 CONCLUDING REMARKS

For the last three decades, intensive research on the GI delivery of macromolecules has been conducted. Various approaches have been tested to promote the absorption of high MW drugs. One of the earliest strategies, which is still under investigation, is the use of surfactants. Surfactants have been demonstrated in numerous studies to efficiently enhance the enteral bioavailability of polypeptides, polysaccharides, and nucleotides. However, a major limiting factor in the application of these promoters is the potential toxicity of the surfactants

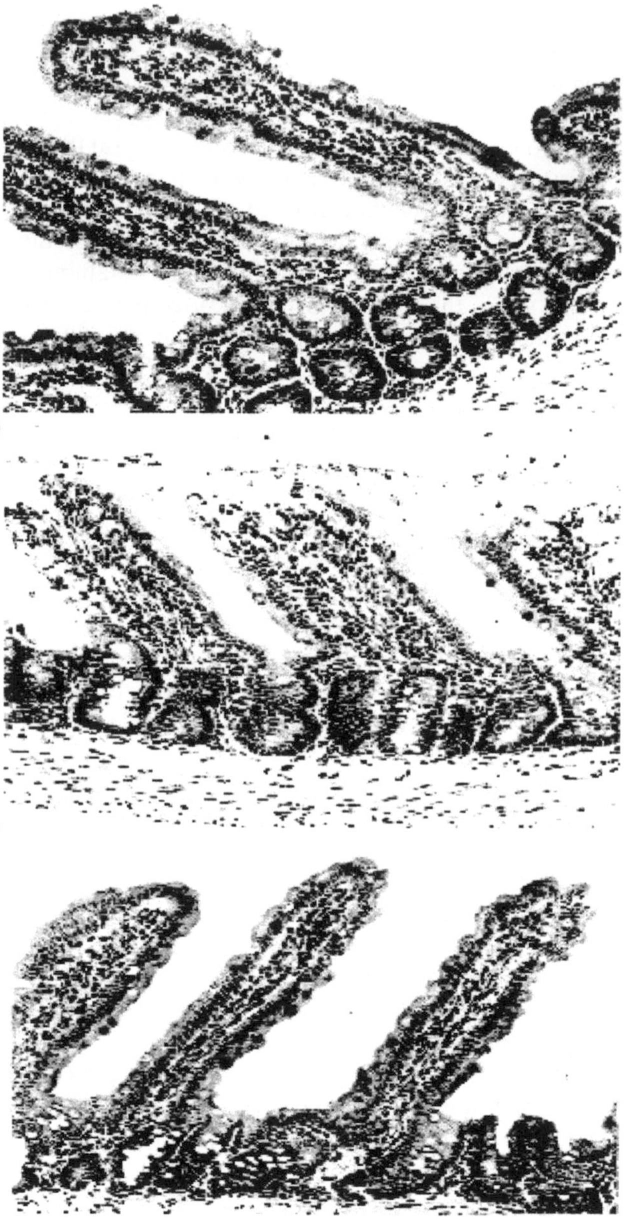

(a)

(b)

(c)

FIGURE 2.7 Light micrographs of rat intestinal villi after (a) 4 h perfusion with phenol red; (b) 1 h perfusion with phenol red plus 1% NP-POE 10.5; and (c) 1 h perfusion with phenol red plus 1% NP-POE 10.5 followed by a 3 h perfusion with phenol red alone. (From Swenson, E.S., Milisen, W.B., and Curatolo,W., *Pharm. Res.*, 11, 1132, 1994. With permission from Springer Science and Business Media.)

themselves. Although nonionic surfactants were reported to be less toxic than the ionic ones, it appears that the membrane changes caused by surface-active agents in the GI tract generally could not be separated from their absorption-enhancing ability.

Other approaches attempted to overcome the obstacle of poor GI absorption of macromolecules include the use of enzyme intestinal inhibitors, chemical derivatization, prodrugs,

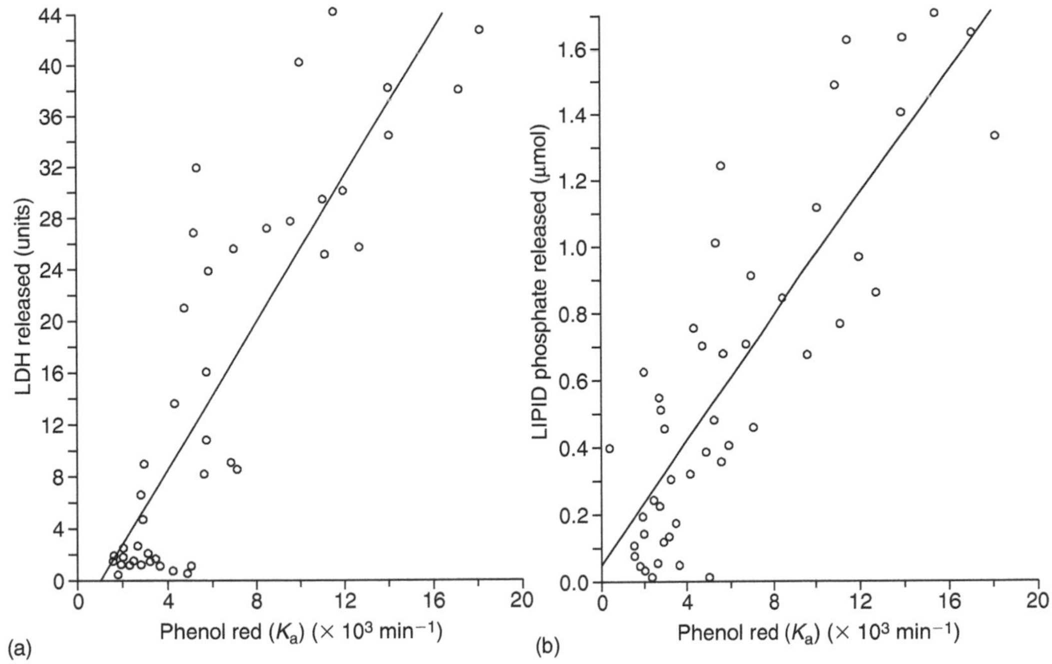

FIGURE 2.8 Relationship between phenol red (K_a) and LDH release (a) or liquid phosphate release (b) into the lumen for 1 h rat intestinal perfusion with phenol red and 1% w/v NP-POEs. The NP-POE consisted of fractions with 9, 10.5, 20, 30, 50, or 100 POE units or mixtures of NP-POE 10.5 and NP-POE 50. (From Swenson, E.S., Milisen, W.B., and Curatolo,W., *Pharm. Res.*, 11, 1501, 1994. With permission from Springer Science and Business Media.)

site-specific drug delivery, encapsulation of the drug in nano- and microparticulate carriers, selective opening of tight junctions, and the use of specially designed molecular carriers. Some of these approaches seem to be promising for the development of future pharmaceutical products. Phase II clinical study results on oral heparin absorption with the SNAC carrier have proven that the oral delivery of macromolecules has graduated from the proof-of-concept stage to the product-development stage.

REFERENCES

1. Salamat-Miller, N., and T.P. Johnston. 2005. Current strategies used to enhance the paracellular transport of therapeutic polypeptides across the intestinal epithelium. *Int J Pharm* 294:201.
2. Turner, J.R. 2000. Show me the pathway! Regulation of paracellular permeability by Na^+–glucose cotransport. *Adv Drug Deliv Rev* 41:265.
3. Nellans, H.N. 1991. Mechanism of peptide and protein absorption, paracellular intestinal transport: Modification of absorption. *Adv Drug Deliv Rev* 7:339.
4. Fasano, A., and S. Uzzau. 1997. Modulation of intestinal tight junctions by zonula occludens toxin permits enteral administration of insulin and other macromolecules in an animal model. *J Clin Invest* 99:1158.
5. Lane, M.E., C.M. O'Driscoll, and O.I. Corrigan. 1996. The relationship between rat intestinal permeability and hydrophilic probe size. *Pharm Res* 13:1554.
6. Tanaka, Y., et al. 1995. Characterization of drug transport through tight-junctional pathway in Caco-2 monolayer: Comparison with isolated rat jejunum and colon. *Pharm Res* 12:523.

7. Bak, A., et al. 1999. Acyloxyalkoxy-based cyclic prodrugs of opioid peptides: Evaluation of the chemical and enzymatic stability as well as their transport properties across Caco-2 cell monolayers. *Pharm Res* 16:24.
8. Salamat-Miller, N., et al. 2005. A randomly-coiled, high-molecular-weight polypeptide exhibits increased paracellular diffusion *in vitro* and *in situ* relative to the highly-ordered alpha-helix conformer. *Pharm Res* 22:245.
9. Bock, U., et al. 1998. Transport of proteolytic enzymes across Caco-2 cell monolayers. *Pharm Res* 15:1393.
10. Ward, P.D., T.K. Tippin, and D.R. Thakker. 2000. Enhancing paracellular permeability by modulating epithelial tight junctions. *Pharm Sci Technol Today* 3:346.
11. Sood, A., and R. Panchagnula. 2001. Peroral route: An opportunity for protein and peptide drug delivery. *Chem Rev* 101:3275.
12. Tomita, M., M. Hayashi, and S. Awazu. 1996. Absorption-enhancing mechanism of EDTA, caprate, and decanoylcarnitine in Caco-2 cells. *J Pharm Sci* 85:608.
13. Rosenblatt, D.E., D.G. Doyle, and A.L. Aronson. 1978. Calcium ethyl-enediaminetetraacetate (Ca EDTA) toxicity: Time- and dose-dependent studies on intestinal morphology in the rat. *Exp Mol Pathol* 28:215.
14. Junginger, H.E., and J.C. Verhoef. 1998. Macromolecules as safe penetration enhancers for hydrophilic drug-A fiction? *Pharm Sci Technol Today* 1:370.
15. Fasano, A. 1998. Novel approaches for oral delivery of macromolecules. *J Pharm Sci* 87:1351.
16. Fasano, A., et al. 1995. Zonula occludens toxin modulates tight junctions through protein kinase C-dependent actin reorganization, *in vitro*. *J Clin Invest* 96:710.
17. Fasano, A., et al. 1991. *Vibrio cholerae* produces a second enterotoxin, which affects intestinal tight junctions. *Proc Natl Acad Sci USA* 88:5242.
18. Salama, N.N., et al. 2004. The effect of delta G on the transport and oral absorption of macromolecules. *J Pharm Sci* 93:1310.
19. Wang, W., et al. 2000. Human zonulin, a potential modulator of intestinal tight junctions. *J Cell Sci* 113:4435.
20. Fasano, A. 2000. Regulation of intercellular tight junctions by zonula occludens toxin and its eukaryotic analogue zonulin. *Ann NY Acad Sci* 915:214.
21. Kotze, A.F., et al. 1999. Enhancement of paracellular drug transport with highly quaternized *N*-trimethyl chitosan chloride in neutral environments: *In vitro* evaluation in intestinal epithelial cells (Caco-2). *J Pharm Sci* 88:253.
22. Dodane, V., M. Amin Khan, and J.R. Merwin. 1999. Effect of chitosan on epithelial permeability and structure. *Int J Pharm* 182:21.
23. Yen, W.C., and V.H.L. Lee. 1995. Penetration enhancement effect of Pz-peptide, a paracellularly transported peptide, in rabbit intestinal segments and Caco-2 cell monolayers. *J Control Release* 36:25.
24. Bernkop-Schnurch, A., C.E. Kast, and D. Guggi. 2003. Permeation enhancing polymers in oral delivery of hydrophilic macromolecules: Thiomer/GSH systems. *J Control Release* 93:95.
25. Wong, V., and B.M. Gumbiner. 1997. A synthetic peptide corresponding to the extracellular domain of occludin perturbs the tight junction permeability barrier. *J Cell Biol* 136:399.
26. Mathiowitz, E., et al. 1997. Biologically erodable microspheres as potential oral drug delivery systems. *Nature* 386:410.
27. Carino, G.P., and E. Mathiowitz. 1999. Oral insulin delivery. *Adv Drug Deliv Rev* 35:249.
28. Ford, J., J. Woolfe, and A.T. Florence. 1999. Nanospheres of cyclosporin A: Poor oral absorption in dogs. *Int J Pharm* 183:3.
29. Lee, J.H., T.G. Park, and H.K. Choi. 1999. Development of oral drug delivery system using floating microspheres. *J Microencapsul* 16:715.
30. Damge, C., et al. 1997. Poly(alkyl cyanoacrylate) nanospheres for oral administration of insulin. *J Pharm Sci* 86:1403.
31. Pinto-Alphandary, H., et al. 2003. Visualization of insulin-loaded nanocapsules: *In vitro* and *in vivo* studies after oral administration to rats. *Pharm Res* 20:1071.

32. Peppas, N.A. 2004. Devices based on intelligent biopolymers for oral protein delivery. *Int J Pharm* 277:11.
33. Mesiha, M.S., M.B. Sidhom, and B. Fasipe. 2005. Oral and subcutaneous absorption of insulin poly(isobutylcyanoacrylate) nanoparticles. *Int J Pharm* 288:289.
34. Morcol, T., et al. 2004. Calcium phosphate–PEG–insulin–casein (CAPIC) particles as oral delivery systems for insulin. *Int J Pharm* 277:91.
35. Morishita, M., et al. 2006. Novel oral insulin delivery systems based on complexation polymer hydrogels: Single and multiple administration studies in type 1 and 2 diabetic rats. *J Control Release* 110:587.
36. Peppas, N.A., K.M. Wood, and J.O. Blanchette. 2004. Hydrogels for oral delivery of therapeutic proteins, *Expert Opin Biol Ther* 4:881.
37. Zhang, N., et al. 2005. Investigation of lectin-modified insulin liposomes as carriers for oral administration. *Int J Pharm* 294:247.
38. Al-Achi, A., and R. Greenwood. 1998. Erythrocytes as oral delivery systems for human insulin. *Drug Dev Ind Pharm* 24:67.
39. Katayama, K., et al. 2003. Double liposomes: Hypoglycemic effects of liposomal insulin on normal rats. *Drug Dev Ind Pharm* 29:725.
40. Shao, Z., et al. 1994. Cyclodextrins as mucosal absorption promoters of insulin. II. Effects of beta-cyclodextrin derivatives on alpha-chymotryptic degradation and enteral absorption of insulin in rats. *Pharm Res* 11:1174.
41. Knipp, G.T., et al. 1997. The effect of β-turn structure on the passive diffusion of peptides across Caco-2 cell monolayers. *Pharm Res* 14:1332.
42. Werner, U., T. Kissel, and W. Stuber. 1997. Effects of peptide structure on transport properties of seven thyrotropin releasing hormone (TRH) analogues in a human intestinal cell line (Caco-2). *Pharm Res* 14:246.
43. Camenisch, G., et al. 1998. Estimation of permeability by passive diffusion through Caco-2 cell monolayers using the drugs' lipophilicity and molecular weight. *Eur J Pharm Sci* 6:317.
44. Leone-Bay, A., et al. 1996. 4-[4-[(2-Hydroxybenzoyl)amino]phenyl]butyric acid as a novel oral delivery agent for recombinant human growth hormone. *J Med Chem* 39:2571.
45. Leone-Bay, A., et al. 2001. Oral delivery of biologically active parathyroid hormone. *Pharm Res* 18:964.
46. Leone-Bay, A., et al. 1998. Synthesis and evaluation of compounds that facilitate the gastrointestinal absorption of heparin. *J Med Chem* 41:1163.
47. Leone-Bay, A., et al. 1996. Oral delivery of rhGH: Preliminary mechanistic considerations. *Drug News Perspect* 9:586.
48. Emisphere technologies official Web site: http://www.emisphere.com.
49. Gibaldi, M., and S. Feldman. 1970. Mechanisms of surfactant effects on drug absorption. *J Pharm Sci* 59:579.
50. Feldman, S., M. Salvino, and M. Gibaldi. 1970. Physiologic surface-active agents and drug absorption. VII. Effect of sodium deoxycholate on phenol red absorption in the rat. *J Pharm Sci* 59:705.
51. Gouda, M.W., N. Khalafalah, and S.A. Khalil. 1977. Effect of surfactants on absorption through membranes V: Concentration-dependent effect of a bile salt (sodium deoxycholate) on absorption of a poorly absorbable drug, phenolsulfonphthalein, in humans. *J Pharm Sci* 66:727.
52. Malik, S.N., D.H. Canaham, and M.W. Gouda. 1975. Effect of surfactants on absorption through membranes III: Effects of dioctyl sodium sulfosuccinate and poloxalene on absorption of a poorly absorbable drug, phenolsulfonphthalein, in rats. *J Pharm Sci* 64:987.
53. Khalafallah, N., M.W. Gouda, and S.A. Khalil. 1975. Effect of surfactants on absorption through membranes IV: Effects of dioctyl sodium sulfosuccinate on absorption of a poorly absorbable drug, phenolsulfonphthalein, in humans. *J Pharm Sci* 64:991.
54. Utsumi, I., K. Kono, and Y. Takeuchi. 1973. Effect of surfactants on drug absorption. IV. Mechanism of the action of sodium glycocholate on the absorption of benzoylthiamine disulfide in the presence of sodium laurylsulfate and polysorbate 80. *Chem Pharm Bull* (*Tokyo*) 21:2161.

55. Levy, G., and A. Perala. 1970. Effect of polysorbate 80 and oleic acid on drug absorption from the rat intestine. *J Pharm Sci* 59:874.
56. Engel, R.H., and M.J. Fahrenbach. 1968. Intestinal absorption of heparin in the rat and gerbil. *Proc Soc Exp Biol Med* 129:772.
57. Engel, R.H., and S.J. Riggi. 1969. Effect of sulfated and sulfonated surfactants on the intestinal absorption of heparin. *Proc Soc Exp Biol Med* 130:879.
58. Engel, R.H., and S.J. Riggi. 1969. Intestinal absorption of heparin facilitated by sulfated or sulfonated surfactants. *J Pharm Sci* 58:706.
59. Touitou, E., M. Donbrow, and E. Azaz. 1978. New hydrophilic vehicle enabling rectal and vaginal absorption of insulin, heparin, phenol red and gentamicin. *J Pharm Pharmacol* 30:662.
60. Shichiri, M., et al. 1978. Increased intestinal absorption of insulin: An insulin suppository. *J Pharm Pharmacol* 30:806.
61. Whitmore, D.A., L.G. Brookes, and K.P. Wheeler. 1979. Relative effects of different surfactants on intestinal absorption and the release of proteins and phospholipids from the tissue. *J Pharm Pharmacol* 31:277.
62. Anderberg, E.K., C. Nystrom, and P. Artursson. 1992. Epithelial transport of drugs in cell culture. VII: Effects of pharmaceutical surfactant excipients and bile acids on transepithelial permeability in monolayers of human intestinal epithelial (Caco-2) cells. *J Pharm Sci* 81:879.
63. Touitou, E., M. Donbrow, and A. Rubinstein. 1980. Effective intestinal absorption of insulin in diabetic rats using a new formulation approach. *J Pharm Pharmacol* 32:108.
64. Touitou, E., and M. Donbrow. 1983. Promoted rectal absorption of insulin: Formulative parameters involved in the absorption from hydrophilic bases. *Int J Pharm* 15:13.
65. Bar-On, H., et al. 1981. Enteral administration of insulin in the rat. *Br J Pharmacol* 73:21.
66. Rubinstein, A., et al. 1981. Increase of the intestinal absorption of gentamicin and amikacin by a nonionic surfactant. *Antimicrob Agents Chemother* 19:696.
67. Touitou, E. 1992. Enhancement of intestinal peptide absorption. *J Control Release* 21:139.
68. Guzman, A., and R. Garcia. 1990. Effects of fatty ethers and stearic acid on the gastrointestinal absorption of insulin. *P R Health Sci J* 9:155.
69. Prasad, Y.V., et al. 2003. Enhanced intestinal absorption of vancomycin with Labrasol and D-alpha-tocopheryl PEG 1000 succinate in rats. *Int J Pharm* 250:181.
70. Rama Prasad, Y.V., et al. 2003. Evaluation of oral formulations of gentamicin containing Labrasol in beagle dogs. *Int J Pharm* 268:13.
71. Rama Prasad, Y.V., et al. 2004. *In situ* intestinal absorption studies on low molecular weight heparin in rats using Labrasol as absorption enhancer. *Int J Pharm* 271:225.
72. Hu, Z., et al. 2002. Diethyl ether fraction of Labrasol having a stronger absorption enhancing effect on gentamicin than Labrasol itself. *Int J Pharm* 234:223.
73. Eaimtrakarn, S., et al. 2002. Absorption enhancing effect of Labrasol on the intestinal absorption of insulin in rats. *J Drug Target* 10:255.
74. Mori, S., et al. 2004. Studies on the intestinal absorption of low molecular weight heparin using saturated fatty acids and their derivatives as an absorption enhancer in rats. *Biol Pharm Bull* 27:418.
75. Ito, Y., et al. 2005. Effect of adsorbents on the absorption of lansoprazole with surfactant. *Int J Pharm* 289:69.
76. Venkatesan, N., et al. 2005. Liquid filled nanoparticles as a drug delivery tool for protein therapeutics. *Biomaterials* 26:7154.
77. Venkatesan, N., et al. 2006. Gastro-intestinal patch system for the delivery of erythropoietin. *J Control Release* 111:19.
78. Hu, Z., et al. 2001. A novel emulsifier, Labrasol, enhances gastrointestinal absorption of gentamicin. *Life Sci* 69:2899.
79. Sechter, I., E. Touitou, and M. Donbrow. 1989. The influence of a non-ionic surfactant on rectal absorption of virus particles. *Arch Virol* 106:141.
80. Hofmann, A.F., and A. Roda. 1984. Physicochemical properties of bile acids and their relationship to biological properties: An overview of the problem. *J Lipid Res* 25:1477.

81. Swenson, E.S., and W.J. Curatolo. 1992. Means to enhance penetration: (2) Intestinal permeability enhancement for proteins, peptides and other polar drugs: Mechanisms and potential toxicity. *Adv Drug Deliv Rev* 8:39.
82. Chadwick, V.S., et al. 1979. Effect of molecular structure on bile acid-induced alterations in absorptive function, permeability and morphology in the perfused rabbit colon. *J Lab Clin Med Sci* 94:661.
83. Gullikson, G.W., et al. 1977. Effects of anionic surfactants on hamster small intestinal membrane structure and function: Relationship to surface activity. *Gastroenterology* 73:501.
84. Guarini, S., and W. Ferrari. 1985. Olive oil-provoked bile-dependent absorption of heparin from gastro-intestinal tract in rats. *Pharmacol Res Commun* 17:685.
85. Guarini, S., and W. Ferrari. 1985. Sodium deoxycholate promotes the absorption of heparin administered orally, probably by acting on gastrointestinal mucosa, in rats. *Experientia* 41:350.
86. Guarini, S., and W. Ferrari. 1984. Structural restriction in bile acids and non-ionic detergents for promotion of heparin absorption from rat gastro-intestinal tract. *Arch Int Pharmacodyn Ther* 271:4.
87. Hosny, E.A., H.I. Al-Shora, and M.M. Elmazar. 2001. Effect of different bile salts on the relative hypoglycemia of witepsol W35 suppositories containing insulin in diabetic beagle dogs. *Drug Dev Ind Pharm* 27:837.
88. Hosny, E.A., et al. 1997. Hypoglycemic effect of oral insulin in diabetic rabbits using pH-dependent coated capsules containing sodium salicylate without and with sodium cholate. *Pharm Acta Helv* 72:203.
89. Hosny, E.A., et al. 1998. Hypoglycemic effect of oral insulin in diabetic rabbits using pH-dependent coated capsules containing sodium salicylate without and with sodium cholate. *Drug Dev Ind Pharm* 24:307.
90. Hosny, E.A., et al. 2003. Evaluation of efficiency of insulin suppository formulations containing sodium salicylate or sodium cholate in insulin dependent diabetic patients. *Boll Chim Farm* 142:361.
91. Small, D.M. 1970. The formation of gallstones. *Adv Intern Med* 16:243.
92. Lowes, S., and N.L. Simmons. 2001. Human intestinal cell monolayers are preferentially sensitive to disruption of barrier function from basolateral exposure to cholic acid: Correlation with membrane transport and transepithelial secretion. *Pflugers Arch* 443:265.
93. Ward, P.D., T.K. Tippin, and D.R. Thakker. 2000. Enhancing paracellular permeability by modulating epithelial tight junctions. *Pharm Sci Technol Today* 3:346.
94. Yamamoto, A., E. Hayakawa, and V.H.L. Lee. 1990. Insulin and proinsulin proteolysis in mucosal homogenates of the albino rabbit: Implications in peptide delivery from nonoral routes. *Life Sci* 47:2465.
95. Morishita, M., et al. 1993. Site-dependent effect of aprotinin, sodium caprate, Na2EDTA and sodium glycocholate on intestinal absorption of insulin. *Biol Pharm Bull* 16:68.
96. Touitou, E., and A. Rubinstein. 1986. Targeted enteral delivery of insulin to rats. *Int J Pharm* 30:95.
97. Ziv, E., et al. 1994. Oral administration of insulin in solid form to nondiabetic and diabetic dogs. *J Pharm Sci* 83:792.
98. Ziv, E., O. Lior, and M. Kidron. 1987. Absorption of protein via the intestinal wall. A quantitative model. *Biochem Pharmacol* 36:1035.
99. Raoof, A.A., et al. 2002. Effect of sodium caprate on the intestinal absorption of two modified antisense oligonucleotides in pigs. *Eur J Pharm Sci* 17:131.
100. Raoof, A.A., et al. 2004. Oral bioavailability and multiple dose tolerability of an antisense oligonucleotide tablet formulated with sodium caprate. *J Pharm Sci* 93:1431.
101. Lennernas, H., et al. 2002. The influence of caprate on rectal absorption of phenoxymethylpenicillin: Experience from an *in-vivo* perfusion in humans. *J Pharm Pharmacol* 54:499.
102. Takatsuka, S., et al. 2006. Enhancement of intestinal absorption of poorly absorbed hydrophilic compounds by simultaneous use of mucolytic agent and non-ionic surfactant. *Eur J Pharm Biopharm* 62:52.

103. Muranishi, S. 1990. Absorption enhancers. *Crit Rev Ther Drug Carrier Syst* 7:1.
104. Swenson, E.S., W.B. Milisen, and W. Curatolo. 1994. Intestinal permeability enhancement: Efficacy, acute local toxicity, and reversibility. *Pharm Res* 11:1132.
105. Nissim, J.A. 1960. Action of some surface-active compounds on the gastro-intestinal mucosa. *Nature* 187:305.
106. Feldman, S., and M. Reinhard. 1976. Interaction of sodium alkyl sulfates with everted rat small intestinal membrane. *J Pharm Sci* 65:1460.
107. Whitmore, D.A., L.G. Brookes, and K.P. Wheeler. 1979. Relative effects of different surfactants on intestinal absorption and the release of proteins and phospholipids from the tissue. *J Pharm Pharmacol* 31:277.
108. Potter, G.D. 1998. Bile acid diarrhea. *Dig Dis* 16:118.
109. Casellas, F., et al. 1991. Human jejunal LTC4 response to irritant bile-acids. *Eur J Gastroenterol Hepatol* 3:393.
110. Quist, R.G., et al. 1991. Activation of mast cells by bile-acids. *Gastroenterology* 101:446.
111. Shah, R.B., A. Palamakula, and M.A. Khan. 2004. Cytotoxicity evaluation of enzyme inhibitors and absorption enhancers in Caco-2 cells for oral delivery of salmon calcitonin. *J Pharm Sci* 93:1070.
112. Yamamoto, A., et al. 1996. Effectiveness and toxicity screening of various absorption enhancers in the rat small intestine: Effects of absorption enhancers on the intestinal absorption of phenol red and the release of protein and phospholipids from the intestinal membrane. *J Pharm Pharmacol* 48:1285.
113. Uchiyama, T., et al. 1999. Enhanced permeability of insulin across the rat intestinal membrane by various absorption enhancers: Their intestinal mucosal toxicity and absorption-enhancing mechanism of *n*-lauryl-beta-D-maltopyranoside. *J Pharm Pharmacol* 51:1241.
114. Swenson, E.S., W.B. Milisen, and W. Curatolo. 1994. Intestinal permeability enhancement: Structure–activity and structure–toxicity relationships for nonylphenoxypolyoxyethylene surfactant permeability enhancers. *Pharm Res* 11:1501.
115. Oberle, R.L., T.J. Moore, and D.A. Krummel. 1995. Evaluation of mucosal damage of surfactants in rat jejunum and colon. *J Pharmacol Toxicol Meth* 33:75.
116. Xia, W.J., and H. Onyuksel. 2000. Mechanistic studies on surfactant-induced membrane permeability enhancement. *Pharm Res* 17:612.

3 Improvement of Oral Drug Absorption by Chitosan and Its Derivatives

Hiraku Onishi and Yoshiharu Machida

CONTENTS

3.1 INTRODUCTION

Chitin is a natural polysaccharide, which is widely and abundantly distributed among living organisms on the Earth. In particular, it is found in the exoskeleton of crustaceans, insects, and some fungi. Chitosan is a product of the alkaline hydrolysis of chitin, and a copolymer of *N*-acetyl-D-glucosamine and D-glucosamine. Generally, chitosan is 70%–100% deacetylated. It is soluble in acidic aqueous solutions, but insoluble in neutral and alkaline aqueous solutions [1,2]. These properties are due to the cationization of its amino groups, that is, the D-glucosamine residues exhibit a pK_a value of 6–7 [3]. Chitosan randomly deacetylated by 50%–60%, dissociated from rigid intra- or intermolecular hydrogen bonding, is soluble even in neutral and weakly alkaline aqueous solutions, and is known as water-soluble chitosan [4,5]. However, water-soluble chitosan is seldom used. Therefore, chitosan with 70%–100% deacetylation and a molecular weight (MW) of 10,000–1,000,000 is described in this chapter.

Chitosan has been the focus of reasearch as a pharmaceutical excipient due to its specific chemically and biologically favorable features [6–8]. As chitosan is soluble in acidic aqueous solutions, it can be processed under acidic conditions. By contrast, as the product made by chitosan is insoluble at neutral or basic pH, it behaves as a delivery system under such conditions. These chemical properties allow chitosan to control drug delivery. Further,

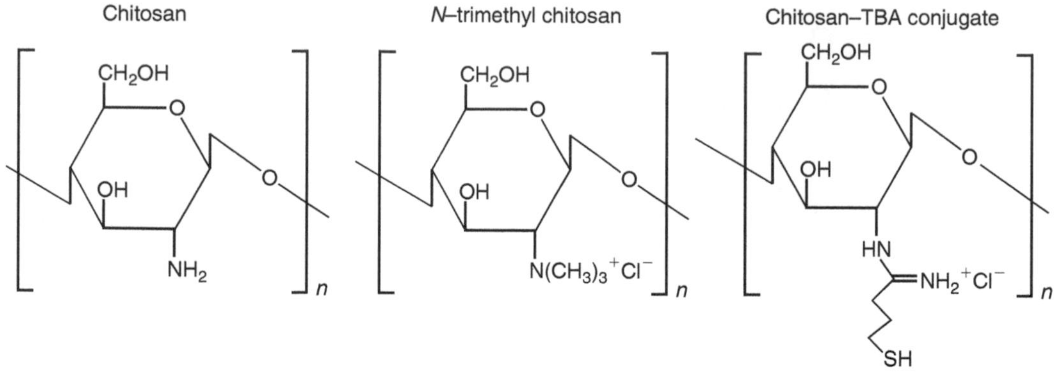

FIGURE 3.1 Chemical structures of chitosan, *N*-trimethyl chitosan hydrochloride, and chitosan–TBA conjugate.

chitosan is biocompatible, biodegradable, and mucoadhesive [4,9–12]. Recently, it was found that chitosan interacts with cells, which is related to changes in the characteristics of the cell membrane or uptake by cells performing phagocytosis [13–16]. Actually, chitosan is a safe polymer for oral administration. These chemical and biological features of chitosan have been utilized to develop drug delivery systems and enhance drug absorption. Furthermore, some chitosan derivatives have been found to be useful as modifiers or enhancers of drug absorption. Chitosan and its derivatives, in particular, *N*-trimethyl chitosan (TMC) and chitosan-TBA (4-thiobutylamidine) conjugate, are reviewed for their physicochemical and biopharmaceutical characteristics and their application to oral dosage forms (Figure 3.1).

3.2 ENHANCEMENT OF ABSORPTION OF SMALL MOLECULES

The controlled release of drugs has been developed using various excipients. Prolonged, time-controlled or site-specific drug release systems are useful for improvement of drug action [17–20]. Chitosan is recognized as a useful excipient for controlled drug release as a tablet, film, or granule formulation [21–24]. These formulations are useful for the control of the plasma concentration–time profiles of drugs. When the formulation allows a sustained drug release, the plasma concentration can be maintained longer, leading to prolonged pharmacological action. However, enhancement of drug absorption and elevation of bioavailability have not been major concerns in most of these studies. Chitosan has been utilized as a modifier of drug release. Actually, oral dosage forms with a simple sustained release cannot always improve drug absorption or prolong drug action. To achieve improvement or enhancement of drug action, it is essential to make dosage forms by taking into account the drug absorption site, gastrointestinal (GI) transition rate, interaction with the GI site, drug stability, etc. Because of their specific physicochemical and biological features (Table 3.1), chitosan and its derivatives allow us to produce such sophisticated formulations.

Drug release and retention at the site of absorption are very important for the enhancement of the bioavailability, particularly when the absorption sites are localized to a certain area of the GI route. This control of the transition of the drug can be achieved using bioadhesive excipients. Chlorothiazide (CT), a diuretic and antihypertensive drug, was better orally absorbed when administered with mucoadhesive polymers [25]. The absorption of CT is considered to be saturable and site-specific, because a low dose is better absorbed, and a decreased stomach emptying rate and slow GI transition rate are better for increased absorption. As chitosan is a mucoadhesive polymer, the absorption of CT is expected to be enhanced

TABLE 3.1
Specific Physicochemical and Biological Characteristics of Chitosan Useful for Improvement of Drug Action or Enhancement of Drug Absorption

Characteristics	Utility
Soluble in acidic pH	Process
Complexation with anions	Carrier, film, matrix
Insoluble in neutral and basic pHs	Carrier, film, matrix
Swelling	Controlled release
Biocompatibility	Safety for administration
Biodegradability	Safety for administration
Mucoadhesion	Localization, transit rate control
Interaction with cell surface	Enhancement of transport

by keeping the drug around the absorption site for a prolonged period. It is considered favorable for the drug to be released in a sustained manner whereas the formulation is located at the absorption site.

3.2.1 Poly(vinyl alcohol)-Gel Spheres with Chitosan

Poly(vinyl alcohol)-gel spheres with chitosan (PVA-GS/Ch) or without chitosan (PVA-GS) were prepared to control the GI transit time of drugs, and their particles were 5–10 μm [26]. PVA-GS/Ch displayed a longer small-intestinal transit time than PVA-GS. The transit rate was considered to decrease by the adhesion of chitosan to the intestinal mucus layer. PVA-GS/Ch and PVA-GS were loaded with theophylline and ampicillin. These released the drugs in a similar manner. The drugs were released almost completely at 4 h after the start of the release test. Both the gel spheres containing theophylline exhibited a bioavailability similar to that of the theophylline solution. Also, the bioavailability of ampicillin was greater in PVA-GS/Ch than in the PVA-GS and ampicillin solution; the bioavailability of PVA-GS/Ch was approximately 150% of that of ampicillin solution (Table 3.2). As theophylline is rapidly absorbed in

TABLE 3.2
Comparison of Pharmacokinetic Profiles after Oral Administration of Ampicillin (50 mg/kg) to Rats or After Intraduodenal Administration of Cephradine (100 mg/kg) to Rats

Drug	Dosage Form	AUC ($0-\infty$) (μgh/mL)	MRT ($0-\infty$) (h)	B.A. (%)	Comparison of Plasma Concentration at 4 and 8 h (for Ampicillin) and 12 and 24 h (for Cephradine)
Ampicillin	Solution	6.91 ± 0.45	1.77 ± 0.07	31.6	
	PVA-GS	8.60 ± 0.46	4.08 ± 0.11	39.3	PVA-GS/Ch > PVA-GS > Solution
	PVA-GS/Ch	10.63 ± 0.63	4.79 ± 0.05	48.5	
Cephradine	Suspension	207.2 ± 29.5	5.6 ± 1.5	65	
	Plain MP	55.0 ± 9.8	4.5 ± 1.3	17.2	MP/Ch > Suspension > Plain MP
	MP/Ch	161.1 ± 55.1	12.7 ± 2.6	50.5	

all regions of the intestine, transit time and release rate minimally influenced the extent of drug absorption. On the other hand, the absorption extent of ampicillin, not absorbed well below the lower intestine, was considered to be affected by the transit time and drug release rate. Further, it is important for the maintenance of drug action to sustain the plasma concentration longer over the minimum effective concentration [27–29]. These suggested chitosans should be available for the control of transit rate of the microparticulate dosage forms, which are based on the mucoadhesive properties of chitosan.

3.2.2 Chitosan-Coated Ethylcellulose Microparticles

As described above, the absorption of β-lactam antibiotics is influenced by the GI transit time and drug release rate. The GI transit time can be controlled by chitosan included in the dosage form due to its mucoadhesion. When the microparticles display a sustained release of drugs with a short biological half-life, their sustained action is expected. However, for drugs having a limited area of absorption in the GI tract, such a prolonged release system is considered insufficient. This is because the sustained release system passes through such absorption areas before releasing a drug sufficiently. As the microparticles move fairly fast through the GI tract, a chitosan coating of the microparticles is considered to be useful to retain the particles longer in the upper part of the GI tract, leading to improved absorption profiles compared to the plain microparticles. It is well known that ethylcellulose microparticles (MP) are useful to achieve the sustained release of many drugs [30,31]. Here, MP-containing cephradine and ketoprfofen, the biological half-lives of which are short, were coated with chitosan [32,33]. The MP with chitosan (MP/Ch) and without chitosan (plain MP) were compared on the drug absorption profiles after intraduodenal administration.

MP/Ch moved fairly slowly through the GI tract after intraduodenal administration; even at 8 h postadministration, more than half of the dose remained in the upper or middle part of the small intestine. MP/Ch and plain MP loaded with cephradine showed a similar release profile, though the drug release rate was slow; i.e., only 15% of the dose was released at 4 h after the start of the release test. As a result, the plasma area under curve (AUC) (0–∞) values of MP/Ch and plain MP were lower than that of drug suspension. The plain MP achieved a lower plasma concentration than the drug suspension through the observation period of 24 h. On the other hand, MP/Ch maintained a fairly high plasma concentration after the plasma level reached a maximum. Actually, the plasma concentration was higher for MP/Ch than for the drug suspension at 12 h or more after the administration (Table 3.2). As the plasma concentration rather than the area under the plasma concentration–time curve is important for the pharmacological effect of antibiotics, MP/Ch is proposed as a good sustained release dosage form of β-lactam antibiotic [27–29]. It was found here that chitosan-coated MP was superior to the plain MP for absorption of cephradine.

When ketoprofen was loaded with chitosan-coated ethylcellulose microparticles and the plain ethylcellulose microparticles, similar findings were obtained. Although ketoprofen is well absorbed from the GI tract, chitosan appears to influence the absorption profile. Chitosan allowed the particles to make contact with the mucosal membrane and be retained at adhesive sites, which means that the drug remains in the small intestine longer. This facilitated the absorption of ketoprofen in chitosan-coated microparticles loaded with ketoprofen. Chitosan is considered to improve the absorption properties of plain microparticles [33].

3.3 ENHANCEMENT OF PEPTIDE ABSORPTION BY CHITOSAN

Chitosan not only displays mucoadhesive properties, but also acts as an absorption enhancer in the intestinal cell layer membrane [34,35]. The mucoadhesion raises the concentration of the drug around the adhesion site, and leads to the higher concentration gradient, which increases the absorption potential. Also, the resistance of the intestinal cell layer decreases with the addition of chitosan under weakly acidic conditions. This is based on the loosening of the tight junctions of the intestinal cell layer membrane. Further, the protection of the peptides from enzymatic hydrolysis in the GI tract is very important for the enhancement of the bioavailability. Adhesion of peptides to the mucosal membrane helps them to avoid exposure to the GI fluid, which is filled with hydrolytic enzymes. These characteristics of chitosan can facilitate the absorption of peptide drugs, which are generally less well absorbed from the intestinal membrane.

3.3.1 Chitosan-Coated Liposomes Containing Peptide Drugs

Multilamellar liposomes composed of dipalmitoylphosphatidylcholine (DPPC) and dicetyl phosphate (DCP) were prepared by the formation of a thin lipid film and subsequent sonication, and coated with chitosan (Ch) [36]. Liposomes with a size of approximately 5 μm were used in the experiment. The Ch-coated and plain liposomes were compared in terms of mucoadhesion to the rat stomach and intestinal parts. Although both the liposomes were less adhesive to the stomach, Ch-coated liposomes displayed much higher mucoadhesion to all the intestinal parts *in vitro* than the plain liposomes. The intestinal adhesion of the plain liposomes were minimal. Further, Ch-coated liposomes showed a great mucoadhesion to the intestine at acidic and neutral pH values. This was also confirmed by fluorescence microscopy when pyrene-loaded Ch-coated liposomes were used in the mucoadhesion test.

The insulin solution and Ch-coated liposomes and plain liposomes containing insulin were tested for their influence on the blood glucose level after intragastric administration to normal rats. When compared with the change in the blood glucose level caused by the subcutaneous administration of insulin solution, Ch-coated liposomes had a pharmacological effect of approximately 5% (Figure 3.2). The effect of the plain liposomes and insulin solution was

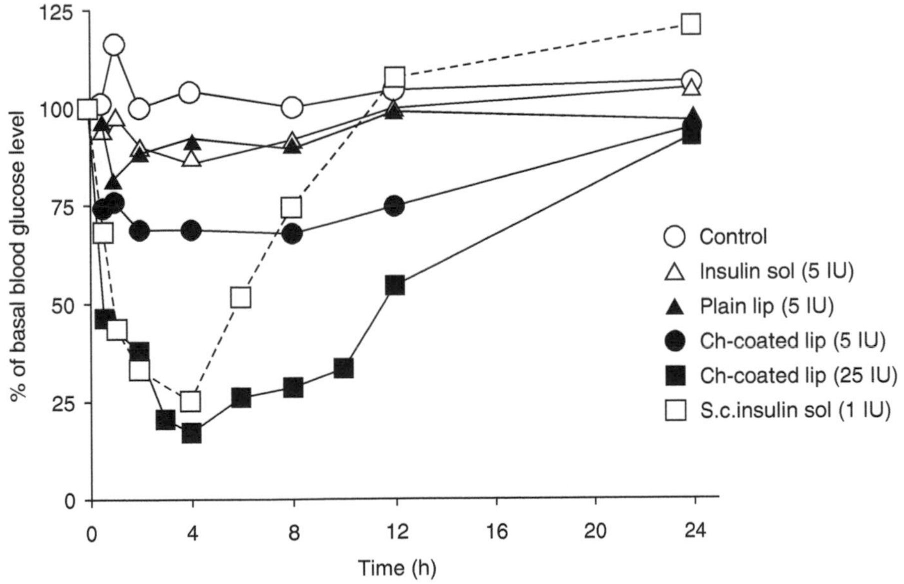

FIGURE 3.2 Blood glucose concentration–time profiles in normal rats after oral administration of insulin-containing chitosan-coated liposomes, insulin-containing plain liposomes and insulin solution, and subcutaneous administration of insulin solution. Sol, solution; Lip, liposomes; Ch, chitosan; s.c., subcutaneous. The results are expressed as the mean values.

less than 1%. Chitosan was found to effectively enhance the absorption of insulin from the liposomes. The mucoadhesive properties of chitosan were considered to facilitate the absorption through intestinal cell layer and prevent digestion by the hydrolytic enzymes.

In addition, insulin-containing Ch-coated liposomes at a size of approximately 100 nm were reported for their hypoglycemic potential after oral administration to normal mice [37]. The liposomes were prepared by the reverse phase evaporation technique. The coating of chitosan was obtained by incubating the liposomes with solutions of chitosan of various MWs at different concentrations. The decrease in the plasma blood glucose level after intragastrical administration was the greatest when the insulin-loaded liposomes were coated with 0.2% chitosan having an MW of 1000 kDa. Although the uncoated liposomes barely protected insulin from being digested by pepsin and trypsin, Ch-coated liposomes protected insulin effectively from proteolytic digestion. The degree of protection from the enzymatic hydrolysis of insulin was consistent with the hypoglycemic effect of Ch-coated liposomes containing insulin. This showed that the MW and concentration of chitosan had marked effects on the hypoglycemic potential of Ch-coated liposomes after oral administration.

3.3.2 Chitosan-Coated Nanoparticles Containing Peptide Drugs

Elcatonin is used as a peptide drug for the treatment of osteoporosis. The nanoparticles containing elcatonin were prepared using poly(DL-lactic acid-*co*-glycolic acid [3:1, mol/mol]) (PLGA) copolymer with an MW of 20,000 as a matrix polymer by the emulsion solvent diffusion method in oil. The nanoparticles were coated with Ch by the adsorption method. The nanoparticles prepared by this method had fairly high peptide levels (more than 1%) and were of nano-order size (ca. 650 nm) [38]. Ch-coated nanoparticles were efficiently coated with chitosan from the zeta potential. Ch-coated and uncoated nanoparticles showed a similar prolonged release of elcatonin. Therefore, Ch-coated nanoparticles were expected to sustain the actions of elcatonin through adhesion to the intestinal mucosa and prolonged drug release when administered orally. The pharmacological effects of Ch-coated and uncoated nanoparticles containing elcatonin and elcatonin solution were compared after intragastrical administration at a dose of 125–500 IU/kg to fasted rats. The elcatonin solution minimally reduced the blood calcium concentration, and uncoated nanoparticles lowered the blood calcium level significantly only at 500 IU/kg. On the other hand, Ch-coated nanoparticles reduced the blood calcium level significantly at 125–500 IU/kg over 24 h after the administration; in particular, at 500 IU/kg, the reduction in the plasma calcium level continued for over 36 h after administration. The Ch-coated nanoparticles showed good mucoadhesion to the rat intestine, and because of their small size are expected to penetrate the mucus layer. Further, the prolonged release properties of elcatonin were considered to cause the sustained action. In addition, Pan et al. reported that chitosan nanoparticles containing insulin, prepared by ionic gelation of chitosan with tripolyphosphate, were useful to improve the oral absorption of insulin [39].

3.4 ENHANCEMENT OF INTESTINAL ABSORPTION BY CHITOSAN DERIVATIVES

As described above, the bioadhesive characteristics of chitosan are readily available for the enhancement of intestinal absorption in combination with micro- and nanoparticulate dosage forms. In addition, chitosan lowers the resistance of the intestinal cell layer membrane. It is shown that the intercellular tight junction is loosened by the interaction of chitosan with the cell membrane. The attachment of chitosan to the intestinal cell surface appears to act on the cells from the outside and weaken the tight junction. The effect of chitosan on the cell surface was found to be reversible, namely, the barrier function of the intestinal layer membrane was recovered by washing chitosan from the intestinal surface. These characteristics of chitosan have been examined *in vitro* using Caco-2 cell monolayers as a model of the intestinal membrane [35]. Calcium chelators and surfactants have been applied as enhancers of intestinal absorption. However, they induce great change in conditions around the cells, leading to damage or an irreversible change in the cells. In addition, small molecular enhancers may enter the cells. Therefore, the application of these small molecular enhancers is considered to be limited. On the other hand, chitosan is a safe macromolecular enhancer of intestinal absorption as it is biocompatible and its effect on the integrity of tight junctions is reversible.

3.4.1 *N*-Trimethylchitosan as an Enhancer of Intestinal Absorption

Although chitosan enhances intestinal permeability, this can only be achieved under acidic pH conditions. At neutral or basic pH values, chitosan is not able to open the tight junctions of the intestinal cell layer membrane. Chitosan loses its cationic charge density at neutral or basic pH values, which causes precipitation within itself, resulting in a lack of interaction with the cell surface. TMC is synthesized by the reaction of chitosan and iodomethane under strong basic conditions at 60°C. The low (12.3%) and high (61.2%) trimethylated chitosans, named TMC-L and TMC-H, respectively, and chitosan hydrochloride (Ch-HCl) were compared with regard to their potential to enhance intestinal permeability [40]. In experiments on transepithelial electric resistance (TEER) and permeability to mannitol, Caco-2 cell monolayers were used. At pH 6.2, all the polymers were effective at reducing TEER and enhancing the rate of mannitol permeation. However, at pH 7.4, only TMC-H caused a reduction of TEER and enhancement of permeation (Figure 3.3). Therefore, the degree of N-trimethylation in TMC affected the function, and only the highly substituted TMC could act efficiently as an enhancer. In these experiments, no polymers induced the deterioration of the cells. These results suggested that the highly substituted TMC should be useful for the improvement of the intestinal absorption of poorly absorbed drugs.

The *in vivo* effect of TMC on intestinal absorption was investigated using the peptide drug buserelin (MW 1240), a metabolically stable LH–RH analogue. A 40% and a 60%

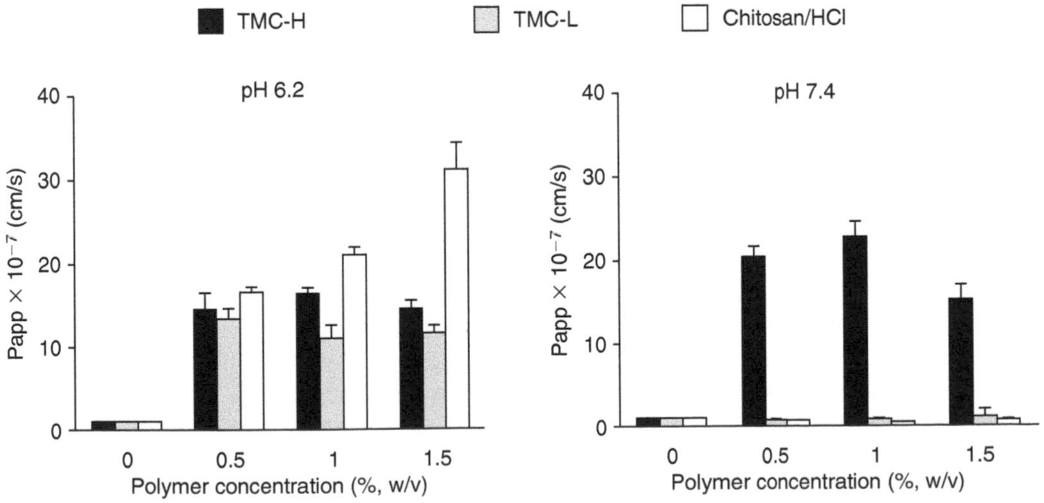

FIGURE 3.3 Effect of TMC-H, TMC-L, and chitosan hydrochloride on the permeability of [^{14}C] mannitol across Caco-2 cell monolayers. The results are expressed as the mean ± SD (n = 3).

trimethylated chitosan, named TMC40 and TMC60, respectively, were used in the *in vitro* and *in vivo* experiments at pH 7.2 [41]. In the *in vitro* permeability experiment at pH 7.2 using Caco-2 cell monolayers, the permeability to buserelin was very low, and chitosan barely altered the permeability. TMC40 and TMC60 gave enhancement ratios of 21 and 60 for the transport of buserelin as compared with the control (buserelin alone). After the intraduodenal administration (500 μg buserelin/animal) of buserelin alone and coadministration of buserelin with TMC40 and TMC60 (1%) to rats, the serum concentration of buserelin was analyzed by radioimmunoassay. Absorption was most enhanced by TMC60, followed by TMC40. The absolute bioavailability of buserelin alone was 0.8%, whereas it was 6% and 13% for buserelin with TMC40 and TMC60, respectively. No damage was observed to the intestinal cell layer membrane after the *in vitro* and *in vivo* experiments. These results suggested that highly substituted TMC should be useful for the enhancement of the intestinal absorption of peptide drugs.

3.4.2 Thiolated Chitosan as a Strongly Mucoadhesive Polymer and Its Applications

Many anionic and catonic mucoadhesive polymers exhibit adherence to the mucosal membrane via hydrogen bonding and ionic interaction. Recently, polymers immobilized with thiol groups have been shown to display an improved mucoadhesion. These polymers interact with mucus glycoproteins in the mucosal layers via a covalent disulfide bond. These characteristics have been drawing much attention for application to drug delivery systems. The polycarbophil–cysteine conjugates were prepared with water-soluble carbodiimide. The mucoadhesive properties were analyzed using porcine intestinal mucosa [42]. The mucoadhesive potential was strengthened by the more covalently attached thiol groups, and the total work of adhesion (TWA) of the conjugate with highly attached thiols was approximately twice that of unmodified polycarbophil. These results were caused by the formation of disulfide bridges between polymers and mucus glycoproteins. Furthermore, polycarbophil–cysteine conjugates displayed an enhancement of mucoadhesive absorption in combination with glutathione (GSH), when sodium fluorescein (FS) was used as a paracellular marker [43]. GSH alone enhanced the transport of FS markedly, but the oxidized form of GSH (GSSG) was less effective. The absorption-enhancing effects of GSH and GSSG increased in combination with polycarbophil–cysteine conjugate. The conjugate reduced GSSG to GSH. GSH inhibits protein tyrosine phosphatase (PTP) through the formation of a disulfide bond with the active site cysteine of the enzyme, resulting in a loosening of the tight junction. This is considered to be a mechanism for the enhancement of the paracellular transport of poorly permeating hydrophilic agents. The polycarbophil–cysteine conjugate adhered well to mucosa, leading to more GSH at the cell surface, which will promote paracellular absorption.

Chitosan was also thiolated in an attempt to enhance the mucoadhesion and intestinal absorption of poorly permeating drugs. Chitosan–TBA (4-thiobutylamidine) conjugates can be produced by simply mixing chitosan and 2-iminothiolane at pH 7. The mucoadhesive properties were investigated with intestinal porcine mucosa. Chitosan–TBA conjugates showed higher TWA values as more 2-iminothiolane was immobilized to the chitosan. The maximum detachment force (MDF) was much higher in chitosan–TBA conjugates than unmodified chitosans [44]. The conjugate with chitosan of MW 400 kDa and 264 μmol thiols/g showed the best mucoadhesion: its mucoadhesive potential indicated by TWA has been improved 100 times more as compared with that of unmodified chitosan (Table 3.3). When the permeation studies were performed using intestinal mucosa from guinea pigs, chitosan–TBA conjugate enhanced the permeation of rhodamine 123 more than unmodified chitosan. In particular, the effect was increased in combination with GSH [45].

TABLE 3.3
Effect of Molecular Weight of Chitosan and Its TBA Conjugate on Mucoadhesive Potentials

	Chitosan (MW, kDa)			Chitosan–TBA Conjugate (MW, kDa)		
Adhesive Parameter	**150**	**400**	**600**	**150**	**400**	**600**
TWA (mJ)	1.5 ± 0.8	4.0 ± 2.5	1.3 ± 0.0	37.2 ± 13.4	467.7 ± 107.1	45.2 ± 22.1
MDF (mN)	0.1 ± 1.3	6.0 ± 1.0	2.3 ± 1.5	26.6 ± 7.8	255.9 ± 29.3	55.6 ± 30.4

TWA, total work of adhesion; MDF, maximum detachment force.
Substitution degrees of chitosan–TBA conjugates: 270, 264, and 235 μmol/g for MW of 150, 400, and 600 kDa, respectively.
The results are expressed as the mean ± SD (n = 3–5).

Chitosan–TBA conjugate was further applied to the oral absorption of calcitonin via the stomach [46]. Drug delivery to the stomach offers several advantages because pepsin is the only major secreted protease and an enteric coating is not needed. Minitablets (5 mg) (Tablet A) containing 50 μg of calcitonin were prepared by a direct compression of salmon calcitonin, chitosan–TBA conjugate, chitosan–pepstatin A conjugate, and GSH. Chitosan–pepstatin A conjugate was prepared by using water-soluble carbodiimide to couple chitosan with a selective and strong pepsin inhibitor, pepstatine A. The minitablets (5 mg) (Tablet B) composed of unmodified chitosan, chitosan–pepstatin A conjugate, and salmon calcitonin (50 μg), and the minitablets (5 mg) (Tablet C) containing unmodified chitosan and salmon calcitonin (50 μg) were obtained in a similar manner. Tablet A and Tablet B released calcitonin fairly quickly, with more than 90% released at 4 h in the release test. Although unmodified chitosan did not protect calcitonin in the pepsin solution, chitosan–pepstatin A conjugate suppressed the degradation of calcitonin completely under similar conditions. The rats received each dosage form orally, and i.v. injection of salmon calcitonin (0.5 μg) was given to rats as a positive control. Tablet C and calcitonin solution (50 μg salmon calcitonin) caused no reduction in the plasma calcium level, whereas Tablets A and B reduced the plasma calcium concentration. The pharmacological efficacy was calculated by comparison of the area under the reduction in plasma calcium level with that given by i.v. injection of calcitonin. Tablet A and Tablet B displayed pharmacological efficacies of 1.35% and 0.41%, respectively; notably, Tablet A reduced the plasma calcium level significantly for more than 12 h after the administration. These tablets existed in the stomach at 4 h after administration, which was confirmed anatomically. The results suggested that chitosan–pepstatin A conjugate could enhance absorption to some extent, and that chitosan–TBA conjugate and GSH, because of their permeation-enhancing effect, should promote the gastric absorption of calcitonin.

3.5 PERSPECTIVES FOR THE USE OF CHITOSAN AND ITS DERIVATIVES FOR ORAL ADMINISTRATION

Chitosan is obtained from the abundant natural polymer chitin by simple alkaline treatment, and its biocompatible and biodegradable characteristics make it a useful medical and pharmaceutical excipient. As described above, chitosan adheres to the mucosal membrane and reduces the resistance of the tight junctions of the cell layer membrane [35,36]. These advantageous properties have been analyzed and applied to oral drug delivery systems. Because of its high MW, chitosan does not permeate cells or enter the systemic circulation, and its interaction with cells, inducing the loosening of tight junctions, is reversible.

Therefore, the enhancement of transport is not accompanied by damage to the cells, which is superior to the absorption-enhancing effects of other small molecules, which lead to irreversible changes in the cell membrane and damage to the cell. Thus, much attention has been paid to chitosan as a polymeric substance that enhances GI absorption.

Recently, some chitosan derivatives were found to act efficiently as enhancers of GI absorption. TMC improves the absorption-enhancing effect of chitosan [40,41]. This polymer overcomes the disadvantage of chitosan, i.e., its inability to enhance the paracellular transport at neutral or alkaline pH. TMC allows the intestinal absorption rate to increase, independent of the pH conditions, and does not damage cells. In addition, although the bioadhesive force of chitosan is not great, it is much improved by the immobilization of thiol groups to chitosan. Chitosan–TBA (4-thiobutylamidine) conjugate is one of the thiol-group introduced chitosans. Chitosan–TBA conjugate induces not only good mucoadhesion but also enhancement of paracellular drug transport in combination with GSH [44–46]. As these derivatives are safe polymers, they are expected to be examined further. Detailed pharmacological and pharmacokinetic analyses may lead to their clinical use. Chitosan, orally administered, could also be of use for targeted delivery to the colon or Peyer's patches [47,14–16]. These areas may not be related directly to the enhancement of GI absorption, but they are very important for the utilization of chitosan. The applicability of chitosan and its derivatives to oral administration including the enhancement of oral absorption will be pursued actively for practical use, and available oral dosage forms are expected to be produced in the near future.

REFERENCES

1. Wang, W., et al. 1991. Determination of the Mark–Houwink equation for chitosans with different degrees of deacetylation. *Int J Biol Macromol* 13:281.
2. Baxter, A., et al. 1992. Improved method for i.r. determination of the degree of N-acetylation of chitosan. *Int J Biol Macromol* 14:166.
3. Li, J., J.F. Revol, and R.H. Marchessault. 1996. Rheological properties of aqueous suspensions of chitin crystallites. *J Colloid Interface Sci* 183:365.
4. Onishi, H., and Y. Machida. 1999. Biodegradation and distribution of water-soluble chitosan in mice. *Biomaterials* 20:175.
5. Cho, Y.W., et al. 2000. Preparation and solubility in acid and water of partially deacetylated chitins. *Biomacromolecules* 1:609.
6. Illum, L. 1998. Chitosan and its use as a pharmaceutical excipient. *Pharm Res* 15:1326.
7. Singla, A.K., and M. Chawla. 2001. Chitosan: some pharmaceutical and biological aspects—an update. *J Pharm Pharmacol* 53:1047.
8. Hejazi, R., and M. Amiji. 2003. Chitosan-based gastrointestinal delivery systems. *J Control Release* 89:151.
9. Nakamura, F., et al. 1992. Lysozyme-catalyzed degradation properties of the conjugates between chitosans having some deacetylation degrees and methotrexate. *Yakuzaigaku* 52:59.
10. Ichikawa, H., et al. 1993. Evaluation of the conjugate between N4-(4-carboxybutyryl)-1-beta-D-arabinofuranosylcytosine and chitosan as a macromolecular prodrug of 1-beta-D-arabinofuranosylcytosine. *Drug Des Discov* 10:343.
11. Lehr, C.M., et al. 1992. *In vitro* evaluation of mucoadhesive properties of chitosan and some other natural polymers. *Int J Pharm* 78:43.
12. Soane, R.J., et al. 1999. Evaluation of the clearance characteristics of bioadhesive systems in humans. *Int J Pharm* 178:55.
13. Carreno-Gomez, B., and R. Duncan. 1997. Evaluation of the biological properties of soluble chitosan and chitosan microspheres. *Int J Pharm* 148:231.
14. Van der Lubben, I.M., et al. 2001. Chitosan microparticles for oral vaccination: preparation, characterization and preliminary *in vivo* uptake studies in murine Peyer's patches. *Biomaterials* 22:687.

15. Van Der Lubben, I.M., et al. 2001. *In vivo* uptake of chitosan microparticles by murine Peyer's patches: visualization studies using confocal laser scanning microscopy and immunohistochemistry. *J Drug Target* 9:39.
16. Van der Lubben, I.M., et al. 2002. Transport of chitosan microparticles for mucosal vaccine delivery in a human intestinal M-cell model. *J Drug Target* 10:449.
17. El Khodairy, K.A., et al. 1992. Preparation and *in vitro* evaluation of slow release ketoprofen microcapsules formulated into tablets and capsules. *J Microencapsul* 9:365.
18. Katikaneni, P.R., et al. 1995. Ethylcellulose matrix controlled release tablets of a water-soluble drug. *Int J Pharm* 123:119.
19. Parikh, N.H., S.C. Porter, and B.D. Rohera. 1993. Aqueous ethylcellulose dispersion of ethylcellulose. I. Evaluation of coating process variables. *Pharm Res* 10:525.
20. Shen, Z., and S. Mitragotri. 2002. Intestinal patches for oral drug delivery. *Pharm Res* 19:391.
21. Kawashima, Y., et al. 1985. Novel method for the preparation of controlled-release theophylline granules coated with a polyelectrolyte complex of sodium polyphosphate-chitosan. *J Pharm Sci* 74: 264.
22. Kawashima, Y., et al. 1985. Preparation of a prolonged release tablet of aspirin with chitosan. *Chem Pharm Bull* 33:2107.
23. Inouye, K., Y. Machida, and T. Nagai. 1987. Sustained release tablets based on chitosan and carboxymethylcellulose sodium. *Drug Des Deliv* 1:297.
24. Miyazaki, S., et al. 1988. Sustained release of indomethacin from chitosan granules in beagle dogs. *J Pharm Pharmacol* 40:642.
25. Longer, M.A., H.S. Ch'ng, and J.R. Robinson, 1985. Bioadhesive polymers as platforms for oral controlled drug delivery III: oral delivery of chlorothiazide using a bioadhesive polymer. *J Pharm Sci* 74:406.
26. Sugimoto, K., et al. 1998. Evaluation of poly(vinyl alcohol)-gel spheres containing chitosan as dosage form to control gastrointestinal transit time of drugs. *Biol Pharm Bull* 21:1202.
27. Singhvi, S.M., A.F. Heald, and E.C. Schreiber. 1978. Pharmacokinetics of cephalosporin antibiotics: protein-binding considerations. *Chemotherapy* 24:121.
28. Soback, S., et al. 1987. Clinical pharmacokinetics of five oral cephalosporins in calves. *Res Vet Sci* 43:166.
29. Hoffman, A., et al. 1998. Pharmacodynamic and pharmacokinetic rationales for the development of an oral controlled-release amoxicillin dosage form. *J Control Release* 54:29.
30. Uchida, T., et al. 1987. Preparation and evaluation of ethyl cellulose microcapsule containing cefadroxil or cephradine. *Yakuzaigaku* 47:254.
31. Yamada, T., H. Onishi, and Y. Machida. 2001. Sustained release ketoprofen microparticles with ethylcellulose and carboxymethylethylcellulose. *J Control Release* 75:271.
32. Takishima, J., H. Onishi, and Y. Machida. 2002. Prolonged intestinal absorption of cephradine with chitosan-coated ethylcellulose microparticles in rats. *Biol Pharm Bull* 25:1498.
33. Yamada, T., H. Onishi, and Y. Machida. 2001. *In vitro* and *in vivo* evaluation of sustained release chitosan-coated ketoprofen microparticles. *Yakugaku Zasshi* 121:239.
34. Lehr, C.M., et al. 1992. Effects of the mucoadhesive polymer polycarbophil on the intestinal absorption of a peptide drug in the rat. *J Pharm Pharmacol* 44:402.
35. Kotze, A.F., et al. 1999. Chitosan for enhanced intestinal permeability: prospects for derivatives soluble in neutral and basic environments. *Eur J Pharm Sci* 7:145.
36. Takeuchi, H., et al. 1996. Enteral absorption of insulin in rats from mucoadhesive chitosan-coated liposomes. *Pharm Res* 13:896.
37. Wu, Z.H., et al. 2004. Hypoglycemic efficacy of chitosan-coated insulin liposomes after oral administration in mice. *Acta Pharmacol Sin* 25:966.
38. Kawashima, Y., et al. 2000. Mucoadhesive DL-lactide/glycolide copolymer nanospheres coated with chitosan to improve oral delivery of elcatonin. *Pharm Dev Technol* 5:77.
39. Pan, Y., et al. 2002. Bioadhesive polysaccharide in protein delivery system: chitosan nanoparticles improve the intestinal absorption of insulin *in vivo*. *Int J Pharm* 249:139.
40. Kotze, A.F., et al. 1999. Enhancement of paracellular drug transport with highly quaternized *N*-trimethyl chitosan chloride in neutral environments: *in vitro* evaluation in intestinal epithelial cells (Caco-2). *J Pharm Sci* 88:253.

41. Thanou, M., et al. 2000. N-trimethylated chitosan chloride (TMC) improves the intestinal permeation of the peptide drug buserelin *in vitro* (Caco-2 cells) and *in vivo* (rats). *Pharm Res* 17:27.
42. Bernkop-Schnurch, A., V. Schwarz, and S. Steininger. 1999. Polymers with thiol groups: a new generation of mucoadhesive polymers? *Pharm Res* 16:876.
43. Clausen, A.E., C.E. Kast, and A. Bernkop-Schnurch. 2002. The role of glutathione in the permeation enhancing effect of thiolated polymers. *Pharm Res* 19:602.
44. Roldo, M., et al. 2004. Mucoadhesive thiolated chitosans as platforms for oral controlled drug delivery: synthesis and *in vitro* evaluation. *Eur J Pharm Biopharm* 57:115.
45. Bernkop-Schnurch, A., D. Guggi, and Y. Pinter. 2004. Thiolated chitosans: development and *in vitro* evaluation of a mucoadhesive, permeation enhancing oral drug delivery system. *J Control Release* 94:177.
46. Guggi, D., A.H. Krauland, and A. Bernkop-Schnurch. 2003. Systemic peptide delivery via the stomach: *in vivo* evaluation of an oral dosage form for salmon calcitonin. *J Control Release* 92:125.
47. Tozaki, H., et al. 1997. Chitosan capsules for colon-specific drug delivery: improvement of insulin absorption from the rat colon. *J Pharm Sci* 86:1016.

4 Targeted GI Delivery

David R. Friend

CONTENTS

4.1 INTRODUCTION

The administration of drugs to the gastrointestinal tract (GIT) normally involves an immediate release formulation, typically a tablet or a capsule. Although such formulations are still preferred for their relative simplicity and low cost, formulations that address specific issues in oral drug delivery require more sophisticated attributes. Extended release dosage forms are now used in a number of products. These products permit reduced dosing frequency, leading to improved patient compliance and, in some instances, improved pharmacologic response.

The optimal site of absorption and the mechanism of a drug's action often suggest that something other than immediate or extended release formulations is required to develop the best possible product. For instance, some drugs demonstrate a window of absorption (e.g., they are absorbed relatively rapidly in the upper small intestine but relatively slowly or incompletely from the distal small intestine or large intestine). Delivering drugs to the large intestine to treat local diseases such as inflammatory bowel disease (IBD) represents another example. Addressing these needs optimally requires a targeted delivery system to bring the drug to the correct site and at a controlled dissolution rate. This chapter covers a number of drug delivery solutions currently under investigation to target delivery of drugs in the GIT.

4.2 BACKGROUND

The delivery of drugs to the GIT should be approached in a rational manner taking into account a range of variables and design considerations [1]. Conditions along the GIT vary widely. These variations present opportunities as well as obstacles to effective drug delivery. Transit times, pH, enzymatic activity (both endogenous and derived from gut microflora), permeability barriers, and effect of disease activity (if any) can impact targeted oral drug delivery systems. Most of the following examples are experimental. Some systems have been tested on humans, but many are part of research programs that will probably not lead to approved pharmaceutical products although they will add to the body of scientific drug delivery knowledge. However, it is possible that such research will be used in the future as new biologically active agents are discovered (or invented). As with many other areas of drug delivery, the approaches being investigated are coming to rely more heavily on exploiting the biology of the gut.

4.3 STOMACH

The delivery of drugs in the stomach (gastric drug delivery) is of interest in the treatment of gastric diseases and for drugs demonstrating reduced absorption from the lower small intestine and large intestine [2]. One specific application of gastroretentive drug delivery systems is treatment of *Helicobacter pylori* infections [3]. Typically, gastric residence time is a variable depending on the size and density of the formulation and whether the stomach is in the fed or fasted state [4,5]. Emptying of solid dosage forms ranges from several minutes to 2 h in the fasted state; emptying of dosage forms during the fed state is more complex but, in general, it is slower compared with the fasted state [6]. Set against these physiologic considerations is the goal of gastric retentive dosage forms—increase the duration and consistency of residence time of a dosage form in the stomach while controlling drug dissolution, which could be immediate or extended.

Approaches, used either singly or in combination, to increase gastric residence time include (1) low density formulations allowing the dosage form to float on gastric fluids, (2) high-density dosage forms that are retained at the bottom of the stomach, (3) expandable or swellable dosage forms that are retained due to their inability to pass through the pyloric sphincter, (4) bio- or mucoadhesive formulations, and (5) coadministration of excipients or drugs that slow gastric motility [7]. Some recent formulation efforts to increase gastric retention based on these approaches are presented herein.

4.3.1 Floating Dosage Forms

Floating drug delivery systems for gastric retention have been studied for some time [8]. A common feature of floating dosage forms is their density of <1.0 g/mL. These dosage forms are designed to float on stomach contents while releasing the drug. The size of a meal (on which the dosage form floats) can influence retention time. Ideally, a dosage form that presents consistent gastroretentive properties despite meal size is preferred and will have the broadest application across patients and their eating habits [9]. Subject posture and size of the floating dosage form also influence gastroretentive properties [8].

Most floating gastroretentive dosage forms under investigation are composed of polysaccharides or other polymeric materials. For instance, hydroxypropyl methylcellulose (HPMC) and Carbopol 934P have been used to create floating gastroretentive dosage forms [10]. Chitosan [11] and alginate [12] are polysaccharides tested extensively as both floating and mucoadhesive (*vide infra*) delivery systems. Dosage forms prepared from various polymeric materials include tablets [13,14], capsules [15], microcapsules [12], microspheres or microbeads

[11,16–18], and microballoons (MBs) [19]. In general, formulation approaches relying on many particles, as opposed to a single unit such as a tablet, are preferred to maximize the time of retention of some portion of the dose in the stomach.

The principles of gastric retentive dosage forms can be illustrated using recent data collected with MBs (floating dosage forms). Sato et al. [20–22] published several papers describing the use of MBs in fasted and fed human volunteers. The compound used in these studies was riboflavin, which is absorbed primarily from the proximal small intestine. Riboflavin therefore represents drugs demonstrating a window of absorption. The gastroretentive properties of these dosage forms were followed pharmacokinetically and visually using gamma scintigraphy. In the scintigraphy study, ^{99m}Tc-labeled floating MBs and nonfloating microspheres (NFMs) containing riboflavin were tested in both fed and fasted human volunteers [22]. The MBs were composed of Eudragit S100 with small amounts of HPMC and monostearin. The particle size of the MBs ranged from 500 to 1000 μm and the MBs were buoyant (defined as percent of particles that float relative to those that sink under defined conditions).

In the fed state, the MBs were retained in the stomach for up to 300 min post-dosing whereas the NFMs descended into the lower portion of the stomach within 90 min of administration [22]. The MBs floated for approximately 60 min in the fasted state at which time they cleared rapidly from the stomach during the housekeeper wave (i.e., the interdigestive migrating motor complex). The NFMs cleared more rapidly than did the MBs in the fasted state [22]. The gastroretention data, expressed as normalized radioactivity remaining in the stomach (GRTn), are shown in Table 4.1. GRTn is the ratio of radioactivity from ^{99m}Tc remaining in the stomach to that of ^{111}In, which was administered as an aqueous solution to measure inherent gastric emptying of a liquid. Thus, the higher the GRTn, the more prolonged the residence time of the formulation in the stomach. Pharmacokinetic data were consistent with the imaging data—longer gastric residence time of the formulation increased bioavailability of riboflavin [22].

These data are consistent with some general findings concerning gastroretentive dosage forms. The administration of dosage forms in the fed state usually increases gastric residence time as compared with the fasted state. Whereas the MBs gave higher GRTn than the NFMs in either the fed or the fasted state, the NFMs were retained longer in the fed stomach than

TABLE 4.1
Normalized Activity Remaining in the Stomach of Volunteers Dosed with MBs or NFMs in Both the Fed and Fasted States

	^{99m}Tc		^{111}In		GRTn
Volunteers	**$T20\%^{a}$ (min)**	**$T80\%^{b}$ (min)**	**$T20\%$ (min)**	**$T80\%$ (min)**	**[$T20\%$–$T80\%$ (^{99m}Tc)/$T20\%$–$T80\%$ (^{111}In)]**
A (MB, fed)	195	50	90	20	2.07
B (MB, fasted)	90	70	30	5	0.80
C (NFM, fed)	90	40	35	5	1.67
D (NFM, fasted)	55	40	45	10	0.43

Source: From Sato, Y. et al., *J. Control. Release*, 98, 75, 2004. With permission.
[a] $T20\%$ = time when 20% of the activity remains in the stomach.
[b] $T80\%$ = time when 80% of the activity remains in the stomach.

were the MBs in the fasted state. Thus, most efforts to increase gastric residence time require administration following ingestion of a meal.

4.3.2 Swellable and Expandable Dosage Forms

The second approach used to increase gastric retention is the swellable or expandable systems that delay passage through the pyloris. Data suggest that a solid object of length more than 5 cm or with a diameter larger than 3 cm is retained in the stomach [23]. Dosage forms can be tailored in size to enhance retention in the stomach. Since these formulations must be swallowed, their size at the time of administration must be physiologically relevant. Once in the stomach, they increase in size either by swelling through the absorption of moisture or by the generation of gas (e.g., CO_2) and expand by unfolding to generate an extended object that resists passage through the pyloris.

An example of an expandable (or swellable) dosage form capable of being retained in the stomach is shown in Figure 4.1 [24]. A key element of this formulation is the envelope composed of an elastic, nondissolvable but water permeable polymer. The envelope contains an agent capable of swelling due to the ingress of water. Once sufficient moisture dissolves the drug in the reservoir, the envelope controls drug release. This system has demonstrated extended gastric retention in dogs but its performance in humans is unknown.

The ability of various swellable dosage forms, composed of silicone in various shapes, to be retained in the stomach was examined in human volunteers using gamma scintigraphy [25]. Three shapes were examined: rods, slabs, and minimatrices. These materials were administered in hard-shell gelatin capsules with 100 mL water following the ingestion of a standardized meal. The rod (22 × 6 mm) was retained in the fed stomach for about 4 h 20 min. The slab (20 × 30 × 1 mm, rolled up into the gelatin capsule) was retained for 4 h 40 min on an average, whereas five minimatrices (6 × 4 mm) gave roughly 3 h retention. Only gastric emptying of the slab was correlated with emptying of the meal [25].

Chitosan in combination with poly(acrylic acid) as a polyionic complex was studied recently as a swellable hydrogel in a gastric retention study [26]. Like a number of gastroretentive dosage forms, this one was designed to deliver one or more antibiotics locally for the treatment of *H. pylori* infection. Gastric retention was assessed by a [^{14}C]octanoic acid breath test [27]. When the formulation was administered following a standardized meal in healthy volunteers, the mean gastric half-emptying time of the polyionic complex was 164 min as

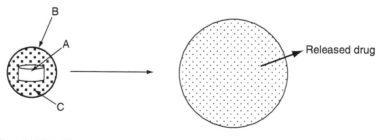

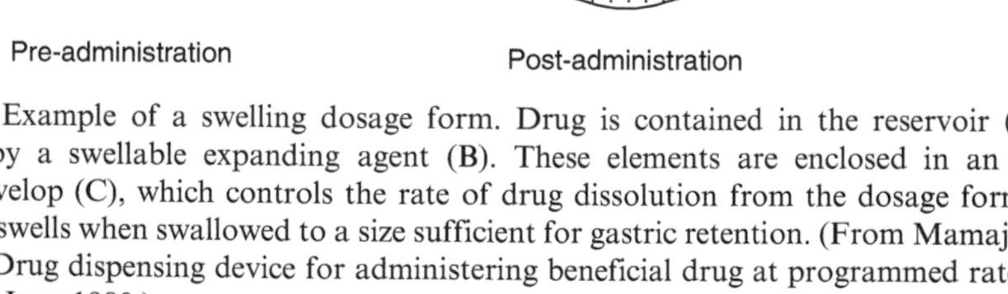

FIGURE 4.1 Example of a swelling dosage form. Drug is contained in the reservoir (A), which is surrounded by a swellable expanding agent (B). These elements are enclosed in an elastic outer polymeric envelop (C), which controls the rate of drug dissolution from the dosage form. A resin or hydrocolloid swells when swallowed to a size sufficient for gastric retention. (From Mamajek, R.C., and E.S. Moyer, Drug dispensing device for administering beneficial drug at programmed rate, US Patent, 4,207,890, 17 June 1980.)

compared with a reference dosage form (nongastroretentive), which gave a mean half-emptying time of 65 min (this difference was statistically significant).

Another example of gastric retention dosage forms is the superporous hydrogel (SPH). These materials are designed to swell rapidly within a few minutes. For example, their diameter can reach 25 mm when starting from an unhydrated diameter of 9 mm [28]. These materials can be compressed while maintaining their ability to swell rapidly [29,30]. Whereas many studies on the synthesis and physical properties of SPHs have been published, few *in vivo* data are available on these interesting materials. One example from a dog study indicated that gastric retention times of 2–3 h were possible in the fasted state and extended retention times of greater than 24 h were observed in the fed state [31].

The fed state, and the known ability to retain solid oral dose forms in this state, has been exploited using a relatively traditional approach of swelling tablets [32]. These dosage forms, based on the swelling of hydrocolloids such as hydroxypropylcellulose, are retained in the stomach for several hours and are capable of releasing ciprofloxacin or metformin over an extended period of time whereas the dosage forms are retained in the fed stomach [33].

There are safety considerations when developing a swellable or expandable gastroretentive dosage form. They should pass through the esophagus without expansion. Once in the stomach, the expanded dosage form should not alter gastric motility. The dosage form should have dull or rounded edges to prevent GI damage and it should be safe with respect to local irritation or damage when retained in the stomach for prolonged periods of time. Collapse or degradation of the dosage form should be controlled and reproducible. There should also be no accumulation of units during multiple dosings.

4.3.3 Mucoadhesive Dosage Forms

A third approach investigated to override normal gastric emptying is the use of materials that promote adherence to the mucosa that is lining the stomach. These dosage forms are often called mucoadhesives or bioadhesives [34]. This approach has been under investigation for many years with limited success. Several recent studies on mucoadhesive dosage forms are presented here.

Chitosan is a cationic polysaccharide investigated extensively as a mucoadhesive substance [35]. There is evidence suggesting that this polymer interacts electrostatically with negatively charged mucus that is lining the epithelial surfaces, including the stomach [36]. ^{152}Sm-labeled microcyrstalline chitosan (MCCh) granules were examined in fasted human volunteers [37]. MCCh prolonged gastric emptying time as compared with reference granules (lactose). Maximum individual half-emptying times ($t_{50\%}$) of greater than 2 h were recorded as compared with the control group $t_{50\%}$ of 0.5 ± 0.3 h. However, retention of the MCCh formulations was erratic and hence too unreliable for further gastroretentive dosage form development. These results generally confirmed earlier studies in human volunteers [38].

A number of other polymer systems have been tested over a number of years for their ability to adhere to gastric mucosa. Primary among these is poly(acrylic acid) and related compositions. While animal studies have shown positive gastric retention results, supporting human data are in short supply. Nonetheless, efforts at exploiting the mucosadhesive properties of these materials to retard gastric emptying continue. For example, retention of ^{14}C-labeled poly(acrylic acids) on gastric and esophageal mucosa was studied under *in vitro* conditions to facilitate local drug delivery [39]. The data demonstrated the ability of these polymers to be retained on mucosal samples obtained from pigs with the highest molecular weight fractions showing the greatest retention. These studies confirmed earlier work by the same group [40].

Another material under investigation as a gastroretentive material is cholestyramine. This substance is an anionic exchange resin capable of coating the gastric mucosa and providing

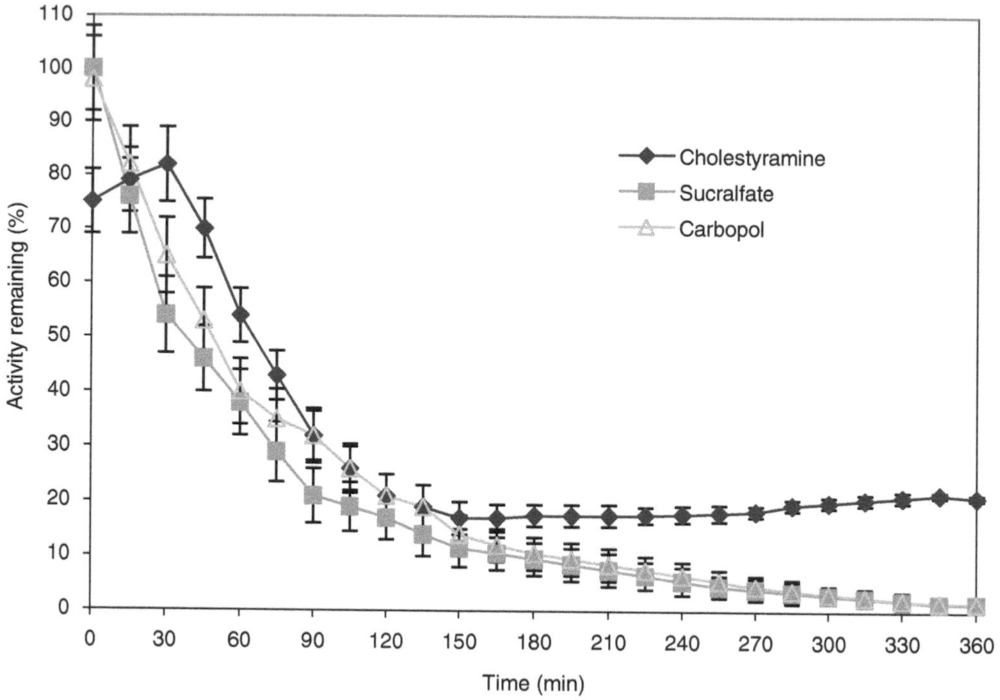

FIGURE 4.2 Mean gastric emptying curves from radiolabeled formulations of cholestyramine, sucralfate, and Carbopol ($n = 10$; mean $\pm$ S.E.M.). The cholestyramine formulation was labeled with ^{99m}Tc-sodium pertechnetate, the sucralfate formulation with ^{99m}Tc-diethylenetriaminepentaacetic acid, and Carbopol with ^{99m}Tc-tin colloid. (From Jackson, S.L., Bush, D., and Perkins, A.C., *Int. J. Pharm.*, 212, 55, 2001. With permission).

extended gastric retention [41,42]. Recently, the intragastric distribution and residence of cholestyramine was compared with that of formulations containing Carbopol 934P or sucralfate [43]. This study was conducted in humans using gamma scintigraphy. The subjects were dosed in the fasted state and then fed 4 h post-dose. Cholestyramine slowed gastric emptying during the later phases of the emptying process leading to an overall increase in total gastric residence time. The mean gastric emptying curves of the three materials studied are shown in Figure 4.2. Also, the cholestyramine preparation spread evenly throughout the entire stomach whereas the other two materials were concentrated in the body and the antrum. While these data, which were consistent with previously published numbers [44], are not compelling in demonstrating that gastric retention is capable of increasing drug absorption from the proximal small intestine, there is the possibility of enhanced topical delivery of antibiotics to gastric sites infected with *H. pylori*.

Amidated gelatin has been formulated into microspheres in an attempt to create a mucoadhesive delivery system [45]. The microspheres were capable of sustaining release of amoxicillin and, in isolated rat stomach, the positively charged microspheres demonstrated increased mucoadhesion as compared with control microspheres. Another approach used to control *H. pylori* infection is based on lipobeads coated with acetohydroxamic acid [46]. This system is designed as a receptor-mediated delivery system for blocking adhesion of *H. pylori* and thus preventing the initiation of gastric infection. The beads were composed of polyvinyl alcohol xerogel coated with phosphatidylcholine ethanolamine. *In situ* adherence studies using human stomach and KATO-III cells demonstrated effective inhibition of *H. pylori* binding in both models.

4.4 SMALL INTESTINE

The small intestine is the primary site of absorption when the drug is administered orally. Most drugs formulated for administration via the oral route are released in the stomach and upper small intestine provide maximize drug absorption. In some instances, this delivery profile is inadequate. For example, some drugs are actively transported across the gut wall only in the proximal small intestine and thus demonstrate a window of absorption as mentioned previously. Other compounds are poorly absorbed from the small intestine and require a targeted mechanism to enhance absorption. The latter issue is encountered when attempting to deliver biotechnology products such as peptides, proteins, genes, oligonucleotides, and vaccines. In addition to poor absorption, these compounds are unstable in the gastrointestinal environment. Oral targeted formulations can be used to address some of these problems.

4.4.1 Biospecific Mucoadhesive Approaches

The localized or targeted delivery of drugs in the small intestine has been approached through mucoadhesion. Unlike most gastric retentive dosage forms, targets in the small intestine are typically molecular rather than nonspecific. For instance, lectins (also called agglutinins) are proteins known to bind to specific sugar residues. As such, these molecules have been investigated as muco- (or bio-) adhesive agents in the GIT [47–49]. Lectins are proteins and are therefore themselves susceptible to proteolytic degradation. They are also potentially immunogenic and cytotoxic. Nonetheless, numerous attempts have been made to use lectins to target and retain drugs, usually conjugated onto particles, in the small intestine. For example, poly(lactide) microspheres have been coated with lectins from *Lycopersicon esculentum* (tomato lectin) and *Lotus tetragonabolis* (asparagus pea lectin) [50]. Transit through the rat GIT was delayed mainly due to gastric retention when the microspheres were conjugated with lectins. The target sugar of tomato lectin is *N*-acetyl-D-glucosamine and that of asparagus pea lectin is L-fucose. These sugar groups are located predominantly in the small intestine [51,52]. Despite this binding specificity, microspheres with conjugated lectins appeared to be retained through nonspecific interactions in the stomach [50]. Additional information on lectin-mediated oral drug delivery can be found in recent reviews [49,53].

An innovative use of lectins to mediate small intestinal drug delivery is based on microfabricated dosage forms. This approach to drug formulation has some advantages over traditional approaches—size and shape can be accurately controlled and bioactive materials (e.g., lectins) can be conjugated to predetermined surfaces. To this end, microfabrication was used to create dose units with multiple reservoirs and bioadhesive molecules (i.e., lectins) conjugated to one side of the particle. These particles are designed to adhere to intestinal mucosa and deliver their contents unidirectionally to the mucosal surface [54,55]. The lectins (tomato lectin and *Artocarpus hirusta* or peanut lectin) were conjugated on the flat side of the microparticles through amino groups using carbodiimide chemistry. When these particles were incubated with Caco-2 monolayers, the tomato-lectin coated microparticles were bound more efficiently than the peanut-lectin or the unmodified particles (see Figure 4.3). Additional studies using this approach in animals are awaited.

4.4.2 Receptor-Mediated Targeting and Absorption

Targeting of drugs in the GIT using liposomes has been examined with little success [56]. To be successful, liposomal preparations need to demonstrate greater intestinal stability, greater intestinal wall affinity, and higher drug loadings than have been possible in the past [57]. To address intestinal wall affinity, folic acid has been investigated as a targeting moiety as it is

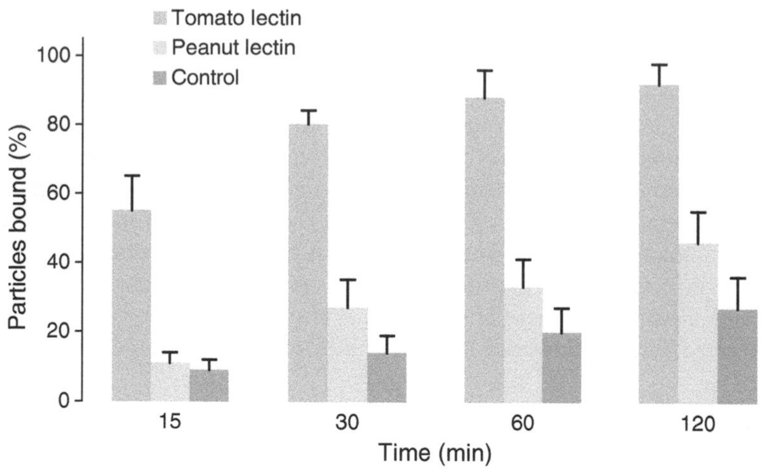

FIGURE 4.3 Proportion of poly(methyl methacrylate) microparticles attached to Caco-2 cells after 15, 30, 60, and 120 min incubation periods. (From Tao, S.I., Lubeley, M.W., and Desai, T.A., *J. Control. Release*, 88, 215, 2003. With permission).

actively transported across the gut wall [58], even when bound to other molecules or particles [59]. To that end, folate was bound to the surface of liposomes and tested for its ability to enhance the uptake of the glycopeptide antibiotic vancomycin. Compared with uncoated liposomes or a simple oral solution, these folate-coated liposomes substantially increased oral absorption in the rat [57]. Other receptors, including those on intestinal M cells, have been studied as a route for effective oral vaccine delivery. Recent reviews have been published on this topic [60,61].

Another target in the small intestine is epithelial tight junctions. Transport of large, polar molecules across cellular membranes (transcellular uptake) is effectively blocked due to the biophysical (lipophilic) nature of cell membranes. Transport around intestinal epithelial cells (the paracellular route) is a more realistic proposition, but this space is blocked by intercellular tight junctions. Certain bacteria, such as *Clostridium* spp., *Escheria coli, Bacteroides fragilis, Vibrio cholerae*, have developed mechanisms to breach these tight junctions [62]. Through the identification and study of the mechanism of action of these bacteria, it should be possible to create delivery systems that effectively mimic the invasive properties of these infectious microorganisms.

A biologically active fragment of zonula occludens toxin (ZOT) obtained from *V. cholerae* [62] reversibly opens tight junctions of intestinal epithelial cells (as well as endothelial cells in the brain). This fragment, called DeltaG, has been found to enhance the oral absorption of certain macromolecules under *in vitro* and *in vivo* conditions [63]. DeltaG has also demonstrated absorption-enhancing properties for low molecular weight drugs that are poorly absorbed from the small intestine [64]. The drugs tested were cyclosporine, ritonavir, saquinavir, and acyclovir. Both C_{max} and area under the curve (AUC) were significantly increased relative to intraduodenal administration of the drug without DeltaG. DeltaG (and its parent ZOT) are biolabile due to their protein structure. Its use *in vivo* will therefore require a method to protect the enhancer from metabolic deactivation [64].

4.4.3 Polymer-Based Targeting

Drug targeting throughout the GIT has traditionally been accomplished, with variable success, using approaches based on standard pharmaceutical excipients. This approach can also be used to assess regional gastrointestinal drug absorption in humans [65]. Advances in polymer chemistry are being applied to oral delivery. One example of a new polymer system for oral drug delivery to the small intestine is SPHs [66]. These gels, which are based on Carbopol, have a high swelling ratio and are reportedly mucoadhesive. Novel mucoadhesive polymers using polymer-tethered structures are also under development [67]. These structures are designed to exhibit reversible, pH-dependent swelling behavior due to the formation of interpolymer complexes between protonated pendent acid groups and etheric groups on the graft chains. Compositionally, these materials are copolymer networks of poly(methacrylic acid) grafted with poly(ethylene glycol). While *in vitro* studies appear promising, these agents must function in the variable and often harsh confines of the GIT, where peptide and protein delivery has proven generally unsuccessful.

4.4.4 Electrically Based Drug Delivery

The final topic covered under small intestinal targeted delivery is taken from a technique applied previously to enhancing skin permeability and in other non-GI *in vitro* and *ex vivo* situations. Electroporation is a technique whereby short electrical pulses are applied transiently to disrupt cell membranes leading to transfection with exogenous molecules such as genes or proteins [68]. Typically, cells are studied in suspension, or when the method is used in transdermal studies, on the skin. In an attempt to extend its application, electroporation was examined for its ability to deliver molecules into model intestinal epithelia [69]. In this work, Caco-2 and T84 monolayers were used to evaluate intracellular delivery of two compounds (calcein- and fluorscein-labeled bovine serum albumin). Delivery into cells was uniform in both cell lines and was increased as a function of voltage, pulse length, and pulse number. The barrier properties were compromised for a time after exposure to electroporation but recovered to normal values within 24 h. These observations are interesting and need to be extended into an *in vivo* situation. The technique could be useful for targeted drug delivery of macromolecules to treat intestinal inflammation and other intestinal disorders. To accomplish this goal, a minimally invasive approach is required. Thus, delivery to the colon is probably more realistic, but reaching the small intestine is feasible [69].

4.5 COLONIC DELIVERY

Delivery of drugs to the colon has attracted interest primarily for the treatment of diseases affecting the colon (e.g., inflammatory bowel disease). It has also been proposed that the colon may be a better site than the small intestine to promote oral macromolecule uptake. The colon is also typically a site of drug absorption from extended release preparations that deliver a substantial portion of the drug in the colon. A variety of approaches are used to deliver drugs locally into the colon. These approaches include time-based systems, pH-based systems, combination of time- and pH-based systems, enzyme-triggered systems (for prodrugs and coated or matrix-based formulations), and pressure-based systems [70]. In addition, lectins, glycoconjugates, and recombinant adenoviruses have been used to deliver drugs into the colon [71,72]. These latter methods, which rely on biologic principles rather than physical pharmacy, will be discussed herein.

4.5.1 Receptor-Mediated Colon Targeting

The use of lectins for colonic delivery is not as well developed as for the small intestine. Nonetheless, the principles are the same in both cases. There are glycoproteins and glycolipids lining the intestine. It is the sugar moieties that are important in lectin-mediated drug delivery.

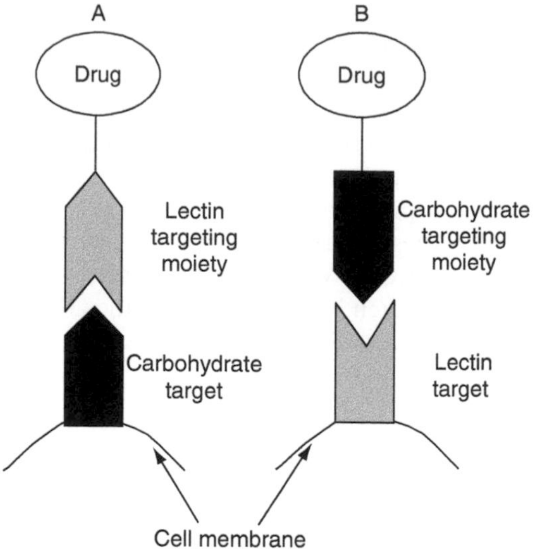

FIGURE 4.4 Schematic representation of direct (A) and reverse (B) lectin–drug mediated targeting. (From Minko, T., *Adv. Drug Deliv. Rev.*, 56, 491, 2004. With permission).

Like the lectin-based systems described above, there are lectins that can be administered to bind to sugars expressed on colonic epithelial cells. Conversely, there are lectins expressed by cells lining the colon, as well as other cells in the body [73]. Lectins expressed on normal colon cells are galectins-1 and -3; these lectins are overexpressed in colon cancer cells [74]. Exploiting these lectins (also referred to as reverse lectins) and direct (exogenous) lectins for colonic drug delivery is currently underway [71]. Direct and reverse lectin–drug mediated targeting is shown schematically in Figure 4.4.

Neoglycoconjugates are synthetic, polyvalent macromolecules with specific molecular weight, solubility, chemical and enzymatic stability, and molecular configuration [75]. An example of a colon-targeted neoglycoconjugate is a copolymer of *N*-(2-hydroxypropyl)methacrylamide (HPMA) bound to wheat germ agglutinin (WGA) or peanut agglutinin (PNA) [76]. Findings from this work indicated that (1) lectins bound to HPMA copolymers did not affect lectin binding activity, (2) WGA-conjugate bound to goblet cells of normal colon tissue whereas PNA-conjugate was detected only in supranuclear (Golgi) regions of these cells, and (3) a PNA-binding glycoprotein sequence was identified on neonatal and diseased but not normal tissue. This difference in binding patterns between healthy and diseased tissues may provide the basis of anti-inflammatory or anticancer agents with greater site-specificity than is currently possible using more traditional approaches to colon targeting.

In the case of colon cancer, aberrant glycosylation patterns are commonly found during neoplastic growth and are an attribute of invasive growth and metastasis [77,78]. Specifically, galectin-1 and -3 are expressed on human colon carcinoma cells [79]. Table 4.2 lists structures of neoglycoconjugates tested for their ability to target colonic tumors. Some of these neoglycoconjugates do not bind (i.e., there are no binding sites) in the liver so that nontarget retention is minimized [80].

A related approach to colon targeting using the neoglycoconjugate approach is based on HPMA containing saccharide epitopes galactosamine (P-Gal), lactose (P-Lac), or triantenary galactose (P-TriGal). These structures are shown schematically in Figure 4.5 along with two other structures that rely on lectin [76] or hyaluronic acid (HA) [81,82] targeting moieties. The structures composed of the saccharide epitopes demonstrated a higher degree of binding

TABLE 4.2
Neoglycoconjugates Tested for Colon Tumor Targeting

Neoglycoconjugate	Trivial Name
Glcβ-PAA[a]	—
Fucaα-PAA	—
Fucaα-2 Glcβ-PAA	H-disaccharide
Fucaα-2 Glcβ1-3GlcNAcβ-PAA	H-type-1
Fucaα-2 Glcβ1-4GlcNAcβ-PAA	H-type-2
Galβ1-3[Fucα1-4]GlcNAcβ-PAA	Lewis a
GalNAcα1-3[Fucα1-2]Galβ-PAA	A-trisaccharide
Galα-3[Fucα1-2]Galβ-PAA	B-trisaccharide

Source: Sato, Y., et al., *J. Control. Release*, 98, 75, 2004. With permission.
Note: Neoglycoconjugates are formed by coupling of mono- or oligosaccharides to PAA via a propyl linking arm.
[a]PAA, polyacrylamide.

when higher sugar content copolymers were studied in human adenocarcinoma cells (see Figure 4.6) [83].

Delivery of genetic material to the GIT may be useful in the treatment of inflammatory bowel disease. Efficient delivery of genes, even to relatively accessible tissue sites such as the colon, presents a significant challenge to formulators. Recombinant adenoviruses have been used to transfer genes *in vivo*. This approach was studied for local gene delivery to normal and inflamed colonic epithelial and subepithelial cells [72]. The ability of two different adenoviruses to effectively introduce genetic information *in vitro* and *in vivo* was followed by the expression of β-galactosidase activity as measured by specific chemiluminescent reporter gene assay. Local (rectal) delivery of the adenoviruses into healthy Balb/c mice resulted in high reporter gene expression in colonic epithelial cells and lamina propria mononuclear cells. In mice with induced experimental colitis (trinitrobenzenesulfonic acid), rectal administration of one of the adenoviruses (standard) studied led to higher β-galactosidase activity in isolated lamina propria cells compared with control mice. The other adenovirus, which incorporated a

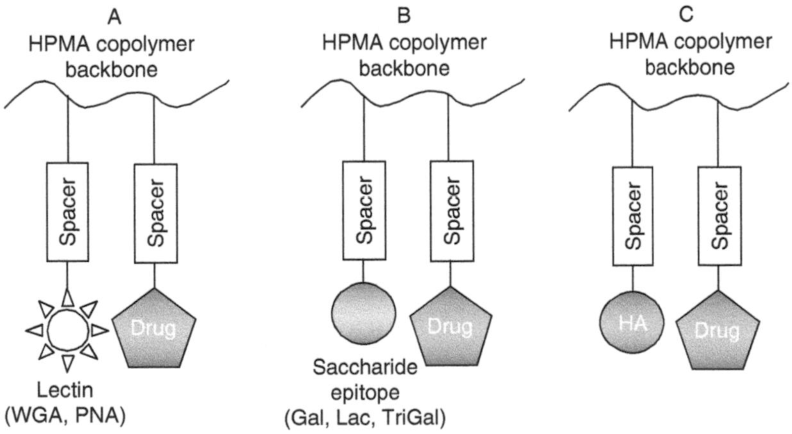

FIGURE 4.5 Schematic representations of HPMA copolymers designed for colon-specific drug targeting. Key to abbreviations is found in text. (From Minko, T., *Adv. Drug Deliv. Rev.*, 56, 491, 2004. With permission.)

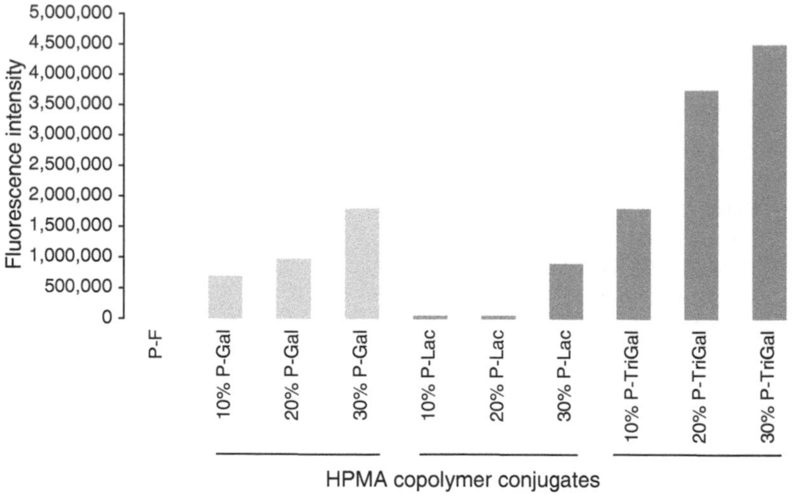

FIGURE 4.6 Quantification of fluorescence intensity (expressed in arbitrary units per square micron) of the confocal fluorescence images of SW-480 cells after incubation with various polysaccharide containing HPMA copolymer-FITC conjugates and the control polymer (P-F) after 24 h at 37°C. (From David, A. et al., *Pharm. Res.*, 19, 1114, 2002. With permission.)

modified fiber structure, led to relatively high reporter gene activity in spleen T cells and lamina propria mononuclear cells of colitic mice compared with the standard vector tested. These data suggest the possibility of using these vectors to locally deliver signal transduction proteins with therapeutic potential into inflamed colonic tissue [72].

4.6 CONCLUSIONS

Targeting of drugs in the GIT remains a challenging goal. Some progress has been made in delivering drugs to the distal intestine using delayed release preparations. These preparations typically rely on enteric coatings on tablets or multiparticulate formulations. The approaches described herein are experimental. Many of the systems are complex, and maintaining stability of the delivery system in the GIT will be a significant hurdle to overcome. Also, some of the approaches described will permit delivery of only the most potent drugs because a relatively large percentage of the mass required for effective delivery is derived from the delivery system (often called the excipients).

Improved understanding of intestinal biology will continue to offer opportunities for targeted drug delivery in the GIT. Active transport processes, including efflux systems, will provide means of improving oral bioavailability. A lectin-based polymer system (neoglycoconjugates) is moving into clinical testing. Thus, novel and sophisticated drug targeting principles are being pursued commercially. If more systems of similar nature reach testing in patients, the field of GIT targeting will continue to receive justified scientific attention.

REFERENCES

1. Wilding, I.R., S.S. Davis, and T.O. O'Hagan. 1994. Targeting of drugs and vaccines. *Pharmcol Ther* 62:97.
2. Hou, S.Y., V.E. Cowles, and B. Berner. 2003. Gastric retentive dosage forms: a review. *Crit Rev Ther Drug Carrier Syst* 20:459.

3. Shah, S., et al. 1999. Evaluation of the factors influencing stomach-specific delivery of antibacterial agents for *Helicobacter pylori* infection. *J Pharm Pharmacol* 51:667.
4. Davis, S.S., J.G. Hardy, and J.W. Fara. 1986. Transit of pharmaceutical dosage forms through the small intestine. *Gut* 27:886.
5. Kaus, K.C., et al. 1984. On the intestinal transit of a single non-disintegrating object. *Int J Pharm* 20:315.
6. Read, N.W., and K. Sugden. 1987. Gastrointestinal dynamics and pharmacology for the optimum design of controlled-release oral dosage forms. *Crit Rev Ther Drug Carrier Syst* 4:221.
7. Klausner, E.A., et al. 2003. Expandable gastroretentive dosage forms. *J Control Release* 90:143.
8. Singh, B.N., and K.H. Kim. 2000. Floating drug delivery systems: an approach to oral controlled drug delivery via gastric retention. *J Control Release* 63:235.
9. Whitehead, L., et al. 1998. Floating dosage forms: an *in vivo* study demonstrating prolonged gastric retention. *J Control Release* 55:3.
10. Li, S., et al. 2003. Effect of HPMC and Carbopol on the release and floating properties of gastric floating drug delivery systems using factorial design. *Int J Pharm* 253:13.
11. El-Gibaly, I. 2002. Development and *in vitro* evaluation of novel floating chitosan microcapsules for oral use: comparison with non-floating chitosan microspheres, *Int J Pharm* 249:7.
12. Whitehead, L., J.H. Collet, and J.T. Fell. 2000. Amoxycillin release from a floating dosage form based on alginates. *Int J Pharm* 210:45.
13. Wei, Z., D. Yu, and D. Bi. 2001. Design and evaluation of a two-layer floating tablet for gastric retention using cisapride as a model drug. *Drug Dev Ind Pharm* 27:469.
14. Wie, Z., et al. 2004. Preparation of a 5-FU floating sustained release tablet for gastric retention. *Beijing Da Xue Xue Bao* 36:439.
15. Menon, A., W.A. Ricschel, and A. Sakr. 1994. Development and evaluation of a monolithic floating dosage form for furosemide. *J Pharm Sci* 83:239.
16. Umamaheshwari, R.B., et al. 2002. Floating-bioadhesive microspheres containing acetohydroxamic acid for clearance of *Helicobacter pylori. Drug Deliv* 9:223.
17. Umamaheshwari, R.B., S. Jain, and N.K. Jain. 2003. A new approach in gastroretentive drug delivery system using cholestyramine. *Drug Deliv* 10:151
18. Soppimath, K.S. 2001. Microspheres as floating drug-delivery systems to increase gastric retention of drugs. *Drug Metab Rev* 33:149.
19. Kawashima, Y., et al. 1991. Preparation of multiple unit hollow microspheres (microballons) with acrylic resin containing tranilast and their drug release characteristics (*in vitro*) and floating behavior (*in vivo*). *J Control Release* 16:279
20. Sato, Y., et al. 2004. *In vitro* and *in vivo* evaluation of riboflavin-containing microballoons for a floating controlled drug delivery system in healthy humans. *Int J Pharm* 275:97.
21. Sato, Y., et al. 2004. *In vivo* evaluation of riboflavin-containing microballoons for floating controlled drug delivery system in healthy human volunteers. *J Control Release* 93:39.
22. Sato, Y., et al. 2004. Pharmacoscintigraphic evaluation of riboflavin-containing microballoons for a floating controlled drug delivery system in healthy humans. *J Control Release* 98:75.
23. Hamilton, J.K., and D.E. Polter. 1993. Gastrointestinal foreign bodies. In *Gastrointestinal Disease*, 286. Philadelphia: W.B. Saunders Co.
24. Mamajek, R.C., and Moyer, E.S. 1980. Drug dispensing device for administering beneficial drug at programmed rate, US Patent 4,207,890, 17 June 1980.
25. Kedzierewicz, F., et al. 1999. Evaluation of peroral silicone dosage forms in humans by gamma scintigraphy. *J Control Release* 58:195.
26. Torrado, S., et al. 2004. Chitosan-poly(acrylic) polyionic complex: *in vivo* study to demonstrate prolonged gastric retention. *Biomaterials* 25:917.
27. Ghoos, Y.F., et al. 1993. Measurement of gastric emptying rate of solid by means of a carbon-labeled octanoic acid breath test. *Gastroenterology* 104:1640.
28. Dorkoosh, F.A., et al. 2000. Preparation and NMR characterization of superporous hydrogels (SPH and SPH composites). *Polymer* 41:8213.
29. Gemeinhart, R.A., H. Park, and K. Park. 2001. Effect of compression on fast swelling of poly(acrylamide-co-acrylic acid) superporous hydrogels. *J Biomed Mater Res* 55:54.

30. Qui, Y., and K. Park. 2003. Superporous IPN hydrogels having enhanced mechanical properties. *AAPS Pharm Sci Technol* 4:406
31. Chen, J., et al. 2000. Gastric retention properties of superporous hydrogel composites. *J Control Release* 64:39.
32. Shell, J.W., J. Louie-Helme, and M. Markey. 2002. Extending the duration of drug release within the stomach during the fed mode. US Patent 6,340,475, 22 January 2002.
33. Gusler, G., et al. 2001. Pharmacokinetics of metformin gastric-retentive tables in healthy volunteers. *J Clin Pharmacol* 41:655.
34. Dodou, D., Breedveld, and P.A. Weiringa. 2005. Mucoadhesives in the gastrointestinal tract: revisiting the literature for novel applications. *Eur J Pharm Biopharm* 60:1.
35. Hejazi, R., and M. Amiji. 2003. Chitosan-based gastrointestinal delivery systems. *J Control Release* 89:151.
36. Lehr, C.M., et al. 1992. *In vitro* evaluation of mucoadhesive properties of chitosan and some other natural polymers. *Int J Pharm* 78:43.
37. Säkkinen, M., et al. 2004. Gamma scintigraphic evaluation of the fate of microcrystalline chitosan granules in human stomach. *Eur J Pharm Biopharm* 57:133.
38. Säkkinen, M., et al. 2003. Evaluation of microcrystalline chitosans for gastro-retentive drug delivery. *Eur J Pharm Sci* 19:345.
39. Smart, J.D., et al. 2003. The retention of ^{14}C-labelled poly(acrylic acids) on gastric and esophageal mucosa: an *in vitro* study. *Eur J Pharm Sci* 20:83.
40. Riley, R.G., et al. 2002. An *in vitro* model for investigating the gastric mucosal retention of ^{14}C-labelled poly(acrylic acid) dispersions. *Int J Pharm* 236:87.
41. Burton, S., et al. 1995. Intragastric distribution of ion-exchange-resins: a drug delivery system for topical treatment of gastric mucosa. *J Pharm Pharmacol* 47:901.
42. Thairs, S., et al. 1998. Effect of dose size, food and surface coating on the gastric residence and distribution of an ion exchange resin. *Int J Pharm* 176:47.
43. Jackson, S.L., D. Bush, and A.C. Perkins. 2001. Comparative scintigraphic assessment of the intragastric distribution and residence of cholestyramine, Carbopol 934P and sucralfate. *Int J Pharm* 212:55.
44. Jackson, S.J., et al. 2000. Effect of resin surface charge on gastric mucoadhesion and residence of cholestyramine. *Int J Pharm* 205:173.
45. Wang, J., et al. 2000. Positively charged gelatin microspheres as gastric mucoadhesive drug delivery system for eradication of *H. pylori. Drug Deliv* 7:237.
46. Umamaheshwari, R.B., and N.K. Jain. 2004. Receptor-mediated targeting of lipobeads bearing acetohydroxamic acid for eradication of *Helicobacter pylori. J Control Release* 99:27.
47. Ponchel, G., and J.M. Irache. 1998. Specific and non-specific bioadhesive particulate systems for oral delivery to the gastrointestinal tract. *Adv Drug Deliv Rev* 34:191.
48. Lehr, C.M. 1994. Bioadhesion technologies for delivery of peptide and protein drugs to the gastrointestinal tract. *Crit Rev Ther Drug Carrier Syst* 11:119.
49. Gabor, F., et al. 2004. The lectin-cell interaction and its implications to intestinal lectin-mediated delivery. *Adv Drug Deliv Rev* 56:459.
50. Montisci, M.J., et al. 2001. Gastrointestinal transit and mucoadhesion of colloidal suspensions of *Lycopersicon esculentum L.* and *Lotus tetragonabolus* lectin-PLA microsphere conjugates in rats. *Pharm Res* 18:829.
51. Pusztai, A., et al. 1993. Antinutritive effects of wheat-germ agglutinin and other *N*-acetylglycosamine-specific lectins. *Br J Nutr* 70:313.
52. Jepson, M.A., et al. 1996. Targeting to intestinal M-cells. *J Anat* 189:507.
53. Bies, C., C.M. Lehr, and J.F.Woodley. 2004. Lectin mediated drug delivery: history and applications. *Adv Drug Deliv Rev* 56:425.
54. Tao, S.I., M.W. Lubeley, and T.A. Desai. 2003. Bioadhesive poly(methyl methacrylate) microdevices for controlled drug delivery. *J Control Release* 88:215.
55. Ahmed, A., C. Bonner, and T.A. Desai. 2002. Bioadhesive microdevices with multiple reservoirs: a new platform for oral drug delivery. *J Control Release* 81:291.

56. Rogers, J.A., and K.E. Anderson. 1998. The potential of liposomes in oral drug delivery. *Crit Rev Ther Drug Carrier Sys* 15:465.
57. Anderson, K.E., et al. 2001. Formulation and evaluation of a folic acid receptor-targeted oral vancomycin liposomal dosage form. *Pharm Res* 18:316.
58. Rosenburg, I. 1990. Herman award lecture. Folate absorption: clinical questions and metabolic answers. *Am J Clin Nutr* 51:531.
59. Leamon, C.P., and P.S. Low. 1991. Delivery of macromolecules into living cells: A method that exploits folate receptor endocytosis. *Proc Natl Acad Sci USA* 88:5572.
60. Foster, N., and B.H. Hirst. 2005. Exploiting receptor biology for oral vaccination with biodegradable polymers. *Adv Drug Deliv Rev* 57:431.
61. Brayden, D.J. and A.W. Baird. 2004. Apical membrane receptors on intestinal M cells: potential targets for vaccine delivery. *Adv Drug Deliv Rev* 56:721.
62. Fasano, A., and J.P. Nataro. 2004. Intestinal epithelial tight junctions for enteric bacteria-derived toxins. *Adv Drug Deliv Rev* 56:795.
63. Salama, N.N., et al. 2004. The effect of DeltaG on the transport and oral absorption of macromolecules. *J Pharm Sci* 93:1310.
64. Salama, N.N., et al. 2005. The impact of DetlaG on the oral bioavailability of low bioavailable therapeutic agents. *J Pharmacol Exp Ther* 312:199.
65. Basit, A.W., et al. 2004. The use of formulation technology to assess regional gastrointestinal drug absorption in humans. *Eur J Pharm Sci* 21:179.
66. Tang, C., et al. 2005. New superporous hydrogels composites based on aqueous Carbopol® solution (SPHCcs): synthesis, characterization and *in vitro* bioadhesive force studies. *Eur Polym J* 41:557.
67. Peppas, N.A. 2004. Devices based on intelligent biopolymers for oral protein delivery. *Int J Pharm* 277:11.
68. Heller, R., R. Gilbert, and M.J. Jaroszeski. 1999. Clinical applications of electrochemotherapy. *Adv Drug Deliv Rev* 35:119.
69. Gharty-Tagoi, E.B., et al. 2004. Electroporation-mediated delivery of molecules to model intestinal epithelia. *Int J Pharm* 270:127.
70. Friend, D.R. 2005. New oral delivery systems for treatment of inflammatory bowel disease. *Adv Drug Deliv Rev* 57:247.
71. Minko, T. 2004. Drug targeting to the colon with lectins and neoglycoconjugates. *Adv Drug Deliv Rev* 56:491.
72. Wirtz, S., P.R. Galle, and M.F. Neurath. 1999. Efficient gene delivery to the inflamed colon by local administration of recombinant adenoviruses with normal and modified fibre structure. *Gut* 44:800.
73. Perillo, N.L., M.E. Marcus, and L.G. Baum. 1998. Galectins: versatile modulators of cell adhesion, cell proliferation, and cell death. *J Mol Med* 76:402.
74. Schoeppner, H.L., et al. 1995. Expression of an endogenous galactose-bining lectin correlates with neoplastic progression in the colon. *Cancer* 75:2818.
75. Bovin, N.V. 1998. Polyacylamide-based glycoconjugates as tools in glycobiology. *Glycoconj J* 15:431.
76. Wroblewski, S., et al. 2000. The influence of a colonic microbiota on HPMA copolymer lectin conjugates binding to rodent intestine. *J Drug Target* 8:173.
77. Remmelink, M., et al. 1999. *In vitro* influence of lectins and neoglycoconjugates on the growth of three human sarcoma cell lines. *J Cancer Res Clin Oncol* 125:275.
78. Kayser, K., et al. 2002. Primary colorectal carcinomas and their intrapulmonary metastases: clinical, glyco-, immuno- and lectin histochemical, nuclear, and syntactic structure analysis with emphasis on correlation with period of occurrence of metastases and survival. *Acta Pathol Microbiol Immunol Scand* 110:435.
79. Ohannesian, D.W., et al. 1994. Concomitant increases in galectin-1 and its glycoconjugate ligands (carcinoembryonic antigen, lamp-1, and lamp-2) in cultured human colon carcinoma cells by sodium butyrate. *Cancer Res* 54:5992.
80. Galanina, O., et al. 1998. Detection of a potential receptor for the H-blood group antigen on rat colon carcinoma cells and normal tissues. *Int J Cancer* 76:136.

81. Luo, Y., and G.D. Prestwich. 1999. Synthesis and selective cytotoxicity of a hyaluronic acid-antitumor bioconjugate. *Bioconjug Chem* 10:755.
82. Luo, Y., et al. 2002. Targeted delivery of doxorubicin by HPMA copolymer-hyaluronan bioconjugates. *Pharm Res* 19:396.
83. David, A., et al. 2002. The role of galactose, lactose, and galactose valency in the biorecognition of *N*-(2-hydroxypropyl)methacrylamide copolymers by human adenocarcinoma cells. *Pharm Res* 19:1114.

5 Inhibition of Enzymes and Secretory Transport

Martin Werle and Andreas Bernkop-Schnürch

CONTENTS

5.1 INTRODUCTION

Efflux systems and presystemic metabolism are, beside other reasons, responsible for low bioavailability of orally administered drugs. Presystemic metabolism itself can be divided into three subtypes: luminal metabolism, first-pass intestinal metabolism, and first-pass hepatic metabolism. Luminal metabolism is caused mainly by secreted enzymes. Most important in this context are proteases secreted from the pancreas such as trypsin and chymotrypsin, esterases, and miscellaneous nucleases. For first-pass intestinal metabolism, brush border membrane-bound enzymes including aminopeptidases and carboxypeptidases, as well as intracellular occurring enzymes like esterases and above all CYP3A4 are responsible. Absorbed drugs reach the liver through the portal circulation, where a certain amount of administered drug is biotransformed before it reaches the systemic circulation. This is known as the first-pass hepatic metabolism [1].

Membrane proteins play a crucial role in the transfer of drugs across the intestinal membrane. On one hand these transport systems can positively influence drug uptake by acting as absorptive systems, on the other hand, they can build a barrier to transport by causing drug secretion back across the intestinal membrane. Two very important classes of membrane transporters located in the intestine are the solute carrier proteins (SLC family) that include peptide transporters like PepT1 and PepT2 and the ATP-binding cassette (ABC) family comprising the most important efflux pumps like P-glycoprotein (P-gp) and the multidrug resistance-associated proteins (MRP 1–5). By inhibiting either presystemic metabolism or efflux systems, higher bioavailability can be achieved.

5.2 ENZYME INHIBITION

Attempts to reduce enzymatic degradation include the use of analogs, prodrugs, and formulations such as nanoparticles, microparticles, and liposomes that shield therapeutic agents from luminal enzymatic attack. The design of delivery systems targeting the colon where proteolytic activity is relatively low [2] is also an important approach to circumvent excessive enzymatic degradation. Moreover, the co-administration of enzyme inhibitors has become of increasing interest, since due to such excipients a significantly improved bioavailability of drugs after oral dosing could be achieved [3–6].

Nevertheless, the coadministration of enzyme inhibitors—especially for drugs, which are used in long-term therapy—remains questionable because of side effects caused by these agents. Beside systemic toxic side effects and intestinal mucosal damage, enzyme inhibitors can also lead to a disturbed digestion and an inhibitor-induced enzyme secretion stimulation caused by feedback regulation [7]. Numerous studies have investigated this regulation with agents such as Bowman–Birk inhibitor, soybean trypsin inhibitor (Kunitz trypsin inhibitor), and camostat in rats and mice. It was observed that this feedback regulation rapidly leads to both hypertrophy and hyperplasia of the pancreas. A prolonged administration of the Bowman–Birk inhibitor and soybean trypsin inhibitor even led to the development of numerous neoplastic foci, frequently progressing to invasive carcinoma [8–11]. However, by utilizing proper formulations, side effects can be minimized or even excluded.

5.2.1 Nuclease Inhibition

In recent years, the administration of plasmid DNA (p-DNA), short interference RNA (siRNA), and (antisense) oligonucleotides (ON) has become an interesting subject. However, regarding the poor oral bioavailability of these compounds, in the range of only a few percent [12,13] nucleotide administration is currently limited to parenteral routes. Another promising approach for the use of oral gene delivery besides reaching systemic circulation through the oral route is the targeting of intestinal cells. Transfection of M-cells of Peyer's patches would even allow the establishment of a barrier against pathogens entering the body through intestinal membranes mediated by a triggering of mucosal immunity. The oral delivery of nucleotides aiming the transfection of Peyer's patches would therefore be a beneficial system for vaccination. For both attempts, oral nucleotide delivery to reach systemic circulation as well as oral nucleotide delivery to transfect Peyer's patches, sufficient protection against nucleases is crucial.

Chemical modification of the molecule [14,15], interaction with polycations [16], liposomes [17,18], polymeric carrier systems [19,20], and nuclease inhibitors [21–25] have already been used to enhance the cellular uptake and to protect nucleotides from enzymatic degradation. Although an improvement in the stability of nucleotide agents by the addition of nuclease inhibitors as auxiliary agents can be anticipated, nucleotide agents have, to our knowledge, not been evaluated for oral gene delivery so far. Recently, enhanced protection of p-DNA against gastrointestinal enzymatic attack by utilizing a chitosan conjugate comprising the covalently bound nuclease inhibitor EDTA compared to unmodified chitosan was demonstrated [19]. Regardless of toxic risks, various nuclease inhibitors are listed in Table 5.1.

5.2.2 Protease Inhibition

Therapeutic peptides, proteins, and peptidomimetics are getting increasingly important. These compounds are degraded in the first instance by luminally secreted and brush border membrane-bound proteases. Because of their hydrophilic character and comparatively large size, they are usually uptaken via the paracellular route after oral administration.

Regardless of toxic risks, various inhibitors of pancreatic and brush border membrane-bound proteases are listed in Table 5.2 and Table 5.3, respectively. Beside this overview, a classification of inhibitory agents based on their chemical structure is shown below.

5.2.2.1 Inhibitors That Are Not Based on Amino Acids

The high toxicity of most of these inhibitory agents makes them useless as agents in human therapeutics. Hence, apart from few exceptions, this class of inhibitors is more of

TABLE 5.1
Nuclease Inhibitors

Nuclease Inhibitor	Annotations	Ref.
β-Mercaptoethanol		[136]
Ammonium sulfate		[136]
Bentonite		[136]
Aurintricarboxylic acid	No tissue toxicity	[21,137]
Chitosan–EDTA	Inhibits furthermore various GI proteases	[19]
Divalent cations	Mg^{+2}, Mn^{+2}, Zn^{+2}, Fe^{+2}, Ca^{+2}, Cu^{+2}	[136]
DMI-2	Polyketide metabolite of Streptomyces	[22]
Ethylenediaminetetraacetic acid (EDTA)	Peptide DNase II inhibitor	[136]
Guanidinium thiocyanate		[136,138]
Heparin		[136]
Human placentyl ribonuclease inhibitor	Narrow spectrum of specificity; commercially available	[139]
Hydroxylamine–oxygen–cupric ion		[136]
ID2 peptides		[24]
Macaloid		[136]
PRIME inhibitor	Narrow spectrum of specificity; commercially available	[25]
Protein kinase K		[136]
Sodium dodecyl sulfate (SDS)		[136]
Vanadyl ribonucleoside complexes (VRC)		[136,137,140]

TABLE 5.2
Inhibitors of Luminally Secreted Proteases

Recommended Name (EC Number)	Synonyms	Cofactor	Inhibitors
Carboxypeptidase A (3.4.17.1)	Carboxypeptidase A3 MC-CPA, RMC-CP	Zinc	EDTA, chitosan–EDTA conjugates [64], poly(acrylate) derivatives [49], polycarbophil–cysteine [67]
Carboxypeptidase B (3.4.17.2)	CPB, pancreas specific protein, PASP, protaminase, ZAP 47	Zinc	EDTA, chitosan–EDTA conjugates, poly(acrylate) derivatives [89], polycarbophil–cysteine [67]
Chymotrypsin (3.4.21.1)	Avazyme, Chymar, Enzeon, Quimar	Calcium	β-Phenylpropionate [26], aprotinin [5,44,45,141–144], Bowman–Birk inhibitor [7,141,145], benzyloxycarbonyl-Pro-Phe-CHO [43], chicken ovoinhibitor [47], chymostatin [3,141], DFP [26], FK-448 [3], PMSF [26], polycarbophil–cysteine [67], soybean trypsin inhibitor [5,7,51], sugar biphenylboronic acids complexes [146], poly(acrylate) derivatives [49]
Pancreatic elastase (3.4.21.36)	Pancreatopeptidase E, Elaszym, serine elastase	Calcium	Bowman–Birk inhibitor [7], chicken ovoinhibitor [47], DFP [26], elastatinal [147], methoxysuccinyl-Ala-Ala-Pro-Val-chloromethylketone (MPCMK) [148], PMSF [26], soybean trypsin inhibitor [7,51]
Trypsin (3.4.21.4)	Parenzyme, parenzymol, tryptase	Calcium	AEBSF [26], antipain [76], APMSF [26], aprotinin [5,44,45,141–144], Bowman–Birk inhibitor [7,141,145], camostat mesilate (= FOY-305) [3,5,76], chicken ovoinhibitor [47], chicken ovomucoid [46], DFP [26], flavonoid inhibitors [149], human pancreatic trypsin inhibitor [48], leupeptin [3], *p*-aminobenzamidine [150], PMSF [151], poly(acrylate) derivatives [152], soybean trypsin inhibitor [5,7,46,51,78,141], TLCK (tosyllysine chloromethlyketone) [26]

theoretical interest than of practical interest. Nevertheless, they are generally very potent inhibitors [26] and the development of slightly chemically modified analogs or the immobilization of these auxiliary agents to unabsorbable matrices might reduce or even eliminate this drawback.

Diisopropylfluorophosphate (DFP) and phenylmethanesulfonyl fluoride (PMSF), both organophosphorus inhibitors, are potent irreversible inhibitors of serine proteases. However, because of their additional inhibition of acetylcholinesterase these compounds are highly toxic [26]. Another toxic but potent trypsin inhibitor is (4-aminophenyl)-methanesulfonyl

TABLE 5.3
Inhibitors of Membrane-Bound Proteases

Recommended Name (EC Number)	Synonyms	Cofactor	Inhibitors
γ-Glutamyl transferase (2.3.2.2)	Gamma-GPT, GGT, glutamyl transpeptidase	Magnesium	Acivicin (amino-(3-chloro-4, 5-dihydro-isoxazol-5-yl)-acetic acid) [153], L-serine-borate [154]
Carboxypeptidase M (3.4.17.12)	CPM, M14006 (Merops-ID)	Zinc	DL-2-mercaptomethyl-3-guanidinoethylthiopropanoic acid [155]
Dipeptidyl peptidase IV (3.4.14.5)	ACT 3, glycoprotein GP110, TP103, WC10, X-PDAP	Zinc	*N*-Peptidyl-*O*-acylhydroxylamines boronic acid analogs of proline and alanine [156], DFP [157]
Glutamyl aminopeptidase (3.4.11.7)	Aminopeptidase A, aspartate aminopeptidase, glutamyl peptidase, membrane aminopeptidase II	Zinc, calcium	1,10-Phenantroline [158], α-aminoboronic acid derivatives [29], EDTA [158], phosphinic acid dipeptide analogs [37], puromycin [158]
Leucyl aminopeptidase (3.4.11.1)	Aminopeptidase I, aminopeptidase II, aminopeptidase III, leucinamide aminopeptidase	Zinc, magnesium, manganese	α-Aminoboronic acid derivatives [29], amastatin [83], bestatin [27], flavonoid inhibitors [149], phosphinic acid dipeptide analogs [37,159]
Membrane alanyl aminopeptidase (3.4.11.2)	Aminopeptidase M, aminopeptidase M II, aminopeptidase N, aminopeptidase microsomal, membrane protein p161, microsomal aminopeptidase, particle-bound aminopeptidase	Zinc, cobalt	1,10-Phenanthroline [28,160,161], 8-hydroxyquinoline [28], α-aminoboronic acid derivatives [29], acitonin [161], amastatin [51,83,162], amino acids [28], bacitracin [5,33,44,159], bestatin [27,163–165], Carbopol(974P)-cysteine [68], chitosan–EDTA conjugates [52], di- and tripeptides [33], EDTA, Na-glycocholate [5,27], phosphinic acid dipeptide analogs [37], polycarbophil-cysteine [69], puromycin [27,33,163]
Membrane Pro-X carboxypeptidase (3.4.17.16)	Carboxypeptidase P, microsomal carboxypeptidase	Zinc, manganese	1,10-Phenanthroline, EDTA, Zn^{2+} [166]

(*Continued*)

TABLE 5.3 (Continued)
Inhibitors of Membrane-Bound Proteases

Recommended Name (EC Number)	Synonyms	Cofactor	Inhibitors
Neprilysin (3.4.24.11)	CALLA, CD 10, kidney-brush-border neutral peptidase, neutral endopeptidase	Zinc	1,10-Phenanthroline [167], phosphoramidon [10,52,166], SQ 28,603 (N-[2-(mercaptomethyl)-1-oxo-3-phenylpropyl]-β-alanine) [167], thiorphan ((2-mercaptomethyl-3-phenyl-propionylamino)-acetic acid) [69]
Peptidyl dipeptidase A (3.4.15.1)	ACE, angiotensin-converting enzyme, dipeptidyl carboxypeptidase	Zinc	1,10-Phenanthroline [168], 2-mercaptoethanol [168], 8-hydroxyquinoline [169], angiotensin II [170], dithiothreitol [168,169], EDTA [1,2]
X-Pro aminopeptidase (3.4.11.9)	Aminopeptidase P, membrane-bound AmP	Zinc, manganese	α-Aminoboronic acid derivatives [161], bestatin [83], phosphinic acid dipeptide analogs [160], PMSF [83]
X-Trp aminopeptidase (3.4.11.16)	Aminopeptidase W	Zinc	α-Aminoboronic acid derivatives [161], phosphinic acid dipeptide analogs [160]

fluoride hydrochloride (APMSF). 4-(2-Aminoethyl)-benzenesulfonyl fluoride (AEBSF) displays comparable inhibitory activity to DFP and PMSF; however, it is markedly less toxic.

Contrary to the above-mentioned inhibitors, FK-448 (4-(4-isopropylpiperadinocarbonyl) phenyl 1,2,3,4,-tetrahydro-1-naphthoate methanesulfonate) is a low toxic as well as a potent and specific inhibitor of chymotrypsin. The effectiveness of this substance as an intestinal absorption enhancer has already been demonstrated in rats as well as in dogs. Coadministration of FK-448 led to an enhanced absorption of insulin, which was monitored by a decrease in blood glucose level. The inhibition of chymotrypsin was found to be mainly responsible for the enhanced bioavailability [3]. Camostat mesilate (*N*,*N*′-dimethyl carbamoylmethyl-*p*-(*p*′-guanidino-benzoyloxy)phenylacetate methanesulfonate) [5] and Na-glycocholate [5,27] are further representatives of this class, exhibiting low toxicity.

5.2.2.2 Amino Acids and Modified Amino Acids

Although amino acids as well as modified amino acids are generally low in toxicity or nontoxic and can be produced at comparatively low cost, their small molecular size and good solubility lead to an extensive dilution during gastrointestinal passage and to a quick absorption so that unrealistically high amounts of these inhibitors are necessary to achieve an inhibitory effect toward luminal proteases. For these reasons, this class of inhibitors has so far gained more interest for other transmucosal routes, e.g., nasal where these effects are relatively lower. However, if reduction or even exclusion of extensive dilution effects can be guaranteed, the use of such inhibitors for oral drug administration seems to be feasible and will mainly depend on delivery systems that are able to overcome the above-mentioned problem.

Amino acids can act as reversible, competitive inhibitors of exopeptidase such as aminopeptidase N [28] but due to their low inhibitory activity, their practical use is quite questionable. In contrast, modified amino acids display a much stronger inhibitory activity. Transition-state inhibitors are reversible, competitive inhibitors, which exhibit strong inhibitory activity. The mechanism of inhibition can be explained by the hypothesis that an inhibitor, which resembles the geometry of a substrate in its transition state, has a much higher affinity to the active site of the enzyme than the substrate itself. An example for an aminopeptidase transition-state inhibitor is boroleucine, which displays a 100-fold higher enzyme inhibition than bestatin and a 1000-fold higher enzyme inhibition than puromycin [29]. Other α-aminoboronic acid derivatives acting as aminopeptidase transition-state inhibitors are borovaline and boroalanine. Up to date, these inhibitors have been evaluated only for nasal delivery so that their potential for oral administration has to be investigated. However, problems connected to their chemical liability are likely to occur.

Also *N*-acetylcysteine is capable of inhibiting aminopeptidase N [30]. Beside its low toxicity, this compound displays mucolytic properties that lead to a reduction of the diffusion barrier [31].

5.2.2.3 Peptides and Modified Peptides

The cyclic dodecapeptide bacitracin A has a molecular mass of 1423 Da and shows remarkable resistance against the action of proteolytic enzymes including trypsin, pepsin and aminopeptidase N [32]. So far, it has been exclusively used in veterinary medicine and as a topical antibiotic in the treatment of infections in human. Bacitracin can inhibit bacterial peptidoglycan synthesis, mammalian transglutaminase activity, and proteolytic enzymes and has therefore been used to inhibit the degradation of various therapeutic (poly)peptides, such as insulin, metkephamid, luteinizing hormone-releasing hormone (LH-RH), and buserelin [5,33,34]. Moreover, bacitracin also displays absorption-enhancing effects without leading to serious intestinal mucosal damage [35].

Unfortunately, bacitracin causes nephrotoxicity [36] and so its use as a suitable adjuvant to overcome the enzymatic barrier is quite questionable. Interestingly, it was demonstrated recently that bacitracin, when covalently linked to a mucoadhesive polymer, still displays its inhibitory activity [30]. The immobilization to an unabsorbable drug carrier matrix is believed to lead to an exclusion of systemic toxic side effects, but detailed toxicological studies are necessary to prove this hypothesis.

Dipeptides as well as tripeptides display a weak and unspecific inhibitory activity toward some exopeptidases [33] but analogous to amino acids, their inhibitory activity can be improved by chemical modifications. Phosphinic acid dipeptide analogs also belong to the group of transition-state inhibitors and exhibit strong inhibitory activity toward aminopeptidases. For example, the phosphinate inhibitor VI is a 10- and 100-times stronger inhibitor than bestatin and puromycin, respectively [37]. The potential of phosphinic acid dipeptide analogs for stabilizing nasally administered leucine enkephalin has already been demonstrated.

Another example of a transition-state analog is the modified pentapeptide pepstatin [38]. It is a very potent inhibitor of pepsin. Although an enzymatic attack of orally administered peptide and protein drugs in the stomach can be excluded by coating the dosage form with a gastric fluid resistant layer, the inhibition of pepsin is nevertheless of practical relevance, especially in cases where pepsin-digested therapeutic drugs should be liberated in the stomach, for example, epidermal growth factor in the treatment of gastric ulcer [39]. However, coadministration of pepstatin can cause several side effects by inhibiting physiologically essential enzymes [40–42].

Another group of modified peptides acting as inhibitors are peptides, which display a terminally located aldehyde function in their structure. The sequence benzyloxycarbonyl-Pro-Phe-CHO fulfills the known primary and secondary specificity requirements of chymotrypsin and has been found to be a potent reversible inhibitor of this target proteinase [43]. Further inhibitors exhibiting terminally located aldehyde function are antipain, leupeptin, chymostatin, and elastatinal. Furthermore, phosphoramidon, bestatin, puromycin, and amastatin represent modified peptides acting as reversible inhibitors.

5.2.2.4 Polypeptide Protease Inhibitors

The comparably high molecular mass of polypeptide protease inhibitors allows to keep them concentrated in delivery systems based on a drug carrier matrix. Kimura et al. [44] already demonstrated the advantage of the slow release of an inhibitor from a delivery system. Within this study, a mucoadhesive delivery system was designed, which exhibited a release rate of the protease inhibitor aprotinin of only approximately 10% per hour. This release rate was almost synchronous with the release rate of a polypeptide drug. *In vivo* studies carried out with this delivery system showed an improved oral drug bioavailability [44]. Due to their low toxicity as well as strong inhibitory activity, polypeptide protease inhibitors have so far been used to the highest extent as auxiliary agents to overcome the enzymatic barrier of perorally administered therapeutic peptides and proteins.

Aprotinin is a polypeptide consisting of 58 amino acid residues derived from bovine lung tissues and shows inhibitory activity toward various proteolytic enzymes including chymotrypsin, kallikrein, plasmin, and trypsin. It was also one of the first enzyme inhibitors used as an auxiliary agent for oral (poly)peptide administration. The co-administration of aprotinin led to an increased bioavailability of peptide and protein drugs [5,44,45]. The Bowman–Birk inhibitor (71 amino acids, 8 kDa) and the Kunitz trypsin inhibitor (184 amino acids, 21 kDa) belong to the soybean trypsin inhibitors. Both are known to inhibit trypsin, chymotrypsin, and elastase, whereas carboxypeptidase A and B cannot be inhibited [7,46].

Further inhibitors of this class are the chicken egg white trypsin inhibitor (=chicken ovomucoid; 186 amino acids) [46], the chicken ovoinhibitor (449 amino acids) [47], and the human pancreatic trypsin inhibitor (56 amino acids), which can already be produced in *Escherichia coli* [48].

5.2.2.5 Complexing Agents

Divalent cations are essential cofactors for many proteases. Complexing agents display an inhibitory activity, as they are capable of removing these cations from the enzyme structure. Ions, which are such cofactors of luminally secreted and membrane-bound proteases, are therefore listed in Table 5.2 and Table 5.3, respectively. In many cases this inhibitory effect can only be achieved at very high concentrations of complexing agents. For example, a concentration of 7.5% (w/v) of the chelating agent EDTA is not sufficient to inhibit trypsin activity *in vitro* [49]. Moreover, taking into consideration that calcium ion concentration in gastric and intestinal fluids was determined to be 0.4–0.7 mM [50], it is very likely that calcium ions will even worsen the inhibition effect *in vivo*. Hence, complexing agents do not seem to be the proper compounds to successfully inhibit endoproteases such as trypsin, chymotrypsin, and elastase. Contrary, the efficacy of complexing agents to inhibit various Zn^{2+}-dependent exoproteases including carboxypeptidase A and B as well as aminopeptidase N has already been demonstrated [51–53]. However, because the inhibition rate correlates strongly with the inhibitor concentration, extensive dilution effects during the gastrointestinal passage have to be avoided. Representatives of this class of inhibitory agents are EDTA, EGTA, 1,10-phenantroline, and hydroxychinoline [51,54–56]. Beside enzyme inhibition, complexation with divalent cations can moreover lead to permeation-enhancing effects [57].

5.2.2.6 Mucoadhesive Polymers Exhibiting Enzyme Inhibitory Activity

Dilution effects during the GI passage as well as systemic side effects can be excluded by the use of mucoadhesive polymers exhibiting enzyme-inhibiting properties. Poly(acrylate) derivatives such as poly(acrylic acid) and polycarbophil can affect the activity of various luminal proteases. It is supposed that the inhibitory effect of these polymers is based on the complexation of divalent cations such as Ca^{2+} and Zn^{2+} [58]. Luessen et al. [49] has done detailed analysis of the inhibitory effect toward luminally secreted and brush border membrane-bound proteases. Within this study it could be shown that poly(acrylate) derivatives inhibit trypsin, chymotrypsin, and carboxypeptidase A and B. Whether the protective effect of these polymers is sufficient to prevent luminal enzymatic degradation of polymer-embedded peptide and protein drugs will depend mainly on the dosage form used. In spite of these inhibitory effects, simple formulations of poly(acrylate) derivatives are not believed to display a sufficient protective effect [59]. Nevertheless, because they are generally known to be safe and furthermore exhibit additional advantages for the oral administration of drugs including mucoadhesive [60,61] as well as permeation-enhancing properties [62,63], these polymers are of high practical relevance. The above-mentioned advantages provide a prolonged residence time of the dosage form in the intestine and enhanced drug absorption. Moreover, beside inhibitory and permeation-enhancing properties, mucoadhesive polymers are also able to reduce the enzymatic barrier by sticking to the mucus layer at the site of drug absorption. This effect leads to reduced drug metabolism by luminally secreted proteases because of the reduced distance between the released therapeutic drug from the dosage form and the absorptive tissue.

Chitosan–EDTA was developed to improve the inhibitory activity of mucoadhesive polymers. These conjugates exhibit high binding capacity as well as high binding affinity toward multivalent cations and are capable of inhibiting zinc-dependent proteases such as aminopeptidase A, N, P, and W as well as carboxypeptidase A, B, M, and P. Although their ability to bind calcium ions is higher than the one of carbomer and carbophil [64], chitosan–EDTA conjugates do not display any inhibitory effects toward calcium-dependent proteases such as elastase, chymotrypsin, and trypsin. However, due to their mucoadhesive properties as well as their potential as carrier matrices for controlled drug release, chitosan–EDTA conjugates are promising novel compounds for the design of oral delivery systems.

Mucoadhesive polymers exhibiting strong complexing properties are capable of inhibiting intestinal brush border membrane-bound proteases through a far distance inhibitory effect [65]. *In vivo*, the mucoadhesive polymer is separated from the brush border membrane by a mucus layer [30]. Although there is no direct contact between polymer- and membrane-bound enzymes, it could be shown that inhibition takes place. The exploitation of this far distance effect seems to be a very promising alternative to small molecular mass inhibitors, which are currently used as inhibitors of brush border membrane-bound proteases.

As demonstrated in a number of studies, the inhibitory activity of mucoadhesive polymers can be extensively improved by the immobilization of enzyme inhibitors onto the polymer without markedly influencing the mucoadhesive properties of the polymer. Such conjugates, comprising a mucoadhesive polymer and a covalently bound enzyme inhibitor exhibit diverse advantages. First of all, due to the mucoadhesive properties of the polymer the distance between the delivery system and the absorption membrane is limited, so that the drug is less affected by luminal enzymes as well as nutritive proteins. Furthermore, the enzyme inhibitor remains at a restricted intestinal area, which provides sufficient inhibitor concentration at the absorption site. Mediated by the covalent attachment of the inhibitor, dilution effects are negligible so that the total amount of inhibitor in such delivery systems is much lower compared to formulations containing unbound inhibitor. Due to the comparatively low inhibitor concentrations and the unlikeliness of absorption of the inhibitor when bound to an unabsorbable matrix, local and systemic toxicity is minimized. Beside the inhibition of luminally secreted enzymes, brush border membrane-bound enzymes can be inhibited through a far distance inhibitory effect mentioned above.

A novel approach for improving oral bioavailability of hydrophilic drugs is the use of thiolated polymers or designated thiomers. By immobilization of thiol groups on various polymers such as chitosan or poly(acrylate) derivatives, significant improvement of mucoadhesion as well as permeation-enhancing properties toward the unmodified polymer could be demonstrated in numerous studies [64,66]. Furthermore, recent studies demonstrated the enhanced enzyme inhibition properties of thiomers in comparison to the corresponding unmodified polymers. Polycarbophil–cysteine exhibited significantly improved inhibitory effects against carboxypeptidase A and B as well as chymotrypsin [67]. Moreover, this thiomer was able to efficiently protect leucine enkephalin from degradation caused by aminopeptidase N as well as porcine intestinal mucosa and showed strongly improved inhibitory effects toward unmodified polycarbophil [68]. Valenta et al. [69] evaluated the properties of the thiomer Carbopol(974 P)-cysteine, another poly(acrylate) derivative, which was found to exhibit significantly higher enzyme inhibitory properties against aminopeptidase N as well as vaginal mucosa compared to the unmodified polymer. Within this study it could also be demonstrated that enzyme inhibition was strongly dependent on the amount of immobilized thiol groups. The more thiol groups available, the higher was the inhibitory effect [69]. So far evaluated thiomers and mucoadhesive polymer–inhibitor conjugates exhibiting enzyme inhibitory activity are listed in Table 5.4.

TABLE 5.4
Mucoadhesive Polymers with Enzyme Inhibitory Properties

Inhibitory Activity toward	Polymer or Polymer–Inhibitor Conjugate
Aminopeptidase N	Polycarbophil–cysteine [67,68], Carbopol(974 P)–cysteine [69], poly(acrylic acid)–bacitracin conjugate [52], chitosan–EDTA conjugate [52,64], chitosan–EDTA–antipain, chymostatin, and elastatinal conjugate [171]
Carboxypeptidase A	Polycarbophil–cysteine [67,68], chitosan–EDTA conjugate [52,64], chitosan–EDTA–antipain, chymostatin, and elastatinal conjugate [171]
Carboxypeptidase B	Polycarbophil–cysteine [67,68], chitosan–EDTA–antipain, chymostatin, and elastatinal conjugate [171]
Chymotrypsin	Poly(acrylic acid)-Bowman–Birk inhibitor conjugate [59], poly(acrylic acid)–chymostatin conjugate [172]
Elastase	Poly(acrylic acid)–elastatinal conjugate [173], polycarbophil–elastatinal conjugate [173], carboxymethylcellulose–elastatinal conjugate [173]
Trypsin	Polycarbophil–cysteine [67,68], chitosan–antipain conjugate [76]

5.2.3 Esterase Inhibition

A multitude of therapeutic drugs bears ester moieties. It is known that various esterases are secreted by the pancreas into the intestine and they are bound to the brush border membrane, or occur in the cytosol [70]. Accordingly, esterases play a crucial role in the presystemic metabolism of many drugs exhibiting ester substructures. A specific inhibition of the responsible esterase may consequently be useful for improving oral bioavailability. According to the Enzyme Commission (EC), esterases can be divided into eight groups, which are listed in Table 5.5.

Pancreatic cholesterol esterase (3.1.1.3.) aids in transporting cholesterol to the enterocyte. By utilizing a selective and potent cholesterol esterase inhibitor 6-chloro-3-(1-ethyl-2-cyclohexyl)-2-pyrone, the absorption of cholesterol in hamsters could be reduced [71]. Wadkins et al. [72] synthesized novel sulfonamide derivatives, which demonstrated greater than 200-fold selectivity for human intestinal carboxylesterase compared with the human liver carboxylesterase hCE1, and none of them was an inhibitor of human acetylcholinesterase or butyrylcholinesterase. Maybe these agents can serve as lead compounds for the development of effective, selective carboxylesterase inhibitors for clinical applications. Also the potent P-gp inhibitor verapamil [73] as well as S,S,S-tributylphosphortrithionate (DEF) [74] may exhibit carboxylesterase inhibitory properties. Various other inhibitors of human esterases are listed in Table 5.6.

5.2.4 Methods to Evaluate the Protective Effect of Inhibitory Agents and Dosage Forms toward Luminal Enzymatic Attack

5.2.4.1 *In Vitro* Test Models for Luminally Secreted Enzymes

The potential of a compound to inhibit gastrointestinal enzymes can be evaluated using different methods. On the one hand, the (model)-drug can be incubated with or without inhibitor in artificial gastric or intestinal juice comprising a mixture of certain degrading enzymes at 37°C. Experiments should be carried out under permanent shaking or stirring in an appropriate vessel. Samples can be withdrawn at predetermined time-points to observe the degradation curve by determining the amount of undegraded (model)-drug [75–77]. For formulations delivering not only the drug but also the inhibitor, the influence of dilution effects of the auxiliary agent on the protective effect can be studied using a flow-through cell [75]. The study can also be carried out with collected gastric or intestinal juice from animals [78,79]. An advantage compared to *in vivo* studies is that the inhibitory effect can also be tested in the presence of isolated enzymes so that more detailed knowledge can be gained [75,76].

TABLE 5.5
EC Esterase Classification

EC Number	Esterase
3.1.1.1	Carboxylesterase
3.1.1.2	Arylesterase
3.1.1.3	Triacylglycerol lipase
3.1.1.4	Phospholipase A2
3.1.1.5	Lysophospholipase
3.1.1.6	Acetylesterase
3.1.1.7	Acetylcholinesterase
3.1.1.8	Cholinesterase

TABLE 5.6
Inhibitors of Human Esterases

Inhibitor	3.1.1.1	3.1.1.2	3.1.1.3	3.1.1.4	3.1.1.5	3.1.1.6	3.1.1.7	3.1.1.8
2-Mercaptoethanol [174]				•				
2-Tetradecanoylaminohexanol-1-phosphocholine [174]				•				
3,4-Dichlorisocoumarin [175]				•				
Acetylthiocholine iodide [176]							•	
Ankyrin-iPLA2s [177]				•				
Bile salts [178]			•					
Bis(4-nitrophenyl)phosphate [179]	•							
Bromoenol lactone [180]				•				
Ca^{2+} [181]				•				
Choline [176]							•	
$CuSO_4$ [182]								•
DFP [183]	•							
Dibucaine [184]								•
Diethyl *p*-nitrophenyl phosphate [175]				•				
Diisopropylfluorophosphate [175,185–187]		•		•			•	
Dimethylsulfoxide [188]	•							
Diphenyl carbonate [188]	•							
Dithiothreitol [189]				•				
DTNB [180,190]	•			•				
EDTA [174,191,192]		•		•				
EGTA [174]				•				
HI 6 [193]								•
Methyl arachidonyl fluorophosphonate [194]					•			
n-Butanol [188]	•							
NEM [179,190]	•							
Neostigmine [195]	•							
p-Bromophenacyl bromide [192]				•				
p-Hydroxymercuribenzoate [196]		•						
PAM-2 [193]								•
Paraoxon [188,190]	•							
PCMB [179]	•							
PD-11612 [197]		•						
Phenyl acetate [198]		•						
Phosphostigmine [193]								•
PMSF [188,195]	•							
Prostigmin [182]								•
Quinidine [199]							•	
Ro 02-0683 [193]								•
Sodium chloride [182]								•
Sodium deoxycholate [180,181]				•				
Sodium fluoride [184,188]	•							•
Tetramethyl ammonium [176]							•	
Triphenyl phosphate [188]	•							
Tris/maleate buffer [200]				•				
Triton X-100 [180,181]				•				

5.2.4.2 *In Vitro* Test Models for Brush Border Membrane-Bound Enzymes

As described earlier, the efficacy of an inhibitor or a drug delivery system containing such an auxiliary agent against membrane-bound enzymes can be evaluated by incubation with either a mixture or single isolated membrane-bound enzymes [30]. Those tests should only be considered as preliminary studies, because of the different *in vivo* situation. Asada et al. [79] and Yamamoto et al. [80] described a method by utilizing mucosal homogenate. This method may be used to determine the degradation of lipophilic drugs, which are absorbed through the transcellular way, whereas it is not the method of choice for hydrophilic drugs. These compounds are transported mainly through the paracellular route and are therefore not affected by cytosolic enzymes, which occur in homogenates. By utilizing homogenates for enzymatic degradation studies, hydrophilic drugs will be digested to a much higher extent compared to test models using intact mucosal tissues. Accordingly, more detailed information will be obtained by incubation with brush border membrane vesicles instead of mucosal homogenates [33,81].

Taking into account the important influence of the diffusion barrier caused by the mucus layer, which covers the brush border membrane and its enzymes, *in vitro* evaluations can be performed with a Franz diffusion cell using an artificial membrane. This membrane can be prepared by immobilization of the pure enzyme of interest on a nitrocellulose membrane, coating remaining free-binding sites with a protein, which will not be digested by the degrading enzyme, and covering the membrane with a mucus layer. The inhibitor or dosage form containing a substrate or model drug is added to the donor chamber, and the amount of remaining undegraded substrate or model drug can be determined in the donor and acceptor chamber. The test model also allows investigations of far distance acting inhibitory agents as well as of the influence of diffusion effects of inhibitors on the inhibition of membrane-bound enzymes [30,31]. Also freshly excised mucosa from sacrificed animals can be used for membrane-bound enzyme degradation tests. Studies can be performed, for example, with Ussing-type chambers in parallel to permeation studies or by using only the donor chamber of a Franz diffusion cell or an Ussing chamber, respectively. Results gained by this technique are commonly reproducible due to the defined surface area and offer the advantage of using mucosa from different intestinal segments [52].

Another method to evaluate the protective effect of an enzyme inhibitor is to incubate the (model)-drug with or without inhibitor in buffer solution containing everted intestinal rings of sacrificed animals [81,82].

5.2.4.3 *In Situ* Test Models

In situ studies can be performed by utilizing perfusion techniques. Test solutions containing the (model)-drug and the enzyme inhibitor are delivered continuously through intestinal segments of an anesthetized animals using a single-pass perfusion technique and the amount of undegraded (model)-drug in the intestinal fluid and blood can be determined in samples taken from perfusion solutions and different veins of animals, respectively [4,81,83].

5.2.4.4 *In Vivo* Test Models

Some model drugs offer the advantage of evaluating an enhanced bioavailability by a biological response, e.g., the decrease in blood glucose after insulin administration [84], the antidiuretic response of vasopressin derivatives [45], or the increase in blood total leukocyte counts after administration of human granulocyte colony-stimulating factor [46]. The time-dependent plasma level of numerous other perorally administered compounds can be evaluated by HPLC analysis, RIA, or ELISA [4,81,85–87]. However, due to the permeation-enhancing effect of some inhibitors including bacitracin, EDTA, and poly(acrylate) derivatives [35,86,88,89] it can be difficult to exclusively evaluate the inhibitory effect *in vivo*.

5.3 SECRETORY TRANSPORT

Beside membrane transporters such as PepT1 and PepT2, which act as absorptive systems, there are transporters like P-gp and the MRP 1–5, which transport certain drugs actively back into the intestinal lumen. These efflux pumps are located in several tissues including liver, kidney, brain, and intestine [90,91]. In the intestine, efflux systems are predominantly located at the apical side of the epithelial cells. Lipophilic drugs are usually absorbed by the transcellular route so that they are mostly affected by these systems. Interestingly, the intracellular occurring CYP3A metabolizes compounds to substrates that are eliminated by P-gp [92].

5.3.1 Inhibition of Efflux Systems

Although much work has been pursued in the process of establishing the role of efflux pumps—first of all P-gp—in multidrug resistance (MDR) in cancer cells, the importance of inhibiting those transport systems to improve bioavailability has been increasingly recognized as a major drug delivery opportunity only in the last few years. Beside antibiotics and anticancer drugs, various other agents are secreted back into the lumen by membrane transport systems.

Several compounds are known to inhibit P-gp. Beside their use in cancer therapy to minimize MDR, P-gp inhibitors have already been used in oral drug delivery to improve bioavailability. Also the intake of miscellaneous liquids such as grapefruit juice [93–95] or green tea [96] can have enormous influence on efflux pumps. The P-gp inhibitors can be divided into three generations. First-generation P-gp inhibitors are compounds that are in clinical use for other indications but show inhibitory properties. Because of their pharmacological activity, which would lead to unwanted side effects, these first-generation inhibitors are not the compounds of choice. Second-generation modulators lack the pharmacological activity of the first-generation agents and have improved inhibitory properties. Third-generation modulators are the most potent and selective inhibitors (Table 5.7).

According to Varma et al. [97], three ways of P-gp inhibition can be distinguished, where the first one is the most important:

TABLE 5.7
Efflux Pump Inhibitors

Generation	Pharmacological Category	Examples
First	Antiarrhythmic drugs	Amiodarone [201,202], quinidine [109,203], verapamil [204]
	Antifungal drugs	Itraconazole [202], ketoconazole [111,113]
	Antihypertensive drugs	Felodipine [205], nifedipine [206], nitrendipine [205],
	Miscellaneous	Chlorpromazine [206], cyclosporine A [110,112], lidocaine [206], mibefradil [207], progesterone [208], RU 486 [209], tamoxifen derivates [210], terfenadine [204],
Second		Biricodar [211], GF120918 [116], KR30031 [113], MS-209 [115], PSC 833 [116]
Third		LY335979 [116], OC144093 [119], XR9576 [212]

- Inhibition by blocking drug-binding sites either competitively or allosterically
- Inhibition by interfering with ATP hydrolysis
- Inhibition by altering integrity of cell membrane lipids

While several P-gp inhibitors have entered clinical trials, the development of specific MRP inhibitors is still in its infancy. Similar to P-gp, various MRP inhibitors are already known—among others MK-571 [98–100], probenecid [101], ritonavir [102], leukotriene C4 [103], gemfibrozil [104], pyrrolpyrimidines [105], sulindac [106], and certain isothiocyanates [107], but their potential as auxiliary agents for oral drug delivery has to be further investigated.

5.3.2 Efflux Pump Inhibition to Improve Oral Bioavailability

5.3.2.1 Improving Oral Bioavailability by Using First-Generation Inhibitors

As mentioned above, first-generation efflux pump inhibitors are, because of their pharmacological effects, not the agents of choice to improve oral bioavailability. Nevertheless, they have already been used in this field. Coadministration of quinidine increased etoposide absorption in everted gut sacs prepared from rat jejunum and ileum [108]. Malingre et al. [109] showed a strong enhancement of oral docetaxel bioavailability when coadministrating cyclosporine. Also the oral availability of tacrolimus [110] was doubled when coadministrating ketoconazole, and digoxin bioavailability was improved *in vivo* when coadministrating atorvastatin [111].

5.3.2.2 Improving Oral Bioavailability by Using Second- and Third-Generation Inhibitors

KR30031, a verapamil analog with fewer cardiovascular effects, improved paclitaxel bioavailability 7.5-fold after oral administration in rats [112]. Oral bioavailability of paclitaxel has also already been improved by coadministrating flavone [113], MS-209 [114], GF120918 [115,116], SDZ PSC 833 [117] as well as PSC 833, LY335979, and R101933 [115]. The third-generation inhibitor OC144-093 (ONT-093) has been used to improve the oral uptake of docetaxel [118]. Verstuyft et al. [119] showed that dipyridamole is an *in vitro* and *in vivo* P-gp inhibitor that increases intestinal digoxin absorption and digoxin plasma concentration.

5.3.2.3 Improving Oral Bioavailability by Using Polymeric Agents

Polymeric agents are also important P-gp inhibitory agents, which have been used to improve oral bioavailability. Poly(ethylene glycol) PEG-300 inhibited efflux transporter activity in Caco-2 cell monolayers, which is probably mediated by P-gp or MRP [120,121]. Pluronic P85 increased apical to basolateral permeability in Caco-2 monolayers with respect to a broad panel of structurally diverse compounds [122]. Also microgels composed of cross-linked copolymers of poly(acrylic acid) and Pluronics enhanced the overall cell absorption of doxorubicin by inhibiting P-gp-mediated efflux in Caco-2 monolayers [123]. The copolymer CRL-1605 has been used successfully as a vehicle to significantly improve the bioavailability of the aminoglycosides tobramycin as well as amikacin, after feeding FVB mice with a liquid formulation [124,125]. Surfactants like the water-soluble derivative of vitamin E, D-alpha-tocopherol polyethylene glycol 1000 succinate, or a polyethoxylated derivative of 12-hydroxy-stearic acid, turned out to be the suitable formulation ingredients for improving systemic exposure from the intestine [126,127]. Other nonionic surfactants including Tween 80 and Cremophor EL increased apical to basolateral permeation of the P-gp substrate rhodamine 123 [128]. Recently, it was demonstrated by our research group that thiolated chitosan can inhibit P-gp [129]. These results were supported by Föger et al., who observed an increased oral bioavailability of the P-gp substrate rhodamine 123 in rats when co-administrating a delivery system comprising of thiolated chitosan and glutathione [130].

A problem, which is always connected to the coadministration of auxiliary agents, is that the drug and the auxiliary agent have to be released at the same time and at the same intestinal segment. Dilution effects during passage through the GI tract play an important role in this context. To avoid this problem and to minimize side effects that are caused by the administration of too high doses of auxiliary agent, synchronized oral delivery systems should be considered. Baluom et al. [131], for example, prepared matrix tablets containing hydroxypropylmethylcellulose (HPMC), the P-gp inhibitor quinidine as well as the drug sulpiride. After intraintestinal administration in rats, the synchronous release increased sulpiride bioavailability. The increase varied from 2.6- to 3.9-fold when using different HPMC ratios compared to a control administration of a powdered mixture of sulpiride and quinidine.

5.3.3 Methods to Evaluate Efflux Pump Inhibition

5.3.3.1 *In Vitro* Evaluation

To verify if a certain drug is affected by efflux systems, permeation studies with freshly excised intestinal mucosa at physiological conditions (37°C) can be performed and compared to results gained at 4°C, where activities of most enzymes as well as efflux pumps are strongly decreased. Permeation studies can be performed with Ussing-like chambers, where the mucosa can be mounted between the donor and the acceptor chamber. Drug permeation should be performed from the apical to the basolateral side as well as from the basolateral to the apical. Efflux pump inhibitor activity can be evaluated by coadministrating the inhibitory agent and using the same method as mentioned above. Apart from utilizing excised intestinal mucosa from wild-type animals, mucosa from P-gp knockout mice can also be used [132]. In addition, efflux pump inhibition studies can be carried out with monolayer cell lines [133]. Furthermore, accumulation assays with substrates such as rhodamine 123 or calcein acetoxymethyl ester or monitoring the ATPase activity are useful methods for the identification of efflux pump substrates and inhibitors.

5.3.3.2 *In Vivo* Evaluation

In vivo studies are commonly the best way to verify the effectiveness of a drug delivery system. Beside conventional *in vivo* studies, where wild-type animals are fed with formulations containing the drug as well as the inhibitory agent, another approach is the use of P-gp knockout mice [134,135].

5.4 CONCLUSION AND FUTURE TRENDS

Although several studies clearly indicate the efficacy of inhibitors to overcome the enzymatic barrier and therefore the feasibility to improve oral drug bioavailability by coadministrating enzyme inhibitors, the practical use of such inhibitors in formulations still remains questionable. One problem, which is always connected to auxiliary agents, is their toxicity. As discussed above, also if the inhibitor itself is not toxic, disturbed nutrition can lead to unwanted side effects. Another important factor is the dilution during the gastrointestinal passage, which makes high concentrations of inhibitors necessary to guarantee sufficient inhibitor concentrations at the absorption site. Novel applications should overcome these problems. Very promising in this context are mucoadhesive multifunctional polymers such as thiomers, which offer many advantages. Because of their mucoadhesive properties, the way of the drug from the formulation to the absorption site is limited and therefore the drug is less affected by luminal degradation. Some of these polymers also exhibit enzyme inhibitory activity against various secreted and membrane-bound enzymes. By immobilization of the required inhibitors onto a polymer with or without inhibitory properties, certain intestinal enzymes can be inhibited and therefore the drug can be protected against enzymatic degradation. Due to the character of mucoadhesive polymers, which target only a restricted intestinal area, only little concentrations of inhibitors are necessary. These little inhibitor concentrations as well as the unlikeliness of being absorbed when bound to unabsorbable matrices minimize the possibility of toxic side effects. Moreover, by utilizing mucoadhesive polymers exhibiting inhibitory activity themselves or mucoadhesive polymers with covalently bound inhibitors, dilution effects are negligible compared to formulations with simply coadministered enzyme inhibitors. Furthermore, some multifunctional polymers such as thiomers also exhibit permeation-enhancing effects. In conclusion, the most promising way of the utilization of enzyme inhibitors so as to improve oral drug bioavailability currently seems to be connected to delivery systems based on unabsorbable, multifunctional polymers.

REFERENCES

1. Swarbrick, J., and J.C. Boylan. 2002. *Encyclopedia of pharmaceutical technology*, 2nd ed. New York: Marcel Dekker.
2. Bernkop-Schnürch, A. 1997. Strategies for the peroral administration of therapeutic peptides and proteins. *Sci Pharm* 65:61.
3. Fujii, S., et al. 1985. Promoting effect of the new chymotrypsin inhibitor FK-448 on the intestinal absorption of insulin in rats and dogs. *J Pharm Pharmacol* 37:545.
4. Langguth, P., H.P. Merkle, and G. Amidon. 1994. Oral absorption of peptides—The effect of absorption side and enzyme inhibition on the systemic availability of metkephamid. *Pharm Res* 11:528.
5. Yamamoto, A., et al. 1994. Effects of various protease inhibitors on the intestinal absorption and degradation of insulin in rats. *Pharm Res* 11:1496.
6. Morishita, I., et al. 1992. Hypoglycemic effect of novel oral microspheres of insulin with protease inhibitor in normal and diabetic rats. *Int J Pharm* 78:9.
7. Reseland, J.E., et al. 1996. Proteinase inhibitors induce selective stimulation of human trypsin and chymotrypsin secretion. *J Nutr* 126:634.
8. Otsuki, M., et al. 1987. Effect of synthetic protease inhibitor camostat on pancreatic exocrine function in rats. *Pancreas* 2:164.
9. Melmed, R.N., A.A. El-Aaser, and S.J. Holt. 1976. Hypertrophy and hyperplasia of the neonatal rat exocrine pancreas induced by orally administered soybean trypsin inhibitor. *Biochim Biophys Acta* 421:280.
10. McGuinness, E.E., D. Hopwood, and K.G. Wormsley. 1982. Further studies of the effects of raw soya flour on the rat pancreas. *Scand J Gastroenterol* 17:273.
11. Ge, Y.C., and R.G.H. Morgan. 1993. The effect of trypsin inhibitor on the pancreas and small intestine of mice. *Br J Nutr* 70:333.
12. Raoof, A.A., et al. 2002. Effect of sodium caprate on the intestinal absorption of two modified antisense oligonucleotides in pigs. *Eur J Pharm Sci* 17:131.
13. Geary, R.S., et al. 2001. Absolute bioavailability of 2′-*O*-(2-methoxyethyl)-modified antisense oligonucleotides following intraduodenal instillation in rats. *J Pharmacol Exp Ther* 293:898.
14. Henry, S.P., et al. 2001. Drug properties of second-generation antisense oligonucleotides: How do they measure up to their predecessors? *Curr Opin Investig Drugs* 2:1444.
15. Gewirtz, A.M. 1996. Perturbing gene expression with oligodeoxynucleotides: Research and potential therapeutic applications. *Mt Sinai J Med* 63:372.
16. Haensler, J., and F.C.J. Szoka. 1993. Polyamidoamine cascade polymers mediate efficient transfection of cells in culture. *Bioconjug Chem* 4:372.
17. Niedzinski, E.J., et al. 2000. Gastroprotection of DNA with a synthetic cholic acid analog. *Lipids* 35:721.
18. Huang, L., and S. Li. 1997. Liposomal gene delivery: A complex package. *Nat Biotechnol* 15:620.
19. Loretz, B., et al. 2006. Oral gene delivery: Strategies to improve stability of pDNA towards intestinal digestion. *J Drug Targ* in press.

20. Brus, C., et al. 2004. Efficiency of polyethylenimines and polyethylenimine-graft-poly(ethylene glycol) block copolymers to protect oligonucleotides against enzymatic degradation. *Eur J Pharm Biopharm* 57:427.
21. Glasspool-Malone, J., et al. 2002. DNA transfection of macaque and murine respiratory tissue is greatly enhanced by use of a nuclease inhibitor. *J Gene Med* 4:323.
22. Ross, G.F., et al. 1998. Enhanced reporter gene expression in cells transfected in the presence of DMI-2, an acid nuclease inhibitor. *Gene Ther* 5:1244.
23. Ciftci, K., and R.J. Levy. 2001. Enhanced plasmid DNA transfection with lysosomotropic agents in cultured fibroblasts. *Int J Pharm* 218 (1–2):81.
24. Sperinde, J.J., S.J. Choi, and F.C.J. Szoka. 2001. Phage display selection of a peptide DNase II inhibitor that enhances gene delivery. *J Gene Med* 3:101.
25. Murphy, N.R., S.S. Leinbach, and R.J. Hellwig. 1995. A potent, cost-effective RNase inhibitor. *Biotechniques* 18:1068.
26. Stryer, L. 1988. *Biochemistry*, 3rd ed., 226. New York: W.H. Freeman.
27. Okagawa, T., et al. 1994. Susceptibility of ebiratide to proteolysis in rat intestinal fluid and homogenates and its protection by various protease inhibitors. *Life Sci* 55:677.
28. McClellan, J.B.J., and C.W. Garner. 1980. Purification and properties of human intestine alanine aminopeptidase. *Biochim Biophys Acta* 613:160.
29. Hussain, M.A., et al. 1989. The use of alpha-aminoboronic acid derivatives to stabilize peptide drugs during their intranasal absorption. *Pharm Res* 6:186.
30. Bernkop-Schnürch, A., and M.K. Marschütz. 1997. Development and *in vivo* evaluation of systems to protect peptide drugs from aminopeptidase N. *Pharm Res* 14:181.
31. Bernkop-Schnürch, A., and R. Fragner. 1996. Investigations into the diffusion behaviour of polypeptides in native intestinal mucus with regard to their peroral administration. *Pharm Sci* 2:361.
32. Hickey, R.J. 1964. Bacitracin, its manufacture and uses. *Prog Ind Microbiol* 5:93.
33. Langguth, P., et al. 1994. Metabolism and transport of the pentapeptide metkephamid by brush-border membrane vesicles of rat intestine. *J Pharm Pharmacol* 46:34.
34. Raehs, S.C., et al. 1988. The adjuvant effect of bacitracin on nasal absorption of gonadorelin and buserelin in rats. *Pharm Res* 5:689.
35. Gotoh, S., et al. 1995. Does bacitracin have an absorption-enhancing effect in the intestine? *Biol Pharm Bull* 18:794.
36. Drapeau, G., et al. 1992. Dissociation of the antimicrobial activity of bacitracin USP from its renovascular effects. *Antimicrob Agents Chemother* 36:955.
37. Hussain, M.A., et al. 1992. A phosphinic acid dipeptide analogue to stabilize peptide drugs during their intranasal absorption. *Pharm Res* 9:626.
38. McConnell, R.M., et al. 1991. New pepstatin analogues: Synthesis and pepsin inhibition. *J Med Chem* 34:2298.
39. Itoh, M., and Y. Matsuo. 1994. Gastric ulcer treatment with intravenous human epidermal growth factor: A double-blind controlled clinical study. *J Gastroenterol Hepatol* 9:78.
40. Plumpton, C., et al. 1994. Effects of phosphoramidon and pepstatin A on the secretion of endothelin-1 and big endothelin-1 by human umbilical vein endothelial cells—Measurement by two-site enzyme-linked immunosorbent assay. *Clin Sci* 87:245.
41. McCaffrey, G., and J.C. Jamieson. 1993. Evidence for the role of a cathepsin D-like activity in the release of Gal beta 1–4 GlcNAc alpha 2–6 sialyltransferase for rat and mouse liver in whole-cell systems. *Comp Biochem Phys C* 104:91.
42. Carmel, R. 1994. *In vitro* studies of gastric juice in patients with food-cobalamin malabsorption. *Dig Dis Sci* 39 (12):2516.
43. Walker, B., et al. 1993. Peptide glyoxals—A novel class of inhibitor for serine and cysteine proteinases. *Biochem J* 193:321.
44. Kimura, T., et al. 1996. Oral administration of insulin as poly(vinyl alcohol)-gel spheres in diabetic rats. *Biol Pharm Bull* 19:897.
45. Saffran, M., et al. 1988. Vasopressin: A model for the study of effects of additives on the oral and rectal administration of peptide drugs. *J Pharm Sci* 77:33.

46. Ushirogawa, Y. 1992. Effect of organic acids, trypsin inhibitors and dietary protein on the pharmacological activity of recombinant human granulocyte colony-stimulating factor (rhG-CSF) in rats. *Int J Pharm* 81:133.
47. Scott, M.J., et al. 1987. Ovoinhibitor introns specify functional domains as in the related and linked ovomucoid gene. *J Biol Chem* 262:5899.
48. Kanamori, T., et al. 1988. Expression and excretion of human pancreatic secretory trypsin inhibitor in lipoprotein-deletion mutant of *Escherichia coli. Gene* 66:295.
49. Luessen, H.L., et al. 1996. Mucoadhesive polymers in peroral peptide drug delivery. I. Influence of mucoadhesive excipients on the proteolytic activity of intestinal enzymes. *Eur J Pharm Sci* 4:117.
50. Lindahl, A., et al. 1997. Characterization of fluids from the stomach and proximal jejunum in men and women. *Pharm Res* 14:497.
51. Ikesue, K., P. Kopecková, and J. Kopecek. 1993. Degradation of proteins by guinea pig intestinal enzymes. *Int J Pharm* 95:171.
52. Bernkop-Schnürch, A., C. Paikl, and C. Valenta. 1997. Novel bioadhesive chitosan–EDTA conjugate protects leucine enkephalin from degradation by aminopeptidase N. *Pharm Res* 14:917.
53. Sanderink, G.-J., Y. Artur, and G. Siest. 1988. Human aminopeptidase: A review of the literature. *J Clin Chem Clin Biochem* 26:795.
54. Garner, C.W.J., and F.J. Behal. 1974. Human liver aminopeptidase. Role of metal ions in mechanism of action. *Biochemistry* 13:3227.
55. Sangadala, S., et al. 1994. A mixture of *Manduca sexta* aminopeptidase and phosphatase enhances *Bacillus thuringiensis* insecticidal CryIA(c) toxin binding and 86Rb(+)-K^+ efflux *in vitro*. *J Biol Chem* 269:10088.
56. Mizuma, T., A. Koyanagi, and S. Awazu. 1997. Intestinal transport and metabolism of analgesic dipeptide, kyotorphin: Rate-limiting factor in intestinal absorption of peptide as drug. *Biochim Biophys Acta* 1335:111.
57. Lee, V.H.L. 1990. Protease inhibitors and permeation enhancers as approaches to modify peptide absorption. *J Control Release* 13:213.
58. Luessen, H.L., et al. 1995. Mucoadhesive polymers in peroral peptide drug delivery. II. Carbomer and polycarbophil are potent inhibitors of the intestinal proteolytic enzyme trypsin. *Pharm Res* 12:1293.
59. Bernkop-Schnürch, A., and N.C. Göckel. 1997. Development and analysis of a polymer protecting from luminal enzymatic degradation caused by α-chymotrypsin. *Drug Dev Ind Pharm* 23:733.
60. Lehr, C.-M. 1994. Bioadhesion technologies for the delivery of peptide and protein drugs to the gastrointestinal tract. *Crit Rev Ther Drug* 11:119.
61. Junginger, H.E. 1990. Bioadhesive polymer systems for peptide delivery. *Acta Pharm Technol* 36:110.
62. Luessen, H.L., et al. 1996. Mucoadhesive polymers in peroral peptide drug delivery. VI. Carbomer and chitosan improve the intestinal absorption of the peptide drug buserelin *in vivo*. *Pharm Res* 13:1668.
63. Brochard, G., et al. 1996. The potential of mucoadhesive polymers in enhancing intestinal peptide drug absorption. III. Effects of chitosan–glutamate and carbomer on epithelial tight junctions *in vitro*. *J Control Release* 39:131.
64. Bernkop-Schnürch, A., and M.E. Krajicek. 1998. Mucoadhesive polymers as platforms for peroral peptide delivery and absorption: Synthesis and evaluation of different chitosan–EDTA conjugates. *J Control Release* 50:215.
65. Luessen, H.L., et al. 1996. Mucoadhesive polymers in peroral peptide drug delivery. V. Effect of poly(acrylates) on the enzymatic degradation of peptide drugs by intestinal brush border membrane vesicles. *Int J Pharm* 141:39.
66. Bernkop-Schnürch, A., C.E. Kast, and D. Guggi. 2003. Permeation enhancing polymers in oral delivery of hydrophilic macromolecules: Thiomer/GSH systems. *J Control Release* 93:95.
67. Bernkop-Schnürch, A., and S. Thaler. 2000. Polycarbophil–cysteine conjugates as platforms for oral (poly)peptide delivery systems. *J Pharm Sci* 89:901.
68. Bernkop-Schnürch, A., G. Walker, and H. Zarti. 2001. Thiolation of polycarbophil enhances its inhibition of intestinal brush border membrane bound aminopeptidase N. *J Pharm Sci* 90:1907.

69. Valenta, C., et al. 2002. Evaluation of the inhibition effect of thiolated poly(acrylates) on vaginal membrane bound aminopeptidase N and release of the model drug LH-RH. *J Pharm Pharmacol* 54:603.
70. Spilburg, C.A., et al. 1995. Identification of a species specific regulatory site in human pancreatic cholesterol esterase. *Biochemistry* 34:15532.
71. Heidrich, J.E., et al. 2004. Inhibition of pancreatic cholesterol esterase reduces cholesterol absorption in the hamster. *BMC Pharmacol* 4:5.
72. Wadkins, R.M., et al. 2004. Discovery of novel selective inhibitors of human intestinal carboxylesterase for the amelioration of irinotecan-induced diarrhea: Synthesis, quantitative structure–activity relationship analysis, and biological activity. *Mol Pharmacol* 65:1336.
73. Prueksaritanont, T., et al. 1998. *In vitro* and *in vivo* evaluations of intestinal barriers for the zwitterion L-767,679 and its carboxyl ester prodrug L-775,318. Roles of efflux and metabolism. *Drug Metab Dispos* 26:520.
74. White, R.D., D.L. Earnest, and D.E. Carter. 1983. The effect of intestinal esterase inhibition on the *in vivo* absorption and toxicity of di-*n*-butyl phthalate. *Food Chem Toxicol* 21:99.
75. Bernkop-Schnürch, A., and K. Dundalek. 1996. Novel bioadhesive drug delivery system protecting (poly)peptides from gastric enzymatic degradation. *Int J Pharm* 138:75.
76. Bernkop-Schnürch, A., I. Bratengeyer, and C. Valenta. 1997. Development and *in vitro* evaluation of a drug delivery system protecting from trypsinic degradation. *Int J Pharm* 157:17.
77. Akiyama, Y., et al. 1996. Novel peroral dosage forms with protease inhibitory activities. 2. Design of fast dissolving poly(acrylate) and controlled drug-releasing capsule formulations with trypsin inhibiting properties. *Int J Pharm* 138:13.
78. Takada, K., et al. 1991. Effect of pH, dietary proteins and trypsin inhibitors on hydrolytic rate of human granulocyte colony-stimulating factor (G-CSF) by rat digestive enzymes. *J Pharmacobiodyn* 14:363.
79. Asada, H., et al. 1994. Stability of acyl derivatives of insulin in the small intestine: Relative importance of insulin association characteristics in aqueous solution. *Pharm Res* 11:1115.
80. Yamamoto, A., E. Hayakawa, and V.H.L. Lee. 1990. Insulin and proinsulin proteolysis in mucosal homogenates of the albino rabbit: Implications in peptide delivery from nonoral routes. *Life Sci* 47:2465.
81. Langguth, P., et al. 1997. The challenge of proteolytic enzymes in intestinal peptide delivery. *J Control Release* 46:39.
82. Heizmann, J., et al. 1996. Enzymatic cleavage of thymopoietin oligopeptides by pancreatic and intestinal brush-border enzymes. *Peptides* 17:1083.
83. Friedman, D.L., and G.L. Amidon. 1991. Oral absorption of peptides: Influence of pH and inhibitors on the intestinal hydrolysis of leu-enkephalin and analogues. *Pharm Res* 8:93.
84. Krauland, A.H., D. Guggi, and A. Bernkop-Schnürch. 2004. Oral insulin delivery: The potential of thiolated chitosan-insulin tablets on non-diabetic rats. *J Control Release* 95:547.
85. Lehr, C.M., et al. 1992. Effects of the mucoadhesive polymer polycarbophil on the intestinal absorption of a peptide drug in the rat. *J Pharm Pharmacol* 44:402.
86. Geary, R.S., and H.W. Schlameus. 1993. Vancomycin and insulin used as models for oral delivery of peptides. *J Control Release* 23:65.
87. New, R.R., et al. 1985. Liposomal immunisation against snake venoms. *TOXICON* 23:215.
88. Aungst, B.J., and N.J. Rogers. 1988. Site dependence of absorption-promoting actions of Laureth-9, sodium salicylate, disodium EDTA, and aprotinin on rectal, nasal, and buccal insulin delivery. *Pharm Res* 5:305.
89. Luessen, H.L., et al. 1997. Mucoadhesive polymers in peroral peptide drug delivery. IV. Polycarbophil and chitosan are potent enhancers of peptide transport across intestinal mucosae *in vitro*. *J Control Release* 45:15.
90. Thiebaut, F., et al. 1987. Cellular localization of the multidrug-resistance gene product P-glycoprotein in normal human tissues. *Proc Natl Acad Sci USA* 84:7735.
91. Cordon-Cardo, C., et al. 1990. Expression of the multidrug resistance gene product (P-glycoprotein) in human normal and tumor tissues. *J Histochem Cytochem* 38:1277.
92. Friend, D.R. 2004. Drug delivery to the small intestine. *Curr Gastroenterol Rep* 6:371.

93. Romiti, N., et al. 2004. Effects of grapefruit juice on the multidrug transporter P-glycoprotein in the human proximal tubular cell line HK-2. *Life Sci* 76:293.
94. Honda, Y., et al. 2004. Effects of grapefruit juice and orange juice components on P-glycoprotein- and MRP2-mediated drug efflux. *Br J Pharmacol* 143:856.
95. Bailey, D.G., and G.K. Dresser. 2004. Natural products and adverse drug interactions. *CMAJ* 170:1531.
96. Jodoin, J., M. Demeule, and R. Beliveau. 2002. Inhibition of the multidrug resistance P-glycoprotein activity by green tea polyphenols. *Biochim Biophys Acta* 1542:149.
97. Varma, M.V., et al. 2003. P-glycoprotein inhibitors and their screening: A perspective from bioavailability enhancement. *Pharmacol Res* 48:347.
98. Khan, S.I., et al. 2003. Transport of parthenolide across human intestinal cells (Caco-2). *Planta Med* 69:1009.
99. Walgren, R.A., et al. 2000. Efflux of dietary flavonoid quercetin 4′-beta-glucoside across human intestinal Caco-2 cell monolayers by apical multidrug resistance-associated protein-2. *J Pharmacol Exp Ther* 294:830.
100. Lowes, S., M.E. Cavet and N.L. Simmons. 2003. Evidence for a non-MDR1 component in digoxin secretion by human intestinal Caco-2 epithelial layers. *Eur J Pharmacol* 458:49.
101. Potschka, H., S. Baltes, and W. Loscher. 2004. Inhibition of multidrug transporters by verapamil or probenecid does not alter blood–brain barrier penetration of levetiracetam in rats. *Epilepsy Res* 58:85.
102. Olson, D.P., et al. 2002. The protease inhibitor ritonavir inhibits the functional activity of the multidrug resistance related-protein 1 (MRP-1). *AIDS* 16:1743.
103. Makhey, V.D., et al. 1998. Characterization of the regional intestinal kinetics of drug efflux in rat and human intestine and in Caco-2 cells. *Pharm Res* 15:1160.
104. Seral, C., et al. 2003. Influence of P-glycoprotein and MRP efflux pump inhibitors on the intracellular activity of azithromycin and ciprofloxacin in macrophages infected by *Listeria monocytogenes* or *Staphylococcus aureus*. *J Antimicrob Chemother* 51:1167.
105. Wang, S., et al. 2004. Studies on pyrrolopyrimidines as selective inhibitors of multidrug-resistance-associated protein in multidrug resistance. *J Med Chem* 47:1329.
106. O'Connor, R., et al. 2004. Increased anti-tumour efficacy of doxorubicin when combined with sulindac in a xenograft model of an MRP-1-positive human lung cancer. *Anticancer Res* 24:457.
107. Hu, K., and M.E. Morris. 2004. Effects of benzyl-, phenethyl-, and alpha-naphthyl isothiocyanates on P-glycoprotein- and MRP1-mediated transport. *J Pharm Sci* 93:1901.
108. Leu, B.L., and J.D. Huang. 1995. Inhibition of intestinal P-glycoprotein and effects on etoposide absorption. *Cancer Chemother Pharmacol* 35:432.
109. Malingre, M.M., et al. 2001. Pharmacokinetics of oral cyclosporin A when co-administered to enhance the absorption of orally administered docetaxel. *Eur J Clin Pharmacol* 57:305.
110. Floren, L.C., et al. 2001. Tacrolimus oral bioavailability doubles with coadministration of ketoconazole. *Clin Pharmacol Ther* 62:41.
111. Lennernas, H. 2003. Clinical pharmacokinetics of atorvastatin. *Clin Pharmacokinet* 42:1141.
112. Woo, J.S., et al. 2003. Enhanced oral bioavailability of paclitaxel by coadministration of the P-glycoprotein inhibitor KR30031. *Pharm Res* 20:24.
113. Choi, J.S., H.K. Choi, and S.C. Shin. 2004. Enhanced bioavailability of paclitaxel after oral coadministration with flavone in rats. *Int J Pharm* 275:165.
114. Kimura, Y., et al. 2002. P-glycoprotein inhibition by the multidrug resistance-reversing agent MS-209 enhances bioavailability and antitumor efficacy of orally administered paclitaxel. *Cancer Chemother Pharmacol* 49:322.
115. Bardelmeijer, H.A., et al. 2004. Efficacy of novel P-glycoprotein inhibitors to increase the oral uptake of paclitaxel in mice. *Invest New Drugs* 22:219.
116. Bardelmeijer, H.A., et al. 2000. Increased oral bioavailability of paclitaxel by GF120918 in mice through selective modulation of P-glycoprotein. *Clin Cancer Res* 6:4416.
117. van Asperen, J., et al. 1997. Enhanced oral bioavailability of paclitaxel in mice treated with the P-glycoprotein blocker SDZ PSC 833. *Br J Cancer* 76:1181.

118. Kuppens, I.E., et al. 2005. Oral bioavailability of docetaxel in combination with OC144-093 (ONT-093). *Cancer Chemother Pharmacol* 55:72.
119. Verstuyft, C., et al. 2003. Dipyridamole enhances digoxin bioavailability via P-glycoprotein inhibition. *Clin Pharmacol Ther* 73:51.
120. Hugger, E.D., et al. 2003. Automated analysis of polyethylene glycol-induced inhibition of P-glycoprotein activity *in vitro*. *J Pharm Sci* 92:21.
121. Hugger, E.D., K.L. Audus, and R.T. Borchardt. 2002. Effects of poly(ethylene glycol) on efflux transporter activity in Caco-2 cell monolayers. *J Pharm Sci* 91:1980.
122. Batrakova, E.V., et al. 1999. Pluronic P85 increases permeability of a broad spectrum of drugs in polarized BBMEC and Caco-2 cell monolayers. *Pharm Res* 16:1366.
123. Bromberg, L., and V. Alakhov. 2003. Effects of polyether-modified poly(acrylic acid) microgels on doxorubicin transport in human intestinal epithelial Caco-2 cell layers. *J Control Release* 88:11.
124. Banerjee, S.K., et al. 2000. Bioavailability of tobramycin after oral delivery in FVB mice using CRL-1605 copolymer, an inhibitor of P-glycoprotein. *Life Sci* 67:2011.
125. Jagannath, C., et al. 1999. Significantly improved oral uptake of amikacin in FVB mice in the presence of CRL-1605 copolymer. *Life Sci* 64:1733.
126. Dintaman, J.M., and J.A. Silverman. 1999. Inhibition of P-glycoprotein by D-alpha-tocopheryl polyethylene glycol 1000 succinate (TPGS). *Pharm Res* 16:1550.
127. Bittner, B., et al. 2002. Improvement of the bioavailability of colchicine in rats by co-administration of D-alpha-tocopherol polyethylene glycol 1000 succinate and a polyethoxylated derivative of 12-hydroxy-stearic acid. *Arzneimittelforschung* 52:684.
128. Rege, B.D., J.P. Kao, and J.E. Polli. 2002. Effects of nonionic surfactants on membrane transporters in Caco-2 cell monolayers. *Eur J Pharm Sci* 16:237.
129. Werle, M., and M. Hoffer. 2006. Glutathione and thiolated chitosan inhibit multidrug resistance P-glycoprotein activity in excised small intestine. *J Control Rel* 111:41.
130. Foger, F., T. Schmitz, and A. Bernkop-Schnürch. 2006. In vivo evaluation of an oral delivery system for P-gp substrates based on thiolated chitosan. *Biomaterials* 27:4250.
131. Baluom, M., M. Friedman, and A. Rubinstein. 2001. Improved intestinal absorption of sulpiride in rats with synchronized oral delivery systems. *J Control Release* 70:139.
132. Stephens, R.H., et al. 2002. Resolution of P-glycoprotein and non-P-glycoprotein effects on drug permeability using intestinal tissues from mdr1a (-/-) mice. *Br J Pharmacol* 135:2038.
133. Walle, U.K., and T. Walle. 1998. Taxol transport by human intestinal epithelial Caco-2 cells. *Drug Metab Dispos* 26:343.
134. Fromm, M.F. 2000. P-glycoprotein: A defense mechanism limiting oral bioavailability and CNS accumulation of drugs. *Int J Clin Pharmacol Ther* 38:69.
135. Fromm, M.F. 2003. Importance of P-glycoprotein for drug disposition in humans. *Eur J Clin Invest* 33:6.
136. Kudlicki, W.A., M.M. Winkler, and B.L. Pastoske. 2003. Nuclease inhibitor cocktail. December 16, 2003, USA.
137. Talib, S., and J.E. Hearst. 1983. Initiation of RNA synthesis *in vitro* by vesicular stomatitis virus: Single internal initiation in the presence of aurintricarboxylic acid and vanadyl ribonucleoside complexes. *Nucleic Acids Res* 11:7031.
138. Dellundé, S., et al. 2002. A fast and sensitive nucleic acid extraction method for the detection of *Cryptosporidium* by PCR in environmental water samples. *Water Supply* 2:95.
139. Blackburn, P., G. Wilson, and S. Moore. 1977. Ribonuclease inhibitor from human placenta: Purification and properties. *J Biol Chem* 252:5904.
140. Berger, S.L., and C.S. Birkenmeier. 1979. Inhibition of intractable nucleases with ribonucleoside–vanadyl complexes: Isolation of messenger ribonucleic acid from resting lymphocytes. *Biochemistry* 18:5143.
141. Morishita, M., et al. 1992. Novel oral microspheres of insulin with protease inhibitor protecting from enzymatic degradation. *Int J Pharm* 78:1.
142. Murakawa, Y., et al. 1996. Gastrointestinal absorption of recombinant human insulin-like growth factor-I in rats. *Proc Int Symp Controlled Release Bioact Mater* 23:589.

143. Fjellestad-Paulsen, A., et al. 1996. Bioavailability of 1-deamino-8-D-arginine vasopressin with an enzyme inhibitor (aprotinin) from the small intestine in healthy volunteers. *Eur J Clin Pharmacol* 50:491.
144. Trenktrog, T., et al. 1996. Enteric coated insulin pellets-development, drug release and *in vivo* evaluation. *Eur J Pharm Sci* 4:323.
145. Birk, Y. 1985. The Bowman–Birk inhibitor. *Int J Pept Protein Res* 25:113.
146. Suenaga, H., et al. 1995. Strong inhibitory effect of sugar biphenylboronic acid complexes on the hydrolytic activity of alpha-chymotrypsin. *J Chem Soc* 13:1733.
147. Abd El-Hameeda, M.D., and I.W. Kellaway. 1997. Preparation and *in vitro* characterisation of mucoadhesive polymeric microspheres as intra-nasal delivery systems. *Eur J Pharm Biopharm* 44:53.
148. Powers, J. C., et al. 1977. Specificity of porcine pancreatic elastase, human leukocyte elastase and cathepsin G. Inhibition with peptide chloromethyl ketones. *Biochim Biophys Acta* 485:156.
149. Parellada, J., and M. Guinea. 1995. Flavonoid inhibitors of trypsin and leucine aminopeptidase: A proposed mathematical model for IC50 estimation. *J Nat Prod* 58:823.
150. Manjabacas, M.C., et al. 1995. Kinetic analysis of an autocatalytic process coupled to a reversible inhibition: The inhibition of the system trypsinogen-trypsin by *p*-aminobenzamidine. *Biol Chem Hoppe Seyler* 376:577.
151. Artursson, P., et al. 1994. Effect of chitosan on the permeability of monolayers of intestinal epithelial cells (Caco-2). *Pharm Res* 11:1358.
152. Luessen, H.L., et al. 1995. Carbomer and polycarbophil are potent inhibitors of the intestinal proteolytic enzyme trypsin. *Pharm Res* 12:1293.
153. Smith, T.K., et al. 1995. Different sites of acivicin binding and inactivation of gamma-glutamyl transpeptidases. *Proc Natl Acad Sci USA* 92:2360.
154. Diaz-Flores, M., G. Duran-Reyes, and J.J. Hicks. 1994. Arrest of rat embryonic development by the inhibition of gamma-glutamyl transpeptidase. I. Intrauterine administration of L-serine–borate complex. *Int J Fertil Menopausal Stud* 39:234.
155. McGwire, G.B., and R.A. Skidgel. 1995. Extracellular conversion of epidermal growth factor (EGF) to des-Arg53-EGF by carboxypeptidase M. *J Biol Chem* 270:17154.
156. Borloo, M., and I. De Meester. 1994. Dipeptidyl peptidase IV: Development, design, synthesis and biological evaluation of inhibitors. *Verh K Acad Geneeskd Belg* 56:57.
157. Kenny, A.J., and S. Maroux. 1982. Topology of microvillar membrane hydrolases of kidney and intestine. *Physiol Rev* 62:91.
158. Auricchio, S., et al. 1978. Dipeptidylaminopeptidase and carboxypeptidase activities of the brush border of rabbit small intestine. *Gastroenterology* 75:1073.
159. Giannousis, P.P., and P.A. Bartlett. 1987. Phosphorus amino acid analogues as inhibitors of leucine aminopeptidase. *J Med Chem* 30:1603.
160. Luciani, N., et al. 1998. Characterization of Glu350 as a critical residue involved in the N-terminal amine binding site of aminopeptidase N (EC 3.4.11.2): Insights into its mechanism of action. *Biochemistry* 37:686.
161. Huang, K., et al. 1997. Alanyl aminopeptidase from human seminal plasma: Purification, characterization, and immunohistochemical localization in the male genital tract. *J Biochem* (*Tokyo*) 122:779.
162. Miller, B.C., et al. 1994. Methionine enkephalin is hydrolyzed by aminopeptidase N on $CD4^+$ and $CD8^+$ spleen T cells. *Arch Biochem Biophys* 311:174.
163. Taki, Y., et al. 1995. Gastrointestinal absorption of peptide drug: Quantitative evaluation of the degradation and permeation of metkephamid in rat small intestine. *J Pharm Exp Therapeut* 274:373.
164. Kramer, W., et al. 1990. Intestinal uptake of dipeptides and beta-lactam antibiotics. I. The intestinal uptake system for dipeptides and beta-lactam antibiotics is not part of a brush border membrane peptidase. *Biochim Biophys Acta* 1030:41.
165. Inoue, T., et al. 1994. Bestatin, a potent aminopeptidase-N inhibitor, inhibits *in vitro* decidualization of human endometrial stromal cells. *J Clin Endocrinol Metab* 79:171.
166. Hedeager-Sorensen, S., and A.J. Kenny. 1985. Proteins of the kidney microvillar membrane. Purification and properties of carboxypeptidase P from pig kidneys. *Biochem J* 229:251.

167. Bunnett, N.W., et al. 1988. Isolation of endopeptidase-24.11 (EC 3.4.24.11, "enkephalinase") from the pig stomach. Hydrolysis of substance P, gastrin-releasing peptide 10, [Leu5] enkephalin, and [Met5] enkephalin. *Gastroenterology* 95:952.
168. Ryan, J.W. 1988. Angiotensin-converting enzyme, dipeptidyl carboxypeptidase I, and its inhibitor. *Methods Enzymol* 163:194.
169. Takada, Y., K. Hiwada, and T. Kokubu. 1981. Isolation and characterization of angiotensin converting enzyme from human kidney. *J Biochem* 90:1309.
170. Soffer, R.L. 1976. Angiotensin-converting enzyme and the regulation of vasoactive peptides. *Annu Rev Biochem* 45:73.
171. Bernkop-Schnürch, A., and A. Scerbe-Saiko. 1998. Synthesis and *in vitro* evaluation of chitosan–EDTA–protease-inhibitor conjugates which might be useful in oral delivery of peptides and proteins. *Pharm Res* 15:263.
172. Bernkop-Schnürch, A., and I. Apprich. 1997. Synthesis and evaluation of a modified mucoadhesive polymer protecting from α-chymotrypsinic degradation. *Int J Pharm* 146:247.
173. Bernkop-Schnürch, A., G.H. Schwarz, and M. Kratzel. 1997. Modified mucoadhesive polymers for the peroral administration of mainly elastase degradable therapeutic (poly)peptides. *J Control Release* 47:113.
174. Rehfeldt, W., K. Resch, and M. Goppelt-Struebe. 1993. Cytosolic phospholipase A2 from human monocytic cells: Characterization of substrate specificity and Ca^{2+}-dependent membrane association. *Biochem J* 293:255.
175. Akiba, S., et al. 1998. Characterization of acidic Ca^{2+}-independent phospholipase A2 of bovine lung. *Comp Biochem Physiol B* 120:393.
176. Ciliv, G., and P.T. Oezand. 1972. Human erythrocyte acetylcholinesterase; purification, properties and kinetic behavior. *Biochim Biophys Acta* 284:136.
177. Larsson Forsell, P.K.A., B.P. Kennedy, and H.E. Claesson. 1999. The human calcium-independent phospholipase A2 gene multiple enzymes with distinct properties from a single gene. *Eur J Biochem* 262:575.
178. Yang, Y., and M.E. Lowe. 1998. Human pancreatic triglyceride lipase expressed in yeast cells: Purification and characterization. *Protein Expr Purif* 13:36.
179. Okada, Y., and K. Wakabayashi. 1988. Purification and characterization of esterases D-1 and D-2 from human erythrocytes. *Arch Biochem Biophys* 263:130.
180. Yang, H.C., A.A. Farooqui, and L.A. Horrocks. 1996. Characterization of plasmalogen-selective phospholipase A2 from bovine brain. *Adv Exp Med Biol* 416:309.
181. Kawauchi, Y., J. Takasaki, and Y.M. Matsuura. 1994. Preparation and characterization of human rheumatoid arthritic synovial fluid phospholipase A2 produced by recombinant baculovirus-infected insect cells. *J Biochem* 116:82.
182. Brown, S.S., et al. 1981. The plasma cholinesterase: A new perspective. *Adv Clin Chem* 22:1.
183. Lombardo, D., O. Guy, and C. Figarella. 1978. Purification and characterization of a carboxyl ester hydrolase from human pancreatic juice. *Biochim Biophys Acta* 527:142.
184. Primo-Parmo, S.L., H. Lightstone, and B.N. La Du. 1997. Characterization of an unstable variant (BChE115D) of human butyrylcholinesterase. *Pharmacogenetics* 7:27.
185. Gan, K.N., et al. 1991. Purification of human serum paraoxonase/arylesterase. Evidence for one esterase catalyzing both activities. *Drug Metab Dispos* 19:100.
186. Smolen, A., et al. 1991. Characteristics of the genetically determined allozymic forms of human serum paraoxonase/arylesterase. *Drug Metab Dispos* 19:107.
187. Ravazzolo, R., et al. 1988. Characterization, localization, and biosynthesis of acetylcholinesterase in K562 cells. *Arch Biochem Biophys* 267:245.
188. Saboori, A.M., and D.S. Newcombe. 1990. Human monocyte carboxylesterase. Purification and kinetics. *J Biol Chem* 265:19792.
189. Lamura, E., et al. 1997. Compartmentalisation and characteristics of a Ca^{2+}-dependent phospholipase A2 in human colon mucosa. *Biochem Pharmacol* 53:1323.
190. Hojring, N., and O. Svensmark. 1988. Molecular and catalytic properties of a butyrylesterase from human red cells and brain. *Arch Biochem Biophys* 260:351.

191. Gonzalo, M.C., et al. 1998. Human liver paraoxonase (PON1): Subcellular distribution and characterization. *Biochem Mol Toxicol* 12:61.
192. Franken, P.A., et al. 1992. Purification and characterization of a mutant human platelet phospholipase A2 expressed in *Escherichia coli. Eur J Biochem* 203:89.
193. Reiner, E., V. Simeon-Rudolf, and M. Skrinjaric-Spoljar. 1995. Catalytic properties and distribution profiles of paraoxonase and cholinesterase phenotypes in human sera. *Toxicol Lett* 82/83, 447, 1995.
194. Wang, A., et al. 1999. A specific human lysophospholipase: cDNA cloning, tissue distribution and kinetic characterization. *Biochim Biophys Acta* 1437:157.
195. Tsujita, T., and H. Okuda. 1983. Human liver carboxylesterase. Properties and comparison with human serum carboxylesterase. *J Biochem* 94:793.
196. Hernandez, A.F., et al. 1993. Characterization of paraoxonase activity in pericardial fluid: Usefulness as a marker of coronary disease. *Chem Biol Interact* 87:173.
197. Aviram, M., et al. 1998. Paraoxonase inhibits high-density lipoprotein oxidation and preserves its functions. A possible peroxidative role for paraoxonase. *J Clin Invest* 101:1581.
198. Haagen, L., and A. Brock. 1992. A new automated method for phenotyping arylesterase (EC 3.1.1.2) based upon inhibition of enzymatic hydrolysis of 4-nitrophenyl acetate by phenyl acetate. *Eur J Clin Chem Clin Biochem* 30:391.
199. Sihotang, K. 1974. Properties of human erythrocyte acetylcholinesterase. *J Biochem* 75:939.
200. Hoffmann, G.E., et al. 1992. Characterization of a phospholipase A2 in human serum. *Eur J Clin Chem Clin Biochem* 30:111.
201. Wigler, P.W. and F.K. Patterson. 1994. Reversal agent inhibition of the multidrug resistance pump in human leukemic lymphoblasts. *Biochim Biophys Acta* 1189 (1):1.
202. Kodawara, T., et al. 2002. Organic anion transporter oatp2-mediated interaction between digoxin and amiodarone in the rat liver. *Pharm Res* 19 (6):738.
203. Baluom, M., D.I. Friedman, and A. Rubinstein. 1997. Absorption enhancement of calcitonin in the rat intestine by carbopol-containing submicron emulsions. *Int J Pharm* 154:235.
204. Perloff, M.D., et al. 2003. Rapid assessment of P-glycoprotein inhibition and induction *in vitro. Pharm Res* 20:1177.
205. Hollt, V., M. Kouba, M. Dietel, and G. Vogt. 1992. Stereoisomers of calcium antagonists which differ markedly in their potencies as calcium blockers are equally effective in modulating drug transport by P-glycoprotein. *Biochem Pharmacol* 43:2601.
206. Barancik, M., et al. 1994. Reversal effects of several Ca(2^+)-entry blockers, neuroleptics and local anaesthetics on P-glycoprotein-mediated vincristine resistance of L1210/VCR mouse leukaemic cell line. *Drugs Exp Clin Res* 20:13.
207. Wandel, C., et al. 2000. Mibefradil is a P-glycoprotein substrate and a potent inhibitor of both P-glycoprotein and CYP3A *in vitro. Drug Metab Dispos* 28:895.
208. Aebi, S., et al. 1999. A phase II/pharmacokinetic trial of high-dose progesterone in combination with paclitaxel. *Cancer Chemother Pharmacol* 44:259.
209. Gruol, D.J., and S. Bourgeois. 1994. Expression of the mdr1 P-glycoprotein gene: A mechanism of escape from glucocorticoid-induced apoptosis. *Biochem Cell Biol* 72:561.
210. Kirk, J., et al. 1994. Reversal of P-glycoprotein-mediated multidrug resistance by pure anti-oestrogens and novel tamoxifen derivatives. *Biochem Pharmacol* 48:277.
211. Peck, R.A., et al. 2001. Phase I and pharmacokinetic study of the novel MDR1 and MRP1 inhibitor biricodar administered alone and in combination with doxorubicin. *J Clin Oncol* 19:3130.
212. Walker, J., C. Martin, and R. Callaghan. 2004. Inhibition of P-glycoprotein function by XR9576 in a solid tumour model can restore anticancer drug efficacy. *Eur J Cancer* 40:594.

6 Enhanced Gastrointestinal Absorption of Lipophilic Drugs

Arik Dahan and Amnon Hoffman

CONTENTS

6.1 INTRODUCTION

The oral route is the preferred mode of drug administration, mainly due to comfort and patient compliance. As a consequence of modern drug discovery techniques (i.e., advances in *in-vitro* screening methods, the introduction of combinatorial chemistry), the number of poor water-soluble drug compounds is constantly increasing, and to date, more than 40% of new chemical entities are lipophilic and exhibit poor water solubility [1]. These molecules suffer from low oral bioavailability, and despite their pharmacological activity, fail to proceed to the advanced stages of research and development. A great challenge facing the pharmaceutical community is to turn these molecules into orally administered medications with sufficient bioavailability. This chapter specifies the barriers that lipophilic drugs have to traverse along the intestinal absorption process, as well as the means that can be utilized to overcome these barriers and enhance the bioavailability of the lipophilic drug. For a comprehensive view, the barriers and the corresponding possible solutions are presented according to their physiological order, i.e., preenterocyte, enterocyte, and postenterocyte barriers (Figure 6.1).

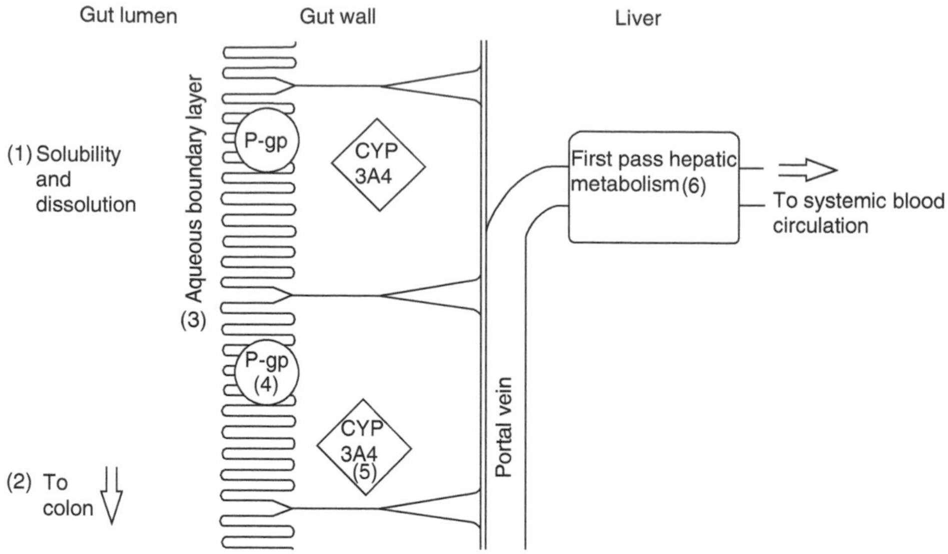

FIGURE 6.1 The barriers that a lipophilic drug has to transverse along the intestinal absorption process: (1) dissolution and solubilization in the intestinal milieu, (2) narrow absorption window, (3) unstirred water layer, (4) efflux pumps, (5) intra-enterocyte metabolism, and (6) first pass hepatic metabolism.

6.2 BARRIERS FACING LIPOPHILIC DRUG ABSORPTION

6.2.1 Preenterocyte Barriers

6.2.1.1 Solubility

Following oral administration, dissolution of the drug molecule in the intestinal milieu is a prerequisite for the absorption process. According to the Biopharmaceutical Classification System (BCS) [2], poor water-soluble compounds (i.e., aqueous solubility less than 100 μg/mL) are class 2 or class 4 compounds. For class 2 compounds, the absorption level is dictated by the dissolution properties of the molecule in the gastrointestinal (GI) fluids. BCS class 4 compounds, which are characterized by both low solubility and poor intestinal wall permeability, are generally poor drug candidates (unless the dose is very low).

As the water solubility of a lipophilic molecule is poor, the aid of surfactants provided by biliary secretions is necessary to solubilize the lipophilic drug and enable its absorption. The bile fluid is secreted by the liver, and is stored and concentrated in the gall bladder. The major organic solutes of the bile are bile acids, phospholipids (particularly lecithin), and cholesterol. The bile acids are derivatives of cholesterol in which hydroxyl and carboxylic acids are attached to the steroid moiety, converting it into a powerful natural surfactant. These biliary secretions aid in the solubilization of the lipophilic drug by forming submicron mixed micelles in which the lipophilic molecule is solubilized and reaches the absorptive membrane of the enterocyte. This process is limited in its capacity and varies considerably in different situations. Also, as it is the rate-limiting step in the absorption of lipophilic drugs, it often becomes a significant absorption barrier.

6.2.1.2 Limited Absorption Site

The solubilization of the lipophilic molecule occurs mostly at the upper part of the GI tract, where pancreatic fluids and biliary lipids (including bile salts, phospholipid and cholesterol ester) are secreted and aid the solubilization process. Therefore, the absorption of these molecules usually takes place along the small intestine. The residence time of a molecule in the upper GI is limited, and the transit time along the small intestine is 3.5–4.5 h in healthy volunteers. Fat can cause a modest extension of the small intestinal transit time (30–60 min), however, this effect is not thought to be significant in drug delivery. The limited transit time along the small intestine may limit the absorption of a lipophilic compound, and if the lipophilic molecule reaches the colon before solubilization its bioavailability is expected to be low.

6.2.1.3 Unstirred Water Layer

The layer of water adjacent to the absorptive membrane of the enterocyte is essentially unstirred. It can be visualized as a series of water lamellas, each progressively more stirred from the gut wall toward the lumen bulk. For BCS class 2 compounds the rate of permeation through the brush border is fast and the diffusion across the unstirred water layer (UWL) is the rate-limiting step in the permeation process. The thickness of the UWL in human jejunum was measured and found to be over 500 μm [3]. Owing to its thickness and hydrophilicity, the UWL may represent a major permeability barrier to the absorption of lipophilic compounds. The second mechanism by which the UWL functions act as a barrier to drug absorption is its effective surface area. The ratio of the surface area of the UWL to that of the underlying brush border membrane is at least 1:500 [4], i.e., this layer reduces the effective surface area available for the absorption of lipophilic compounds and hence impairs its bioavailability.

6.2.2 Enterocyte Barriers

6.2.2.1 Cytochrome P450 3A4 and P-glycoprotein

Once a drug molecule enters the enterocyte, it faces biochemical barriers that affect the magnitude of its absorption. The enterocyte cytochrome P450 3A4 (CYP3A4) enzymes are located in the endoplasmic reticulum of the enterocyte and are responsible for most drug metabolism in the intestinal wall. This isoenzyme accounts for more than 70% of all small intestinal CYP450s [5], and many studies have established its role as a major barrier to the absorption of lipophilic drugs, which are the most likely molecules to undergo oxidative metabolism.

Whereas some transporters located in the apical wall of the enterocyte facilitate absorption, there are others that serve as efflux transporters. These are considered as the multiple drug resistance (MDR) transporters, and they play a major role in the disposition of many drugs. The most extensively studied MDR transporter is the apical P-glycoprotein (P-gp) efflux pump that reduces the fraction of drug absorbed by transporting the drug from the enterocyte back to the intestinal lumen [6].

There is an interplay between the activity of the metabolic CYP3A4 enzymes and the P-gp system. A drug molecule that escapes the intraenterocyte metabolism may either reach the blood circulation or be effluxed back to the GI lumen, and then may be reabsorbed [7].

These two prehepatic systems have been shown to contribute to the limited oral bioavailability of many lipophilic drugs including cyclosporine [8], terfenadine [9], saquinavir [10], midazolam [11], tacrolimus [12], atorvastatin [13], and others.

6.2.3 Postenterocyte Barriers

A drug molecule that manages to escape the intraenterocyte metabolism and the MDR efflux systems can diffuse across the cell and be secreted from the basolateral membrane of the enterocyte into the lamina proporia, where the drug is usually absorbed into the portal blood (unless incorporated into a chylomicron, as further detailed in Section 6.3.2.1). Following absorption into the blood capillaries and before reaching the systemic blood circulation, the drug molecules pass through the liver, and hence are exposed to metabolic enzymes. This first pass hepatic metabolism has been shown to be a major barrier to the absorption of lipophilic drugs, which are the most likely molecules to undergo oxidative metabolism.

6.3 TECHNIQUES FOR ENHANCING LIPOPHILIC DRUG ABSORPTION

To enhance absorption, it is important to identify the rate-limiting step in this process and to counter the relevant barrier in each case. The possible solutions for the absorption barriers facing lipophilic drug absorption are presented according to their physiological order, i.e., issues concerning the GI lumen (preenterocyte), followed by issues associated with the enterocyte and onward. However, it should be noted that a few concepts affect more than one step of the absorption process.

6.3.1 Preenterocyte Associated Solutions

6.3.1.1 Lipid-Based Formulation Approach

The ability of lipid vehicles (either in the pharmaceutical formulation or in food) to enhance the absorption of lipophilic drugs has been well known for many years. Recently, successful bioavailability enhancement utilizing lipid-based formulations has been accomplished with the immunosuppressive agent cyclosporine A (Neoral, Novartis Pharmaceuticals Corporation, East Hanover, NJ), and for the two HIV protease inhibitors ritonavir (Norvir, Abbott Laboratories, IL) and saquinavir (Fortovase, Roche Pharmaceuticals, Nutley, NJ). Consequently, considerable interest in lipid-based formulations has been aroused.

The mechanisms by which lipid-based formulations enhance the absorption of lipophilic drugs are:

1. Enhanced dissolution/solubilization
   The presence of lipids in the GI tract stimulates gall bladder contracts and biliary and pancreatic secretions, including bile salts, phospholipids, and cholesterol. These products, along with the gastric shear movement, form a crude emulsion, which promotes the solubilization of the coadministered lipophilic drug. Exogenous surface-active agents incorporated into the formulation may further stimulate the solubilization of the lipophilic compound.
2. Prolongation of gastric residence time
   Lipids in the GI tract provoke delay in gastric emptying, i.e., gastric transit time is increased. As a result, the residence time of the coadministered lipophilic drug in the small intestine increases. This enables better dissolution of the drug at the absorptive site, and thereby improves absorption [14].
3. Stimulation of lymphatic transport
   Bioavailability of lipophilic drugs may be enhanced also by the stimulation of the intestinal lymphatic transport pathway. This issue will be addressed separately (Section 6.3.2.1).
4. Affecting intestinal permeability
   A variety of lipids have been shown to change the physical barrier function of the gut wall, and hence, to enhance permeability. For BCS class 2 compounds, permeability

through the GI wall is not a limiting step toward absorption, and hence, this mechanism is not thought to be a major contributor for the absorption enhancement of lipophilic drugs.

5. Reduced metabolism and efflux activity
Recently, certain lipids and surfactants have been shown to reduce the activity of efflux transporters in the GI wall, and hence, to increase the fraction of drug absorbed. Because of the interplay between P-gp and CYP3A4 activity this mechanism may reduce intraenterocyte metabolism as well. This issue will be further explained in detail separately (Section 6.3.2.2).

Lipid-based formulations offer a large variety of optional systems. They can be made as solutions, suspensions, emulsions, self-emulsifying systems and microemulsions. Moreover, it is possible to form blends that are composed of several excipients: they can be pure triglyceride (TG) oils or blends of different TG, diglyceride (DG) and monoglyceride (MG). In addition, different types of surfactants (lipophilic and hydrophilic) can be added, as well as hydrophilic co-solvents. Lack of enhanced absorption when one of the above key formulations is tested does not necessarily indicate the effectiveness of alternative lipid-based formulations, and their suitability has to be examined.

The type of oil that is incorporated into the formulation has a major influence on the capability of the formulation to enhance absorption. Nondigestible lipids, including mineral oil, sucrose polyesters, and others, are not absorbed from the gut lumen. They remain in the GI lumen, retain the lipophilic drug within the oil, and hence, limit the absorption of the drug [15]. Digestible lipids, including TG, DG, phospholipids, fatty acids (FA), cholesterol, and other synthetic derivatives, are suitable oils for oral formulations of lipophilic drugs. These lipids are usually defined according to their carbon chain length, i.e., long-chain triglyceride (LCT) or medium-chain triglyceride (MCT); lipid class, i.e., TG, DG, MG, or FA; degree of saturation and their interaction with water. In general, rate and extent of digestion, amount of digestion products formed during the digestion process and degree of dispersion of these products are the factors governing the formulation potential to enhance absorption. However, thus far, a correlation between these parameters and the magnitude of *in vivo* bioavailability enhancement has met with only little success.

The effect of different lipid vehicles on the absorption of the hypocholesterolemic lipophilic (log $P \sim 10$) drug probucol was investigated [16]. Peanut oil (LCT), miglyol (MCT), and paraffin oil probucol solutions were administered to conscious rats. AUC value for probucol following the peanut oil solution was higher than the miglyol solution, whereas no absorption of probucol following the paraffin oil solution was observed (Figure 6.2).

The degree of dispersion of the lipid-based formulation has a marked influence on its ability to augment absorption. Increased degree of dispersion leads to enlarged surface area that is available for lipid digestion and drug absorption. Porter et al. [17] investigated the absorption and lymphatic transport of the lipophilic antimalarial halofantrine in the rat following oral administration in four lipid vehicles with different degrees of dispersion: two oil solutions, emulsion and micellar system. The rank-order effect of the vehicles for the promotion of halofantrine absorption and lymphatic transport was micellar > emulsion > lipid solution. These results suggest that increased dispersion of the oily formulation facilitates the absorption of the coadministered lipophilic drug. Another example of improved oral bioavailability through increased degree of dispersion of the formulation is cyclosporine A [18].

The drug particle size is also an important parameter [19]. Scholz et al. [20] studied the influence of drug particle size on felodipine absorption in dogs. Micronized (8 μm) or coarse (125 μm) suspensions of felodipine were orally administered to dogs, and absorption parameters were studied. The reduction in particle size led to a 22-fold increase in C_{max} and to a

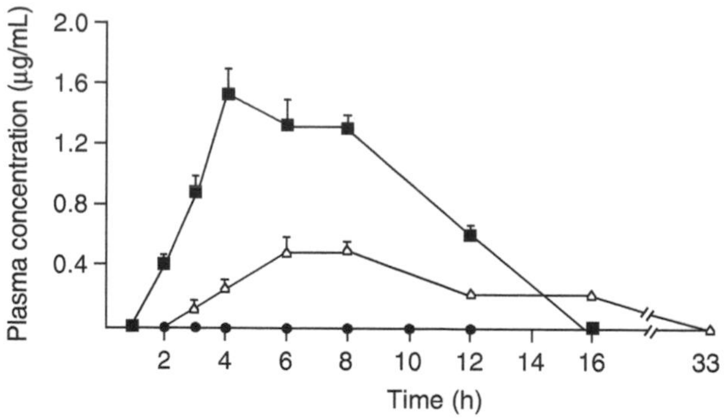

FIGURE 6.2 Probucol plasma profiles following oral administration (100 mg/kg) to rats in three different oily vehicles: peanut oil (LCT, ■), miglyol (MCT, Δ), or paraffin oil (●). (From Palin, K.J. and Wilson, C.G., *J. Pharm. Pharmacol.*, 36, 641, 1984. With permission.)

14-fold increase in AUC. Moreover, the absorption of felodipine from the coarse suspension was sensitive to hydrodynamic changes in the GI tract, whereas the reduction in particle size cancelled this impact.

Further examples for of importance of particle size on the absorption of lipophilic drugs have been described for griseofulvin [21] and progesterone [22].

6.3.1.1.1 Different Types of Lipid-Based Formulations

6.3.1.1.1.1 Solutions and Suspensions The most straightforward approach in developing a lipid-based formulation is to dissolve or suspend the drug in the chosen vehicle. This option may be suitable only for highly lipophilic drugs (log $P > 4$) or for the cases where the drug is highly potent and requires a low dose, due to insufficient solubility of the drug in the lipid vehicle. The addition of DG and MG to the oil can help in overcoming this problem.

Hargrove et al. [22] administered progesterone (200 mg) to seven subjects in different formulations: suspension in LCT, micronized suspension in the same oil and micronized drug in nonoily formulation. Peak plasma concentration following the micronized oily suspension was 30 ng/mL, whereas the nonoily micronized formulation C_{max} was only 13 ng/mL, hence, the oil had an enhancing effect on the oral bioavailability of the progesterone. On the other hand, the nonmicronized oily suspension of progesterone had no effect on progesterone absorption (C_{max} value of 11 ng/mL), highlighting the importance of the particle size of the dispersed drug.

Further cases in which simple oily solutions or suspensions contributed to higher absorption are cinnarizine [23], phenytoin [24], and lipophilic steroid [25].

6.3.1.1.1.2 Emulsions Emulsification of the lipid-based formulation holds a few important advantages: the degree of dispersion is elevated and the particle size decreased, leading to increased surface area available for the drug to be released from the vehicle and absorbed thereafter. The increased surface area also permits an easier access for lipase–colipase complex, whereby the rate and extent of lipid digestion in the lumen is augmented. Another advantage of the emulsion is the presence of exogenous surfactants in the formulation. These surfactants reduce surface tension and thereby contribute to the increased surface area of the emulsified lipid. Exogenous surfactants may aid the solubilization process of the lipophilic drug, and increase its penetration through the aqueous boundary layer. Moreover, a few surfactants may inhibit the activity of efflux transporter systems in the GI wall, leading to enhanced bioavailability, as further detailed separately (Section 6.3.2.2).

Hauss et al. [26] investigated the absorption of ontazolast, a poor water-soluble (0.14 μg/mL) LTB_4 inhibitor, following oral administration of a few formulations to rats. All the oil-containing formulations markedly improved ontazolast bioavailability, with the most profound effect observed following the emulsion administration. Bioavailability following a nonoily suspension of ontazolast was 0.5%, whereas an oily solution improved bioavailability to 5.3%, and the emulsion formulation caused total bioavailability of 9.6%.

Carrigan and Bates examined the absorption of micronized griseofulvin from an oil/water (o/w) emulsion, an oil suspension, and an aqueous suspension. Both oily formulations improved oral griseofulvin bioavailability, with the emulsion having the greatest effect. The administration of the emulsion produced a 2.5-fold greater AUC value than the aqueous suspension, and 1.6-fold greater than the oil suspension (Figure 6.3) [27]. Subsequent study evaluated the same corn o/w griseofulvin emulsion in comparison to two different commercial tablets and an aqueous suspension in humans. A twofold enhancement in griseofulvin bioavailability was observed following administration of the emulsion in comparison to the three other formulations [28].

Other examples in which emulsion formulation has been shown to enhance bioavailability are danazol [29] and penclomedine [30].

6.3.1.1.1.3 Self-Emulsifying Systems Emulsion systems have the disadvantage of being physically unstable, and over time a separation between the oil and water phases of the emulsion will occur. The use of conventional emulsions is also less attractive due to poor precision of the taken dose and the relatively large volume that has to be administered. To overcome these limitations, self-emulsifying drug delivery systems (SEDDS) have been developed. The

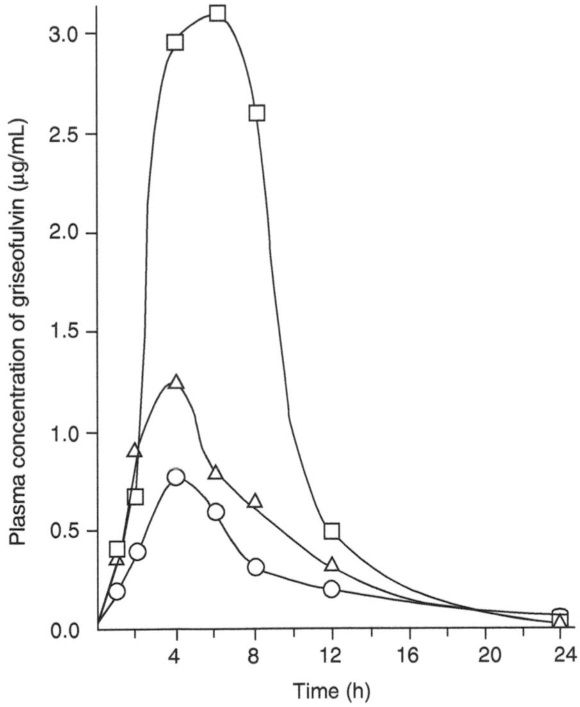

FIGURE 6.3 Micronized griseofulvin plasma profiles following oral administration (50 mg/kg) to rats in three different vehicles: aqueous suspension (○), corn oil suspension (△), or corn oil/water emulsion (□). (From Carrigan, P.J. and Bates, T.R., *J. Pharm. Sci.*, 62, 1476, 1973. With permission.)

principal characteristic of these systems is their ability to form a fine o/w emulsion upon mild agitation following dilution by the aqueous milieu of the GI [31]. SEDDS are homogenous mixtures of oil, surfactants, and drug, where hydrophilic cosolvents may also be introduced. They are thermodynamically stable systems, and upon dilution in aqueous phase form an emulsion with a particle size that is significantly smaller than in conventional emulsions. Moreover, SEDDS formulations can be compounded in soft capsules, providing a precise and convenient unit dose form.

The bioavailability of vitamin E from a self-emulsifying preparation was evaluated in comparison to a commercially available soybean oil solution in soft gelatin capsule. Vitamin E administered as a self-emulsifying preparation had markedly higher plasma levels compared to the oil solution formulation, and the extent of absorption was increased almost threefold. Moreover, variability between subjects was markedly reduced in the case of the self-emulsifying preparations [32].

6.3.1.1.1.4 Microemulsions and Self-Microemulsifying Systems Further developments in lipid-based formulation technologies include microemulsions and self-microemulsifying drug delivery systems (SMEDDS). Microemulsion particle size is usually less than 50 nm; it shows preferred thermodynamic stability and can hold a large dose of the drug. For these reasons, microemulsifying systems are thought to be superior to other formulations.

Cyclosporine is a lipophilic ($\log P = 2.92$) potent immunosuppressive agent. The standard oral formulation of cyclosporine (Sandimmun, Novartis Pharmaceuticals Corporation, East Hanover, NJ) is an oil-based solution that contains the drug together with LCT, ethanol, and polyglycolized LCT. Following oral administration of this formulation, a crude o/w droplet mixture is formed, and emulsification of this mixture by bile salts is required before digestion of the oil and release of the drug can occur. After this emulsification stage, the digestion process is continued, a mixed micellar phase with cyclosporine solubilized in MG and bile salts is formed, and only then the drug molecules can be absorbed [33]. Cyclosporine bioavailability following this process varies considerably, with values from 1% to 89% reported in the literature, indicating a nonlinear and unpredictable relationship between dose and blood concentration. Moreover, the rate and extent of absorption are highly dependent on food intake, bile secretion, and GI motility [34].

Studies demonstrating that cyclosporine absorption is enhanced through reduction in the formulation particle size [18] have established that the bioavailability is improved by microemulsion formulation [35,36].

The new cyclosporine formulation (Sandimmun Neoral, Novartis Pharmaceuticals Corporation, East Hanover, NJ) is a self-microemulsifying drug delivery system, which consists of the drug in a lipophilic solvent (corn oil), hydrophilic cosolvent (propylene glycol) surfactant and an antioxidant [37]. Upon contact with GI fluids, Sandimmun Neoral readily forms a homogenous, monophasic microemulsion, which allows the absorption of the drug molecules. Unlike Sandimmun, the formation of this microemulsion is independent of bile salt activity, and indeed, studies have shown that the absorption of cyclosporine from the new formulation is much less dependent on bile flow [38] and is unaffected by food intake [39].

Improved pharmacokinetic behavior was also seen with the microemulsion in many other studies. Inter- and intraindividual variabilities of cyclosporine primary pharmacokinetic parameters (C_{max}, t_{max}, AUC, and terminal half-life) were compared following the administration of the two formulations to 24 healthy volunteers. Both inter- and intraindividual variabilities were significantly reduced with the microemulsion: intraindividual variability ranged between 9% and 22% for the microemulsion, compared with 19%–41% for the standard formulation, and the interindividual variability values were 3%–22% and 20%–34%, respectively [40].

The pharmacokinetic dose proportionality and relative bioavailability of cyclosporine from the microemulsion formulation were compared to those of the regular formulation over the dosage range of 200 to 800 mg. The AUC for Sandimmun increased in a less than proportional manner with respect to dose, whereas that for Neoral was consistent with linear pharmacokinetics. In addition, the relative bioavailability of cyclosporine from the microemulsion formulation ranged from 174% at the 200 mg dose to 239% at the 800 mg dose compared to the regular formulation [41].

Cyclosporine A absorption was evaluated following oral administration of the regular Sandimmun soft gelatin capsule and the Neoral microemulsion formulation to patients with psoriasis. The Neoral formulation showed 32% increased mean AUC value, increased C_{max}, decreased t_{max} and reduced variability between patients (Figure 6.4) [42].

The pharmacokinetic properties of the microemulsion were compared with those of the standard formulation in 39 liver-transplanted patients. AUC and C_{max} values were significantly increased, and t_{max} was decreased following the microemulsion formulation, in both fasted and postprandial state. Differences between the microemulsion and the regular formulation in the fasting state and after high-fat meal, respectively, were +64% and +38% for AUC, +119% and +53% for C_{max}, and −21% and −59% for t_{max} [43].

6.3.1.1.2 Summary of the Lipid-Based Formulation Approach

The knowledge and rationale accumulated thus far regarding the lipid-based formulation approach for enhancing oral absorption of lipophilic compounds have been presented in this section. Additional comprehensive review articles are also recommended [44–46].

As presented in this section, the success of the lipid-based delivery systems evolves from the suitable selection of the vehicle composition and rational formulation design for the drug candidate. The lipid-based formulation needs to maximize the rate and extent of drug dissolution and maintain the drug in solution during its transit throughout the GI tract. Hence, methods for tracking the solubilization state of the drug after the dispersion of the

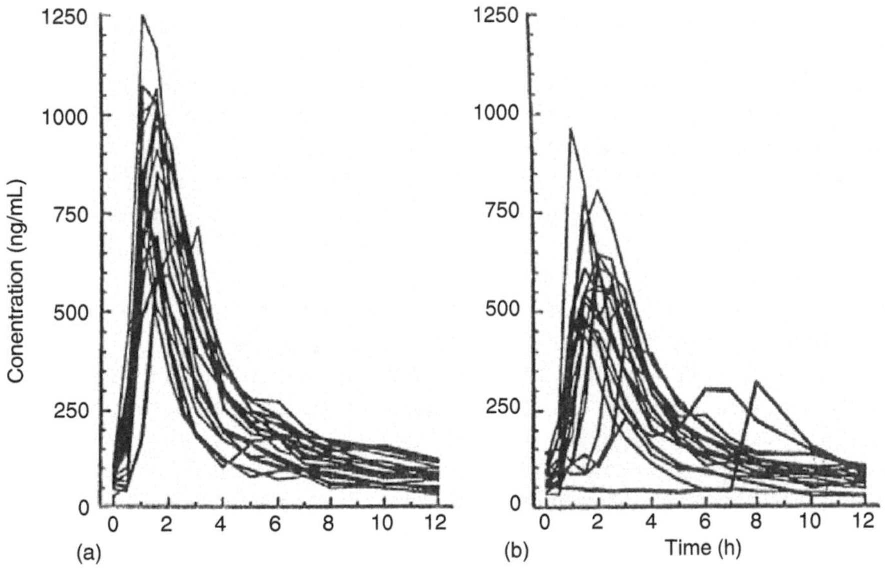

FIGURE 6.4 Individual plasma cyclosporine A profiles following oral administration of Neoral microemulsion formulation (a) and the regular Sandimmun soft gelatin capsule (b) in psoriasis patients. (From Erkko, P., et al., *Br. J. Dermatol.*, 136, 82, 1997. With permission.)

lipid-based formulation in the GI tract are needed. Such methods may better demonstrate the rationale for the design of improved lipid-based formulations to provide better absorption properties of lipophilic compounds.

6.3.1.2 Prodrug Approach

Another approach to overcome the absorption problems of lipophilic molecules is the synthesis of a prodrug: a derivative of an active parent compound, which undergoes transformation in the body to form the free drug. Two diverse rationales can be applied in this approach: (1) to form a prodrug complex that has decreased lipophilicity in comparison to the free drug, or (2) to make a prodrug with elevated lipophilicity.

6.3.1.2.1 Water-Soluble Prodrug

In this strategy, a water-soluble prodrug is synthesized, usually by adding to the drug moiety functional groups that are ionized at physiological pH. The prodrug must undergo reconversion at the intestinal mucosal membrane or at post absorption stages, to form the free drug moiety [47]. The increased solubility of the prodrug and high membrane permeability of the parent compound provide the driving force for increased absorption properties of the drug [48]. Another mechanism by which prodrugs may improve absorption via increased aqueous solubility is by reducing intermolecular hydrogen bonding, without affecting the log octanol–water partition coefficient [49].

Taxol (paclitaxel) is a cytotoxic drug that has been shown to have potent antileukemic and tumor inhibitory properties. Taxol suffers from very poor water solubility (0.25 μg/mL), and hence, a few phosphate ester prodrugs of taxol have been synthesized. While great improvement of water solubility was achieved (>10 mg/mL), these prodrugs were resistant to degradation by alkaline phosphatase, probably due to hindered enzyme access, and no free taxol was formed [50].

Phosphonoxyphenyl propionate esters of taxol were developed to reduce this steric hindrance induced by the phosphate group [51]. These prodrugs were readily converted to taxol upon incubation with alkaline phophatase *in vitro*, but no conversion to taxol occurred following incubation with plasma, probably due to the significant protein binding of the prodrug to albumin. These data indicate that this prodrug may be effective following oral administration, as it has high solubility on one hand, and good cleavage by the brush border membrane enzyme alkaline phosphatase on the other [52].

Celecoxib is a COX-2 selective inhibitor, used in the treatment of rheumatoid arthritis, osteoarthritis, and pain management. Celecoxib comprises very poor water solubility (<50 μg/mL), and hence, suffers from low oral bioavailability (15%). Three aliphatic acyl water-soluble prodrugs of celecoxib were synthesized: celecoxib–acetyl, celecoxib–propionyl, and celecoxib–butyryl, all sharing a similar water solubility of 15 mg/mL. Following oral administration of these prodrugs to rats, the butyryl and propionyl prodrugs improved celecoxib bioavailability by fivefold, whereas the acetyl prodrug showed no improvement over the administration of free celecoxib. Following incubation for 1 h with gastric mucosal suspension, celecoxib–butyryl was completely hydrolyzed to the free celecoxib, whereas the propionyl and acetyl prodrugs released only 56% and 14%, respectively. Similarly, in liver homogenates, the butyryl prodrug hydrolyzed rapidly and completely within 30 min incubation, whereas the propionyl and acetyl prodrugs hydrolysis were only 88% and 38%, respectively. Systemic exposure of the celecoxib–propionyl as the whole complex was fourfold greater than the systemic exposure of the celecoxib–butyryl complex. These findings further illustrate the superiority of the butyryl prodrug over the other two. Nevertheless, further evaluation of celecoxib–butyryl is required before its possible usage in the clinic [53].

Pharmacokinetic studies in dogs demonstrated a 3.5-fold increase in oral bioavailability of phenytoin when administered as the disodium phosphate prodrug versus sodium phenytoin [54]. A few additional examples of investigations with this strategy include the steroids betamethasone [55] and hydrocortisone [56], HIV protease inhibitors [57], and the anticancer drug etoposide [58,59]. Comprehensive reviews of this strategy are also available [52,60,61].

To date, the success of the water-soluble prodrug strategy to improve the absorption of lipophilic drugs has been limited, and frequently, the enhanced solubility of the prodrug has failed to translate into augmented bioavailability of the drug. The need to balance between the hydrophilicity of the prodrug for solubilization and the lipophilicity of the drug for gut wall permeation, as well as between the prodrug chemical stability and the enzymatic cleavage, are some of the reasons this strategy is so challenging.

6.3.1.2.2 Lipidic Prodrug

Lipidic prodrugs are chemical entities with two distinct parts: a drug covalently bound to a lipid moiety, i.e., a fatty acid, a glyceride, or a phospholipid. As a result, the lipophilicity of the prodrug compared to the drug is increased.

The rationale for these prodrugs is to affect their absorption path. By attaching a lipid moiety to the drug the complex may confer the attached drug to the physiologic absorption pathway of natural lipids, e.g., enhancing the lymphatic transport of the molecule. The lymphatic absorption pathway is further detailed in Section 6.3.2.1.

To succeed in manipulating the prodrug to the lymphatic absorption pathway, the prodrug must bear a few essential characteristics. A log P value of >5 appears to be necessary for lymphatic transport. The log P value should be determined at the physiological pH of 7.4 (actually defined as log $D_{7.4}$), where many acidic–basic molecules may be ionized and thereby lowering the effective log P value of the prodrug.

Log P is not the only parameter to determine the ability of a molecule to be transported via the lymph. As the lymphatic absorption is associated with chylomicron formation, which is composed mainly from a triglycerides core (85%–92%) [62], the solubility of the prodrug in triglycerides is an important indicator for lymphatic transport capacity potential. Charman and Stella suggested that solubility in triglycerides of at least 50 mg/mL is required before lymphatic transport is likely to occur [63].

To work well, the prodrug must be designed to be stable within the GI lumen and the gut wall, and to be cleaved only after reaching the systemic circulation in the plasma or other peripheral tissue.

The strategy of making the prodrug more lipophilic than the parent compound has the possible disadvantage of having a complex that has extremely poor solubility. It would have very limited solubilization and dissolution properties, and hence, poor absorption and bioavailability. For this reason the prodrug complex must have an adequate lipophilicity, with a balance between encouraging lymphatic transport and avoiding over-lipophilicity. Figure 6.5 presents the illustration of Stelle and Pochopin that clarifies this concept [64].

These prerequisites for the prodrug to follow have made its task to effectively promote the delivery of drugs through the lymph quite complex, and various attempts in the past have met with only little success.

Testosterone is used to treat various androgen deficiency syndromes. Following oral administration, free testosterone is effectively metabolized and inactivated in the liver before it reaches the systemic circulation [65], and unless very high doses are administered, no testosterone is detected in the plasma [66]. Testosterone undecanoate is a prodrug of testosterone esterified in the 17β-position with undecanoic acid. Following oral administration, this prodrug is metabolized only partly in the intestinal wall [67], and the remaining fraction of

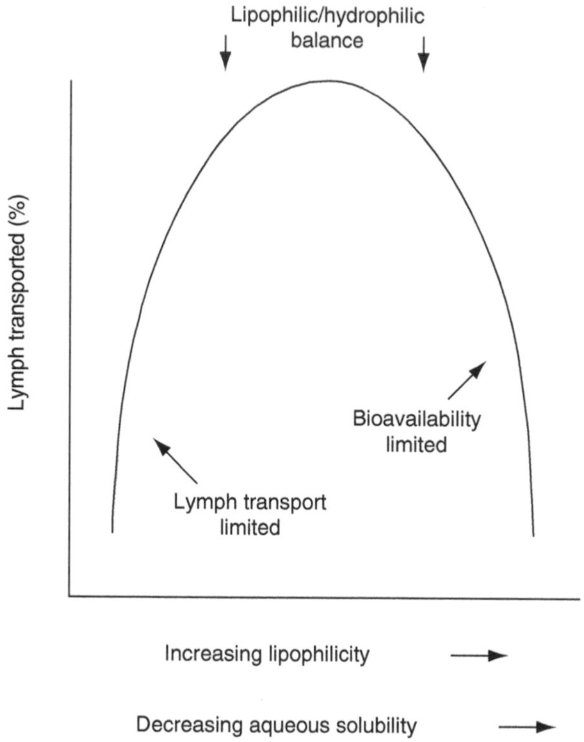

FIGURE 6.5 Relationship between the extent of intestinal lymphatic transport of a drug and lipophilicity after oral administration. (From Stella, V.J. and Pochopin, N.L. *Lymphatic Transport of Drugs*, Charman, W.N. and Stella, V.J., Eds., CRC Press, Boca Raton, 1992, p. 181. With permission.)

testosterone undecanoate is absorbed via the lymphatic system, and hence, avoids first pass hepatic metabolism. After reaching the systemic circulation, testosterone undecanoate rapidly converts to free testosterone. The absolute bioavailability of testosterone following oral administration of testosterone undecanoate is estimated to be 7% [68]. Recent studies in lymph duct cannulated dogs revealed that all of the testosterone undecanoate absorbed into the systemic circulation is due to lymphatic absorption of the prodrug, and more than 80% of the free testosterone in the plasma is contributed by systemic hydrolysis of the lymphatically transported prodrug [69].

The same prodrug approach was used to synthesize testosterone palmitate. Testosterone palmitate has a higher log *P* value than testosterone undecanoate (9 vs. 6.5), so that the comparison between these two prodrugs is very interesting. Although a larger amount of this prodrug was transported through the lymph in comparison to testosterone undecanoate, the undecanoate ester produced a higher amount of free testosterone. No free testosterone was observed following oral administration of testosterone palmitate [70]. This example demonstrates the importance of the knowledge about each and every step in the absorption process of the prodrug, and its degradation to the free drug, with little ability to predict the prodrug performance in advance.

Another strategy in this area is "triglyceride like" prodrugs, in which the drug is attached to a triglyceride replacing one of its fatty acids. The hypothesis for designing this prodrug is that the complex should behave as a triglyceride and therefore be transported via the lymph. The influence of the fatty acid length chain on the absorption properties has been investigated [71]. A fatty acid containing more than 14 carbon atoms was found necessary to obtain lymphatic

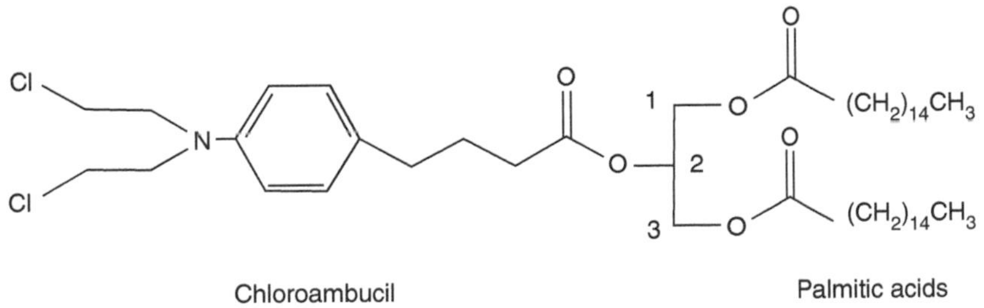

FIGURE 6.6 A triglyceride derivative of chlorambucil, where the chlorambucil attached to the 2-position, and palmitic acid to positions 1 and 3 of the triglyceride. (From Stella, V.J. and Pochopin, N.L. *Lymphatic Transport of Drugs*, Charman, W.N. and Stella, V.J., Eds., CRC Press, Boca Raton, 1992, p. 181. With permission.)

transport of the prodrug. An example for the utilization of this strategy is the case of chlorambucil.

Chlorambucil is an anticancer chemotherapeutic agent, used in the treatment of lymphomas. The targeting of this drug to the lymphatic system may improve the treatment of lymphatic cancers. Hence, a prodrug of chlorambucil was synthesized, with the parent drug attached to the 2-position of a triglyceride with palmitic acid in the 1 and 3 positions (Figure 6.6).

Following oral administration of the drug or the prodrug to mice with leukemic cells disseminating along the lymphatics, an increased life span of the mice and reduced toxicity was observed following the prodrug treatment in comparison to free chlorambucil. Lymph concentrations following oral administration of chlorambucil or its triglyceride analogue showed an increase from 3% of dose transported via the lymph for free chlorambucil up to 26% for the prodrug. Nevertheless, no evidence of free chlorambucil in plasma was observed following the administration of the triglyceride derivative, implying that the triglyceride analogue did not behave as a prodrug, and did not release the chlorambucil molecule. Thus, the improved pharmacology was attributed to the triglyceride–chlorambucil complex itself [72].

This example highlights the need to evaluate the issue of the pharmacological activity of the prodrug itself.

A triglyceride derivative of L-dopa was studied in rats and found to be transported through the lymph (23% of the dose). The prodrug doubled plasma and brain L-dopa AUC, and improved dopamine to L-dopa ratio in comparison to administration of L-dopa itself [73]. There are a few additional examples of investigations utilizing this strategy, including GABA and γ-vinyl-GABA [74,75], the alkylating agent melphalan [76], and fluorouridine attached to a phospholipid moiety [77].

In conclusion, an adequate selection of both the drug and the lipid moiety during the prodrug development may improve the outcomes that can be obtained by this drug delivery approach, and allow the utilization of its advantages.

6.3.2 Enterocyte-Associated Solutions

6.3.2.1 Lymphatic Transport

The majority of orally administered drugs gain access to the systemic circulation by direct absorption into the portal blood. However, highly lipophilic compounds may reach the systemic blood circulation via the intestinal lymphatic system. The overall bioavailability of these drugs is composed of the portion absorbed through the portal blood plus the component absorbed via the lymphatic system. A recent study from our laboratory investigated the absorption characteristics of vitamin D_3 following oral administration in rats. Collecting the lymph through a cannula implanted in the mesenteric lymph duct reduced absolute bioavailability of the vitamin from 52% to 12.6%, hence, 75% of the absorbed vitamin D_3 was associated with lymphatic transport pathway, whereas only 25% was absorbed directly to the portal blood [78]. This example shows the important contribution of the lymphatic transport pathway to the overall absorption of lipophilic drugs.

In addition to increased overall bioavailability of lipophilic molecules, lymphatic transport of a drug provides further advantages, including avoidance of hepatic first pass metabolism, a potential to target specific disease states known to spread via the lymphatics, and improved plasma profile of the drug.

Following oral administration of a lipophilic drug, the main route for the drug to access into the intestinal lymphatics is transcellular, by tracking the same pathway as the lipidic nutrients in food, which use the physiological intestinal lipid transport system. Hence, a brief description of this process is described.

Lingual lipase secreted by the salivary glands, together with gastric lipase that is secreted from the gastric mucosa are responsible for the initiation of hydrolysis of limited amounts of TG, to form the corresponding DG and FA within the stomach. These products, along with the shear movement of the stomach and the passage through the pyloric sphincter into the duodenum, cause the formation of a crude emulsion. Lipids present in the duodenum facilitate the secretion of bile salts, biliary lipids (phospholipid and cholesterol ester) and pancreatic fluids into the duodenum. These agents absorb to the o/w interface and produce a more stabilized emulsion with a reduced droplet size [79]. The enzymatic hydrolysis is completed by the action of pancreatic lipase, which upon complexation with colipase acts at the surface of the emulsified TG droplets to produce the corresponding 2-MG and two FA. These amphiphilic lipid digestion products arranged in bile salt micelles solubilize the coadministered lipophilic drug, and deliver it through the UWL adjacent to the gut wall. Micelles are not thought to be absorbed intact, and the drug dissociates from the micelles prior to absorption into the enterocyte [80].

Once inside the enterocyte, the long-chain fatty acids (more than 12 carbons) migrate to the endoplasmic reticulum, where they reesterify to form TG, and are arranged as large TG-rich droplets. These droplets than assemble as the lipidic core of a lipoprotein called chylomicron. It is this TG core of the chylomicron that upon association with the lipophilic drug carries the drug into the lymphatic system. After its synthesis, the chylomicron is packaged in the golgi and secreted from the basolateral membrane of the enterocyte to the intracellular space.

Because of its large size (200–800 μm), the chylomicron cannot permeate the blood capillaries, and as a result, is absorbed into a porous mesenteric lymph vessel called lacteal, and travels with the lymph until drainage into the systemic blood circulation.

As the association between the lipophilic drug and the lipidic core of the chylomicron is the step dictating the degree of lymphatic absorption of the drug, augmentation of this association can increase the lymphatic transport of the drug. One approach to achieve this is to increase the lipophilicity of the drug via synthesis of a prodrug, as discussed above (Section 6.3.1.2.2). Another approach is to use lipid-based formulations: the coadministered lipids stimulate lipid turnover through the enterocyte, enhancing chylomicron synthesis, and thereby, increasing the lymphatic transport pathway capacity.

In many lymphatic absorption studies a complex lipidic vehicle has been used, and the role of an individual component in the increased lymphatic transport is difficult to evaluate. However, some guidance regarding the nature of the lipid preferable for lymphatic transport enhancement can be drawn.

The impact of the FA chain length of the TG in the formulation on the lymphatic transport of the lipophilic antimalarial drug halofantrine was investigated by Caliph et al. [81]. Both lymphatic transport and total systemic exposure of halofantrine were enhanced by an increase in the FA chain length of the coadministered lipid (Figure 6.7). Halofantrine lymphatic transport was 15.8% of dose following C_{18}-based lipid formulation administration, 5.5% after C_{8-10}-based vehicle, and only 2.2% when C_4 base formulation was used. This work shows that formulating with long-chain lipids, rather than shorter lipids, provokes an increased lymphatic absorption of the coadministered lipophilic drug.

The effect of degree of unsaturation of the administered lipid on the degree of lymphatic transport has also been studied. In general, lipids with increasing degrees of unsaturation preferentially promote lymphatic transport, probably due to improved digestion and increased fluidity of the lipid [82,83]. This trend was further supported by subsequent investigations in caco-2 cells [84,85]. Lipoprotein secretion and TG output were found to be doubled following incubation with linoleic acid (18:2) or oleic acid (18:1) in comparison to stearic acid (18:0). Moreover, the lipoproteins secreted following incubation with $C_{18:1}$ or $C_{18:2}$ FA were similar to chylomicrons, whereas saturated FA (16:0, 18:0) caused secretion of more dense lipoproteins [86].

The class of the lipid in the formulation may also influence the extent and rate of lymphatic absorption of the coadministered drug. Charman and Stella examined the lymphatic transport of DDT following oral administration in different lipidic formulations. Cumulative DDT transported via lymph was twofold greater when administered in a FA (oleic acid)-based vehicle in comparison to a TG-based formulation. In addition, a faster appearance of the DDT in the lymph was observed with the FA formulation [87].

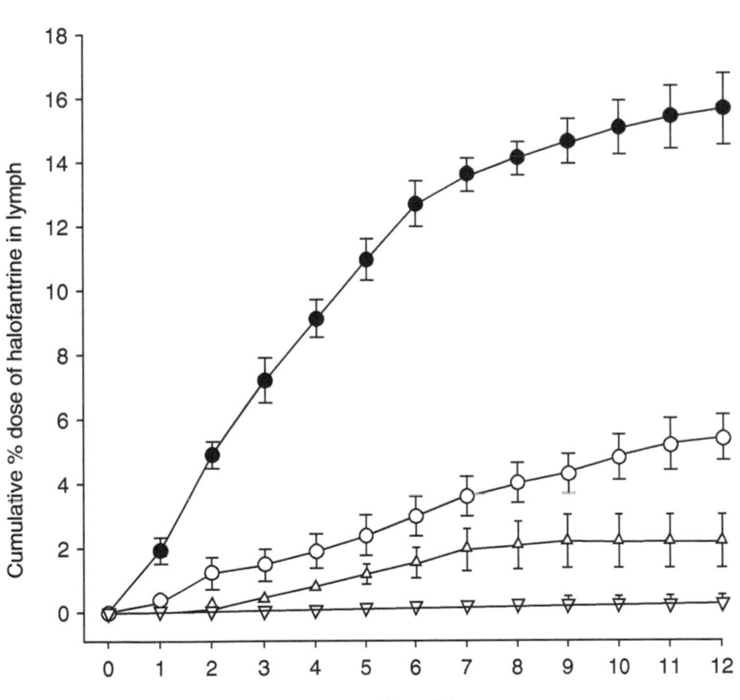

FIGURE 6.7 Cumulative lymphatic transport of halofantrine after oral administration in long-chain, C_{18} (●), medium-chain, C_{8-10} (○), and short-chain, C_4 (△) triglyceride lipid-based formulations, and a lipid-free (▽) formulation. (From Caliph, S.M., et al., *J. Pharm. Sci.*, 89, 1073, 2000. With permission.)

In contrast, no significant difference in the degree of lymphatic absorption of halofantrine was detected following administration in TG or in FA–MG blend [88]. Hence, definite conclusions regarding the effect of lipid class utilized on the extent of lymphatic transport cannot yet be drawn.

Although the strategies for the selection of the proper formulation need to be optimized, there has been substantial progress in the field of drug delivery to the lymphatic system. Khoo et al. [89] investigated the lymphatic absorption of halofantrine in dogs. More than 50% of the administered dose of halofantrine was transported via the lymph in the postprandial state. Ichihashi et al. [90] investigated the lymphatic absorption of the lipophilic anticancer drug mepitiostane following oral administration of different oily vehicles to rats. More than 40% of the mepitiostane dose was transported in the lymph when administered in long-chain triglyceride. These findings illustrate that absorption via the intestinal lymph is clinically relevant, and is accountable for the therapeutic concentration of certain drugs following oral administration.

Intestinal lymphatic transport has also been shown to substantially contribute to the absorption of fat-soluble vitamins [91], ontazolast [26], probucol [16] and others.

A number of animal models have been proposed for the investigation of intestinal lymphatic drug transport, based on an indwelling cannula in a lymphatic vessel [92]. This provides the ability to collect the lymph fluid and indicates the amount of drug in the lymph following the absorption process. However, lymphatic cannulation is time consuming and demands a high level of surgical skill. Furthermore, the success rate is limited. Simplified investigational models developed in our laboratory enable easier and faster study of the oral lymphatic absorption route. A chemical blockage of chylomicron flow *in vivo* was shown to cancel the lymphatic transported component without affecting the nonlymphatic absorption component of a lipophilic drug. This blockage enabled the determination of lymphatic transport degree without the need for lymphatic cannulation [78]. More recently, the uptake of nine lipophilic molecules by chylomicron emulsion was investigated, and a linear correlation was found between this uptake and the degree of lymphatic transport reported in rats. This correlation enables the prediction of the lymphatic transport of a given molecule by a fast *ex vivo* method [93].

In conclusion, drug delivery to the lymph may provide a number of significant advantages, including an increase in the overall bioavailability of a lipophilic drug. The simplified investigational models introduced may assist in the investigation of lymphatic absorption and enable a faster screening of candidate molecules for oral lymphatic transport.

6.3.2.2 Cytochrome P450 3A4 and P-glycoprotein Inhibitors

An increasing body of evidence has shown that certain lipids and excipients found in many lipid-based formulations are capable of inhibiting both P-gp mediated drug efflux and presystemic metabolism in the enterocyte. This inhibition may increase the bioavailability of a drug coadministered with such lipid-based vehicle.

The permeability of two peptides, P-gp substrate and non P-gp substrate, across caco-2 cells in the presence or absence of polysorbate 80 and cremophor EL, commonly used surfactants in pharmaceutical formulations, was investigated. The permeability of the P-gp substrate peptide across caco-2 cells was enhanced in the presence of polysorbate 80 and cremophor EL, whereas the non-P-gp substrate peptide was not affected by these surfactants [94]. Another commonly used lipidic excipient that has been shown to inhibit P-gp mediated efflux is D-α-tocopheryl polyethylene glycol 1000 succinate (TPGS) [95]. The insertion of a known CYP3A4 and P-gp inhibitor to the formulation is another approach to elevate bioavailability.

Paclitaxel is an antitumor drug, with a low oral bioavailability (6%) [96] due to poor water solubility, hydrophobicity (log $P = 4$), and its high affinity to P-gp and CYP3A4. A novel SMEDDS of paclitaxel increased the bioavailability by up to 1.5-fold in comparison to the commercial micellar solution taxol. Concomitant intake of cyclosporine A, a known P-gp inhibitor, further increased paclitaxel bioavailability (1.8-fold in comparison to taxol) [97].

Grapefruit juice has also been shown to inhibit CYP3A4 and P-gp in the small intestine, and hence, can elevate the oral bioavailability of lipophilic drugs [98,99]. Concomitant intake of grapefruit juice with the lipophilic cholesterol-lowering medication simvastatin caused a 16-fold increase in the AUC of simvastatin [100]. The oral bioavailability of the lipophilic HIV protease inhibitor saquinavir was doubled following concomitant administration with grapefruit juice [101].

Although coadministration of P-gp and CYP3A4 inhibitors with the lipophilic drug can improve drug bioavailability, other aspects of drug–drug interaction makes this strategy unsafe. As for grapefruit juice, until the active components of the juice are fully isolated and defined, and the intensity of the bioavailability increment is well predicted, this exploitation might be unsafe as well.

6.4 CONCLUSIONS

Looking into the future, more lipophilic drugs are likely to be produced, and the delivery of these molecules through the oral route is expected to be a most important and relevant issue.

As reviewed in this chapter, certain means can be utilized to improve the bioavailability of lipophilic drugs, whether by formulative approach or molecular changes strategies. These means present a number of attractive propositions to the scientist, ranging from an enhancement of drug dissolution and solubilization by lipid-based formulation, increased solubility via the synthesis of a prodrug, specific delivery to the intestinal lymphatics, and reduction in enterocyte–hepatic presystemic metabolism and efflux systems.

It is also clear from the data presented in this chapter that care must be taken when applying any of the above-proposed solutions. Logical selection of excipients and rational design of lipid-based formulations or prodrug synthesis tailored for drug candidates are key factors for the success of these delivery systems.

REFERENCES

1. Lipinski, C.A., et al. 2001. Experimental and computational approaches to estimate solubility and permeability in drug discovery and development settings. *Adv Drug Deliv Rev* 46:3.
2. Amidon, G.L., et al. 1995. A theoretical basis for a biopharmaceutic drug classification: The correlation of *in vitro* drug product dissolution and *in vivo* bioavailability. *Pharm Res* 12:413.
3. Read, N.W., et al. 1977. Unstirred layer and kinetics of electrogenic glucose absorption in the human jejunum *in situ*. *Gut* 18:865.
4. Westergaard, H., and J.M. Dietschy. 1974. Delineation of the dimensions and permeability characteristics of the two major diffusion barriers to passive mucosal uptake in the rabbit intestine. *J Clin Invest* 54:718.
5. Wacher, V.J., et al. 1998. Role of P-glycoprotein and cytochrome P450 3A in limiting oral absorption of peptides and peptidomimetics. *J Pharm Sci* 87:1322.
6. Gottesman, M.M., I. Pastan, and S.V. Ambudkar. 1996. P-glycoprotein and multidrug resistance. *Curr Opin Genet Dev* 6:610.
7. Benet, L.Z., and C.L. Cummins. 2001. The drug efflux-metabolism alliance: Biochemical aspects. *Adv Drug Deliv Rev* 50:S3.

8. Kolars, J.C., et al. 1991. First-pass metabolism of cyclosporin by the gut. *Lancet* 338:1488.
9. Yun, C., R. Okerholm, and F. Guengerich. 1993. Oxidation of the antihistaminic drug terfenadine in human liver microsomes. Role of cytochrome P-450 3A(4) in *N*-dealkylation and C-hydroxylation. *Drug Metab Dispos* 21:403.
10. Fitzsimmons, M.E., and J.M. Collins. 1997. Selective biotransformation of the human immunodeficiency virus protease inhibitor saquinavir by human small-intestinal cytochrome P4503A4. Potential contribution to high first-pass metabolism. *Drug Metab Dispos* 25:256.
11. Thummel, K.E., et al. 1996. Oral first-pass elimination of midazolam involves both gastrointestinal and hepatic CYP3A-mediated metabolism. *Clin Pharmacol Ther* 59:491.
12. Hebert, M.F. 1997. Contributions of hepatic and intestinal metabolism and P-glycoprotein to cyclosporine and tacrolimus oral drug delivery. *Adv Drug Deliv Rev* 27:201.
13. Lennernas, H. 2003. Clinical pharmacokinetics of atorvastatin. *Clin Pharmacokinet* 42:1141.
14. Hunt, J.N., and M.T. Knox. 1968. Control of gastric emptying. *Am J Dig Dis* 13:372.
15. Rozman, K., L. Ballhorn, and T. Rozman. 1983. Mineral oil in the diet enhances fecal excretion of DDT in the rhesus monkey. *Drug Chem Toxicol* 6:311.
16. Palin, K.J., and C.G. Wilson. 1984. The effect of different oils on the absorption of probucol in the rat. *J Pharm Pharmacol* 36:641.
17. Porter, C.J.H., S.A. Charman, and W.N. Charman. 1996. Lymphatic transport of halofantrine in the triple-cannulated anesthetized rat model: Effect of lipid vehicle dispersion. *J Pharm Sci* 85:351.
18. Tarr, B.D., and S.H. Yalkowsky. 1989. Enhanced intestinal absorption of cyclosporine in rats through the reduction of emulsion droplet size. *Pharm Res* 6:40.
19. Kaneniwa, N., and N. Watari. 1974. Dissolution of slightly soluble drugs. I. Influence of particle size on dissolution behavior. *Chem Pharm Bull (Tokyo)* 22:1699.
20. Scholz, A., et al. 2002. Influence of hydrodynamics and particle size on the absorption of felodipine in Labradors. *Pharm Res* 19:42.
21. Atkinson, R.M., et al. 1962. The effect of griseofulvin particle size on blood levels in man. *Antibiot Chemother* 12:232.
22. Hargrove, J.T., W.S. Maxson, and A.C. Wentz. 1989. Absorption of oral progesterone is influenced by vehicle and particle size. *Am J Obstet Gynecol* 161:948.
23. Tokumura, T., et al. 1987. Enhancement of the oral bioavailability of cinnarizine in oleic acid in beagle dogs. *J Pharm Sci* 76:286.
24. Chakrabarti, S., and F.M. Belpaire. 1978. Biovailability of phenytoin in lipid containing dosage forms in rats. *J Pharm Pharmacol* 30:330.
25. Abrams, L.S., et al. 1978. Comparative bioavailability of a lipophilic steroid. *J Pharm Sci* 67:1287.
26. Hauss, D.J., et al. 1998. Lipid-based delivery systems for improving the bioavailability and lymphatic transport of a poorly water-soluble LTB_4 inhibitor. *J Pharm Sci* 87:164.
27. Carrigan, P.J., and T.R. Bates. 1973. Biopharmaceutics of drugs administered in lipid-containing dosage forms. I. GI absorption of griseofulvin from an oil-in-water emulsion in the rat. *J Pharm Sci* 62:1476.
28. Bates, T.R., and J.A. Sequeria. 1975. Bioavailability of micronized griseofulvin from corn oil-in-water emulsion, aqueous suspension, and commercial tablet dosage forms in humans. *J Pharm Sci* 64:793.
29. Charman, W., et al. 1993. Effect of food and a monoglyceride emulsion formulation on danazol bioavailability. *J Clin Pharmacol* 33:381.
30. Myers, R.A., and V.J. Stella. 1992. Systemic bioavailability of penclomedine (NSC-338720) from oil-in-water emulsions administered intraduodenally to rats. *Int J Pharm* 78:217.
31. Gershanik, T., and S. Benita. 2000. Self-dispersing lipid formulations for improving oral absorption of lipophilic drugs. *Eur J Pharm Biopharm* 50:179.
32. Julianto, T., K.H. Yuen, and A.M. Noor, 2000. Improved bioavailability of vitamin E with a self-emulsifying formulation. *Int J Pharm* 200:53.
33. Vonderscher, J., and A. Meinzer, 1994. Rationale for the development of Sandimmune Neoral. *Transplant Proc* 26:2925.
34. Fahr, A. 1993. Cyclosporin clinical pharmacokinetics. *Clin Pharmacokinet* 24:472.
35. Ritschel, W.A., et al. 1990. Improvement of peroral absorption of cyclosporine A by microemulsions. *Methods Find Exp Clin Pharmacol* 12:127.

36. Drewe, J., et al. 1992. Enhancement of the oral absorption of cyclosporin in man. *Br J Clin Pharmacol* 34:60.
37. Levy, G., et al. 1994. Cyclosporine Neoral in liver transplant recipients. *Transplant Proc* 26:2949.
38. Trull, A.K., et al. 1995. Absorption of cyclosporin from conventional and new microemulsion oral formulations in liver transplant recipients with external biliary diversion. *Br J Clin Pharmacol* 39:627.
39. Mueller, E.A., et al. 1994. Influence of a fat-rich meal on the pharmacokinetics of a new oral formulation of cyclosporine in a crossover comparison with the market formulation. *Pharm Res* 11:151.
40. Kovarik, J.M., et al. 1994. Reduced inter- and intraindividual variability in cyclosporine pharmacokinetics from a microemulsion formulation. *J Pharm Sci* 83:444.
41. Mueller, E.A., et al. 1994. Improved dose linearity of cyclosporine pharmacokinetics from a microemulsion formulation. *Pharm Res* 11:301.
42. Erkko, P., et al. 1997. Comparison of cyclosporin A pharmacokinetics of a new microemulsion formulation and standard oral preparation in patients with psoriasis. *Br J Dermatol* 136:82.
43. Freeman, D., et al. 1995. Pharmacokinetics of a new oral formulation of cyclosporine in liver transplant recipients. *Ther Drug Monit* 17:213.
44. Porter, C.J., and W.N. Charman. 2001. Lipid-based formulations for oral administration: Opportunities for bioavailability enhancement and lipoprotein targeting of lipophilic drugs. *J Recept Signal Transduct Res* 21:215.
45. Humberstone, A.J., and W.N. Charman. 1997. Lipid-based vehicles for the oral delivery of poorly water soluble drugs. *Adv Drug Deliv Rev* 25:103.
46. Pouton, C.W. 2000. Lipid formulations for oral administration of drugs: Non-emulsifying, self-emulsifying and self-microemulsifying drug delivery systems. *Eur J Pharm Sci* 11:S93.
47. Amidon, G.L., G.D. Leesman, and R.L. Elliott. 1980. Improving intestinal absorption of water-insoluble compounds: A membrane metabolism strategy. *J Pharm Sci* 69:1363.
48. Chan, O.H., et al. 1998. Evaluation of a targeted prodrug strategy to enhance oral absorption of poorly water-soluble compounds. *Pharm Res* 15:1012.
49. Amidon, G.L. 1981. Drug derivatization as a means of solubilization: Physicochemical and biochemical strategies. In *Techniques of solubilization of drugs*, ed. S.H. Yalkowsky, 183. New York: Marcel Dekker.
50. Vyas, D.M., et al. 1993. Synthesis and antitumor evaluation of water soluble taxol phosphates. *Bioorg Med Chem Lett* 3:1357.
51. Ueda, Y., et al. 1993. Novel water soluble phosphate prodrugs of taxol(R) possessing *in vivo* antitumor activity. *Bioorg Med Chem Lett* 3:1761.
52. Fleisher, D., R. Bong, and B.H. Stewart. 1996. Improved oral drug delivery: Solubility limitations overcome by the use of prodrugs. *Adv Drug Deliv Rev* 19:115.
53. Mamidi, R.N., et al. 2002. Pharmacological and pharmacokinetic evaluation of celecoxib prodrugs in rats. *Biopharm Drug Dispos* 23:273.
54. Varia, S.A., and V.J. Stella. 1984. Phenytoin prodrugs V: *In vivo* evaluation of some water-soluble phenytoin prodrugs in dogs. *J Pharm Sci* 73:1080.
55. Loo, J.C., et al. 1981. Pharmacokinetic evaluation of betamethasone and its water soluble phosphate ester in humans. *Biopharm Drug Dispos* 2:265.
56. Fleisher, D., et al. 1986. Oral absorption of 21-corticosteroid esters: A function of aqueous stability and intestinal enzyme activity and distribution. *J Pharm Sci* 75:934.
57. Vierling, P., and J. Greiner. 2003. Prodrugs of HIV protease inhibitors. *Curr Pharm Des* 9:1755.
58. Doyle, T.W., and D.M. Vyas. 1990. Second generation analogs of etoposide and mitomycin C. *Cancer Treat Rev* 17:127.
59. Sessa, C., et al. 1995. Phase I clinical and pharmacokinetic study of oral etoposide phosphate. *J Clin Oncol* 13:200.
60. Beaumont, K., et al. 2003. Design of ester prodrugs to enhance oral absorption of poorly permeable compounds: Challenges to the discovery scientist. *Curr Drug Metab* 4:461.
61. Ettmayer, P., et al. 2004. Lessons learned from marketed and investigational prodrugs. *J Med Chem* 47:2393.
62. Hussain, M.M. 2000. A proposed model for the assembly of chylomicrons. *Atherosclerosis* 148:1.

63. Charman, W.N., and V.J. Stella. 1986. Estimating the maximal potential for intestinal lymphatic transport of lipophilic drug molecules. *Int J Pharm* 34:175.
64. Stella, V.J., and N.L. Pochopin. 1992. Lipophilic prodrugs and the promotion of intestinal lymphatic drug transport. In *Lymphatic transport of drugs*, eds. W.N. Charman, and V.J. Stella, 181. Boca Raton: CRC Press.
65. Daggett, P.R., M.J. Wheeler, and J.D. Nabarro. 1978. Oral testosterone, a reappraisal. *Horm Res* 9:121.
66. Nieschlag, E., et al. 1975. Plasma androgen levels in men after oral administration of testosterone or testosterone undecanoate. *Acta Endocrinol* (*Copenh*) 79:366.
67. Horst, H.J., et al. 1976. Lymphatic absorption and metabolism of orally administered testosterone undecanoate in man. *Klin Wochenschr* 54:875.
68. Tauber, U., et al. 1986. Absolute bioavailability of testosterone after oral administration of testosterone-undecanoate and testosterone. *Eur J Drug Metab Pharmacokinet* 11:145.
69. Shackleford, D.M., et al. 2003. Contribution of lymphatically transported testosterone undecanoate to the systemic exposure of testosterone after oral administration of two andriol formulations in conscious lymph duct-cannulated dogs. *J Pharmacol Exp Ther* 306:925.
70. Noguchi, T., W.N. Charman, and V.J. Stella. 1985. The effect of drug lipophilicity and lipid vehicles on the lymphatic absorption of various testosterone esters. *Int J Pharm* 24:173.
71. Lambert, D.M. 2000. Rationale and applications of lipids as prodrug carriers, *Eur J Pharm Sci* 11:S15.
72. Garzon-Aburbeh, A., et al. 1983. 1,3-dipalmitoylglycerol ester of chlorambucil as a lymphotropic, orally administrable antineoplastic agent. *J Med Chem* 26:1200.
73. Garzon-Aburbeh, A., et al. 1986. A lymphotropic prodrug of L-dopa: Synthesis, pharmacological properties, and pharmacokinetic behavior of 1,3-dihexadecanoyl-2-[(S)-2-amino-3-(3,4-dihydroxyphenyl)propanoyl]propane-1,2,3-triol. *J Med Chem* 29:687.
74. Deverre, J.R., et al. 1989. *In-vitro* evaluation of filaricidal activity of GABA and 1,3-dipalmitoyl-2-(4-aminobutyryl)glycerol HCl: A diglyceride prodrug. *J Pharm Pharmacol* 41:191.
75. Jacob, J.N., G.W. Hesse, and V.E. Shashoua. 1990. Synthesis, brain uptake, and pharmacological properties of a glyceryl lipid containing GABA and the GABA-T inhibitor gamma-vinyl-GABA. *J Med Chem* 33:733.
76. Loiseau, P.M., et al. 1994. Study of lymphotropic targeting and macrofilaricidal activity of a melphalan prodrug on the Molinema dessetae model. *J Chemother* 6:230.
77. Sakai, A., et al. 1993. Deacylation-reacylation cycle: A possible absorption mechanism for the novel lymphotropic antitumor agent dipalmitoylphosphatidylfluorouridine in rats. *J Pharm Sci* 82:575.
78. Dahan, A., and A. Hoffman. 2005. Evaluation of a chylomicron flow blocking approach to investigate the intestinal lymphatic transport of lipophilic drugs. *Eur J Pharm Sci* 24:381.
79. Carey, M.C., D.M. Small, and C.M. Bliss. 1983. Lipid digestion and absorption. *Annu Rev Physiol* 45:651.
80. Westergaard, H., and J.M. Dietschy. 1976. The mechanism whereby bile acid micelles increase the rate of fatty acid and cholesterol uptake into the intestinal mucosal cell. *J Clin Invest* 58:97.
81. Caliph, S.M., W.N. Charman, and C.J.H. Porter. 2000. Effect of short-, medium-, and long-chain fatty acid-based vehicles on the absolute oral bioavailability and intestinal lymphatic transport of halofantrine and assessment of mass balance in lymph-cannulated and non-cannulated rats. *J Pharm Sci* 89:1073.
82. Cheema, M., K.J. Palin, and S.S. Davis. 1987. Lipid vehicles for intestinal lymphatic drug absorption. *J Pharm Pharmacol* 39:55.
83. Bergstedt, S.E., et al. 1990. A comparison of absorption of glycerol tristearate and glycerol trioleate by rat small intestine. *Am J Physiol Gastrointest Liver Physiol* 259:G386.
84. Field, F., E. Albright, and S. Mathur. 1988. Regulation of triglyceride-rich lipoprotein secretion by fatty acids in Caco-2 cells. *J Lipid Res* 29:1427.
85. van Greevenbroek, M.M., D.W. Erkelens, and T.W. de Bruin. 2000. Caco-2 cells secrete two independent classes of lipoproteins with distinct density: Effect of the ratio of unsaturated to saturated fatty acid. *Atherosclerosis* 149:25.

86. van Greevenbroek, M.M., et al. 1996. Effects of saturated, mono-, and polyunsaturated fatty acids on the secretion of apo B containing lipoproteins by Caco-2 cells. *Atherosclerosis* 121:139.
87. Charman, W.N., and V.J. Stella. 1986. Effects of lipid class and lipid vehicle volume on the intestinal lymphatic transport of DDT. *Int J Pharm* 33:165.
88. Porter, C.J.H., et al. 1996. Lymphatic transport of halofantrine in the conscious rat when administered as either the free base or the hydrochloride salt: Effect of lipid class and lipid vehicle dispersion. *J Pharm Sci* 85:357.
89. Khoo, S.M., et al. 2001. A conscious dog model for assessing the absorption, enterocyte-based metabolism, and intestinal lymphatic transport of halofantrine. *J Pharm Sci* 90:1599.
90. Ichihashi, T., et al. 1992. Effect of oily vehicles on absorption of mepitiostane by the lymphatic system in rats. *J Pharm Pharmacol* 44:560.
91. Kuksis, A. 1987. Absorption of fat soluble vitamins. In *Fat absorption*, ed. A. Kuksis, 65. Boca Raton: CRC Press.
92. Edwards, G.A., et al. 2001. Animal models for the study of intestinal lymphatic drug transport. *Adv Drug Deliv Rev* 50:45.
93. Gershkovich, P., and A. Hoffman. 2005. Uptake of lipophilic drugs by plasma derived isolated chylomicrons: Linear correlation with intestinal lymphatic bioavailability. *Eur J Pharm Sci* 26:394.
94. Nerurkar, M.M., P.S. Burton, and R.T. Borchardt. 1996. The use of surfactants to enhance the permeability of peptides through caco-2 cells by inhibition of an apically polarized efflux system. *Pharm Res* 13:528.
95. Dintaman, J.M., and J.A. Silverman. 1999. Inhibition of P-glycoprotein by D-α-tocopheryl polyethylene glycol 1000 succinate (TPGS). *Pharm Res* 16:1550.
96. Malingre, M.M., J.H. Beijnen, and J.H.M. Schellens. 2001. Oral delivery of taxanes. *Invest New Drugs* 19:1552.
97. Yang, S., et al. 2004. Enhanced oral absorption of paclitaxel in a novel self-microemulsifying drug delivery system with or without concomitant use of P-glycoprotein inhibitors. *Pharm Res* 21:261.
98. Bailey, D.G., et al. 1998. Grapefruit juice–drug interactions. *Br J Clin Pharmacol* 46:101.
99. Dahan, A., and H. Altman. 2004. Food–drug interaction: Grapefruit juice augments drug bioavailability—Mechanism, extent and relevance. *Eur J Clin Nutr* 58:1.
100. Lilja, J.J., K.T. Kivisto, and P.J. Neuvonen. 1998. Grapefruit juice-simvastatin interaction: Effect on serum concentrations of simvastatin, simvastatin acid, and HMG-CoA reductase inhibitors. *Clin Pharmacol Ther* 64:477.
101. Kupferschmidt, H.H., et al. 1998. Grapefruit juice enhances the bioavailability of the HIV protease inhibitor saquinavir in man. *Br J Clin Pharmacol* 45:355.

Part II

Promoted Rectal Absorption

7 Permeation Pathways in Rectal Absorption

Yoshiteru Watanabe

CONTENTS

The rectal route of drug administration has been used for many years because drugs may be readily introduced and retained in the rectal cavity. Rectal administration may be a practical alternative to oral administration when patients are prone to nausea, vomiting, convulsion, and, in particular, disturbances of consciousness. Therefore, rectal administration has been used to deliver many kinds of drugs such as anticonvulsants, analgesics (including narcotics), antiemetics, antibacterial agents, anesthetics for children, and some anticancer agents. Rectal drug delivery, on the one hand, is effective because of the extensive rectal vasculature and the presence of lymphatic vessels in the rectal region. On the other hand, patient acceptability of rectal administration is poor and drug absorption may be affected by defecation. Further development and optimization of rectal drug formulations for clinical use may be expected in the near future. Several reviews concerning rectal drug delivery and absorption have already been published [1–4].

This chapter focuses on rectal drug administration, which represents one of the most common routes of transmucosal drug delivery, and some aspects of rectal drug absorption, including enhancement strategies.

7.1 GENERAL CONSIDERATIONS

7.1.1 Anatomy and Physiology of the Rectum and Its Role in Drug Absorption

The rectum is the terminal part of the digestive tract that proceeds from the sigmoid colon and ends at the anus. In humans, the rectum is formed by the end of the large intestine and is 10–15 cm in length and 1.5–3.5 cm in width. Usually, the rectal cavity is expanded and is then called the rectal ampulla. The rectum of a man is generally larger than the rectum of a woman. Three rectal valves are clearly observed inside the rectum. When a suppository, the most common dosage form, is administered in the rectum, it becomes liquefied. Consequently, the drugs contained in the suppository are absorbed from the luminal surface and transferred to the systemic circulation.

The human colon and rectum, in the lower part of digestive tract, do not have functions in the absorption of nutritional substances ingested through the mouth. The absorption of water, sodium, and chloride and the secretion of potassium and bicarbonate takes place in the human colon, where active transports of glucose and amino acids are lacking. In the rectum, absorption of sodium and water is negligible [5]. The structural features of the mucous membrane on the luminal surfaces in the colon and rectum do not differ from those in the stomach or the small intestine. The luminal surface of epithelial cells in the small intestine contains microvilli; however, villi are absent in the large intestine including the rectum. The luminal surfaces of the colon and rectum are somewhat wrinkled. In the proximal part of the anal canal, formed by the last 4–5 cm of the rectum, a transition from columnar epithelium to stratified squamous epithelium occurs [6]. The entire surface area of the rectum is approximately 200–400 cm^2. When compared with the surface area of a segment of small intestine having the same volume, the rectum surface area is only 0.2%. In addition, when compared with the absorption surface area of the entire small intestine (20×10^7 cm^2), the rectum surface area is only 0.01%. From its anatomical features, we cannot conclude that the rectum is an effective organ for drug absorption in comparison with the upper alimentary tract. However, previous studies have shown that the difference in surface areas is not a discriminatory factor in drug absorption.

Compared to the mucous membrane of the sigmoid colon, the rectal mucous membrane is deeper red because as the rectum approaches the anus, it becomes richer in blood vessels. Regarding the arterial system that leads to the rectum, the superior rectal artery, the middle rectal artery, and the inferior rectal artery supply blood to the rectal mucous membrane, the muscular tuft of the anus, and the skin. The venous system consists of the superior rectal vein, the middle rectal vein, and the inferior rectal vein (Figure 7.1). An important aspect of rectal administration in drug therapy is the possibility of partial elimination of hepatic first-pass metabolism or first-pass effect. The superior rectal vein perfuses the upper part of the rectum, drains into the portal vein, and subsequently empties into the liver. Therefore, drugs encounter an extensive first-pass effect when they are absorbed from the upper part of the rectum. The middle and inferior rectal veins drain the lower part of the rectum and venous blood is returned to the inferior vena cava. Drugs absorbed in the latter system will be delivered preferentially to the systemic circulation, bypassing the liver and avoiding first-pass metabolism [7,8]. As mentioned above, the presence of extensive anastomosis of blood vessels may decrease this effect.

The epithelial mucous membrane surface has a specific bottom layer that contains capillary vessels and lymphatic vessels for absorption, with smooth muscle and the neuroplexus. Large numbers of lymphatic nodules, derived from the reticular tissue of the mucous membrane, are localized in this layer. The blood vessel system is extensively related to absorption and secretion of water and substances in the intestinal tract. However, we cannot ignore the simultaneous contributions of the lymphatic and blood vessel systems. The lymphatic vessel

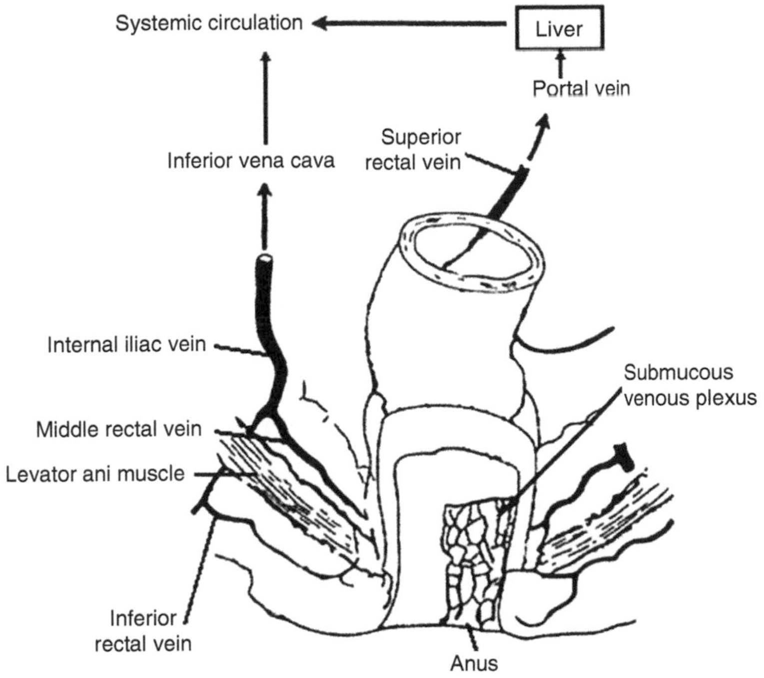

FIGURE 7.1 Schematic illustration of the venous drainage of the human rectum. Drugs absorbed in the inferior and middle rectal veins that drain the lower part of the rectum will be delivered preferentially to the systemic circulation, bypassing the liver and avoiding first-pass metabolism. (Illustrated with modification from Nishimura, M., et al. (Eds.), *Shin-Gekakgaku*, 5th ed., Nanzando, Tokyo, 1979, 583. With permission.)

system in the colon and rectum contributes to the absorption of water, the circulation of protein in the extracellular fluid, the movement of lymphocytes, and the metastasis of cancer from these parts. The relatively large pore radius of a lymphatic capillary, compared with a blood capillary, reportedly establishes the lymphotropic delivery of compounds of high molecular weight (10 to 500 kDa) in rats [9,10]. However, the total amount of drug delivered via the lymphatic system appears to be quite small, possibly because of the low lymph flow compared with the blood flow.

In humans, the luminal surfaces of the rectum are covered by a membrane formed by one layer of columnar epithelial, endocrine, and goblet cells. These cells related to absorption have histological features that closely resemble those of the epithelium cells of the small intestine. Mechanistically, the rectal absorption of drugs is not dissimilar to their absorption in the upper gastrointestinal tract. With rectal administration, molecules may be absorbed across the epithelial cells (transcellular) or via the tight junctions interconnecting the mucosal cells (intercellular, paracellular) [2]. Several potential barriers may influence the passage of a dissolved drug from luminal fluid into venous blood or lymph [8,11]. These barriers may be located in an unstirred water layer, the mucous layer, the apical cell membranes, the basal cell membrane, the tight junctions, the basement membranes, and the walls of lymph vessels or blood capillaries.

Concerning the absorption theory for organic compounds in the alimentary tract, the pH-partition hypothesis has been accepted. Under normal physiological conditions, drug absorption from the lower alimentary tract is well described by the pH-partition hypothesis. Therefore, the passive transport mechanism is dominant in drug absorption from rectal mucous membranes. On the other hand, does any specific transport mechanism, such as active transport, for some substances, such as monosaccharides, amino acids, and vitamins, exist in the rectum? No experimental results have suggested the existence of a carrier-mediated drug transport mechanism, which includes active transport, in the colon and rectum.

Many physiological aspects affect drug absorption from the rectum (Table 7.1). Influential factors include the pH of the rectal contents, state of the mucus layer, volume and viscosity of rectal fluid, luminal pressure from the rectal wall on the dosage form, enzymatic and microbacterial degradation by rectal epithelium, presence of stools, and venous drainage differences within the rectosigmoid regions.

7.1.2 Effect of Dosage Form and Formulation on Drug Dissemination from Rectum to Colon

The physicochemical properties of the drug molecules and the formulation can also influence rectal drug absorption (Table 7.1). Crucial parameters in this regard include drug concentration, molecular weight, solubility, lipophilicity, pK_a, surface properties, and particle size [12]. In addition, formulation properties such as the nature of the suppository base material may play a critical role in regulating drug absorption.

Although several drug formulations (suppositories, gelatin capsules, and enemas of various volumes) may be rectally administered in clinical use, the suppository is the most widely used dosage form applied to the rectum. When a suppository is inserted through the anus, it settles in the rectal ampulla before becoming deformed by melting. After a suppository is inserted into the rectum, its liquefying behavior may differ from that of other suppositories because various types of base materials are used to give the suppository its solid form. When oleaginous base materials such as cacao butter are used, these suppositories gradually melt at the rectal temperature (approximately 37°C). On the other hand, when water-soluble base materials such as polyethylene glycol and gelatin capsule are used, these suppositories dissolve

TABLE 7.1
Influential Factors on Rectal Drug Absorption from Suppository

Drug	pK_a Lipophilicity Solubility Concentration Molecular weight (size) Particle diameter (size) Surface characteristics of particle
Base material	Composition Fusibility Surface tension Rheological characteristics
Rectal fluid	Volume pH Buffering ability Surface tension Viscosity Composition Luminal pressure from rectal wall

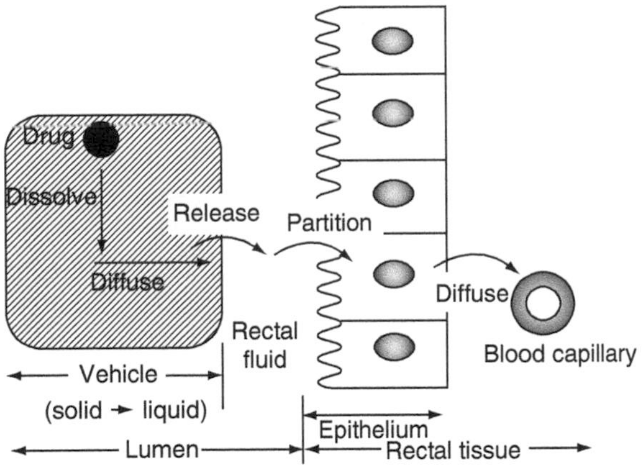

FIGURE 7.2 Process of the drug movement from rectal dosage vehicles to blood capillary in the rectum. The drug must be released from the base materials after liquefaction into the rectal fluid before drug is absorbed.

in the rectal fluid, gradually losing their shape. Drugs in suppository form are not limited to being dissolved in base materials; they may also be suspensions or emulsions. Therefore, drugs in suppository form have to be dissolved in the rectal fluid before permeation in the rectal epithelial cells. For most rectal dosage forms, the drug must be released from the base material after liquefaction into the rectal fluid before the drug is absorbed (Figure 7.2).

About 3 mL of rectal fluid exists in the human rectum. This is very little when compared with the amounts of gastric juice and intestinal fluid in oral drug administration. Before transmucosal passage can occur, the released drug must dissolve into the limited volume of rectal fluid. The scant amount of fluid secreted inside the rectum suggests that the drug release and dissolution are very important in the process of drug absorption when a suppository is administered. The pH value of rectal fluid is usually 7–8, but the buffering ability of this fluid is low. The disintegration and dissemination of an inserted suppository may be influenced by the luminal pressure from the rectal wall on the dosage form. However, it is unclear whether the surface tension and viscosity of the rectal fluid affect the drug release from the suppository form.

Dissemination characteristics influence the effectiveness of drugs administered for local action, and they are decisive for the fraction of the dose bypassing first-pass metabolism by the liver [7,13]. Thus, the migration of drugs delivered by suppositories and enemas in the human rectum has been studied. On the one hand, the migration of drugs administered in suppository bases appears to be limited to the rectal area [14–16]. On the other hand, the drug dissemination from enemas may be significant and administration of volumes of 80–100 mL may result in migration into the ascending colon and even the ileum [17,18].

7.2 RECTAL PERMEABILITY IN DISEASES

7.2.1 Inflammatory Diseases

The gastrointestinal tract has a limited number of responses to pathogens and entirely different mechanisms may lead to similar histopathologic responses. The expression of food protein allergy in humans is heterogeneous, varies with the age of the subject, and is, to a certain extent, genetically determined. Food allergy may involve the entire gut, from mouth to rectum, including the esophagus. Abnormalities in intestinal permeability are the hallmarks of an inflamed gut, occurring in subjects with celiac disease as well as in those with Crohn's disease, and may contribute to a diagnosis of food-induced enteropathy. The characteristics of the intestinal inflammatory response are largely determined by the cytokine responses triggered by a pathologic mechanism, regardless of origin, in the stomach, the small intestine, and the colon [19]. A so-called T-helper (Th) type 2 response is characteristic of the allergic subject and may be a predominant response in subjects with ulcerative colitis.

Thorn and coworkers [20] investigated whether basic fibroblast growth factor (bFGF) could be detected in the intraluminal secretions of the small intestine, sigmoid colon, and rectum in healthy individuals and in patients with ulcerative colitis. These investigators used endoscopic perfusion techniques to obtain samples from well-defined intestinal segments and measured the concentrations of bFGF and the biochemical markers of inflammation (myeloperoxidase [MPO] and interleukin-6 [IL-6]) in the perfusion fluid. Also, the permeability was determined using albumin. There were strong correlations between the concentration of bFGF and the concentrations of MPO and IL-6. The bFGF concentration varies in different intestinal anatomical locations and increases significantly in patients with ulcerative colitis in close relationship with biochemical markers of inflammation and permeability [20].

Evidence of local eosinophil activation and altered mucosal permeability in collagenous colitis has been reported [21]. The local release of the inflammatory mediators eosinophil cationic protein and MPO, and the permeability marker albumin was studied in collagenous colitis using a technique for segmental perfusion of the rectum and the descending colon of patients. A significant correlation was found between the increased concentrations of eosinophil cationic protein and albumin, indicating a possible relationship between eosinophil activation and disturbed mucosal permeability in collagenous colitis. Patients with collagenous colitis have increased colorectal mucosal secretion of vascular endothelial growth factor (VEGF), which may lead to albumin leakage and promote fibrosis with deposition of collagen [22].

Membrane permeability and antipyrine absorption were investigated in a rat model of ischemic colitis in which ischemia and associated inflammation were induced by marginal vessel ligation of the distal rectum [23]. Vessel ligation caused some sloughing of epithelial cells and elevated the MPO level, a key indicator of inflammation. The results suggested that the absorption kinetics of antipyrine depends on blood flow changes in the large intestine that occur with inflammation.

Although in recent years more data have become available, the pathogenesis of inflammatory bowel disease (IBD) remains unknown. The contribution of genetic and environmental factors is evident and luminal bacterial flora plays a major role in the irritation and perpetuation of chronic IBD [24]. In human IBD, inflammation is present in parts of the gastrointestinal tract containing the highest bacterial concentrations. Furthermore, the terminal ileum, caecum, and rectum are areas of relative stasis, providing prolonged mucosal contact with luminal contents. Due to a genetic predisposition or because of direct contact with bacteria or their products, enhanced mucosal permeability may play a pivotal role in maintaining a chronic inflammatory state. A defective epithelial barrier may cause a loss of tolerance to the normal enteric flora. Moreover, increased mucosal absorption of viable bacteria and products is found in IBD [24]. Therefore, much attention should be paid to rectal administration of drugs in cases of IBD. Recently, elevated levels of substance P (SP), which has proinflammatory effects on immune and epithelial cells and participates in inflammatory diseases of the gastrointestinal, respiratory, and musculoskeletal systems, and upregulated neurokinin-1 receptor (NK-1R) expression have been reported in the rectum and colon of patients with IBD. Furthermore, their levels correlated with the disease activity [25].

7.2.2 Other Pathophysiological Conditions

Intestinal permeability in humans is reportedly increased after radiation therapy [26]. The mucosal permeability in the irradiated rectum and nonirradiated sigmoid colon of patients subjected to radiation therapy before surgical treatment of rectal cancer was investigated. The mucosa-to-serosa passage of all marker molecules (^{14}C-mannitol, fluorescein isothiocyanate-dextran (FITC-dextran) 4400, and ovalbumin) was increased in the irradiated rectum compared with the nonirradiated sigmoid colon, whereas in specimens from nonirradiated patients, there were no differences in marker molecule passage between the rectum and sigmoid colon. Kierbel and coworkers [27] reported effects of medium hypertonicity on water permeability in the mammalian rectum. Minute-by-minute net water fluxes (J_w) were measured across the isolated rectal epithelium in rats and rabbits. Immunohistochemical studies showed the expression of aquaporin 3 (AQP3) at the basolateral membrane of epithelial cells in the rat. They concluded that the epithelium of the mammalian rectum is a highly polarized, AQP3-containing, water permeability structure. A second conclusion is that serosal hypertonicity induces an increase in transcellular water permeability in both rat and rabbit rectums. On the other hand, colonic surface hydrophobicity reportedly modulates permeability to hydrophilic molecules and protects against toxins [28]. In control rats, surface hydrophobicity (contact angle measurement) was low on the caecal mucosa and high in the colon and rectum. Detergent treatment reduced surface hydrophobicity and increased colonic permeability to mannitol and dextran. Conversely, treatment with lipids increased surface hydrophobicity and reduced colonic permeability. Surface hydrophobicity of the rat colonic mucosa is a defensive barrier against macromolecules and toxins.

7.3 ENHANCEMENT OF RECTAL ABSORPTION

7.3.1 Rate-Controlled Rectal Drug Delivery

Recently, various studies concerning the application of devices to perform rectal drug administration with controlled delivery rate have been reported. Generally, the aim of these investigations was to evaluate the ability to achieve relatively constant drug concentrations for longer periods by using new systems [2,3]. Using an osmotic device with zero-order drug release characteristics, steady-state concentrations in plasma were obtained with drugs [29,30]. As alternative formulations with sustained drug release properties, hydrogel preparations of the matrix type and subsequent near zero-order delivery characteristics may be used rectally, e.g., those containing theophylline [31] or morphine [32]. Rate-controlled drug delivery was employed to modulate systemic drug concentrations and thereby influence drug effects. In addition, mucosal dissemination can be reduced using osmotic delivery systems, resulting in rate-controlled and site-specific rectal drug delivery, and consequently resulting in enhancement of the absorption profiles. Furthermore, the results of these studies indicate that rate-controlled and site-specific rectal delivery makes partial avoidance of first-pass metabolism possible, although the extent depends on the site of rectal absorption in humans [33]. Local effects may also be influenced by the rate of delivery; the delivery rate affected the local action of absorption-enhancing agents such as sodium salicylate and sodium octanoate. Yet, cefoxitin, a poorly absorbed drug, was efficiently absorbed from the rectum in human volunteers [34]. This delivery-rate dependent effect was interpreted in terms of amount of absorption enhancing agent delivered per unit of mucosal area and per unit of time.

7.3.2 Absorption-Enhancing Agents

In the past two decades, many studies have tested adjuvants that act by either permeabilizing the rectal mucosa or inhibiting drug degradation. Oral and rectal routes of drug administration are unsuitable for adequate absorption of various compounds with a peptide or protein structure and of several hydrophilic antibiotics. The use of absorption enhancers, e.g., salicylates, enamines, surfactants, and straight-chain fatty acids, has gained wide interest

[11]. For example, the incorporation of sodium caprate and Witepsol H-15 (triacylglycerols in an oleaginous base) can increase the bioavailability of ampicillin and ceftizoxime in animal models [35,36]. Suppositories containing each antibiotic (ampicillin or ceftizoxime) in combination with sodium caprate are clinically used in patients, particularly children. Interestingly, the absorption-enhancing effect of sodium salicylate and sodium octanoate on rectal cefoxitin absorption was shown to be delivery-rate dependent [34]. Lindmark and coworkers [37] reported the mechanism of absorption enhancement in humans after rectal administration of ampicillin in suppositories containing sodium caprate. They also elucidated the mechanism of absorption enhancement by medium-chain fatty acids in intestinal epithelial Caco-2 cell monolayers [38]. Concerning medium-chain fatty acids as an absorption enhancer in rectal dosage forms, several studies in formulation and mechanism of action have been reported in the past two decades [35,39–42]. Other chemical enhancers include medium-chain monoglycerides [43,44], sodium tauro-24,25-dihydrofusidate (STDHF) [45], and cyclodexitrins [46,47]. Furthermore, in recent years, the effects of new types of absorption enhancers (Table 7.2), such as *S*-nitroso-*N*-acetyl-DL-penicillamine (SNAP) as a nitric oxide (NO) donor [48], α-cyprinol sulfate [49], and glycyrrhizin monoammonium salt [50] on drug permeability in the rectum have been reported.

Enzyme inhibitors represent another class of enhancers. Examples of these include aprotinin, amastatin, and salicylate [4]. Sayani and coworkers [51] reported that EDTA enhanced the permeation of leucine enkephalin (Tyr-Gly-Gly-Phe-Leu; Leu-Enk) through the rectal mucous membrane. EDTA had a good stabilizing effect on Leu-Enk degradation. A combination of amastatin, EDTA, and thimerosal had the greatest stabilizing effect on Leu-Enk and its degradation intermediates.

Many studies have focused on the development of rectal formulations for the administration of peptide and protein drugs such as insulin. In 1992, De Boer and coworkers [52] reviewed rate-controlled rectal peptide absorption enhancement as part of penetration enhancement for polypeptides through the epithelia. According to a previous study, an insulin solution (100–150 U) or suspension (45–60 U) in a triacylglycerol base containing lecithin (10% w/v) increased plasma insulin levels and reduced blood glucose levels in healthy volunteers and in patients with diabetes [53]. Interestingly, highly purified docosahexaenoic acid (polyunsaturated fatty acid) was investigated as a potential absorption enhancer for the rectal delivery of insulin, using a water-in-oil-in-water (W/O/W) multiple emulsion [54]. The

TABLE 7.2
New Types of Absorption Enhancing Systems Used in Rectal Drug Administration

Poorly Absorbed Drug	Absorption Enhancer	Reference
Insulin	Nitric oxide (NO) donor	
	S-nitroso-*N*-acetyl-DL-penicillamine (SNAP)	48
	(±)-(*E*)-4-methyl-2-[(*E*)-hydroxyimino]-5-nitro-6-methoxy-3-hexenamide (NOR1)	48
	(±)-*N*-[(*E*)-4-ethyl-2-[(*Z*)-hydroxyimino]-5-nitro-hexen-1-yl]3-pyridine carboxamide (NOR4)	48
Ampicillin sodium salt	α-Cyprinol sulfate	49
Glycyrrhizin monoammonium salt	High concentrated solution without additive	50
Insulin	Docosahexaenoic acid	54
Insulin	Sodium laurate and acrylic hydrogel	55
Human chorionic gonadotropin (hCG)	α-Cyclodextrin	58
Rebamipide	Sodium laurate and taurine	65

results showed that eicosapentaenoic acid (DHA) has a potential enhancement effect on insulin permeability in the rectum. On the other hand, insulin-loaded acrylic hydrogels containing some absorption enhancers were also effective for the rectal delivery of insulin [55]. In more recent years, we have demonstrated that several peptide and protein drugs such as insulin [46], recombinant human granulocyte colony-stimulating factor (rhG-CSF) [56], elcatonin [57], and human chorionic gonadotropin (hCG) [58] can be efficiently absorbed from rabbit rectal mucosa using the hollow-type suppository developed by Watanabe and coworkers. [59], in combination with cyclodextrins as absorption enhancers. Recently, Liu and coworkers [60] reported that one of the recombinant hirudin variants, rHV2, a polypeptide used clinically as an anticoagulant agent, was more rapidly absorbed in anesthetized rats after both intratracheal and rectal administration, compared with its absorption after nasal administration.

An important issue that has been recognized recently concerns the potential adverse effect of absorption enhancers on the rectal mucosa, as shown in rats after a single application [61,62]. Safety evaluation of the applicability of absorption enhancers is imperative. In clinical application, medium-chain fatty acids such as sodium caprate are used only for suppositories containing antibiotics [36].

To reduce irritation problems when an absorption enhancer is administered with poorly absorbed drugs, we have recently demonstrated a novel absorption-enhancing system based on the function of physiologically active internal substances in humans. In 1998, we reported [48] that NO from a donor compound, such as *S*-nitroso-*N*-acetyl-DL-penicillamine (SNAP), significantly increased the rectal absorption of fluorescein isothiocyanate-dextrans (FDs), particularly one with an average molecular weight 4000 (FD-4), a model of a poorly absorbed drug, with very little mucosal damage in rabbits. As shown in Table 7.3, these absorption-enhancing effects of NO significantly decreased after simultaneous administration with an NO scavenger such as 2-(4-carboxyphenyl)-4,4,5,5-tetramethyimidazole-1-oxyl-3-oxide sodium salt (carboxy-PTIO). Also, we found the same enhancing effect of SNAP and other

TABLE 7.3
Pharmacokinetic and Pharmacodynamic Parameters for Insulin and Glucose Following Rectal Administration of Insulin, SNAP, and Carboxy-PTIO

	SNAP (mg)	Carboxy-PTIO (mg)	Insulin $AUC_{0\to6}$ (h μIU/mL)	Insulin C_{max} (μIU/mL)	Glu (h mg/dL)
Control	0	0	337±75	168±16	78±19
SNAP	0.25	0	432±95	227±15[a]	112±30
	1.0	0	1059±155[a]	457±105[a]	233±22[b]
	4.0	0	1466±174[b]	797±50[b]	279±23[b]
Carboxy-PTIO	4.0	30	506±79[c]	260±72[c]	117±70[d]

Note: Glu is the decrease in the plasma glucose concentration calculated from the area under the decreased plasma glucose level versus time curve from 0 to 6 h after rectal administration of hollow-type suppository [59] using the linear trapezoidal rule. Each value represents the mean ± SE for 3–5 experiments.

[a] $P < 0.01$ compared to control.
[b] $P < 0.001$ compared to control.
[c] $P < 0.001$ compared to SNAP 4.0 mg.
[d] $P < 0.005$ compared to SNAP 4.0 mg.

NO donors on the absorption of macromolecules such as FD and rhG-CSF from the jejunum of rats [63] and the nasal mucous membrane of rabbits [64]. These findings suggest that an NO donor can improve the absorption of macromolecular drugs from the intestinal tract without mucosal damage. Recently, it was reported that fatty acid suppository employing the combinatorial use of sodium laurate (C12) with taurine (Tau), an ajuvant exerting the cytoprotective action, could be a promising formulation for effective and safe administration of poorly absorbable drugs [65].

7.4 CONCLUSIONS

Many drugs can now be delivered rectally instead of by parenteral injection (intravenous route) or oral administration. Generally, the rectal delivery route is particularly suitable for pediatric and elderly patients who experience difficulty ingesting medication or who are unconscious. However, rectal bioavailabilities tend to be lower than the corresponding values of oral administration. The nature of the drug formulation has been shown to be an essential determinant of the rectal absorption profiles. The development of novel absorption enhancers with potential efficacy without mucosal irritation (low toxicity) is very important. The delivery of peptide and protein drugs by the rectal route is currently being explored and seems to be feasible.

REFERENCES

1. De Boer, A.G., L.G.J. De Leede, and D.D. Breimer. 1984. Drug absorption by sublingual and rectal routes. *Br J Anaesth* 56:69.
2. Muranishi, S. 1984. Characteristics of drug absorption via the rectal route. *Methods Find Exp Clin Pharmacol* 6:763.
3. Van Hoogdalem, E.J., A.G. De Boer, and D.D. Breimer. 1991. Pharmacokinetics of rectal drug administration, Part I. General considereations and clinical applications of centrally acting drugs. *Clin Pharmacokinet* 21:11.
4. Song, Y., et al. 2004. Mucosal drug delivery: Membranes, methodologies, and applications. *Crit Rev Ther Drug Carrier Syst* 21:195.
5. Binder, H.J., and G.I. Sandle. 1987. Electrolyte absorption and secretion in mammalian colon. In *Physiology of the gastrointestinal tract*, 2nd ed., ed. L.R. Johnson, 1389. New York: Raven Press.
6. Netter, F.H. 1973. *The Ciba collection of medical illustrations: digestive system: lower digestive tract*, vol. 3, ed. E. Oppenheimer, 57. Summit: Ciba Pharmaceutical Co.
7. De Boer, A.G., et al. 1982. Rectal drug administration: Clinical pharmacokinetic consideration. *Clin Pharmacokinet* 7:285.
8. Müller, B.W., ed. 1986. *Suppositorien*. Stuttgart: Wissenschaftliche Verlagsgesellschaft mbH.
9. Kaji, Y., et al. 1985. Selective transfer of 1-hexylcarbamoyl-5-fluorouracil into lymphatics by combination of β-cyclodextrin polymer complexation and absorption promoter in the rat. *Int J Pharm* 24:79.
10. Yoshikawa, H., et al. 1985. Comparison of disappearance from blood and lymphatic delivery of human fibroblast interferon in rat by different administration route. *J Pharmacobiodyn* 8:206.
11. Van Hoogdalem, E.J., A.G. De Boer, and D.D. Breimer. 1989. Intestinal drug absorption enhancement: an overview. *Pharmacol Ther* 44:407.
12. Lee, V.H.L. 1991. *Peptide and protein drug delivery*. New York: Marcel Dekker.
13. De Leede, et al. 1984. Rectal and intravenous propranolol infusion to steady state kinetics and β-receptor blockade. *Clin Pharamcol Ther* 35:148.
14. Jay, M., et al. 1985. Disposition of radiolabelled suppositories in humans. *J Pharm Pharmacol* 37:266.
15. Hardy, J.G., et al. 1987. The application of gamma-scintigraphy for the evaluation of the relative spreading of suppository bases in rectal hard gelatin capsules. *Int J Pharm* 38:103.

16. Sugito, K., et al. 1988. The spreading of radiolabelled fatty suppository bases in human rectum. *Int J Pharm* 47:157.
17. Vitti, R.A., et al. 1989. Quantitative distribution of radiolabelled 5-aminosalicylic acid enemas in patients with left-sided ulcerative colitis. *Dig Dis Sci* 34:1792.
18. Van Buul, M.M., et al. 1989. Retrograde spread of therapeutic enemas in patients with inflammatory bowel disease. *Hepatogastroenterology* 36:199.
19. Dupont, C., and M. Heyman. 2000. Food protein-induced enterocolitis syndrome: Laboratory perspectives. *J Pediatr Gastroenterol Nutr* Suppl. no. 30:S57.
20. Thorn, M., et al. 2000. Intestinal mucosal secretion of basic fibroblast growth factor in patients with ulcerative colitis. *Scand J Gastroenterol* 35:501.
21. Taha, Y., et al. 2001. Evidence of local eosinophil activation and altered mucosal permeability in collagenous colitis. *Dig Dis Sci* 46:888.
22. Taha, Y., et al. 2004. Vascular endothelial growth factor (VEGF)—a possible mediator of inflammation and mucosal permeability in patients with collagenous colitis. *Dig Dis Sci* 49:109.
23. Koga, K., et al. 2004. Membrane permeability and antipyrine absorption in a rat model of ischemic colitis. *Int J Pharm* 286:41.
24. Linskens, R.K., et al. 2001. The bacterial flora in inflammatory bowel disease: Current insights in pathogenesis and the influence of antibiotics and probiotics. *Scand J Gastroenterol* Suppl. no. 234:29.
25. O'Connor T.M., et al. 2004. The role of substance P in inflammatory disease. *J Cell Physiol* 201:167.
26. Nejdfors, P., et al. 2000. Intestinal permeability in humans is increased after radiation therapy. *Dis Colon Rectum* 43:1582.
27. Kierbel, A., et al. 2000. Effects of medium hypertonicity on water permeability in the mammalian rectum: Ultrastructural and molecular correlates. *Pflugers Arch* 440:609.
28. Lugea, A., et al. 2000. Surface hydrophobicity of the rat colonic mucosa is a defensive barrier against macromolecules and toxins. *Gut* 46:515.
29. De Leede, L.G.J., et al. 1982. Zero-order rectal delivery of theophylline in man with an osmotic system, *J Pharmacokinet Biopharm* 10:525.
30. De Leede, L.G.J., et al. 1986. Rate-controlled rectal delivery in man with a hydrogel preparation. *J Control Rel* 4:17.
31. Breimer, D.D., et al. 1985. Rate controlled rectal drug delivery. In *Rate control in drug therapy*, eds. L.F. Prescott, and W.S. Nimmo, 54. Edinburgh: Churchill Livingstone.
32. Hanning, C.D., et al. 1988. The morphine hydrogel suppository. *Br J Anaesth* 61:221.
33. De Leede, L.G.J., et al. 1984. Site specific rectal drug administration in man with an osmotic system: Influence on "first-pass" elimination of lidocaine. *Pharm Res* 1:129.
34. Van Hoogdalem, E.J., et al. 1989. Rate-controlled rectal absorption enhancement of cefoxitin by coadministeration of sodium salicylate or sodium octanoate in healthy volunteers. *Br J Clin Pharmacol* 27:75.
35. Nishimura, K., et al. 1985. Studies on the promoting effects of carboxylic acid derivatives on the rectal absorption of β-lactam antibiotics in rats. *Chem Pharm Bull* 33:282.
36. Bergogne-Bérézin, E., and A. Bryskier. 1999. The suppository form of antibiotic administration: Pharmacokinetics and clinical application. *J Antimicrob Chemother* 43:177.
37. Lindmark, T., et al. 1997. Mechanism of absorption enhancement in humans after rectal administration of ampicillin in suppositories containing sodium caprate. *Phram Res* 14:930.
38. Lindmark, T., T. Nikkilä, and P. Artursson. 1995. Mechanisms of absorption enhancement by medium chain fatty acids in intestinal epithelial Caco-2 cell monolayers. *J Pharmacol Exp Ther* 275:958.
39. Van Hoogdalem, E.J., et al. 1988. Absorption enhancement of rectally infused cefoxitin sodium by medium-chain fatty acids in conscious rats: Concentration–effect relationship. *Pharm Res* 5:453.
40. Watanabe, Y., et al. 1994. Bioavailability of gentamicin from a new rectal dosage vehicle in rabbtis. *Yakuzaigaku* 54:122.
41. Takahashi, H., et al. 1997. The enhancing mechanism of capric acid (C10) from a suppository on rectal drug absorption through a paracellular pathway. *Biol Pharm Bull* 20:446.
42. Lennernas, H., et al. 2002. The influence of caprate on rectal absorption of phenoxymethylpenicillin: Experience from an *in-vivo* perfusion in humans. *J Pharm Pharmacol* 54:499.

43. Watanabe, Y., et al. 1988. Absorption enhancement of rectally infused cefoxitin by medium chain monoglycerides in conscious rats. *J Pharm Sci* 77:847.
44. Van Hoogdalem, E.J., et al. 1989. Rectal absorption enhancement of des-enkephalin-γ-endorphin (DeγE) by medium-chain glycerides and EDTA in conscious rats. *Pharm Res* 6:91.
45. Van Hoogdalem, E.J., et al. 1990. Absorption enhancement of rectally infused insulin by sodium tauro-24,25-dihydrofusidate (STDHF) in rats. *Pharm Res* 7:180.
46. Watanabe, Y., et al. 1992. Absorption enhancement of polypeptide drugs by cyclodextrins. I. Enhanced rectal absorption of insulin from hollow-type suppositories containing insulin and cyclodextrins in rabbits. *Chem Pharm Bull* 40:3042.
47. Matsuda, H., and Arima, H. 1999. Cyclodextrins in transdermal and rectal delivery. *Adv Drug Deliv Rev* 36:81.
48. Utoguchi, N., et al. 1998. Nitric oxide donors enhance rectal absorption of macromolecules in rabbits. *Pharm Res* 15:870.
49. Murakami, T., et al. 2000. Enhancing effect of 5 α-cyprinol sulfate on mucosal membrane permeability to sodium ampicillin in rats. *Eur J Pharm Biopharm* 49:111.
50. Koga, K., et al. 2003. Preparation and rectal absorption of highly concentrated glycyrrhizin solution. *Biol Pharm Bull* 26:1299.
51. Sayani, A.P., I.K. Chun, and Y.W. Chien. 1993. Transmucosal delivery of leucine enkephalin: Stabilization in rabbits enzyme extracts and enhancement of permeation through mucosae. *J Pharm Sci* 82:1179.
52. De Boer, A.G., E.J. Van Hoogdalem, and D.D. Breimer. 1992. (D) Routes of delivery: Case studies, (4) rate-controlled rectal peptide drug absorption enhancement. *Adv Drug Deliv Rev* 8:237.
53. Nishihata, T., et al. 1989. Effectiveness of insulin suppositories in diabetic patients. *J Pharm Pharmacol* 4:799.
54. Onuki, Y., et al. 2000. *In vivo* effects of highly purified docosahexaenoic acid on rectal insulin absorption. *Int J Pharm* 198:147.
55. Uchida, T., et al. 2001. Preparation and characterization of insulin-loaded acrylic hydrogels containing absorption enhancers. *Chem Pharm Bull* 49:1261.
56. Watanabe, Y., et al. 1996. Pharmacokinetics and pharmacodynamics of recombinant human granulocyte colony-stimulating factor (rhG-CSF) after administration of a rectal dosage vehicle. *Biol Pharm Bull* 19:1059.
57. Watanabe, Y., et al. 1998. Studies of drug delivery systems for a therapeutic agent used in osteoporosis. I. Pharmacodynamics (hypocalcemic effect) of elcatonin in rabbits following rectal administration of hollow-type suppositories containing elcatonin. *Biol Pharm Bull* 21:1187.
58. Kowari, K., et al. 2002. Pharmacokinetics and pharmacodynamics of human chorionic gonadotropin (hCG) after rectal administration of hollow-type suppositories containing hCG. *Biol Pharm Bull* 25:678.
59. Watanabe, Y., et al. 1986. Pharamaceutical evaluation of hollow type suppositories. IV. Improvement of bioavailability of propranolol in rabbits after rectal administration. *J Pharmacobiodyn* 9:526.
60. Liu, Y., et al. 2005. Pharmacodynamics and pharmacokinetics of recombinant hirudin via four nonparenteral routes. *Peptides* 26:243.
61. Van Hoogdalem, E.J., et al. 1990. Topical effects of absorption enhancing agents on the rectal mucosa of rats *in vivo. J Pharm Sci* 79:866.
62. Swenson, E.S., W.B. Milisen, and W. Curatolo. 1994. Intestinal permeability enhancement efficacy, acute local toxicity, and reversibility. *Pharm Res* 11:1132.
63. Numata, N., et al. 2000. Improvement of intestinal absorption of macromoleclules by nitric oxide donor. *J Pharm Sci* 89:1296.
64. Watanabe, Y., et al. 2000. Absorption enhancement of a protein drug by nitric oxide donor: Effect on nasal absorption of human granulocyte colony-stimulating factor. *J Drug Target* 8:185.
65. Miyake, M., et al. 2004. Development of suppository formulation safely improving rectal absorption of rebamipide, a poorly absorpbable drug, by utilizing sodium laurate and taurine. *J Control Rel* 99:63.

8 Cyclodextrins and Other Enhancers in Rectal Delivery

Hidetoshi Arima and Kaneto Uekama

CONTENTS

8.1 INTRODUCTION

The rectum has historically been an accepted site of drug delivery [1]. Its principal applications have been for local therapy, e.g., hemorrhoids, and for systemic delivery of drugs, e.g., fever and pain. The rectal route can be an extremely useful route for delivery of drugs to infants, young children, and patients where difficulties can arise from oral administration because of swallowing, nausea, and vomiting [2]. Also, this route offers several potential opportunities for drug delivery, including the avoidance of hepatic first-pass elimination, absorption enhancement, and the possibility of rate-controlled drug delivery. However, the rectal route has some potential disadvantages: (1) poor or erratic absorption across the rectal mucosa of many drugs, (2) a limiting absorbable surface area, (3) a dissolution problem due to the small fluid content in the rectum. To overcome these problems, many attempts to enhance rectal absorption of various drugs have been made. For example (1) increased drug dissolution rate, (2) increased drug release rate from vehicles, (3) drug stabilization, (4) enhanced drug membrane permeability, and (5) increased viscosity to escape the first-pass effect. Actually, various drug delivery systems in the rectal route have been developed, e.g., prodrugs, liposomes, micro- and nanocapsules, chemical or biotechnological modifications and their combinatorial uses of absorption enhancer, prodrug, carriers, and enzymatic inhibitors [3–5].

In this chapter, we especially focus on the strategies for enhancement of rectal absorption of various drugs including peptides and proteins from rectal mucosa using pharmaceutically useful excipients, cyclodextrins (CyDs), and the other absorption enhancers.

8.2 CYCLODEXTRINS AS MULTIFUNCTIONAL ENHANCERS FOR RECTAL DRUG ABSORPTION

CyDs are known to alter various properties of drugs, pharmaceutical formulations, and biomembranes, resulting in enhancement and modulation of rectal drug absorption. CyDs, cyclic oligosaccharides consisting of several glucopyranose units, are host molecules, which form inclusion complexes. So far, the usefulness of three parent CyDs (α-CyD, β-CyD, and γ-CyD, Figure 8.1) in rectal drug delivery has been reported with respect to drug stabilization, improvement in drug release and bioavailability, and alleviation of local irritation [6–8]. In the last decade, different kinds of chemically modified CyD derivatives have been prepared to extend the physicochemical properties and inclusion capacity of parent CyDs [9–11]. Table 8.1 summarizes the pharmaceutically useful β-CyD derivatives, classified into hydrophilic, hydrophobic, and ionizable derivatives. Among these compounds, hydrophilic CyDs such as hydroxypropyl-β-CyD (HP-β-CyD) and branched β-CyD have received special attention, because their toxicity is extremely low and aqueous solubility is very high, promising a parenteral use. The hydrophobic CyDs include ethylated CyDs such as 2,6-di-*O*-ethyl-β-CyD (DE-β-CyD), which can retard the dissolution rate of water-soluble drugs. In addition, the ionizable CyDs include sulfobutyl ether β-CyD (SBE-β-CyD), which can realize an improvement in inclusion capacity, a modification of dissolution rate, and the alleviation of local irritation of drugs.

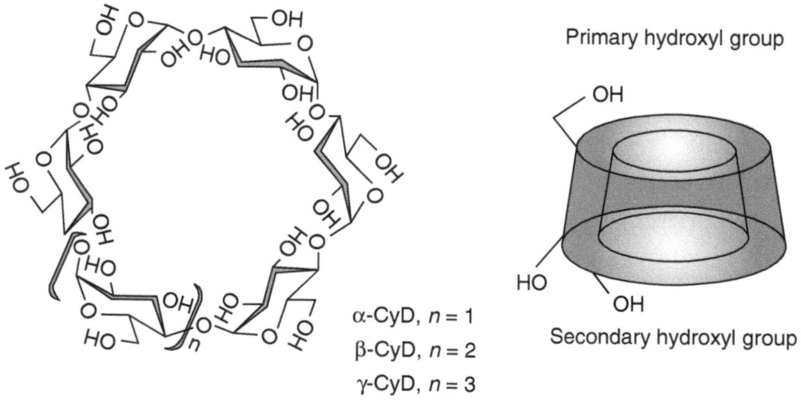

FIGURE 8.1 Chemical structure of CyD.

TABLE 8.1
Pharmaceutically Useful CyDs

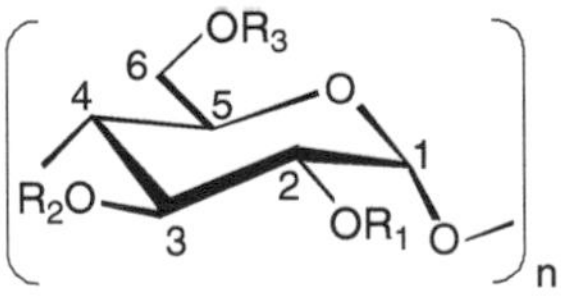

CyD	Abbreviation	Substitution Group	Possible Use
α-Cyclodextrin	α-CyD	R_1, R_2, R_3, = H	Oral, parenteral, local
β-Cyclodextrin	β-CyD	R_1, R_2, R_3 = H	Oral, local
γ-Cyclodextrin	γ-CyD	R_1, R_2, R_3 = H	Oral, parenteral, local
Dimethyl-α-cyclodextrin	DM-α-CyD	$R_1 = CH_3$, $R_2 = H$, $R_3 = CH_3$	Oral, local
Dimethyl-β-cyclodextrin	DM-β-CyD	$R_1 = CH_3$, $R_2 = H$, $R_3 = CH_3$	Oral, local
2-Hydroxypropyl-α-cyclodextrin	HP-α-CyD	R_1, R_2, R_3 = H or $CH_2CH(OH)CH_3$	Oral, parenteral, local
2-Hydroxypropyl-β-cyclodextrin	HP-β-CyD	R_1, R_2, R_3 = H or $CH_2CH(OH)CH_3$	Oral, parenteral, local
2-Hydroxypropyl-γ-cyclodextrin	HP-γ-CyD	R_1, R_2, R_3 = H or $CH_2CH(OH)CH_3$	Oral, parenteral, local
Gluculonylglucosyl-β-cyclodextrin	GUG-β-CyD	$R_1 = H$, $R_2 = H$, $R_3 = H$ or gluculonylglucose	Oral, parenteral, local
Maltosyl-β-cyclodextrin	G_2-β-CyD	$R_1 = H$, $R_2 = H$, $R_3 = H$ or maltose	Oral, parenteral, local
Randomly methyl-β-cyclodextrin	RM-β-CyD	R_1, R_2, R_3 = H or CH_3	Oral, local
Sulfobutyl ether-β-cyclodextrin	SBE-β-CyD	R_1, R_2, R_3 = H or $(CH_2)_4SO_3Na$	Oral, parenteral, local
Sulfobutyl ether-γ-cyclodextrin	SBE-γ-CyD	R_1, R_2, R_3 = H or $(CH_2)_4SO_3Na$	Oral, parenteral, local

Meanwhile, the hydrophobic CyD derivatives may modulate the release of drugs from the vehicles.

Hydrophilic CyDs' possible enhancing mechanisms on the bioavailability of drugs in various administration routes are summarized as follows: (1) increase the solubility, dissolution rate, and wettability of poorly water-soluble drugs, (2) prevent the degradation or disposition of chemically unstable drugs in gastrointestinal (GI) tracts as well as during storage, (3) perturb the membrane fluidity to lower the barrier function, which consequently enhances the absorption of drugs including peptide and protein drugs through the rectal mucosa, (4) release the included drug by competitive inclusion complexation with third components (bile acid, cholesterol, lipids, etc.), (5) inhibit P-glycoprotein-mediated efflux of drug from intestinal epithelial cells [12]. Thus, CyDs would be useful for drug carriers in the rectal routes and have been mostly applied to optimizing the rectal delivery of drugs intended for a systemic use. The representative examples of this application of parent CyDs and hydrophilic CyD derivatives to the rectal delivery are summarized in Table 8.2.

Many reports have indicated the findings that the effects of CyDs on the rectal delivery of drugs depend markedly on vehicle type (hydrophilic or oleaginous), physicochemical properties of the complexes, and an existence of tertiary excipients such as viscous polymers. The enhancing effects of CyDs on the rectal absorption of lipophilic drugs are generally based on the improvement of the release from vehicles and the dissolution rates in rectal fluids, whereas those of CyDs on the rectal delivery of poorly absorbable drugs such as antibiotics, peptides,

TABLE 8.2
The Use of CyDs in Rectal Delivery

CyD	Drug	Ref.
α-CyD	Cefmetazole	[35]
	Chorionic gonadotropin	[33]
	Granulocyte colony-stimulating factor	[32]
	Morphine hydrochloride	[26,37]
	Sulfanilic acid	[34]
β-CyD	Acetaminophen	[38]
	AD1590	[24]
	Carmofur	[36]
	4-Biphenylacetic acid	[18,21]
	Ethyl 4-biphenylyl acetate	[23]
	Naproxen	[14]
	Phenobarbital	[15]
	Piroxicam	[44]
	Sulfanilic acid	[34]
γ-CyD	Diazepam	[24]
	Flurbiprofen	[16]
	Sulfanilic acid	[34]
DM-β-CyD	4-Biphenyacetic acid	[21]
	Carmofur	[17,36]
	Diazepam	[24]
	Ethyl 4-biphenylyl acetate	[23]
	Flurbiprofen	[16,20]
	Insulin	[31]
TM-β-CyD	Carmofur	[36]
	Diazepam	[24]
	Flurbiprofen	[16]
HP-β-CyD	4-Biphenylacetic acid	[21]
	Diazepam	[24]
	Ethyl 4-biphenylyl acetate	[23]
β-CyD polymer	Carmofur	[36]

and proteins are based on the direct action of CyDs to rectal epithelial cells as described later. On the other hand, the prolonging effects of CyDs on the drug levels in blood are caused by the sustained release from the vehicles, slower dissolution rates in the rectal fluid, or the retardation in the rectal absorption of drugs by a poorly absorbable complex formation. Thus, the effects of CyDs on the rectal drug delivery seem to be associated with several physiological and physicochemical factors.

8.2.1 Improvement of Drug Release from Suppositories and Drug Absorption by Cyclodextrins

There are many reports that CyDs enhance the release of drugs from the suppository [7,13]. Most of these examples have been observed when the physicochemical characteristics of suppository base and drugs used were oleaginous and lipophilic, respectively. For example, the complexation with parent CyDs augmented the release of hydrophobic drugs such as naproxen [14], phenobarbital [15], flurbiprofen [16], carmofur [17], and 4-biphenylacetic acid (BPAA) [18] from oleaginous suppository bases. So far, it has been proposed that the enhancing effects of parent CyDs on the release of lipophilic drugs from the oleaginous suppository bases could be attributed to the formation of more hydrophilic complexes of these drugs with parent CyDs, because the complexes have low affinity with the bases and rapidly dissolve into the rectal fluids. In addition, Frijlink et al. [19] reported that the complexation with CyDs enhanced the dissolution of lipophilic drugs at an interface between the molten base and the surrounding fluid, and inhibited the reverse diffusion of the drug into the vehicle.

In comparison with parent CyDs, DM-β-CyD and HP-β-CyD enhance the rectal absorption of lipophilic drugs such as carmofur [17], flurbiprofen [20], BPAA [18], ethyl 4-biphenylyl acetate (EBA) [21], and *n*-butyl-*p*-aminobenzoate [22], to a great extent. These superior effects of both β-CyD derivatives may be explained by the high aqueous solubility and faster release of the complex together with the lowering of affinity of complexed drug to the oleaginous suppository base. For instance, Uekama et al. [18] demonstrated that the complexation with DM-β-CyD and HP-β-CyD significantly enhanced the rectal bioavailability of BPAA after rectal administration of Witepsol H-5 suppository containing BPAA or its β-CyD complexes to rats through increasing drug release from vehicles, compared with BPAA alone and β-CyD complexes (Figure 8.2). In addition, Arima et al. [23] revealed that HP-β-CyD enhanced the release of EBA from the oleaginous suppository base in comparison with that of EBA alone, β-CyD and DM-β-CyD complexes. The superior effects of HP-β-CyD to DM-β-CyD on the drug release could be ascribed to lower dissociation of the complex in the vehicle and lower viscosity of the base in comparison with DM-β-CyD complex. The *in situ* recirculation study revealed that these complexes of EBA were less absorbable from the rectal lumen in solution state, but this disadvantageous effect of β-CyDs was compensated in part by the inhibition of the bioconversion of EBA to BPAA. The enhancing effects of HP-β-CyD on rectal absorption of EBA were the only ones observed *in vivo*, reflecting *in vitro* data. Interestingly, rather high amounts of HP-β-CyD and DM-β-CyD, compared with β-CyD, were absorbed from the rat rectum, when β-CyDs were coadministered with EBA to the rat rectum *in vivo*. Thus, the enhancement of the *in vivo* rectal absorption of EBA can be due to the fact that β-CyDs increased the release rate of EBA from the vehicle and stabilized EBA in the rectal lumen, and in part the absorption of EBA in the form of complex from the rectum.

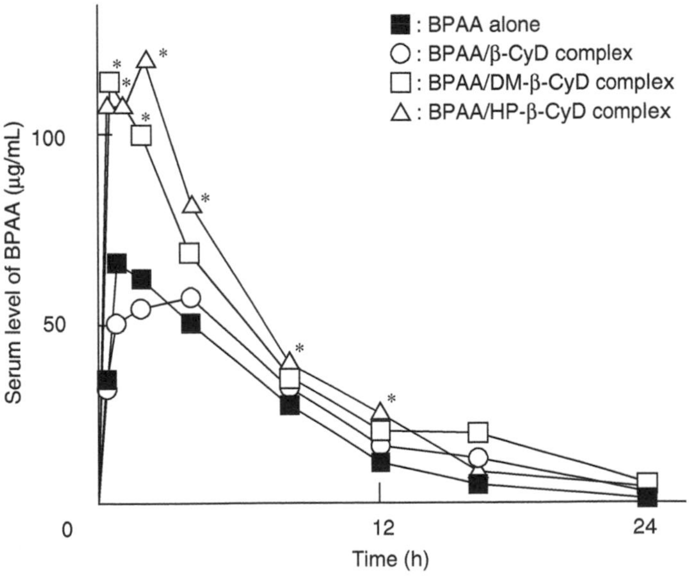

FIGURE 8.2 Serum levels of 4-biphenylacetic acid (BPAA) after rectal administrations of Witepsol H-5 suppositories containing BPAA or its CyD complexes to rats. Each point represents the mean of five rats. *$p < 0.05$, compared with BPAA alone.

Differences in the effects of CyDs on the rectal absorption of drugs were reported in clinical studies using oleaginous suppositories containing diazepam, ibuprofen, or prednisolone [24]. When diazepam was complexed with γ-CyD, the drug release from the suppositories and the rectal absorption of the drug increased compared with suppositories containing the drug alone. However, the complexation of ibuprofen with β-CyD decreased the rectal absorption due to the prevention of the spreading of the suppository and dissolution of the complex. In addition, the bioavailability of prednisolone from the suppositories was not increased by complexation with β-CyD [24]. Thus, the hydrophilic CyDs may increase the rectal absorption of lipophilic drugs from oleaginous suppositories, but more attention should be paid to oleaginous suppositories containing β-CyD complex with somewhat low aqueous solubility is administered to the rectum.

8.2.2 Stabilization of Drugs by Cyclodextrins in Rectal Delivery

The complexation of drugs with CyDs can improve the chemical stability in suppository bases and bioconversion of the drugs to pharmacologically inactive metabolites in the rectum.

Takahashi et al. [25] reported that the autoxidation of AD-1590, an acid nonsteroidal anti-inflammatory drug (NSAID), was improved by complexation with β-CyD, although the autoxidation could not be inhibited by hydroquinone or similar antioxidants. Likewise, Kikuchi et al. [17] reported that carmofur, a 1-hexylcarbamoyl-5-fluorouracil, a masked compound of 5-fluoroyracil (5-FU), which is more likely to hydrolyze to 5-FU, is stabilized by the complexation with DM-β-CyD and 2,3,6-tri-*O*-methyl-β-cyclodextrin (TM-β-CyD) in an oleaginous suppository base. These stabilizing effects of β-CyDs seem to be attributable to the decrease in solubility of these drugs in the oleaginous suppository base, which may lead to a lesser interaction of drugs with the base.

As described above, the bioconversion of EBA, a prodrug of BPAA, in the rectal lumen of rats was markedly inhibited by the complexation with β-CyD, DM-β-CyD, and HP-β-CyD using an *in situ* recirculation technique [23]. These inhibitory effects may also lead to the alleviation of the rectal irritancy of the drugs. In addition, the inhibitory effects of CyDs on the bioconversion of morphine were reported by Kondo et al. [26]: the two types of morphine glucuronide, one being less and another being more pharmacologically active metabolites, in plasma were found to be reduced by the combined use of α-CyD and xanthan gum, a polysaccharide-type polymer with high swelling capacity, after rectal administration of hollow-type oleaginous suppositories in rabbits. The inhibitory effects of α-CyD on the glucuronate conjugation of morphine could be ascribed to the inhibition of the upward movement of morphine from areas, which are impacted by first-pass metabolism, indicating that the combination of α-CyD and viscous polymer such as xanthan gum is beneficial to the stability of drug in suppositories and rectum. Thus, it is clear that CyDs are useful for the stabilization of drugs in suppositories and rectal mucosa.

8.2.3 Cyclodextrins as Absorption Enhancers

It is well known that CyDs, especially methylated β-CyDs, release membrane components such as phospholipids and cholesterol from biomembranes, resulting in enhancement of drug absorption in various routes such as rectal, nasal, ophthalmic, and transdermal [8,27–30]. Hence, CyDs have the potency of the absorption enhancer in rectum as well. There are some reports that CyDs enhance the permeability of drugs through the rectal epithelium cells. For example, Kondo et al. [26] demonstrated that α-CyD and β-CyD, but not γ-CyD, enhanced the rate and extent of bioavailability of morphine after the rectal administration of hollow-type oleaginous suppositories containing morphine hydrochloride, the former being more effective (Figure 8.3, probably because of an increase in the permeability of poorly absorbable drugs, although these parent CyDs did not alter the release rate of morphine from the vehicle.

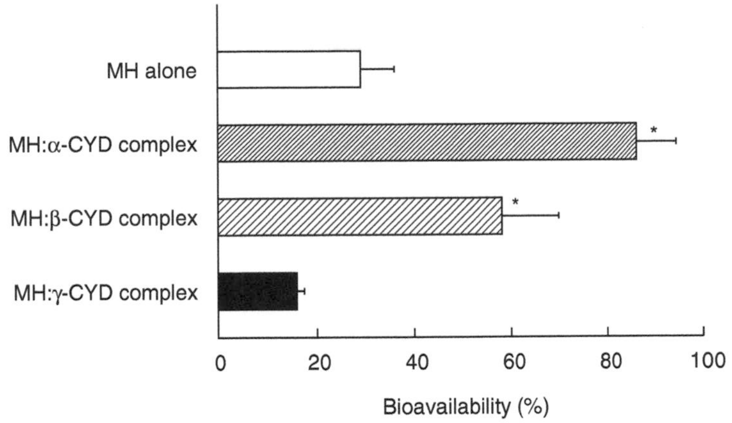

FIGURE 8.3 Bioavailability of morphine after rectal administrations of Witepsol H-15 hollow-type suppositories containing morphine hydrochloride (MH) and CyDs in rabbits. Each value represents the mean ±SE of three rabbits. *$p < 0.05$, compared with MH alone.

CyDs were found to enhance the permeability of proteins such as insulin [31], recombinant human granulocyte colony-stimulating factor (rhG-CSF) [32], and human chorionic gonadotropin (hCG) [33] through the rectal epithelium cells of rabbits. With respect to insulin suppository, the absorption of insulin from the rectum of rabbits after the administration of hollow-type suppositories containing insulin and CyDs significantly increased with a marked decrease in the glucose levels. The enhancing effects of CyDs on rectal insulin absorption by CyDs, i.e., DM-β-CyD and HP-β-CyD, were higher than those by natural CyDs (α-, β-, and γ-CyD) [31]. The absorption-enhancing effect disappeared 24 h after preadministration, suggesting that CyDs enhance insulin absorption from the rectum, and that attenuation of the membrane transport barrier function in the rectum recovered at a maximum of 24 h after administration of CyDs. Thus, α-CyD and DM-β-CyD, possibly as well as HP-β-CyD, act as absorption enhancers in the rectal route, although this depends on the type of peptides and proteins.

On the contrary, Nakanishi et al. [34] reported that mucin layer works as a barrier to the increased absorption of sulfanilic acid, a nonabsorbable drug and noninteracted drug with CyDs, by β-CyD because enhanced absorption of sulfanilic acid from the rat rectum was not induced by β-CyD even with pretreatment with *N*-acetyl-L-cysteine (NAC), sodium deoxycholate (NaDC), or sodium lauryl sulfate (SLS). Thus, β-CyD may have a moderate absorption-enhancing effect in rectum.

8.2.4 Cyclodextrins as Absorption Coenhancers

CyDs are known to be able to solubilize lipophilic drugs as well as lipophilic absorption enhancers, leading to the improvement of the enhancer's efficiency. There are some reports on the use of CyDs as a candidate for a coenhancer. For example, Yanagi et al. [35] reported that CyDs may promote the potency of absorption enhancers in rectum of rabbits. Inclusion complex of decanoic acid (C10), an absorption enhancer, with α-CyD was prepared as an additive of cefmetazole sodium suppository and rectally administered to rabbits. Plasma concentration and area under the curve (AUC) of cefmetazole sodium after rectal administration of a suppository containing C10-α-CyD complex to rabbits increased more significantly than those with no additive.

The combinatorial effects of CyDs and absorption enhancers on the selective transfer of antitumor drug into lymphatic have been demonstrated [36]. When the complex of carmofur with β-CyD polymer was administered together with mixed micelles, an absorption enhancer, into the lumen of rat large intestine of rats, the selective transfer of the drug into the lymphatic was observed [36].

8.2.5 Long-Acting Absorption-Enhancing Effects of Cyclodextrins

A need has arisen to develop rectal long-active preparations for clinical practice, such as in the treatment of intractable chronic pain in advanced cancer patients. An attempt to optimize the rectal delivery of morphine was performed by the addition of CyDs as an absorption enhancer and xanthan gum as a swelling hydrogel in oleaginous hollow-type suppositories [37]. As a result, α-CyD enhanced the rate and extent of bioavailability after the rectal administration as described above, whereas xanthan gum retarded the plasma morphine levels after the rectal administration (Figure 8.4), reflecting *in vitro* slow release characteristics. In addition, the retarding effects of xanthan gum were superior to the other viscous polysaccharides such as guar gum, arabinogalactan, cardlan, and hydroxypropyl cellulose-H (HPC-H). Consequently, a combination of α-CyD and xanthan gum produced sustained plasma profiles of morphine along with an increased rectal bioavailability (more than four times) [37].

However, some negative effects of the combination of CyDs and polysaccharide on the rectal drug delivery were reported. Lin et al. [38] demonstrated that the mixture of β-CyD and hydroxypropylmethylcellulose (HPMC) markedly reduced the *in vivo* bioavailability of acetaminophen from both aqueous solution and hydrogels. Not only the lower partition coefficient but also the higher hydrophilic property of the β-CyD complex and the higher viscosity of HPMC hydrogel matrix might be responsible for the decrease in the *in vitro* permeation rate and depression of *in vivo* rectal absorption of acetaminophen.

8.2.6 Enhancing Mechanism of Cyclodextrins for Rectal Absorption

CyDs are known to induce shape change of membrane invagination on human erythrocytes, which induce hemolysis at higher concentrations. The hemolytic activity of CyDs is in the order of DM-β-CyD > TM-β-CyD > β-CyD > G_2-β-CyD > HP-β-CyD >

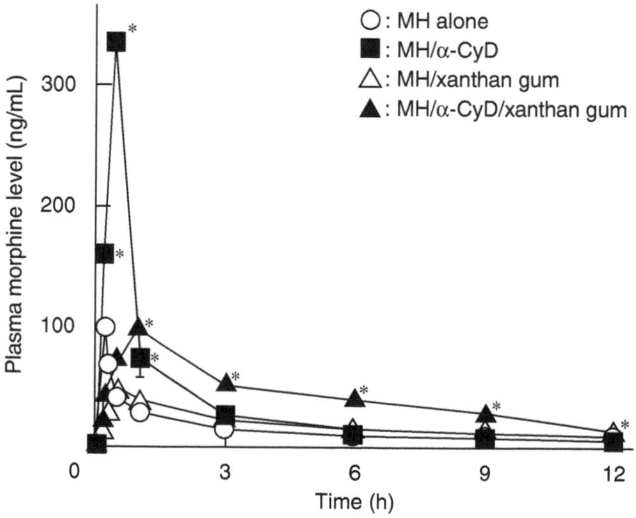

FIGURE 8.4 Plasma levels of morphine after rectal administrations of Witepsol H-15 hollow-type suppositories containing morphine hydrochloride, α-CyD and xanthan gum in rabbits. Each point represents the mean ±SE of three rabbits. *$p < 0.05$, compared with MH alone.

α-CyD > SBE7-β-CyD > γ-CyD > α-CyD > δ-CyD > DMA-β-CyD [28,39,40]. These differences are ascribed to the differential solubilization rates of membrane components, such as cholesterol and phospholipids by each CyD. Likewise, the solubilizing effects of CyDs on membrane lipids may induce changes in cellular function in higher concentrations [41]. Figure 8.5 shows the possible enhancing mechanism of CyD for rectal bioavailability of water-insoluble drugs and poorly absorbable drugs including peptide and protein. As shown in Figure 8.5a, water-insoluble drugs have a rate-limiting step of release from vehicles and dissolution in rectal fluid; the complexation with hydrophilic CyDs improves the low release and dissolution of drugs, resulting in enhancement of rectal absorption. In general, CyD itself

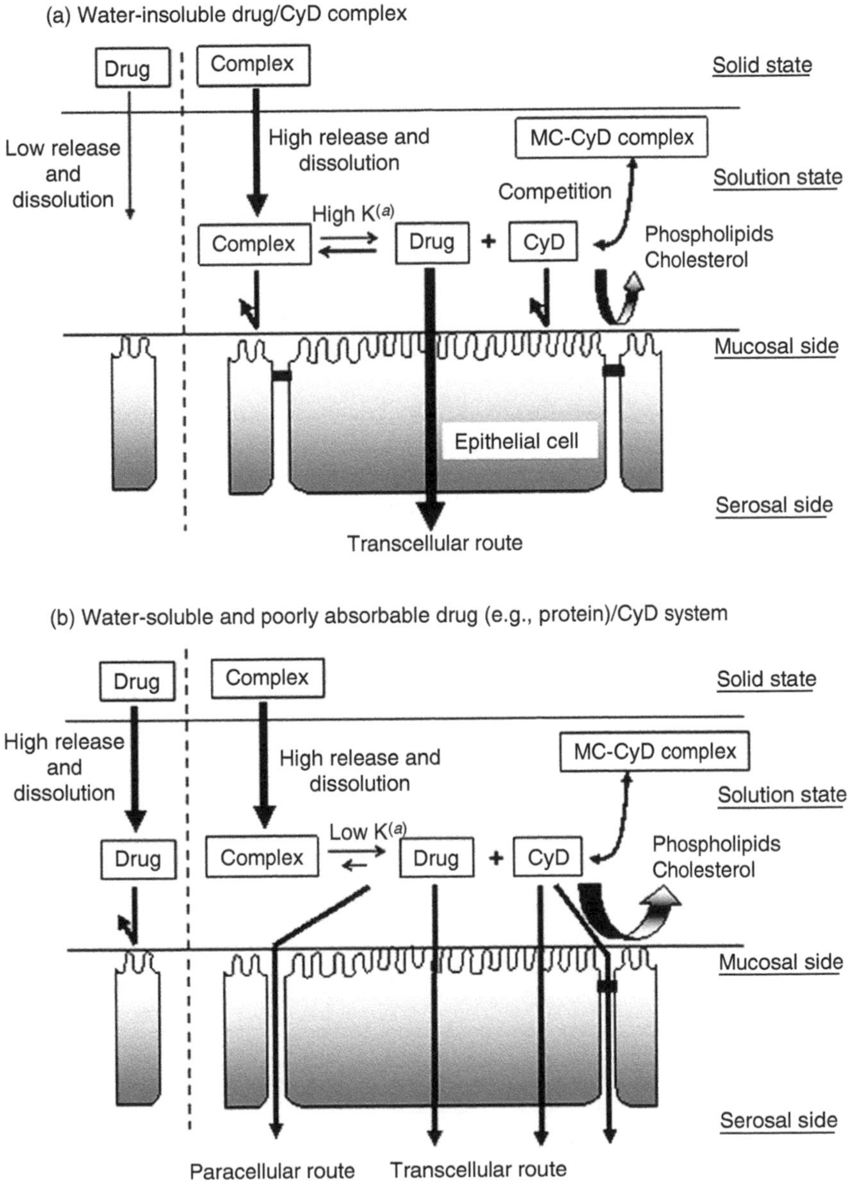

FIGURE 8.5 Possible enhancing mechanism of CyD for rectal bioavailability of water-insoluble drugs (a) and poorly absorbable drugs including peptide and protein (b). *a*, Stability constant.

and CyD complex are known to be hardly absorbed from intestinal tracts because of its hydrophilicity and high molecular weight. In most cases, the enhancing effect of CyDs on the drug release and dissolution could, however, surpass the negative effect of complexation against the drug absorption in solution. In water-soluble and poorly absorbable drug, e.g., peptide and protein as shown in Figure 8.5b, CyDs are unable to make complexes with drugs in solution (low stability constant) and predominantly include membrane components such as phospholipids and cholesterol, but not drugs, depending on their cavity sizes. Consequently, CyDs are likely to increase membrane fluidity and loose tight junction, leading to the enhanced permeation of drugs as well as CyDs. Evidently, DM-β-CyD is the most powerful in all aspects and causes an increase in the permeability of the cytoplasm membrane in a concentration-dependent manner. Actually, it was possible to increase the overall transport of the macromolecular pore marker polyethylene glycol 4000 (PEG4000) 10-fold by the use of DM-β-CyD in low concentrations where the toxic effects on Caco-2 cell monolayers were insignificant [42]. DM-β-CyD was able to produce an absorption-enhancing effect on PEG4000 in concentrations where the toxic effects on Caco-2 monolayers were low. Thus, it may be worth discussing DM-β-CyD as an absorption enhancer in rectum. Additionally, DM-β-CyD is likely to act in both paracellular and transcellular routes as described below.

So far it has been believed that CyDs cannot be absorbed from the rectum and only free drug can be permeated through epithelial monolayers (Figure 8.5). However, other studies may contradict the accepted opinion [23]. Rather high amounts of HP-β-CyD (~26% of dose) and DM-β-CyD (~21% of dose), compared with β-CyD (~5% of dose), were absorbed from the rat rectum, when β-CyDs were coadministered with EBA in the rectum of rats. These results suggest that not only the free form but also the complexed form of EBA can be absorbed in the cases of HP-β-CyD and DM-β-CyD. As described below, the unexpected rectal absorption may be explained in the presence of NSAIDs and triglycerides, which act as absorption enhancers. A similar permeation behavior of α-CyD was observed in the *in vitro* study using the isolated rabbit rectal mucosa (Figure 8.6). However, there is still less information about the rectal absorption of CyDs. We hope that there will be further studies regarding the rectal absorption of CyDs. Thus, it is apparent that CyDs can improve the pharmacokinetics behavior in the rectal delivery with the modification of the release of drugs from the vehicles and permeability of drugs through the rectal epithelium cells. However, the reasons for the superior effects of CyDs seem to be unclear, although probably due to the rather low aqueous solubility.

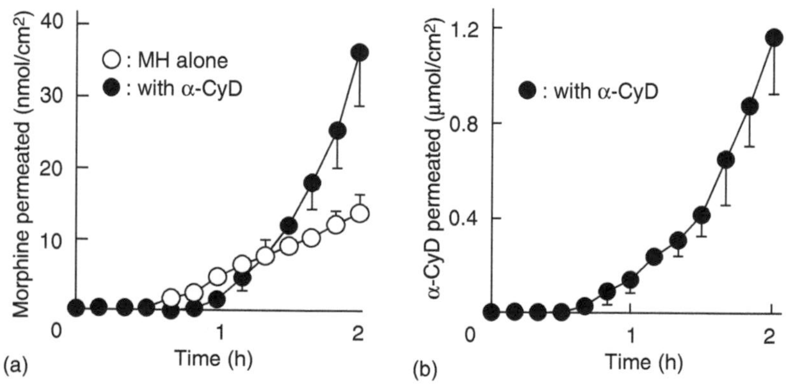

FIGURE 8.6 Permeation profiles of morphine (a) and α-CyD (b) through rabbit rectal mucosa after applications of morphine hydrochloride and α-CyD in Ringer's solution (pH 7.4) at 37°C. Each point represents the mean ±SE of four experiments. *$p < 0.05$, compared with MH alone.

8.2.7 Inhibitory Effects of Cyclodextrins of Drug Irritancy in Rectal Delivery

It is a well-established fact that CyDs reduce the irritation caused by NSAIDs to GI mucosa [7]. Arima et al. [21,23] revealed that HP-β-CyD significantly reduced the irritation of the rectal mucosa caused by BPAA and EBA after single and multiple administrations of the oleaginous suppositories to rats. Additionally, gross and microscopic observations of rectal membranes 12 h after rectal administration of hollow-type suppositories containing morphine hydrochloride, α-CyD, and xanthan gum indicated that the morphine preparation containing α-CyD and xanthan gum caused less irritation to the rectal mucosa, compared with the morphine preparation containing α-CyD alone [37]. In a clinical study, the suppository containing piroxicam-β-CyD complex was shown to produce an earlier onset of pain relief with a longer-lasting effect and tolerability [43]. On the other hand, there are some reports on the cytotoxicity and local irritation of CyD itself, even DM-β-CyD [39,44]. However, the rectal irritancy of CyD itself seems to be less serious [31].

8.3 OTHER ABSORPTION ENHANCERS

The most efficient rectal absorption enhancers, which have been studied, include surfactants, bile acids, sodium salicylate (NaSA), medium-chain glycerides (MCG), NaC10, enamine derivatives, EDTA, and others [45–47]. Transport from the rectal epithelium primarily involves two routes, i.e., the paracellular route and the transcellular route. The paracellular transport mechanism implies that drugs diffuse through a space between epithelial cells. On the other hand, an uptake mechanism which depends on lipophilicity involves a typical transcellular transport route, and active transport for amino acids, carrier-mediated transport for β-lactam antibiotics and dipeptides, and endocytosis are also involved in the transcellular transport system, but these transporters are unlikely to express in rectum (Figure 8.7). Table 8.3 summarizes the typical absorption enhancers in rectal routes.

8.3.1 Surfactants

Surfactants are capable of improving various pharmaceutical properties such as wettability, solubility, dissolution rate, and miscibility, and are useful for improving the rectal bioavailability

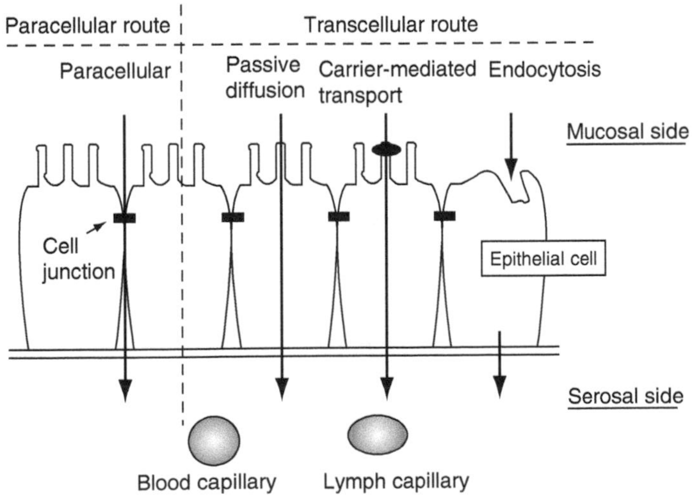

FIGURE 8.7 Transport mechanisms for drugs across rectal epithelia.

TABLE 8.3
Various Absorption Enhancers for Rectal Route

Enhancer	Drug	Ref.
Surfactants	[Asul,7]-eel calcitonin	[55]
	Cefoxitin	[56]
	Insulin	[51–53]
	p-Aminobenzoic acid	[49]
	Sulfaguanidine	[48]
	Sulfanilic acid	[48]
Bile acids	Antipyrine	[58,59]
	Ampicillin	[60]
	Cefoxitin	[65]
	Heparin	[61]
	Insulin	[62–64,66,67]
	Vasopressin derivative	[65]
Enamine derivatives	β-Lactum antibiotics	[69,70]
	Calcitonin	[75]
	Heparin	[76]
	Insulin	[71–74]
	Lysozyme	[76]
Salicylate, 5-methoxysalicylate, other salicylate derivatives	Ara-C	[87]
	Ampicillin	[81]
	Cefmetazole	[79,80]
	Erythropoietin	[84]
	Gastrin	[85]
	Heparin	[86]
	Insulin	[51,64,88–91]
	Levodopa	[79,80]
	Lidocaine	[79,80]
	Pentagastrin	[85]
	Phenylalanine	[82]
	Phenylglycine	[82]
	Theophylline	[78,79,94]
NSAIDs	Insulin	[95,96]
	Latamoxef	[97]
	Sulfanilic acid	[95]
Caprate, fatty acids	Acyclovir	[106]
	Ampicillin	[99]
	Cefoxitin	[104,109]
	Gentamicin	[102]
	Glycyrrhizin	[100,101]
	Insulin	[113]
	Phenoxymethylpenicillin	[103]
	Propranolol	[112]
Mixed micelles	Cefmetazole	[121]
	Cis-diamminedichloroplatinum (II)	[122]
EDTA	Cefoxitin	[132]
EGTA	Des-enkephalin-γ-endorphin	[130]
APD	Insulin	[68,129]
Glycyrrhetinic acid derivative	Ampicillin	[133]

TABLE 8.3 (Continued)
Various Absorption Enhancers for Rectal Route

Enhancer	Drug	Ref.
Saponin	β-lactam antibiotics	[134]
Nitric oxide donor	Insulin	[135]
N-acylcollagen peptide	Ampicillin	[136]
Concanovalin A	Cefoxitin	[137]
Phenothiazines	Cefmetazole	[138]
Diethylmalate	Cefmetazole, lysozyme	[139]
Glycerin ester	Insulin	[140]
Phosphate derivative	Cefoxitin	[141]

of impermeable drugs across biomembranes, including *p*-aminobenzoic acid (PABA), sulfaguanidine, and proteins [48,49].

Sakai et al. [50] reported the enhanced absorption of PABA from the colon by polyoxyethylated nonionic surfactants using an *in situ* perfusion technique. The order of their absorption-enhancing effect was as follows: polyoxyethylene lauryl ether > polyoxyethylene sorbitan fatty acid esters = polyoxyethylene fatty acid esters. The enhancing effects of surfactants on rectal absorption of proteins such as insulin and calcitonin in various formulations have been reported: polyoxyethylene-9-lauryl ether (BL-9) enhanced the rectal insulin absorption in suppository (Witepsol W35) and solution [51,52]. Recently, the use of Tween 60 as an absorption enhancer for insulin suppository in rabbits was demonstrated [53]. Additionally, Morimoto et al. [54] reported that polyacrylic acid (PAA) gel itself has an absorption-enhancing effect on various drugs in rats in a concentration-dependent manner, and that the further incorporation of BL-9 in the PAA gel base significantly enhanced the rectal absorption of [Asu1,7]-eel calcitonin, a calcitonin analog, in rats [55]. However, the irritation of surfactants to rectal mucosa and viral absorption from rectum to bloodstream should be noted [56]. The mechanisms by which surfactants enhance rectal absorption of poorly membrane permeable drugs are likely to be Ca^{2+} sequestration, membrane lipids solubilization, and protein release from rectal mucosa.

8.3.2 Bile Acids

Bile acids are surface-active compounds that consist of a facially amphiphilic steroid nucleus with a hydrophobic α-side and a hydrophilic β-side, which are known to enhance impermeable drugs in the rectal routes [57]. In particular, sodium taurocholate (NaTC), sodium glycocholate (NaGC), and sodium cholate (NaC) enhanced rat rectal absorption of antipyrine [58,59], sodium ampicillin [60], and heparin [61]. Moreover, there are some papers regarding the enhancing effects of bile acids on insulin and peptides. Namely, insulin suppositories including NaDC plus NaC, sodium taurodeoxycholate (NaTDC), or NaTC showed a large decrease in plasma glucose levels in alloxan-induced hyperglycemic rabbits [62] as well as in diabetic beagle dogs [63]. Recently, Hosny et al. [64] reported that the use of insulin–NaC suppositories can serve as effective buffer against meal-related hyperglycemia patients. Besides, van Hoogdalem et al. [65,66] reported the potential use of the other novel bile acid derivative, sodium tauro-24,25-dihydrofusidate (STDHF) on the rectal absorption of cefoxitin, desglycinamide arginine vasopressin, and insulin. The enhancing mechanisms of bile acids for rectal absorption seem to involve Ca^{2+} sequestration, an increase in solvent drag and water channel, dissociation of protein oligomers, and solubilization of membrane lipids [67,68]. The example of the enhancing effects of mixed micelle composed of lipids and bile acids will be shown below.

8.3.3 Enamine Derivatives

Amino acid enamines (phenylalanine and phenylglycine) of β-diketones (ethylacetoacetate) are known to be novel absorption enhancers in the rectal routes. Murakami et al. [69,70] reported that these enamine derivatives enhanced the rectal absorption of β-lactam antibiotics in rabbits. In addition, enamine derivatives of phenylglycine [71] and the enamine of sodium DL-phenylalanate or leucin are likely to be useful as adjuvants for the rectal absorption of insulin [72,73]. Likewise, Nishihata et al. [74] reported that rectal absorption of insulin in depancreatized dogs was significantly enhanced by the coadministration of enamine as a suppository adjuvant. Furthermore, the enhancing effects of enamines on the rectal absorption of other macromolecules such as calcitonin, lysozyme, and heparin were also reported [75,76]. However, enamine derivatives are somewhat labile in aqueous solution. Therefore, they may have the absorption-enhancing effect with a very short duration, and may be more useful in prodrug development rather than as absorption enhancers [77].

8.3.4 Salicylate and Its Derivatives

There are many reports on the utility of salicylic acid (SA) and its derivatives, e.g., sodium 5-methoxysalicylate (NaMSA), as adjuvants in promoting rectal absorption of theophylline [78,79], lidocaine [79,80], levodopa [79,80], cefmetazole [79,80], ampicillin [81], phenylalanine [82], phenylalanylglycine [82], insulin [83], recombinant human erythropoietin [84], gastrin [85], pentagastrin [85], heparin [86], and AraC [87]. It has also been reported that SA enhanced rectal absorption of dextran with molecular weights of more than 100,000 in rats [88]. Hosny et al. [51] demonstrated the absorption-promoting effect of NaSA on insulin suppository in dogs [51] and in normal volunteers and insulin-dependent diabetic patients [64,89]. Thus, SA may be beneficial to enhancement in rectal absorption of various drugs including macromolecules. However, we must pay particular attention to their pharmacological activity. On the other hand, Hauss et al. [90,91] reported the potential effect of other SA derivatives such as 3,5-diiodoSA sodium (DIS), a highly lipophilic SA and the sodium salts of 3,5-dichloro of SA for rectal insulin delivery. These enhancing mechanisms of NaSA may relate to the membrane perturbation by the interaction with membrane proteins, but not membrane lipids, and a decrease in intracellular glutathione levels [92,93]. However, van Hoogdalem reported the deficient effect of NaSA and anesthesia on the rectal absorption of theophylline in rats [94]. Thus, a thoughtful approach to the practical use of NaSA as an absorption enhancer is suggested.

8.3.5 Nonsteroidal Anti-Inflammatory Drugs

NSAIDs such as indomethacin, phenylbutazone, diclofenac sodium, and aspirin have been reported to act as absorption enhancers in the rectal route [95]. For example, NSAIDs enhanced rectal absorption of sulfanilic acid and creatinine in the *in situ* single perfusion experiment in rats [95]. The enhancing effects of NSAIDs increased in the order of phenylbutazone < diclofenac sodium < indomethacin [95]. Similar enhancing effects were observed in inulin (MW = 5500), insulin (MW = 6000), polyvinyl pyrrolidone (MW = 35,000), and albumin (MW = 69,000) [96]. Nakanishi et al. [97] reported the enhancing effect of diclofenac sodium for the rectal absorption of Latamoxef sodium, an antibiotic, in rats [97]. The enhancing effect of NSAIDs could be attributed to change in membrane permeability of poorly permeable drugs owing to the accumulation of the drugs in rectal mucosa [98]. Similar to NaSA, one must be careful of local irritation induced by the pharmacological activity of NSAIDs.

8.3.6 Caprate and Fatty Acids

NaC10 is an acknowledged novel absorption enhancer for ampicillin sodium [99], glycyrrhizin [100,101], gentamicin [102], phenoxymethyl penicillin [103], cefoxitin sodium [104,105], and acyclovir [106]. Takahashi et al. [107] reported that the enhanced membrane permeability of phenolsulfonphthalein depends on the disappearance kinetics of C10 from the loop and its calcium ion sequestration capacity. The enhancing mechanisms of NaC10 are proposed to be involved in (1) Ca^{2+} sequestration, (2) increase in pore size and solvent drag, (3) interaction with membrane proteins and lipids, and (4) increase in the intracellular calcium level [104,105,108–111].

Other fatty acids as absorption enhancers have been reported. Ogiso et al. [112] demonstrated that lauric acid (C12) produced the largest increase in permeation rate, penetration coefficient, and partition coefficient of propranolol. Onuki et al. [113] reported that docosahexaenoic acid (DHA) has a strong insulin permeability enhancement effect and little toxicity, compared to oleic acid and eicosapentaenoic acid (EPA) using a water-in-oil-in-water (W/O/W) multiple emulsion with no or little mucosal damage.

8.3.7 Mixed Micelles

Mixed micelles are composed of lipids and bile acids or surfactants. These micelles are known to enhance poorly absorbable drugs in small intestines as well as large intestines. Muranishi et al. reported the use of mixed micelles as an absorption enhancer for various drugs including gentamicin, 5-fluorouracil, streptomycin, heparin, and human interferon in the large intestinal route [46,57,114–120]. However, there are only a few reports regarding the use of mixed micelles in the rectal route. That is, MCG and triglyceride (Vosco S-55 and Vosco S-55 + methylcellulose) enhanced rectal absorption of cefmetazole and acyclovir in rats, respectively [121,122], and then the promoting effect of MCG on the rectal absorption of cefmetazole was significantly increased by further addition of nonionic surfactants such as ether-type (Brij 30 and Brij 35) and ester-type (Nikkol MYL-10) in dogs [121]. Wakatsuki et al. [123] also reported the use of mixed micelles for an absorption enhancer; tumor growth ratios were reduced significantly in mixed micelles plus *cis*-diamminedichloroplatinum (II) suppository plus irradiation groups. On the other hand, it should be noted that mixed micelles enhance transfer of human fibroblast interferon-β, cyclosporine A, and the other drugs in lymph rather than in blood after absorption from a GI tract, which applies to cancer therapy and immunotherapy [119,123–128].

8.3.8 Ethylenediamine Tetraacetic Acid and Egtazic Acid

Strong chelating agents such as EDTA and EGTA have been employed for the study of tight junctions between intestinal and rectal epithelial cells. In fact, these chelating agents increased the rectal absorption of insulin and des-enkephalin-γ-endorphin [68,129,130]. On the other hand, the extent of rectal cefoxitin absorption was enhanced by 3-amino-1-hydroxypropylidene-1, 1-diphosphonate disodium salt (APD), the calcium-binding agent, on rectal infusion as well as on bolus delivery, the latter regimen tending to result in lower bioavailabilities in rats [131]. Regarding the enhancing mechanisms, Tomita et al. [132] revealed that EDTA activates protein kinase C by depletion of extracellular calcium through chelation, resulting in expansion of the paracellular route in Caco-2 cell monolayers.

8.3.9 Other Absorption Enhancers

Mishima et al. [133] reported on the enhancing effect on rectal absorption of sodium ampicillin by the glycyrrhetinic acid derivative disodium glycyrrhetinic acid 3β-*O*-monohemiphthalate (GA MHPh) in rats. Interestingly, GA MHPh was more effective as an absorption promoter than either C10 or glycyrrhetinic acid. Yata et al. [134] reported that monodesmosides, saponin A, B, and C, isolated from pericarps of *Sapindus mukurossi* (Enmei-hi) were shown to promote the rectal absorption of β-lactam antibiotics in rats. Saponin is known to be surface-active, and may be involved in the enhancing effects. Utoguchi et al. [135] reported that the nitric oxide (NO) donor *S*-nitroso-*N*-acetyl-DL-penicillamine (SNAP) induced a significant increase in the absorption of insulin and FITC-dextran (MW 4000) from the rabbit's rectum. Other NO donors, NOR1 and NOR4, also induced increases in the insulin absorption. The absorption enhancing effect of SNAP was inhibited by coadministration of the NO scavenger carboxy-PTIO. Thus, NO donors can act as potent absorption enhancers. Moreover, Yata et al. [136] reported the promoting effects of *N*-acyl derivatives of a collagen peptide, which was fractionated from the hydrolysates of collagen on the rectal absorption of sodium ampicillin. The adjuvant interaction with the calcium ions located in the rectal membrane may be involved in the enhanced absorption of sodium ampicillin. Nishihata and Higuchi [137] reported that concanavalin A, a lectin, enhanced the rat rectal absorption of phenol red and cefoxitin. The enhancing action of concanavalin A is sodium ion-dependent and inhibited by the presence of 4,4′-diisothiocyano-2,2′-disulfonate stilbene and phlorizin, thus suggesting the involvement of the membrane protein fraction. In addition, phenothiazines, calmodulin antagonist tranquilizers are reported to enhance the rectal absorption of cefmetazole, suggesting the involvement of calmodulin in the absorption-enhancing mechanism [138]. Other absorption enhancers such as diethylmalate (DEM) [139], diethylethoxy methylene malonate (DEEMM) [139], glycerin ester [140], and phosphate derivative [141] have been reported as well. Thus, a number of absorption enhancers have been extensively developed for their potential use. Of these enhancers, NaC10 and CyD have been successfully included in commercially available suppositories. To extend the practical use of the other enhancers, pivotal issues such as their local irritation and safety as well as variability of the efficacy should be addressed.

8.4 ENHANCING MECHANISM OF ABSORPTION ENHANCERS

Clarification of the enhancing mechanism of various absorption enhancers is beneficial for optimal selection of the enhancers and the prediction of combinatorial effects of absorption enhancers. In fact, promising enhancing mechanisms of various absorption enhancers in the rectal route have been proposed. Transcellular and paracellular fluxes are tightly controlled by membrane pumps, ion channels, and tight junctions, thus adapting permeability to physiological needs [142]. The possible enhancing mechanisms of absorption enhancers are briefly summarized in Table 8.3.

8.4.1 Effect of Absorption Enhancers on Paracellular Routes

Tight junctions form selective barriers that regulate paracellular transport across epithelia and endothelia. Tight junctions are composed of transmembrane proteins (occludin, claudins, and JAMs) linked to the actin cytoskeleton through cytoplasmic ZO proteins, whereas adherence junctions are composed of the nectin–afadin system and the E-cadherin–catenin system [143]. These cell–cell adhesion proteins create the barrier and regulate electrical resistance, size, and ionic charge selectivity [144]. Several experimental procedures have been invented to disrupt tight junctions and adherent junctions in cultured epithelial cells, as well as primary cells obtained from animals. These methods are Ca^{2+}-chelating methods, ATP depletion, oxidative stress-induced disruption, and scratch- or wound-healing assays [143]. In terms of the enhancing mechanism of absorption enhancers, the Ca^{2+} sequestration capacity of the absorption enhancers such as surfactants, bile acids, enamine derivatives, NaC10, saponin, and collagen peptides results in their ability to enhance absorption of drugs [131].

Tomita et al. [145] reported that NaC10, NaC12, and mixed micelles caused the colonic pore radius, determined from the equivalent pore theory using an everted sac procedure, to increase significantly, thus making it possible for inulin to permeate the everted sac from the mucosal to the serosal side. In addition, the promoting effects of bile salts on rectal absorption clearance of antipyrine might be due to the increase in solvent drag [58]. With respect to water movement, at least six aquaporin (AQP) isoforms, AQP1, AQP3, AQP4, AQP5, AQP8, and AQP9, which are water transporters, have recently been expressed in the digestive system [146]. However, it is still unclear if AQPs are active in rectum. Besides it is likely that transepithelial water transfer can occur not only through AQP water channels but also through the intercellular route and other transcellular pathways employing other channels and transporters [146]. Hence, water transfer may occur through paracellular route as well as transcellular routes. Therefore, further studies focused not only on AQPs but also on other possible water transfer systems and will help us to understand the mechanism of absorption enhancers in a GI tract as well as rectum. Additionally, CyDs are likely to act in paracellular routes as shown in the findings that mannitol permeation and transepithelial electrical resistance were changed by adding CyDs [147,148]. In fact, recent findings that lipid rafts may play an important role in the spatial organization of tight junctions and in the regulation of the intestinal barrier function [143] may support the opening effects of CyDs on tight junction, especially β-CyDs, which release cholesterol from plasma membranes.

8.4.2 Effect of Absorption Enhancers on Transcellular Routes

In the bilayer membrane model of the 1980s, cell membranes were based largely on a fluid lipid bilayer in which proteins were embedded [149,150]. The bilayer was highly dynamic; lipids and proteins could flex, rotate, and diffuse laterally in a two-dimensional fluid. Based on this, the enhancing mechanisms of absorption enhancers on transcellular routes have been clarified. In summary, most of the mechanisms are strongly associated with membrane fluidity. The fluidity is likely to be changed by the following factors.

The apparent correlation with partition coefficient indicates that the uptake of some absorption enhancers into rectal tissue must be a key factor in their potency as absorption-promoting adjuvants. For example, SA binding to some feature of the rectal membrane appears to be important in the enhanced absorption of drugs from the rectum [79]. NaC10 is also known to have direct interaction not only with proteins but also with lipids [151]. Meanwhile, the involvement of interaction of enhancers with membrane components has been demonstrated. Nishihata et al. [152] reported that both polyoxyethylene-23-lauryl ether (POE) and EDTA increased the release of protein from the rectal mucosa, although treatment of rat's rectal mucosa with SA resulted in slightly less protein release than treatment with NaCl. Similar to POE and EDTA, the enhancing mechanism of CyDs, especially α-CyD and methylated CyDs, for rectal absorption of poorly absorbable drugs, may be due to the interaction of membrane lipids as well, although the cytotoxic effects of CyDs are much lower than POE and EDTA, causing membrane fluidity.

Other enhancing mechanisms have been proposed. Tomita et al. [153] reported that NaC10 enhances permeability through the transcellular route through membrane perturbation [153]. Similarly, Kajii et al. [92] demonstrated that NaSA caused a significant decrease in the fluorescence polarization of 1,6-diphenyl-1,3,5-hexatriene (DPH) and a slight increase in the fluorescence polarization of 8-anilino-1-naphthalene sulfonic acid (ANS) in the isolated rat small intestinal epithelial cell suspension, resulting in an increase in the membrane fluidity of epithelial cells [92]. Nishihata et al. [77,139,154] found that protein thiol and nonprotein thiols play an important role in the permeability of the epithelial membranes of rectum; the decrease in thiols may lead to altered glutathione metabolism in cells, which results in attenuation of membrane integrity. Likewise, Murakami et al. [155] reported that the absorption-enhancing action of fatty acids was suppressed by pretreatment with *N*-ethylmaleimide, a sulfhydryl modifier.

The contribution of osmolarity to the enhancing mechanism of the enhancers is still controversial. Yata et al. [136] reported that the promoting effect of *N*-acyl derivatives of a collagen peptide on the rectal absorption of ampicillin sodium was markedly increased in the hypertonic solution. Likewise, Nishihata et al. [137] reported that the enhancing effect of concanavalin A on the rat rectal absorption of phenol red and cefoxitin was augmented by the addition of NaCl, but not by KCl, demonstrating a sodium ion dependency. Furthermore, the enhancing effect of SA or EDTA on the absorption of cefoxitin from the rectum was also enhanced by NaCl, whereas the activity of POE was not [152]. However, they reported unlike the case of *N*-acyl derivatives of collagen peptides, no influence of osmolarity of the administered solution on the absorption-promoting action was observed [134]. Thus, the involvement of osmotic pressure in the enhancing effects of absorption enhancers may be dependent on the enhancer's type.

As described above, versatile enhancing mechanisms of absorption enhancers in the colorectal route have been proposed. In fact, it is unlikely that only one individual mechanism acts in the enhancing effect, it is rather reasonable to assume that more than two mechanisms play an important role in the enhancing effects. Studies are needed to clarify the detailed mechanism of the absorption enhancers.

8.5 PEPTIDASE INHIBITORS

As mentioned above, the rectal route is very attractive for systemic delivery of peptide and protein drugs, but rectal administration of peptides often results in very low bioavailability due to not only poor membrane penetration characteristics (transport barrier) but also due to hydrolysis of peptides by digestive enzymes of the GI tract (enzymatic barrier). Of these two barriers, the latter is of greater importance for certain unstable small peptides, as these peptides, unless they have been degraded by various proteases, can be transported across the intestinal membrane. Therefore, the use of protease inhibitors is one of the most promising approaches to overcome the delivery problems of these peptides and proteins. Many compounds have been used as protease inhibitors for improving the stability of various peptides and proteins. These include aprotinin, trypsin inhibitors, bacitracin, puromycin, bestatin, and bile salts such as NaCC and are frequently used with absorption enhancers for improvement in rectal absorption.

Readers should refer to the excellent reviews [4,5,156].

8.6 COMBINATORIAL EFFECTS OF ABSORPTION ENHANCERS AND OTHER MATERIALS

It may be possible to achieve significant penetration enhancement by using a combination of other penetration enhancers and peptidase inhibitors.

Combinations of various absorption enhancers are known to increase drug absorption in rectal routes as follows: α-CyD and C10 [35], NaSA and EDTA [157], NaGC and EDTA [68], drug solubilizer (HPMC acetate succinate) and absorption enhancer (polyoxyethylene (23) cetylether (BC-23) or NaCl) [158], acrylic (Eudispert) hydrogel and NaC12 [159], NaC10 + taurin [160], Tween 80 and diclofenac sodium [97], EDTA + NaC10 [161], NaC10 + CMC-Na [162], peptidase inhibitor (Na_2EDTA) and MCG [130], polysorbate 80 (PS-80) micelle and esterase for enzymatic degradation of PS-80 [163], peptidase inhibitor+enhancer [130], and NaMSA+enamine+NSAIDs [164]. Thus, a wide variety of combinations have been attempted, and the suitable combination in terms of both efficacy and safety should be selected.

8.7 PROLONGED RECTAL DRUG DELIVERY

The rectal milieu is quite constant as its pH is about 7.5, and the temperature is usually 37°C. It is normally empty and the pressure varies between 0 and 50 cm. This makes this route suitable for the (controlled) delivery of drugs by applying adequate (controlled release) dosage forms such as osmotic pumps and hydrogels, since the classical suppositories are, in general, not the most suitable dosage form to achieve a reproducible rate and extent of drug absorption.

We have already mentioned the potential use of sustained-release suppositories including morphine/α-CyD/xanthan gum [37]. Likewise, Morgan et al. [165] reported a prolonged release of morphine from a lipophilic suppository base *in vitro* and *in vivo*. In addition, Nakajima et al. [166] and Watanabe et al. [167] reported the use of indomethacin sustained-release suppositories containing sugar ester in polyethylene glycol base and of the sustained-release hydrogel suppositories, including indomethacin prepared with water-soluble dietary fibers, xanthan gum, and locust bean gum, respectively. Interestingly, thermally reversible gels of the block copolymer, Synperonic T908, have been recently evaluated as vehicles for rectal administration of indomethacin [168]. Thus, the rectal sustained-release formulations are suitable for some drugs such as morphine and indomethacin, especially when the oral route is no longer tolerable.

8.8 CONCLUSION

The effects of CyDs and absorption enhancers on the release rate, rectal bioavailability, and pharmacological effects of drugs applied into the rectum change considerably with vehicle type and drug type. Rectal preparations containing drug-β-CyD complex are commercially available, i.e., piroxicam–β-CyD complex. In fact, NaC10 is in clinical use as an absorption enhancer for suppositories including ampicillin sodium or ceftizoxime sodium. On the other hand, recent advancements in biotechnology, a molecular biology and a computer-assisted drug design technique, have been producing larger, labile, complex, insoluble, and poorly absorbable drugs. Hence, further investigation for CyDs and absorption enhancers will overcome some of these problems. Various mechanisms for the enhancing effects of CyDs and absorption enhancers on rectal absorption of drugs have been proposed. Recent progresses in the molecular biology and cell biology fields have elicited a new paradigm of cell membrane structure (lipid raft model) and the molecular machinery of cell–cell interaction [150]. However, there seem to have been few attempts to reveal the mechanism recently. Thereby, these enhancing mechanisms of CyDs and absorption enhancers will be reelucidated by further study, based on the new membrane model and new findings. We hope that further investigations on CyDs and absorption enhancers will lead to their widespread use in rectal formulation and delivery.

REFERENCES

1. de Boer, A.G., et al. 1982. Rectal drug administration: Clinical pharmacokinetic considerations. *Clin Pharmacokinet* 7:285.
2. van Hoogdalem, E., A.G. de Boer, and D.D. Breimer. 1991. Pharmacokinetics of rectal drug administration. Part I. General considerations and clinical applications of centrally acting drugs. *Clin Pharmacokinet* 21:11.
3. Song, Y., et al. 2004. Mucosal drug delivery: Membranes, methodologies, and applications. *Crit Rev Ther Drug Carrier Syst* 21:195.
4. Sayani, A.P., and Y.W. Chien. 1996. Systemic delivery of peptides and proteins across absorptive mucosae. *Crit Rev Ther Drug Carrier Syst* 13:85.
5. Lee, V.H. 1988. Enzymatic barriers to peptide and protein absorption. *Crit Rev Ther Drug Carrier Syst* 5:69.
6. Uekama, K., and M. Otagiri. 1987. Cyclodextrins in drug carrier systems. *Crit Rev Ther Drug Carrier Syst* 3:1.
7. Szejtli, J. 1994. Medicinal applications of cyclodextrins. *Med Res Rev* 14:353.
8. Matsuda, H., and H. Arima. 1999. Cyclodextrins in transdermal and rectal delivery. *Adv Drug Deliv Rev* 36:81.
9. Thompson, D.O. 1997. Cyclodextrins-enabling excipients: Their present and future use in pharmaceuticals. *Crit Rev Ther Drug Carrier Syst* 14:1.
10. Albers, E., and B.W. Muller. 1995. Cyclodextrin derivatives in pharmaceutics. *Crit Rev Ther Drug Carrier Syst* 12:311.
11. Stella, V.J., and R.A. Rajewski. 1997. Cyclodextrins: Their future in drug formulation and delivery. *Pharm Res* 14:556.
12. Uekama, K. 2004. Design and evaluation of cyclodextrin-based drug formulation. *Chem Pharm Bull* 52:900.
13. Szente, L., et al. 1985. Suppositories containing cyclodextrin complexes. Part 2: Dissolution and absorption studies. *Pharmazie* 40:406.
14. Celebi, N., M. Iscanoglu, and T. Degim. 1991. The release of naproxen in fatty suppository bases by β-cyclodextrin complexation. *Pharmazie* 46:863.
15. Iwaoku, R., et al. 1982. Enhanced absorption of phenobarbital from suppositories containing phenobarbital-β-cyclodextrin inclusion complex. *Chem Pharm Bull* 30:1416.
16. Otagiri, M., et al. 1983. Improvements to some pharmaceutical properties of flurbiprofen by β- and γ-cyclodextrin complexations. *Acta Pharm Suec* 20:1.
17. Kikuchi, M., F. Hirayama, and K. Uekama. 1987. Improvement of oral and rectal bioavailability of carmofur by methylated β-cyclodextrin complexations. *Int J Pharm* 38:191.
18. Uekama, K., et al. 1986. Possible utility of β-cyclodextrin complexation in the preparation of biphenyl acetic acid suppositories. *Yakugaku Zasshi* 106:1126.
19. Frijlink, H.W., A.J.M. Schoonen, and C.F. Lerk. 1980. The cyclodextrins on drug release from fatty suppository bases. *In-vitro* observations, ed. H.W. Frijlink, 77. Groningen: Drukkerij van denderen.
20. Uekama, K., et al. 1985. Improvement of dissolution and suppository release characteristics of flurbiprofen by inclusion complexation with heptakis(2,6-di-*O*-methyl)-β-cyclodextrin. *J Pharm Sci* 74:841.
21. Arima, H., T. Kondo, and T. Irie. 1992. Use of water-soluble β-cyclodextrin derivatives as carriers of anti-inflammatory drug biphenyl-acetic acid in rectal delivery. *Yakugaku Zasshi* 112:65.
22. Frijlink, H.W., et al. 1992. The effects of cyclodextrins on drug release from fatty suppository bases. III. Application of cyclodextrin derivatives. *Eur J Pharm Biopharm* 38:174.
23. Arima, H., et al. 1992. Enhanced rectal absorption and reduced local irritation of the anti-inflammatory drug ethyl 4-biphenylylacetate in rats by complexation with water-soluble β-cyclodextrin derivatives and formulation as oleaginous suppository. *J Pharm Sci* 81:1119.
24. Frijlink, H.W., et al. 1991. The effects of cyclodextrins on drug release from fatty suppository bases. II. *In-vivo* observations. *Eur J Pharm Biopharm* 37:183.
25. Takahashi, T., et al. 1986. Stabilization of AD-1590, a non-steroidal antiinflammatory agent, in suppository bases by β-cyclodextrin complexation. *Chem Pharm Bull* 34:1770.

26. Kondo, T., T. Irie, and K. Uekama. 1996. Combination effects of α-cyclodextrin and xanthan gum on rectal absorption and metabolism of morphine from hollow-type suppositories in rabbits. *Biol Pharm Bull* 19:280.
27. Rajewski, R.A., and V.J. Stella. 1996. Pharmaceutical applications of cyclodextrins. 2. *In vivo* drug delivery. *J Pharm Sci* 85:1142.
28. Irie, T., and K. Uekama. 1999. Cyclodextrins in peptide and protein delivery. *Adv Drug Deliv Rev* 36:101.
29. Loftsson, T., and J.H. Olafsson. 1998. Cyclodextrins: New drug delivery systems in dermatology. *Int J Dermatol* 37:241.
30. Loftsson, T., and T. Jarvinen. 1999. Cyclodextrins in ophthalmic drug delivery. *Adv Drug Deliv Rev* 36:59.
31. Watanabe, Y., et al. 1992. Absorption enhancement of polypeptide drugs by cyclodextrins. I. Enhanced rectal absorption of insulin from hollow-type suppositories containing insulin and cyclodextrins in rabbits. *Chem Pharm Bull* 40:3042.
32. Watanabe, Y., et al. 1996. Pharmacodynamics and pharmacokinetics of recombinant human granulocyte colony-stimulating factor (rhG-CSF) after administration of a rectal dosage vehicle. *Biol Pharm Bull* 19:1059.
33. Kowari, K., et al. 2002. Pharmacokinetics and pharmacodynamics of human chorionic gonadotropin (hCG) after rectal administration of hollow-type suppositories containing hCG. *Biol Pharm Bull* 25:678.
34. Nakanishi, K., et al. 1990. Effect of cyclodextrins on biological membrane. I. Effect of cyclodextrins on the absorption of a non-absorbable drug from rat small intestine and rectum. *Chem Pharm Bull* 38:1684.
35. Yanagi, H., et al. 1991. Effect of inclusion complexation of decanoic acid with α-cyclodextrin on rectal absorption of cefmetazole sodium suppository in rabbits. *Yakugaku Zasshi* 111:65.
36. Kaji, Y., et al. 1985. Selective transfer of 1-hexylcarbamoyl-5-fluorouracil into lymphatics by combination of β-cyclodextrin polymer complexation and absorption promotor in the rat. *Int J Pharm* 24:79.
37. Uekama, K., et al. 1995. Modification of rectal absorption of morphine from hollow-type suppositories with a combination of α-cyclodextrin and viscosity-enhancing polysaccharide. *J Pharm Sci* 84:15.
38. Lin, S.Y., and J.C. Yang. 1990. Effect of β-cyclodextrin on the *in vitro* permeation rate and *in vivo* rectal absorption of acetaminophen hydrogel preparations. *Pharm Acta Helv* 65:262.
39. Irie, T., and K. Uekama. 1997. Pharmaceutical applications of cyclodextrins. III. Toxicological issues and safety evaluation. *J Pharm Sci* 86:147.
40. Miyazaki, I., et al. 1995. Physicochemical properties and inclusion complex formation of δ-cyclodextrin. *Eur J Pharm Sci* 3:153.
41. Rothblat, G.H., et al. 1999. Cell cholesterol efflux: Integration of old and new observations provides new insights. *J Lipid Res* 40:781.
42. Hovgaard, L., and H. Brondsted. 1995. Drug delivery studies in Caco-2 monolayers. IV. Absorption enhancer effects of cyclodextrins. *Pharm Res* 12:1328–1332.
43. Costa, S., G. Zinelli, and L. Bufalino. 1990. Piroxicam-β-cyclodextrin in the treatment of dysmenorrhoea. *Drug Invest* 4:73.
44. Totterman, A.M., et al. 1997. Intestinal safety of water-soluble β-cyclodextrins in paediatric oral solutions of spironolactone: Effects on human intestinal epithelial Caco-2 cells. *J Pharm Pharmacol* 49:43.
45. Muranishi, S. 1984. Characteristics of drug absorption via the rectal route. *Methods Find Exp Clin Pharmacol* 6:763.
46. Muranishi, S. 1990. Absorption enhancers. *CRC Crit Rev Ther Drug Carrier Syst* 7:1.
47. de Boer, A.G., E.J. van Hoogdalem, and D.D. Breimer. 1990. Improvement of drug absorption through enhancers. *Eur J Drug Metab Pharmacokinet* 15:155.
48. Nakanishi, K., M. Masada, and T. Nadai. 1983. Effect of pharmaceutical adjuvants on the rectal permeability of drugs. II. Effect of Tween-type surfactants on the permeability of drugs in the rat rectum. *Chem Pharm Bull* 31:3255.

49. Nakanishi, K., et al. 1982. Effect of pharmaceutical adjuvants on the rectal permeability of drugs. I. Effect of pharmaceutical additives on the permeability of sulfaguanidine through the rat rectum. *Yakugaku Zasshi* 102:1133.
50. Sakai, K., et al. 1986. Contribution of calcium ion sequestration by polyoxyethylated nonionic surfactants to the enhanced colonic absorption of *p*-aminobenzoic acid. *J Pharm Sci* 75:387.
51. Hosny, E.A., H.I. Al-Shora, and M.M. Elmazar. 2001. Relative hypoglycemic effect of insulin suppositories in diabetic beagle dogs: Optimization of various concentrations of sodium salicylate and polyoxyethylene-9-lauryl ether. *Biol Pharm Bull* 24:1294.
52. Ichikawa, K., et al. 1980. Rectal absorption of insulin suppositories in rabbits. *J Pharm Pharmacol* 32:314.
53. Kosior, A. 2002. Investigation of physical and hypoglycaemic properties of rectal suppositories with chosen insulin. *Acta Pol Pharm* 59:353.
54. Morimoto, K., T. Iwamoto, and K. Morisaka. 1987. Possible mechanisms for the enhancement of rectal absorption of hydrophilic drugs and polypeptides by aqueous polyacrylic acid gel. *J Pharmacobiodyn* 10:85.
55. Morimoto, K., et al. 1985. Effect of non-ionic surfactants in a polyacrylic acid gel base on the rectal absorption of [Asu1,7]-eel calcitonin in rats. *J Pharm Pharmacol* 37:759.
56. Sechter, I., E. Touitou, and M. Donbrow. 1989. The influence of a non-ionic surfactant on rectal absorption of virus particles. *Arch Virol* 106:141.
57. Muranishi, S. 1985. Modification of intestinal absorption of drugs by lipoidal adjuvants. *Pharm Res* 2:108.
58. Hirasawa, T., et al. 1985. Promoting mechanism by bile salt related to water absorption in drug rectal absorption. *J Pharmacobiodyn* 8:211.
59. Shiga, M., et al. 1985. The promotion of drug rectal absorption by water absorption. *J Pharm Pharmacol* 37:446.
60. Murakami, T., et al. 1984. Effect of bile salts on the rectal absorption of sodium ampicillin in rats. *Chem Pharm Bull* 32:1948.
61. Ziv, E., et al. 1983. Bile salts facilitate the absorption of heparin from the intestine. *Biochem Pharmacol* 32:773.
62. Hosny, E.A. 1999. Relative hypoglycemia of rectal insulin suppositories containing deoxycholic acid, sodium taurocholate, polycarbophil, and their combinations in diabetic rabbits. *Drug Dev Ind Pharm* 25:745.
63. Hosny, E.A., H.I. Al-Shora, and M.M. Elmazar. 2001. Effect of different bile salts on the relative hypoglycemia of Witepsol W35 suppositories containing insulin in diabetic Beagle dogs. *Drug Dev Ind Pharm* 27:837.
64. Hosny, E.A., et al. 2003. Evaluation of efficiency of insulin suppository formulations containing sodium salicylate or sodium cholate in insulin dependent diabetic patients. *Boll Chim Farm* 142:361.
65. van Hoogdalem, E.J., et al. 1989. Rectal absorption enhancement of cefoxitin and desglycinamide arginine vasopressin by sodium tauro-24,25-dihydrofusidate in conscious rats. *J Pharmacol Exp Ther* 251:741.
66. van Hoogdalem, E.J., et al. 1990. Absorption enhancement of rectally infused insulin by sodium tauro-24,25-dihydrofusidate (STDHF) in rats. *Pharm Res* 7:180.
67. Karino, A., et al. 1982. Solvent drag effect in drug intestinal absorption. I. Studies on drug and D_2O absorption clearances. *J Pharmacobiodyn* 5:410.
68. Yamamoto, A., et al. 1992. A mechanistic study on enhancement of rectal permeability to insulin in the albino rabbit. *J Pharmacol Exp Ther* 263:25.
69. Murakami, T., et al. 1981. Studies on absorption promoters for rectal delivery preparations. I. Promoting efficacy of enamine derivatives of amino acids for the rectal absorption of β-lactam antibiotics in rabbits. *Chem Pharm Bull* 29:1998.
70. Murakami, T., et al. 1982. Studies of absorption promoters for rectal delivery preparations. II. A possible mechanism of promoting efficacy of enamine derivatives in rectal absorption. *Chem Pharm Bull* 30:659.
71. Kamada, A., et al. 1981. Study of enamine derivatives of phenylglycine as adjuvants for the rectal absorption of insulin. *Chem Pharm Bull* 29:2012.

72. Kim, S., et al. 1983. Effect of enamine derivatives on the rectal absorption of insulin in dogs and rabbits. *J Pharm Pharmacol* 35:100.
73. Yagi, T., et al. 1983. Insulin suppository: Enhanced rectal absorption of insulin using an enamine derivative as a new promoter. *J Pharm Pharmacol* 35:177.
74. Nishihata, T., et al. 1985. Enhanced bioavailability of insulin after rectal administration with enamine as adjuvant in depancreatized dogs. *J Pharm Pharmacol* 37:22.
75. Miyake, M., et al. 1985. Rectal absorption of [Asu1,7]-eel calcitonin in rats. *Chem Pharm Bull* 33:740.
76. Miyake, M., et al. 1984. Rectal absorption of lysozyme and heparin in rabbits in the presence of non-surfactant adjuvants. *Chem Pharm Bull* 32:2020.
77. Nishihata, T., and J.H. Rytting. 1997. Absorption-promoting adjuvants: Enhancing action on rectal absorption. *Adv Drug Deliv Rev* 28:205.
78. Nishihata, T., J.H. Rytting, and T. Higuchi. 1981. Effects of salicylate on rectal absorption of theophylline. *J Pharm Sci* 70:71.
79. Nishihata, T., J.H. Rytting, and T. Higuchi. 1982. Enhanced rectal absorption of theophylline, lidocaine, cefmetazole, and levodopa by several adjuvants. *J Pharm Sci* 71:865.
80. Nishihata, T., J.H. Rytting, and T. Higuchi. 1982. Effect of salicylate on the rectal absorption of lidocaine, levodopa, and cefmetazole in rats. *J Pharm Sci* 71:869.
81. Nishihata, T., et al. 1984. Enhancement of rectal absorption of ampicillin by sodium salicylate in rabbits. *Chem Pharm Bull* 32:2433.
82. Nishihata, T., et al. 1984. The effects of salicylate on the rectal absorption of phenylalanine and some peptides, and the effects of these peptides on the rectal absorption of cefoxitin and cefmetazole. *J Pharm Sci* 73:1326.
83. Nishihata, T., et al. 1983. Enhancement of rectal absorption of insulin using salicylates in dogs. *J Pharm Pharmacol* 35:148.
84. Mizuno, A., M. Ueda, and G. Kawanishi. 1992. Effects of salicylate and other enhancers on rectal absorption of erythropoietin in rats. *J Pharm Pharmacol* 44:570.
85. Yoshioka, S., L. Caldwell, and T. Higuchi. 1982. Enhanced rectal bioavailability of polypeptides using sodium 5-methoxysalicylate as an absorption promoter. *J Pharm Sci* 71:593.
86. Nishihata, T., et al. 1981. Enhanced rectal absorption of insulin and heparin in rats in the presence of non-surfactant adjuvants. *J Pharm Pharmacol* 33:334.
87. Nishihata, T., et al. 1986. Rectal absorption and lymphatic uptake of Ara-C in rats. *Int J Pharm* 31:185.
88. West, G.B. 1982. Rectal absorption of dextran in rats in the presence of salicylates. *Int Arch Allergy Appl Immunol* 68:283.
89. Hosny, E., et al. 1994. Effect of sodium salicylate on insulin rectal absorption in humans. *Arzneimittelforschung* 44:611.
90. Hauss, D.J., and H.Y. Ando. 1988. The influence of concentration of two salicylate derivatives on rectal insulin absorption enhancement. *J Pharm Pharmacol* 40:659.
91. Hauss, D.J., et al. 1991. The relationship of salicylate lipophilicity to rectal insulin absorption enhancement and relative lymphatic uptake. *J Pharmacobiodyn* 14:139.
92. Kajii, H., et al. 1985. Fluorescence study on the interaction of salicylate with rat small intestinal epithelial cells: Possible mechanism for the promoting effects of salicylate on drug absorption *in vivo*. *Life Sci* 37:523.
93. Kajii, H., et al. 1986. Effects of salicylic acid on the permeability of the plasma membrane of the small intestine of the rat: A fluorescence spectroscopic approach to elucidate the mechanism of promoted drug absorption. *J Pharm Sci* 75:475.
94. van Hoogdalem, E.J., A.G. de Boer, and D.D. Breimer. 1986. Influence of salicylate and anaesthesia on the rectal absorption of theophylline in rats. *Pharm Weekbl Sci* 8:281.
95. Nakanishi, K., et al. 1984. Effect of nonsteroidal anti-inflammatory drugs on the permeability of the rectal mucosa. *Chem Pharm Bull* 32:1956.
96. Nakanishi, K., M. Masada, and T. Nadai. 1986. Effect of nonsteroidal anti-inflammatory drugs on the absorption of macromolecular drugs in rat rectum. *Chem Pharm Bull* 34:2628.
97. Nakanishi, K., et al. 1994. Improvement of the rectal bioavailability of latamoxef sodium by adjuvants following administration of a suppository. *Biol Pharm Bull* 17:1496.

98. Nakanishi, K., et al. 1984. Mechanism of the enhancement of rectal permeability of drugs by nonsteroidal anti-inflammatory drugs. *Chem Pharm Bull* 32:3187.
99. van Hoogdalem, E.J., A.G. de Boer, and D.D. Breimer. 1988. Rectal absorption enhancement of rate-controlled delivered ampicillin sodium by sodium decanoate in conscious rats. *Pharm Weekbl Sci* 10:76.
100. Fujioka, T., et al. 2003. Efficacy of a glycyrrhizin suppository for the treatment of chronic hepatitis C: A pilot study. *Hepatol Res* 26:10.
101. Sasaki, K., et al. 2003. Improvement in the bioavailability of poorly absorbed glycyrrhizin via various non-vascular administration routes in rats. *Int J Pharm* 265:95.
102. Matsumoto, Y., et al. 1989. Rectal absorption enhancement of gentamicin in rabbits from hollow type suppositories by sodium salicylate or sodium caprylate. *Drug Des Deliv* 4:247.
103. Lennernas, H., et al. 2002. The influence of caprate on rectal absorption of phenoxymethylpenicillin: Experience from an *in-vivo* perfusion in humans. *J Pharm Pharmacol* 54:499.
104. Takahashi, H., et al. 1997. The enhancing mechanism of capric acid (C10) from a suppository on rectal drug absorption through a paracellular pathway. *Biol Pharm Bull* 20:446.
105. van Hoogdalem, E.J., et al. 1988. Absorption enhancement of rectally infused cefoxitin sodium by medium-chain fatty acids in conscious rats: Concentration–effect relationship. *Pharm Res* 5:453.
106. Yamazaki, M., et al. 1990. The effect of fatty acids on the rectal absorption of acyclovir in rats. *J Pharm Pharmacol* 42:441.
107. Takahashi, K., et al. 1994. Pharmacokinetic analysis of the absorption enhancing action of decanoic acid and its derivatives in rats. *Pharm Res* 11:388.
108. Lindmark, T., et al. 1997. Mechanism of absorption enhancement in humans after rectal administration of ampicillin in suppositories containing sodium caprate. *Pharm Res* 14:930.
109. Lindmark, T., et al. 1998. Absorption enhancement in intestinal epithelial Caco-2 monolayers by sodium caprate: Assessment of molecular weight dependence and demonstration of transport routes. *J Drug Target* 5:215.
110. Hayashi, M., M. Tomita, and S. Awazu. 1997. Transcellular and paracellular contribution to transport processes in the colorectal route. *Adv Drug Deliv Rev* 28:191.
111. Hayashi, M., et al. 1999. Physiological mechanism for enhancement of paracellular drug transport. *J Control Release* 62:141.
112. Ogiso, T., et al. 1991. Enhancement by fatty acids of the rectal absorption of propranolol: *In vitro* evaluation in the rat. *J Pharmacobiodyn* 14:385.
113. Onuki, Y., et al. 2000. *In vivo* effects of highly purified docosahexaenoic acid on rectal insulin absorption. *Int J Pharm* 198:147.
114. Muranishi, S., N. Muranishi, and H. Sezaki. 1979. Improvement of absolute bioavailability of normally poorly absorbed drugs: Inducement of the intestinal absorption of streptomycin and gentamicin by lipid-bile salt mixed micelles in rat and rabbit. *Int J Pharm* 2:101.
115. Muranishi, N., et al. 1980. Mechanism for the inducement of the intestinal absorption of poorly absorbed drugs by mixed micelles. I. Effects of various lipid-bile salt mixed micelles on the intestinal absorption of streptomycin in rat. *Int J Pharm* 4:271.
116. Muranishi, S., H. Yoshikawa, and H. Sezaki. 1979. Absorption of 5-fluorouracil from various regions of gastrointestinal tract in rat. Effect of mixed micelles. *J Pharmacobiodyn* 2:286.
117. Tokunaga, Y., S. Muranishi, and H. Sezaki. 1978. Enhanced intestinal permeability to macromolecules. I. Effect of monoolein-bile salts mixed micelles on the small intestinal absorption of heparin. *J Pharmacobiodyn* 1:28.
118. Taniguchi, K., S. Muranishi, and H. Sezaki. 1980. Enhanced intestinal permeability to macromolecules. II. Improvement of the large intestinal absorption of heparin by lipid-surfactant mixed micelles in rat. *Int J Pharm* 4:219.
119. Yoshikawa, H., H. Sezaki, and S. Muranishi. 1983. Mechanism for selective transfer of bleomycin into lymphatics by a bifunctional delivery system via the lumen of the large intestine. *Int J Pharm* 13:321.
120. Yoshikawa, H., et al. 1984. A method to potentiate enteral absorption of interferon and selective delivery into lymphatics. *J Pharmacobiodyn* 7:59.

121. Sekine, M., et al. 1985. Improvement of bioavailability of poorly absorbed drugs. V. Effect of surfactants on the promoting effect of medium chain glyceride for the rectal absorption of β-lactam antibiotics in rats and dogs. *J Pharmacobiodyn* 8:653.
122. Ogiso, T., et al. 1993. Rectal absorption of acyclovir in rats and improvement of absorption by triglyceride base. *Biol Pharm Bull* 16:315.
123. Wakatsuki, K., et al. 2005. Effects of irradiation combined with *cis*-diamminedichloroplatinum (CDDP) suppository in rabbit VX2 rectal tumors. *World J Surg* 29:388.
124. Yoshikawa, H., K. Takada, and S. Muranishi. 1984. Molecular weight dependence of permselectivity to rat small intestinal blood–lymph barrier for exogenous macromolecules absorbed from lumen. *J Pharmacobiodyn* 7:1.
125. Yoshikawa, H., et al. 1985. Comparison of disappearance from blood and lymphatic delivery of human fibroblast interferon in rat by different administration routes. *J Pharmacobiodyn* 8:206.
126. Takada, K., et al. 1986. Effect of administration route on the selective lymphatic delivery of cyclosporin A by lipid-surfactant mixed micelles. *J Pharmacobiodyn* 9:156.
127. Supersaxo, A., W.R. Hein, and H. Steffen. 1991. Mixed micelles as a proliposomal, lymphotropic drug carrier. *Pharm Res* 8:1286.
128. Yoshikawa, H., K. Takada, and S. Muranishi. 1992. Molecular weight-dependent lymphatic transfer of exogenous macromolecules from large intestine of renal insufficiency rats. *Pharm Res* 9:1195.
129. Aungst, B.J., and N.J. Rogers. 1988. Site dependence of absorption-promoting actions of laureth-9, Na salicylate, Na_2EDTA, and aprotinin on rectal, nasal, and buccal insulin delivery. *Pharm Res* 5:305.
130. van Hoogdalem, E.J., et al. 1989. Rectal absorption enhancement of des-enkephalin-γ-endorphin (DE γ E) by medium-chain glycerides and EDTA in conscious rats. *Pharm Res* 6:91.
131. van Hoogdalem, E.J., et al. 1989. 3-Amino-1-hydroxypropylidene-1,1-diphosphonate (APD): A novel enhancer of rectal cefoxitin absorption in rats. *J Pharm Pharmacol* 41:339.
132. Tomita, M., M. Hayashi, and S. Awazu. 1996. Absorption-enhancing mechanism of EDTA, caprate, and decanoylcarnitine in Caco-2 cells. *J Pharm Sci* 85:608.
133. Mishima, M., et al. 1995. Promotion of rectal absorption of sodium ampicillin by disodium glycyrrhetinic acid 3β-*O*-monohemiphthalate in rats. *Biol Pharm Bull* 18:566.
134. Yata, N., et al. 1985. Enhanced rectal absorption of β-lactam antibiotics in rat by monodesmosides isolated from pericarps of *Sapindus mukurossi* (Enmei-hi). *J Pharmacobiodyn* 8:1041.
135. Utoguchi, N., et al. 1998. Nitric oxide donors enhance rectal absorption of macromolecules in rabbits. *Pharm Res* 15:870.
136. Yata, N., et al. 1985. Enhanced rectal absorption of sodium ampicillin by *N*-acyl derivatives of collagen peptide in rabbits and rats. *J Pharm Sci* 74:1058.
137. Nishihata, T., and T. Higuchi. 1984. Promoting effect of concanavalin A on transport of sodium cefoxitin and phenol red from rat rectal compartment. *Life Sci* 34:419.
138. Suzuka, T., et al. 1986. Effects of phenothiazines and *N*-(6-aminohexyl)-5-chloro-1-naphthalenesulfonamide on rat colonic absorption of cefmetazole. *J Pharmacobiodyn* 9:460.
139. Nishihata, T., M. Miyake, and A. Kamada. 1984. Study on the mechanism behind adjuvant action of diethylethoxymethylene malonate enhancing the rectal absorption of cefmetazole and lysozyme. *J Pharmacobiodyn* 7:607.
140. Nishihata, T., et al. 1983. Adjuvant effects of glyceryl esters of acetoacetic acid on rectal absorption of insulin and inulin in rabbits. *J Pharm Sci* 72:280.
141. Nishihata, T., et al. 1984. Possible mechanism behind the adjuvant action of phosphate derivatives on rectal absorption of cefoxitin in rats and dogs. *J Pharm Sci* 73:1523.
142. Baumgart, D.C., and A.U. Dignass. 2002. Intestinal barrier function. *Curr Opin Clin Nutr Metab Care* 5:685.
143. Miyoshi, J., and Y. Takai. 2005. Molecular perspective on tight-junction assembly and epithelial polarity. *Adv Drug Deliv Rev* 57:815.
144. Van Itallie, C.M., and J.M. Anderson. 2004. The molecular physiology of tight junction pores. *Physiology* 19:331.
145. Tomita, M., et al. 1988. Enhancement of colonic drug absorption by the paracellular permeation route. *Pharm Res* 5:341.

146. Matsuzaki, T., et al. 2004. Aquaporins in the digestive system. *Med Electron Microsc* 37:71.
147. Ono, N., et al. 2001. A moderate interaction of maltosyl-α-cyclodextrin with Caco-2 cells in comparison with the parent cyclodextrin. *Biol Pharm Bull* 24:395.
148. Marttin, E., J.C. Verhoef, and F.W. Merkus. 1998. Efficacy, safety and mechanism of cyclodextrins as absorption enhancers in nasal delivery of peptide and protein drugs. *J Drug Target* 6:17.
149. Singer, S.J., and G.L. Nicolson. 1972. The fluid mosaic model of cell membranes. *Science* 175:720.
150. Edidin, M. 2003. Lipids on the frontier: A century of cell-membrane bilayers. *Nat Rev Mol Cell Biol* 4:414.
151. Kajii, H., et al. 1986. Effects of sodium salicylate and caprylate as adjuvants of drug absorption on isolated rat small intestinal epithelial cells. *Int J Pharm* 33:253.
152. Nishihata, T., et al. 1985. Comparison of the effects of sodium salicylate, disodium ethylenediaminetetraacetic acid and polyoxyethylene-23-lauryl ether as adjuvants for the rectal absorption of sodium cefoxitin. *J Pharm Pharmacol* 37:159.
153. Tomita, M., et al. 1988. Enhancement of colonic drug absorption by the transcellular permeation route. *Pharm Res* 5:786.
154. Nishihata, T., et al. 1987. Influence of diethyl maleate-induced loss of thiol on cefmetazole uptake into isolated epithelical cells and on cefmetazole absorption from ileal loop of rats. *Chem Pharm Bull* 35:2914.
155. Murakami, M., et al. 1988. Intestinal absorption enhanced by unsaturated fatty acids. Inhibitory effect of sulfhydryl modifiers. *Biochim Biophys Acta* 939:238.
156. Yamamoto, A., and S. Muranishi. 1997. Rectal drug delivery systems—Improvement of rectal peptide absorption by absorption enhancers, protease inhibitors and chemical modification. *Adv Drug Deliv Rev* 28:275.
157. Nishihata, T., et al. 1987. The synergistic effects of concurrent administration to rats of EDTA and sodium salicylate on the rectal absorption of sodium cefoxitin and the effects of inhibitors. *J Pharm Pharmacol* 39:180.
158. Takeichi, Y., et al. 1990. Combinative improving effect of increased solubility and the use of absorption enhancers on the rectal absorption of uracil in beagle dogs. *Chem Pharm Bull* 38:2547.
159. Uchida, T., et al. 2001. Preparation and characterization of insulin-loaded acrylic hydrogels containing absorption enhancers. *Chem Pharm Bull* 49:1261.
160. Miyake, M., et al. 2004. Development of suppository formulation safely improving rectal absorption of rebamipide, a poorly absorbable drug, by utilizing sodium laurate and taurine. *J Control Release* 99:63.
161. Tomita, M., M. Hayashi, and S. Awazu. 1994. Comparison of absorption-enhancing effect between sodium caprate and disodium ethylenediaminetetraacetate in Caco-2 cells. *Biol Pharm Bull* 17:753.
162. Murakami, T., et al. 1991. Enhanced rectal and nasal absorption of human epidermal growth factor by combined use of the absorption promoter and the synthetic polymer in rats. *Eur J Drug Metab Pharmacokinet* Spec No 3:125.
163. Yamamoto, K., A.C. Shah, and T. Nishihata. 1994. Enhanced rectal absorption of itazigrel formulated with polysorbate 80 micelle vehicle in rat: Role of co-administered esterase. *J Pharm Pharmacol* 46:608.
164. Nishihata, T., et al. 1984. Effect of adjuvants on the rectal absorption and lymphatic uptake of pepleomycin in rats. *J Pharmacobiodyn* 7:278.
165. Morgan, D.J., et al. 1992. Prolonged release of morphine alkaloid from a lipophilic suppository base *in vitro* and *in vivo*. *Int J Clin Pharmacol Ther Toxicol* 30:576.
166. Nakajima, T., et al. 1990. Indomethacin sustained-release suppositories containing sugar ester in polyethylene glycol base. *Chem Pharm Bull* 38:1680.
167. Watanabe, K., et al. 1993. Investigation on rectal absorption of indomethacin from sustained-release hydrogel suppositories prepared with water-soluble dietary fibers, xanthan gum and locust bean gum. *Biol Pharm Bull* 16:391.
168. Miyazaki, S., et al. 1995. Thermally gelling poloxamine Synperonic T908 solution as a vehicle for rectal drug delivery. *Biol Pharm Bull* 18:1151.

Part III

Enhancement of Buccal and Sublingual Absorption

9 Basic Biopharmaceutics of Buccal and Sublingual Absorption

Priya Batheja, Rashmi Thakur, and Bozena Michniak

CONTENTS

9.1 INTRODUCTION

The quest for alternative modes of drug delivery to the ever-popular oral route of administration has led to exploration of the various mucosae as possible delivery routes. Since the invention of the nitroglycerin sublingual tablet, the oral mucosal route has generated interest as a substitute delivery approach. Recent developments in the design of dosage forms range from tablets and patches to "lollipops" and insertable chips. One of the leading examples of successful buccal delivery includes the oral transmucosal fentanyl citrates (OTFCs). The OTFC in the form of a lollipop, Actiq® (Cephalon, Inc., Frazer, PA) has revolutionized the treatment for breakthrough cancer pain relief (BTCP) [1], providing better pain relief as compared to morphine sulfate immediate release (MSIR) tablets. Another novel dosage form is the Periochip® (Dexcel-Pharma Ltd., UK), composed of chlorhexidine gluconate in a biodegradable matrix of hydrolyzed gelatin and cross-linked with glutaraldehyde, which is inserted into the intraperiodontal pocket. This has proven to be an excellent treatment for periodontitis, and is often given in conjunction with other dental procedures such as scaling and root planing [2]. Recent investigations have also concentrated on delivery of macromolecules such as peptides using the buccal route. The avoidance of first-pass metabolism and intermediate permeation properties make it attractive for drugs which are sensitive to pH and enzymatic degradation. This chapter aims to provide an overview of the physiology and biopharmaceutics of the oral mucosal route and present an insight into the delivery systems designed for this route.

9.2 PHYSIOLOGY OF THE ORAL MUCOSA

9.2.1 STRUCTURE

The cheeks, lips, hard and soft palates and tongue form the oral cavity. The main difference between the oral mucosa and skin as compared to the gastrointestinal (GI) tract lining lies in the organization of the different epithelia. While the latter has a single layer of cells forming the simple epithelium, the skin and the oral cavity have several layers of cells with various degrees of differentiation.

Within the oral cavity, the masticatory mucosa has a keratinized or cornified epithelium, and covers the stress-enduring regions such as the gingival and the hard palate, providing chemical resistance and mechanical strength. It is divided into four layers: keratinized, granular, prickle-cell, and basal layer (Figure 9.1). The lining mucosa, which provides elasticity, in contrast, is comprised of noncornified surface epithelium covering the rest of the regions including the lips, cheeks, floor of the mouth, and soft palate. It also can be further divided into superficial, intermediate, prickle-cell, and basal layers. The third type of mucosa is the specialized mucosa consisting of both keratinized and nonkeratinized layers, and is restricted to the dorsal surface of the tongue. The intercellular spaces contain water, lipids, and proteins.

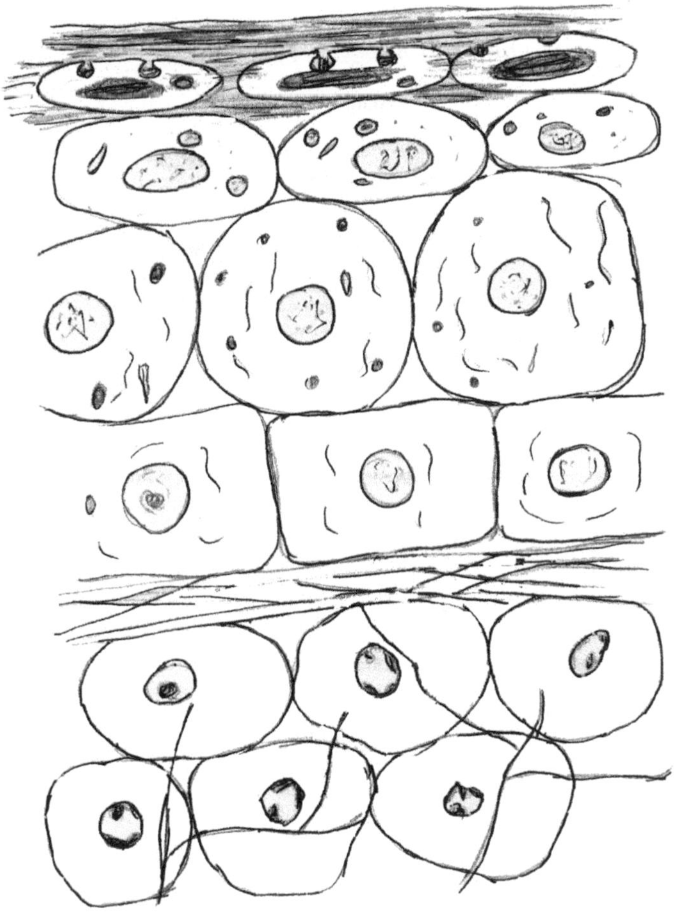

FIGURE 9.1 Structure of the mucosa.

9.2.1.1 Physiological Importance of Mucins and Saliva

The mucosal tissues are further covered with mucus, which is negatively charged, and contains large glycoproteins termed mucins. These are thought to contribute significantly to the viscoelastic nature of saliva, and maintain a pH of 5.8–7.4 [3]. Mucin consists of a protein core, rich in *O*-glycosylated serine and threonine, containing many helix-breaking proline residues. The salivary glands secreting mucus also synthesize saliva, which offers protection to the soft tissues from chemical and mechanical abrasions. The average thickness of the salivary film in the mouth varies between 0.07 and 0.10 mm.

Sustained adhesion of the dosage form (tablet, patch) to the mucosa is an important first step to successful buccal delivery. The mucus plays an important role during this mucoadhesive process by buccal drug delivery systems. The interaction between the mucus and mucoadhesive polymers generally used in most dosage forms can be explained by theories summarized in Table 9.1.

The mean total surface area of the mouth has been calculated to be 214.7 ± 12.9 cm^2 [4]. The teeth, keratinized epithelium, and nonkeratinized epithelium occupy about 20%, 50%, and 30% of this surface area, respectively.

Drug delivery through the oral mucosa can be achieved via different pathways: sublingual (floor of the mouth), buccal (lining of the cheeks), and gingival (gums). The sublingual

TABLE 9.1
Postulated Mechanisms for Polymer–Mucosal Adhesive Properties

Theory of Adhesion	Mechanism of Adhesion
Adsorption	Secondary chemical bonds such as van der Waals forces, hydrophobic interactions, electrostatic attractions, and hydrogen bonds between mucus and polymer [107]
Diffusion	Entanglements of the polymer chains into mucus network [108]
Electronic	Attractive forces across electrical double layer formed due to electron transfer across polymer and mucus [109]
Wetting	Analyzes the ability of a paste to spread over a biological surface and calculates the interfacial tension between the two [110]. The tension is considered proportional to $X^{1/2}$, where X is the Flory polymer–polymer interaction parameter. Low values of this parameter correspond to structural similarities between polymers and an increased miscibility
Fracture	Relates the force necessary to separate two surfaces to the adhesive bond strength and is often used to calculate fracture strength of adhesive bonds [111]

mucosa is the most permeable followed by the buccal and then the palatal. This is due to the presence of neutral lipids such as ceramides and acylceramides in the keratinized epithelia present on the palatal region, which are impermeable to water. The nonkeratinized epithelia contain water-permeable ceramides and cholesterol sulfate. A comparison of the various mucosae is provided in Table 9.2.

The thickness of the buccal epithelium varies from 10 to about 50 cell layers in different regions because of serrations in connective tissue. In fact, the thickness of buccal mucosa has been observed to be 580 μm, the hard palate 310 μm, the epidermis 120 μm, and the floor of mouth mucosa 190 μm.

9.2.2 Tissue Permeability

In comparison to the skin, the buccal mucosa offers higher permeability and faster onset of drug delivery, whereas the key features which help it score over the other mucosal route, the nasal delivery system, include robustness, ease of use, and avoidance of drug metabolism and degradation. The buccal mucosa and the skin have similar structures with multiple cell layers at different degrees of maturation. The buccal mucosa, however, lacks the intercellular lamellar bilayer structure found in the stratum corneum, and hence is more permeable. An additional factor contributing to the enhanced permeability is the rich blood supply in the

TABLE 9.2
Suitability of Various Regions of the Oral Mucosa for Transmucosal Drug Delivery based on Various Tissue Properties

	Permeability	Blood Flow	Residence Time
Buccal	+	++	+
Sublingual	++	−−	−−
Gingival	−−	+	+
Palatal	−−	−−	++

Source: From de Vries, M.E. et al., *Crit. Rev. Ther. Drug Carrier Syst.*, 8, 271, 1991. With permission.
Note: ++ means very suitable; −− means least suitable.

oral cavity. The lamina propia, an irregular dense connective tissue, supports the oral epithelium. Though the epithelium is avascular, the lamina propia is endowed with the presence of small capillaries. These vessels drain absorbed drugs along with the blood into three major veins—lingual, facial, and retromandibular, which open directly into the internal jugular vein, thus avoiding first-pass metabolism. Numerous studies have been conducted comparing the blood supply of the oral cavity to the skin in animals [5–7]. A thicker epithelium has been associated with a higher blood flow probably due to the greater metabolic demands of such epithelia. Gingiva and anterior and posterior dorsum of tongue have significantly higher blood flows than all other regions; skin has a lower flow than the majority of oral regions; and palate has the lowest of all regions. In fact, the mean blood flow to the buccal mucosa in the rhesus monkey was observed to be 20.3 mL/min/100 g tissue as compared to 9.4 mL/min/100 g in the skin.

9.2.3 Barriers to Permeation

The main resistance to drug permeation is caused by the variant patterns of differentiation exhibited by the keratinized and nonkeratinized epithelia. As mucosal cells leave the basal layer, they differentiate and become flattened. Accumulation of lipids and proteins also occurs. This further culminates in a portion of the lipid that concentrates into small organelles called membrane-coating granules (MCGs).

In addition, the cornified cells also synthesize and retain a number of proteins such as profillagrin and involucrin, which contribute to the formation of a thick cell envelope. The MCGs then migrate further and fuse with the intercellular spaces to release the lipid lamellae. The lamellae then fuse from end to end to form broad lipid sheets in the extracellular matrix, forming the main barrier to permeation in the keratinized regions in the oral cavity. These lamellae were first observed in porcine buccal mucosa [8], and have been recently identified in human buccal mucosa [9]. Though the nonkeratinized epithelia also contain a small portion of these lamellae, the random placement of these lamellae in the noncornified tissue vis-a-vis the organized structure in the cornified tissue makes the former more permeable. Also, the nonkeratinized mucosa does not contain acylceramides, but has small amounts of ceramides, glucosylceramides, and cholesterol sulfate. The lack of organized lipid lamellae and the presence of other lipids instead of acylceramides make the nonkeratinized mucosa more water permeable as compared to the keratinized mucosa.

9.3 ADVANTAGES AND DISADVANTAGES OF BUCCAL AND SUBLINGUAL ROUTES OF ADMINISTRATION

As an alternative method of delivery to the GI route for several classes of pharmaceutical agents, the mucosal route offers protection against low gastric pH, proteases, and first-pass degradation. Proteins and peptides as well as hormones, which are highly pH sensitive, cannot be delivered via the oral route. Avoidance of presystemic metabolism and the near-neutral pH of the oral cavity as compared to the highly acidic pH in the stomach make the buccal route promising for these entities. Peptides can be rapidly metabolized by proteolysis at most sites of administration in the body. Among the mucosal routes, peptide transport through the buccal mucosa was found to be much less sensitive to degrading enzymes as compared to nasal, vaginal, and rectal administration [10]. However, not all peptides escape hydrolysis. Though the buccal mucosa seems to be deficient in proteinases such as pepsin, trypsin, and chymotrypsin present in gastric and intestinal secretions, it is still known to include exo- and endopeptidases, amino-peptidases, carboxypeptidases, deamidases (human, pig, monkey, rat, rabbit, and cultured hamster buccal cells). In addition, saliva also contains esterases and carboxylesterases.

The highly vascularized nature of the buccal mucosa offers the added advantage of fast onset of action. Additionally, patients experiencing difficulty in swallowing, or experiencing nausea, or vomiting can be treated via this route. The noninvasiveness and low level of irritation expected on application also makes the route less intimidating for patients as compared to injections. The lack of need for technical expertise for administration of drugs via this route also makes it a cost-effective alternative to other routes. A relatively high patient compliance due to reduced frequency of dosage administration is another benefit of this route.

The major limitation is that oral mucosa suffers from a relatively small surface area available for drug administration. As mentioned previously, the surface area of the oral mucosa is 200 cm^2 as compared to the GI tract (350,000 cm^2) and skin (20,000 cm^2) [11]. Thus, lower amounts of drug are absorbed, rendering achievement of therapeutically efficacious drug concentrations a challenge. Variability in absorption through this route also leads to inconsistent systemic levels. These factors together with peptidases and proteases present in the mucosa lead to systemic drug bioavailabilities of less than 5% of administered dose [12]. Furthermore, the barrier properties of oral mucosa make it a less popular choice as compared to the nasal route. This combined with the small surface area available makes it a satisfactory choice only in the case of highly potent drugs.

Like with any other route of drug delivery, pathological conditions influencing the regular physiology, and functioning of the mucosa will affect the extent and success of the route. Mucosal irritation is another major impediment to the use of buccal route. Such irritation is common because buccal delivery involves applications of drugs, excipients, and enhancers to the mucosal lining for extended periods of time in the form of patches or buccoadhesive tablets.

Other factors which limit mucosal absorption include environmental factors such as the exposure of oral mucosa to salivary flow and the production of shearing forces due to tongue movement and swallowing. Hence in most cases, the actual dose available for buccal absorption is reduced since a high proportion of drug ends up swallowed by the patient.

9.4 PHYSICOCHEMICAL PROPERTIES AND ROUTES OF PERMEATION

Small molecules can be transported across buccal mucosa through two routes: the transcellular (intracellular) route and the paracellular (intercellular) route. Molecules can use either of these two routes or a combination, and the physicochemical properties of the molecule and the mucosal membrane determine the final route of permeation. Hydrophilic molecules can pass through the aqueous pores adjacent to the polar head groups of the lipids in the membrane or through the hydrophilic intercellular cytoplasm. At the same time, lipophilic molecules are likely to use the transcellular route through the lipophilic cell membrane, where the permeation will depend on the partition coefficient. They can also pass through the lipophilic lipid lamellae present between the cells. The intercellular route is a tortuous route and presents a lesser area for the drug to permeate, but has been found to be a predominant route for absorption of many drugs.

Figure 9.2 shows the two routes of permeation that can be used by drugs to pass through the buccal mucosa.

9.5 BIOPHARMACEUTICS OF BUCCAL AND SUBLINGUAL ABSORPTION

9.5.1 Principles of Drug Absorption

The oral mucosa contains both hydrophilic and hydrophobic components and a combination of both keratinized and nonkeratinized epithelia. Passive diffusion is the most common route

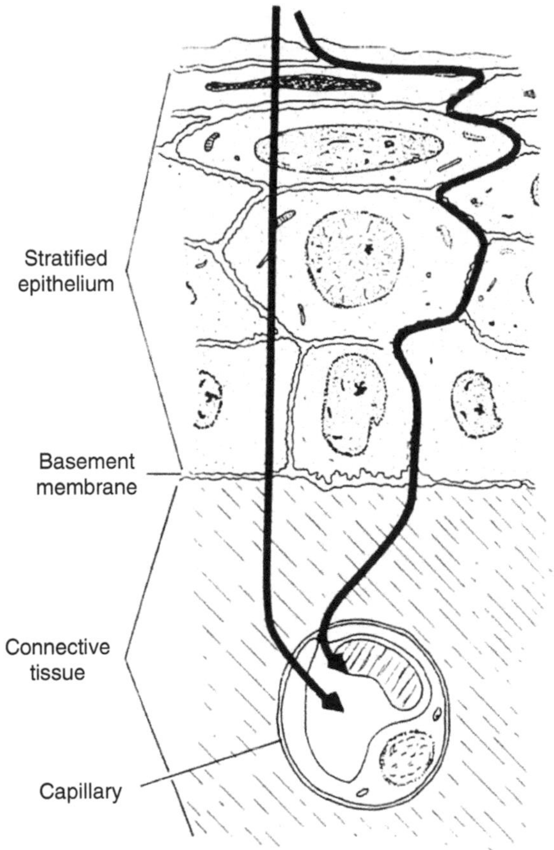

FIGURE 9.2 Routes of transepithelial penetration: transcellular route versus paracellular route. (From Wertz, P.W. and Squier, C.A., *Crit. Rev. Ther. Drug Carrier Syst.*, 8, 237, 1991. With permission.)

of permeation through the oral mucosa, and uses the Fick's first law of diffusion given by the general equation [11]

$$P = \frac{DK_p}{h} \tag{9.1}$$

The amount of drug absorbed A is given by

$$A = PCSt = \frac{DK_p}{h} CSt \tag{9.2}$$

where P is the permeability coefficient, C is the free drug concentration in the delivery medium, D is the diffusion coefficient of the drug in the oral mucosa, K_p is the partition coefficient of the drug between the delivery medium and the oral mucosa, h is the thickness of the oral mucosa, S is the surface area of the delivery or the absorption site on the mucosa, and t is the duration of time the drug stays in contact with the mucosa.

The thickness of the tissue, partition coefficient, and the diffusion coefficient are properties of the mucosa and cannot be altered. Designing appropriate formulations that heed the necessary conditions can vary the surface area for delivery of the drug, time of contact, and the free drug concentration. The partitioning of the drug into the membrane will depend on its ratio of hydrophilicity and lipophilicity. Studies performed with amines and acids showed that their absorptions were proportional to their partition coefficients, thus also establishing the fact that the transcellular route was the primary route of absorption of these drugs [13]. Similar results were obtained for β-adrenoreceptor-blocking drugs [14]. Since the drug will face different barriers through the paracellular and the transcellular routes, the flux of drug permeation through these routes will differ to some extent. The equation above can be modified to account for this difference.

Hydrophilic compounds will tend to use the paracellular route and permeate through the intercellular spaces, which present a smaller surface area. The flux of drug permeation through this pathway can be described as [15]

$$J_{\mathrm{H}} = \frac{D_{\mathrm{H}}\varepsilon}{h_{\mathrm{H}}} C_{\mathrm{D}} \tag{9.3}$$

where D_{H} is the diffusion coefficient, h_{H} is the length of the tortuous path followed in the paracellular route, C_{D} is the concentration of the drug on the donor side, and ε is the fraction of the surface area of the paracellular route.

A lipophilic drug will preferably use the transcellular route since it will be easier for it to partition into the lipophilic cell membrane. The path length here is shorter than for the paracellular route but the drug has to move through several types of barriers (cell membrane, the cytoplasm, as well as intercellular spaces). Thus the equation for flux through the transcellular route is given as

$$J_{\mathrm{L}} = \frac{(1-\varepsilon)D_{\mathrm{L}}K_{\mathrm{p}}}{h_{\mathrm{L}}} C_{\mathrm{D}} \tag{9.4}$$

where K_{p} is the partition coefficient between the lipophilic regions (cell membrane) and the hydrophilic regions (cytoplasm, formulation vehicle, and the intercellular space).

9.5.2 Factors Affecting Drug Absorption

Besides the biochemical characteristics of the buccal and sublingual membranes, which are responsible for the barrier function and permeability, various factors of the drug molecule influence the extent of permeation through the membranes.

The lipid solubility, degree of ionization, pK_{a} of the drug, pH of the drug solution, presence of saliva and the membrane characteristics, molecular weight and size of the drug, various physicochemical properties of the formulation, and the presence or absence of permeation enhancers, all affect the absorption and the permeation of drugs through the oral mucosa.

9.5.2.1 Degree of Ionization, pH, and Lipid Solubility

The permeability of unionizable compounds is a function of their lipid solubilities, determined by their oil–water partition coefficients. Squier et al. [16] demonstrated this dependence of water permeability on the lipid contents of keratinized and nonkeratinized epithelia. The lipids present however contribute to this effect more in the keratinized epithelia (more total lipid content, nonpolar lipids, ceramides) than in the nonkeratinized epithelia where permeability seems to be related to the amount of glycosylceramides present.

The absorption of drug through a membrane depends upon its lipophilicity, which in turn depends on its degree of ionization and partition coefficient. The higher the unionized fraction of a drug, the greater is its lipid solubility. The degree of ionization in turn depends on the pH of the mucosal membrane and the pK_{a} of the drug. Beckett and Triggs [17] studied the buccal absorption of basic drugs over a range of concentration, pH, and the use of different drug combinations (alone and mixtures). The resultant pH–absorption curves showed that the percentage of drug absorbed increased as the concentration of drug in the unionized form increased. Also, the shapes of the absorption curves were a function of the pK_a values and the lipid solubility of their unionized form. A study conducted with fentanyl [17], a weak base with a pK_a of 8.2, further demonstrated the relationship between the pH and the absorption across oral mucosa. When the pH of the delivery solution was increased, more of the drug was present in the unionized form, with the drug being 2.45% unionized at pH 6.6, 9.1% unionized at pH 7.2, and 24% unionized at pH 7.7. The fentanyl solutions with a pH range of 6.6 to 7.7 showed a three- to fivefold increase in peak plasma concentration, bioavailability, and permeability coefficients. Similar studies conducted with sublingual administration of opioids such as buprenorphine, methadone, and fentanyl showed increased absorption with increase in pH, where the drug was predominantly present in the unionized form [18]. However, absorption of other opioids such as levorphanol, hydromorphone, oxycodone, and heroin under similar conditions did not improve. These drugs, however, were more hydrophilic as compared to the earlier set of opioids. Thus pH modifiers can be used to adjust the pH of the saliva prior to drug administration to increase the absorption of such drugs through the mucosal membranes.

However, the nature of the buccal and sublingual membrane complicates the above condition since the pH may vary depending on the area of the membrane and also on the layer of the membrane that is considered. The pH of the mucosal surface may be different from that of buccal and sublingual surfaces throughout the length of the permeation pathway [19]. Thus the drug in its unionized form may be well absorbed from the surface of the membrane, but the pH in the deeper layers of the membrane may change the ionization and thus the absorption. Also, the extent of ionization of a drug reflects the partitioning into the membrane, but may not reflect the permeation through the lipid layers of the mucosa. Henry et al. [20] studied the buccal absorption of propranolol followed by repeated rinsing of the mouth with buffer solutions and recovered much of this drug in the rinsing. In addition, the effect of lipophilicity, pH, and pK_a will depend on the transport pathway used by the drug. Studies conducted with busiprone [21] showed that the unionized form of the drug used the more lipophilic pathway, the transcellular route, but an increase in the pH increased the ionization of the drug and subsequently the absorption. It was concluded that this transport of the ionized form of the drug was through the more hydrophilic paracellular pathway. Therefore, at neutral pH the preferred pathway was found to be transcellular, but at acidic pH, the ionized species of the drug also contributed to the absorption across the membrane.

9.5.2.2 Molecular Size and Weight

The permeability of a molecule through the mucosa is also related to its molecular size and weight, especially for hydrophilic substances. Molecules that are smaller in size appear to traverse the mucosa rapidly. The smaller hydrophilic molecules are thought to pass through the membrane pores, and larger molecules pass extracellularly. Increases in molar volume to greater than 80 mL/mol produced a sharp decrease in permeability [22,23].

Due to the advantages offered by the buccal and the sublingual route, delivery of various proteins and peptides through this route has been investigated. It is difficult for the peptide molecules with high molecular weights to make passage through the mucosal membrane. Also, peptides are usually hydrophilic in nature. Thus they would be traversing the membrane by the paracellular route, between cells through the aqueous regions next to the intercellular lipids. In addition, peptides often have charges associated with their molecules, and thus their absorption would depend on the amount of charge associated with the peptide, pH of the formulation and the membrane, and their isoelectric point (p*I*).

9.5.2.3 Permeability Coefficient

To compare the permeation of various drugs, a standard equation calculating the permeability coefficient can be used. One form of this equation is [19]

$$P = \frac{\%\ \text{permeated} \times V_\text{d}}{A \times t \times 100} \tag{9.5}$$

where P is the permeability coefficient (cm/s), A is the surface area for permeation, V_d is the volume of donor compartment, and t is the time. This equation assumes that the concentration gradient of the drug passing through the membrane remains constant with time, as long as the percent of drug absorbed is small. Another approach to determine the permeability coefficient has been described by Dowty et al. [24]. In their studies of the transport of thyrotropin-releasing hormone (TRH) in rabbit buccal mucosa, they incorporated the metabolism of TRH in the rabbit buccal mucosa into the equation, which was given as

$$P = \frac{\left[\frac{\text{d}(Q - M)}{\text{d}t}\right]}{AC_\text{d}} \tag{9.6}$$

where $\text{d}(Q-M)/\text{d}t$ is the change in quantity of the solute Q minus its metabolites M with time t. A is the exposed area to solute transport and C_d is the concentration of solute in the donor.

9.5.2.4 Formulation Factors

The permeation of drugs across mucosal membranes also depends to an extent on the formulation factors. These will determine the amount and rate of drug released from the formulation, its solubility in saliva, and thus the concentration of drug in the tissues. In addition, the formulation can also influence the time the drug remains in contact with the mucosal membrane. After release from the formulation, the drug dissolves in the surrounding saliva, and then partitions into the membrane, thus the flux of drug permeation through the oral mucosa will depend on the concentration of the drug present in the saliva. This concentration can be manipulated by changing the amount of drug in the formulation, its release rate, and its solubility in the saliva. The first two factors vary in different types of formulations, and the last can be influenced by changing the properties of the saliva that affect the solubility (e.g., pH).

9.5.3 Permeation Enhancement

Enhancers have been used to increase the permeation of drugs through the membrane, and thus increase the subsequent bioavailability. These should be pharmacologically inert and nontoxic, and should have reversible effects on the physicochemical properties of the oral mucosa.

Penetration enhancers have different mechanisms of action depending on their physicochemical properties. Some examples of penetration enhancers and their mechanisms are bile salts (micellization and solubilization of epithelial lipids), fatty acids such as oleic acid (perturbation of intracellular lipids) [25,26], azone (1-dodecylazacycloheptan-2-one) (increasing fluidity of intercellular lipids), and surfactants such as sodium lauryl sulfate (expansion of intracellular spaces). The complete list of enhancers and their mechanism of actions are discussed in detail in Chapter 10.

9.6 *IN VITRO* AND *IN VIVO* STUDY METHODS

9.6.1 Animal Models for Studies

The limited available tissue area in the human buccal cavity has encouraged the use of animal models that may mimic human oral mucosal absorption. Rats, hamsters, dogs, rabbits, guinea pigs, and rhesus monkeys have all been used in buccal studies [27–31]. As with any animal model, these all have their advantages and disadvantages. Almost all animals have a completely keratinized epithelium. The hamster cheek pouch offers a large surface area but is not flushed with saliva. The oral mucosa of the monkey, a primate, has been widely used but the high cost of procurement as well as challenging handling are disadvantages when it comes to selecting these animals. Rabbit mucosa is similar to human mucosa since it has regions of nonkeratinized tissue. However, the small surface area and difficulty in accessing the required tissue make it an impractical choice. The animal of choice remains the pig because of comparable permeability to human buccal mucosa and a large surface area enabling reduced variability in the data [32].

The methods used for measuring the amount of drug absorbed have to be designed in such a way as to account for local delivery of the drug to the mucosa as well as systemic delivery through the mucosa into the circulation. A selection of *in vivo* and *in vitro* techniques has been developed and tested over the years.

9.6.2 *In Vivo* Methods

Both human and animal models have been used for *in vivo* testing of oral mucosal drug delivery. Choices of animal models depend on how closely the mucosal membrane reflects the structure and properties of human mucosa. An important *in vivo* technique using human test subjects, the "buccal absorption test" was developed and established by Beckett and Triggs [17]. They adjusted solutions of several basic drugs to various pH values with buffer, and placed the solution in the subject's mouth. The solution was circulated about 300–400 times by the movement of the cheeks and tongue for a contact time of 5 min. The solution was then expelled, and the subject's mouth was rinsed with 10 mL distilled water for 10 s. The rinsing was collected, and combined with the earlier expelled solution, and the fraction of the drug remaining in this solution was measured by gas–liquid chromatography. It was observed that the absorption of drug from the oral cavity was dependent on pH. Though this technique is easy to perform, noninvasive, and gives relatively consistent results with little intra- and intersubject variation, limitations for the method do exist [19,33]. It does not provide information concerning the varying permeabilities of different regions in the oral cavity. Also, the continuous flow of saliva affects the pH of the applied solution as well as the overall volume. In addition, the test analyzes the amount of drug that has been transported from the sample into the oral cavity and does not provide information on the actual systemic absorption of the drugs. Some of the drugs could be swallowed or accumulated, and redistributed into the epithelium or biotransformed in the mucosa [34]. Simultaneous measurement of appearance of the drug in the systemic circulation could further validate this test.

"Disk methods" for assessing absorption have also been studied where the drug-loaded disk is kept in contact with certain area of the mucosal membrane to allow for absorption. One such polytef disk was used by Anders et al. [35] for the buccal absorption of protirelin. The disk had an area of ~10 cm^2 and a central circular depression containing the drug. It was removed after 30 min of contact with the buccal mucosa, and blood samples were taken to determine the amount of drug absorbed from the mucosa.

The disk method provides information about absorption from a specific area of the mucosa. Interference from salivary secretions, difficulties in keeping the disk adhered, and loss of drug permeating due to leakage of the drug from the disk are some disadvantages with this method.

Another method has been the use of perfusion cells [36,37]. These cells have certain specific area and can contain a drug solution that is stirred continuously. The closed cell isolates the solution from the surroundings, thus negating the effects of the environmental factors such as saliva and pH. Solution under test can be passed through the mucosal membrane once or it can be recirculated. The solution in the cells is then analyzed for drug content. However, the surface area for absorption is low, and the tissue has a tendency to become erythematous [38]. This method, like the buccal absorption test, measures the loss of drug from the cell, but not the actual absorption of the drug through the buccal mucosa. These methods have been used to analyze different types of dosage forms (composite films, patches, and bioadhesive tablets) and their mucosal drug absorption and have been used to assess both buccal and sublingual absorptions across the respective mucosa [39].

A glass perfusion cell was developed and used by Yamahara et al. [40] for the measurement of drug absorption through mucosal membranes of anesthetized male beagle dogs. The cell contained a biocompatible bioadhesive polymer O-ring that adhered the cell to the oral mucosal membrane. This type of cell can be used to measure buccal and sublingual absorption as well as perfusion through the surface of the tongue.

9.6.3 *In Vitro* Methods

These methods have proven to be important tools in the study of transmucosal absorption, since they can facilitate studies of drug permeation under controlled experimental conditions. Oral mucosal tissue can be surgically removed from the oral cavity of animals. These tissues contain a fair amount of connective tissue, which is separated from the mucosal membrane. This connective tissue, if not removed, may contribute to the permeability barrier. This separation can be carried out with the aid of heat where tissues are separated at 60°C, or chemically by the use of various enzymes or EDTA [41,42]. These tissues are then stored in buffer solution (usually Krebs). This storage step is important in preserving the viability and integrity of the tissue. The tissue is then placed in a side-by-side diffusion cell, where the placement of the tissue is in between the donor and the receptor chambers. The donor contains the drug solution, whereas the receptor usually contains a buffer solution to emulate the body fluids. The chambers can be stirred continuously to ensure even distribution of the drug and are maintained at a desired temperature. The epithelial side of the tissue faces the donor chamber, allowing the drug to pass from the donor chamber through the tissue into the receptor chamber from where samples can be withdrawn at specific time intervals and replaced with fresh receptor solution. A detailed experiment is described in Junginger et al. [43] where transport of fluorescein isothiocyanate (FITC)-labeled dextrans of different molecular weights through porcine buccal mucosa are studied.

Different kinds of diffusion cell apparatus have been used in such *in vitro* experiments. Some of these are small volume diffusion cells as described by Grass and Sweetana [44], Using chambers [45] and Franz diffusion cells [46].

The *in vitro* methods, though relatively simple have various disadvantages: (a) The conditions of tissue separation, preparation, and storage may affect the viability, integrity, and therefore their barrier function. Tests assessing the ATP levels have been used to analyze the viability and integrity of tissue. A method for ATP extraction using perchloric acid and subsequent analysis of ATP in nanomoles per gram of tissue has been described by Dowty et al. [24]. (b) Human oral mucosa is relatively expensive and available in limited amounts. Therefore, animal mucosae which have to be chosen carefully in order to resemble the human mucosa as closely as possible are used. (c) A specific complication occurs in cases of sublingual mucosa. Various ducts from the submandibular and the sublingual salivary glands open into the mucosal surface, and thus a sufficiently large piece of mucosa that is not perforated by these ducts is difficult to obtain [19]. Also, the presence of enzymes in the tissue indicates that there is a high probability of the drugs being metabolized during transport across the mucosa and therefore appropriate metabolism studies and drug-stabilizing efforts should be undertaken. Studies reported by Dowty et al. [24] measured the extent of metabolism of TRH in rabbit buccal mucosa *in vitro*.

9.6.4 Cell and Tissue Culture Systems

The advantages of the *in vitro* approaches described above also apply to buccal cell culture systems. In addition, other aspects such as cell growth and differentiation can be studied in these systems in detail. Also, once the source is established, a continuous supply of cell lines can be obtained, which obviates the need for expensive animal or human tissues that are often difficult to obtain in large quantities.

On the other hand, the established cell line must simulate, as closely as possible, the physical and biochemical properties of the buccal or sublingual tissues *in vivo*. These properties such as the growth, differentiation, biological barrier effectiveness, permeability levels, and metabolic pathways are crucial to the permeation studies.

Of the different types of oral mucosal cell cultures that have been used [47,48], the most commonly used ones are explants of primary cultures. Small pieces of excised buccal or sublingual tissue are placed in a support system and fed with culture medium. The outgrowths obtained from these tissue explants are then transferred and grown in appropriate media. For example, outgrowths of fibroblasts [49] thus obtained have been described. Gibbs and Ponec [50] reconstructed the epithelium of mucosal tissue by placing a tissue biopsy (with the epithelial side upwards) onto a fibroblast-populated collagen gel. The explants obtained were cultured immediately at the air–liquid interface until the epithelium had expanded over the gel (2–3 weeks). These explant cultures may retain many of the *in vivo* tissue characteristics.

Freshly excised buccal or sublingual tissues have also been used to generate dissociated cells. Hedberg et al. [49] used one such culture to measure the expression of alcohol dehydrogenase-3 in cultured cells from human oral mucosal tissue. Human buccal tissue was incubated with 0.17% trypsin in phosphate-buffered saline (PBS) at 4°C for 18 to 24 h to obtain dissociated primary keratinocytes, and subsequently these keratinocytes were seeded onto fibronectin and collagen-coated dishes in serum-free epithelial medium.

Various buccal epithelial cell lines have also been established. The biochemical properties of these cell lines depend greatly upon the growth media and other conditions used during culturing. Hennings et al. [51] showed that the amount of calcium present in the media affects the differentiation of epithelial cells in culture. Different types of cell lines are used for different applications. The TR146 cell line that originated from human neck metastasis of a buccal carcinoma [52] was used as an *in vitro* model of human buccal mucosa to study and compare the enzyme activity with respect to human and porcine buccal epithelium [53]. This cell line has also been used to study and compare the permeability of drugs across cell monolayers, and human buccal tissue to assess the effect of pH and concentration on the permeability [53,54]. The SqCC/Y1 cell line (a squamous epithelial cell line derived from buccal carcinoma) was used to characterize the expression and function of cytochrome P-450s in human buccal epithelium [55].

9.7 DOSAGE FORMS

A wide range of formulations have been developed and tested for buccal and sublingual administration. Various advances have been made over the years, which counteract the problems faced in delivering drugs through the sublingual and buccal mucosae to the systemic circulation. The primary challenges for these routes of delivery are:

1. The varying structure of the mucosal membrane in different parts of the oral cavity and the reduced permeation due to the barrier presented by the mucosal epithelial layers
2. The constant presence of saliva, which prevents the retention of the formulation in one area of the oral cavity leading to shorter contact time
3. Person to person variability caused by differences in tongue movements, saliva amounts, and saliva content
4. The limited surface area available for absorption
5. Ensuring patient comfort with a dosage form small and flexible enough to fit comfortably in the oral cavity, easy to install and remove, and not causing any local reactions, discomfort, or erythema

Buccal and sublingual deliveries have been used in various clinical applications such as cardiovascular, smoking cessation, sedation, analgesia, antiemesis, diabetes, and hormonal therapy. The specific drugs will be discussed in relation to the dosage form category.

Buccal delivery has also been actively researched for the delivery of peptides, since these molecules are sensitive to the acidic and proteolytic environment of the GI tract and are subjected to first-pass metabolism. The application of this field to peptide delivery will be discussed later in this chapter.

9.7.1 Chewing Gums

Gums are now considered pharmaceutical dosage forms, and have been used to deliver drugs for buccal absorption. These formulations consist of a gum base, which primarily consists of resins, elastomers, waxes, and fats. Emulsifiers such as glycerol monostearate and lecithin are added to facilitate and enhance the uptake of saliva by the gum. Resin esters and polyvinyl acetate (PVA) are added to improve texture and decrease sticking of the gum to teeth. Additives such as sweeteners, glycerol (to keep the gum soft and flexible), and flavors can be added as desired [56]. These chewing gums move about in the oral cavity, and the process of chewing mixes it with the saliva where the drug is rapidly released, partitioned, and then absorbed into the mucosal membrane. Thus, the solubility of the drug in saliva is an important factor in increasing the amount of drug released and absorbed. Intersubject variation such as the intensity of chewing, amount of saliva produced, and inconsistent dilution of the drug influence the amount of drug released. Also, the saliva can be swallowed, leading to disappearance of an often unknown amount of drug.

Gum formulations containing caffeine showed rapid release and absorption of the agent with comparable bioavailability to the capsule form [57]. Various gum formulations with vitamin C [56], diphenhydramine [58], methadone [59], and verapamil [60] have been developed and tested.

Recently, sustained release of catechins from chewing gums has been achieved by using a special procedure involving granulation of the active principles with PVA followed by coating of the pellets with acrylic insoluble polymer [61]. One of the most important and successful applications for chewing gum as a dosage form is that for nicotine replacement therapy (NRT) [62]. Nicorette (GlaxoSmithKline, USA), a chewing gum containing nicotine, is available in regular strength (2 mg) and extra strength (4 mg) and has a specially recommended chewing technique to maximize efficacy.

9.7.2 Lozenges

Lozenges can be used as an alternative dosage form to tablets and capsules when patients are unable to swallow. The use of lozenges has been reported for systemic drug delivery but it is more usual to see this dosage form used to bathe the oral cavity or the throat areas.

While sublingual lozenges may be impractical due to their size, buccal lozenges have been extensively used, and are kept between the cheek and the gums. Though the lozenge usually dissolves in about 30 min, the patient controls the rate of dissolution and absorption because the patient sucks on the lozenge until it dissolves. This process can result in high variability of amounts delivered each time the lozenge is administered. Increases in the amount of sucking and production of saliva may also lead to increased dilution of the drug and often accidental swallowing. In a study conducted by de Blaey and de Haseth [63], there was a noticeable intrasubject variation in residence time (from 2 to 10 min) of unflavored buccal lozenges. They also found that stronger lozenges prolonged the buccal residence time, a factor which can be used as an advantage in local delivery of agents from lozenges.

Despite their drawbacks and an additional requirement of palatability, lozenges have had considerable success in the market. For example, zinc lozenges have been studied and used extensively in the treatment of common colds [64]. A study utilizing NRT was conducted with 2 and 4 mg lozenges. It was found that the lozenges achieved better abstinence from smoking in low- and high-dependent smokers compared to those patients receiving an identical dose in a chewing gum [65]. Transmucosal administration of fentanyl citrate, a medication for breakthrough pain, resulted in a bioavailability substantially greater than oral administration and led to faster achievement of peak plasma concentration [66].

9.7.3 Buccal and Sublingual Tablets

These tablets are placed and held between the cheek and gum or the lip and gum (buccal) or under the tongue (sublingual) until they dissolve. Nitroglycerin tablets have been used extensively in the form of buccal and sublingual tablets for the fast onset and quick relief from angina [67,68]. Similarly isosorbide dinitrate is available in the form of sublingual tablets to be placed under the tongue or chewable tablets where the tablet has to be chewed in the mouth for 2 min before swallowing, and the drug is adsorbed through the oral mucosa [69]. Other formulations that have been used are nifedipine (sublingual capsules) [70], sublingual misoprostol for labor induction [71], methyl testosterone (buccal and sublingual tablets), buprenorphine (sublingual and buccal) [34], and selegiline (Zydis selegiline, RP Scherer Corporation, Troy, MI, USA) for monoamine oxidase-B inhibition [72].

9.7.4 Mucoadhesive Systems

One of the primary problems in oral mucosal drug delivery is the retention of the device on the desired area of the membrane for a sufficiently long period of time to allow for absorption of the drug and hence achievement of the desired blood levels. To assist in this, bioadhesive systems have been designed to stay and maintain intimate contact with the mucous membrane that covers the epithelium. These systems are referred to as "mucoadhesive," and they isolate the delivery of the drug from environmental factors in the cavity and allow the drug to be absorbed only from a specific (buccal or sublingual) region. This results in prolonged contact, and these systems can also be designed to control the release rate of the drug.

Mucoadhesives are generally macromolecular organic polymers made from natural (gelatin, agarose, chitosan, hyaluronic acid) or synthetic polymers (polyvinylpyrrolidone (PVP), polyacrylates, polyvinyl alcohol, cellulose derivates). They possess hydrophilic groups that can form hydrogen bonds such as carboxyl, hydroxyl, amide, and amine groups. These mucoadhesives are called "wet" adhesives and need to be in the presence of water in order to hydrate and swell. The amount of water uptake by the system depends on the number of hydrophilic groups in the polymer, and the degree of adhesion in turn depends on the amount of hydration [73]. Upon hydration and swelling, they adhere nonspecifically to the mucosal surfaces. Mucoadhesives can also be used in the dry or partially hydrated forms. Hypotheses have described the mucoadhesion process as initial establishment of contact with the substrate and the subsequent formation of chemical bonds. The attachment to the substrate can be governed by covalent interaction, electrostatic interaction, hydrogen bonding, or hydrophobic interactions. The result is the formation of a tight and intimate contact between the mucosal surface and the polymeric chains of the mucoadhesive, and this "intertangling" between the two surfaces leads to adhesiveness. The mucoadhesion achieved depends on various polymer properties, such as molecular weight, chain length, conformation, and chain flexibility.

Effective mucoadhesion has been used to design different formulations, some of which are discussed below.

9.7.4.1 Films and Patches

Patches are flexible dosage forms that adhere to a specific region of the mucosa and provide either a unidirectional flow or a bidirectional flow of drug, depending on the type of delivery intended (local or systemic). The permeation of the drug into the membrane will depend on the surface area of the patch. Different patches are designed to achieve objectives such as local and systemic drug delivery, varying duration of action and varying rates of release. In general, most patches contain either a "matrix system" in which the drug is dispersed along with excipients or the mucoadhesive, or a "reservoir system." The mucoadhesive can be dispersed in the drug matrix as described above or as a separate layer. The patches may incorporate a backing layer that protects it from the surrounding oral cavity if strictly transmucosal delivery is required. Otherwise, the backing layer is omitted. The polymer within the mucoadhesive layer swells, and a network is produced through which the drug diffuses into the membrane [74]. Combinations of the above factors have been used to design and develop three kinds of patches: patches with a dissolvable matrix, patches with a nondissolvable backing, and patches with a dissolvable backing. Patches with a dissolvable matrix release the drug into the entire oral cavity, but the presence of a mucoadhesive layer prolongs this release. Patches with a nondissolvable backing provide a unidirectional flow of the drug through the mucosa for a long period of time, whereas patches with a dissolvable backing are short acting as the backing layer dissolves fairly rapidly in the oral cavity [11].

Figure 9.3 shows the two kinds of patch system designs. The patches should be comfortable for the patients to wear for a long period of time, should not hinder or obstruct day-to-day activities, should be easy to attach and remove, and should not cause any local irritation.

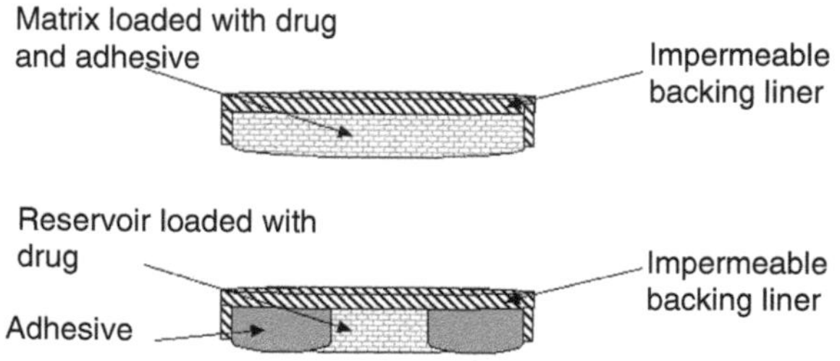

FIGURE 9.3 Alternative matrix and reservoir patch designs. (Modified from Rathbone, M.J., *Oral Mucosal Drug Delivery*, Marcel Dekker, New York, 1996.)

Flexible buccal patches for the controlled delivery of metoprolol, a selective β_1-adrenergic antagonist, which is widely used to treat essential hypertension, were developed using water-insoluble Eudragit® (Röhm GmbH & Co., Darmstadt, Germany) NE40D as the base matrix. Eudragit® NE40D is a neutral poly(ethylacrylate methylmethacrylate) copolymer, and is widely used in the development of controlled release delivery systems and film-coating technology [75]. Various hydrophilic polymers, namely Methocel K4M, Methocel K15M, SCMC 400, Cekol 700, Cekol 10000, CP934P, CP971P, and CP974P, were incorporated into the Eudragit® patches to modify the drug-release profile and the bioadhesiveness of the buccal patch. Incorporation of the hydrophilic polymers was found to alter both the amount of bioadhesion as well as the drug release [76].

The oral mucosa has also been investigated as a site for immunization, and bilayer films have been developed and administered to rabbits. The films were prepared using different ratios of Noveon and Eudragit® S-100 for the mucoadhesive layer and a pharmaceutical wax as the impermeable backing layer. Noveon is a cross-linked mucoadhesive polyacrylate polymer and Eudragit S-100 is an anionic pH-sensitive copolymer of polymethacrylic acid-*co*-methylmethacrylate. The films were pre- or postloaded with 100 g of plasmid DNA expressing β-galactosidase (CMV-β-gal) or β-galactosidase. The films were then applied to the buccal pouch of rabbits and immunological responses were measured. It was found that the weight ratio of Noveon and Eudragit® S-100 had a significant effect on adhesion time of the bilayer films. Postloaded films were observed to release 60%–80% of both plasmid DNA and β-galactosidase in 2 h. It was found that this technique of buccal immunization led to comparable antigen-specific IgG titer to that of subcutaneous protein injection [77].

The delivery of buprenorphine, a partial opioid agonist, has been extensively studied using the buccal and sublingual routes since the oral dosage form results in poor bioavailability. In order to increase the retention time on the sublingual membrane, a thin polymeric film consisting of mucoadhesive polymers Carbopol 934P, Carbopol 974P, and the polycarbophil (PCP) Noveon AA-1 was prepared, and polyethylene glycol (PEG) was used as a plasticizer to make the films flexible [78].

A novel buccal delivery system Striant® (Columbia Laboratories, Inc., Livingston, NJ) approved by the Food and Drug Administration (FDA) in 2003 is a controlled and sustained release buccal mucoadhesive system, containing 30 mg of testosterone and bioadhesive excipients [79]. The patch contains the bioadhesive polymer PCP, along with other inert ingredients including hydroxypropylcellulose, monohydrated lactose, and cornstarch. After the patch was placed on the gum above the right or left canine, testosterone was slowly released from the matrix. The system was left on for 12 h, then slid out and replaced by another system for the next dosing interval. The testosterone concentrations obtained from the buccal system were found to be within the physiological range for a significantly greater portion of the 24 h treatment period as compared to a marketed testosterone transdermal patch.

9.7.4.2 Tablets

Buccal and sublingual tablets are compressed dosage forms, and like patches can provide either unidirectional flow of drug through the mucosa if they contain a backing layer or bidirectional flow into the oral cavity if no backing is present. The basic formulation is similar to that of patches with a matrix containing the drug, a bioadhesive polymer either in a separate layer or incorporated into the matrix, and the presence or absence of an impermeable backing film.

Recently a study investigated different types of mucoadhesive polymers for buccal tablet formation [80]. The polymers used were Carbopol (CP934 and CP940), PCP, sodium carboxymethyl cellulose (SCMC) and pectin, all anionic-type polymers, chitosan (cationic type), and hydroxypropyl methylcellulose (HPMC) as a nonionic polymer. These polymers were used alone or in combination to form compressed bioadhesive tablets that were tested for bioadhesion and swelling. Also, residence time *in vitro* was tested using a locally modified USP disintegration apparatus. The polyacrylic acid (PAA) derivatives (CP934, CP940, PCP) showed the highest bioadhesion force and prolonged residence time. While HPMC and pectin demonstrated weaker bioadhesion, SCMC and chitosan showed stronger bioadhesive properties. Among the combinations, a mixture of 5% CP934, 65% HPMC, and 30% spray-dried lactose or 2% PCP, 68% HPMC, and 30% mannitol showed optimal bioadhesion and good residence time.

Bioadhesive tablets can be made by the compression of polymers or can consist of a matrix base or bilayers, with an impermeable backing layer covering the layer with the drug and the mucoadhesion polymer. Examples of these systems are discussed below.

Buccoadhesive-controlled release tablets for delivery of nifedipine were prepared by direct compression of carboxymethyl cellulose (CMC) with carbomer (CP) and compared to those prepared with PVP, PVA, HPMC, and acacia by a modified tensiometry method *in vitro*. It was found that the adhesion force was significantly affected by the mixing ratio of CP:CMC in the tablets. CMC is necessary for controlling the release rate, whereas CP is important in providing bioadhesion. The tablets containing 15% CMC and 35% CP were found to have optimum drug release rate and bioadhesion [81].

Miyazaki et al. [82] designed and evaluated both single and bilayer tablets of pectin and HPMC in the ratio of 1:1 for the sublingual delivery of diltiazem. Bilayer tablets consisted of a backing layer and an adhesive, drug reservoir layer, and were made by covering one side of the single-layer tablet with an inert ethylcellulose layer. The plasma concentration curves for both single-layer and bilayer sublingual tablets showed evidence of a sustained release of diltiazem, with the bilayer tablets with backing layer having a significantly more prolonged effect when compared with single-layer tablets. Bioavailability of diltiazem was 2.5 times that achieved by oral administration for single-layer tablets and 1.8 times for the bilayered tablets.

Biphasic buccal adhesive tablets have also been used for smoking cessation therapy with nicotine [83].

In order to improve the mucosal absorption of poorly absorbed drugs such as peptides and proteins, newer delivery systems with higher mucoadhesive and permeation-enhancing polymers have been developed. While the first generation of mucoadhesive polymers provided adhesion to the mucus gel layer via secondary bonds, the new generation of mucoadhesive polymers is able to form covalent bonds with the mucous layer. The immobilization of thiol groups on mucoadhesive polymers results in thiolated polymers or thiomers that can form disulfide bonds with cysteine-rich subdomains of mucus glycoproteins [84,85].

Langoth et al. [86] studied the properties of matrix-based tablets containing the novel pentapeptide leu-enkephalin (Tyr-Gly-Gly-Phe-Leu) that has been shown to have pain-modulating properties. The matrix-based tablets were made with the thiolated polymer PCP. The covalent attachment of cysteine to the anionic polymer PCP leads to an improvement of the stability of matrix tablets, enhances the mucoadhesive properties, and increases the inhibitory potency of PCP towards buccal enzymes. All these factors lead to stability of the peptide and a controlled drug release for the peptide was obtained for more than 24 h. Also, the tablets based on thiolated PCP remained attached on freshly excised porcine mucosa 1.8 times longer than the corresponding unmodified polymer.

Solubilization of poorly water-soluble drugs by complexation with cyclodextrins and then delivery via the buccal or sublingual mucosa has been studied as an additional strategy for increasing drug absorption. Cyclodextrins are able to form inclusion complexes with drugs, and can increase the aqueous solubility, dissolution rate, and bioavailability. Jug and Becirevic-Lacan [87] studied the drug carrier system of a molecular complex of piroxicam with hydroxypropyl β-cyclodextrin incorporated in a hydrophilic matrix. The buccal tablets were prepared by a direct compression of HPMC and Carbopol 940 (C940). The *in vitro* release results demonstrated that complexed matrix tablets displayed faster piroxicam release compared to those containing free drug. The combination of HPMC and C940 was shown to demonstrate good bioadhesion properties.

Buprenorphine films prepared with the polymers Carbopol 934P, Carbopol 974P, and PCP Noveon AA-1 were compared to similar mucoadhesive sublingual tablets by Das and Das [78]. The tablets were prepared with or without excipients, and the mucoadhesive properties were studied. It was found that the mucoadhesive tablet formulations produced overall superior results compared to the mucoadhesive film formulations, and optimum results were reported in the case of high lactose, low mucoadhesive polymer, Carbopol 974P- and PEG 3350-containing tablet formulations. These formulations provide a sustained release profile of the drug without producing any sudden "burst release" effects. Also, the tablets were capable of releasing their entire drug content within 2 h, which is optimal for sublingual administration.

9.7.4.3 Hydrogels

Hydrogels are three-dimensional, hydrophilic, polymeric networks that can take up large amounts of water or other biological fluids. The networks consist of homopolymers or copolymers having physical or chemical cross-links that make them insoluble, which are responsible for the integrity of the network.

Depending on their chemical side groups, hydrogels can be neutral or ionic. For a hydrogel to possess mucoadhesive properties, the polymer chains have to be mobile to facilitate the interpenetration into the mucous layer and formation of bonds leading to mucoadhesion. Absorption of water by the hydrogel results in lowering of the glass transition temperature (T_g), and the gel becomes more rubbery. This leads to increased mobility of the polymer chains and establishment of mucoadhesion. The swelling of a hydrogel depends on the properties of the hydrogel itself or properties of the changing external environment. The cross-linking ratio (the ratio of the moles of cross-linking agent to the moles of polymer-repeating units) is one of the primary factors affecting the swelling [88]. The higher the amount of cross-linking agent, the greater is the ratio, thus leading to a tighter structure which leads to less mobility of the polymer and lesser swelling. Also, gels containing more hydrophilic groups will swell more as compared with those containing more hydrophobic groups.

Swelling of physiologically responsive hydrogels is affected by various external factors such as pH, ionic strength, temperature, and electromagnetic radiation [89].

The drug can be either present in a matrix core anchored by a hydrogel to the mucosa or it can be dispersed into the mucoadhesive matrix. In the second case, swelling will play a primary role in the release of the drug from the system.

de Vries et al. [90] determined the adhesiveness of the copolymer hydrogels made of acrylic acid (polar) and butyl acrylate (apolar) in different molar ratios to porcine oral mucosa. Azo-*bis*-isobutylonitrile was used as the polymerization initiator, and ethylene glycol dimethacrylate was used as the cross-linker in varying concentrations. The glass transition temperatures and the water contact angles were measured to indicate the mobility of the polymer chain and the extent of surface polarity of the hydrogel, respectively. The peel and shear detachment forces from the mucosa were determined for the copolymers, which are directly related to the extent of adhesiveness. It was found that the contact angle maximized at

50% butyl acrylate content, whereas the glass transition temperatures decreased as the concentration of butyl acrylate was changed from 0% to 100%. The data indicated that not only a low T_g, but also an optimal number of polar groups, are necessary for optimal adhesion to the mucosal surface.

A study compared the buccal mucoadhesive properties for different polymeric films that differed in their cross-linking status [91]. Synthetic (Carbopol 971P, PCP), semisynthetic (SCMS), and natural carrageenan (λ-type) were analyzed for their mucoadhesive properties using a TA-XT2i texture analyzer.

The texture analyzer gave detachment profiles of these polymers from bovine sublingual mucosa after mucoadhesion under a force of 0.5 N for periods of 0.5, 2, 15, and 30 min, with the polymeric film of PVP K-90 used as a control. Rheological examinations, torque sweep, frequency sweep, and oscillatory examinations were also conducted. In addition, swelling properties were determined with weight measurement before and after wetting with saliva. After a contact time of 2 min, the strength of mucoadhesion was established as CMC > PCP > Carbopol 971P > Carageenan. But after a contact time of 15 min, the order was reversed to Carbopol 971P > PCP > Carageenan > CMC. The swelling of the polymers at 2 and 15 min showed the same reversal of order. Thus as compared to CMC, the other three polymers were found to have good mucoadhesive and swelling properties. The study also emphasized the importance of the composition of the chains, the charge density, and the molecular weight to form a network that is capable of forming relatively strong links with the mucous membrane.

Copolymers of acrylic acid and poly(ethylene glycol) monomethylether monomethacrylate (PEGMM) were used to design a buccal delivery system for the systemic delivery of the antiviral agent, acyclovir [92]. The system consisted of the copolymer, an adhesive, and an impermeable backing layer to allow strictly unidirectional flow of the drug. The drug was loaded by equilibrium swelling of the copolymeric films in isotonic buffer (pH 6.8) solutions at 37°C for 24 h. Permeation studies through porcine buccal mucosa were carried out using side-by-side flow through diffusion cells (Crown Glass Co., NJ). It was found that buccal permeation of acyclovir from the mucoadhesive delivery system was controlled for up to 20 h with a time lag of 10.4 h and a steady-state flux of 144.2 $\mu g/cm^2/h$. With the incorporation of sodium glycocholate (NaGC) as a penetration enhancer, the lag time was decreased to 5.6 h, and the steady-state flux increased to 758.7 $\mu g/cm^2/h$.

Hydrogels have also been used to deliver drugs *in vivo* through the oral mucosa. One such example is the preparation of a hydrogel containing 17-β-estradiol, which is administered for osteoporosis, but has very poor oral bioavailability [93]. The hydrogels were prepared by mixing an ethanolic solution containing the drug and an absorption enhancer with an aqueous solution of carboxyvinyl polymer and triethanolamine to produce an ointment. The buccal administration of the hydrogel formulation containing the estradiol in 40% (w/w) ethanol and using 2% (w/w) LAU (glyceryl monolaurate) as the absorption enhancer allowed the maintenance of the plasma level at above 300 $ng/(mL\ cm^2)$ for 7 h.

9.7.5 Other Dosage Forms

9.7.5.1 Sprays

These can be sprayed orally onto the buccal or the sublingual membrane to achieve a local or a systemic effect. One such spray called insulin buccal spray (IBS) was developed with soybean lecithin and propanediol [94]. Soybean lecithin has high affinity for biomembranes but does not enhance the transport of drugs due to low solubility. Propanediol can improve the solubility of soybean lecithin, and act as an enhancer. IBS was administered to diabetic rabbits, and the hypoglycemic effect of this formulation was investigated. The results show that when the diabetic rabbits were administrated with IBS in dosages of 0.5, 1.5, and 4.5 U/kg, the blood glucose level decreased significantly compared with that of the control group, and the hypoglycemic effect lasted over 5 h. To investigate the transport route for insulin through the buccal mucosa, penetration of FITC-labeled insulin was studied by scanning the distribution of the fluorescent probe in the epithelium using confocal laser scanning microscopy. The results revealed that FITC–insulin can pass through the buccal mucosa promoted by the enhancer and the passage of insulin across the epithelium involved both intracellular and paracellular routes.

9.7.5.2 Carrier-Associated Suspensions

Another novel approach to buccal administration of insulin involves using insulin associated with a carrier, namely erythrocyte ghosts (EG) [95]. The insulin was administered either free or attached to carrier systems (erythrocyte ghosts–insulin, EG–INS) to streptozocin diabetic rats by instilling the dose in the mouth cavity using a syringe. To prevent swallowing of the dose, the rats were anesthetized, and blood samples were collected from the tail over 5 h. The magnitude of blood glucose level decline was found to be at its maximum of 39.53 mg/dL (at 2 h) for free insulin and 26.23 mg/dL (at 4 h) for EG–INS insulin, showing that the carrier-associated system was significantly effective at decreasing the blood glucose levels.

9.7.5.3 Liposomes

Liposomes have been used in the local delivery of drugs to the oral mucosa. Farshi et al. [96] studied the biodistribution of dexamethasone sodium phosphate (DSP) encapsulated in multilamellar vesicle (MLV) liposomes labeled with 99mTc in ulcerated and intact oral mucosae of rats. The liposomes were found to localize the drug in the ulcerated area and increase local drug concentration while decreasing systemic concentration.

Yang et al. [97] investigated the effect of deformable lipid vesicles as compared to conventional vesicles for delivering insulin to the buccal mucosa. The deformable lipid vesicles also called "transfersomes" contain at least one inner aqueous compartment, which is surrounded by a lipid bilayer. It has been postulated that these vesicles respond to changing external environments by shape transformations, and this deformation enables them to release the drug across various barriers. Surfactants such as sodium deoxycholate are used to render these vesicles deformable. Conventional vesicles and deformable vesicles (with sodium deoxycholate) containing insulin were administered using a buccal spray to male rabbits and blood samples were taken. These data were compared to subcutaneous administration of insulin. The results showed that the entrapment efficiencies of the deformable and conventional vesicles were $18.87\% \pm 1.78\%$ and $22.07\% \pm 2.16\%$, respectively. The relative bioavailability of the insulin-deformable vesicles group was 19.78% as compared to subcutaneous administration. This bioavailability was found to be higher than that from conventional insulin vesicles.

9.7.5.4 Nanoparticles

In an effort to develop an effective bioadhesive system for buccal administration, insulin was encapsulated into polyacrylamide nanoparticles by the emulsion solvent evaporation method [98]. Though nanoparticle formation ensures even distribution of the drug, pelleting of the nanoparticles was performed to obtain three-dimensional structural conformity. In addition, it was hypothetized that the pelletized particles will remain adhered to the mucosa, leading to good absorption. While studying bioadhesion and drug release profiles, it was found that the system showed a sustained drug release profile that was mainly governed by polymer concentration. A significant and nonfluctuating hypoglycemic response with this formulation was observed after 7 h in diabetic rats.

9.7.5.5 Microparticulate Delivery Systems

Microparticulate delivery systems containing piroxicam in amorphous form were designed to improve the drug dissolution rate via the sublingual route [99]. Two low-swellable mucoadhesive methacrylic copolymers, namely Eudragit® L sodium salt (EuLNa) and Eudragit® S sodium salt (EuSNa), were chosen as carriers for the preparation of the microparticles. Two series of microparticles containing piroxicam and EuLNa or EuSNa in ratios ranging from 15:85 to 85:15 (m/m) were prepared by spray drying. The effect of the different compositions on the dissolution profile of piroxicam was determined. In addition, the mucoadhesive properties were also assessed. The microparticles of piroxicam and the copolymer improved the piroxicam dissolution rate in comparison with that of micronized piroxicam in cubic form. Also, the drug released from the microparticles reached a plateau within 12 min, and the concentrations were always higher than the maximum solubility of piroxicam in the cubic form.

9.8 IONTOPHORESIS

Iontophoresis is the process of delivering drugs or other charged molecules across a membrane using a small electrical charge. The "like-repels-like" phenomenon is applied here to drive charged molecules that are repelled by similarly charged electrodes into a tissue. Besides its use in transdermal delivery, this method has also been used to enhance oral mucosal drug delivery. Jacobsen [100] used iontophoresis to enhance the absorption of atenelol into porcine buccal mucosa. A newly designed *in vitro* three-chamber iontophoretic permeation cell was used to measure the permeability of the drug over a period of 8 h. High enhancement ratios were obtained, and were found to be a factor of the electric current rather than the concentration gradient.

Though this method can be used to increase the penetration of drugs, the inconvenience and accessibility issues faced in administration to the oral mucosa limit its applications.

9.9 BUCCAL AND SUBLINGUAL DELIVERY OF PEPTIDES AND PROTEINS

Proteins and peptides have emerged as an important class of therapeutic agents. The advances in biotechnology, proteomics, and increasing clinical applications have resulted in an increase in the number of formulations that are developed and introduced into the market. The buccal route has been researched for peptide delivery to overcome the disadvantages of the oral and parenteral routes. With oral delivery, peptides are quickly degraded in the GI tract since they are susceptible to degradation by the acidic pH of the stomach and metabolism by the peptidases present in the luminal, brush border, and cytosolic membranes. Also their large size, associated charge, and hydrophilicity hinder absorption through the intestinal epithelium. Most importantly, they undergo hepatic first-pass metabolism, which further reduces the bioavailability [101,102]. The parenteral route has also been extensively used for the delivery of peptides. This route, however, necessitates frequent injections to maintain therapeutically significant levels of the drugs due to short biological half-lives of the molecules leading to irritation at the site of delivery and reduced patient comfort and compliance.

In order to overcome these issues, various noninvasive routes are tested for the delivery of peptides. The oral mucosa due to its high vascularity, avoidance of hepatic first-pass metabolism, and the absence of degradative enzymes normally present in the GI tract has been explored as a suitable route for peptide delivery. Several studies of peptide absorption through the oral mucosa have been conducted, and the results have been impressive in some cases, and not in the others. The development of mucoadhesive systems for buccal and sublingual delivery has increased the absorption and bioavailability of peptides, and various formulations have been developed using these systems.

The factors that hinder the absorption of peptides through the intestinal epithelium, namely high molecular weight, charge, and hydrophilicity also affect their absorption through the oral mucosa. Combinations of mucoadhesive systems, absorption enhancers, and enzyme inhibitors have enabled better absorption.

A mucoadhesive buccal patch was evaluated for transmucosal delivery of oxytocin (OT) [103]. OT was incorporated with coformulations of Carbopol 974P and silicone polymer. The plasma concentrations of OT remained 20- to 28-fold greater than levels obtained from placebo patches for a period of 0.5 to 3.0 h.

Transmucosal delivery of salmon calcitonin (sCT) via the buccal route was studied using a mucoadhesive bilayer thin-film composite (TFC) [104]. *In vitro* studies showed that over 80% of sCT was released from the TFCs within 240 min. The relative bioavailability for rabbits treated with the film composites was 43.8% ± 10.9% as compared to intravenous injection.

Buccal delivery for insulin has been investigated using different formulations such as buccoadhesive tablets [105], deformable vesicles [97], and pelleted bioadhesive polymeric nanoparticles [98].

Generex Biotechnology (Toronto, Ontario, Canada) markets a spray for delivery of insulin through the buccal mucosa [106]. The spray called Oralin, uses the RapidMist™ technology, has been also developed by Generex Biotechnology. The device sprays a high-velocity, fine-particle aerosol into the patient's mouth, which results in an increased deposition of the particles over the mucosa. Since the particles are very fine and move fast, the insulin molecules delivered through this system traverse the topmost layers of the epithelial membrane, pass through the other layers, and are absorbed into the blood stream with the aid of absorption enhancers. Oralin has been found to produce rapid absorption and metabolic control comparable to subcutaneously injected insulin.

9.10 CONCLUSION

Despite various disadvantages, the oral mucosal route might be the potential option for drug delivery and for macro- and micromolecular deliveries. While buccal sprays, tablets, lozenges, and patches for smaller molecules have already been commercialized, not many buccal peptide formulations have been marketed. Administered peptides still remain susceptible to the permeability and enzymatic barrier of the buccal mucosa, and in many studies only moderate bioavailability has been observed. The advent of techniques like enzyme inhibitors, effervescent tablets, mucoadhesive devices, and absorption enhancers along with other advantages such as patient acceptability and low degradation have initiated numerous studies for delivery of proteins and peptides by this route. The development and evaluation of thiomers (thiolated polymers) for buccal delivery of peptides discussed earlier in this chapter provide many advantages in one system [86]. The applications of chemical enhancers as a promising technique for improved buccal and sublingual delivery are discussed in the next chapter.

REFERENCES

1. Coluzzi, P.H., et al. 2001. Breakthrough cancer pain: A randomized trial comparing oral transmucosal fentanyl citrate (OTFC) and morphine sulfate immediate release (MSIR). *Pain* 91:123.
2. Heasman, P.A., et al. 2001. Local delivery of chlorhexidine gluconate (PerioChip) in periodontal maintenance patients. *J Clin Periodontol* 28:90.
3. Wu, A.M., et al. 1994. Structure, biosynthesis, and function of salivary mucins. *Mol Cell Biochem* 137:39.
4. Collins, L.M., and C. Dawes. 1987. The surface area of the adult human mouth and thickness of the salivary film covering the teeth and oral mucosa. *J Dent Res* 66:1300.
5. Canady, J.W., et al. 1993. Measurement of blood flow in the skin and oral mucosa of the rhesus monkey (*Macaca mulatta*) using laser Doppler flowmetry. *Comp Biochem Physiol Comp Physiol* 106:61.
6. Johnson, G.K., et al. 1987. Blood flow and epithelial thickness in different regions of feline oral mucosa and skin. *J Oral Pathol* 16:317.
7. Squier, C.A., and D. Nanny. 1985. Measurement of blood flow in the oral mucosa and skin of the rhesus monkey using radiolabelled microspheres. *Arch Oral Biol* 30:313.
8. Wertz, P.W., et al. 1996. Biochemical basis of the permeability barrier in skin and oral mucosa. In *Oral mucosal drug delivery*, ed. M.J. Rathbone, 27. New York: Marcel Dekker.
9. Garza, J., et al. 1998. Membrane structure in human buccal epithelium. *J Dent Res* 77:293.
10. Veuillez, F., et al. 2001. Factors and strategies for improving buccal absorption of peptides. *Eur J Pharm Biopharm* 51:93.
11. Zhang, H., et al. 2002. Oral mucosal drug delivery: Clinical pharmacokinetics and therapeutic applications. *Clin Pharmacokinet* 41:661.
12. Sayani, A.P., and Y.W. Chien. 1996. Systemic delivery of peptides and proteins across absorptive mucosae. *Crit Rev Ther Drug Carrier Syst* 13:85.
13. Beckett, A.H., and A.C. Moffat. 1969. Correlation of partition coefficients in *n*-heptane-aqueous systems with buccal absorption data for a series of amines and acids. *J Pharm Pharmacol* 21 (Suppl.):144S$^+$.
14. Le Brun, P.P.H., et al. 1989. *In vitro* penetration of some beta-adrenoreceptor blocking drugs through porcine buccal mucosa. *Int J Pharm* 49:141.
15. Rathbone, M.J. 1996. *Oral mucosal drug delivery*. New York: Marcel Dekker.
16. Squier, C.A., et al. 1991. Lipid content and water permeability of skin and oral mucosa. *J Invest Dermatol* 96:123.
17. Beckett, A.H., and E.J. Triggs. 1967. Buccal absorption of basic drugs and its application as an *in vivo* model of passive drug transfer through lipid membranes. *J Pharm Pharmacol* 19 (Suppl.):31S.
18. Weinberg, D.S., et al. 1988. Sublingual absorption of selected opioid analgesics. *Clin Pharmacol Ther* 44:335.
19. Harris, D., and J.R. Robinson. 1992. Drug delivery via the mucous membranes of the oral cavity. *J Pharm Sci* 81 (1):1.
20. Henry, J.A., et al. 1980. Drug recovery following buccal absorption of propranolol. *Br J Clin Pharmacol* 10:61.
21. Birudaraj, R., et al. 2005. Buccal permeation of buspirone: Mechanistic studies on transport pathways. *J Pharm Sci* 94:70.
22. Siegel, I.A., et al. 1981. Mechanisms of non-electrolyte penetration across dog and rabbit oral mucosa *in vitro*. *Arch Oral Biol* 26:357.
23. Siegel, I.A. 1984. Permeability of the rat oral mucosa to organic solutes measured *in vivo*. *Arch Oral Biol* 29:13.
24. Dowty, M.E., et al. 1992. Transport of thyrotropin releasing hormone in rabbit buccal mucosa *in vitro*. *Pharm Res* 9:1113.
25. Turunen, T.M., et al. 1994. Effect of some penetration enhancers on epithelial membrane lipid domains: Evidence from fluorescence spectroscopy studies. *Pharm Res* 11:288.
26. Oh, C.K., and W.A. Ritschel. 1990. Absorption characteristics of insulin through the buccal mucosa. *Methods Find Exp Clin Pharmacol* 12:275.

27. Ritschel, W.A., et al. 1985. Disposition of nitroglycerin in the beagle dog after intravenous and buccal administration. *Methods Find Exp Clin Pharmacol* 7:307.
28. Mehta, M., et al. 1991. *In vitro* penetration of tritium-labelled water (THO) and [3H]PbTx-3 (a red tide toxin) through monkey buccal mucosa and skin. *Toxicol Lett* 55:185.
29. Squier, C.A., and B.K. Hall. 1985. *In-vitro* permeability of porcine oral mucosa after epithelial separation, stripping and hydration. *Arch Oral Biol* 30:485.
30. Hussain, M.A., et al. 1986. Buccal and oral bioavailability of nalbuphine in rats. *J Pharm Sci* 75:218.
31. Singh, B.B., et al. 1988. *In vivo* effects of *sn*-1,2-dioctanoylglycerol, TPA and A23187 on hamster cheek pouch epithelium. *J Oral Pathol* 17:517.
32. Song, Y., et al. 2004. Mucosal drug delivery: Membranes, methodologies, and applications. *Crit Rev Ther Drug Carrier Syst* 21:195.
33. Rathbone, M.J., and J. Hadgraft. 1991. Absorption of drugs from the human oral cavity. *Int J Pharm* 74:9.
34. Hoogstraate, A.J., and P.W. Wertz. 1998. Drug delivery via the buccal mucosa. *PSTT* 1:309.
35. Anders, R., et al. 1983. Buccal absorption of protirelin: An effective way to stimulate thyrotropin and prolactin. *J Pharm Sci* 72:1481.
36. Rathbone, M.J. 1991. Human buccal absorption. I. A method for estimating the transfer kinetics of drugs across the human buccal membrane. *Int J Pharm* 69:103.
37. Rathbone, M.J. 1991. Human buccal absorption. II. A comparative study of the buccal absorption of some parahydroxybenzoic acid derivatives using the buccal absorption test and a buccal perfusion cell. *Int J Pharm* 74:189.
38. Rathbone, M.J., and I.G. Tucker. 1993. Mechanisms, barriers and pathways of oral mucosal drug permeation. *Adv Drug Del Rev* 12:41.
39. Aungst, B.J., et al. 1988. Comparison of nasal, rectal, buccal, sublingual and intramuscular insulin efficacy and the effects of a bile salt absorption promoter. *J Pharmacol Exp Ther* 244:23.
40. Yamahara, H., et al. 1990. *In situ* perfusion system for oral mucosal absorption in dogs. *J Pharm Sci* 79:963.
41. de Vries, M.E., et al. 1991. Localization of the permeability barrier inside porcine buccal mucosa: A combined *in vitro* study of drug permeability, electrical resistance and tissue morphology. *Int J Pharm* 76:25.
42. Garren, K.W., and A.J. Repta. 1989. Buccal drug absorption. II: *In vitro* diffusion across the hamster cheek pouch. *J Pharm Sci* 78:160.
43. Junginger, H.E., et al. 1999. Recent advances in buccal drug delivery and absorption—*In vitro* and *in vivo* studies. *J Control Release* 62:149.
44. Grass, G.M., and S.A. Sweetana. 1988. *In vitro* measurement of gastrointestinal tissue permeability using a new diffusion cell. *Pharm Res* 5:372.
45. Artusi, M., et al. 2003. Buccal delivery of thiocolchicoside: *In vitro* and *in vivo* permeation studies. *Int J Pharm* 250:203.
46. Senel, S., et al. 1998. *In vitro* studies on enhancing effect of sodium glycocholate on transbuccal permeation of morphine hydrochloride. *J Control Release* 51 (2–3):107.
47. Autrup, H., et al. 1985. Metabolism of benzo[*a*]pyrene by cultured rat and human buccal mucosa cells. *Carcinogenesis* 6:1761.
48. Sundqvist, K., et al. 1991. Growth regulation of serum-free cultures of epithelial cells from normal human buccal mucosa. *In Vitro Cell Dev Biol* 27A:562.
49. Hedberg, J.J., et al. 2000. Expression of alcohol dehydrogenase 3 in tissue and cultured cells from human oral mucosa. *Am J Pathol* 157:1745.
50. Gibbs, S., and M. Ponec. 2000. Intrinsic regulation of differentiation markers in human epidermis, hard palate and buccal mucosa. *Arch Oral Biol* 45:149.
51. Hennings, H., et al. 1980. Calcium regulation of growth and differentiation of mouse epidermal cells in culture. *Cell* 19:245.
52. Rupniak, H.T., et al. 1985. Characteristics of four new human cell lines derived from squamous cell carcinomas of the head and neck. *J Natl Cancer Inst* 75:621.

53. Nielsen, H.M., and M.R. Rassing. 2000. TR146 cells grown on filters as a model of human buccal epithelium. V. Enzyme activity of the TR146 cell culture model, human buccal epithelium and porcine buccal epithelium, and permeability of leu-enkephalin. *Int J Pharm* 200:261.
54. Nielsen, H.M., and M.R. Rassing. 2002. Nicotine permeability across the buccal TR146 cell culture model and porcine buccal mucosa *in vitro*: Effect of pH and concentration. *Eur J Pharm Sci* 16:151.
55. Vondracek, M., et al. 2001. Cytochrome P450 expression and related metabolism in human buccal mucosa. *Carcinogenesis* 22:481.
56. Lingstrom, P., et al. 2005. The release of vitamin C from chewing gum and its effects on supragingival calculus formation. *Eur J Oral Sci* 113:20.
57. Rowe, R.C. 2003. By gum—A buccal delivery system. *Drug Discov Today* 8:617.
58. Valoti, M., et al. 2003. Pharmacokinetics of diphenhydramine in healthy volunteers with a dimenhydrinate 25 mg chewing gum formulation. *Methods Find Exp Clin Pharmacol* 25:377.
59. Christrup, L.L., et al. 1990. Relative bioavailability of methadone hydrochloride administered in chewing gum and tablets. *Acta Pharm Nord* 2:83.
60. Christrup, L.L., et al. 1990. Relative bioavailability of (+/−)-verapamil hydrochloride administered in tablets and chewing gum. *Acta Pharm Nord* 2:371.
61. Yang, X., G. Wang, and X. Zhang. 2004. Release kinetics of catechins from chewing gum. *J Pharm Sci* 93:293.
62. Russell, M.A., et al. 1980. Clinical use of nicotine chewing-gum. *Br Med J* 280:1599.
63. de Blaey, C.J., and C.P. de Haseth. 1977. Buccal residence time of lozenges. *Pharm Acta Helv* 52:116.
64. Eby, G.A., et al. 1984. Reduction in duration of common colds by zinc gluconate lozenges in a double-blind study. *Antimicrob Agents Chemother* 25:20.
65. Shiffman, S., et al. 2002. Efficacy of a nicotine lozenge for smoking cessation. *Arch Intern Med* 162:1267.
66. Streisand, J.B., et al. 1991. Absorption and bioavailability of oral transmucosal fentanyl citrate. *Anesthesiology* 75:223.
67. Lahiri, A., et al. 1986. Buccal nitroglycerin tablets in heart failure. *Ann Intern Med* 105:141.
68. Abrams, J. 1987. Glyceryl trinitrate (nitroglycerin) and the organic nitrates. Choosing the method of administration. *Drugs* 34:391.
69. Bomber, J.W., and P.L. De Tullio. 1995. Oral nitrate preparations: An update. *Am Fam Physician* 52:2331.
70. Save, T., et al. 1994. Comparative study of buccoadhesive formulations and sublingual capsules of nifedipine. *J Pharm Pharmacol* 46:192.
71. Wolf, S.B., et al. 2005. Sublingual misoprostol for labor induction: A randomized clinical trial. *Obstet Gynecol* 105:365.
72. Clarke, A., et al. 2003. A new formulation of selegiline: Improved bioavailability and selectivity for MAO-B inhibition. *J Neural Transm* 110:1241.
73. Gu, J.M., J.R. Robinson, and S.H. Leung. 1988. Binding of acrylic polymers to mucin/epithelial surfaces: Structure–property relationships. *Crit Rev Ther Drug Carrier Syst* 5 (1):21.
74. Gandhi, R.B., and J.R. Robinson. 1994. Oral cavity as a site for bioadhesive drug delivery. *Adv Drug Del Rev* 13:43.
75. Lehman, K.O.R. 1989. Chemistry and application properties of polymethacrylate coating systems. In *Aqueous polymeric coatings for pharmaceutical applications*, ed. J.W. McGinity, 153. New York: Marcel Dekker.
76. Wong, C.F., et al. 1999. Formulation and evaluation of controlled release Eudragit buccal patches. *Int J Pharm* 178:11.
77. Cui, Z., and R.J. Mumper. 2002. Bilayer films for mucosal (genetic) immunization via the buccal route in rabbits. *Pharm Res* 19:947.
78. Das, N.G., and S.K. Das. 2004. Development of mucoadhesive dosage forms of buprenorphine for sublingual drug delivery. *Drug Deliv* 11:89.
79. Korbonits, M., et al. 2004. A comparison of a novel testosterone bioadhesive buccal system, striant, with a testosterone adhesive patch in hypogonadal males. *J Clin Endocrinol Metab* 89:2039.

80. Nafee, N.A., et al. 2004. Mucoadhesive delivery systems. I. Evaluation of mucoadhesive polymers for buccal tablet formulation. *Drug Dev Ind Pharm* 30:985.
81. Varshosaz, J., and Z. Dehghan. 2002. Development and characterization of buccoadhesive nifedipine tablets. *Eur J Pharm Biopharm* 54:135.
82. Miyazaki, S., et al. 2000. Oral mucosal bioadhesive tablets of pectin and HPMC: *In vitro* and *in vivo* evaluation. *Int J Pharm* 204:127.
83. Park, C.R., and D.I. Munday. Development and evaluation of a biphasic buccal adhesive tablet for nicotine replacement therapy. *Int J Pharm* 237:215.
84. Bernkop-Schnurch, A., et al. 1999. Polymers with thiol groups: A new generation of mucoadhesive polymers? *Pharm Res* 16:876.
85. Bernkop-Schnurch, A., et al. 2004. Thiolated chitosans: Development and *in vitro* evaluation of a mucoadhesive, permeation enhancing oral drug delivery system. *J Control Release* 94:177.
86. Langoth, N., et al. 2003. Development of buccal drug delivery systems based on a thiolated polymer. *Int J Pharm* 252:141.
87. Jug, M., and M. Becirevic-Lacan. 2004. Influence of hydroxypropyl-beta-cyclodextrin complexation on piroxicam release from buccoadhesive tablets. *Eur J Pharm Sci* 21:251.
88. Peppas, N.A., et al. 2000. Hydrogels in pharmaceutical formulations. *Eur J Pharm Biopharm* 50:27.
89. Peppas, N.A. 1991. Physiologically responsive gels. *J Bioact Compat Polym* 6:241.
90. de Vries, M.E., et al. 1988. Hydrogels for buccal drug delivery: Properties relevant for mucoadhesion. *J Biomed Mater Res* 22:1023.
91. Eouani, C., et al. 2001. *In-vitro* comparative study of buccal mucoadhesive performance of different polymeric films. *Eur J Pharm Biopharm* 52:45.
92. Shojaei, A.H., et al. 1998. Transbuccal delivery of acyclovir (II): Feasibility, system design, and *in vitro* permeation studies. *J Pharm Pharm Sci* 1:66.
93. Kitano, M., et al. 1998. Buccal absorption through golden hamster cheek pouch *in vitro* and *in vivo* of 17β-estradiol from hydrogels containing three types of absorption enhancers. *Int J Pharm* 174:19.
94. Xu, H.B., et al. 2002. Hypoglycaemic effect of a novel insulin buccal formulation on rabbits. *Pharmacol Res* 46:459.
95. Al-Achi, A., and R. Greenwood. 1993. Buccal administration of human insulin in streptozocin-diabetic rats. *Res Commun Chem Pathol Pharmacol* 82:297.
96. Farshi, F.S., et al. 1996. *In-vivo* studies in the treatment of oral ulcers with liposomal dexamethasone sodium phosphate. *J Microencapsul* 13:537.
97. Yang, T.Z., et al. 2002. Phospholipid deformable vesicles for buccal delivery of insulin. *Chem Pharm Bull* (*Tokyo*) 50:749.
98. Venugopalan, P., et al. 2001. Pelleted bioadhesive polymeric nanoparticles for buccal delivery of insulin: Preparation and characterization. *Pharmazie* 56:217.
99. Cilurzo, F., et al. 2005. Fast-dissolving mucoadhesive microparticulate delivery system containing piroxicam. *Eur J Pharm Sci* 24:355.
100. Jacobsen, J. 2001. Buccal iontophoretic delivery of atenolol. HCl employing a new *in vitro* three-chamber permeation cell. *J Control Release* 70:83.
101. Lee, V.H.L., and A. Yamamoto. 1989. Penetration and enzymatic barriers to peptide and protein absorption. *Adv Drug Del Rev* 4:171.
102. Aungst, B.J. 1993. Novel formulation strategies for improving oral bioavailability of drugs with poor membrane permeation or presystemic metabolism. *J Pharm Sci* 82:979.
103. Li, C., et al. 1997. Transmucosal delivery of oxytocin to rabbits using a mucoadhesive buccal patch. *Pharm Dev Technol* 2:265.
104. Cui, Z., and R.J. Mumper. 2002. Buccal transmucosal delivery of calcitonin in rabbits using thin-film composites. *Pharm Res* 19:1901.
105. Hosny, E.A., et al. 2002. Buccoadhesive tablets for insulin delivery: *In-vitro* and *in-vivo* studies. *Boll Chim Farm* 141:210.
106. Guevara-Aguirre, J., et al. 2004. Oral spray insulin in treatment of type 2 diabetes: A comparison of efficacy of the oral spray insulin (Oralin) with subcutaneous (SC) insulin injection, a proof of concept study. *Diabetes Metab Res Rev* 20:472.

107. Kaelble, D.H. 1977. A surface energy analysis of bioadhesion. *Polymer* 18:475.
108. Voyutskii, S.S. 1963. Autohesion and adhesion of high polymers. In *Polymer reviews*, 140. eds. Mark, H.F. and Immergut, E.H. New York: John Wiley & Sons.
109. Deryaguin, B.V., 1997. On the relationship between the electrostatic and molecular component of the adhesion of elastic particles to a solid surface. *J Colloid Interface Sci* 58:528.
110. Helfand, E., and Y. Tagami. 1972. Theory of the interface between immiscible polymers. *J Chem Phys* 57:1812.
111. Chickering, D.E., III, and E. Mathiowitz, eds. 1999. Definitions, mechanisms and theories of bioadhesion. In *Bioadhesive drug delivery systems: Fundamentals, novel approaches and development*, 1. New York: Marcel Dekker.
112. de Vries, M.E., et al. 1991. Developments in buccal drug delivery. *Crit Rev Ther Drug Carrier Syst* 8:271.
113. Wertz, P.W., and C.A. Squier. 1991. Cellular and molecular basis of barrier function in oral epithelium. *Crit Rev Ther Drug Carrier Syst* 8:237.

10 Chemical Enhancers in Buccal and Sublingual Absorptions

John D. Smart

CONTENTS

10.1 INTRODUCTION

One of the major issues for drug delivery scientists is the delivery of the new products generated by the genomics and proteomic revolutions [1–4]. The pharmaceutical sciences now have to consider new strategies to effectively deliver these new "biopharmaceutical" products (typically large, hydrophilic and unstable proteins, oligonucleotides, and polysaccharides), as well as conventional small drug molecules; the oral mucosa provides one potential route for achieving this. Oral mucosae are more permeable than skin [5], although less permeable (but more accessible) than most other mucosal routes [6]. Buccal formulations have been developed to allow prolonged localized therapy and enhanced systemic delivery whereas the sublingual route (considered to be the more permeable [5]) is usually used when a rapid onset of action is required. The oral mucosa, however, while avoiding first pass effects and the high levels of enzymes present in the middle gastrointestinal tract, is a formidable barrier to drug absorption, especially for larger molecules. For this reason, chemical/permeation enhancers that can be included in a formulation to allow improved drug absorption have been developed.

10.2 BUCCAL AND SUBLINGUAL DRUG DELIVERY

To understand the mechanism of action of permeation enhancers, the routes by which drugs are absorbed first need to be considered. Chapter 9 reviews the basic biopharmaceutics of buccal and sublingual drug delivery, so only a very brief discussion of this in the context of permeation enhancement will be given below.

The major barrier for drug absorption is considered to be the squamous stratified epithelium of the oral mucosae. There are two possible routes of drug absorption through this: transcellular (intracellular, passing through the cell) and paracellular (intercellular, passing around the cell). Permeation across the buccal mucosa has been reported to be mainly by the paracellular route through the intercellular lipids produced by the membrane coating granules [7–9]. It has been argued however that the route taken depends on the physicochemical properties of the drug [10,11].

Generally small molecules that are predominantly lipophilic, with a log P of 1.6–3.3, are absorbed most rapidly; above 3.3, limited water solubility restricts their absorption [12]. Most drugs delivered successfully via the buccal or sublingual route are therefore small and lipophilic (such as glyceryl trinitrate and nicotine), whereas large hydrophilic molecules are in general poorly absorbed. However, it has been proposed that in the nonkeratinized buccal and sublingual mucosae, the hydrophilic nature of the lipids means that this is the predominant route for the absorption of hydrophilic molecules, whereas lipophilic molecules pass through the cell membranes and are absorbed by the transcellular route [10,11]. The amphiphilic nature of the intercellular lipids suggests that both a hydrophobic and a hydrophilic pathway through the paracellular route are likely to exist [13] so the situation may be more complex than the relatively simple models sometimes described. Although passive diffusion is the main mechanism of drug absorption [10,11], Veuillez et al. [13] have reported specialized transport mechanisms, surprising for an epithelium of this nature, for a few drugs and nutrients. For example, Kurosaki et al. [14] reported that the amino-β-lactam antibiotic, cefadroxil, was absorbed by active transport in the oral cavity, based on demonstrating saturation absorption kinetics which could be partially inhibited by the presence of a similar molecule.

In order to address the challenge of delivering poorly absorbed drugs via this route, especially those large hydrophilic molecules, permeation enhancers have become of increasing interest in recent years.

10.3 MECHANISMS OF ABSORPTION ENHANCEMENT

Permeation enhancers in general disrupt the integrity of the epithelial barrier, and have been said to act in the following ways [10,13,15] (Figure 10.1):

- Increasing the fluidity of the cell membrane
- Extracting inter- and intracellular lipids
- Disrupting lipid structure, e.g., solubilization by formation of micelles to create aqueous channels
- Altering cellular proteins
- Increasing the thermodynamic activity of the drug (promoting passive diffusion)
- Overcoming enzymatic barriers, particularly for peptide and protein drugs
- Altering surface mucin rheology

They have been shown to be clearly effective in promoting the absorption of large molecules; the *in vitro* penetration of some proteins was typically 1%–3% but the addition of an appropriate absorption enhancer increased this to ca. 10% [13].

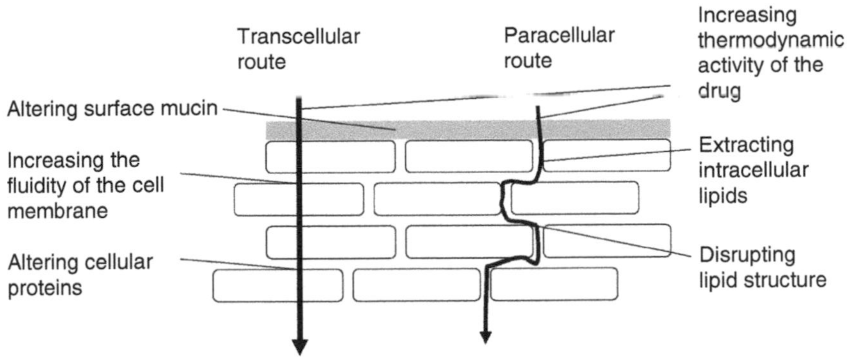

FIGURE 10.1 Methods by which a permeation enhancer can enhance drug absorption through buccal and sublingual mucosae.

10.4 TYPES OF PERMEATION ENHANCERS

The ideal permeation enhancer is safe, effective, pharmacologically inactive, chemically inert, and has a reversible effect (does not compromise the mucosal barrier function for extended periods) [10]. It is noticeable that the majority of the most widely investigated permeation enhancers have surfactant-like properties, and those that are water soluble seem to be most active at concentrations above the critical micelle concentration. The following have been investigated as a means of enhancing buccal permeability.

10.4.1 Bile Salts

Bile salts are steroids with surfactant-like properties that form associations in water. Their physiological role is to emulsify lipids in foodstuff passing through the intestine to enable fat digestion and absorption through the intestinal wall. It is perhaps not surprising, therefore, that they can be used as permeation enhancers and have been extensively employed to enhance the absorption of drugs through various epithelia. They are believed to act on both the transcellular and paracellular routes by a variety of mechanisms including solubilization and micellar entrapment of intercellular lipids, denaturation and extraction of proteins, enzyme inactivation, tissue swelling, the extraction of lipids or proteins from the cell wall, membrane fluidization, and reverse membrane micellation [13,15]. One *in vitro* study particularly indicated that sodium glycodeoxycholate acts in the intercellular lipid domain at lower concentrations (2 mM), apparently reducing the amount of polar lipids, whereas disorganizing cell membrane lipids at higher concentrations [16]. Generally they are believed to act reversibly without producing major damage to the mucosa (although currently there is a lack of chronic safety studies on their use).

Bile salts used in permeation enhancement studies include the trihydroxy salts sodium cholate, sodium glycocholate, and sodium taurocholate (Figure 10.2); and the dihydroxy salts sodium deoxycholate, sodium glycodeoxycholate, and sodium taurodeoxycholate. Several *in vitro* permeation studies carried out in isolated animal buccal mucosa and *in vivo* bioavailability studies conducted in animals and human subjects have proven their potential as effective buccal permeation enhancers.

The dihydroxy salts have been reported to be more active permeation enhancers than the trihydroxy salts [9,17], probably related to their increased lipophilicity, although interestingly other studies appear to show equal effectiveness between the two types of salts [18]. The permeation enhancing effect of dihydroxy bile salts seems to be more pronounced at or above

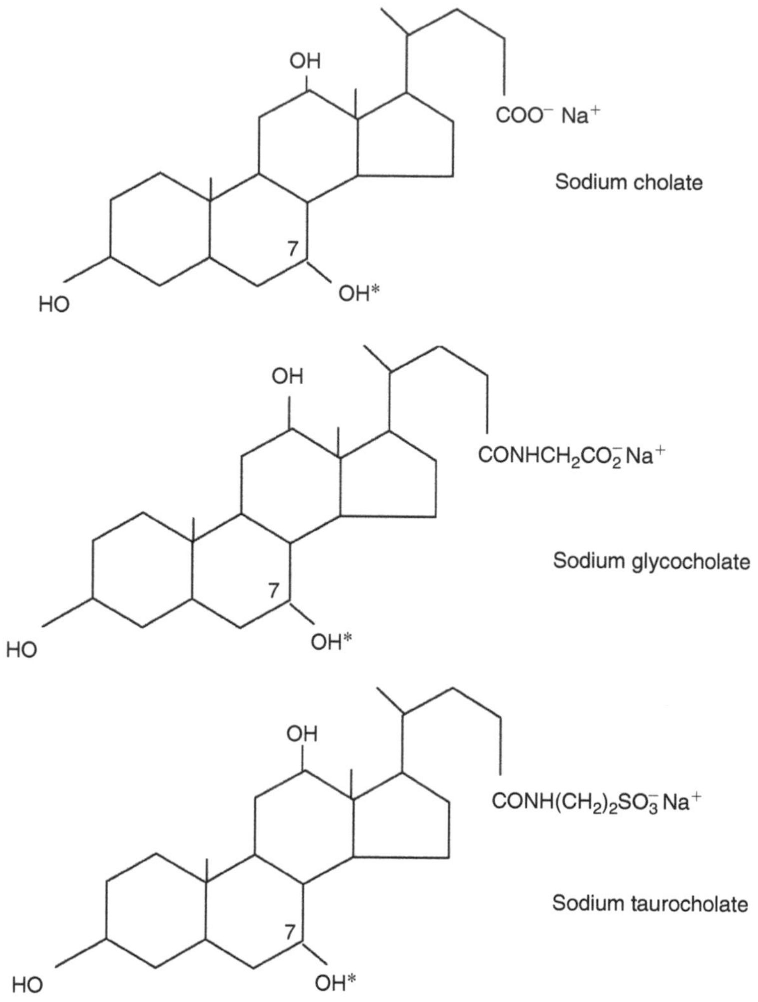

FIGURE 10.2 Bile salts used as absorption enhancer (deoxy-forms of these bile salts do not have the position 7 hydroxyl (OH*) group).

the critical micelle concentration [19]. In a cell culture model the dihydroxy bile salt sodium glycodeoxycholate has been reported to possess a better enhancing effect for mannitol than trihydroxy bile salts such as sodium glycocholate and sodium taurocholate [19]. Other studies have shown a 32-fold enhancement in the permeability of 2′,3′-dideoxycytidine in the presence of 4 mM sodium glycodeoxycholate across porcine buccal mucosa [19]. The absolute bioavailability of buserelin and fluorescein isothionate dextran in pigs has been enhanced by 5–7-fold when administered with sodium glycodeoxycholate [20,21]. It also enhanced the buccal permeation of morphine sulfate by approximately 5 times at 100 mM concentration [22] and the bioavailability of some proteins [9]. A correlation between the *in vitro* permeation across the nonkeratinized porcine buccal mucosa and *in vivo* bioavailability in rabbits (with largely keratinized mucosa) was observed with triamcinolone acetonide gel containing 5% sodium deoxycholate [23,24].

Trihydroxy bile salts have also been proven to be effective. Enhanced permeation of mannitol in TR 146 cells (a model for oral cavity mucosa) was noticed when the concentration of sodium taurocholate and sodium glycocholate was about 2- to 3-fold higher than their critical micelle concentration values [17]. At higher concentrations (100 mM) the flux and the morphological changes caused by sodium glycocholate were comparable with sodium glycodeoxycholate. Sodium glycocholate was found to be an effective enhancer at 2–100 mM concentration for a model antisense oligonucleotide [25]. Using dogs and an *in vivo* absorption cell, sodium taurocholate and sodium glycocholate were found to enhance the absorption of insulin while not damaging the oral mucosae; the latter bile salt was found to have a prolonged effect [26]. A 100-fold enhancement in the permeability of the ionized form of flecainide in the presence of 1% sodium glycocholate at pH 5.8 has been reported [27] (although it had less effect on the permeation of sotalol), and the permeation of morphine hydrochloride has been enhanced by a factor of 9.3 [28]. The ability of sodium cholate to almost triple the absorption of calcitonin through rat buccal mucosa has been reported [6] (although the buccal route was the least effective relative to other routes of administration).

10.4.2 Fatty Acids and Their Salts and Esters

Fatty acids include oleic acid, lauric acid, and cod liver oil extract, whereas fatty acid salts include sodium laurate and sodium caprate, and esters include glyceryl monostearate, diethylene glycol monoethyl ether, and various sucrose fatty acid esters (Figure 10.3) (related surfactants like the Laureth series will be dealt with in the next section). These are generally lipophilic in nature with limited water solubility, which restricts the delivery systems in which they can be applied. The unsaturated fatty acids such as oleic acid are believed to act by reducing lipid order and increasing fluidity in the skin due to their "kinked" molecular conformation arising from the double bond in the hydrocarbon chain [29], and they should have a similar effect on oral mucosa. Oleic acid has been reported to be a good adsorption enhancer for insulin [30]. Unionized ergotamine absorption was enhanced significantly in the presence of 5% cod liver oil extract, which contained oleic acid as one of its major components [31]. The distribution of the drug within the lipid-rich region of the buccal mucosa and the resultant reduction in the barrier structure were correlated with its permeation enhancing effects (although at higher concentrations [7%–10%] permeation was seen to apparently decrease).

Fatty acid esters would be predicted to have little irritation or toxic effects. *Ex vivo* permeability studies conducted in porcine buccal mucosa showed significant permeation enhancement of an enkephalin from liquid crystalline phases of glycerine monooleate [32]. These were reported to enhance peptide absorption by a cotransport mechanism. Diethylene glycol monoethyl ether was reported to enhance the permeation of essential oil components of *Salvia desoleana* through porcine buccal mucosa from a topical microemulsion gel formulation [33]. Some sucrose fatty acid esters, namely, sucrose laurate, sucrose oleate, sucrose palmitate, and sucrose stearate, were investigated on the permeation of lidocaine hydrochloride [34], with 1.5% w/v sucrose laurate showing a 22-fold increase in the enhancement ratio.

$CH_3\,[CH_2]_7\,CH{:}CH\,[CH_2]_7\,CO_2H$	Oleic acid
$CH_3\,[CH_2]_{10}\,CO_2H$	Lauric acid
$CH_3\,[CH_2]_7\,CO_2H$	Capric acid
$CH_3\,[CH_2]_{16}\,CO\,OCH_2\,CHOH\,CH_2OH$	Glyceryl monooleate

FIGURE 10.3 Some fatty acid and fatty acid esters used as permeation enhancers.

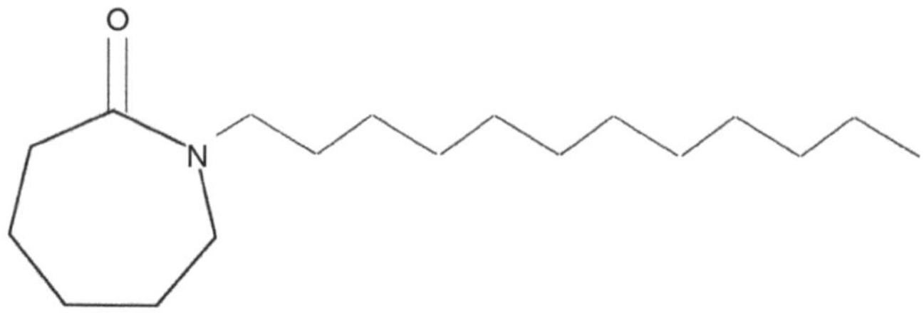

FIGURE 10.4 Azone (laurocapram).

10.4.3 Azone

Azone (laurocapram) is used extensively as a transdermal permeation enhancer, and has also found use in buccal drug delivery. It is a lipophilic surfactant in nature (Figure 10.4). Permeation of salicylic acid was enhanced by the pre-application of an Azone emulsion *in vivo* in a keratinized hamster cheek pouch model [35]. Octreotide and some hydrophobic compounds' absorption have also been improved by the use of Azone [36]. Azone was shown to interact with the lipid domains and alter the molecular moment on the surface of the bilayers [37]. In skin it has been proposed that Azone was able to form ion pairs with anionic drugs to promote their permeation [38].

10.4.4 Other Surfactants

The other surfactants include sodium dodecyl (lauryl) sulfate, the polysorbates, the laureths, Brijs and benzalkonium chloride (Figure 10.5). These are predominantly water soluble and can form associations (micelles) in aqueous solution.

They are believed to enhance the transbuccal permeation by a mechanism that is similar to that of bile salts, namely, extraction of lipids, protein denaturation, inactivation of enzymes, and swelling of tissues [39]. Sodium dodecyl sulfate is reported to have a significant absorption enhancing effect but may also produce damage to the mucosa [13]. The effect of sodium

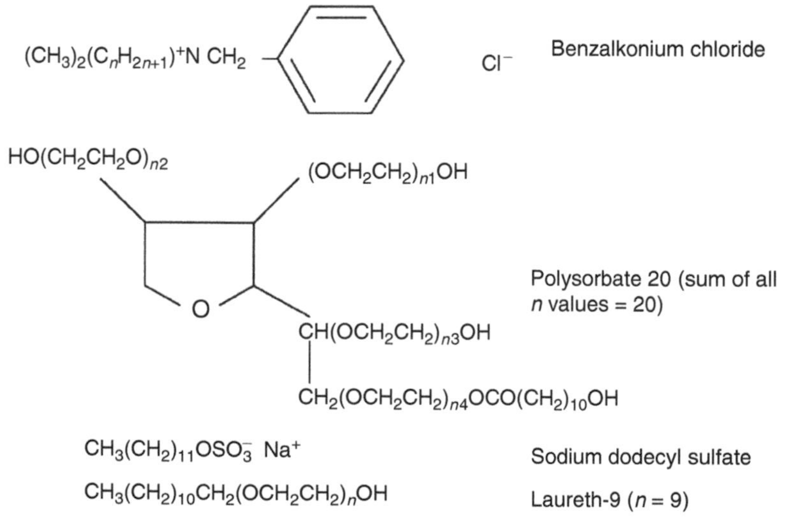

FIGURE 10.5 Some examples of surfactants used as permeation enhancers.

dodecyl sulfate on the *in vitro* buccal permeability of caffeine and estradiol has been evaluated using porcine buccal tissue [40]. With caffeine, sodium dodecyl sulfate (0.05%–1% w/v) enhanced the flux (enhancement ratios ranging from 1.6 to 1.8) while having the opposite effect with estradiol, a lipophilic drug. The significant reduction in flux was partly attributed to the micellar entrapment of the lipophilic drug and the resultant poor permeation of the complex. The permeation enhancing property of laureth-9, a nonionic surfactant, from the oral cavity of anaesthetized rats [41] has also been reported. A 5% solution produced insulin activity at 25%–33% of the intramuscular (i.m.) dose. In an *in vivo* rabbit model using a buccal cell to administer solutions to the mucosa, the surfactant Brij 35 (polyoxyethylene-23-lauryl ether) was seen to be more effective than sodium lauryl sulfate or several bile salts at a concentration of 1 mM in enhancing the absorption of insulin from solution [42]. This effect was markedly increased when the concentration of Brij was above the critical micelle concentration; it then reached a plateau at concentrations above 1%. The effect was considered to be due to a combination of the prevention of insulin aggregation in solution, permeation enhancement, and protease inhibition. However, the rabbit oral epithelium is keratinized, so it differs a little from that of the human buccal and sublingual mucosa. A similar consideration occurs for an *in vivo* study in rats [41]. Insulin in a 5% solution of octoxynol-9 has ca. 15% availability, and in a pH 8.9 solution containing lauric acid has 22.4% availability via the buccal route, relative to an i.m. injection.

10.4.5 Complexing Agents

The complexing agents include cyclodextrins and sodium edetate. Cyclodextrins are enzymatically modified starches, forming rings of 6–8 units. The outer surface of the ring is polar whereas the internal surface is nonpolar. Hence, the center of the cyclodextrin can be used to carry water-insoluble molecules in an aqueous environment by forming inclusion complexes. Effective buccal absorption of steroidal hormones using two different hydrophilic cyclodextrin derivatives, namely, 2-hydroxypropyl betacyclodextrin and poly betacyclodextrin, was reported [43]. The effect of cyclodextrins (5%) on the buccal absorption of interferon has also been described [44]. Chelators such as EDTA, sodium citrate, and also the polyacrylic acids are reported to have an absorption enhancing effect by interfering with calcium ions [15].

10.4.6 Cosolvents

Cosolvents include water-miscible solvents such as ethanol and propylene glycol. The use of vehicles that enhance absorption has been considered in transdermal drug delivery, and would also be of use in buccal delivery. They work by changing the thermodynamic activity of the drug in solution, increasing its concentration and facilitating partition of the drug into the membrane, and promoting passive diffusion. As ethanol and propylene glycol penetrate into mucosa, drugs dissolved in these cosolvents are expected to be carried with them [45]. In most studies, the vehicle is used in combination with a permeation enhancer to further increase absorption.

A combination of oleic acid (1%) and polyethylene glycol 200 (PEG, 5% and 10%) appreciably enhanced the *ex vivo* permeation of a model peptide across porcine buccal mucosa [46]. A hydrogel formulation containing glyceryl monolaurate (2%) and alcohol (40%) effectively enhanced the permeability of 17β-estradiol across hamster cheek pouch buccal mucosa with no morphological changes evident in the mucosa 7 h after application. Permeation enhancement was also observed when sodium caprate and alcohol or propylene glycol were used in combination [47]. The inclusion of 10% lauric acid in propylene glycol produced almost 30% of the i.m. dose of insulin [41].

10.4.7 Others

Lecithin (phosphatidylcholine) is a phospholipid, which may be isolated from either egg yolk or soybeans. It is commercially available in high purity for medical uses and has been used to enhance the absorption of insulin *in vivo* [26]. The antibiotic sodium fusidate, a steroid similar in molecular structure to bile salts has also been shown to have permeation enhancing properties for insulin *in vitro* [41].

Chitosan, a polysaccharide containing glucosamine and acetyl glucosamine units, has been shown to have permeation-enhancing activity [48]. Solutions and gels of chitosan were found to be effective absorption enhancers by their transient widening of the tight junctions within the mucosa [49]. It was found to promote the transport of mannitol and fluorescent-labeled dextrans across a tissue culture model of the buccal epithelium [50], chitosan glutamate being particularly effective. Chitosan has been shown to be an effective permeation enhancer for peptide absorption across porcine buccal mucosa [51] without producing any histological evidence of tissue damage [15].

Nicolazzo et al. [52] considered the use of the lipophilic skin penetration enhancers, octisalate and padimate (both used in sunscreens), in comparison to Azone on the buccal absorption of various drugs *in vitro*. They were found to have limited effect in enhancing the permeation of triamcinolone acetonide (although some increase in tissue uptake was proposed in some cases) relative to Azone, while reducing the penetration of estradiol and caffeine. One interesting report is that of the effect of capsaicin from capsicum, a commonly used food ingredient, which has been reported to enhance the permeability of sulfathiazole in human volunteers [53] presumably by a direct irritation effect on the mucosa. This raised an interesting issue of the effect of diet on oral mucosal permeability.

10.5 ENZYME INHIBITION

The inhibition of enzymes present in the oral mucosa [54] that may degrade protein drugs can also enhance absorption, and protease inhibitors such as aprotinin [55] and puromycin [56] have been used to affect this. Although not strictly permeation enhancers, they will help to increase the transport of intact drugs across oral mucosae. Bile salts have also been proposed as having an inhibitory effect on enzymes, sodium glycocholate inhibited insulin metabolism in a range of mucosae [56], whereas the primary reason given for bile-salt-induced enhanced calcitonin absorption in rat oral mucosa was that of inhibition of degradation [57].

10.6 TOXICITY

Local irritation will need to be avoided if a permeation enhancer is to find routine use. Intuitively, the mechanism of action of permeation enhancers would suggest that tissue damage would occur. For surfactants like sodium dodecyl sulfate, mucosal irritation would be expected to be an issue, but its widespread use in oral healthcare products such as toothpastes suggests that this is not of major importance within the oral cavity. No evidence of toxicity was observed when the buccal mucosa of dogs were exposed *in situ* to sodium glycocholate, sodium taurocholate, and lysophosphatidylcholine, although the enhancing effect of the glycocholate was seen to persist [26]. However, in an *in vitro* study using porcine buccal tissue, histological changes indicating tissue damage were evident relative to controls when the tissue was exposed to 100 mM solutions of di- and trihydroxy bile salts over a 4 h period [18]. Histological changes from loss of upper cell layers to separation of the epithelium from the underlying connective tissue were also evident in mucosal tissues exposed to a range of bile salt solutions at a concentration of 100 mM for 4 h periods [15].

10.7 CONCLUSIONS

The buccal route is currently limited to drugs with the appropriate physicochemical properties to allow effective absorption. The use of permeation enhancers to enhance the buccal and sublingual absorptions of drugs would be required, especially if larger molecules are to be delivered by this route. Their major mechanism of action is via the disruption of cell membranes, so it is not surprising that irritancy, toxicity, and compromising of the mucosal barrier remain issues, and long-term toxicity studies will be required before their widespread usage. Ensuring that the optimum concentration of the enhancer is employed to limit toxicity while facilitating an enhancing effect which is reproducible (with limited inter-subject variability) will be challenges for future developments.

The physicochemical properties of the permeation enhancers considered would appear to provide challenges to the pharmaceutical scientist in their formulation. Many are predominantly oil soluble, whereas others are charged surfactants that can readily interact with oppositely charged molecules, thereby limiting their compatibility with many drugs and excipients. It would appear that many of the permeation enhancers need to be presented to the absorbing surface in the form of a solution and then given time to penetrate into the mucosa in order to produce their effect. This would clearly limit the type of formulation that could be used to retentive (bioadhesive) liquids, semisolids, and gelling matrices, and suggests that the sublingual route, with all the challenges of locating and retaining a formulation within this region of the oral cavity, would probably present the greatest challenge for the use of this technology.

REFERENCES

1. Tyers, M., and M. Mann. 2003. From genomics to proteomics. *Nature* 422:193.
2. Betz, S.F., S.M. Baxter, and J.S. Fetrowm. 2002. Function first: A powerful approach to post genomic drug discovery. *Drug Discov Today* 7:8765.
3. Meyer, J.M., and G.S. Ginsburg. 2002. The path to personalised medicine. *Cur Opin Chem Biol* 6:434.
4. Gurwitz, D., A. Wiezman, and M. Rehavi. 2003. Teaching pharmacogenomics to prepare future physicians and researchers for personalised medicine. *Trends Pharmacol Sci* 24:122.
5. Kurosaki, Y., and T. Kimura. 2002. Regional variation in oral mucosal drug permeability. *Crit Rev Ther Drug Carrier Syst* 17:467.
6. Yamamoto, A., et al. 2001. Absorption of water soluble compounds with different molecular weights and [ASU1.7]-eel calcitonin from various mucosal administration sites. *J Control Release* 76:363.
7. Squier, C.A., and C.A. Lesch. 1988. Penetration pathways of different compounds through epidermis and oral epithelia. *J Oral Pathol* 17:512.
8. Squier, C.A., and P.W. Wertz. 1993. Permeability and pathophysiology of oral mucosa. *Adv Drug Deliv Rev* 12:13.
9. Junginger, H., J. Hoogstraate, and J. Verhoef. 1999. Recent advances in buccal drug delivery and absorption in vitro and in vivo studies. *J Control Release* 62:149.
10. Song, Y., et al. 2004. Mucosal drug delivery: Membranes, methodologies and applications. *Crit Rev Ther Drug Carrier Syst* 21(3): 195–256.
11. Shojaei, A.M. 1998. Buccal mucosa as a route for systemic drug delivery, a review. *J Pharm Sci* 1:15.
12. Florence A.T., and D.A. Attwood. 1998. Buccal and sublingual absorption. In *Physicochemical principles of pharmacy*. 3rd ed., 392. Basingstoke, UK: MacMillan Press.
13. Veuillez, F., et al. 2001. Factors and strategies for improving buccal absorption of peptides. *Eur J Pharm Biopharm* 51:93.
14. Kurosaki, Y., et al. 1992. Existence of a specialised absorption mechanism for cefadroxil, an aminocephalosporin antibiotic, in the human oral cavity. *Int J Pharm* 82:165.

15. Senel, S., and A.A. Hincal. 2001. Drug permeation enhancement via buccal route: Possibilities and limitations. *J Control Release* 72:133.
16. Hoogstraate A.J., et al. 1997. Effects of the penetration enhancer glycodeoxycholate on the lipid integrity in porcine buccal epithelium in-vitro. *Eur J Pharm Sci* 5:189.
17. Hanne, M.N., and R.R. Margrethe. 1999. TR 146 cells grown on filters as a human buccal epithelium: III. Permeability enhancement by different pH values, different osmolarity values, and bile salts. *Int J Pharm* 185:215.
18. Senel, S., et al. 1994. Enhancement of in-vitro permeability of porcine buccal mucosa by bile salts: Kinetic and histological studies. *J Control Release* 32:45.
19. Jun, X., F. Xiaoling, and L. Xiaoling. 2002. Transbuccal delivery of 2′,3′-dideoxycytidine: In vitro permeation study and histological investigation. *Int J Pharm* 231:57.
20. Hoogstraate, A., et al. 1996. In vivo buccal delivery of the peptide drug buserelin with glycodeoxycholate as an absorption enhancer in pigs. *Pharm Res* 13:1233.
21. Hoogstraate, A., et al. 1996. In-vivo buccal delivery of fluorescein isothiocyanate-dextran 4400 with glycodeoxycholate as an absorption enhancer in pigs. *J Pharm Sci* 85:457.
22. Senel, S., Y. Capan, and M.F. Sargon. 1997. Enhancement of transbuccal permeation of morphine sulfate by sodium glycodeoxycholate in vitro. *J Control Release* 45:153.
23. Shin, S., J. Bum, and J. Choi. 2000. Enhanced bioavailability by buccal administration of triamcinolone acetonide from the bioadhesive gels in rabbits. *Int J Pharm* 209:37.
24. Shin, S., and J. Kim. 2000. Enhanced permeation of triamcinolone acetonide through the buccal mucosa. *Eur J Pharm Biopharm* 50:217.
25. Jasti, B., et al. 2000. Permeability of antisense oligonucleotide through porcine buccal mucosa. *Int J Pharm* 208:35.
26. Zhang, J., et al. 1994. An in-vivo dog model for studying recovery kinetics of the buccal mucosa permeation barrier after exposure to permeation enhancers; Apparent evidence of effective enhancement without tissue damage. *Int J Pharm* 101:15.
27. Deneer, V., et al. 2002. Buccal transport of flecainide and sotalol: Effect of a bile salt and ionization state. *Int J Pharm* 241:127.
28. Senel, S., et al. 1998. In vitro studies on enhancing effect of sodium glycocholate on transbuccal permeation of morphine hydrochloride. *J Control Release* 51:107.
29. Ogiso, T., M. Iwaki, and T. Paku. 1995. Effect of various enhancers on transdermal penetration of indomethacin and urea, and relationship between penetration parameters and enhancement factors. *J Pharm Sci* 84:482.
30. Morishita, M., et al. 2001. Pluronic F-127 gels incorporating highly purified unsaturated fatty acids for buccal delivery of insulin. *Int J Pharm* 212:289.
31. Tsutsumi, K., et al. 1998. Effect of cod-liver oil extract on the buccal permeation of ergotamine tartrate. *Drug Dev Ind Pharm* 24:757.
32. Lee, J., and I.W. Kellaway. 2000. Buccal permeation of [D-Ala(2), D-Leu(5)]enkephalin from liquid crystalline phases of glyceryl monooleate. *Int J Pharm* 195:35.
33. Ceschel, G.C., et al. 2000. In vitro permeation through porcine buccal mucosa of Salvia desoleana Atzei & Picci essential oil from topical formulations. *Int J Pharm* 195:171.
34. Quintanar, A., et al. 1998. *Ex vivo* oral mucosal permeation of lidocaine hydrochloride with sucrose fatty acid esters as absorption enhancers. *Int J Pharm* 173:203.
35. Kurosaki, Y., et al. 1989. Enhancing effect of 1-dodecylazacycloheptan-2-one (Azone) on the absorption of salicylic acid from keratinized oral mucosa and the duration of enhancement in vivo. *Int J Pharm* 51:47.
36. Araki, M., et al. 1992. Interaction of percutaneous absorption enhancer with stratum corneum of hamster cheek pouch; An electrophysiological study. *Int J Pharm* 81:39.
37. Turunen, T., et al. 1994. Effect of some penetration enhancers on epithelial membrane lipid domains: Evidence from fluorescence spectroscopy studies. *Pharm Res* 11:288.
38. Hadgraft, J., D.G. Williams, and G. Allan. 1993. Azone mechanism of action and clinical effect. In *Pharmaceutical Skin Penetration Enhancement*, ed. Walters, K.A., and J. Hadgraft, 175. New York: Marcel Dekker.
39. Quintanar, A., et al. 1997. Mechanisms of oral permeation enhancement. *Int J Pharm* 156:127.

40. Nicolazzo, J.A., B.L. Reed, and B.C. Finnin. 2004. Assessment of the effects of sodium dodecyl sulfate on the buccal permeability of caffeine and estradiol. *J Pharm Sci* 93:431.
41. Aungst, B., and N. Rogers. 1989. Comparison of the effects of various transmucosal absorption promoters on buccal insulin delivery. *Int J Pharm* 53:227.
42. Oh, C.K., and W.A. Ritschel. 1990. Biopharmaceutic aspects of buccal absorption of insulin. *Methods Find Exp Clin Pharmacol* 12:205.
43. Pitha, J., S. Harma, and M. Michel. 1986. Hydrophilic cyclodextrin derivatives enable effective oral administration of steroidal hormones. *J Pharm Sci* 75:165.
44. Stewart, A., D. Bayley, and C. Howes. 1994. The effect of enhancers on the buccal absorption of hybrid (BDBB) alpha-interferon. *Int J Pharm* 104:145.
45. Bendas, B., U. Schmalfub, and R. Nuebert. 1995. Influence of propylene glycol as cosolvent on mechanism of drug transport from hydrogels. *Int J Pharm* 116:19.
46. Lee, J., and I.W. Kellaway. 2000. Combined effect of oleic acid and polyethylene glycol 200 on buccal permeation of [D-ala2, D-leu5] enkephalin from a cubic phase of glyceryl monooleate. *Int J Pharm* 204:137.
47. Kitano, M., et al. 1998. Buccal absorption through golden hamster cheek pouch in vitro and in vivo of 17-estradiol from hydrogels containing three types of absorption enhancers. *Int J Pharm* 174:19.
48. Dodane, V., M. Amin Khan, and J. Merwin. 1999. Effect of chitosan on epithelial permeability and structure. *Int J Pharm* 182:21.
49. Portero, A., C. Remunan-Lopez, and H.M. Nielsen. 2002. The potential of chitosan in enhancing peptide and protein absorption across the TR146 cell culture model—an in vitro model of the buccal epithelium. *Pharm Res* 19:169.
50. Senel, S., et al. 2000. Enhancing effect of chitosan on peptide drug delivery across buccal mucosa. *Biomaterials* 21:2067.
51. Sandri, G., et al. 2004. Assessment of chitosan derivatives as buccal and vaginal penetration enhancers. *Eur J Pharm Sci* 2:351.
52. Nicolazzo, J.A., B.L. Reed, and B.C. Finnin. 2004. Modification of buccal delivery following pre-treatment with skin penetration enhancers. *J Pharm Sci* 93:2054.
53. Raouf Hamid, M., F. Shmela, and S.A. Metwally. 1985. Influence of capsaicin on drug absorption and transport across biological membranes. *J Drug Res Egypt* 16:67.
54. Tavakoli-Saberi, M.R., A. Williams, and K.L. Audus. 1991. Aminopeptidase activity in human buccal epithelium and primary cultures of hamster buccal epithelium. *Pharm Res* 6:S197.
55. Aungst, B.J., and N.J. Rogers. 1988. Site dependence of absorption-promoting actions of laureth-9, Na salicylate, Na_2EDTA, and aprotinin on rectal, nasal, and buccal insulin delivery. *Pharm Res* 5:305.
56. Yamamoto, A., E. Hayakawa, and V.H.L. Lee. 1990. Insulin and proinsulin proteolysis in mucosal homogenates of the albino rabbit: Implications in peptide drug delivery from non-oral routes. *Life Sci* 47:2465.
57. Nakada, Y., et al. 1988. The effect of additives on the oral mucosal absorption of human calcitonin in rats. *J Pharmacobiodyn* 11:395.

Part IV

Transdermal Enhanced Delivery

11 The Lipid Organization in Stratum Corneum and Model Systems Based on Ceramides

Miranda W. de Jager, Maja Ponec, and Joke A. Bouwstra

CONTENTS

11.1 INTRODUCTION

Microscopically, the skin is a multilayered organ composed of many histological layers. It is generally subdivided into three layers: the epidermis, the dermis, and the hypodermis [1]. The uppermost nonviable layer of the epidermis, the stratum corneum, has been demonstrated to constitute the principal barrier to percutaneous penetration [2,3]. The excellent barrier properties of the stratum corneum can be ascribed to its unique structure and composition. The viable epidermis is situated beneath the stratum corneum and responsible for the generation of the stratum corneum. The dermis is directly adjacent to the epidermis and composed of a matrix of connective tissue, which renders the skin its elasticity and resistance to deformation. The blood vessels that are present in the dermis provide the skin with nutrients and oxygen [1]. The hypodermis or subcutaneous fat tissue is the lowermost layer of the skin. It supports the dermis and epidermis and provides thermal isolation and mechanical protection of the body.

The outer layer of the skin forms an effective barrier to retain water within the body and keep exogenous compounds out of the body. As a result, the major problem in dermal and transdermal drug deliveries is the low penetration of drug compounds through the stratum corneum. Dermal drug delivery comprises the topical application of drugs for the local treatment of skin diseases. It requires the permeation of a drug through the outer skin layers to reach its site of action within the skin, with little or no systemic uptake. The application of drugs to the skin for systemic therapy is referred to as transdermal drug delivery. Hence, it is required that a pharmacologically potent drug reaches the dermis where it can be taken up by the systemic blood circulation. In either case, the drug has to cross the outermost layer of the skin, the stratum corneum.

In the first part of this chapter, the formation and structure of the stratum corneum will be discussed. The second part describes the composition and organization of the intercellular stratum corneum lipids *in vivo* and *in vitro*.

11.2 KERATINOCYTE TERMINAL DIFFERENTIATION

11.2.1 The Viable Epidermis

The epidermis is approximately 100 to 150 μm thick and consists of various layers, characterized by different stages of differentiation. Figure 11.1 shows a schematic representation of

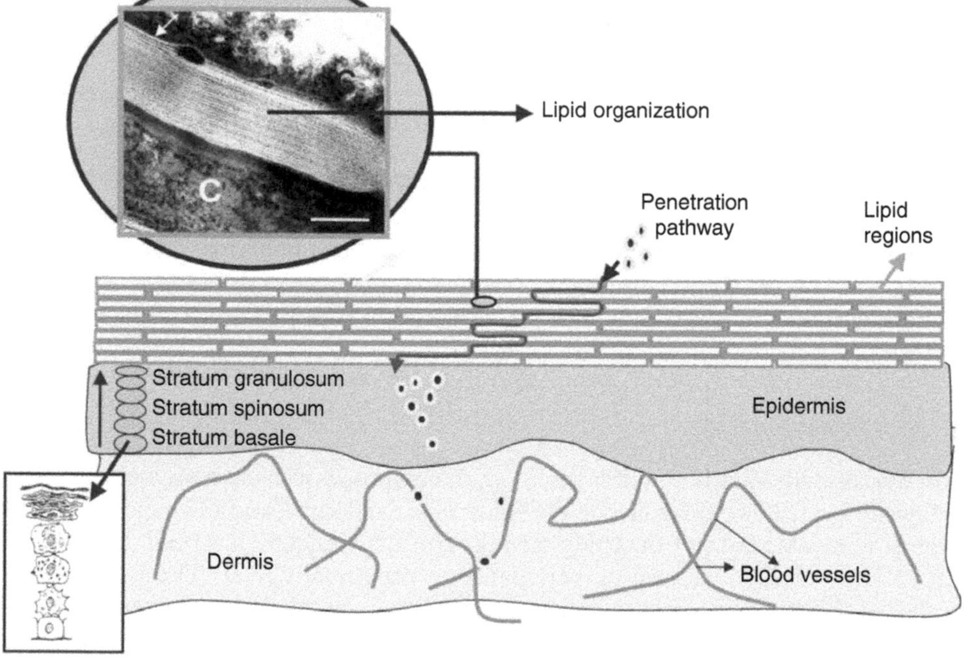

FIGURE 11.1 A schematic drawing of a skin cross section. The skin is composed of a dermis and an epidermis. In the basal layer of the epidermis cells proliferate. Upon leaving the basal layer cells start to differentiate and migrate in the direction of the skin surface. At the interface between stratum granulosum and stratum corneum final differentiation occurs, during which the viable cells are transformed into dead keratin filled cells (corneocytes). The corneocytes are surrounded by a cell envelope composed of cross-linked proteins and a covalently bound lipid envelope (see arrow). In stratum corneum the corneocytes are embedded in lipid lamellar regions, which are orientated parallel to the corneocyte surface. Substances permeate mainly along the tortuous pathway in the intercellular lamellar regions. C = corneocyte filled with keratin. Bar = 100 nm.

the four layers present in the epidermis: stratum basale (or basal layer), stratum spinosum (or spinous layer), stratum granulosum (or granular layer), and stratum corneum (or cornified layer). The main cell type in the viable epidermis is the keratinocyte, which contains keratin filaments and constitutes approximately 90% of the tissue [4]. Other more sparingly distributed cells in the viable epidermis are melanocytes for pigment formation, Merkel cells for sensory reception, and the antigen-presenting Langerhans cells.

The innermost layer of the epidermis, the stratum basale, consists of a single layer of columnar-shaped, undifferentiated stem cells. Mitosis of these cells constantly renews the epidermis and this proliferation compensates for the loss of dead stratum corneum cells (corneocytes) from the skin surface. As the cells produced by the basal layer move upward, they alter morphologically as well as histochemically to form the outermost layer, the stratum corneum. Over a 4- to 5-week period the entire epidermis is renewed [5].

At the spinous layer, the cells appear to be nearly round. They still contain a nucleus and organelles, but contain more keratin filament bundles and are connected by more desmosomes than the basal cells. Desmosomes are specialized structures that are involved in intercellular adhesion between adjacent keratinocytes. They create a transcellular network of keratin filaments and are therefore crucial for tissue integrity [6,7]. From the basal side of the stratum spinosum to the stratum granulosum, the keratinocytes become more flattened and some cell organelles disappear. In the upper spinous regions, two types of intracellular granules are formed: keratohyalin granules and membrane-bound granules. Keratohyalin granules are electron dense, irregularly shaped granules, which are predominantly composed of profilaggrin, loricrin, and keratin [8,9]. Membrane-bound granules, often referred to as lamellar bodies or membrane-coated granules, are round to ovoid, measure about 0.2 μm, and contain flattened lamellar disks. They were first observed by Selby in the late 1950s [10] and later described in detail by others [11–13]. Lipid analysis of isolated lamellar bodies revealed that these organelles are enriched mainly in polar lipids, including glucosylsphingolipids, phospholipids, free sterols, and cholesterol sulfate, which are present as lipid stacks. Furthermore, they contain catabolic enzymes, like acid hydrolases, sphingomyelinase, and phospholipase A_2 [14–16].

The stratum granulosum is the most superficial cell layer of the viable epidermis and contains highly differentiated keratinocytes. The lamellar bodies, which have been formed in the stratum spinosum, migrate to the apical periphery of the uppermost granular cells and eventually fuse with the membrane of the keratinocyte. Via exocytosis their content is extruded into the intercellular spaces at the stratum granulosum–stratum corneum interface. The lipids derived from the lamellar bodies are essential for the formation of the stratum corneum barrier.

11.2.2 The Stratum Corneum

The outermost layer of the skin, the cornified layer or stratum corneum, has been identified as the principal diffusion barrier for substances, including water [2,3]. It is approximately 10 to 20 μm thick when dry but swells to several times this thickness when fully hydrated [17]. It contains 10 to 25 layers lying parallel to the skin surface of nonviable cells, the corneocytes, which are surrounded by a cell envelope and imbedded in a lipid matrix. This architecture is often modeled as a wall-like structure, with the corneocytes as protein bricks embedded in a lipid mortar [18]. Similarly to the viable epidermis, desmosomes (corneodesmosomes) contribute to the cell cohesion.

During the transition of the mature keratinocyte into the corneocyte, profilaggrin that is released from the keratohyalin granules is dephosphorylated and proteolytically processed to filaggrin monomers. Filaggrin is responsible for the formation of extensive disulfide bonds between keratin fibers. This aggregation results in a macrostructure of keratin fibers, which ultimately fill the interior of the corneocytes. Subsequently, filaggrin is degraded into free amino acids and their derivatives, which contribute to the hydration of the stratum corneum (reviewed in Ref. [19]).

The corneocytes are entirely enveloped in a uniform 12 nm thick proteinaceous layered structure. This cornified envelope is formed via a complex, but well-organized process during terminal differentiation. Several precursor proteins, including involucrin, loricrin, and cornifine [20,21], are cross-linked by the action of calcium-dependent transglutaminases, resulting in a very rigid and stable structure. The protein envelope has a lipoidal exterior formed by a monolayer of lipids. These lipids are covalently bound to the cornified envelope proteins, most abundantly to involucrin, and mainly consist of long chain (C30–C34) ω-hydroxy fatty acids, linked to sphingosine and 6-hydroxysphingosine, respectively [22,23]. A number of possible roles for the covalently bound lipids have been hypothesized [24]: (i) they are assumed to play an important role in the organization of the intercellular lipids by acting as a substrate that facilitates the orientation of the lamellae parallel to the corneocyte surface; (ii) they may facilitate the interaction of the hydrophilic interior of corneocytes with the intercellular lipid domain; (iii) the covalently bound lipids may stabilize the stratum corneum structure and the cohesiveness with the intercellular lipids; (iv) they may provide a permeability barrier around each corneocyte to impede diffusion of substances across the envelope.

The cornified envelope has an unusually high resistance to proteolytic enzymes and organic solvents. Lipid extraction causes collapse of the intercellular spaces, but does not result in the disintegration of the cells. In a few studies it is suggested that the covalently bound lipids interdigitate to close the intercellular space and hold together the cell layers [22–24]. Abnormalities in the process of envelope formation strongly influence the structural and functional integrity of the cornified envelope. For instance, in lamellar ichthyosis, a genetic deficiency of transglutaminase 1 activity is observed [25,26]. Transglutaminase 1 is not only involved in the cross-linking process of the precursor proteins, but is the only transglutaminase that is capable of covalently attaching the lipids onto the proteins. Hence, mutations in transglutaminase 1 result in severe defects in the formation of the skin barrier. The amount of protein-bound ω-hydroxyceramides is also significantly reduced in atopic dermatitis [27].

Just prior to the formation of the cornified envelope, the content of the lamellar bodies is discharged into the intercellular space. After extrusion, the polar glucosylceramides are enzymatically converted into ceramides, whereas the phospholipids are catabolized into saturated fatty acids. The stacks of disks rearrange parallel to the corneocytes and join edge-to-edge to form multiple, continuous intercellular lipid sheets or lamellae [24]. These lamellae have been visualized by electron microscopy using ruthenium postfixation and have a unique structure of alternating broad–narrow–broad sequences of electron-lucent bands. The lipids from which the intercellular lamellae are composed are highly unusual. The major lipid classes present are ceramides, cholesterol, and free fatty acids. In addition, minor amounts of cholesterol sulfate are present. The composition and molecular organization of the intercellular lipids will be discussed in more detail in the next section.

11.3 STRATUM CORNEUM BARRIER LIPIDS

11.3.1 Penetration Pathway through the Stratum Corneum

A compound may use two diffusional routes to penetrate normal intact human skin: the transappendageal route and the transepidermal route. The transappendageal route involves transport via the sweat glands or the pilosebaceous units (hair follicles with their associated sebaceous glands). This route circumvents penetration through the stratum corneum and is therefore known as the shunt route. The transappendageal route is considered to be of less importance than the transepidermal route because of its relatively small area, approximately 0.1% of the total skin area [28]. However, recent studies have demonstrated the possibility of specifically targeting certain compounds to the pilosebaceous structures [29,30]. The rate of success largely depends on the lipophilicity of the permeant and the composition of the vehicle. The appendageal route is of further importance during electrically enhanced transport such as iontophoresis [31].

Compounds that penetrate the stratum corneum via the transepidermal route may follow a transcellular (or intracellular) or intercellular pathway (see Figure 11.1). Because of the highly impermeable character of the cornified envelope (see previous section), the tortuous intercellular pathway has been suggested to be the route of preference for most drug molecules [32]. This is confirmed by several microscopic transport studies, in which compounds have been visualized in the intercellular space of the stratum corneum [33–35]. Moreover, it has been demonstrated that drug permeation across stratum corneum increases many folds after lipid extraction [36]. Hence, knowledge of the structure and physical properties of the intercellular lipids is crucial to broaden our insight into the skin barrier function.

11.3.2 Composition of the Intercellular Lipids in the Stratum Corneum

The lipid composition changes dramatically during terminal differentiation. After extrusion from the lamellar bodies, the polar lipid precursors are enzymatically converted into more hydrophobic lipids. As a result, phospholipids are almost absent in the stratum corneum. The lipid lamellae surrounding the corneocytes are predominantly composed of ceramides, cholesterol, and free fatty acids. It is generally assumed that these lipids are present in nearly equimolar ratios. However, inspection of literature data shows that there is a high interindividual variability in the lipid composition [37].

Figure 11.2 illustrates the various ceramide classes present in human and pig stratum corneum. Ceramides are structurally heterogeneous and consist of two long saturated hydrocarbon chains and a small polar head group with several functional groups. Each of the nine ceramides (CER1–CER9) identified in human stratum corneum [38–41] contains a sphingoid base and a fatty acid, which are linked by an amide bond between the carboxyl group of the fatty acid and the amino group of the base. The sphingoid moiety can be sphingosine (S), phytosphingosine (P), or 6-hydroxysphingosine (H), whereas the fatty acid moiety is nonhydroxylated (N) or α-hydroxylated (A) with chain lengths of predominantly 24 to 26 hydrocarbon atoms. The most remarkable ceramides are the acylceramides (CER1, CER4, and CER9). These ceramides consist of an unusually long ω-hydroxy fatty acid of 30 to 34 hydrocarbon atoms (O) to which unsaturated linoleic acid is ester-linked (E). Essential fatty acids, such as linoleate have to be derived from our diet, but may also be recycled within the epidermis [42,43].

The composition of the free fatty acids is also unique. In both human and pig stratum cornea, the free fatty acid fraction consists mainly of long and saturated hydrocarbon chains [44,45]. Oleic and linoleic acid are the only unsaturated free fatty acids detected in the stratum corneum. There are various sterols present in human stratum corneum, of which cholesterol predominates. Cholesterol is the only major lipid class that is present in both plasma membranes and the intercellular lipid lamellae. Cholesterol is synthesized in the epidermis and this synthesis is independent of the hepatic one. A minor fraction is sulfated to

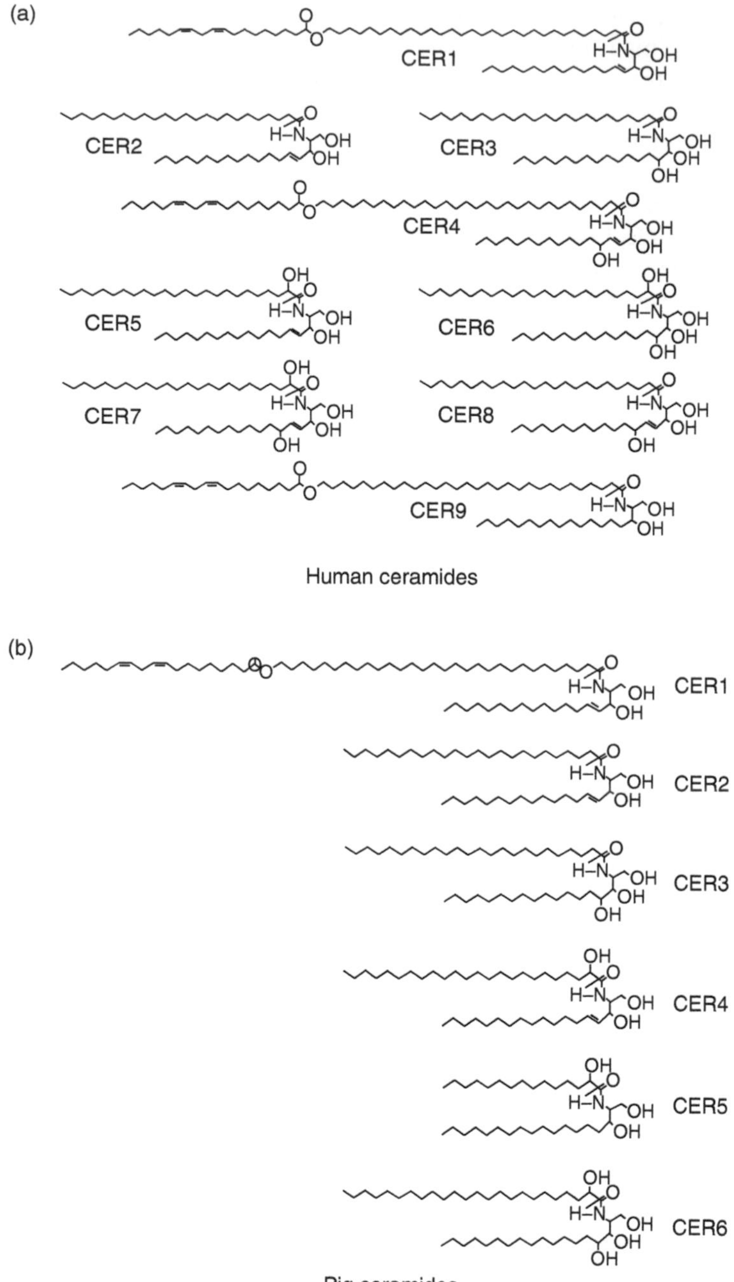

FIGURE 11.2 Molecular structures of the ceramides (CER) present in human stratum corneum (a) and pig stratum corneum (b).

form cholesterol sulfate. Although it is present in only small amounts (typically 2%–5% w/w), cholesterol sulfate is considered to play an important role in the desquamation process of the stratum corneum.

11.4 LIPID ORGANIZATION OF STRATUM CORNEUM LIPIDS

11.4.1 LIPID ORGANIZATION IN STRATUM CORNEUM OF NORMAL SKIN

The intercellular lipid lamellae were first visualized in the early 1970s using freeze fracture electron microscopy [46]. However, they could not be visualized in transmission electron microscopy. It is now well recognized that the saturated nature of the stratum corneum lipids does not permit a chemical reaction with the osmium tetroxide that is routinely used in the preparation of specimens for transmission electron microscopy. However, the use of ruthenium tetroxide as a postfixation agent made it possible to visualize the unique lamellar arrangement of the intercellular lipids [47,48]: multiple lamellae, consisting of a broad–narrow–broad sequence of electron-lucent bands, exist throughout the depth in the stratum corneum. The periodicity of the repeating structural unit could be measured by small angle x-ray diffraction studies on human, pig, and mouse stratum corneum [49–53] Two lamellar phases are present, a 13 nm lamellar phase, indicated as the long periodicity phase, and a second lamellar phase with a periodicity of approximately 6 nm, indicated as the short periodicity phase. In contrast to the short periodicity phase, the long periodicity phase is only identified in the stratum corneum and not in other biological membranes. Due to its characteristic periodicity, it is generally suggested that this phase plays an important role in the skin barrier function.

Besides the lamellar organization, the crystallinity of the intercellular lipids is also of crucial importance for the barrier function of human skin. The packing density decreases in the order orthorhombic > hexagonal > liquid. As a result, the orthorhombic packing is the least permeable structure, whereas the liquid phase is highly permeable. Wide angle x-ray diffraction studies reveal that the lipids in human stratum corneum are predominantly packed in an orthorhombic lattice, although the presence of a coexisting hexagonal packing could not be excluded [54]. In addition, it could not be concluded whether a liquid phase coexists, as its broad reflection in the diffraction pattern was overlapped by the reflections attributed to keratin, which is present in the corneocytes. Electron diffraction and Fourier transformed infrared studies on tape-stripped stratum corneum have confirmed that the bulk of the stratum corneum lipids forms mainly an orthorhombic phase. However, near the skin surface an increased fraction of lipids is in a hexagonal state [55,56]. It is suggested that this transition from an orthorhombic to a hexagonal packing is caused by the interaction of the intercellular lipids with sebum. Sebum is excreted by the sebaceous glands and forms a protective layer covering the skin surface. It consists of neutral, polar lipids that contain primarily triglycerides, short-chain free fatty acids, wax esters, and squalene, as well as smaller amounts of cholesterol and cholesteryl esters [57].

11.4.2 ALTERED LIPID COMPOSITION AND ORGANIZATION IN STRATUM CORNEUM OF DISEASED AND DRY SKIN

There are several genetic skin diseases with known defects in the lipid metabolism. Atopic dermatitis, lamellar ichthyosis, and psoriasis have been the most widely studied with respect to epidermal barrier function and alterations in the lipid profile. Deviations in the lipid profile have been linked with an impaired stratum corneum barrier function. Atopic dermatitis is characterized by inflammatory, dry and easily irritable skin, and overall reduced ceramide levels in the stratum corneum [58–60]. In particular a significant decrease in the ceramide 1 level is observed, whereas the levels of oleate that is esterified to ceramide 1 are elevated [59]. Both aberrations may be responsible for the reduced order of the lamellar phases as observed with freeze fracture electron microscopy [61]. It has further been established that, in comparison to healthy stratum corneum, the fraction of lipids forming a hexagonal packing is increased [61]. A recent study reveals that the level of free fatty acids with more than 24 carbon atoms is remarkably reduced in both lesion and nonlesion parts of atopic skin as compared to healthy skin [27]. Previous x-ray diffraction studies on isolated ceramide mixtures reveal that long-chain fatty acids are required for the formation of the orthorhombic packing. In addition, it was demonstrated that the fraction of lipids that forms a hexagonal packing is increased at reduced ceramide1 levels [62,63]. Both observations may explain the decreased packing density of the lipids in atopic dermatitis.

Lamellar ichthyosis is a rare skin disorder that is characterized by thick plate-like scales, skin dryness, and variable redness. Characteristics of this skin disorder are significantly reduced levels of free fatty acids and an altered overall ceramide profile, without significant changes in the ceramide 1 content [64]. These changes in lipid profile likely explain the altered lamellar organization in lamellar ichthyosis, as observed by x-ray diffraction [64]. Transmission electron microscopy studies using ruthenium tetroxide as a postfixation further showed that in the intercellular space irregularly distributed lipid lamellae are present with areas containing excessive numbers of lamellae [65]. Concerning the lateral lipid organization, it has been demonstrated that the lateral packing is predominantly hexagonal instead of orthorhombic [61]. This latter observation can be associated with reduced levels of free fatty acids.

Psoriasis is a chronic skin disorder, in which an abnormally fast transition of basal cells into corneocytes results in a thickening of the stratum corneum. Transmission electron microscopy studies show an aberrant stratum corneum lipid ultrastructure in psoriatic skin [66], which is expected to be related to abnormalities in the lipid profile. Particularly, a significant reduction in ceramide 1 and a predominance of sphingosine ceramides at the expense of phytosphingosine ceramides are reported in psoriatic stratum corneum [67,68].

In recessive X-linked ichthyosis, the amount of cholesterol sulfate in the stratum corneum is increased due to a deficiency in cholesterol sulfatase deficiency [69,70]. Lipid analysis of scales reveals a nearly 10-fold increase in the cholesterol sulfate to free cholesterol ratio as compared to healthy stratum corneum [71]. Previous x-ray diffraction studies on isolated ceramide mixtures revealed that increased cholesterol sulfate levels induce the formation of a fluid phase, which is likely to reduce the skin barrier function [72].

Abnormalities in the lipid composition and organization have also been established in dry skin. Interestingly, pronounced seasonal changes in the stratum corneum lipid profile have been reported. During the winter months, decreased levels of all major lipid species are observed [73]. In addition, the CER1 linoleate to CER1 oleate ratio dramatically drops from 1.74 in summer to 0.51 in winter. These changes may explain the disorganized lipid lamellae, which are observed in winter xerosis [74]. Similarly to psoriatic skin, dry skin contains reduced levels of phytosphingosine ceramides and increased levels of sphingosine ceramides [75]. One of the suggested pathways for the phytosphingosine biosynthesis involves the addition of water to the corresponding sphingosine double bond. The observed changes in the sphingosine to phytosphingosine ceramide ratio may therefore be caused by disturbed water availability, associated with dry skin [67].

All the above-mentioned changes in lipid composition and organization in diseased and dry skin likely contribute to an impaired stratum corneum barrier function and increased susceptibility to dry skin. However, as previously indicated, abnormalities in the process of envelope formation may also strongly influence the stratum corneum barrier integrity. Therefore, more information is required to elucidate the precise mechanisms by which stratum corneum structure and function are altered.

11.4.3 *In Vitro* Lipid Organization of Lipid Mixtures Prepared with Isolated Ceramides

To further increase our insight into the stratum corneum barrier function, stratum corneum lipid models have been developed. The main advantage of these lipid models is that the composition of the mixtures can be systematically altered, which allows studying the individual role of the various lipid classes in the stratum corneum lipid organization. Furthermore, proteins (keratin) are absent. This considerably facilitates the interpretation of the results. A large number of studies describe the lipid phase behavior of mixtures prepared with ceramides isolated from either pig or human stratum corneum. Hydrated mixtures of cholesterol and purified ceramides isolated from human or pig stratum corneum resemble the characteristic stratum corneum lipid organization. Two lamellar phases with periodicities of 5.2 and 12.2 nm with a hexagonal lateral packing are formed in mixtures of cholesterol and pig ceramides over a wide molar range (0.4 to 2) [62]. Addition of long-chain free fatty acids increases the packing density and induces a hexagonal to orthorhombic transition. In addition, free fatty acids slightly increase the periodicity of the long periodicity phase to 13 nm. Additional studies revealed that the lipid organization in equimolar mixtures of cholesterol and pig ceramides is insensitive to the composition of the ceramides [76]. The only exception is CER1, as the formation of the long periodicity phase is dramatically decreased in the absence of CER1 [63,77,78]. This demonstrates that CER1 is of crucial importance for proper lipid organization in stratum corneum. In an additional study, the role of the fatty acid, which is linked to the ω-hydroxy fatty acid of CER1, on the lipid phase behavior was elucidated [79]. This investigation was performed with equimolar mixtures of cholesterol, free fatty acids, and isolated human ceramides, in which native CER1 was replaced by synthetic CER1 linoleate, CER1 oleate, or CER1 stearate. The results show that the degree of saturation of the fatty acid chain of CER1 highly affects the lamellar and lateral organization. The long periodicity phase is predominantly present in mixtures prepared with CER1 linoleate, to a lesser extent in mixtures prepared with CER1 oleate, and absent in mixtures prepared with CER1 stearate. Concerning the lateral packing, besides the hexagonal and orthorhombic packing, a small fraction of lipids forms a fluid phase. The formation of this fluid phase decreases in the order CER1 oleate > CER1 linoleate > CER1 stearate, indicating that the formation of the long periodicity phase correlates with the presence of a fluid phase and that for the formation of the 13 nm phase a certain optimal amount of lipids should be present in a fluid phase.

Cholesterol sulfate is another intercellular lipid. Addition of low levels of cholesterol sulfate, as observed in normal healthy stratum corneum, to lipid mixtures has little effect on the phase behavior at room temperature. However, addition of high levels of cholesterol sulfate, at levels similar to that observed in the skin disease recessive X-linked ichthyosis, promotes the formation of the long periodicity phase, induces the formation of a fluid phase, and increases the solubility of cholesterol in the lamellar phases [72,80].

Various models have been proposed to describe the molecular organization of intercellular stratum corneum lipids, such as the stacked monolayer model [81], domain mosaic model [82], the single gel phase model [83], and the sandwich model [79,84]. According to the sandwich model (see Figure 11.3) the lipids within the long periodicity phase are organized in a trilayer structure: two broad layers with a crystalline (orthorhombic) structure are separated by a narrow central lipid layer with fluid domains. This broad–narrow–broad pattern of hydrocarbon chains corresponds to the images obtained with electron microscopy of the stratum corneum intercellular lamellae. Cholesterol and the linoleic acid moieties of the acylceramides CER1, CER4, and CER9 are proposed to be located in the central narrow layer, whereas crystalline packed ceramides are present on both sides of this central layer [79]. Due to their unusually long structure, the acylceramides are able to span a layer and extend into another one within the trilayer unit. The acylceramides are therefore thought to contribute to the stability of the 13 nm phase. The central, noncontinuous fluid phase may be of importance for proper elasticity of the lamellae and the enzyme activity in the stratum corneum, as enzymes are unlikely to be active in crystalline phases.

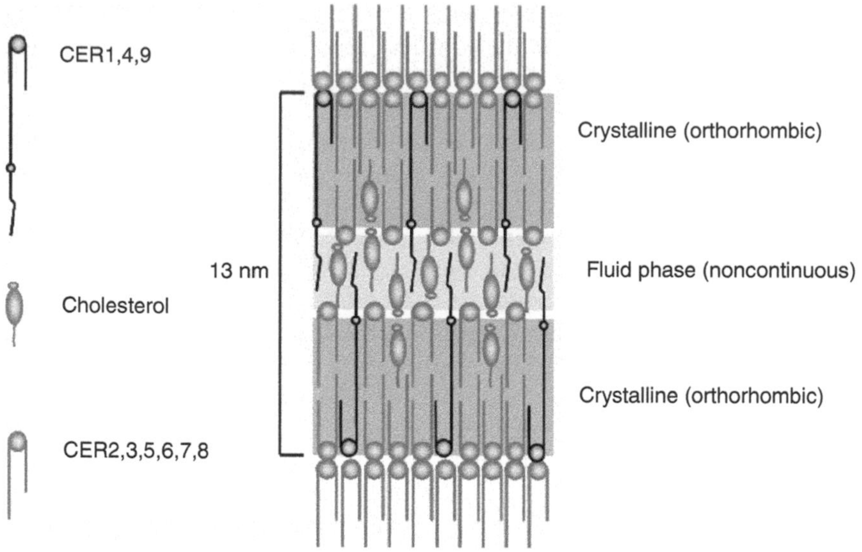

FIGURE 11.3 Lipid organization in the 13 nm lamellar phase according to the sandwich model.

11.4.4 *In Vitro* Lipid Organization of Lipid Mixtures Prepared with Synthetic Ceramides

Various lipid models have been studied that consisted of equimolar mixtures of the commercially available bovine brain ceramide type III (structurally similar to CER2, but with shorter fatty acid chain lengths [85] referred to as ΣCERIII), cholesterol, and palmitic acid. Using NMR, it has been demonstrated that the majority of lipids forms an orthorhombic phase, whereas a small portion forms a more mobile phase [86,87]. FTIR studies on similar stratum corneum lipid models, containing bovine brain ceramide type IV, synthetic CER2, or synthetic CER5, confirm the presence of orthorhombic lattices [88–90]. However, small angle x-ray diffraction studies reveal that mixtures prepared with these (semi)synthetic ceramides do not form the characteristic long periodicity phase [91]. This is likely caused by the fact that the composition of these lipid mixtures does not mimic that of native stratum corneum with respect to the choice of the ceramide mixtures not containing ceramide1 and the chain length of the free fatty acids.

In additional studies four commercially available ceramides were selected, namely CER3 with a chain length of 24 carbon atoms (CER3(C24)), CER3 with a chain length of 16 carbon atoms (CER3(C16)), bovine brain ceramide type III (ΣCERIII), and bovine brain ceramide type IV (ΣCERIV). Synthetic CER1(C30) and ΣCERIII and ΣCERIV contain a sphingosine backbone to which fatty acids with varying chain lengths are linked. The effects of compositional changes, including the addition of synthetic CER1(C30), on the phase behavior of lipid mixtures prepared with these ceramides were systematically examined [92]. It was demonstrated that both the presence of synthetic CER1 as well as a proper composition of the other ceramides in the lipid mixture are crucial for the formation of a phase with a long periodicity. Lipid mixtures prepared with a ceramide comprising CER1 and ΣCERIII or CER1 and ΣCERIV do not form the LPP. Only a mixture containing synthetic CER1, CER3(C24), cholesterol, and free fatty acids revealed the formation of a lamellar phase with a long periodicity. However, the number and intensities of the reflections attributed to this phase were rather low, which indicates that only a small fraction of lipids participates in the formation of this phase. Moreover, the periodicity of 11.6 nm considerably differs from that observed in stratum corneum or lipid mixtures prepared with isolated ceramides. To increase the repeat distance and the fraction of lipids forming the LPP, both the preparation method and the composition of the lipid mixtures were optimized. Firstly, the influence of the equilibration temperature during sample preparation and the fraction of free fatty acids in the lipid mixture on the phase behavior was investigated [93]. All variations in that study were performed with a fixed synthetic ceramide composition, namely CER1, CER3(C24), and ΣCERIV in a 1:7:2 molar ratio (further referred to as synth CER I). ΣCERIV was incorporated into the synthetic ceramide mixture to introduce some variation in acyl chain length and head group architecture. The results demonstrated that an increased equilibration temperature from 60°C to 100°C promotes the formation of the LPP. Furthermore, an increased equilibration temperature enhances also the formation of a phase with a periodicity of 4.3 nm at the expense of a phase with a periodicity of 3.7 nm. Both phases are attributed to phase-separated CER3 in a V-shaped formation and not present in native stratum corneum [94].

The influence of the relative content of free fatty acids on the lipid phase behavior was subsequently investigated by preparing mixtures of cholesterol, CER1, CER3(C24), and ΣCERIV and a varying amount of fatty acids [93]. Up to an equimolar content, systematic increase in the free fatty acid content promotes the formation of both LPP (12.2 nm) and SPP (5.5 nm). However, when the relative free fatty acids content was further raised, the formation of the LPP was reduced and the SPP dominated. The main difference between synthetic and isolated ceramide mixtures or stratum corneum is the presence of either one or two additional phases with repeat distances of approximately 4.3 and 3.7 nm, also observed in the previous study [93]. In a follow-up study [95], the main focus was to establish the optimal CER3 to ΣCERIV molar ratio to minimize the formation of these additional phases and maximize the fraction of lipids forming the LPP. All lipid mixtures in that study were prepared with a fixed CER1 level (10%), whereas the molar ratio between CER3 (uniform chain length) and ΣCERIV (chain length variation) ranged from 1:8 to 8:1. For the formation of the LPP in the absence of free fatty acids, an optimal CER3 to ΣCERIV ratio of 5:4 was found in the absence of free fatty acids, whereas the optimal CER3 to ΣCERIV ratio was 7:2 in the presence of free fatty acids. A further reduction in the CER3 content in favor of the ΣCERIV level reduced the formation of the additional phases. However, it also considerably inhibited the formation of the LPP. This shift toward an increased CER3 content is most likely due to the chain length variation of the free fatty acids (ranging from C16 to C26).

The above results demonstrate that the unique stratum corneum lipid organization can be reproduced *in vitro* with mixtures based on cholesterol, free fatty acids, and a limited number of synthetic ceramides. The results further reveal that the formation of the LPP is rather insensitive toward changes in the total composition of cholesterol, CER, and free fatty acids over a wide range of molar ratios. This is in excellent agreement with the *in vivo* situation, in which a high interindividual variability in stratum corneum lipid composition usually does not lead to substantial changes in the lipid organization.

To assess the effect of acylceramide type and relative content on the lipid phase behavior, synthetic ceramide mixtures were prepared with various amounts of sphingosine-based CER1, phytosphingosine-based CER9, or combinations of CER1 and CER9, thereby maintaining a constant CER3 to ΣCERIV molar ratio [96]. The difference between CER1 and CER9 is the presence of an additional hydroxyl group at the sphingosine base in the latter. The results reveal that CER9 less efficiently enhances the formation of the LPP than CER1.

In contrast to the previous investigation, in which solely mixtures prepared with a limited number of synthetic ceramides were studied, in a very recent study [97] the lipid organization in mixtures prepared with various synthetic ceramides was investigated. The initial composition of this ceramide mixture is 15% CER1(C30), 51% CER2(C24), 16% CER3(C24), 4% CER4(C24), 9% CER3(C16), and 5% CER6(C24), which resembles the composition of the ceramides in pig stratum corneum. Subsequently, the CER mixture was gradually modified to elucidate the role various synthetic ceramides play in the formation of the LPP. As the synthetic counterpart of natural CER5 is not available, the ceramide mixtures were prepared with CER3(C16).

The lipid organization in equimolar mixtures of cholesterol, synthetic CER, and free fatty acids closely resembles that in stratum corneum, as both LPP (12.2 nm) and SPP (5.4 nm) are present, the lateral packing of the lipids is orthorhombic, a minor fraction of cholesterol phase separates into crystalline domains, and no additional phases can be detected. Interestingly, free fatty acids are required for proper lipid organization, as only in their presence a dominant formation of the LPP could be detected. This might be related to the limited acyl chain length distribution present in these CER mixtures.

The results described above unequivocally demonstrate that lipid mixtures prepared with CER1, CER2, CER3, CER4, and CER6 offer an attractive tool to unravel the importance of individual ceramides for proper stratum corneum lipid organization. The results show that the presence of CER1, a proper choice of the equilibration temperature during the preparation of the mixtures, and a variation in chain length distribution are crucial for the formation of the LPP. These mixtures most closely mimic the lipid composition and organization of the intercellular lipids in human stratum corneum, as no additional crystalline phases are present.

11.5 CONCLUSION

The lipid organization in stratum corneum is very unusual. Two lamellar phases are present with periodicities of approximately 6 and 13 nm. This lipid organization can be reproduced *in vitro* with mixtures prepared from either isolated ceramides (from pig and human stratum corneum) or synthetic ceramides. However, for the formation of the long periodicity phase the presence of acyl ceramides in the lipid mixtures is very crucial. The phase behavior of the lipid mixtures provides important information to explain the deviation in phase behavior in diseased skin.

REFERENCES

1. Schaefer, H., and T.E. Redelmeier. 1996. *Skin barrier. Principles of percutaneous absorption*. Basel: Karger.
2. Blank, I.H. 1969. Transport across the stratum corneum. *Toxicol Appl Pharmacol* (Suppl. 3):23.
3. Scheuplein, R.J., and I.H. Blank. 1971. Permeability of the skin. *Physiol Rev* 51:702.
4. Eckert, R.L. 1989. Structure, function, and differentiation of the keratinocyte. *Physiol Rev* 69:1316.
5. Baker, H., and A.M. Kligman. 1967. Technique for estimating turnover time of human stratum corneum. *Arch Dermatol* 95:408.
6. Burdett, I.D. 1998. Aspects of the structure and assembly of desmosomes. *Micron* 29:309.
7. Fuchs, E., and S. Raghavan. 2002. Getting under the skin of epidermal morphogenesis. *Nat Rev Genet* 3:199.
8. Harding, C.R., and I.R. Scott. 1983. Histidine-rich proteins (filaggrins): Structural and functional heterogeneity during epidermal differentiation. *J Mol Biol* 170:651.
9. Steven, A.C., et al. 1990. Biosynthetic pathways of filaggrin and loricrin—two major proteins expressed by terminally differentiated epidermal keratinocytes. *J Struct Biol* 104:150.
10. Selby, C.C. 1957. An electron microscope study of thin sections of human skin. II. Superficial layers of footpad epidermis. *J Invest Dermatol* 29:131.
11. Odland, G.F.A. 1960. Submicroscopic granular component in human epidermis. *J Invest Dermatol* 34:11.

12. Elias, P.M., N.S. McNutt, and D.S. Friend. 1977. Membrane alterations during cornification of mammalian squamous epithelia: A freeze-fracture, tracer and thin-section study. *Anat Rec* 189:577.
13. Landmann, L. 1980. Lamellar granules in mammalian, avian, and reptilian epidermis. *J Ultrastruct Res* 72:245.
14. Freinkel, R.K., and T.N. Traczyk. 1985. Lipid composition and acid hydrolase content of lamellar granules of fetal rat epidermis. *J Invest Dermatol* 85:295.
15. Wertz, P.W., et al. 1984. Sphingolipids of the stratum corneum and lamellar granules of fetal rat epidermis. *J Invest Dermatol* 83:193.
16. Wertz, P.W. 1992. Epidermal lipids. *Semin Dermatol* 11:106.
17. Bouwstra, J. A., et al. 2003. Water distribution and related morphology in human stratum corneum at different hydration levels. *J Invest Dermatol* 120:750.
18. Elias, P.M. 1983. Epidermal lipids, barrier function, and desquamation. *J Invest Dermatol* 80:44s.
19. Harding, C.R., A. Watkinson, and A.V. Rawlings. 2000. Dry skin, moisturization and corneodesmolysis. *Int J Cosmet Sci* 22:21.
20. Harding, C.R., et al. 2003. The cornified cell envelope: An important marker of stratum corneum maturation in healthy and dry skin. *Int J Cosmet Sci* 25:157.
21. Hirao, T. 2003. Involvement of transglutaminase in *ex vivo* maturation of cornified envelopes in the stratum corneum. *Int J Cosmet Sci* 25:245.
22. Swartzendruber, D.C., et al. 1987. Evidence that the corneocyte has a chemically bound lipid envelope. *J Invest Dermatol* 88:709.
23. Wertz, P.W., K.C. Madison, and D.T. Downing. 1989. Covalently bound lipids of human stratum corneum. *J Invest Dermatol* 92:109.
24. Wertz, P.W., and D.T. Downing. 1989. Stratum corneum: Biological and biochemical considerations. In *Transdermal drug delivery: Developmental issues and research initiatives*, eds. J. Hadgraft, and R.H. Guy, 1. New York: Marcel Dekker
25. Huber, M., et al. 1995. Mutations of keratinocyte transglutaminase in lamellar ichthyosis. *Science* 267:525.
26. Hohl, D., M. Huber, and E. Frenk. 1993. Analysis of the cornified cell envelope in lamellar ichthyosis. *Arch Dermatol* 129:618.
27. Macheleidt, O., H.W. Kaiser, and K. Sandhoff. 2002. Deficiency of epidermal protein-bound omega-hydroxyceramides in atopic dermatitis. *J Invest Dermatol* 119:166.
28. Barry, B.W. 1983. Structure, function, diseases, and topical treatment of human skin. In *Dermatological formulations percutaneous absorption*, 1. New York: Marcel Dekker.
29. Rolland, A., et al. 1993. Site-specific drug delivery to pilosebaceous structures using polymeric microspheres. *Pharm Res* 10:1738.
30. Grams, Y.Y., et al. 2003. Permeant lipophilicity and vehicle composition influence accumulation of dyes in hair follicles of human skin. *Eur J Pharm Sci* 18:329.
31. Cullander, C., and R.H. Guy. 1992. Visualization of iontophoretic pathways with confocal microscopy and the vibrating probe electrode. *Solid State Ionics* 53–56:197.
32. Williams, M.L., and P.M. Elias. 1987. The extracellular matrix of stratum corneum: Role of lipids in normal and pathological function. *Crit Rev Ther Drug Carrier Syst* 3:95.
33. Boddé, H.E., et al. 1991. Visualization of *in vitro* percutaneous penetration of mercuric chloride transport through intercellular space versus cellular uptake through desmosomes. *J Control Release* 15:227.
34. Johnson, M.E., D. Blankschtein, and R. Langer. 1997. Evaluation of solute permeation through the stratum corneum: Lateral bilayer diffusion as the primary transport mechanism. *J Pharm Sci* 86:1162.
35. Meeuwissen, M.E.M.J., et al. 1998. A cross-section device to improve visualization of fluorescent probe penetration into the skin by confocal laser scanning microscopy. *Pharm Res* 15:352.
36. Rastogi, S.M., and J. Singh. 2001. Lipid extraction and transport of hydrophilic solutes through porcine epidermis. *Int J Pharm* 225:75.
37. Weerheim, A., and M. Ponec. 2001. Determination of stratum corneum lipid profile by tape stripping in combination with high-performance thin-layer chromatography. *Arch Dermatol Res* 293:191.

38. Wertz, P.W., et al. 1985. The composition of the ceramides from human stratum corneum and from comedones. *J Invest Dermatol* 84:410.
39. Robson, K.J. et al. 1994. 6-Hydroxy-4-sphingenine in human epidermal ceramides. *J Lipid Res* 35:2060.
40. Stewart, M.E., and D.T. Downing. 1999. A new 6-hydroxy-4-sphingenine-containing ceramide in human skin. *J Lipid Res* 4:1434.
41. Ponec, M., et al. 2003. New acylceramide in native and reconstructed epidermis. *J Invest Dermatol* 120:581.
42. Madison, K.C., et al. 1989. Murine keratinocyte cultures grown at the air/medium interface synthesize stratum corneum lipids and "recycle" linoleate during differentiation. *J Invest Dermatol* 93:10.
43. Wertz, P.W., and D.T. Downing. 1990. Metabolism of linoleic acid in porcine epidermis. *J Lipid Res* 31:1839.
44. Wertz, P.W., and D.T. Downing. 1991. In *Physiology, biochemistry and molecular biology of the skin*, ed. Goldsmith L.A., 2nd ed., 205. Oxford: Oxford University Press.
45. Ponec, M., et al. 2001. Barrier function in reconstructed epidermis and its resemblance to native human skin. *Skin Pharmacol Appl Skin Physiol* 14 (Suppl. 1):63.
46. Breathnach, A.S., et al. 1973. Freeze fracture replication of cells of stratum corneum of human epidermis. *J Anat* 114:65.
47. Madison, K.C., et al. Presence of intact intercellular lipid lamellae in the upper layers of the stratum corneum. *J Invest Dermatol* 88:714.
48. Hou, S.Y., et al. 1991. Membrane structures in normal and essential fatty acid-deficient stratum corneum: Characterization by ruthenium tetroxide staining and x-ray diffraction. *J Invest Dermatol* 96:215.
49. White, S.H., D. Mirejovsky, and G.I. King. 1988. Structure of lamellar lipid domains and corneocyte envelopes in murine stratum corneum: An X-ray diffraction study. *Biochemistry* 27:3725.
50. Bouwstra, J.A., et al. 1991. The structure of human stratum corneum as determined by small angle X-ray scattering. *J Invest Dermatol* 96:1006.
51. Bouwstra, J.A., et al. 1991. Structural investigations of human stratum corneum by small angle X-ray scattering. *J Invest Dermatol* 97:1005.
52. Bouwstra, J.A., et al. 1994. The lipid and protein structure of mouse stratum corneum: A wide and small angle diffraction study. *Biochim Biophys Acta* 1212:183.
53. Bouwstra, J.A., et al. 1995. Lipid organization in pig stratum corneum. *J Lipid Res* 36:685.
54. Bouwstra, J.A., et al. 1992. Structure of human stratum corneum as a function of temperature and hydration: A wide-angle X-ray diffraction study. *Int J Pharm* 84:205.
55. Pilgram, G.S.K., et al. 1999. Electron diffraction provides new information on human stratum corneum lipid organization studied in relation to depth and temperature. *J Invest Dermatol* 113:403.
56. Bommannan, D., R.O. Potts, and R.H. Guy. 1990. Examination of stratum corneum barrier function *in vivo* by infrared spectroscopy. *J Invest Dermatol* 95:403.
57. Stewart, M.E., and D.T. Downing. 1991. Chemistry and function of mammalian sebaceous lipids. *Adv Lipid Res* 24:263.
58. Imokawa, G., et al. 1991. Decreased level of ceramides in stratum corneum of atopic dermatitis: An etiologic factor in atopic dry skin? *J Invest Dermatol* 96:523.
59. Yamamoto, A., et al. 1991. Stratum corneum lipid abnormalities in atopic dermatitis. *Arch Dermatol Res* 283:219.
60. Di Nardo, A., et al. 1998. Ceramide and cholesterol composition of the skin of patients with atopic dermatitis. *Acta Derm Venereol* 78:27.
61. Pilgram, G.S.K., et al. 2001. Aberrant lipid organization in stratum corneum of patients with atopic dermatitis and lamellar ichthyosis. *J Invest Dermatol* 117:710.
62. Bouwstra, J.A., et al. 1996. Phase behavior of isolated skin lipids. *J Lipid Res* 37:999.
63. Bouwstra, J.A., et al. 1998. Role of ceramide 1 in the molecular organization of the stratum corneum lipids. *J Lipid Res* 39:186.
64. Lavrijsen, A.P.M., et al. 1995. Reduced skin barrier function parallels abnormal stratum corneum lipid organization in patients with lamellar ichthyosis. *J Invest Dermatol* 105:619.

65. Fartasch, M. 1997. Epidermal barrier in disorders of the skin. *Microsc Res Tech* 38:361.
66. Ghadially, R., J.T. Reed, and P.M. Elias. 1996. Stratum corneum structure and function correlates with phenotype in psoriasis. *J Invest Dermatol* 107:558.
67. Motta, S., et al. 1993. Ceramide composition of the psoriatic scale. *Biochim Biophys Acta* 1182:147.
68. Motta, S., et al. 1994. Abnormality of water barrier function in psoriasis. *Arch Dermatol* 130:452.
69. Zettersten, E., et al. Recessive x-linked ichthyosis: Role of cholesterol-sulfate accumulation in the barrier abnormality. *J Invest Dermatol* 111:784.
70. Rehfeld, S.J., et al. 1988. Calorimetric and electron spin resonance examination of lipid phase transitions in human stratum corneum: Molecular basis for normal cohesion and abnormal desquamation in recessive X-linked ichthyosis. *J Invest Dermatol* 91:499.
71. Williams, M.L., and P.M. Elias. 1981. Stratum corneum lipids in disorders of cornification: Increased cholesterol sulfate content of stratum corneum in recessive x-linked ichthyosis. *J Clin Invest* 68:1404.
72. Bouwstra, J.A., et al. 1999. Cholesterol sulfate and calcium affect stratum corneum lipid organization over a wide temperature range. *J Lipid Res* 40:2303.
73. Rogers, J., et al. 1996. Stratum corneum lipids: The effect of ageing and the seasons. *Arch Dermatol Res* 288:765.
74. Rawlings, A.V., et al. 1994. Abnormalities in stratum corneum structure, lipid composition, and desmosome degradation in soap-induced winter xerosis. *J Soc Cosmet Chem* 45:203.
75. Fulmer, A.W., and G.J. Kramer. 1986. Stratum corneum lipid abnormalities in surfactant-induced dry scaly skin. *J Invest Dermatol* 86:598.
76. Bouwstra, J.A., et al. 1999. The role of ceramide composition in the lipid organisation of the skin barrier. *Biochim Biophys Acta* 1419:127.
77. McIntosh, T.J., M.E. Stewart, and D.T. Downing. 1996. X-ray diffraction analysis of isolated skin lipids: Reconstitution of intercellular lipid domains. *Biochemistry* 35:3649.
78. Bouwstra, J.A., et al. 2001. Phase behavior of lipid mixtures based on human ceramides: Coexistence of crystalline and liquid phases. *J Lipid Res* 42:1759.
79. Bouwstra, J.A., et al. 2002. Phase behavior of stratum corneum lipid mixtures based on human ceramides: The role of natural and synthetic ceramide 1. *J Invest Dermatol* 118:606.
80. Bouwstra, J.A., et al. 1998. pH, cholesterol sulfate, and fatty acids affect the stratum corneum lipid organization. *J Invest Dermatol Symp Proc* 3:69.
81. Swartzendruber, D.C., et al. 1989. Molecular models of the intercellular lipid lamellae in mammalian stratum corneum. *J Invest Dermatol* 92:251.
82. Forslind, B. 1994. A domain mosaic model of the skin barrier. *Acta Derm Venereol* 74:1.
83. Norlen, L. 2001. Skin barrier structure and function: The single gel phase model. *J Invest Dermatol* 117:830.
84. Bouwstra, J.A., et al. 2000. The lipid organisation in the skin barrier. *Acta Derm Venereol Suppl (Stockh)* 208:23.
85. ten Grotenhuis, E., et al. 1996. Phase-behavior of stratum corneum lipids in mixed langmuir-blodgett monolayers. *Biophys J* 71:1389.
86. Fenske, D.B., et al. 1994. Models of stratum corneum intercellular membranes: ^{2}H NMR of microscopically oriented multilayers. *Biophys J* 67:1562.
87. Kitson, N., et al. 1994. A model membrane approach to the epidermal permeability barrier. *Biochemistry* 33:6707.
88. Moore, D.J., M.E. Rerek, and R. Mendelsohn. 1997. Lipid domains and orthorhombic phases in model stratum corneum: Evidence from fourier transform infrared spectroscopy studies. *Biochem Biophys Res Commun* 231:797.
89. Lafleur, M. 1998. Phase behaviour of model stratum corneum lipid mixtures: An infrared spectroscopy investigation. *Can J Chem* 76:1500.
90. Moore, D.J., and M.E. Rerek, 2000. Insights into the molecular organisation of lipids in the skin barrier from infrared spectroscopy studies of stratum corneum lipid models. *Acta Derm Venereol Suppl (Stockh.)* 208:16.
91. Bouwstra, J.A., et al. 1997. A model membrane approach to the epidermal permeability barrier: An X-ray diffraction study. *Biochemistry* 36:7717.

92. de Jager, M.W., et al. 2003. The phase behaviour of skin lipid mixtures based on synthetic ceramides. *Chem Phys Lipids* 124:123.
93. de Jager, M.W., et al. 2004. Novel lipid mixtures based on synthetic ceramides reproduce the unique stratum corneum lipid organization. *J Lipid Res* 45:923.
94. Abrahamsson, S., et al. 1978. Lateral packing of hydrocarbon chains. *Prog Chem Fats Other Lipids* 16:125.
95. de Jager, M.W., et al. 2004. Modelling the stratum corneum lipid organisation with synthetic lipid mixtures: The importance of synthetic ceramide composition. *Biochim Biophys Acta* 30:132.
96. de Jager, M.W., et al. Acylceramide head group architecture affects lipid organization in synthetic ceramide mixtures. *J Invest Dermatol* 123:911.
97. de Jager, M.W., et al. 2005. Lipid mixtures prepared with well-defined synthetic ceramides closely mimic the unique stratum corneum lipid phase behavior. *J Lipid Res* 46:2649.

12 Chemical Permeation Enhancement

Adrian C. Williams and Brian W. Barry

CONTENTS

12.1 INTRODUCTION

Among the myriad strategies employed to increase both the amount of a therapeutic agent traversing the skin and the range of drugs that can be effectively delivered through this route, lies in the application of chemical penetration enhancers. These agents interact with stratum corneum constituents to promote drug flux. Such materials have been used empirically in topical and transdermal preparations for as long as pastes, poultices, creams, and ointments have been applied to skin, though it is only over the last four decades that enhancers have been employed deliberately for this specific purpose. To date, nearly 400 chemicals have been evaluated as penetration enhancers (accelerants, absorption promoters), yet their inclusion into topical or transdermal formulations is limited because the underlying mechanisms of action of these agents are seldom clearly defined and regulatory approval is costly and difficult. Here, we review some applications of the more widely investigated chemical penetration enhancers and consider some of the complex mechanisms by which they may exert their activities.

12.2 BACKGROUND

Penetration enhancers promote drug flux through diverse body membranes, from gastric epithelia to nasal membranes, as discussed in other chapters of this book. Many chemicals have been evaluated for increasing permeant delivery through the stratum corneum but to be of clinical value the penetration enhancer must exert its effects without injuring underlying viable skin cells. In addition, accelerants should act reversibly, i.e., stratum corneum barrier properties should reduce only temporarily. Desirable properties for such an enhancer acting on human skin include the following [1]:

1. It should be pharmacologically inert within the body, either locally or systemically.
2. It should not irritate or induce allergic responses.
3. The enhancer should work rapidly with a predictable onset of action.
4. The operation of enhancement (both in terms of activity and duration of effect) should be predictable and reproducible.
5. When the enhancer leaves the skin, the barrier resistance of the membrane should return rapidly and fully.
6. The penetration enhancer should work unidirectionally, i.e., should allow medicaments to enter the body while preventing the release of endogenous materials.
7. The accelerant should be suitable for formulation into topical and transdermal preparations, being compatible with drugs and excipients and promoting appropriate drug solubility in the formulation.
8. It should be cosmetically acceptable, being odorless, colorless, and with appropriate skin feel.

Many topical and transdermal formulations contain chemicals with promoter activity, such as some surfactants and solvents (e.g., ethanol), even if their primary purpose for inclusion was not originally for their enhancing activity. In research laboratories, an extensive array of chemicals has been evaluated for accelerant activity toward numerous drugs across a plethora of membranes, human, animal, and artificial. Thus, it is not practicable to review in depth all the agents used as putative enhancers of transdermal permeants. This chapter attempts, therefore, to draw out some general trends and uses of different classes of enhancers, and to discuss potential modes of action of these agents in promoting drug flux. Furthermore, accelerants have also been combined with several other strategies for promoting drug delivery through skin, such as iontophoresis. For example, fatty acids have been used synergistically to enhance hormone delivery by iontophoresis. Iontophoresis has also been employed to increase amounts of promoters in the skin. Likewise, chemicals have been employed with electroporation for two principal reasons:

1. To obtain an additional enhancement effect alongside the creation of pores in the stratum corneum bilayers
2. To stabilize and possibly expand the transient pores that the high-voltage pulses create

The literature describing such synergy between electrical enhancement techniques with chemical penetration enhancers was reviewed recently [2] and is beyond the scope of this chapter.

12.3 MAJOR CLASSES OF CHEMICAL PENETRATION ENHANCERS

Accelerants operate in complex, interacting ways to change the intercellular region of the horny layer by fluidization, alteration of polarity, phase separation, or lipid extraction. More drastically, they may form vacuoles within corneocytes, denature their keratin or split squames (Figure 12.1). The following selection of enhancers illustrates examples of all these modes of action.

12.3.1 Water

It has been long known that water often increases transdermal and topical drug deliveries. Typically the water content of human stratum corneum is around 15%–20% of the tissue dry weight, although this varies depending on the temperature and humidity of the external environment. Soaking the skin in water, exposing the membrane to high humidities or, as is more usual under clinical conditions, occluding the tissue so preventing transepidermal water loss, allows the stratum corneum water content to equilibrate with that of the underlying epidermal skin cells. Thus, horny layer water concentration can approach 400% of the tissue dry weight. Many clinically effective preparations such as ointments and patches are occlusive, so illustrating one mechanism of enhanced drug permeation; patch formulations deliver drugs at relatively high rates due at least in part to the raised stratum corneum hydration.

In general, increased tissue wetness promotes transdermal delivery of both hydrophilic and lipophilic permeants. However, Bucks and Maibach [3] cautioned against too wide a generalization, stating that occlusion does not necessarily increase percutaneous absorption and may not always enhance transdermal delivery of hydrophilic compounds. Further, they warned that occlusion could irritate skin with clear implications for the design and clinical application of transdermal and topical preparations.

Considering the heterogeneous nature of human stratum corneum it is not surprising that water within this membrane forms several states. Typically, thermal analysis and spectroscopic methodologies assess that 25%–35% of the water present in stratum corneum is bound, i.e., associated with some structural elements within the tissue [4]. The remaining water is free and is available within the membrane to receive polar permeants. Human skin also contains a hygroscopic humectant mixture of amino acids, amino acid derivatives and salts, termed the natural moisturizing factor (NMF). This material retains water within the stratum corneum and helps to maintain tissue pliability. Furthermore, the keratin-filled corneocytes contain functional groups such as –OH and –COOH that are also expected to bind water molecules. Considering such disparate potential water-binding sites, the absorption (and desorption) of water to and from stratum corneum is complex. However, it is notable that even maintaining a stratum corneum membrane over a strong desiccant such as phosphorous pentoxide will not remove all the water from the tissue—there remains a strongly bound fraction of 5%–10% water.

Despite extensive research in the area, the mechanisms by which water increases transdermal drug delivery are unclear. Free water within the tissue could alter the solubility of a permeant within the stratum corneum and hence could modify partitioning from the vehicle into the membrane. Such a mechanism could partially explain elevated hydrophilic drug fluxes under occlusive conditions but would fail to account for hydration-enhanced delivery for lipophilic permeants such as steroids. As the principle route for transdermal drug delivery usually resides in the horny layer lipids, it might be expected that a high water content, generated by occlusion or soaking, would swell and hence disrupt these domains, possibly by expanding the polar head group regions of the bilayers. However, Bouwstra and coworkers, using x-ray diffractometry, showed that water does not modify the lipid bilayer packing in this way [5]. Such findings raise the question "Where does the water go?" Clearly the corneocytes take up water and engorge. One may expect that such swelling of cells would impact upon the lipid structure between the corneocytes, squeezing and disrupting the bilayer packing. Again, some experimental evidence contradicts this view. Freeze fracture electron

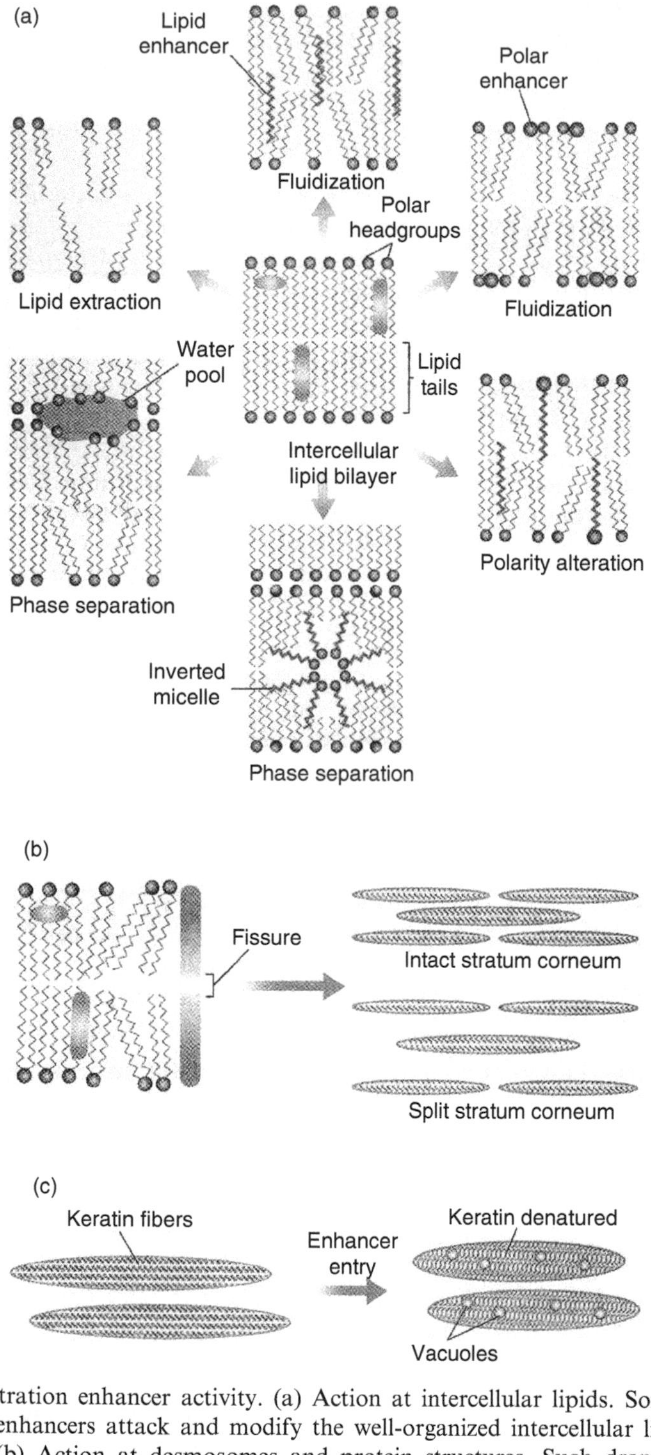

FIGURE 12.1 Penetration enhancer activity. (a) Action at intercellular lipids. Some of the ways by which penetration enhancers attack and modify the well-organized intercellular lipid domain of the stratum corneum. (b) Action at desmosomes and protein structures. Such dramatic disruption by enhancers (particularly potent solvents) as they split the stratum corneum into additional squames and individual cells would be clinically unacceptable. (c) Action within corneocytes. Swelling, further keratin denaturation and vacuolation within individual horny layer cells would not be so drastic but would usually be cosmetically challenging (see Menon and Lee [69] for further details). (Reproduced from Barry, B.W., *Nat. Biotechnol.* 22, 165, 2004. With permission.)

microscopy of fully hydrated stratum corneum shows that the intercellular lipid bilayers contain water pools with occasional vesicle-like structures, but no gross distortion to the lipid regions [6].

Other workers [7] propose that an aqueous pore pathway forms at sites of corneodesmosome degradation (lacunar domains) embedded within the lipid bilayers. They suggest that under high stress conditions (such as extensive hydration, iontophoresis, or ultrasound) the scattered, discontinuous lacunae expand, interconnect, and develop a continuous pore pathway. Drugs would diffuse easily down this route.

When investigating the effects of water on transdermal permeation, animal skin may yield results markedly different to human data. For example, hairless mouse skin is unsuitable for modeling human stratum corneum regarding hydration effects; the murine skin, when hydrated for 24 h, became grossly more permeable than human skin membranes [8]. Thus water effects on skin permeability obtained using animal models need cautious assessment.

12.3.2 Sulfoxides and Similar Chemicals

Dimethylsulfoxide (DMSO), the archetypal penetration enhancer, is a powerful aprotic solvent that is colorless, odorless, and hygroscopic; its value as an enhancer may be predicted from its use chemically as a universal solvent (Figure 12.2).

Extensive investigations on the accelerant activities of DMSO show it to be effective in promoting the flux of both lipophilic and hydrophilic permeants, e.g., antiviral agents,

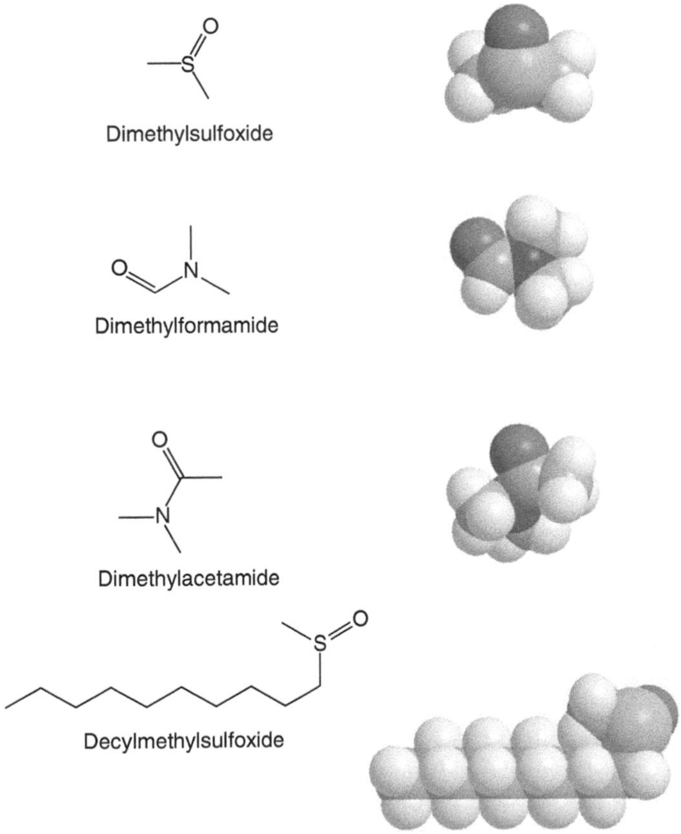

FIGURE 12.2 Aprotic solvents and the derivative, decylmethylsulfoxide.

steroids, and antibiotics. DMSO works rapidly but its effects are markedly concentration dependent and generally cosolvents containing more than 60% DMSO are needed for acceptable enhancement. However, at such concentrations DMSO can incite erythema, raise wheals, and may denature proteins. Thus painting volunteers with 90% DMSO twice daily for 3 weeks provoked erythema, scaling, contact uticaria, stinging, and burning sensations; several subjects developed systemic symptoms [9]. DMSO also produces the metabolite, dimelthylsulfide, raising a foul odor on the breath. When considering DMSO activity it is necessary to note that animal membranes (especially those from rodents) tend to be more fragile than human skin membranes. Thus, the actions of this powerful aprotic solvent on animal tissue may be dramatically greater than the effects seen on a human skin membrane. Such differences between animal and human skin responses are common to many accelerants (see Section 12.3.1).

Once such clinical problems became apparent with the originally much vaunted DMSO, researchers investigated similar, chemically related materials as accelerants. Dimethylacetamide (DMAC) and dimethylformamide (DMF) are also powerful aprotic solvents that illustrate a broad range of penetration-enhancing activities, for example, promoting the flux of hydrocortisone, lidocaine, and naloxone through skin membranes. However, Southwell and Barry [10], although showing a 12-fold increase in caffeine flux across DMF-treated human skin, concluded that the accelerant irreversibly damaged the membrane. DMF has been used *in vivo* when it increased the bioavailability of betamethasone-17-benzoate as judged by the vasoconstrictor assay [11,12]. Further structural analogs have been prepared including alkylmethylsulfoxides such as decylmethylsulfoxide (DCMS). The concept of its synthesis was to try to combine the accelerant activity of DMSO with that of a surfactant alkyl chain. This compound acts reversibly on human skin and most literature shows it to be a potent enhancer for hydrophilic permeants but less effective with lipophilic medicaments.

The mechanisms of action of the sulfoxide enhancers are complex. DMSO denatures proteins and on application to human skin alters the intercellular keratin confirmation, changing it from a α-helical to a β-sheet [13,14]. DMSO also interacts with the intercellular lipid domains of human stratum corneum. Considering its small highly polar nature it is feasible that DMSO interacts with the head groups of some bilayer lipids to distort their packing geometry, as well as dissolving in, and extracting, lipids. Further, DMSO dissolved within skin membranes may alter the polarity and facilitate drug partitioning from a formulation into this universal solvent within the tissue (see Figure 12.1).

12.3.3 Azone

Azone (1-dodecylazacycloheptan-2-one or laurocapram), the most famous modern enhancer, was the first accelerant specifically designed as such (Figure 12.3). It is a hybrid of a cyclic amide, as with pyrrolidone structures (see below), with an alkylsulfoxide; the sulfoxide group presumed to provide some of the disadvantages of DMSO is absent. Azone is a colorless, odorless liquid possessing a smooth, oily but yet nongreasy feel. It is highly lipophilic with a log $P_{\text{octanol/water}}$ around 6.2 and is soluble in, and compatible with, most organic solvents, including propylene glycol (PG) and alcohols. Laurocapram has low irritancy, very low toxicity (oral LD_{50} in rat of 9 g/kg), and little pharmacological activity although there is some evidence for an antiviral effect. Thus, Azone appears to possess many of the desirable qualities for a penetration enhancer as listed above.

Azone increases the permeation of many drugs such as steroids, antibiotics, and antiviral agents. Reports describe its activity in promoting the flux of both hydrophilic and lipophilic medicaments. The efficacy of Azone appears to be strongly concentration dependent and is also influenced by the choice of vehicle in which it is applied. Unlike the aprotic solvents

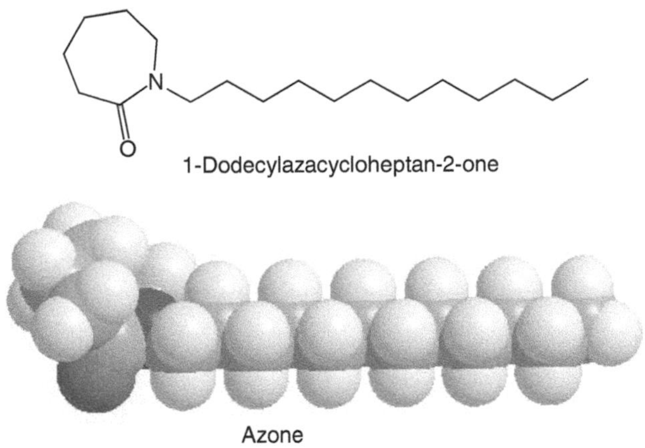

FIGURE 12.3 Azone, the original molecule synthesized specifically to act as a skin penetration enhancer.

DMSO, DMF, and DMAC, Azone is most effective at low concentrations, typically between 0.1% and 5%, often between 1% and 3%. Although Azone has been used for nearly 30 years, workers continue to investigate its mechanism of action. It presumably operates through interactions with the lipid domains of the stratum corneum. Considering the molecule's structure (a large polar head group linked to a lipid alkyl chain) the enhancer would be expected to partition into the bilayer lipids and thus disrupt their organized packing; a homogeneous integration into the lipids is unlikely in the light of the variety of compositional and packing domains within the lipid bilayers. Thus, Azone molecules may disperse within the intercellular lipids or remain in separate domains within the bilayers. A soup-spoon model for Azone's conformation within stratum corneum lipids supports the above hypothesis [15]; many accelerants have such a soup-spoon structure (e.g., see Figure 12.7). Electron diffraction studies using lipids isolated from human stratum corneum provide good evidence that Azone exists (or partially forms) as a distinct phase within the lipids [16]. Extensive controversy concerning the use, metabolism and fate of Azone has been reviewed and the molecule continues to be investigated [17–19].

12.3.4 Pyrrolidones

A range of pyrrolidones and structurally related compounds has been investigated as potential penetration enhancers in human skin. As with azone and many other accelerants, they apparently have greater effects on hydrophilic diffusants than for lipophilic permeants, although this may be attributable to the greater enhancement potential for the poorer hydrophilic permeants. *N*-methyl-2-pyrrolidone (NMP) and 2-pyrrolidone (2P) are the most widely studied enhancers of this group (Figure 12.4). NMP is a polar aprotic solvent being a clear liquid at room temperature and miscible with most common solvents including water and alcohols. Similarly 2P is miscible with most solvents and is a liquid above 25°C.

Pyrrolidones show activity with numerous molecules including hydrophilic (e.g., mannitol, 5-fluorouracil, and sulfaguanidine) and lipophilic (betamethasone-17-benzoate, hydrocortisone, and progesterone) permeants. As with many studies, higher flux enhancements have been reported for the hydrophilic molecules. Recently, NMP was employed with limited success as an absorption promoter for captopril when formulated into a matrix-type transdermal patch [20].

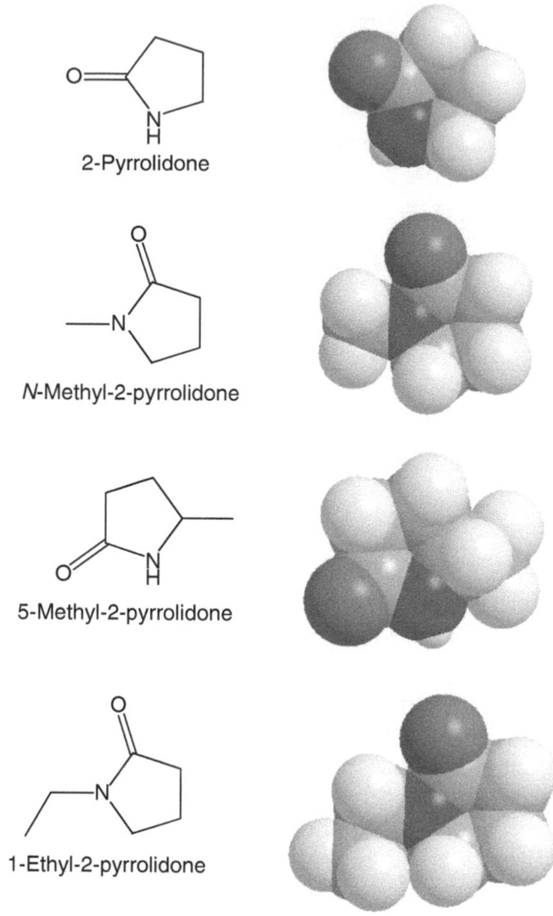

FIGURE 12.4 Various pyrrolidones investigated as skin accelerants.

In terms of the mechanisms of action, the pyrrolidones partition well into human horny layer. They may act by altering the solvent nature of the membrane and pyrrolidones have been used to generate reservoirs within skin membranes. Such a reservoir effect offers potential for sustained release of a permeant from the stratum corneum over extended time periods. However, as with many other potential enhancers, clinical use of pyrrolidones is problematic due to adverse reactions. An *in vivo* vasoconstrictor bioavailability study demonstrated that pyrrolidones caused erythema in some volunteers, although this effect was relatively short-lived. Also, a toxic hygroscopic contact reaction to NMP has been reported recently [21].

12.3.5 Fatty Acids

A wide variety of long-chain fatty acids increase transdermal delivery; the most popular is oleic acid. It is relevant that many penetration enhancers contain saturated or unsaturated hydrocarbon chains and some structure–activity relationships have been drawn from the extensive studies of Aungst et al. [22,23] who employed a range of fatty acids and alcohols, sulfoxides, surfactants, and amides as enhancers for naloxone. From these experiments, it appears that saturated alkyl chain lengths of around C_{10} to C_{12} attached to a polar head group yield the best promoters. In contrast, for accelerants containing unsaturated alkyl chains, C_{18} appears near optimal. For such unsaturated compounds, the bent *cis* configuration is expected to disturb intercellular lipid packing more so than the *trans* arrangement, which differs little from the saturated analog (Figure 12.5).

Numerous reports that fatty acids improve percutaneous absorption of, among others, estradiol, progesterone, acyclovir, 5-fluorouracil, and salicylic acid, indicate that these chemicals can be used to promote delivery of both lipophilic and hydrophilic permeants. Lauric acid in PG increased the delivery of highly lipophilic antiestrogens [24]. Fatty acid effects on drug delivery to and through human skin can vary. For example, Santoyo and Ygartua [25] employed the monounsaturated oleic acid, polyunsaturated linoleic and linolenic acids, and the saturated lauric acid enhancers for promoting piroxicam flux [25]. Pretreating the tissue with the accelerants increased the amount of piroxicam retained within the skin and also decreased the lag-time required to reach pseudosteady state flux. Oleic acid is effective for many drugs, for example increasing the flux of salicylic acid 28-fold and 5-fluorouracil 56-fold through human skin membrane *in vitro* [26]. As with azone, oleic acid is effective at relatively low concentrations (typically less than 10%) and can work synergistically when delivered from vehicles such as PG or ternary systems with dimethyl isosorbide [27]. Various analogs of fatty acids have been researched as penetration enhancers, for example diesters increased the permeation of nonsteroidal antiinflammatory drugs (NSAIDs) through rat skin [28].

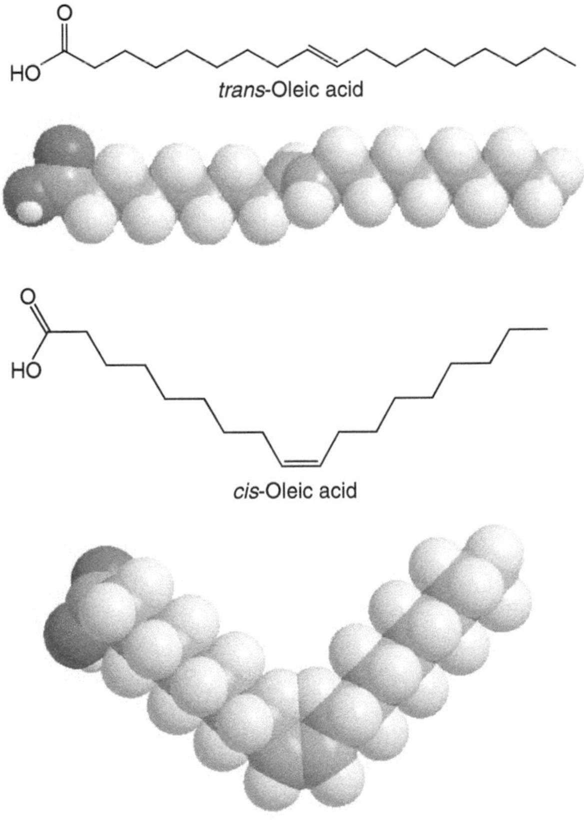

FIGURE 12.5 The fatty acid, oleic acid, illustrating the bent *cis* structure that would disrupt intercellular lipid packing much more than the *trans* form.

Significant efforts have been directed at investigating the mechanisms of action of oleic acid. It is apparent from numerous literature reports that the accelerant interacts with and modifies the lipid domains of the horny layer, as would be expected for a long-chain fatty acid with a *cis* configuration (Figure 12.5). Spectroscopic investigations using deuterated oleic acid in human stratum corneum indicate that at higher concentration it can also exist as a separate phase (or as pools) within the bilayer lipids [29]. More recently, electron microscopic studies showed that oleic acid induces formation of a discreet lipid domain within stratum corneum bilayer lipids [30]. The creation of such pools would provide permeability defects within the intercellular domain thus facilitating penetration of hydrophilic permeants through the membrane.

12.3.6 Alcohols, Fatty Alcohols, and Glycols

Ethanol is used in many transdermal formulations and is often the solvent of choice for incorporation into patches. As with water, ethanol permeates rapidly through human skin with a steady-state flux of approximately 1 $mg/cm^2/h$ [31].

Ethanol enhances the flux of levonorgestrel, estradiol, hydrocortisone, and 5-fluorouracil through rat skin [32] and of estradiol through human skin *in vivo* [33]. However, when using an ethanol:water cosolvent vehicle, the enhancement effect of ethanol is concentration dependent. Salicylate ion permeation across human epidermal membranes was promoted up to an ethanol:water composition of 0.63; higher levels of the alcohol decreased diffusion [34]. Similar results were obtained for glyceryl trinitrate [31], estradiol [35], and zidovudine [36]. It is probable that at higher ethanol levels dehydration of the biological membrane reduced permeation across the tissue.

Ethanol exerts its accelerant activity through various mechanisms. Firstly, as a solvent, it can increase the solubility of the drug in the vehicle—although at steady state the flux of a permeant from any saturated, nonenhancing, vehicle should ideally be equivalent. However, for poorly soluble permeants that are prone to depletion within the donor during a steady-state permeation study, then ethanol can increase permeant solubility in the donor phase, delay depletion, and hence raise the flux [33]. Furthermore, penetration of ethanol into the horny layer can alter the solubility properties of the tissue with a consequent improvement for drug partitioning into the membrane [35]. Additionally, it is also feasible that the rapid permeation of ethanol, or its evaporative loss from the donor phase, could increase the thermodynamic activity of the drug remaining within the formulation. Such an effect is most apparent upon applying a finite dose of a preparation onto the skin surface; as ethanol disappears, the drug concentration may increase beyond saturated solubility, yielding a supersaturated state with a greater driving force for permeation. Such a mechanism may operate for transdermal delivery from patches where ethanol, typically included to solubilize the drug, may traverse the stratum corneum rapidly leaving behind a metastable supersaturated permeant that is inhibited from crystallizing by polymers that are typically incorporated into patches. A further controversial potential mechanism of action arising because of rapid ethanol permeation across the skin has been reported; it has been suggested that solvent drag may carry permeant into the tissue as ethanol traverses it, although such a mechanism has been discounted for morphine hydrochloride permeation from formulations containing ethanol and methanol [37]. In addition, ethanol as a potent solvent may extract some of the lipid fraction from within the stratum corneum when used at high concentration for prolonged times; such a mechanism would clearly improve drug flux through skin.

Fatty alcohols may also have accelerant activity. These molecules are typically applied to the skin in a cosolvent—often PG or ethanol—at concentrations between 1% and 10%. As with fatty acids described above, some structure–activity relationships for fatty alcohol penetration enhancement have been deduced with lower activities reported for branched alkanols, whereas 1-butanol was shown to be the most effective enhancer for levonorgestrel traversing rat skin [32]. Other investigators showed that 1-propanol is effective for salicylic acid and nicotinamide traversing hairless mouse skin. More recent structure–activity relationships have been drawn for fatty alcohols, using melatonin penetrating through porcine and human skin *in vitro* [38]; comparing activities for saturated fatty alcohols from octanol to myristyl alcohol, a parabolic relationship was found with a maximum enhancement effect being given by decanol. Accelerant activity also showed a general increase when adding up to two unsaturated bonds into the alcohols, but activity fell when three double bonds were introduced.

PG is widely used as a vehicle for promoters and shows synergistic action when used with, for example, oleic acid or azone. However, PG has also been applied as an accelerant in its own right. Literature reports concerning the efficacy of simple PG as a permeation enhancer are mixed; evidence suggests at best only a very mild enhancement effect with drugs such as estradiol and 5-fluorouracil. As with ethanol, PG permeates well through human stratum corneum and its mechanisms of action are probably similar but much milder than those suggested above for ethanol. Loss of the solvent into the tissue or through evaporation could alter the thermodynamic activity of the drug in the vehicle, which would in turn modify the driving force for diffusion. Solvent may also partition into the tissue, facilitating uptake of the drug into skin and there may be some minor disturbance to intercellular lipid packing within the stratum corneum bilayers.

12.3.7 Surfactants

As with some of the materials described previously, surfactants are incorporated into many therapeutic, cosmetic, and agrochemical preparations. Usually, they are added to formulations to stabilize emulsions and suspensions or to solubilize lipophilic active ingredients, and so they have the potential to dissolve lipids within the stratum corneum. Typically composed of lipophilic alkyl or aryl fatty chains, together with a hydrophilic head group, surfactants are often described in terms of the nature of the hydrophilic moiety. Anionic surfactants include sodium lauryl sulfate (SLS), cationic surfactants encompass cetyltrimethyl ammonium bromide, the nonoxynol surfactants are nonionic surfactants and zwitterionic surfactants include dodecyl betaine. Anionic and cationic surfactants can potentially damage human skin; SLS is a powerful irritant and increased the transepidermal water loss in human volunteers *in vivo* [39] and both anionic and cationic surfactants swell the stratum corneum and interact with intracellular keratin. Nonionic surfactants are much less damaging, but are also less potent as enhancers. Surfactants generally have low chronic toxicity and most investigations have shown them to enhance the flux of materials permeating through biological membranes.

Many studies evaluating accelerant activity have focused on the use of anionic and nonionic surfactants. Anionic materials themselves tend to penetrate relatively poorly through human horny layer over short-time exposure (for example when investigations mimic occupation exposure) but permeation obviously increases with application time. Fewer studies assessed nonionic surfactant penetration; Watkinson et al. [40] showed that around 0.5% of the applied dose of nonoxynol surfactant traversed human skin after 48 h exposure *in vitro*. Surfactant-facilitated permeation of many materials through skin membranes has been researched, with reports of significant enhancement of drugs such as chloramphenicol through hairless mouse skin by SLS, and acceleration of hydrocortisone and lidocaine permeating across hairless mouse skin by the nonionic surfactant Tween 80 [41,42].

However, as is common with enhancers in general, the choice of model membrane can dramatically affect the scale of permeation promotion. Tween 80 did not accelerate nicardipine or ketorolac permeation in monkeys *in vivo* [43]. Likewise, 5-fluorouracil permeation through human and snakeskin *in vitro* was not improved by 0.1% Tween 20 in normal saline [26,44], whereas the same enhancer formulation increased 5-fluorouracil permeation across hairless mouse skin sixfold [44]. From the literature, it is apparent that, in general terms, nonionic surfactants have only a minor enhancement effect in human skin whereas anionic surfactants can have a more pronounced consequence.

12.3.8 Urea

Urea is a hydrating agent (a hydrotrope) used to treat scaling conditions such as psoriasis, ichthyosis, and other hyperkeratotic skin conditions. Applied in a water-in-oil vehicle, urea alone or in combination with ammonium lactate hydrated stratum corneum and improved barrier function when compared to the vehicle alone in human volunteers *in vivo* [45]. Urea also has keratolytic properties, usually when combined with salicylic acid for keratolysis. The somewhat modest penetration-enhancing activity of urea probably arises from a combination of increasing stratum corneum water content (water is a valuable penetration enhancer) and through the keratolytic activity.

As urea itself possesses only marginal accelerant activity, attempts have been made to synthesize analogs containing more potent-enhancing moieties. Thus Wong et al. prepared cyclic urea analogs and found them to be as powerful as azone for promoting indomethacin penetration across shed snakeskin and hairless mouse skin [46]. A series of alkyl and aryl urea analogs was moderately effective as enhancers for 5-fluorouracil when applied in PG to human skin *in vitro*, though urea itself was ineffectual [47].

12.3.9 Essential Oils, Terpenes, and Terpenoids

Terpenes, found in essential oils, are nonaromatic compounds comprising only carbon, hydrogen, and oxygen. Several terpenes have long been used as medicines, flavorings, and fragrance agents. For example, menthol is traditionally employed for inhalation and has a mild antipruritic effect when incorporated into emollient preparations. It is also used as a fragrance and to flavor toothpastes, peppermint sweets, and mentholated cigarettes.

The essential oils of eucalyptus, chenopodium, and ylang ylang were effective penetration enhancers for 5-fluorouracil traversing human skin *in vivo* [48]. The most potent of these essential oils, eucalyptus, increased the permeability coefficient of the drug 34-fold. The principal terpene element within eucalyptus oil is 1,8-cineole (eucalyptol) and this molecule was one of a series of 17 monoterpenes and terpenoids evaluated as enhancers for the model hydrophilic drug 5-fluorouracil tested in human skin *in vitro* [49] (see Figure 12.6). The data yielded some structure–activity relationships in that hydrocarbon terpenes were less potent accelerants for this hydrophilic drug than were alcohol- or ketone-containing terpenes, and the greatest enhancement activity was shown by the oxide terpenes and terpenoids. Within this oxide subclass, potencies varied with ring-bridged oxides (cyclic ethers) being better than 1,2-oxygen linked (epoxide) molecules; pretreatment of human epidermal membranes with 1,8-cineole increased the permeability coefficient of 5-fluorouracil by nearly 100-fold. However, such tentative structure–activity relationships appear to be drug specific. The same terpenes were employed in an identical protocol for the lipophilic drug estradiol [50]. The results showed that, unlike for 5-fluorouracil where alcohol and ketone terpenes had relatively moderate enhancement activities (10–40-fold increases in permeability coefficient), these agents had no accelerant potential toward the lipophilic model drug and indeed appeared to retard its permeation. The cyclic ethers, so potent for 5-fluorouracil, provided only moderate enhancements for estradiol permeation and, in contrast to the hydrophilic

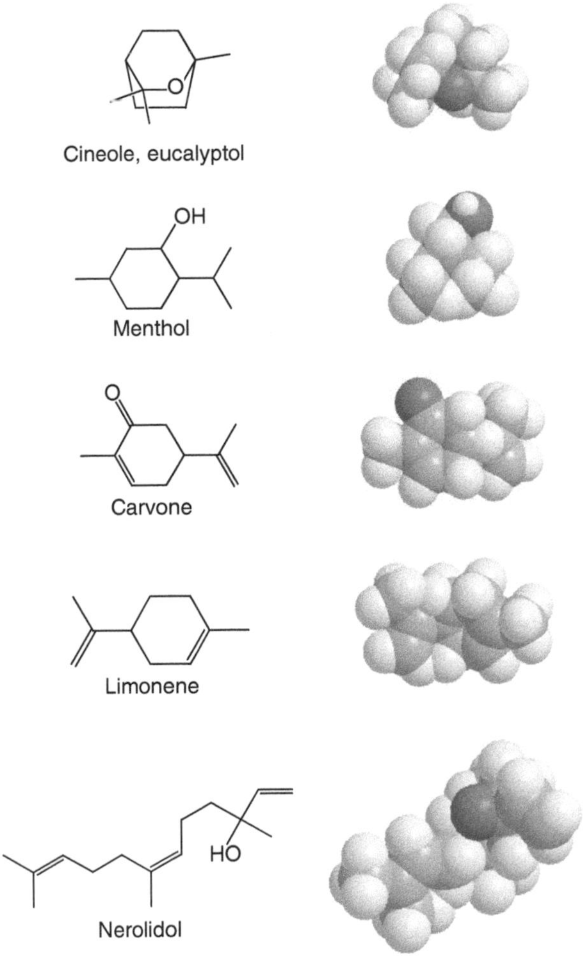

FIGURE 12.6 Four monoterpenes and a sesquiterpene that enhance drug penetration.

drug, hydrocarbon terpenes (such as D-limonene) were generally the most effective terpene enhancers for the steroid. Similar results were deduced for the permeation of another lipophilic molecule, indomethacin, traversing rat skin; hydrocarbon terpenes, especially limonene, were as effective as azone in promoting drug flux and oxygen-containing terpenes (carvone, 1–8 cineole) were ineffective [51,52]. Other hydrophilic drugs, such as propranolol and diazepam, were also enhanced by nonpolar terpenes and cyclic monoterpenes generally showed stronger promotion of curcumin than did other terpenes, flavanoids, and cholestanol [53].

As with many of the enhancers described above, a synergistic effect for terpene efficacy has also been shown when PG was used as the vehicle [54]; with this cosolvent, enhancer activities for carveol, carvone, pulegone, and 1–8 cineole rose approximately fourfold, explained by improved partitioning of the terpene into the stratum corneum. Enhancement using terpenes in PG has recently been revisited for haloperidol delivery through human skin [55,56].

Additionally to the small monoterpenes described above, larger molecules (sesquiterpenes) have also been evaluated as enhancers. Thus, materials such as nerolidol increased 5-fluorouracil permeability over 20-fold through human skin *in vitro* [57]. As larger lipophilic enhancers, these agents exerted their effects over prolonged periods—for up to 5 d—in contrast to the monoterpenes that tended to wash out relatively easily from the stratum corneum. Moderate enhancement activity has also been reported for the cosmetically valuable terpene α-bisabolol [58].

Terpenes continue to be a popular choice as experimental enhancers for delivering materials across skin membranes. For example, L-menthol facilitated *in vitro* permeation of morphine hydrochloride through hairless rat skin [37], imipramine hydrochloride across rat skin [59], and hydrocortisone through hairless mouse skin [60]. Recently, niaouli oil was found to be the most effective of six essential oils in promoting estradiol penetration through hairless mouse skin [61]. It is noteworthy that there is currently little control on the topical use of most terpenes, and many aromatherapy oils and formulations contain appreciable quantities of these chemicals. Their excessive use offers potential for permeation of hazardous compounds from the same formulations into the skin; some terpenes also have pharmacological activity.

From the above, it appears that the smaller terpenes tend to be more active permeation enhancers than the larger sesquiterpenes. Further, it also seems that hydrocarbon or the nonpolar group containing terpenes, such as limonene, provides better enhancement for lipophilic permeants than do the polar terpenes. Conversely, those containing polar groups (such as menthol, 1–8 cineole) provide better enhancement for hydrophilic permeants. Such a relationship implies that one mechanism by which such agents operate is to modify the solvent nature of the stratum corneum, improving drug partitioning into the tissue. Many terpenes permeate human skin well [57], and a matrix-type patch introduced large amounts of terpenes (up to 1.5 $\mu g/cm^2$) into the epidermis [62]. With loss of terpenes, which are generally good solvents, from a formulation the thermodynamic activity of the permeant in the formulation could alter, as was described for ethanol. Terpenes may also increase drug diffusivity through the membrane. During steady-state permeation experiments using such accelerants, the lag-time for drug permeation usually decreased, indicating some increase in diffusivity of the drug through the membrane. Small angle x-ray diffraction studies have also shown that D-limonene and 1–8 cineole disrupt stratum corneum bilayer lipids, whereas nerolidol, a long-chain sesquiterpene, reinforces the bilayers, possibly by orientating alongside the stratum corneum lipids [63]. Spectroscopic evidence has also suggested that, as with azone and oleic acid, terpenes could exist within separate domains in stratum corneum lipids.

12.3.10 Phospholipids

Many studies have employed phospholipids as liposomes (vesicles) to transport drugs into and through human skin. However, a few investigations have also employed phospholipids in a nonvesicular form as penetration enhancers. For example, 1% phosphatidylcholine in PG, a concentration at which liposomes would not form, enhanced theophylline penetration through hairless mouse skin [64]. Similarly, indomethacin flux was enhanced through rat skin by the same phospholipid and hydrogenated soybean phospholipids increased diclofenac permeation through rat skin *in vivo*.

There is no compelling evidence to show that phospholipids interact directly to modify stratum corneum packing, though this might be assumed when considering their physicochemical properties and structures. Additionally, phospholipids can occlude somewhat the skin surface and thus can increase tissue hydration, which, as discussed above, can increase drug permeation. When applied to the stratum corneum as vesicles, phospholipids can sometimes fuse with stratum corneum lipids. This collapse of structure liberates permeant into the vehicle in which the drug may be poorly soluble and hence thermodynamic activity temporarily increases, facilitating drug delivery.

12.3.11 Ceramide Analogs

An interesting novel approach in penetration enhancer science was to synthesize and test a series of ceramide analogs containing eight polar groups and six chain lengths based on L-serine and glycine (ceramides form the main components of the intercellular lipid domains). The concept was that there would be some similarity between an enhancer molecule and ceramides. The results suggested two trends: first, the larger the polar head group of the enhancer, the lower was its activity, and second, the presence of a hydrogen-bonding group decreased the accelerant activity. The authors concluded that the polar head group of the enhancers is responsible for the penetration and the anchoring of the molecule in the stratum corneum lipids, and the length of the hydrophilic chains controls the disordering of the lipids. Surprisingly, the most active substances possessed the most ordered chains [65–68].

12.3.12 Solvents at High Concentrations

In addition to the general activities of accelerants within the intercellular domain, high levels of potent solvents may have more drastic effects. They may damage desmosomes and protein-like bridges, leading to fissuring of the intercellular lipid and splitting of the stratum corneum squames (Figure 12.1). Solvent may enter the corneocyte, drastically disrupting the keratin and even forming vacuoles [69]. A major problem is that such dramatic effects would be even less acceptable to regulatory agencies (and patients) than lesser insults to the intercellular lipid.

12.3.13 Metabolic Interventions

More interventionist approaches to drug delivery through human skin have also been proposed, as reviewed by Elias et al. [70]. Strategies that alter barrier homeostasis by interfering with any or all of the processes of synthesis, assembly, secretion, activation, processing, or assembling and disassembling of the extracellular lamellar membranes, could promote permeation. Such an approach would pose significant regulatory problems, not least of which would be issues related to increased xenobiotic or microbial access. The concept of interfering with barrier homeostasis on a relatively long timescale poses many clinical considerations and objections.

12.3.14 Enhancer Combinations

As detailed above for enhancers such as Azone, oleic acid, urea, and terpenes, it is apparent that the vehicle used to deliver the enhancer may dramatically affect its efficacy, with PG and Transcutol often used to generate synergy of action. Recently, a high-throughput screening approach to the use of enhancer combinations has been developed [71]. This technique claims to be much more efficient than other screening methods and provides what are called SCOPE formulations (Synergistic Combinations of Penetration Enhancers). The authors selected 32 enhancers from chemicals reported in the literature. They then assessed over 5000 binary formulations in 50% ethanol/buffer, four times each, using conductivity measurements *in vitro* with porcine skin. The leading hits were evaluated for their irritation potential and potent, safe enhancer mixtures (SCOPE formulations) were selected for flux measurements with candidate drugs. Finally, the best formulations were assessed for bioavailability and safety *in vivo* in hairless rats.

Ninety-eight percent of formulations were eliminated as poorly potent, 99.5% were discarded after irritation studies, the remaining 0.5% was tested for flux enhancement and 0.02% was finally assessed for bioavailability. The investigators thus revealed rare nonirritant mixtures of enhancers that increased the skin permeability to macromolecules, such as heparin, luteinizing hormone-releasing hormone, and an oligonucleotides, by up to 100-fold. The two most successful SCOPE formulations were a mixture of sodium laureth sulfate with phenyl piperazine (Figure 12.7a) and a combination of *N*-lauroyl sarcosine with sorbitan monolaurate (Figure 12.7b).

Future work may elucidate why the areas of potency hot spots were so restricted, and the fundamental molecular mechanisms producing the enhancement. The molecular structures of the most successful SCOPE mixtures, as illustrated in Figure 12.7, suggest that surface-active phenomena may play a crucial role.

Instead of using a screening approach, with its heavy workload, investigators have tried other techniques. Many studies demonstrated that a rule-based approach to enhancement was fraught with difficulties; enhancer combinations in different vehicles for specific permeants traversing a particular membrane thus tend to be evaluated on a case-by-case basis [72,73]. However, attempts have been made at a more rational approach to enhancer selection, applying quantitative (and qualitative) structure–activity relationships to penetration enhancers [74–77]. Naturally, such models depend upon the quality of data used to obtain the relationship. Hence inclusion of information derived from, for example, different animal models or dosing regimens must be carefully assessed as the generated relationship may only be applicable to the specific conditions used in obtaining the input data.

12.4 GENERAL COMMENTS ON PENETRATION ENHANCERS

The list of materials that have been used as penetration enhancers as discussed above is not exhaustive but is intended to illustrate the range of agents that have been employed for facilitating transdermal drug delivery. Several common themes emerge from these considerations:

1. It is difficult to select rationally a penetration enhancer for a given permeant. Accelerant potencies appear to be drug specific, or at best may be predictive for a series of permeants with similar physicochemical properties (such as similar partition coefficients, molecular weights, and solubilities). Some broad trends are apparent, such as the use of hydrocarbon monoterpenes for lipophilic permeants, but the level of enhancement expected for these agents is unpredictable.
2. Penetration enhancements through animal skins, and rodent tissues in particular, are generally considerably greater than those obtained with human skin, correlating with the increased barrier resistance of human stratum corneum. Hairless mouse skin is particularly fragile and its use may grossly mislead the investigator. Most experiments are performed *in vitro*, although there are exceptions, for example, the use of confocal Raman spectroscopy to monitor the penetration of DMSO through volunteer skin [78].
3. Accelerants tend to work well with cosolvents such as PG or ethanol. Synergistic effects arise enhancers such as azone, oleic acid (and other fatty acids), and terpenes dissolved in, for example, PG.
4. Many enhancers have a complex concentration-dependent effect. This is shown clearly by azone, which is effective in promoting the transdermal flux of many drugs when used at 1% in PG but which is far less potent when applied at higher concentrations or neat.
5. Potential mechanisms of action of enhancers are varied, and can range from direct effects on the skin to modification of the formulation. Thus, directly acting on the skin, enhancers can do the following (see Figure 12.1):
   (i) Modify the intercellular lipid domains to reduce the barrier resistance of the bilayer lipids. Disruption to the lipid bilayers could be homogeneous where the enhancer

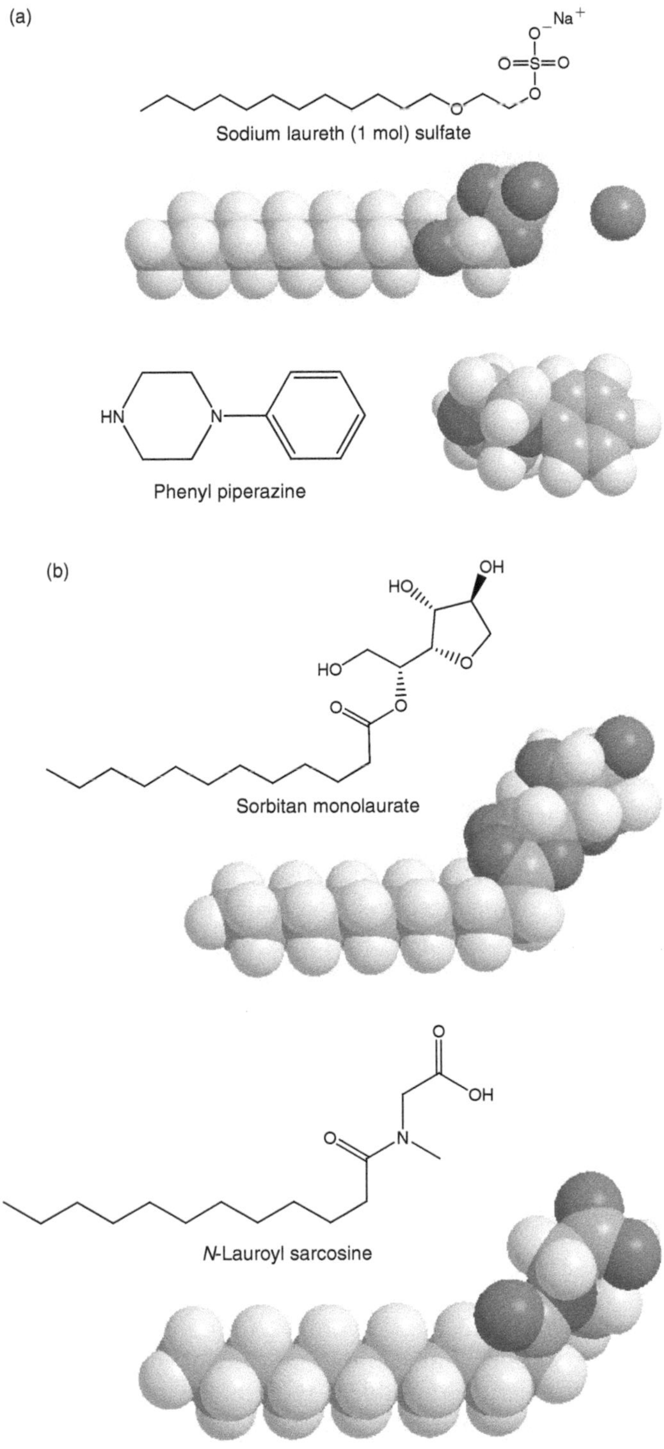

FIGURE 12.7 The two most potent SCOPE formulations. (a) A mixture of sodium laureth sulfate with phenyl piperazine and (b) a combination of sodium monolaurate with *N*-lauroyl sarcosine. (From Karande, P., Jain, A. and Mitragotri, S. *Nat Biotechnol* 22:192, 2004.)

distributes evenly within the complex bilayer lipids, but sometimes the accelerant will be heterogeneously concentrated within domains of the bilayer lipids. Such a pooling phenomenon has been shown for oleic acid, terpenes, and Azone, and is likely to occur for several similar enhancers considering the range of packing and different molecular domains in the stratum corneum lipids. The crystalline/gel/liquid crystal domain may fluidize, alter its polarity, separate phase, or have lipids extracted.

(ii) Alter the solvent nature of the stratum corneum so as to modify partitioning of the drug, coenhancer, or a cosolvent into the tissue. Many enhancers are good solvents and so, for example, the pyrrolidones can increase the amount of permeant within the skin.

(iii) Act on the stratum corneum intracellular keratin, denature it, or modify its conformation causing swelling, increased hydration, and vacuolization.

(iv) Affect the desmosomes that maintain cohesion between corneocytes and other protein structures, leading finally to splitting of the stratum corneum.

The above mechanisms of action have been embraced within a general scheme to explain enhancer effects on stratum corneum, termed the lipid–protein-partitioning concept; enhancers can act by altering skin lipids and proteins or by affecting partitioning behavior [79]. Recent general reviews on skin penetration enhancers have been written (see Refs. [80–83]). A very recent, interesting and novel approach assumes that chemical accelerants perturb the stratum corneum barrier by lipid extraction or fluidization of the lipid bilayers, as assessed through Fourier transform infrared spectroscopy [84]. By analyzing the underlying molecular forces responsible for irritancy and potency, the authors isolated inherent constraints that limit performance. Using this knowledge, they designed more than 300 potential enhancers, which were screened *in silico* for testing *in vitro*. This publication and a prior paper from the same group [71] represent two of the most important texts published in recent years in the area of transdermal drug delivery. In addition, penetration enhancement can be indirect by:

(i) Modification of thermodynamic activity of the vehicle. Rapid permeation of a good solvent such as ethanol from the donor solution, or its evaporation, can leave the permeant in a more thermodynamically active state than when all the solvent was originally present—even to the point of supersaturation.

(ii) It has been suggested that solvent permeating through the membrane could drag the permeant with it, though this concept is somewhat controversial and remains to be proven.

(iii) Solubilizing the permeant in the donor (e.g., with surfactants), especially where solubility is very low as with steroids in aqueous donor solutions, can reduce depletion effects and prolong drug permeation.

6. Many of the chemicals described above are used for alternative reasons within topical and transdermal preparations. For example, a dermatological preparation could contain PG as a vehicle, a surfactant to solubilize the drug or to stabilize a dispersion, and a terpene as a fragrance material. The efficacies of some topical preparations, particularly those long established, are probably due to penetration enhancement by these types of agents, although the commercial preparations are not claimed to incorporate an agent specifically for its enhancing ability.

REFERENCES

1. Barry, B.W. 1983. *Dermatological formulations percutaneous absorption.* New York: Marcel Dekker.
2. Williams, A.C. 2003. *Transdermal and topical drug delivery: From theory to clinical practice.* London: Pharmaceutical Press.
3. Bucks, D., and H.I. Maibach. 1999. Occlusion does not uniformly enhance penetration *in vivo*. In *Percutaneous absorption: Drugs, cosmetics, mechanisms, and methodology*, 3rd ed., eds. R.L. Bronaugh and H.I. Maibach. New York: Marcel Dekker, chap. 4.
4. Walkley, K.1972. Bound water in stratum corneum measured by differential scanning calorimetry. *J Invest Dermatol* 59:225.
5. Cornwell, P.A., et al. 1994. Wide-angle x-ray diffraction of human stratum corneum: Effects of hydration and terpene enhancer treatment. *J Pharm Pharmacol* 46:938.
6. Van Hal, D.A., et al. 1996. Structure of fully hydrated human stratum corneum: A freeze fracture electron microscopy study. *J Invest Dermatol* 106:89.
7. Menon, G.K., and P.M. Elias. 1997. Morphological basis for a pore-pathway in mammalian stratum corneum. *Skin Pharmacol* 10:235.
8. Bond, J.R., and B.W. Barry. 1986. Limitations of hairless mouse skin as a model for *in vitro* permeation studies through human skin: Hydration damage. *J Invest Dermatol* 90:486.
9. Kligman, A.M. 1965. Topical pharmacology and toxicology of dimethylsulfoxide. *JAMA* 193:796.
10. Southwell, D., and B.W. Barry. 1983. Penetration enhancers for human skin: Mode of action of 2-pyrrolidone and dimethylformamide on partition and diffusion of model compounds water, *n*-alcohols and caffeine. *J Invest Dermatol* 80:507.
11. Barry, B.W., D. Southwell, and R. Woodford. 1984. Optimisation of bioavailability of topical steroids: Penetration enhancers under occlusion. *J Invest Dermatol* 82:49.
12. Bennett, S.L., B.W. Barry, and R. Woodford. 1984. Optimisation of bioavailability of topical steroids: Non-occluded penetration enhancers under thermodynamic control. *J Pharm Pharmacol* 37:298.
13. Oertel, R.P. 1997. Protein conformational changes induced in human stratum corneum by organic sulphoxides: An infrared spectroscopic investigation. *Biopolymer* 16:2329.
14. Anigbogu, A.N.C., et al. 1995. Fourier transform Raman spectroscopy of interactions between the penetration enhancer dimethylsulphoxide and human stratum corneum. *Int J Pharm* 125:265.
15. Hoogstraate, A.J., et al. 1991. Kinetics, ultrastructural aspects and molecular modelling of transdermal peptide flux enhancement by *N*-alkylazacycloheptanones. *Int J Pharm* 76:37.
16. Pilgram, G.S.K., et al. 2001. The influence of two azones and sebaceous lipids on the lateral organization of lipids isolated from human stratum corneum. *Biochim Biophys Acta Biomembr* 1511:244.
17. Wiechers, J.W. 1989. Absorption, distribution, metabolism, and excretion of the cutaneous penetration enhancer azone. PhD thesis, University of Groningen.
18. Wiechers, J.W., and R.A. de Zeeuw. 1990. Transdermal drug delivery: Efficacy and potential applications of the penetration enhancer azone. *Drug Design Del* 6:87.
19. Afouna, M.I., et al. 2003. Effect of azone upon the *in vivo* antiviral efficacy of cidofovir or acyclovir topical formulations in treatment/prevention of cutaneous HSV-1 infections and its correlation with skin target site free drug concentration in hairless mice. *Int J Pharm* 253:159.
20. Park, E.S., et al. 2001. Effects of adhesives and permeation enhancers on the skin permeation of captopril. *Drug Dev Ind Pharm* 27:975.
21. Jungbauer, F.H.W., P.J. Coenraads, and S.H. Kardaun. 2001. Toxic hygroscopic contact reaction to *N*-methyl-2-pyrrolidone. *Contact Dermatitis* 45:303.
22. Aungst, B.J., N.J. Rogers, and E. Shefter. 1986. Enhancement of naloxone penetration through human skin *in vitro* using fatty acids, fatty alcohols, surfactants, sulfoxides and amides. *Int J Pharm* 33:2256.
23. Aungst, B.J. 1989. Structure-effects studies of fatty acid isomers as skin penetration enhancers and skin irritants. *Pharm Res* 6:244.
24. Funke, A.P., et al. 2002. Transdermal delivery of highly lipophilic drugs: *In vitro* fluxes of antiestrogens, permeation enhancers, and solvents from liquid formulations. *Pharm Res* 19:661.

25. Santoyo, S., and P. Ygartua. 2000. Effect of skin pre-treatment with fatty acids on percutaneous absorption and skin retention of piroxicam after its topical application. *Eur J Pharm Biopharm* 50:245.
26. Goodman, M., and B.W. Barry. 1989. Lipid-protein-partitioning theory (LPP) theory of skin enhancer activity: Finite dose technique. *Int J Pharm* 57:29.
27. Aboofazeli, R., H. Zia, and T.E. Needham. 2002. Transdermal delivery of nicardipine: An approach to *in vitro* permeation enhancement. *Drug Del* 9:239.
28. Takahashi, K., et al. 2002. Effect of fatty acid diesters on permeation of anti-inflammatory drugs through rat skin. *Drug Dev Ind Pharm* 28:1285.
29. Ongpipattanakul, B., et al. 1991. Evidence that oleic acid exists in a separate phase within stratum corneum lipids. *Pharm Res* 7:350.
30. Tanojo, H., et al. 1997. *In vitro* human skin barrier perturbation by oleic acid: Thermal analysis and freeze fracture electron microscopy studies. *Thermochim Acta* 293:77.
31. Berner, B., et al. 1989. Ethanol: Water mutually enhanced transdermal therapeutic system II: Skin permeation of ethanol and nitroglycerin. *J Pharm Sci* 78:402.
32. Friend, D., et al. 1988. Transdermal delivery of levonorgestrel. I. Alkanols as permeation enhancers. *J Control Release* 7:243.
33. Pershing, L.K., L.D. Lambert, and K. Knutson. 1990. Mechanism of ethanol-enhanced estradiol permeation across human skin *in vivo. Pharm Res* 7:170.
34. Kurihara-Bergstrom, T., et al. 1990. Percutaneous absorption enhancement of an ionic molecule by ethanol–water systems in human skin. *Pharm Res* 7:762.
35. Megrab, N.A., A.C. Williams, and B.W. Barry. 1995. Oestradiol permeation across human skin, silastic and snakeskin membranes: The effects of ethanol/water co-solvent systems. *Int J Pharm* 116:101.
36. Thomas, N.S., and R. Panchagnula. 2003. Transdermal delivery of zidovudine: Effect of vehicles on permeation across rat skin and their mechanism of action. *Pharm Sci* 18:71.
37. Morimoto, H., et al. 2002. *In vitro* skin permeation of morphine hydrochloride during the finite application of penetration-enhancing system containing water, ethanol and l-menthol. *Biol Pharm Bull* 25:134.
38. Andega, S., N. Kanikkannan, and M. Singh. 2001. Comparison of the effect of fatty alcohols on the permeation of melatonin between porcine and human skin. *J Control Release* 77:17.
39. Tupker, R.A., J. Pinnagoda, and J.P. Nater. 1990. The transient and cumulative effect of sodium lauryl sulphate on the epidermal barrier assessed by transepidermal water loss: Inter-individual variation. *Acta Derm Venereol (Stockh)* 70:1.
40. Watkinson, A.C., et al. 1998. Skin penetration of a series of nonoxynol homologues. In *Perspectives in percutaneous penetration*, vol. 5b, eds. K.R. Brain, V.J. James, and K.S. Waters. Cardiff: STS Publishing Ltd.
41. Sarpotdar, P.P., and J.L. Zatz. 1986. Percutaneous absorption enhancement by non-ionic surfactants. *Drug Dev Ind Pharm* 12:1625.
42. Sarpotdar, P.P., and J.L. Zatz. 1986. Evaluation of penetration enhancement of lidocaine by non-ionic surfactants through hairless mouse skin *in vitro. J Pharm Sci* 75:176.
43. Yu, D., et al. 1988. Percutaneous absorption of nicardipine and ketorolac in rhesus monkeys. *Pharm Res* 5:457.
44. Rigg, P.C., and B.W. Barry. 1990. Shed snakeskin and hairless mouse skin as model membranes for human skin during permeation studies. *J Invest Dermatol* 94:235.
45. Gloor, M., et al. 2001. Clinical effect of salicylic acid and high dose urea applied in standardized NRF formulations. *Pharmazie* 56:810.
46. Wong, O., et al. 1989. Unsaturated cyclic ureas as new non-toxic biodegradable transdermal penetration enhancers. II. Evaluation study. *Int J Pharm* 52:191.
47. Williams, A.C., and B.W. Barry. 1989. Urea analogues in propylene glycol as penetration enhancers in human skin. *Int J Pharm* 56:43.
48. Williams, A.C., and B.W. Barry. 1989. Essential oils as novel human skin penetration enhancers. *Int J Pharm* 57:R7.
49. Williams, A.C., and B.W. Barry. 1991. Terpenes and the lipid-protein-partitioning theory of skin penetration enhancers. *Pharm Res* 8:17.

50. Williams, A.C., and B.W. Barry. 1991. The enhancement index concept applied to terpene penetration enhancers for human skin and model lipophilic (oestradiol) and hydrophilic (5-fluorouracil) drugs. *Int J Pharm* 74:157.
51. Nagai, T., et al. 1989. Effect of limonene and related compounds on the percutaneous absorption of indomethacin, *Proceedings of the 16th International Symposium on Controlled Release on Bioactive Materials*, Chicago, USA, 181.
52. Okabe, H., et al. 1989. Effect of limonene and related compounds on the percutaneous absorption of indomethacin. *Drug Design Del* 4:313.
53. Fang, J.Y., et al. 2003. Efficacy and irritancy of enhancers on the *in vitro* and *in vivo* percutaneous absorption of curcumin. *J Pharm Pharmacol* 55:593.
54. Barry, B.W., and A.C. Williams. 1989. Human skin penetration enhancement: The synergy of propylene glycol with terpenes, *Proceedings of the 16th International Symposium on Controlled Release on Bioactive Materials*, Chicago, USA, 33.
55. Vaddi, H.K., P.C. Ho, and S.Y. Chan. 2002. Terpenes in propylene glycol as skin-penetration enhancers: Permeation and partition of haloperidol, Fourier transform infrared spectroscopy, and differential scanning calorimetry. *J Pharm Sci* 91:1639.
56. Vaddi, H.K., et al. 2003. Oxide terpenes as human skin penetration enhancers of haloperidol from ethanol and propylene glycol and their modes of action on stratum corneum. *Biol Pharm Bull* 26:220.
57. Cornwell, P.A., and B.W. Barry. 1994. Sesquiterpene components of volatile oils as skin penetration enhancers for the hydrophilic permeant 5-fluorouracil. *J Pharm Pharmacol* 46:261.
58. Kadir, R., and B.W. Barry. 1991. α-Bisabolol, a possible safe penetration enhancer for dermal and transdermal therapeutics. *Int J Pharm* 70:87.
59. Jain, A.K., N.S. Thomas, and R. Panchagnula. 2002. Transdermal drug delivery of imipramine hydrochloride. I. Effect of terpenes. *J Control Release* 79:93.
60. El-Kattan, A.F., C.S. Asbill, and B.B. Michniak. 2000. The effect of terpene enhancer lipophilicity on the percutaneous permeation of hydrocortisone formulated in HPMC gel systems. *Int J Pharm* 198:179.
61. Monti, D., et al. 2002. Effect of different terpene-containing essential oils on permeation of estradiol through hairless mouse skin. *Int J Pharm* 237:209.
62. Cal, K., S. Janicki, and M. Sznitowska. 2001. *In vitro* studies on penetration of terpenes from matrix-type transdermal systems through human skin. *Int J Pharm* 224:81.
63. Cornwell, P.A., et al. 1996. Modes of action of terpene penetration enhancers in human skin; differential scanning calorimetry, small-angle x-ray diffraction and enhancer uptake studies. *Int J Pharm* 127:9.
64. Kato, A., Y. Ishibashi, and Y. Miyake. 1987. Effect of egg-yolk lecithin on transdermal delivery of bunazosin hydrochloride. *J Pharm Pharmacol* 39:399.
65. Vavrova, K. 2003. Modification of the skin barrier properties. Penetration enhancers and the stratum corneum repair. PhD diss., Charles University in Prague.
66. Vavrova, K., et al. 2003. L-serine and glycine based ceramide analogues as transdermal permeation enhancers: Polar head size and hydrogen bonding. *Bioorg Med Chem Lett* 14:2351.
67. Vavrova, K., et al. 2003. Synthetic ceramide analogues as skin permeation enhancers: Structure–activity relationships. *Bioorg Med Chem* 24:5381.
68. Vavrova, K., et al. 2003. Synthesis and evaluation of the novel ceramide analogue based on L-serine for the skin barrier repair. *J Invest Dermatol* 5:1261.
69. Menon, G.K., and S.H. Lee. 1998. Ultrastructural effects of some solvents and vehicles on the stratum corneum and other skin components: Evidence for an extended mosaic-partitioning model of the skin barrier. In *Dermal absorption and toxicity assessment*, eds. M.S. Roberts and K.A. Walters. New York: Marcel Dekker, chap. 29.
70. Elias, P.M., et al. 2003. Metabolic approaches to transdermal drug delivery. In *Transdermal drug delivery*, 2nd ed., eds. R.H. Guy and J. Hadgraft. New York: Marcel Dekker, chap. 8.
71. Karande, P., A. Jain, and S. Mitragotri. 2004. Discovery of transdermal penetration enhancers by high-throughput screening. *Nat Biotechnol* 22:192.
72. Gwak, H.S., I.S. Oh, and I.K. Chun. 2004. Transdermal delivery of ondansetron hydrochloride: Effects of vehicles and penetration enhancers. *Drug Dev Ind Pharm* 30:187.

73. Wagner, H., et al. 2004. Effects of various vehicles on the penetration of flufenamic acid into human skin. *Eur J Pharm Biopharm* 58:121.
74. Yamashita, F., and M. Hashida. 2003. Mechanistic and empirical modelling of skin permeation of drugs. *Adv Drug Deliv Rev* 55:1185.
75. Ghafourian, T., et al. 2004. The effect of penetration enhancers on drug delivery through skin: A QSAR study. *J Control Release* 99:113.
76. Warner, K.S., et al. 2003. Structure–activity relationship for chemical skin permeation enhancers: Probing the chemical microenvironment of the site of action. *J Pharm Sci* 92:1305.
77. He, N., et al. 2004. Mechanistic study of chemical skin permeation enhancers with different polar and lipophilic functional groups. *J Pharm Sci* 93:1415.
78. Caspers, P.J., et al. 2002. Monitoring the penetration enhancer dimethyl sulfoxide in human stratum corneum *in vivo* by confocal Raman spectroscopy. *Pharm Res* 19:1577.
79. Barry, B.W. 1991. Lipid-protein-partitioning theory of skin penetration enhancement. *J Control Release* 15:237.
80. Bouwstra, J.A., I. van den Bergh, and M. Suhonen. 2001. Topical application of drugs: Mechanisms involved in chemical enhancement. *J Recept Signal Transduct Res* 21:259.
81. Purdon, C.H., et al. 2004. Penetration enhancement of transdermal delivery—current permutations and limitations. *Crit Rev Ther Drug Carrier Syst* 21:97.
82. Ting, W.W., C.D. Vest, and R.D. Sontheimer. 2004. Review of traditional and novel modalities that enhance the permeability of local therapeutics. *Int J Dermatol* 43:538.
83. Barry, B.W. 2001. Novel mechanisms and devices to enable successful transdermal drug delivery. *Eur J Pharm Sci* 14:101.
84. Karande, P., et al. 2005. Design principles of chemical penetration enhancers for transdermal drug delivery. *PNAS* 102:4688.
85. Barry, B.W. 2004. Breaching the skin's barrier to drugs. *Nat Biotechnol* 22:165.

13 Vesicular Carriers for Enhanced Delivery through the Skin

Elka Touitou and Biana Godin

CONTENTS

13.1 INTRODUCTION

Dermal and transdermal delivery requires efficient penetration of compounds through the skin barrier, the bilayer domains of intercellular lipid matrices, and keratin bundles in the stratum corneum (SC). Lipid vesicular systems are a recognized mode of enhanced delivery of drugs into and through the skin. However, it is noteworthy that not every lipid vesicular system has the adequate characteristics to enhance skin membrane permeation. Specially designed lipid vesicles in contrast to classic liposomal compositions could achieve this goal. This chapter describes the structure, main physicochemical characteristics, and mechanism of action of prominent vesicular carriers in this field and reviews reported data on their enhanced delivery performance.

13.2 LIPOSOMES—PHOSPHOLIPID VESICLES FOR TOPICAL ADMINISTRATION OF DRUGS

Liposomes were first proposed for drug topical administration to the skin more than 25 years ago by Mezei and Gulusekharam [1,2]. The basic components of liposomes are phospholipids (phosphatidylcholine, phophatidylethanolamine, phophatidylserine, dipalmitoyl phosphatidylcholine, and others), cholesterol, and water. Liposomes may vary significantly in terms of size (from tens of nm to microns) and structure. In liposomes, one or more concentric bilayers surround an aqueous core generating small or large unilamellar vesicles (SUV, LUV) or multilamellar vesicles (MLV), respectively [3].

In the early 1980s Mezei and Gulusekharam published the results of their investigation on topical administration of triamcinolone acetonide (TA) delivered from liposomes, showing higher drug concentrations in the upper skin strata and reduced drug percutaneous transport, as compared to a conventional formulation [1,2]. This local delivery behavior of liposomes was further confirmed by many studies where various drugs were tested *in vitro* or *in vivo* [4–16]. Investigators have mostly focused on dermal corticosteroid and retinoid liposome products. However, localized effects of liposome-encapsulated proteins, enzymes, tissue growth factors, interferons, and other immunomodulators have also been reported [17–20].

Another study showed that drug delivery to or across the skin could be modulated either by the use of liposomes or permeation enhancers, respectively. In this work, a high flux of caffeine through the skin was achieved from a system containing a combination of the enhancers diethylglycol and oleic acid, whereas the application of caffeine from small unilamellar liposomes resulted in accumulation of large drug quantities within the skin (Figure 13.1). Quantitative skin autoradiography results confirmed these findings, showing that the highest concentration of caffeine was accumulated in epidermis [10].

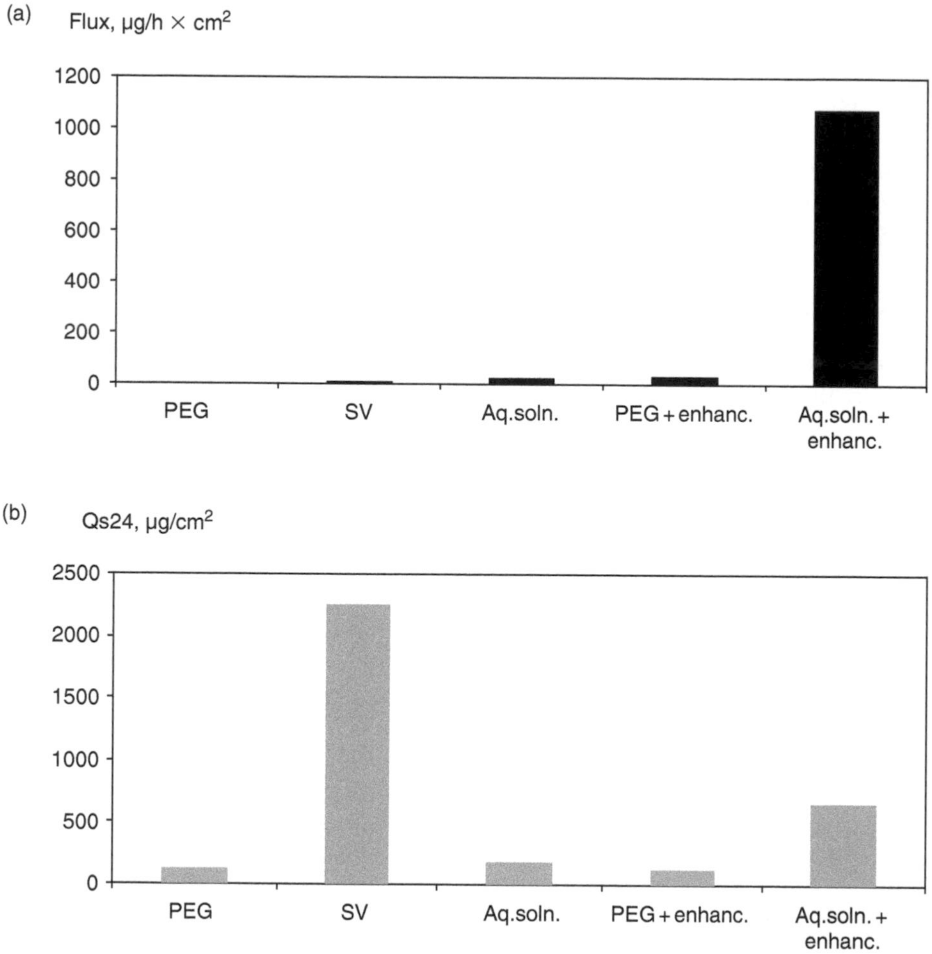

FIGURE 13.1 Modulation of caffeine skin delivery by carrier design. Effect of different formulations on the flux (a) and the skin accumulation (Qs24) (b) of caffeine in *in vitro* permeation experiments through the hairless mouse skin for 24 h. Each system contained 3% of the drug. SV-small vesicle liposomes (mean diameter: 40 nm); PEG, polyethylene glycol base; Aq. soln., aqueous solution; PEG + enhanc., PEG base containing 10% oleic acid and 20% transcutol; Aq. soln. + enhanc., aqueous system containing 10% oleic acid and 20% transcutol.

It is commonly agreed that direct contact between liposomes and skin is necessary for efficient delivery. The work carried out by Hofland et al. resulted in an important contribution to the understanding of the interaction between liposomes and human stratum corneum [21]. In this study, by using freeze fracture electron microscopy, SUV structures were visualized on the stratum corneum surface at the interface of the liposomal dispersion and the stratum corneum (Figure 13.2). Near the interface, the liposomes appear to fuse and adsorb on the stratum corneum, forming clusters of bilayers and rough structures on the top of the outermost corneocytes. No changes in the ultrastructure of the intercellular lipid regions were detected within the stratum corneum, not even in the lipid regions between the first and second layers of corneocytes. The results of this study confirmed the work hypothesis that liposomes remain adsorbed on the skin surface thus producing drug reservoir in the upper cutaneous strata. These findings are in agreement with the results obtained by Egbaria and Weiner who tested the penetration of radiolabeled lipids (^{14}C-dipalmitoylphosphatidylcholine and ^{3}H-cholesterol) from liposomes into various strata of excised hairless mouse using the striping technique [16]. In this work the highest quantities of lipid probes were detected on the stratum corneum surface and no radioactivity was measured in the receiver compartment.

A number of clinical implications of drug reservoir formation in the upper skin layers by delivery from liposomes have been reported [4–6,13]. From these studies it appears that the efficiency of liposomal-incorporated drugs was superior to other formulations in the treatment of disorders, which do not affect the deep layers of the skin. For example, in a double-blind, randomized paired study on patients suffering from atopic eczema or psoriasis vulgaris, a liposomal betamethasone dipropionate was more efficient than a nonliposomal preparation in eczematous but not in psoriatic patients [6].

Tolerability is another aspect that should be considered during the design of topical preparations. In a double-blind study on 20 acne vulgaris patients, who received for 10 weeks liposomal tretinoin (0.01%) on one side of the body and a commercial gel preparation with either 0.025% or 0.05% of the drug on the other, Schafer-Korting et al. evaluated efficacy and local tolerability [13]. The measured clinical parameters included: quantity of comedones/papules/pustules, and rating redness, scaling, and burning on to a 4-point scale. Results indicate that despite the lower tretinoin content, the liposomal preparation was equipotent to the reference gels, but significantly superior in terms of skin tolerability. In a further study, morphological changes following *in vitro* application of a liposomal (0.05% and 0.025%) or conventional (0.05%) tretinoin were investigated on reconstructed epidermis. The findings of this work show that after 24 h structural changes characterizing toxic dermatitis in the epidermis visualized by light and electron microscopies were more pronounced for the conventional formulation tested [22].

Liposomes applied on the skin were also investigated for their delivery proprieties to the pilosebaceous units [15,23–28]. The *in vitro* skin penetration behavior of carboxyfluorescein incorporated in multilamellar liposomes (phosphatidylcholine: cholesterol: phosphatidylserine) and in another four nonliposomal systems (HEPES pH 7.4 buffer; 5% propylene glycol; 10% ethanol; and 0.05% sodium lauryl sulfate) was studied by Lieb et al. [25]. Using two fluorescent techniques the authors found a higher accumulation of the probe within skin follicles when delivered from liposomes [25]. Further, in an interesting setup of *in vitro* and *in vivo* experiments in mice, Hoffman's group observed liposomal delivery of the active Lac-Z gene and its expression mostly in the hair follicles [26,28].

It was shown that liposomes, due to their structure, have a retarding effect on the incorporated drug release. In early studies, Knepp et al. reported that progesterone release from agarose gel was faster than from liposomes embedded in the gel [29]. This retarding release behavior from liposomes was further confirmed by a lower drug transport rate as compared to the gel measured across hairless mouse skin [30]. Another study by Foldvari et al. [8] examined the

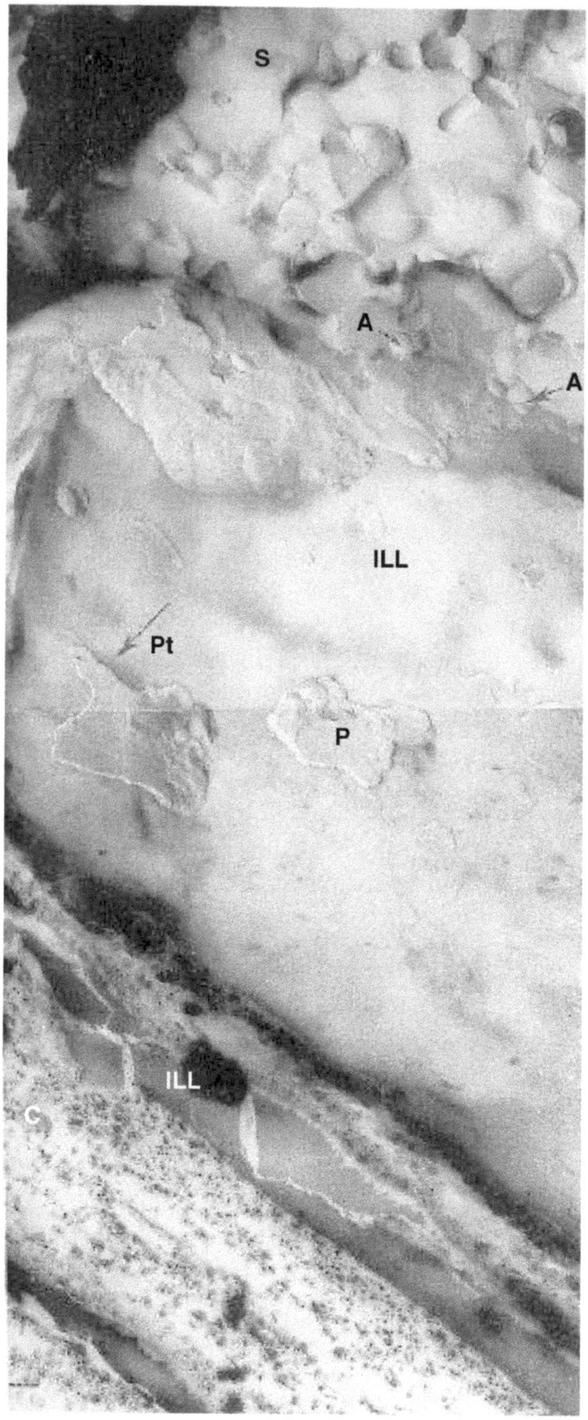

FIGURE 13.2 Stratum corneum incubation with liposome suspension (S). The liposomes appear to adsorb (A) on to the surface of the stratum corneum. The liposomes are fusing edge to edge, and patches of liposomal clusters (P) are formed on the top of the outermost corneocytes (C). No changes in ultrastructure could be found deeper in the stratum corneum. Pt, the direction of platinum shadow scale bar (0.1 μm). (Reproduced from Hofland, H.E. et al., *Br. J. Dermatol.*, 132, 853, 1995. With permission from Blackwell Publishing.)

release and the pharmacodynamic effect of tetracaine delivered from liposomes in the presence of gel-forming polymers including carbopol, veegum, and methylcellulose. Results of this work show that the presence of additives decreased both the total amount released and the amount of drug in peak sample, with the exception of methylcellulose when used at low concentrations. The appearance of peak was also delayed in the presence of most additives.

From a formulative point of view, it is important to take into consideration that the base in which liposomes are incorporated could also affect the drug delivery profile as well as the interactions between the liposomes and skin structures. The effect of formulation bases on dyphylline skin permeation from liposomes was examined by Touitou et al. [9]. In this work the effect of four bases for dyphylline liposomes (polyethylene glycol (PEG), carbopol gel, a PEG-enhancer base, and water) was studied. With these bases, the lowest skin permeation flux and a superior skin partitioning of liposomal dyphylline were reported for the PEG base, suggesting that this base favored dyphylline accumulation within the skin. A number of liposomes are currently marketed by cosmetic raw material companies for incorporation in different bases to obtain final products. It is important to keep in mind that the delivery properties of the final formulation could be altered by the vehicle used and should be tested to confirm any performance characteristics claimed.

One of the major drawbacks of liposomes is related to their preparation methods [3,4]. Liposomes for topical delivery are prepared by the same classic methods widely described in the literature for preparation of these vesicles. The majority of the liposome preparation methods are complicated multistep processes. These methods include hydration of a dry lipid film, emulsification, reverse phase evaporation, freeze–thaw processes, and solvent injection. Liposome preparation is followed by homogenization and separation of unentrapped drug by centrifugation, gel filtration, or dialysis. These techniques suffer from one or more drawbacks such as the use of solvents (sometimes pharmaceutically unacceptable), an additional sizing process to control the size distribution of final products (sonication, extrusion), multiple-step entrapment procedure for preparing drug-containing liposomes, and the need for special equipment.

The stability of liposomes is of great concern during the design of topical formulations. As liposomal preparations contain an aqueous phase, the lipid vesicles may be subject to a series of adverse effects including aggregation, fusion, phospholipid oxidation, and hydrolysis [31–35]. The hydrolysis is dependent on several factors such as pH, temperature, buffer species, ionic strength, acyl chain length and headgroup, and the state of aggregation. Oxidation of phospholipids in liposomes mainly takes place via a free radical chain mechanism in unsaturated fatty acyl chain-carrying phospholipids. Furthermore, cholesterol present in the liposomal vesicles is readily oxidized, creating a stability problem for lipid-based drug products. Some of these oxidation by-products tend to be rather toxic in biological systems. The oxidation products 25-hydroxycholesterol, 7-keto-cholesterol, 7α- and 7β-hydroxycholesterol, cholestane-3β,5α,6β-triol, and the 5- and 7-hydroperoxides, were found in a concentrate, which had activity causing toxic effects on aortic smooth muscle cells [35]. Storage at low temperatures and protection from light and oxygen reduces the chance of oxidation. Further protection can be achieved with the addition of antioxidants. Working under nitrogen or argon also minimizes the oxidation of lipids during preparation [32,33].

13.3 NIOSOMES AND ELASTIC NIOSOMES—NONIONIC SURFACTANT VESICLES

Nonionic surfactant vesicles (niosomes) were first proposed by Handjani-Vila et al. [36] as systems to improve accumulation of the active molecule within the skin and thus benefit cosmetic products. These reports opened the way for an intensive investigation of these vesicles as carriers for skin administration of drugs [37–47].

Niosomes are bilayer structures formed from amphiphiles in aqueous media. Many types of surfactants have been used for formulation of niosomes [41,48–55]. Basically, these vesicles are analogous to liposomes. Niosome formation requires the presence of a particular class of amphiphile and aqueous solvent. Among the surfactants we can enumerate polyoxyethylene alkyl ethers, sorbitan esters, polysorbate–cholesterol mixtures, crown ether derivatives, perfluoroalkyl surfactants, alkyl glycerol ethers, and others. The hydrophobic moiety of the surfactant may contain one or more alkyl (C_{12}–C_{18}) or perfluoroalkyl (C_{10}) groups or a steroidal group. Among the hydrophilic headgroups found in these amphiphils are ethylene oxides, glycerols, crown ether, polyhydroxyls, and sugars. The hydrophilic and the hydrophobic moieties are generally linked by ether, ester, or amide bonds.

In certain cases, cholesterol is required for vesicle formation. It is commonly accepted that the hydrophilic lipophilic balance (HLB) is a parameter that could indicate the vesicle-forming potential of surfactants. For amphiphils such as sorbitan esters and alkyl ethers, low HLB values could predict vesicle formation [52,55]. However, niosomes were obtained from polysorbate 20 (HLB 16.7), a highly hydrophilic molecule, when cholesterol at an appropriate concentration was added to the amphiphil [44]. In this case it could be assumed that a kind of amphiphilic complex with a lower HLB was responsible for the vesicle formation. An excellent review on the structure, characteristics, chemical composition, and mechanism of action of niosomes was published by Uchegbu and Vyas [41].

The methods for preparation of niosomes are similar and as complicated as those used for liposomes. One of the most frequently utilized techniques consists of the hydration of a mixture of the surfactant–lipid at elevated temperature followed by optional size reduction (by sonication, extrusion, homogenization, etc.) to obtain a homogeneous colloidal dispersion and separation of the unentrapped drug [36,40,41,52,55].

A number of works investigated the interaction between niosomes and human skin. With niosomes prepared from C_{12} alcohol polyoxyethylene ether and cholesterol, vesicular structures of about 100 nm size have been observed between the first and second layers of human corneocytes 48 h after incubation as well as in the deeper strata of the skin [37]. The authors concluded that the structures visualized in the deeper regions could be vesicles reorganized from individual molecules that penetrated the skin. In another study, electron micrographs illustrated that niosomes containing surfactants and cholesterol affected only the most superficial corneocytes. Moreover, two-photon fluorescence microscopy confirmed that fluorescent probe encapsulated in niosomes was confined to the intercellular spaces within the apical stratum corneum layers [56].

Niosomes, prepared from polyoxyethylene alkyl ethers (C_nEO_m, $m = 3$ or 7, $n = 12$ or 18) or stearate/palmitate sucrose ester, Wasag-7, were tested for skin delivery of protonated and unprotonated lidocaine. The investigated systems resulted in a relatively low encapsulation efficiency of the drug. No significant differences in the flux through human stratum corneum membrane were obtained when the drug was applied from vesicles or a control solution through an occlusive or nonocclusive mode [45].

Hofland et al. investigated the *in vitro* permeation behavior of estradiol from niosomes (*n*-alkyl polyoxyethylenes/cholesterol) through human stratum corneum. In this study examining drug delivery from multilamellar niosomes, small unilamellar niosomes and a micellar solution, all being saturated systems containing 1.5, 1.5, and 0.75 mM estradiol, very low permeation fluxes were detected (64 ± 17, 45 ± 15, and 42 ± 2 ng/cm^2/h, respectively) [40].

Occlusion is a condition that could affect drug transport from niosomes and through the stratum corneum. Such an effect was reported for saturated estradiol niosomal formulations composed of polyoxyethylene alkyl ether surfactants and sucrose ester surfactants with cholesterol and dicetyl phosphate, for which occlusion enhanced the drug human stratum corneum transport [43].

More recently, Carafa et al. showed that niosomes could be obtained from polyoxyethylene sorbitan monolaurate–cholesterol in aqueous environment. These authors investigated the delivery of lidocaine HCl and lidocaine base from vesicles through silicone membrane and nude mice skin [44]. It was found that only the charged molecule (loading pH 5.5) could be encapsulated within the vesicles (~30%). This behavior was explained by the entrapment ability of the hydrophilic moiety within the aqueous core of the vesicles. The lipophilic unionized form of lidocaine (loading pH 8.6) remained unattached. The amount of lidocaine permeated through nude mice skin from these niosomes was similar to liposomes and only about twofold greater than from a micellar system.

Niosomes were also investigated for their ability to deliver drugs to hair follicles. Niemiec et al. investigated the deposition of two polypeptide drugs, interferon-α (INF-α) and cyclosporin into the pilosebaceous units of the hamster ear [46]. In this work, following 12 h *in vivo* application, only one niosomal system comprised of glyceryl dilaurate/cholesterol/polyoxyethylene-10-stearyl ether improved drug delivery into skin shafts as compared to other niosomal systems, liposomes, hydroethanolic, or aqueous solutions of these drugs.

Van den Bergh et al. [56–58] introduced a new type of niosomes, the elastic surfactant vesicles containing octaoxyethylene laurate-ester (PEG-8-L) and sucrose laurate-ester (L-595). To study the elasticity of the vesicles, fatty acid spin labels were incorporated into the systems [58]. The results of these studies indicate that the molar content of PEG-8-L in vesicles directly affects the bilayers elasticity. The *in vivo* interactions of these elastic vesicles with hairless mouse skin were evaluated following nonocclusive application in comparison to rigid vesicles composed of sucrose stearate-ester (Wasag-7) [57]. Disrupted organization of skin bilayers and increased skin permeability were observed following application of elastic vesicles but not the rigid ones. Thin layer chromatography measurements showed that after 1 h application of elastic and rigid vesicles, a sixfold increase in the amount of elastic vesicle constituents was present within the stratum corneum compared to Wasag-7 rigid vesicles. However, after a longer application time, the quantity of PEG-8-L vesicle constituent in stratum corneum significantly decreased as compared to the rigid vesicle material.

Mechanistic studies on elastic vesicles were further conducted by Honeywell-Nguyen et al. [59–63]. In one of these works, ketorolac-loaded elastic (PEG-8L: L-595) and rigid vesicles (L-595 only) applied nonocclusively to the skin were used to quantify skin distribution profiles of a deuterium-labeled phospholipid (as a marker for the vesicle constituent) and ketorolac [59]. Data obtained by tape-stripping suggested that the lipid probe and the drug from elastic vesicles can enter the deeper layers of the stratum corneum, in contrast to rigid vesicle material. Elastic vesicles also improved ketorolac permeation across the skin (Figure 13.3). In a further study, the effect of elastic and rigid vesicles on the penetration of pergolide across human skin was further assessed *in vitro* using flow-through Franz diffusion cells [60]. For this purpose, various L-595/PEG-8-L/sulfosuccinate vesicles, from very rigid to very elastic, were investigated in comparison to a saturated buffer solution. In this study, skin occlusion enhanced drug permeation from elastic vesicles as well as from the buffer solution. It is noteworthy that the highest pergolide skin permeation was obtained when a simple saturated buffer solution not containing vesicles was applied under occlusion.

The stability of various niosomal formulations depends on factors such as preparation methods, storage temperature, the encapsulated drug, the surfactants, and additive mixture [41,52,64,65]. It may be possible to stabilize niosomes by a variety of methods such as the use of membrane-spanning lipids, the interfacial polymerization of surfactant monomers *in situ*, addition of polymerized surfactants, cholesterol, steric and electrostatic stabilizers to the formulation [41,52]. In general, vesicle aggregation may be prevented by inclusion of

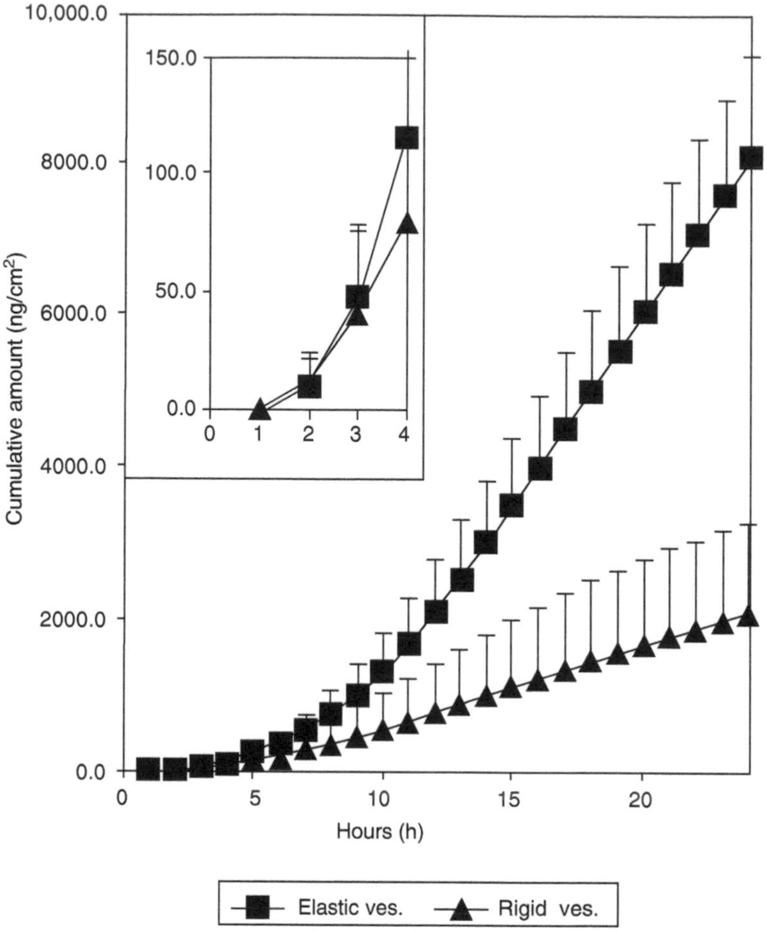

FIGURE 13.3 The cumulative amount of ketorolac as a function of time from elastic and rigid vesicle formulations across human skin *in vitro*. Elastic vesicles were clearly more effective as compared with rigid vesicles in the enhancement of ketorolac transport across the skin. The cumulative amounts found after 1 and 4 h of elastic vesicle treatment (corresponding to the time periods chosen for this study), however, were still very low. (Reproduced from Honeywell-Nguyen, P.L. et al., *J. Invest. Dermatol.*, 123, 902, 2004. With permission from Blackwell Publishing.)

molecules that stabilize the system. Steric and electrostatic stabilization of sorbitan monostearate niosomes could be achieved by inclusion of cholesteryl polyoxyethylene ethers or dicetyl phosphate [54,55].

Skin safety of niosomes was tested in a number of studies. As an example, the toxicity of polyoxyethylene alkyl ether vesicles containing C_{12-18} alkyl chains and 3 and 7 oxyethylene units was assessed by measuring the effect on proliferation of cultured human keratinocytes [47]. It was found that the length of either polyoxyethylene headgroup or alkyl chain had only a minor influence on keratinocyte proliferation. However, the ether surfactants were much more toxic than esters tested in this study. The concentrations of ether surfactants required to inhibit cell proliferation by 50% were 10-fold lower than for ester surfactants. Neither the HLB nor the critical micelle concentration values or cholesterol content affected keratinocyte proliferation.

13.4 TRANFERSOMES—ULTRADEFORMABLE LIPOSOMES

Transfersomes, deformable liposomes, were introduced by Cevc et al. in the early 1990s [66–69]. The main components of these systems are phospholipids, a surfactant edge activator (such as sodium cholate), water and sometimes very low concentrations of ethanol ($\leq$7%) [66]. Transfersomes are prepared by the same methods as liposomes. The preparation process is usually followed by homogenization, sonication, or other mechanical means to reduce the size of the lipid vesicles.

The mechanism of skin permeation by transfersomes as proposed by the authors involves a number of processes, which occur when the system is applied nonoccluded to the skin. Transfersomes dehydrate on the skin surface by evaporation, resulting in an osmotic pressure difference between the region of higher water concentration inside the skin and the nearly dry surface of the skin. It is suggested that these lipid vesicles could avoid the osmotic tension by dehydration, thereby opening narrow intercellular pores in the stratum corneum and penetrating the barrier. Thus, nonoccluded administration is vital to allow for enhanced delivery from transfersomes. The authors claim that due to the presence of the polar surfactant molecule in the lipid phase of the aggregate, the deformable vesicle is able to squeeze and forge through the small pores [66,68,70].

Another possible explanation for the fact that transfersomes are able to deliver molecules only under nonoccluded conditions could be that as a result of water evaporation from the applied sodium cholate–phospholipid aggregates system, concentrated micellar systems of cholates or cholates–phospholipids, or both, are generated. The micelles may further delipidize the stratum corneum creating small pores through which the drug could penetrate.

An early study confirmed that occlusion is detrimental to transfersome penetration enhancement ability. The results of this study clearly demonstrated that, under the occlusive conditions, murine skin permeation of fluorescently labeled lipids from transfersomal suspensions and liposomes is comparable [67]. Furthermore, Guo et al. reported that the vesicles failed to transfer detectable quantities of cyclosporin A through the hydrated abdominal mice skin [71].

In a study carried out in normoglycemic human volunteers, following epicutaneous application of insulin associated with transfersomes (Transferinsulin™) the first signs of hypoglycemia were observed at 90–180 min and a maximum transfersome-mediated decrease in blood glucose levels was about 35% of the effect of the same insulin dose administrated through subcutaneous injection. The cumulative pharmacodynamic response of at least 50% of the subcutaneous dose was obtained after skin application of insulin from transfersomes with systemic normoglycaemia that lasted for 16 h [67,72,73].

Hofer et al. [74–76] designed and characterized transfersomes with immunomodulators interleukin-2 (IL-2) and INF-α and reported that both molecules retained their biological activity and could be efficiently encapsulated in these vesicles.

A study on diclofenac-containing transfersomes (Transfenac) carried out in comparison to a commercial Voltaren emulgel formulation evaluated the distribution of radiolabeled drug and a number of pharmacokinetic parameters in mice, rats, and pigs [70]. The relative drug quantities measured in the mice dorsal muscles epicutaneously treated with 9 mg diclofenac/kg were 6.7 and 3.5 μg/g for Transfenac and Voltaren, respectively. In rats, following application of 0.6 mg drug/kg body weight the ratio of diclofenac quantities delivered from Transfenac vs. Voltaren measured deep in the hind legs was close to 10. Similar results were obtained in pigs (Figure 13.4).

The transfersome technology has been tested for delivery of a number of additional molecules including steroids, vaccines, and adjuvants [67,77–79] and these studies have been recently reviewed by Cevc [80].

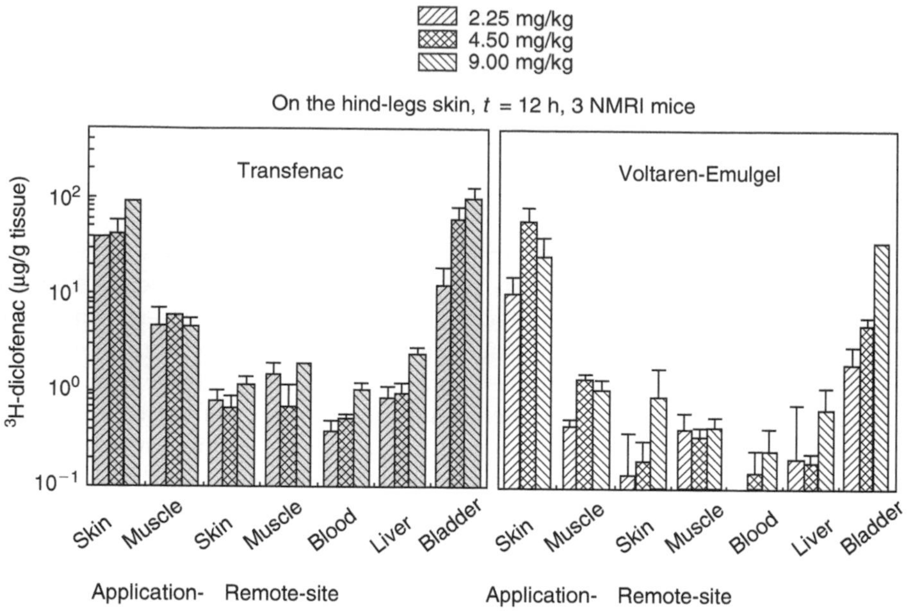

FIGURE 13.4 Left: Tissue concentration of diclofenac-derived ^{3}H-radioactivity at $t = 12$ h after an epicutaneous application of 2.25 mg/kg body weight (right-dashed), 4.5 mg/kg body weight (hatched), and 9 mg/kg body weight (left-dashed) of the drug in ultradeformable vesicles, transfersomes, on the hind thigh of mice. Right: Biodistribution of diclofenac-derived ^{3}H-radioactivity 12 h after an application of 2.25 mg/kg body weight (right-dashed), 4.5 mg/kg body weight (hatched), and 9 mg/kg body weight (left-dashed) in the commercial hydrogel. (Reproduced from Cevc, G. and Blume, G., *Biochim. Biophys. Acta*, 1514, 191, 2001. With permission from Elsevier.)

Various aspects of skin drug delivery from transfersomes were also studied by Barry's group [81–85]. In one work, the authors compared the potential use of ultradeformable and standard liposomes as skin drug delivery systems and found that the ultradeformable formulation was superior to standard liposomes for *in vitro* skin delivery of 5-FU [81]. Interestingly, in these systems the entrapment efficiency was less than 10%. In further work these authors replaced sodium cholate with Span 80 and Tween 80 for preparation of ultradeformable vesicles of estradiol [82]. The maximum fluxes of estradiol through the human epidermal membrane increased by 18-, 16-, and 15-fold for vesicles containing sodium cholate, Span 80, and Tween 80 as compared with aqueous control. The authors concluded that these nonionic surfactants are as efficient edge activators as sodium cholate in the preparation of transfersomes.

13.5 ETHOSOMES—SOFT PHOSPHOLIPID VESICLES FOR ENHANCED DERMAL AND TRANSDERMAL DELIVERY

Although frequently referred to as a kind of liposomes, ethosomes are very different from other lipid vesicles by their composition, structure, mechanism of action, and delivery properties. Ethosomal carriers contain soft lipid vesicles (mainly composed of phospholipids, ethanol, and water) in a hydroethanolic milieu. They have appropriate features, designed to allow for enhanced delivery by passive transport to the deep skin strata and through the skin [86–90]. Their delivery enhancing properties can be modulated by changes in the composition and structure. Ethosomes, invented by Touitou [86–88], were so named to emphasize the presence of ethanol in a vesicular structure. Due to the interdigitation effect of alcohol on lipid bilayers, it was previously thought that the ethanol milieu is destructive to vesicular structures and that vesicles could not coexist with high concentrations of alcohol [91,92]. In contrast to this belief, the existence of vesicles and the structure of ethosomes were evidenced by various methods including ^{31}P-NMR, transmission electron microscopy (TEM), and scanning electron microscopy (SEM) [86]. Phosphorous NMR spectra of ethosomal systems exhibited a solid-state lineshape, the phospholipid bilayer configuration typically observed in phosphatidylcholine vesicles in the aqueous environment. Moreover, the paramagnetic-ion NMR spectra pointed toward a greater ethosomal phospholipid membrane fluidity and permeability to cations, in comparison to liposomes [86]. Using the electron microscopy techniques, the authors were able to visualize the ethosomal vesicles. Ethosomal systems may contain unilamellar [93,94] or multilamellar [86,95] lipid vesicles, with their sizes ranging from 30 nm to microns [96]. Scanning electron micrographs confirmed a three-dimensional nature of ethosomal vesicles, whereas negatively stained transmission electron micrographs showed that multilamellar ethosomes are characterized by a structure with phospholipid bilayers throughout the vesicle [86,93–96] (Figure 13.5a). One of the distinguishing characteristics of ethosomes is their ability to efficiently entrap molecules of various lipophilicities (Figure 13.5b). The presence of ethanol, together with high-vesicle lamellarity, allows for efficient entrapment of hydrophilic, lipophilic, and amphiphilic molecules [86]. For example, ultracentrifugation studies demonstrated that the encapsulation efficiency of ethosomes could be as high as 90% and 83% in the case of the lipophilic drugs testosterone and minoxidil, respectively. This is different from the problematic entrapment of lipophilic compounds in liposomes restricted by a limited number of phospholipid bilayers surrounding the aqueous core.

To gain insights into the characteristics of ethosomes that allow them to efficiently promote drug delivery into and through the skin the transition temperature (T_m) of vesicular lipids and free energy measurements of the vesicle bilayers were assessed by differential scanning calorimetry (DSC) and fluorescence anisotropy [86,94,95]. The T_m values were reported to be 20°C–35°C lower for ethosomes as compared to liposomes composed from the same phospholipids but without ethanol, and are explained by the fluidity of phospholipid bilayers in ethosomes [86,95]. This suggestion was further strengthened by fluorescent anisotropy

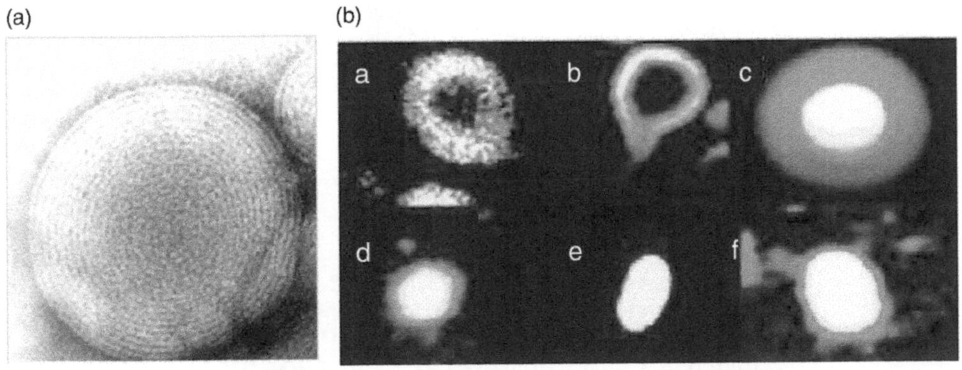

FIGURE 13.5 (a) Visualization of a typical multilamellar ethosome containing 2% PL, 30% ethanol, and water by TEM: (b) Entrapment of fluorescent probes by phopholipid vesicles as visualized by CSLM. Liposomes (a–c) or ethosomes (d–f) were prepared with one of three fluorescent probes: rhodamine red (a, d), D-289 (b, e), or calceine (c, f). White represents the highest concentration of probe. (Reproduced from Touitou, E. et al., *J. Control. Release*, 65, 403, 2000. With permission from Elsevier.)

measurements of AVPC (9-Antrylvinyl labeled analog of phosphatidylcholine) where a 20% lower value was measured in comparison to liposomes [86]. These results evidenced that ethosomal vesicles possess a soft malleable structure, which could be related to the fluidizing effect of ethanol on the phospholipid bilayers [91,92]. In comparison to liposomes, ethosomes are much less rigid.

Besides this important fluidizing effect of ethanol on the phospholipid vesicle bilayers, the alcohol interferes with the lipid organization in stratum corneum. Based on the results obtained in fluorescent anisotropy and DSC experiments as well as in skin permeation studies, a mechanism for enhanced delivery by the ethosomal system was proposed [86] (Figure 13.6). The authors suggest that the fluidizing effect of ethanol on the lipid bilayers of stratum corneum together with the characteristics of ethosomes contribute to the skin permeation enhancement of drug from the ethosomal carrier. The softness of ethosomal vesicles imparts to them the ability to penetrate the disturbed stratum corneum lipid bilayers and promote delivery of active agents into the deep layers of the skin and through the skin. Further, the soft vesicle penetrates the fluidized stratum corneum bilayers forging a pathway through the skin by virtue of its particulate nature and later on fuses with cell membranes in the deeper skin layers releasing the active agent there. Further data, which shed more light on understanding this sequence of synergistic processes, were obtained by researchers from the skin permeation studies. For example, it was found that an important amount (10.5% of initial) of phosphatidylcholine (PL) permeated the skin during a 24 h experiment, suggesting that the vesicles might have traversed the skin strata. In a further confocal laser scanning microscopy (CLSM) study by Godin and Touitou, fluorescently labeled bacitracin

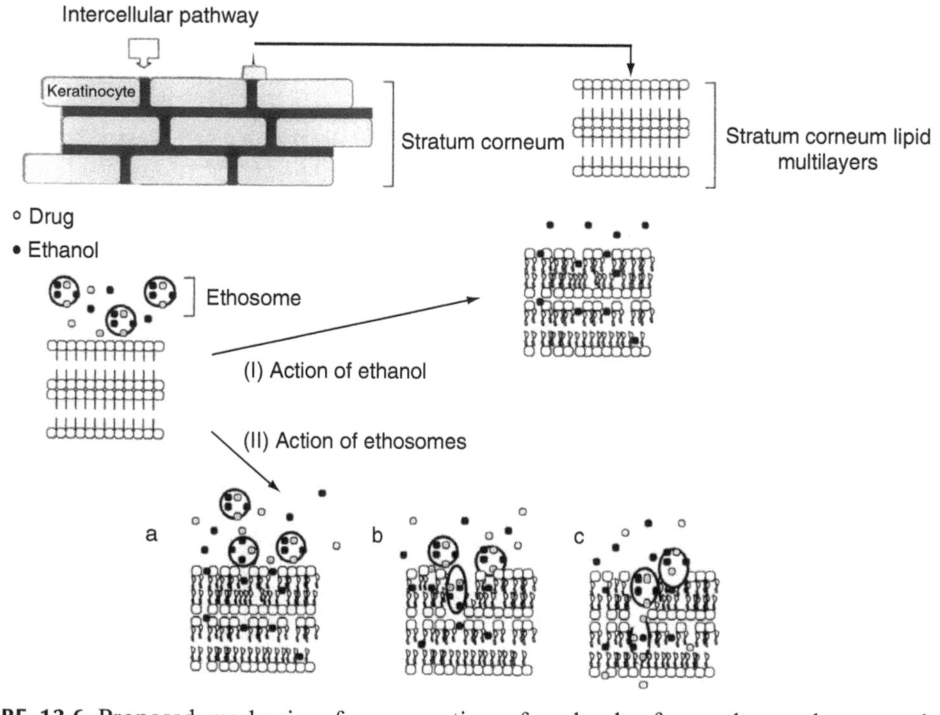

FIGURE 13.6 Proposed mechanism for permeation of molecules from ethosomal system through stratum corneum (SC) lipids. (Reproduced from Touitou, E. et al., *J. Control. Release*, 65, 403, 2000. With permission from Elsevier.)

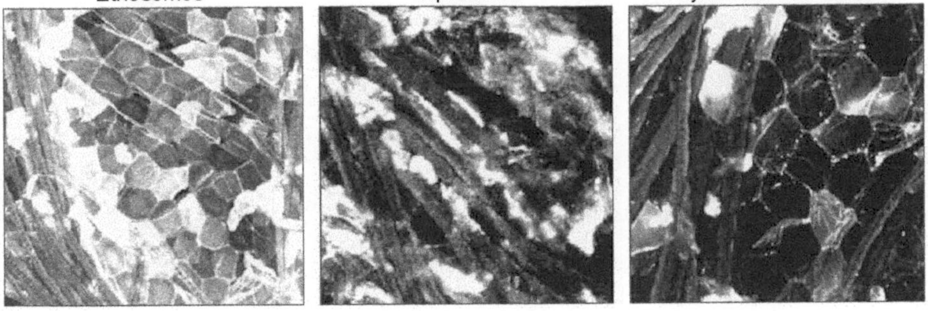

FIGURE 13.7 Reconstituted CLSM optical slices of the stratum corneum of the skin following skin delivery of FITC-Bac *in vivo* in SD rats. Comparison of skin permeation routes from systems containing 0.1% FITC-Bac following an 8 h skin exposure: ethosomes vs. liposomes and hydroethanolic solution. (Reproduced from Godin, B. Touitou, E., *J. Control. Release*, 94, 365, 2004. With permission from Elsevier.)

(FITC-Bac) delivered *in vivo* from ethosomes, penetrated the rat skin through the intercorneocyte pathways, which typically exist along the lipid domain of the stratum corneum [95] (Figure 13.7). In contrast, significantly lower fluorescence staining of the intercellular penetration pathway and no inter- or intracorneocyte fluorescence were observed with FITC-Bac hydroethanolic solution and liposomes, respectively.

An important number of agents from various pharmacological groups and with a variety of physicochemical characteristics were formulated with ethosomes and tested *in vitro*, in animals and clinical studies. When testing the enhanced skin permeation properties of ethosomes, it is important to compare them to hydroethanolic solutions, liposomes, or phospholipid ethanolic solutions, which contain the system ingredients in various combinations. In research using fluorescent probes, the ability of a rhodamine red dihexadecanoyl glycerophosphoethanolamine (RR), which is a phopholipid probe, to penetrate deep into the skin was evaluated by CLSM [86,95,97]. The obtained CLS micrographs of nude mice skin after 8 h application of RR from ethosomes, hydroalcoholic solution, and liposomes illustrate that delivery from ethosomes resulted in the highest intensity of fluorescence up to a depth of 150 μm. In agreement with previously described works on liposomes, no deep penetration of RR from liposomal dispersion was visualized. As RR is used as an indicator of lipid fusion, which does not usually cross lipid bilayers, the results obtained in these experiments suggest that ethosomes traversed the skin strata to a high depth, in contrast to liposomes that remained on the skin surface. In further CLSM studies, delivery of two additional fluorescent probes possessing distinct characteristics, a hydrophilic probe calceine and an amphiphilic cationic probe D-289 (4-(4-diethylamino) styryl-*N*-methylpyridinium iodide) from ethosomes and control systems was assessed [93,96,97]. The authors reported that for calceine, a maximum fluorescence intensity (MaxFI) value of 150 arbitrary units (AU) was obtained when delivered from ethosomes. This value was reached at the skin depth of 20 μm, remained constant throughout approximately 50 μm, and dropped to zero only at 160 μm. In contrast, when calceine was applied from liposomes or from a hydroethanolic solution, lower MaxFI values were obtained followed by a sharp decrease of fluorescent intensity to zero at the depths of 60 and 80 μm, respectively [97]. Similar data were measured for the amphiphilic cationic probe D-289, which was delivered from trihexyphenidyl HCl ethosomes deeply into the skin (170 μm) with a significantly higher intensity than the two control systems [93] (Figure 13.8).

Charged and hydrophilic large molecules (polypeptides, proteins) were challenging compounds tested with ethosomal carriers [93,98–100]. The ability of trihexyphenidyl hydrochloride

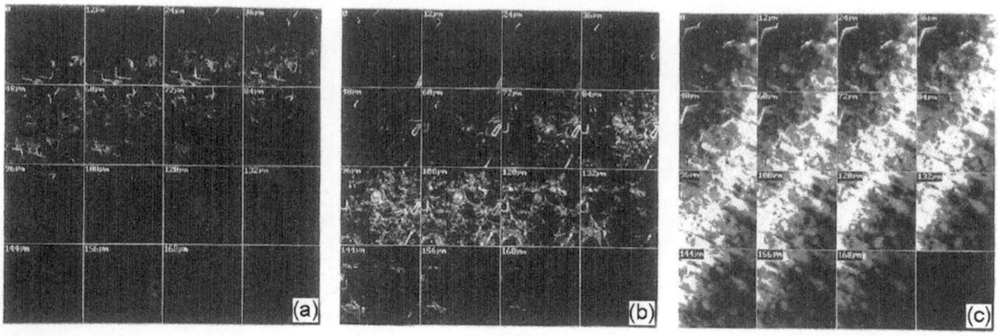

FIGURE 13.8 Visualization of penetration of fluorescent probe D-289 to full-thickness nude mice skin by CLSM, following 8 h application from systems containing 0.03% D-289 and 1% THP: (a) liposomes (containing 2% PL) (b) 30% hydroethanolic solution and (c) ethosomes (containing 2% PL and 30% ethanol). (Reproduced from Dayan, N. and Touitou, E., *Biomaterials*, 21, 1879, 2000. With permission from Elsevier.)

(THP), an anti-Parkinsonian agent, to be delivered across the skin from ethosomes was investigated [93]. In this study, THP fluxes were 87, 51, and 4.5 times higher for ethosomes (0.21 mg/cm^2/h) than for liposomes, phosphate buffer, and hydroethanolic drug solution, respectively. Furthermore, significantly greater quantities of this cationic molecule were detected in the skin following delivery from ethosomes than from any other control system tested. THP ethosomes designed in this work were found to efficiently entrap THP (75%) and be stable for at least 2 years at room temperature. In further work with ethosomal insulin, a significant decrease (up to 60%) in blood glucose levels (BGL) in both normal and diabetic rats was measured (Figure 13.9) [96,99,100]. On the other hand, insulin application from a control nonethosomal formulation was not able to reduce the blood levels of glucose . It is noteworthy that the ethosomal insulin system resulted in a favorable pharmacodynamic profile with the plateau effect lasting for at least 8 h. The possibility to tailor the desirable hypoglycemic effect by modulating system composition presents another advantage reported for ethosomal insulin [99]. The obtained data show that the ethosomal carrier enabled percutaneous absorption of insulin by passive diffusion.

As known, highly lipophilic molecules ($\log P > 5$) exhibit insufficient transdermal absorption. These compounds accumulate within stratum corneum layers and encounter problems in their clearance into the viable epidermis interface where partitioning into a predominantly aqueous environment is required. Lodzki et al. examined the transdermal delivery of cannabidiol (CBD), a molecule with $\log P \sim 8$, from an ethosomal carrier [101–103]. The application of 100 mg ethosomal composition containing 3% CBD on nude mice skin for 24 h resulted in a significant drug amount transported through the skin (559 μg/cm^2) and formation of an important CBD skin reservoir (845 μg/cm^2). The authors further reported that *in vivo* application of CBD ethosomes to the abdominal skin of CD1 nude mice resulted in significant localization of the molecule within the skin (110.07 ± 24.15 μg/cm^2) and in the underlying muscle (11.537 μg CBD/g muscle) [96,101]. Pharmacokinetic study in ICR mice revealed that following system application to the animal abdomen for 72 h, CBD's plasma concentrations reach steady-state levels of 0.67 μg/mL in 24 h and last at least until the end of the experiment. In this work the anti-inflammatory effect of CBD ethosomal systems was also evaluated. Significant differences in the pharmacodynamic profiles between CBD-treated and -untreated animals at all times during the experiment were observed, indicating that the inflammation was prevented by transdermal delivery of ethosomal CBD [102]. Testosterone,

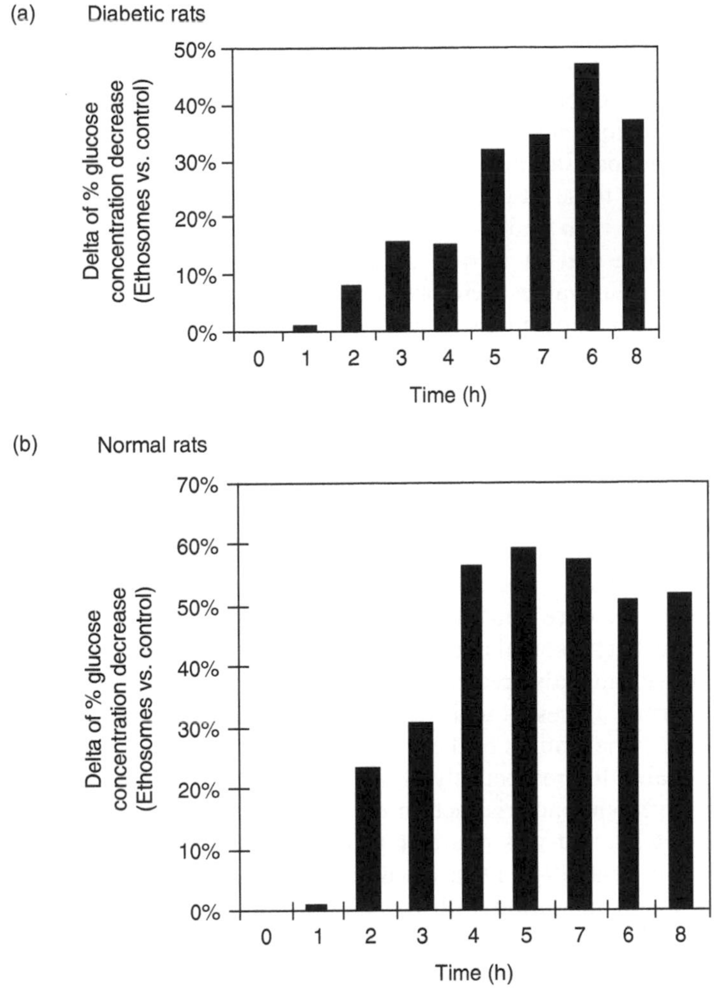

FIGURE 13.9 Delta blood glucose levels measured following *in vivo* application to the abdominal area of ethosomal insulin vs. no treatment in (a) diabetic and (b) normal SD-1 rats.

prescribed as a hormone replacement therapy in primary and secondary hypogonadism, is an additional lipophilic molecule tested with ethosomes. A 30-fold higher quantity of testosterone was found to permeate *in vitro* through rabbit pinna skin from an ethosomal patch, Testosome, (848.16 ± 158.38 μg) as compared to Testoderm (Alza), each containing 0.25 mg hormone/cm^2. AUC and C_{max} values measured in rabbits treated with Testosome were 2.2 and 2.4 times higher than in animals treated with Testoderm [86]. In a more recent study, a nonpatch ethosomal testosterone system was designed and tested vs. AndroGel (Unimed, USA) for systemic absorption *in vivo* in rats and skin permeation through human skin *in vitro* [104]. A single dermal application of 400 mg ethosomal formulation containing 1% testosterone resulted in C_{max} and AUC values of 1970 ± 251 ng/dL and 9313 ± 385 and ng/dL*h, respectively. For this ethosomal testosterone formulation, the estimated application area needed for obtaining physiological testosterone human plasma levels is only 40 cm^2, which is approximately 10-fold lesser than for the currently marketed products. These works demonstrated that ethosomes possess the ability not only to enhance the drug partitioning into the lipophilic layers of the skin, but also to permit its clearance into the hydrophilic environment, leading to transdermal delivery.

Optimization of drug delivery to the target tissue is the important goal in modern therapy. Currently, bacterial infections localized within deep dermal and subdermal tissues are cured only by systemic antibiotic administration. Recently, a novel approach to treat these problematic infections by local application of antibiotic in ethosomal carrier was investigated [94,95,96,105–107]. In these studies, ethosomes enabled the delivery of the polypeptide antibiotic bacitracin *in vitro* and *in vivo* through human cadaver and rat skin, respectively. It is worth mentioning that in *in vitro* experiments occlusion had no effect on drug skin transport and similar flux values were obtained for occlusive and nonocclusive applications. *In vivo* experiments with ethosomal bacitracin demonstrated that the antibiotic was efficiently delivered into deep skin layers from ethosomes but not from liposomes or a hydroethanolic solution [95]. Of particular interest are the results of the *in vivo* study on *Staphylococcus aureus* infected mice, designed to test the hypothesis that ethosomal erythromycin is able to eradicate deep dermal infection [107]. In this work the authors compared the efficiency of ethosomal erythromycin applied to the skin-infected site with parenteral or topical administration of the drug in a hydroethanolic solution. It was found that the therapy with ethosomal erythromycin was as effective as the systemically administered erythromycin. A very efficient healing of *S. aureus*-induced deep dermal infections and zero bacterial skin counts were measured when the mice were treated with ethosomal erythromycin. Histological evaluation of the skin treated with ethosomal antibiotic revealed normal skin structure with no bacterial colonies. In contrast, animals treated with topical hydroethanolic erythromycin solution developed deep dermal abscesses with destroyed dermal structures colonized by *S. aureus*. In these animals, bacterial counts of the infected tissues were 1.06×10^7 and 0.27×10^7 cfu/g tissue on days 7 and 10, respectively. Further work shows that the ethosomal system significantly improved antibacterial action of erythromycin as compared to free drug in *in vitro* susceptibility tests, and was nontoxic to dermal 3T3 cultured fibroblasts in live/dead viability/cytotoxicity assay [94]. Data discussed above suggest the possibility to substitute systemic antibiotic administration with topical ethosomal drug application in clinical use. This new therapeutical mode could result in decreased drug exposure and associated side effects, thereby potentially increasing patient compliance. In addition to these therapeutic benefits, effective bacterial kill with ethosomal erythromycin could ultimately result in minimizing bacterial resistance.

Insufficient drug delivery into the basal epidermis where virus replication occurs is at the origin of the low efficiency of acyclovir (ACV) dermal products for *Herpes simplex* topical treatment [108–110]. Keeping this in mind, a new formulation containing ethosomal ACV (EA) was designed and evaluated in a two-armed, double-blind, randomized study. In this work ACV ethosomal cream was compared to a commercial product (Zovirax, ZC) in 40 subjects (61 herpetic episodes) [111]. The assessed clinical parameters were the time to crust formation, time to loss of crust (the healing period), and the percentage of abortive lesions (lesions not progressive beyond the papular stage). The results clearly showed that EA significantly improved all evaluated clinical parameters in both parallel and crossover arms. In the crossover arm, both the time to crust formation and the healing period were significantly reduced with EA vs. ZC (1.8 vs. 3.5 days and 4.2 vs. 5.9 days, respectively). In the parallel arm 80% of lesions crusted on the third day from the initiation of treatment with EA compared to only 10% in the ZC group. The fraction of abortive herpetic lesions was ~30% in the case of EA vs. only 10% with ZC (Figure 13.10). The results of this clinical trial show that treatment with EA formulation significantly improved all evaluated clinical parameters, which could be explained by the efficient delivery of the drug to its target tissue.

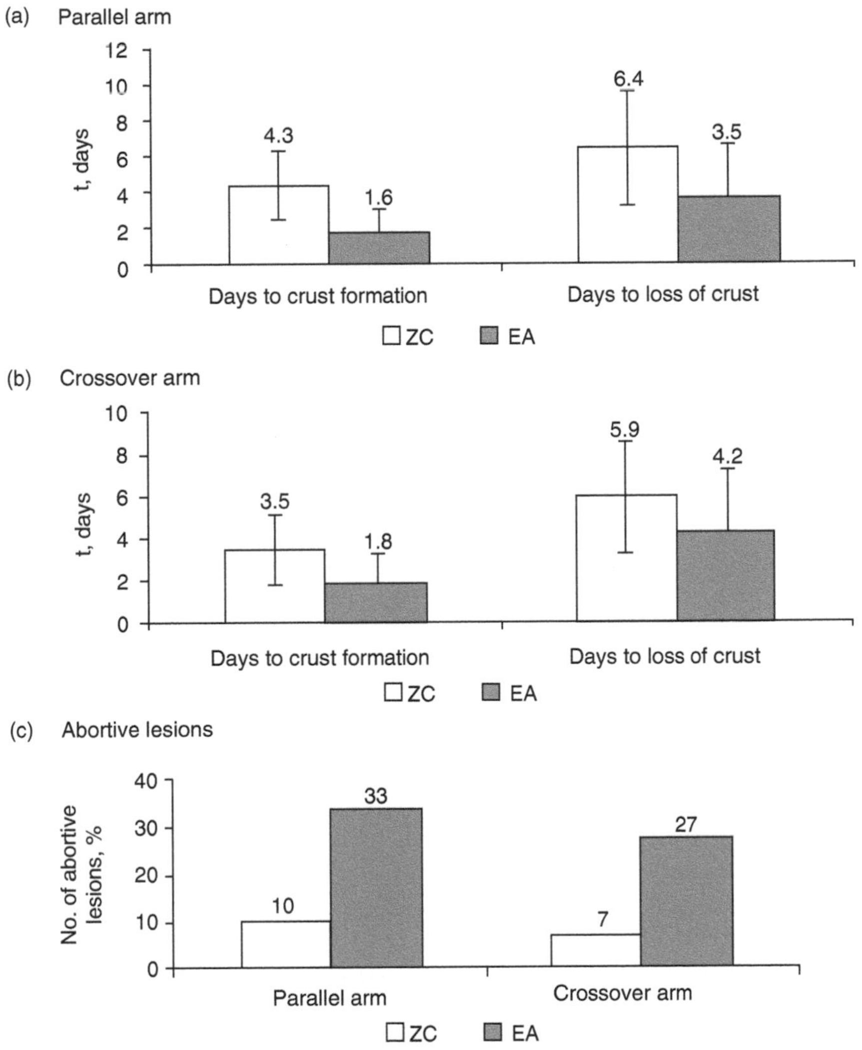

FIGURE 13.10 Parameters assessed in a two-armed, randomized, double-blind clinical study in RHL patients with two formulations containing 5% acyclovir: ethosomal acyclovir (EA) vs. Zovirax cream (ZC). (a) Data obtained in parallel arm; (b) data obtained in crossover arm; (c) number of abortive lesions (%). (Reproduced from Touitou, E. et al., *Drug Dev. Res.*, 50, 406, 2000. With permission from John Wiley & Sons.)

Ethosomal carriers were found to be very efficient in promoting delivery of molecules to pilosebaceous and hair follicular units. Tailored minoxidil ethosomes were investigated *in vivo* for drug localization into the pilosebaceous units as a potential way to improve its therapeutic effect in hair loss disorders [96,112]. The designed minoxidil ethosomes appeared as multilayered vesicles with an encapsulation capacity of $83 \pm 6\%$. *In vivo* application of ethosomes and liposomes containing 0.5% minoxidil and 50 μCi H^3-minoxidil to the dorsal region of hairless rats for up to 24 h resulted in the accumulation of the tritiated drug within the skin follicles (Figure 13.11). Quantitative skin autoradiography [113–115] was further utilized to assess the levels of H^3-minoxidil within pilosebaceous elements of the skin. The obtained results demonstrated that the ethosomal system was 5 times more efficient in

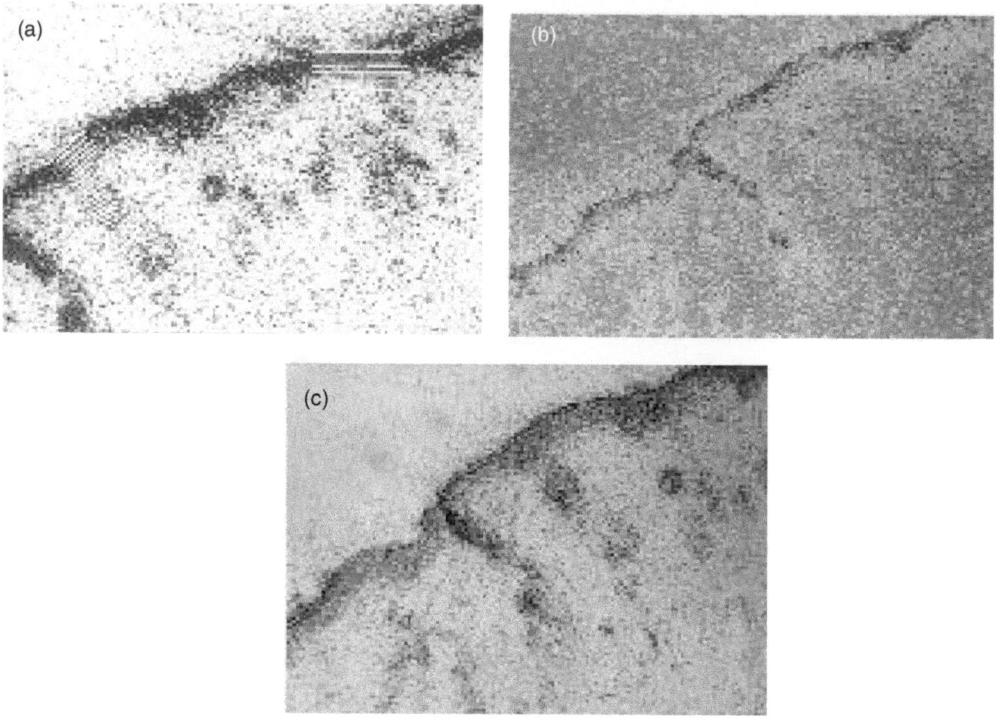

FIGURE 13.11 Localization of H^3-minoxidil within the pilosebaceous units. Skin images were obtained following 12 h application of ethosomal system containing 0.5% minoxidil and 50 μCi tritiated drug to the dorsal region of hairless rat (6 cm^2) *in vivo*. (a) skin autoradiogram; (b) histological image; (c) superimposition of a and b.

targeting minoxidil to the skin shaft than liposomes (22 vs. 4.5 nmol/g tissue, respectively, $p < 0.005$) [90].

The tolerability and safety of ethosomes were tested in cultured skin cells, on animals and in humans. An *in vitro* live/dead viability/cytotoxicity viability test carried out with various vesicular systems and controls indicated that ethosomal carriers were not toxic to 3T3 fibroblasts and cultured cells kept their viability [97]. In animal studies on rabbits no acute skin irritation was observed following a single-dose 48 h occlusive application of patches containing the ethosomal systems. Furthermore, cumulative 14 d repeated ethosomal patch application also did not generate any significant erythema [86]. In another study following 12, 24, or 48 h application of ethosomes containing 2% PL and 45% ethanol on healthy human volunteers, the formulation was very well tolerated and no signs of erythema were detected. In this study no significant difference in erythema index (ΔEI) was measured between skin areas treated with ethosomes and saline [116].

There are a number of methods, which are used to prepare stable ethosomal formulations depending on drug and the target of drug delivery. In contrast to liposomes and other vesicular carriers described in this chapter, methods used for preparation of ethosomes do not require special equipment and can be scaled up without difficulty. Moreover, the produced vesicles are of uniform size and no homogenization/size-reducing steps are required. The specific methods for ethosomes preparation are given in patents published on ethosomes [87,88].

The stability of ethosomal systems incorporating various drugs was assessed in a number of studies by comparing the average diameter and the structure of the vesicles during a 2-year period at room temperature [86,93,94]. The mean size of empty and cationic trihexyphenidyl loaded vesicles remained unaffected during the storage interval. Moreover, visualization by negative stain TEM confirmed that the vesicular structure of the ethosomes persisted after 2 years of storage and no significant structural changes occurred over that time in both systems [86,93]. In another study with ethosomes containing 1% erythromycin no significant variations in the dimensions of ethosomes throughout the storage at room temperature were measured [94]. The initial mean size of the vesicles was 123 ± 15 nm, whereas the diameter of ethosomes following a 1-year interval was 117 ± 18 nm. TEM micrographs confirmed that erythromycin unilamellar ethosomes kept their configuration during the stability evaluation experiments. A number of pharmaceutical and cosmetic products covered by patents on ethosomes are marketed. Data on SupraVir cream (Trima, Israel), an ethosomal formulation of acyclovir, indicate that the formulation and the drug has a long shelf life with no stability problems. Acyclovir in SupraVir cream has been shown by HPLC assay to be stable for at least 3 years at 25°C. Furthermore, skin permeation experiments showed that the cream after 3 years retains its initial penetration enhancing capacity [117].

Studies conducted with ethosomal systems *in vitro*, *in vivo*, and in humans confirm that these soft vesicles enable efficient transport of active agents into the deep strata and across the skin. The above mentioned drugs that have been tested with ethosomal carriers are clearly not exhaustive but intended to illustrate the wide range of molecules that could be tailored with ethosomes for facilitating dermal, transdermal, or follicular delivery.

13.6 CONCLUDING REMARKS

In summary, a classic liposomes remain confined to the upper skin layers, resulting in the formation of drug reservoir mainly in the horny strata and generally do not penetrate into the deeper layers of the skin. Thus, the liposomal carriers could be efficient in local treatment of skin disorders and for cosmetic uses.

Lipid vesicles for enhancing drug skin permeation have been specially designed using various approaches. For this purpose, ethosomes, soft vesicles with fluid bilayers, and the elastic vesicles, transfersomes, and elastic niosomes, have been invented. Although each vesicle type has its own characteristics, their common feature is their ability to improve the delivery of drugs across the skin barrier. The high tolerability and efficiency of vesicular systems, such as ethosomes, open vast potential therapeutic uses. These carriers might offer advanced local and systemic new therapies with agents that are unable to efficiently penetrate the stratum corneum via passive diffusion.

REFERENCES

1. Mezei, M., and V. Gulusekharam. 1980. Liposomes, a selective drug delivery system for the topical route of administration. *Life Sci* 26:1473.
2. Mezei, M., and V. Gulusekharam. 1982. Liposomes, a selective drug delivery system for the topical route of administration: Gel dosage form. *J Pharm Pharmacol* 34:473.
3. New, R.R.C. 1990. *Liposomes—a practical approach.* Oxford: Oxford University Press.
4. Braun-Falco, O., H.C. Korting, and H.I. Maibach. 1992. *Gries Conference: Liposome Dermatics.* Heidelberg: Springer-Verlag.
5. Schmid, M.H., and H.C. Korting. 1994. Liposomes: A drug carrier system for topical treatment in dermatology. *Crit Rev Ther Drug Carrier Syst* 11:97.
6. Korting, H.C., et al. 1990. Liposome encapsulation improves efficacy of betamethasone dipropionate in atopic eczema but not in psoriasis vulgaris. *Eur J Clin Pharmacol* 39:349.
7. Lasch, J., and W. Wohlrab. 1986. Liposome-bound cortisol: A new approach to cutaneous therapy. *Biomed Biochim Acta* 45:1295.

8. Foldvari, M., B. Jarvis, and C.J.N. Ogueijofor. 1993. Topical dosage form of liposomal tetracaine: Effect of additives on the *in vitro* release and *in vivo* efficacy. *J Control Release* 27:193.
9. Touitou, E., et al. 1992. Diphylline liposomes for delivery to the skin. *J Pharm Sci* 81:131.
10. Touitou, E., et al. 1994. Modulation of caffeine skin delivery by carrier design: Liposomes versus permeation enhancers. *Int J Pharm* 103:131.
11. Touitou, E., et al. 1994. Liposomes as carriers for topical and transdermal delivery. *J Pharm Sci* 83:1189.
12. Masini, V., et al. 1993. Cutaneous bioavailability in hairless rats of tretinoin in liposomes or gel. *J Pharm Sci* 82:17.
13. Schafer-Korting, M., H.C. Korting, and E. Ponce-Poschl. 1994. Liposomal tretinoin for uncomplicated acne vulgaris. *Clin Invest* 72:1086.
14. Fresta, M., and G. Puglisi. 1996. Application of liposomes as potential cutaneous drug delivery systems. In vitro and in vivo investigation with radioactively labelled vesicles. *J Drug Target* 4:95.
15. Weiner, N., et al. 1994. Liposomes: A novel topical delivery system for pharmaceutical and cosmetic applications. *J Drug Target* 2:405.
16. Egbaria, K., and N. Weiner. 1992. In *Liposome Dermatics*, eds. O. Braun-Falco, H.C. Korting, and H.I. Maibach, 172. Berlin, Heidelberg: Springer-Verlag.
17. Jeschke, M.G., et al. 2002. Non-viral liposomal keratinocyte growth factor (KGF) cDNA gene transfer improves dermal and epidermal regeneration through stimulation of epithelial and mesenchymal factors. *Gene Ther* 9:1065.
18. Jeschke, M.G., et al. 2001. Therapeutic success and efficacy of nonviral liposomal cDNA gene transfer to the skin *in vivo* is dose dependent. *Gene Ther* 8:1777.
19. Wolf, P., et al. 2000. Topical treatment with liposomes containing T4 endonuclease V protects human skin *in vivo* from ultraviolet-induced upregulation of interleukin-10 and tumor necrosis factor-alpha. *J Invest Dermatol* 114:149.
20. Vorauer-Uhl, K., et al. 2001. Topically applied liposome encapsulated superoxide dismutase reduces postburn wound size and edema formation. *Eur J Pharm Sci* 14:63.
21. Hofland, H.E., et al. 1995. Interactions between liposomes and human stratum corneum *in vitro*: Freeze fracture electron microscopical visualization and small angle x-ray scattering studies. *Br J Dermatol* 132:853.
22. Schaller, M., R. Steinle, and H.C. Korting. 1997. Light and electron microscopic findings in human epidermis reconstructed *in vitro* upon topical application of liposomal tretinoin. *Acta Derm Venereol* 77:122.
23. Ciotti, S.N., and N. Weiner. 2002. Follicular liposomal delivery systems. *J Liposome Res* 12:143.
24. Lieb, L.M., G. Flynn, and N. Weiner. 1994. Follicular (pilosebaceous unit) deposition and pharmacological behavior of cimetidine as a function of formulation. *Pharm Res* 11:1419.
25. Lieb, L.M., et al. 1992. Topical delivery enhancement with multilamellar liposomes into pilosebaceous units: I. *In vitro* evaluation using fluorescent techniques with the hamster ear model. *J Invest Dermatol* 99:108.
26. Hoffman, R.M. 1998. Topical liposome targeting of dyes, melanins, genes, and proteins selectively to hair follicles. *J Drug Target* 5:67.
27. Hoffman, R.M. 2005. Gene and stem cell therapy of the hair follicle. *Methods Mol Biol* 289:437.
28. Li, L., and R.M. Hoffman. 1995. The feasibility of targeted selective gene therapy of the hair follicle. *Nat Med* 1:705.
29. Knepp, V.M., et al. 1988. Controlled drug release from a novel liposomal delivery system. I. Investigations of transdermal potential. *J Control Release* 5:211.
30. Knepp, V.M., F.C. Szoka, and R.H. Guy. 1990. Controlled drug release from a novel liposome delivery system. II. Transdermal delivery characteristics. *J Control Release* 12:25.
31. Ulrich, A.S. 2002. Biophysical aspects of using liposomes as delivery vehicles. *Biosci Rep* 22:129.
32. Brandl, M. 2001. Liposomes as drug carriers: A technological approach. *Biotechnol Annu Rev* 7:59.
33. Grit, M., and D.J. Crommelin. 1993. Chemical stability of liposomes: Implications for their physical stability. *Chem Phys Lipids* 64:3.
34. Anderson, M., and A. Omri. 2004. The effect of different lipid components on the *in vitro* stability and release kinetics of liposome formulations. *Drug Deliv* 11:33.

35. Cox, D.C., K. Comai, and A.L. Goldstein. 1988. Effects of cholesterol and 25-hydroxycholesterol on smooth muscle cell and endothelial cell growth. *Lipids* 23:85.
36. Handjani-Vila, R.M., et al. 1979. Dispersion of lamellar phases of non-ionic lipids in cosmetic products. *Int J Cosmet Sci* 1:303.
37. Junginger, H.E, H.E. Hofland, and J.A. Bouwstra. 1991. Liposomes and niosomes: Interactions with human skin. *Cosmet Toiletries* 106:45.
38. Hofland, H.E., et al. 1995. Interactions between liposomes and human stratum corneum *in vitro*: Freeze fracture electron microscopical visualization and small angle diffraction scattering studies. *Br J Dermatol* 132:853.
39. Schreier, H., and J.A. Bouwstra. 1994. Liposomes and niosomes as topical drug carriers: Dermal and transdermal drug delivery. *J Control Release* 30:1.
40. Hofland, H.E., et al. 1994. Estradiol permeation from nonionic surfactant vesicles through human stratum corneum *in vitro. Pharm Res* 11:659.
41. Uchegbu, I.F., and S.P. Vyas. 1998. Non-ionic surfactant based vesicles (niosomes) in drug delivery. *Int J Pharm* 172:33.
42. Manconi, M., et al. 2002. Niosomes as carriers for tretinoin. I. Preparation and properties. *Int J Pharm* 234:237.
43. Van-Hal, D., et al. 1996. Diffusion of estradiol from nonionic surfactant vesicles through human stratum corneum *in vitro. STP Pharm Sci* 6:72.
44. Carafa, M., E. Santucci, and G. Lucania. 2002. Lidocaine-loaded non-ionic surfactant vesicles: Characterization and *in vitro* permeation studies. *Int J Pharm* 231:21.
45. Van-Hal, D.A., et al. 1996. Encapsulation of lidocaine base and hydrochloride into non-ionic surfactant vesicles (NSVs) and diffusion through human stratum corneum *in vitro. Eur J Pharm Sci* 4:147.
46. Niemiec, S.M., C. Ramachandran, and N. Weiner. 1995. Influence of nonionic liposomal composition on topical delivery of peptide drugs into pilosebaceous units: An *in vivo* study using the hamster ear model. *Pharm Res* 12:1184.
47. Hofland, H.E., et al. 1991. Interactions of non-ionic surfactant vesicles with cultured keratinocytes and human skin in vitro: A survey of toxicological aspects and ultrastructural changes in stratum corneum. *J Control Release* 16:155.
48. Arunothayanun, P., et al. 2000. The effect of processing variables on the physical characteristics of non-ionic surfactant vesicles (niosomes) formed from a hexadecyl diglycerol ether. *Int J Pharm* 201:7.
49. Lawrence, M.J., S.M. Lawrence, and D.J. Barlow. 1997. Aggregation and surface properties of synthetic double-chain-ionic surfactants in aqueous solution. *J Pharm Pharmacol* 49:594.
50. Baillie, A.J., et al. 1985. The preparation and properties of niosomes—non-ionic surfactant vesicles. *J Pharm Pharmacol* 37:863.
51. Paspaleeva-Kuhn, V., and E. Nurnberg. 1992. Participation of Macrogolstearate 400 lamellar phases in hydrophilic creams and vesicles. *Pharm Res* 9:1336.
52. Uchegbu, I.F., and A.T. Florence. 1995. Non-ionic surfactant vesicles (niosomes): Physical and pharmaceutical chemistry. *Adv Colloid Interface Sci* 58:1.
53. Assadullahi, T.P., R.C. Hider, and A.J. McAuley. 1991. Liposome formation from synthetic polyhydroxyl lipids. *Biochim Biophys Acta* 1083:271.
54. Uchegbu, I.F., et al. 1995. Distribution, metabolism and tumoricidal activity of doxorubicin administered in sorbitan monostearate (span-60) niosomes in the mouse. *Pharm Res* 12:1019.
55. Yoshioka, T., B. Sternberg, and A.T. Florence. 1994. Preparation and properties of vesicles (niosomes) of sorbitan monoesters (span-20, span-40, span-60 and span-80) and a sorbitan triester (span-85). *Int J Pharm* 105:1.
56. Van den Bergh, B.A., et al. 1999. Interactions of elastic and rigid vesicles with human skin *in vitro*: Electron microscopy and two-photon excitation microscopy. *Biochim Biophys Acta* 1461:155.
57. Van den Bergh, B.A., et al. 1999. Elasticity of vesicles affects hairless mouse skin structure and permeability. *J Control Release* 62:367.
58. Van den Bergh, B.A., et al. 2001. Elasticity of vesicles assessed by electron spin resonance, electron microscopy and extrusion measurements. *Int J Pharm* 217:13.

59. Honeywell-Nguyen, P.L., G.S. Gooris, and J.A. Bouwstra. 2004. Quantitative assessment of the transport of elastic and rigid vesicle components and a model drug from these vesicle formulations into human skin *in vivo*. *J Invest Dermatol* 123:902.
60. Honeywell-Nguyen, P.L., and J.A. Bouwstra. 2003. The *in vitro* transport of pergolide from surfactant-based elastic vesicles through human skin: A suggested mechanism of action. *J Control Release* 86:145.
61. Honeywell-Nguyen, P.L., S. Arenja, and J.A. Bouwstra. 2003. Skin penetration and mechanisms of action in the delivery of the D2-agonist rotigotine from surfactant-based elastic vesicle formulations. *Pharm Res* 20:1619.
62. Honeywell-Nguyen, P.L., et al. 2003. The *in vivo* transport of elastic vesicles into human skin: Effects of occlusion, volume and duration of application. *J Control Release* 90:243.
63. Honeywell-Nguyen, P.L., et al. 2002. Transdermal delivery of pergolide from surfactant-based elastic and rigid vesicles: Characterization and *in vitro* transport studies. *Pharm Res* 19:991.
64. Udupa, N., et al. 1993. Formulation and evaluation of methotrexate niosomes. *Drug Dev Ind Pharm* 19:1331.
65. Uchegbu, I.F., et al. 1996. Phase-transitions in aqueous dispersions of the hexadecyl diglycerol ether (c(16)g(2)) non-ionic surfactant, cholesterol and cholesteryl poly-24-oxyethylene ether-vesicles, tubules, discomes and micelles. *STP Pharm Sci* 6:33.
66. Cevc, G., and G. Blume. 1992. Lipid vesicles penetrate into intact skin owing to the transdermal osmotic gradients and hydration force. *Biochim Biophys Acta* 1104:226.
67. Cevc, G. 1996. Transfersomes, liposomes and other lipid suspensions on the skin: Permeation vesicle penetration and transdermal drug delivery. *Crit Rev Ther Drug Carrier Syst* 13:257.
68. Cevc, G., A. Schatzein, and G. Blume. 1995. Transdermal drug carriers: Basic properties, optimization and transfer efficiency in the case of epicutaneously applied peptides. *J Control Release* 36:3.
69. Cevc, G., et al. 1998. Ultraflexible vesicles, transfersomes, have an extremely low pore penetration resistance and transport therapeutic amounts of insulin across the intact mammalian skin. *Biochim Biophys Acta* 1368:201.
70. Cevc, G., and G. Blume. 2001. New, highly efficient formulation of diclofenac for the topical, transdermal administration in ultradeformable drug carriers. Transfersomes. *Biochim Biophys Acta* 1514:191.
71. Guo, J., et al. 2000. Lecithin vesicular carriers for transdermal delivery of cyclosporin A. *Int J Pharm* 194:201.
72. Cevc, G. 2003. Transdermal drug delivery of insulin with ultradeformable carriers. *Clin Pharmacokinet* 42:461.
73. Cevc, G., et al. 1998. Ultraflexible vesicles, transfersomes, have an extremely low pore penetration resistance and transport therapeutic amounts of insulin across the intact mammalian skin. *Biochim Biophys Acta* 1368:201.
74. Hofer, C., et al. 1999. Formulation of interleukin-2 and interferon-alpha containing ultradeformable carriers for potential transdermal application. *Anticancer Res* 19:1505.
75. Hofer, C., et al. 2000. New ultradeformable drug carriers for potential transdermal application of interleukin-2 and interferon-alpha: Theoretic and practical aspects. *World J Surg* 24:1187.
76. Hofer, C., et al. 2004. Transcutaneous IL-2 uptake mediated by transfersomes depends on concentration and fractionated application. *Cytokine* 25:141.
77. Gupta, P.N., et al. 2005. Tetanus toxoid-loaded transfersomes for topical immunization. *J Pharm Pharmacol* 57:295.
78. Cevc, G., and G. Blume. 2004. Hydrocortisone and dexamethasone in very deformable drug carriers have increased biological potency, prolonged effect, and reduced therapeutic dosage. *Biochim Biophys Acta* 1663:61.
79. Cevc, G., and G. Blume. 2003. Biological activity and characteristics of triamcinolone-acetonide formulated with the self-regulating drug carriers, transfersomes. *Biochim Biophys Acta* 1614:156.
80. Cevc, G. 2004. Lipid vesicles and other colloids as drug carriers on the skin. *Adv Drug Deliv Rev* 56:675.

81. El Maghraby, G.M., A.C. Williams, and B.W. Barry. 2001. Skin delivery of 5-fluorouracil from ultradeformable and standard liposomes *in vitro*. *Pharm Pharmacol* 53:1069.
82. El Maghraby, G.M., A.C. Williams, and B.W. Barry. 2000. Oestradiol skin delivery from ultra-deformable liposomes: Refinement of surfactant concentration. *Int J Pharm* 196:63.
83. El Maghraby, G.M., A.C. Williams, and B.W. Barry. 2000. Skin delivery of oestradiol from lipid vesicles: Importance of liposome structure. *Int J Pharm* 204:159.
84. El Maghraby, G.M., A.C. Williams, and B.W. Barry. 1999. Skin delivery of oestradiol from deformable and traditional liposomes: Mechanistic studies. *J Pharm Pharmacol* 51:1123.
85. El Maghraby, G.M., A.C. Williams, and B.W. Barry. 2001. Skin hydration and possible shunt route penetration in controlled estradiol delivery from ultradeformable and standard liposomes. *J Pharm Pharmacol* 53:1311.
86. Touitou, E., et al. 2000. Ethosomes-novel vesicular carriers for enhanced delivery: Characterization and skin penetration properties. *J Control Release* 65:403.
87. Touitou, E. 1996. Compositions for applying active substances to or through the skin, US Patent 5,540,934.
88. Touitou, E. 1998. Composition for applying active substances to or through the skin, US Patent 5,716,638.
89. Touitou, E., et al. 1997. Ethosomes: The novel vesicular carriers for enhanced skin delivery. *Pharm Res* 14:S305.
90. Touitou, E., B. Godin, and C. Weiss. 2000. Enhanced delivery of drugs into and across the skin by ethosomal carriers. *Drug Dev Res* 50:406.
91. Chin, J.H., and D.B. Goldstein. 1997. Membrane disordering action of ethanol: Variation with membrane cholesterol content and depth of the spin label probe. *Mol Pharmacol* 13:435.
92. Harris, R.A., et al. 1987. Effect of ethanol on membrane order: Fluorescence studies. *Ann NY Acad Sci* 492:125.
93. Dayan, N., and E. Touitou. 2000. Carriers for skin delivery of trihexyphenidyl HCl: Ethosomes vs. liposomes. *Biomaterials* 21:1879.
94. Godin, B., and E. Touitou. 2005. Erythromycin ethosomal systems: Physicochemical characterization and enhanced antibacterial activity. *Curr Drug Deliv* 2:269.
95. Godin, B., and E. Touitou. 2004. Mechanism of bacitracin permeation enhancement through the skin and cellular membranes from an ethosomal carrier. *J Control Release* 94:365.
96. Godin, B., and E. Touitou. 2003. Ethosomes: New prospects in transdermal delivery. *Crit Rev Ther Drug Carrier Syst* 20:63.
97. Touitou, E., et al. 2001. Intracellular delivery mediated by an ethosomal carrier. *Biomaterials* 22:3053.
98. Dayan, N. 2000. Enhancement of skin permeation of trihexyphenidyl HCl. PhD thesis, The Hebrew University of Jerusalem, Jerusalem, Israel.
99. Dkeidek, I., and E. Touitou. 1999. Transdermal absorption of polypeptides. *AAPS Pharm Sci* 1:S202.
100. Dkeidek, I. 1999. Transdermal transport of macromolecules. MSc thesis, The Hebrew University of Jerusalem, Jerusalem, Israel.
101. Lodzki, M. 2002. Transdermal delivery of cannabidiol by ethosomal carrier. MSc thesis, The Hebrew University of Jerusalem, Jerusalem, Israel.
102. Lodzki, M., et al. 2003. Cannabidiol-transdermal delivery and anti-inflammatory effect in a murine model. *J Control Release* 93:377.
103. Touitou, E., et al. 2002. Transdermal delivery of cannabinoids by ethosomal carriers, 4th World Meeting ADRITELF/APV/APGI Abstracts, Florence.
104. Ainbinder, D., and E. Touitou. 2005. Testosterone ethosomes for enhanced transdermal delivery. *Drug Deliv* 12:1.
105. Godin, B., and E. Touitou. 2002. Intracellular and dermal delivery of polypeptide antibiotic bacitracin, Drug research between information and life sciences, ICCF, 3rd Symposium Abstracts, Bucharest.
106. Godin, B., E. Rubinstein, and E. Touitou. 2004. A new approach to interfere with microorganisms' resistance to antibiotics, 31st Annual Meeting and Exposition of the Controlled Release Society, Honolulu, Hawaii, 354.

107. Godin, B., et al. 2005. A new approach for treatment of deep skin infections by an ethosomal antibiotic preparation: An *in vivo* study. *J Antimicrob Chemother* 55:989.
108. Shaw, M., et al. 1985. Failure of acyclovir cream in the treatment of recurrent herpes labialis. *Br Med J* 291:7.
109. Spruance, S.L., and C.S. Crumpacker. 1982. Topical 5% acyclovir in polyethylene-glycol for herpes simplex labialis: Antiviral effect without clinical benefit. *Am J Med* 315:73A.
110. Huff, J.C., et al. 1981. The histopathologic evolution of recurrent herpes labialis. *J Am Acad Dermatol* 5:550.
111. Horwitz, E., et al. 1999. A clinical evaluation of a novel liposomal carrier for acyclovir in the topical treatment of recurrent herpes labialis. *Oral Surg Oral Med Oral Pathol Oral Radiol Endod* 88:700.
112. Godin, B., M. Alcabez, and E. Touitou. 1999. Minoxidil and Erythromycin targeted to pilosebaceous units by ethosomal delivery systems. *Acta Technol Legis Medicament* 10:107.
113. Meidan, V., and E. Touitou. 2001. Treatments for androgenetic alopecia and alopecia areata: Current options and future prospects. *Drugs* 61:53.
114. Touitou, E., V. Meidan, and E. Horwitz. 1998. Methods for quantitative determination of drug localized in the skin. *J Control Release* 56:7.
115. Fabin, B., and E. Touitou. 1991. Localization of lipophilic molecules penetrating rat skin *in vivo* by quantitative autoradiography. *Int J Pharm* 74:59.
116. Paolino, D., et al. 2005. Ethosomes for skin delivery of ammonium glycyrrhizinate: *In vitro* percutaneous permeation through human skin and *in vivo* anti-inflammatory activity on human volunteers. *J Control Release* 106:99.
117. Trima Israel Pharmaceutical Products Maabarot Ltd., data on SupraVir cream file.

14 Iontophoresis in Transdermal Delivery

Blaise Mudry, Richard H. Guy, and M. Begoña Delgado-Charro

CONTENTS

14.1 INTRODUCTION

Iontophoresis enhances drug transport across biological membranes by the application of a small electric field (for example, transdermal iontophoresis usually applies 0.1–0.5 mA, which, given skin's typical resistance, requires a voltage of less than 10 V). Because of its efficiency, safety, and potential to control transport across biological barriers, iontophoresis has been widely investigated, for example, for the administration of local anesthetics in dentistry [1,2], for drug delivery to the eye in ophthalmology [3–5], and mainly, for transdermal drug delivery and noninvasive monitoring via the skin [6,7]. Despite the advantages of passive transdermal drug delivery, the approach remains limited to a few drugs having adequate pharmacokinetic, pharmacodynamic, and physicochemical properties [8,9]. This can be principally ascribed to the extraordinary barrier properties of the skin. Only potent, small, and lipophilic drugs are currently marketed as transdermal patches. The development of iontophoresis (a method known for over 100 years) as a modern drug delivery technique has expanded the range of drugs that can be efficiently delivered through and into the skin. Most currently marketed drugs are either weak acids (~20%) or weak bases (~75%), and many new drug entities, including peptides and proteins, are also charged. Thus, a considerable number of drugs may benefit from iontophoretic delivery for the treatment of both local and systemic diseases. The first iontophoretic device to be commercialized comprised a power supply and electrodes that were extemporaneously filled with ready-made solutions (Iomed system TransQ Flex, TransQ 1&2GS, Salt Lake City, UT, USA). Nowadays, state-of-the-art devices (LidoSite topical system, Vyteris, Inc, Fair Lawn, NJ, USA; E-TRANS, Alza Corporation, Palo Alto, CA, USA) and the GlucoWatch G2 Biographer (Cygnus, Inc., Redwood City, CA, USA) employ smaller power supplies and integrated drug–electrode single-use units.

Polar, ionizable compounds are obvious candidates for iontophoretic delivery. Charged drugs are driven across the skin through a direct interaction with the applied electric field. This mechanism of transport is called electromigration or electrorepulsion. Further, the electroosmotic, convective solvent flow, which is induced across the net negatively charged skin during transdermal iontophoresis, provides a second mechanism of transport that supplements the electrotransport of cations and permits a significantly enhanced delivery of neutral, polar compounds [10,11]. Of particular importance is the fact that the electric field interacts directly with the drug to "push" it across the barrier (unlike most other enhancement technologies which act to increase the skin's permeability). This means that iontophoretic delivery of a drug is relatively independent of the skin's inherent permeability and can be easily controlled (Table 14.1 and Table 14.2) [12,13].

Over the last 20 years, the mechanisms of iontophoretic drug transport have been elucidated. In the process, several models have been developed which (a) describe drug transport across the skin under the influence of an electric field, (b) permit drug candidates to be selected rationally, and (c) optimize iontophoretic conditions and formulations. In the first part of this chapter, the main contributions of these models to the field are summarized, and

TABLE 14.1
Passive and Iontophoretic Fluxes of Three Model Cationic Drugs across Different Skin Barriers

Drug [Applied Concentration]	Skin Type	Passive Flux (μg/h/cm^2)	Iontophoretic Flux (μg/h/mA)
Pyridostigmine Br [1 M]	Human	0.4 ± 0.6	1610 ± 50
	Porcine	1.2 ± 1.4	1610 ± 50
	Mouse	4.2 ± 1.2	1720 ± 90
LiCl [1 M]	Human	0.2 ± 0.2	83 ± 3.7
	Porcine	6.9 ± 0.8	74 ± 3
	Rabbit	13 ± 1	71 ± 4
Hydromorphone Cl [1 M]	Human	0.25 ± 0.7	1053 ± 172
	Porcine	26 ± 10	1150 ± 159

Source: Data from Phipps, J.B., Padmanabhan, R.V., and Lattin, G.A., *J. Pharm. Sci.*, 78 (5), 365, 1989; Padmanabhan, R.V. et al., *J. Control. Release*, 11 (3), 123, 1990.

Iontophoretic fluxes are significantly enhanced over, and less variable than, those achieved by passive diffusion.

TABLE 14.2
Cumulative (8 h) Passive and Iontophoretic Lidocaine Transport through Porcine Skin as a Function of Barrier Integrity

	Cumulative Lidocaine Delivery (fg/cm^2)	
Barrier Integrity (Relative TEWL)	**Passive**	**Iontophoresis**
Intact (1)	0.7 ± 0.4	1837 ± 583
Intermediate "plus" (2.0–2.7)	6.9 ± 0.6	1837 ± 275
Intermediate "less" (2.9–4.2)	66 ± 24	Not determined
Fully compromised (4.5–8.0)	116 ± 69	1979 ± 364

Source: Data from Sekkat, N., Kalia, Y.N., and Guy, R.H., *Pharm. Res.*, 21, 1290, 2004.
"Relative TEWL" indicates the value of transepidermal water loss across the compromised skin barriers relative to that of normal, intact skin.

the manner in which they may be exploited by the transdermal scientist is discussed. In the second component of this chapter, the major factors which determine iontophoretic drug transport are examined in detail.

14.2 OVERVIEW

Transdermal iontophoresis involves the application of an electric field across the skin to facilitate (primarily) ionic transport across the membrane. Iontophoresis, it is important to point out, is differentiated from electroporation [14], another electrical approach to enhance transdermal transport, by the low fields employed. Whereas iontophoresis has achieved commercialization, there is (to our knowledge) no active development in progress of a transdermal delivery system employing electroporation.

An iontophoretic system (Figure 14.1) works similarly to an electrolytic cell [15]. Electrical energy is supplied from an external voltage source such that oxidation and reduction reactions are driven at the electrodes (usually Ag/AgCl). Oxidation occurs at the positively charged anode whereas reduction takes place at the negatively charged cathode. Electrolytic half-cells are typically connected by a salt bridge through which the ions generated at the electrode reactions are transported to maintain electroneutrality. In transdermal iontophoresis, the circuit is completed via the skin. Thus, some distinctive characteristics of iontophoresis as a transdermal enhancement technique can be deduced:

1. The same amount of charge, which is carried by electrons through the external circuit, will be transported through the skin by ions. The amount of charge is externally controlled by manipulation of the power supply; it follows that the extent of the ionic transport through the skin (including that of the drug) can be precisely determined. Further, on–off current profiles and "delivery" pulses of different magnitude can be used to achieve complex and individualized drug input profiles [16].
2. Ionic transport occurs in both directions across the skin. For example, to preserve electroneutrality at the anode, cations migrate into the body and anions migrate from the body into the electrode chamber. Hence, iontophoresis can be used for both drug delivery and noninvasive sampling [6,7,17].
3. All the ions present in the system (above and below the skin) may contribute to charge transport. The transport number is defined as the fraction of the total charge

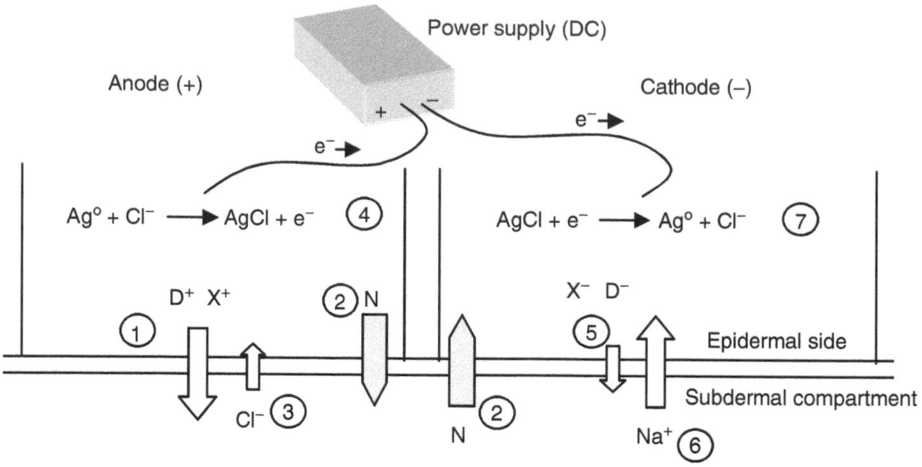

FIGURE 14.1 Schematic representation of the events taking place during iontophoresis. At the anodal compartment: drug (D^+) and competitor (X^+) coions migrate through the skin into the body (1); a convective, electroosmotic solvent flow transports neutral species (N) across the barrier and compliments the cationic flux (2); endogeneous anions, principally chloride, traverse the skin and enter the anodal solution (3); Cl^- are consumed at the Ag/AgCl electrode by oxidation of silver (4). At the cathodal compartment: the anionic drug (D^-) and competitor coions (X^-) migrate into the body (5); endogenous cations, mainly sodium, are extracted (6); neutral species (N) are extracted by electroosmosis (2); Cl^- are released into solution following silver reduction at the cathode (7).

> transported by a specific ion during iontophoresis. Because the sum of all the transport numbers must equal 1, it follows that iontophoretic transport is competitive. The transport number of a drug is related to its effectiveness as a charge carrier, and to the presence of competitor co- and counterions and their corresponding charge-carrying abilities.

In summary, drug flux is directly controlled by the intensity of current passed through the circuit but is limited by the maximum transport number achievable for a given ion. Not surprisingly, optimization of iontophoretic transport requires formulations that maximize the fraction of charge transported by the ion of interest; this can be accomplished in part by rational manipulation of the pH, the drug concentration, and the ionic strength and composition of the driving electrode solution. On the other hand, little can be done about competition from endogenous counterions.

The development of the first transdermal patches in the 1980s generated considerable interest in this route of drug administration. Soon afterwards, iontophoresis was rediscovered and its potential to contribute to the new field of transdermal drug delivery was examined. This work provided the basic principles for modern iontophoretic devices [13,18–21]. Furthermore, and importantly, they demonstrated the existence of a (primarily) electroosmotic, convective solvent flux during transdermal iontophoresis [10,11,22–24], and it was shown that the permselective properties of the skin (a) could be exploited to enhance the transport of neutral, polar species and (b) have a clear impact on ionic transport. Subsequent research has better characterized skin permselectivity and the factors which determine the magnitude of electroosmosis [25–27].

The total iontophoretic flux of a drug ion, therefore, is the combination of passive, electromigration, and electroosmotic contributions. The relative contribution of each mechanism depends on the characteristics of the permeant. For example, lithium, a small, highly mobile cation, is transported primarily by electromigration [17]; glucose, on the other hand, which is uncharged and has a very low passive permeability across the skin, owes its iontophoretic transport exclusively to electroosmosis [28,29]. Nevertheless, the passive contributions to other substances, such as nicotine [30] and urea [31], during iontophoresis cannot be neglected. In general, though, for most drugs (including, for example, propranolol, lidocaine, and many peptides [32–35]), the passive contribution is small compared to electrotransport. It should also be emphasized that the experimental conditions (pH, ionic strength and composition, drug concentration) also influence the relative contributions of electromigration and electroosmosis. Describing and integrating this complexity into one model, and then demonstrating validity, are extremely challenging objectives yet to be completely attained.

14.3 PATHWAYS OF TRANSPORT DURING TRANSDERMAL IONTOPHORESIS

Elucidation of iontophoretic transport pathways through the skin is relevant to the development of useful and predictive models. The skin can be considered, for example, as an electrical circuit having several resistances in parallel [36]; in this case, ions will travel via the pathways of lowest resistance. Initial experiments [37,38] suggested that iontophoretic transport was highly localized in appendageal structures, specifically hair follicles and sweat glands. Certainly, the early clinical applications of iontophoresis, for the diagnosis of cystic fibrosis [39] and the treatment of hyperhidrosis [40], implied that at least some transport was taking place via sweat glands. Indeed, this important role of the appendageal pathway in iontophoresis is a significant differentiating characteristic from passive transport. Clearer elucidation of iontophoretic pathways has been achieved with confocal microscopy [41,42], scanning electrochemical microscopy [43–47], and x-ray microanalysis [48,49]. Overall, the results from this work indicate that both electromigrative and electroosmotic transport are strongly associated with follicular structures, but other pathways (including the sweat glands and the intercellular route [42,49]) also participate. It is likely that the relative importance of each route will depend on the nature (size, lipophilicity, etc.) of the permeant [42,49].

The major conclusions can be summarized as follows: (a) iontophoretic transport is strongly associated with hair follicles and sweat glands, and may be advantageous, therefore, when local delivery to these structures is required (an opportunity yet to be practically realized, however); (b) the intercellular pathway has also been demonstrated, at least for ionic compounds; (c) the current density flowing through the localized transport pathways is probably much higher than that nominally applied onto the skin; (d) passive transport via the appendages may increase when an electric field is imposed across the skin; and (e) the existence of a multiple pathways must be considered when theoretical models are used to fit iontophoretic transport data. Furthermore, the relative contributions of electromigration and electroosmosis through each of these routes are not necessarily the same (in fact, very little is known about the "local" permselectivity built into each of these pathways). Because most experiments provide only global measurements of iontophoretic flux across an excised membrane, considerable caution is warranted when imposing models based on preconceived populations of pores and pathways.

14.4 THEORETICAL MODELS OF IONTOPHORETIC DRUG TRANSPORT

Complex membrane transport phenomena are relevant to several areas in medicine, such as ion transport through neurons, ion exchange, and electrodialysis [50,51]. The Nernst–Planck equation is often the starting point for the interpretation of transport through both artificial and biological membranes [50,51]. Strictly speaking, the Nernst–Planck equation requires the membrane to meet certain criteria and satisfy some demands with respect to its properties. For an artificial membrane, pore distribution, pore size, and charge density are usually known. In contrast, as discussed above, iontophoretic transport takes place simultaneously through different pathways, whose exact dimensions, charge density, and permselective properties have been only partially characterized. Hence, models applied to transdermal iontophoresis are less than satisfactory as they typically regard the skin as an ideal, homogenous and, in some cases, uncharged membrane. Despite this limitation, the development and application of models for iontophoresis has significantly clarified the mechanisms of iontophoretic transport and allowed the identification of the key factors which determine the drug flux.

Before examining the relevant models, a brief word about the practical aspects of iontophoresis is required. The first question to answer is whether drug delivery will be performed under constant current or constant voltage conditions? Ion transport at constant voltage is traditionally described by the Nernst–Planck equation, which has been used to model transport through the gastrointestinal and neuronal membranes [50,51]. A similar approach has been adopted for iontophoresis across the skin (discussed later). However, experiment shows that the skin's resistance decreases dramatically during iontophoresis [52,53]; it follows from Ohm's law (voltage = resistance × current) at constant voltage that the current flowing across the membrane must increase. As a result, the flux of an ion cannot be fully controlled during constant voltage iontophoresis; furthermore, as there is variability in skin resistance both within and between individuals, there will be significant variations in drug flux under these conditions (indeed, it has been proposed that skin electrical resistance be kept constant to decrease variability [54]).

It is therefore more sensible (and probably easier and safer) to perform constant current iontophoresis, and this is by far the most common approach. In this case, the power supply adapts the voltage imposed to the resistance of the circuit to keep the intensity of current (and hence the drug flux) constant. Drug transport can now be modeled with Faraday's equation, which links current intensity to ionic flux.

A comprehensive review of all existing models and their subsequent development is beyond the scope of this chapter. Instead, a succinct discussion of the most representative approaches (Figure 14.2) is presented, with references to the original work containing comprehensive descriptions. Electrodiffusion theory and the Nernst–Planck equation have been applied in two constant voltage scenarios [55–57]: (i) constant field conditions and its development to include electroosmosis [58] and (ii) the electroneutrality case. The application of this model to the single-ion situation is of particular interest with respect to practical applications of iontophoresis. The principal model for electroosmosis [24] is based on irreversible thermodynamics. Models of drug transport under constant current conditions begin at Faraday's law [37], and have been expanded to address the issue of competing co- and counterions [59]. The "ionic pore–mobility" model [60–62] represents an important effort both to include the electroosmosis and to consider the impact of drug and vehicle properties on iontophoresis.

14.4.1 Nernst–Planck Models

The Nernst–Planck equation constitutes the starting point for the electrotransport models [55–57]. The overall flux of the ionic species "i" (J_i) comprises the diffusion term driven by the chemical potential gradient ($\mathrm{d}c_i/\mathrm{d}x$) and the electric transference term due to the electrical potential gradient ($\mathrm{d}\varphi/\mathrm{d}x$):

$$J_i = -D_i \frac{\mathrm{d}c_i}{\mathrm{d}x} - \frac{D_i z_i F c_i}{RT} \frac{\mathrm{d}\varphi}{\mathrm{d}x} \qquad (14.1)$$

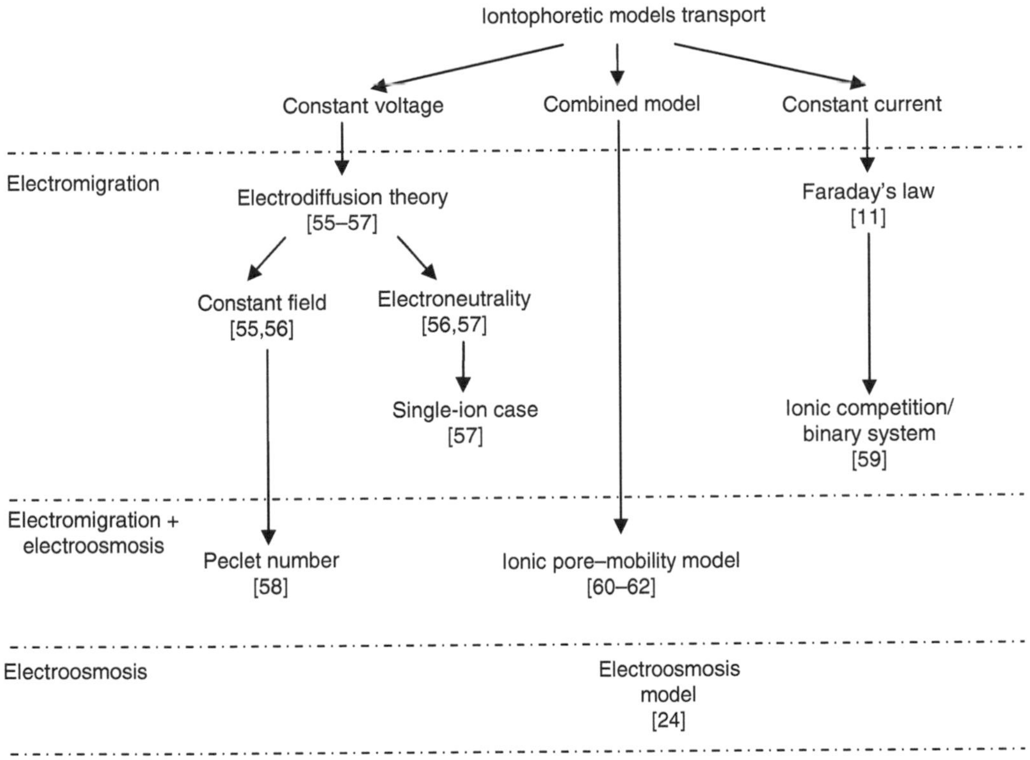

FIGURE 14.2 Relationship between the models, which have been proposed to describe iontophoretic transport.

where D_i, c_i, and z_i are the diffusion coefficient, concentration, and valence of "*i*," F is Faraday's constant, R is the gas constant, and T is the absolute temperature. Equation 14.1 reduces to Fick's first law of diffusion when $z_i = 0$. This equation takes no account of any contribution to the total transport by convection.

14.4.1.1 Constant Field Approximation

Equation 14.1 contains derivatives of both the concentration and electrical potential, and requires both quantities to be known as a function of position in the membrane so as to predict the flux. To simplify matters, Goldman proposed the approximation that the electric field across the membrane be considered constant (i.e., the electric field is not affected by the presence of ions in the membrane). Under these conditions, the electrical potential gradient reduces to $E = -\mathrm{d}\varphi/\mathrm{d}x = \Delta\varphi/h$ [55,56], where $\Delta\varphi$ is the applied potential and h is the membrane thickness. This approximation is appropriate for membranes that are relatively thin compared to the Debye length. This is not the case for the skin, even when heat-separated (~100 μm) or dermatomed (~0.5 mm). This approximation is also reasonable when the total ion concentrations on both sides of the membrane are equal. However, this is rarely the case in iontophoresis for which the applied ionic concentration is typically much less than that subdermally.

This model has been used to predict flux, flux enhancement, and the reduction in the lag time when a voltage is applied [55,56]. An enhancement factor (EF) has been defined as the ratio of the steady-state flux with an applied voltage (J_{ion}) to the corresponding passive flux (J_{pas}):

$$\mathrm{EF} = \frac{J_{\mathrm{ion}}}{J_{\mathrm{pas}}} = \frac{\nu}{1 - \mathrm{e}^{-\nu}} \tag{14.2}$$

where

$$\nu = \frac{ze(\Delta\varphi/h)}{RT}$$

This model estimates the theoretical effect of the electric field on EF and good agreement has been found between prediction and experiment using synthetic cellulose membranes at low voltages (<0.125 V) [63]. In contrast, similar studies using hairless mouse and human skin have shown significant deviations [53,56,58,64,65] from the theory. According to the Goldman approximation, the flux of each ion is independent of the concentration of other ions in the system, a prediction inconsistent with the experimental data. This discrepancy is probably due, at least in part, to (a) the permselectivity and inhomogeneity of the skin, (b) the nonuniformity of the electric field owing to the presence of ionized drug (or other ions) at relatively high concentrations, and (c) the decrease in skin resistance induced by iontophoretic application.

14.4.1.2 Electroneutrality Approximation

As the skin is relatively thick compared to the space–charge layers at its boundaries, the bulk of the membrane may be expected to be electroneutral [56,57]. The Nernst–Planck equation can be solved, therefore, by imposing the electroneutrality condition: $\sum C_j/C = \sum C_k/C$, where the subscripts j and k refer to positive and negative ions, respectively, and C is the average total ion concentration in the membrane. In the case of a homogenous and uncharged membrane bathed by a 1:1 electrolyte, the total ion concentration profile across the membrane is linear and the resulting steady-state flux is described by

$$J_i = -\frac{D_i}{h}\left(1 + \frac{\nu}{\ln \chi}\right)\left(\frac{\chi - 1}{\chi - \mathrm{e}^{-\nu}}\right)(c_{ih} - c_{io} \times \mathrm{e}^{-\nu}) \tag{14.3}$$

where D_i is the diffusivity of ion "i," C_{io} and C_{ih} are its concentrations at the exterior and interior boundaries of the membrane, ν is defined as before ($\nu = ze(\Delta\varphi/h)/RT$), and χ is the ratio of the total ionic concentrations, C_{io}/C_{ih}. This equation assumes that the charge on each ion is identical; in practice, therefore, Equation 14.3 is restricted to monovalent ions (given that Na^+ and Cl^- are the predominant endogenous ions present during iontophoresis *in vivo*).

14.4.1.3 Single Ion and Multiple Competing Coions

The previous model [57] has been further developed for the iontophoretic delivery of a monovalent drug. The first case considered (the single-ion situation) is exemplified by the anodal iontophoresis of a monovalent, cationic drug (m^+), which is the only available charge carrier at the skin surface. The cathodal (subdermal) electrolyte is assumed to be normal saline. Under these circumstances, the transport number of the drug is given by

$$t_m = D_m/(D_m + D_{\mathrm{Cl}}) \tag{14.4}$$

where D_m and D_{Cl} are the diffusion coefficients of the drug and Cl^-, respectively, in the membrane. The efficiency of drug delivery is therefore determined by the ratio of the

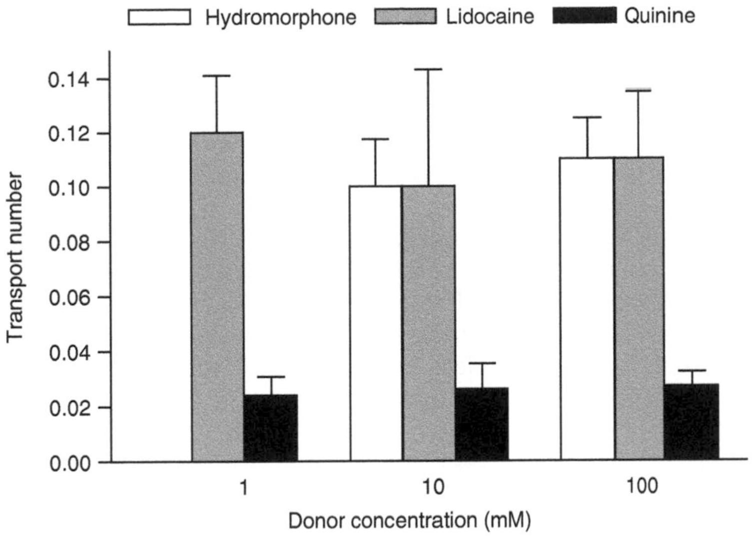

FIGURE 14.3 The single-carrier or single-ion scenario illustrated for hydromorphone, lidocaine, and quinine. Transport numbers (mean $\pm$ SD, $n \geq 3$) are essentially constant for these three drugs despite a significant increase in donor drug concentration. (Data taken from Padmanabhan, R.V. et al., *J. Control. Release*, 11 (3), 123, 1990; Marro, D. et al., *Pharm. Res.*, 18 (12), 1701, 2001.)

diffusivities of the drug and the principal counterion, and is independent of drug concentration. The validity of this equation (during constant current iontophoresis) has been verified *in vitro* and *in vivo* for lidocaine, quinine, and hydromorphone and ropinirole (Figure 14.3) [20,32,66,67]. The second scenario adds a competing coion to the applied drug solution such as Na^+. The drug must now compete with both Na^+ and Cl^- to carry the charge across the skin, and its transport number becomes strongly dependent on concentration. Thus, when iontophoresis is performed in the presence of competing ions, the drug must be concentrated. In addition, ideally, the molecular weight of the drug should be reasonable so that its diffusivity is such that efficient delivery can be achieved. From a drug delivery point of view, therefore, the second situation is much less efficient than the single-carrier scenario. Optimization of the flux of a cationic drug requires, therefore, that (a) the small, mobile coions in the applied drug formulation be minimized and (b) the pH be buffered either by bulky organic ions, or by the drug itself. The validity of these principles has been widely recognized [6,13,16,32].

14.4.2 Expansion of the Nernst–Planck Model to Incorporate Electroosmosis

A significant modification of the model is the addition of an additional term to include the convective contribution to the total iontophoretic transport. This is achieved by adding a linear term ($v \times C$) [58], where v is the average velocity of the solvent and C is the concentration of the drug. Because the skin has a net negative charge, this term is positive for cations and negative for anions. From the constant field approximation, the EF is predicted to be

$$\mathrm{EF} = \frac{Pe}{[1 - \mathrm{e}^{(-Pe)}]} \tag{14.5}$$

The Peclet number, $Pe = (v \times h)/D$, is a function of the solvent velocity (v), the thickness of the membrane (h), and the drug's diffusion coefficient (D). While v and h depend on membrane properties (charge density, thickness) and the experimental conditions imposed (electrolyte, electric field strength), D clearly depends on the nature of the solute. As D is inversely related to molecular size, Pe should increase with drug molecular weight. The validity of Equation 14.5 has been demonstrated for neutral and ionic solutes, permeating nucleopore and human skin membranes [53,68]. In general, the model correctly predicts the asymmetry in EF for cations and anions, and that neutral, polar compounds are more efficiently delivered from the anode [53,58,68]. Further development of the model has taken into account the porosity of the membrane, the size and charge of the permeant, hindrance effects, and the effect of ionic surfactants [69–73]. In general, while the application of this model to iontophoresis through artificial membranes has been straightforward, transdermal electrotransport is more complex (due, for example, to the coexistence of multiple pathways, and induction of skin changes upon current application), and typically requires a more specific approach.

14.4.3 Electroosmosis

Electroosmosis is the bulk fluid flow that occurs when a voltage gradient is imposed across a charged membrane. Transport by convection allows the delivery and extraction of neutral and zwitterionic compounds and plays a major role in the movement of large, poorly mobile cations. Electroosmosis is an electrokinetic phenomenon, which may be described by nonequilibrium thermodynamics [24]:

$$J_i = L_{ve}(-\mathrm{d}\Phi/\mathrm{d}x) \qquad (14.6)$$

where L_{ve} is a coefficient reflecting the magnitude and the direction of the volume flow. Transport of water and neutral solutes across the skin at physiological pH has been typically observed in the anode-to-cathode direction, suggesting that the skin has a net negative charge [10,11,22,23,25,76,77]. One approach to model such behavior is a pore with a fixed negative charge [10,24,74,75]. In this way, a double layer is formed with mobile positive charges to maintain electroneutrality. The ion atmosphere within the pore fluid has a nonzero charge density; in fact, it has a total charge equal to that of the pore wall but opposite in sign. The interaction of the electric field with the volume in the ion atmosphere is the driving force for the electroosmotic flow, which follows the same direction as the counterions. Hence, given that the skin's isoelectric point is between 4 and 5, it follows that the membrane is negatively charged at pH 7.4, the counterions are cations, and electroosmotic flow is from anode to cathode. Electrosomosis has been measured directly using tritiated water [25], or by fluid flow into capillary tubes connected to the electrode compartments [10]. More typically, an indirect estimation of the current-induced water flow (V_w) is deduced from the flux of a neutral polar marker ($J_{eo,i}$), such as mannitol, and the following equation [24,32]:

$$J_{eo,i} = V_w C_i \qquad (14.7)$$

where C_i is the applied concentration of the neutral marker.

14.4.4 Constant Current Iontophoresis: Application of Faraday's Law

Faraday's law links, at steady state, the amount of charge exchanged at the electrodes to the mass transfer of ions through the membrane [37,59]. For example, oxidation occurs at the anode where one electron is "lost"; to maintain electroneutrality, either one cation must leave the anodal compartment and transfer into the skin, or one anion must arrive from beneath the skin. Faraday's law relates the electromigration flux (J_i) to the current applied:

$$J_i = (t_i \times I)/(z_i \times F) \tag{14.8}$$

where

$$t_i = I_i/I \tag{14.9}$$

and

$$\Sigma t_i = 1 \tag{14.10}$$

where I is the intensity of current passed through the circuit, I_i is the current carried by ion "i," t_i and z_i are the transport number and valence of "i" respectively, and F is Faraday's constant.

Faraday's law is particularly interesting for drug delivery indicating that, once the drug's transport number is known for a given set of conditions, the current and time of application can be precisely manipulated to achieve the dose and input rate required.

Transport numbers have values between 0 and 1 and measure the efficacy of drug transport. The transport numbers of $Na^+$and Cl^-, during saline iontophoresis, are ~0.6 and ~0.4, respectively, reflecting the skin's cation permselectivity [23,29,78,79]. As no drug can carry current more efficiently than these two endogenous ions, values of the transport numbers of cationic and anionic species never exceed 0.6 and 0.4, respectively. In fact, the typical range of transport numbers for low molecular weight drugs is approximately 0.05 to 0.20; for larger compounds, such as peptides or proteins, the values are much less [32,34,66,79,80].

Transport numbers can be measured by different methods. The small mono- and divalent inorganic ions have been used to demonstrate skin permselectivity, and have been determined from membrane potential measurements, or by the Hittorf method [10,25,77,79]. The latter method has been frequently used for drugs; alternatively, the transport number can be estimated from the slope of a plot of drug flux as a function of current intensity (Figure 14.4) [18,66].

14.4.4.1 Binary Case

To relate the flux of a drug to its concentration and to that of extraneous coions, Equation 14.8, Ohm's law, and the electroneutrality condition can be combined to give an expression for the transport number of a drug in a binary cation system through a homogeneous nonionic membrane [59]:

$$t_{\mathrm{D}} = \{C_{\mathrm{D}} \times u_{\mathrm{D}} \times z_{\mathrm{D}}\}/\sum C_i \times u_i \times z_i \tag{14.11}$$

where C_i and C_{D} are the concentrations of the ion "i" and the drug D within the membrane, u_i and u_{D} are their mobilities, and z_i and z_{D} are their valences. Note that the denominator includes all ions present on both sides of the skin, i.e., the drug, the coion (e.g., Na^+), and the counterion (Cl^-). Further development of the model yields

$$t_{\mathrm{d}} = \frac{\frac{t_{\mathrm{d}}^{\mathrm{o}}}{1-t_{\mathrm{d}}^{\mathrm{o}}}}{\frac{t_{\mathrm{d}}^{\mathrm{o}}}{1-t_{\mathrm{d}}^{\mathrm{o}}} + BZ_{\mathrm{c}}X_{\mathrm{c}}\frac{1}{1-t_{\mathrm{e}}^{\mathrm{o}}} + 1} \tag{14.12}$$

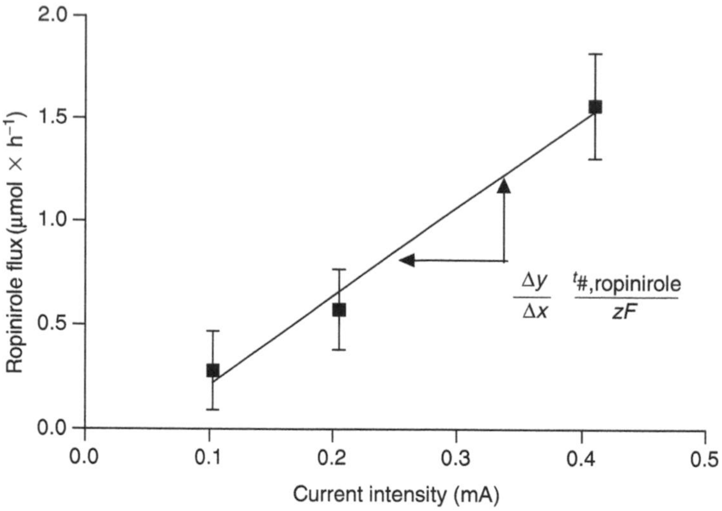

FIGURE 14.4 Linear dependence of ropinirole hydrochloride iontophoretic flux (mean $\pm$ SD, $n \geq 5$) upon the intensity of current applied. The transport number of the drug is estimated from the slope of the line. (Data from Luzardo-Alvarez, A., Delgado-Charro, M.B., and Blanco-Méndez, J., *Pharm. Res.*, 18 (12), 1714, 2001.)

where t_d^o and t_e^o are the drug and coion transport numbers in the absence of competition from each other (single-ion situation), and Z_c and X_c are the valence and the mole fraction ratios, respectively of the cations in the anode compartment. B relates the cation concentration ratio in the solution to that in the membrane. A value of B equal to 1 indicates the same apparent molar ratio of the two coions in the membrane as in the donor solution. When $B = 1$, the ratio of cation concentration in the membrane equals that in the anode formulation; if $B \neq 1$, then the "partitioning" of the drug and the coion into the membrane is different reflecting, presumably, the existence of interactions between the ions and the membrane under the influence of the electric field.

The validity of Equation 14.12 has been demonstrated experimentally for the lithium–hydromorphone and calcium–hydromorphone couples [59]. However, different values of B are required to fit the two data sets. The uncertainty of the significance of the parameter B is the weakest point of this model. In fact, B cannot be predicted and its evaluation requires experimental data.

Application of this model to published data on the iontophoresis of lidocaine [32,78] in the binary system lidocaine–sodium shows a good agreement between experiment and prediction (Figure 14.5). On the other hand, discrepancies have been found for ropinirole [66]; in this case, parallel increments in the concentrations of the drug and sodium were undertaken to maintain their molar fractions constant. While Equation 14.8 predicts that the drug flux should remain constant under these circumstances, ropinirole flux actually decreased as the sodium concentration increased. Clearly, further research is required to optimize the form of this model.

14.4.5 Ionic Mobility–Pore Model

This model attempts to integrate solute properties, vehicle composition, and electroosmosis [60–62], and derives an expression which includes as many as 12 determinants of iontophoretic transport: (i) solute size (molecular weight and molecular volume), (ii) solute mobility,

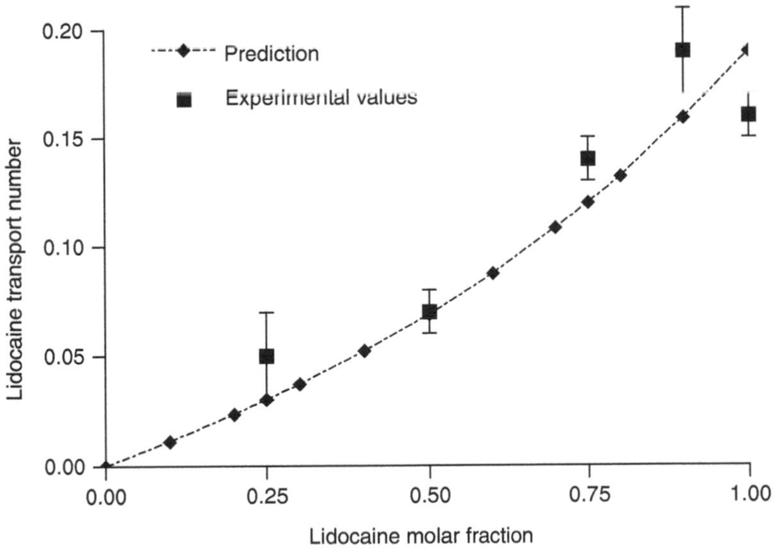

FIGURE 14.5 Prediction of lidocaine transport number in the binary system (lidocaine–sodium) from the drug molar fraction in the anodal solution. The theoretical transport numbers were estimated using Equation 14.12 [59], and are compared with the experimental values from the literature [32]. A value of $B = 1$ was used for the prediction.

(iii) molecular geometry, (iv) solute charge, (v) Debye layer thickness, (vi) total current applied, (vii) solute concentration, (viii) fraction ionized, (ix) presence of extraneous ions (quantified by the conductivity of the solution), (x) epidermal permselectivity, (xi) interaction of unionized and ionized solutes with the pore, and (xii) electroosmosis.

An interesting idea is the use of conductivity for predicting the competing effects of other ions present in both the anodal and cathodal solutions, and for estimating the drug's mobility. The model [61,62] postulates that iontophoretic transport is restricted to pores present in the appendages and in the intercellular regions of the epidermal barrier. Solute transport is therefore size-limited. Two scenarios are envisaged: the free-volume case, in which the molecules jump between a limited number of "holes" in the pathway, and the pore-restriction case where there is steric hindrance both for entry into and to movement through the pore. Application of this model to a series of local anesthetics identified for two major predictors of iontophoretic permeability are ionic mobility (and its determinants: pK_a, molecular weight, and conductivity) and molecular size.

14.5 KEY PARAMETERS IN TRANSDERMAL IONTOPHORESIS

14.5.1 The Drug

Iontophoretic flux is the sum of electromigrative and electroosmotic contributions, the relative importance of which is determined by the physicochemical properties of the drug and the experimental conditions. It follows that adequate structure–activity relationships can only be obtained from experiments performed under the same conditions (for example, not only at the same pH, but also with the same buffer composition). For electromigration, the efficiency of the drug to carry charge is probably the most critical attribute. Thus, small, mobile drug cations are the best candidates for iontophoretic delivery as they have higher transport numbers. Indeed, lithium, which competes quite efficiently with the endogenous chloride ion, is the drug with the best physicochemical properties for electromigration [17,81]. All other drugs, however, are larger and less mobile than lithium, and they are much less efficiently transported. Several attempts have been made to relate transport number to more easily measured properties such as molecular weight [82–87], log P [84,88,89], mobility in capillary electrophoresis [90], and specific conductivity [91]. To date, molecular weight has been the most indicative, and an inverse relationship with iontophoretic flux has been demonstrated [60–62,83,84,86,87,92]. An effective estimation of the iontophoretic charge-carrying efficiency of a drug candidate can be obtained from measurements of iontophoretic flux in the single-ion situation. In this way, confounding effects (i.e., competition) from other coions are eliminated and the flux will reflect only the relative diffusivity of the drug to that of the main endogenous, competing counterion [32,57]. When coions are required in the vehicle for practical purposes (e.g., for buffering, preservation, or stabilization), it is preferable to use either uncharged or large (and therefore less mobile) additives, in small molar concentrations, to minimize coion competition.

The effects of valence, charge distribution, and molecular structure on electromigration have not been systematically studied, but it is expected that these parameters will influence the mobility and diffusivity of the ion. Faraday's equation (Equation 14.8) indicates that iontophoretic flux is inversely proportional to the valence of the drug. For a fixed transport number, the flux of a divalent ion should be one-half of that of a monovalent ion [13,23,59]. To illustrate this point, the transport numbers of lithium and calcium are 0.336 and 0.438, respectively [59]; yet, even though t_{Ca} is higher, its experimentally measured single-ion flux is only 86 μmol/h/cm^2 compared to 142 μmol/h/cm^2 for lithium. However, the valence of the ion also determines the magnitude of transport number (Equation 14.11) and, hence, the precise impact of valence on flux cannot be directly derived. As for polarity, it has been frequently shown [84,93] that the cationic transport is preferred because of the skin's permselectivity. As all other things remain equal, iontophoretic fluxes of cationic drugs are anticipated to be higher than those for anions.

The electroosmotic contribution is allegedly less dependent on the molecular structure of the permeant, which is viewed as simply "carried" along by the solvent flux [24,75]. Indeed, the electroosmotic transport of several compounds spanning a range of molecular weight from 60 to 400 Da is rather similar [31,94,95]. On the other hand, it has also been argued that transport should decrease for larger permeants, and an inverse effect of size on electroosmotic delivery has been observed [86,87]. An interesting experimental observation is that iontophoresis of both small, divalent cations (such as calcium) and lipophilic cationic drugs reduces the magnitude of electroosmotic flow [23,32,33,59,80,96]. This has been attributed to the adsorption of these cations onto the membrane, and the resulting partial neutralization of its negative charge. In support of this hypothesis, electroosmotic flux has been enhanced by delivery of ethylenediaminetetraacetate (EDTA), which scavenges endogenous, divalent ions like Ca^{2+} [27]. The practical relevance of an alteration in the electroosmotic contribution to the total flux depends on the molecule involved. For a small drug like propranolol, for example, experimental conditions can be arranged such that the electromigrative contribution is dominant. Conversely, when electroosmosis is the principal mechanism of transport (e.g., for neutral and less mobile drugs), a change in skin permselectivity may have dramatic consequences on the magnitude of the flux and its reproducibility.

14.5.2 Current Intensity

Faraday's law indicates that electromigrative flux increases proportionally with the intensity of current applied. Figure 14.4 shows that the ropinirole hydrochloride transport number can be estimated from the linear relationship between drug flux and applied current [18,66]. The direct dependence of flux upon intensity has been repeatedly demonstrated and provides the basis for controlled drug delivery by iontophoresis [66,79,97].

Modulation of current profiles can be used to modify and control drug delivery. Pulsatile and different intensity profiles offer the opportunity of individualized dosing. Such condition is the case for the iontophoretic device containing fentanyl (E-Trans, Alza Corp., Palo Alto, CA, USA) that has been developed for the treatment of acute (postoperative) and chronic (cancer) pain. Moreover, *in vitro* experiments have shown that iontophoresis can be used to mimic physiological hormone secretion rhythms [98–101], or to assure rapid nicotine input [30]. Finally, it has been suggested that an alternating current allows for skin depolarization and may be more efficient than direct current [102]. However, there is no experimental evidence to support this hypothesis; the data available indicate that (as predicted) the transport of drug is proportional to the net charge delivered [103].

14.5.3 Current Density

Another significant feature that differentiates iontophoretic transport from passive diffusion is the former's lack of dependence on area [104]. At fixed current, iontophoretic fluxes do not increase proportionally with application area (unlike passive transport). Nevertheless, in the literature, electrotransport rates are frequently normalized per unit area. This can result in misunderstanding because, in iontophoresis, it is the total charge, which determines delivery. Hence, a current of 0.5 mA applied over an area of 5 cm^2 (0.1 mA/cm^2), for example, will deliver the same amount of drug as the same current applied to an area of 10 cm^2 (0.05 mA/cm^2), assuming a negligible passive contribution, which is usually the case. On the other hand, increasing the area of a passive transdermal patch by a factor of 2 will double drug delivery [8,9]. It must be emphasized, however, that current density is a critical, practical factor in that it determines skin irritation and patient discomfort. A value of 0.5 mA/cm^2 has been typically considered as an upper limit [105].

14.5.4 Drug Concentration

The effect of drug concentration on iontophoretic flux has been widely studied. With respect to electromigration, the two principal scenarios are (a) the so-called "single-ion" or "single-carrier" situation and (b) the multiple coions case [57]. In the single-ion configuration (discussed above), the drug is the only charge carrier available in the electrode formulation. Under these circumstances, it has been demonstrated (*in vitro* and *in vivo*) that transport is maximized and is independent of drug concentration. By way of illustration, up to 100-fold increase in the concentrations of hydromorphone [20], lidocaine [32], and ropinirole [66,67] had no significant effect on their flux (Figure 14.3). From a practical point of view, the single-ion situation is clearly advantageous, allowing maximum fluxes at low drug load (an important factor when the active species is expensive).

Despite the clear benefit of the single-ion case, iontophoresis is frequently performed in the presence of multiple competing coions. In the "real world," an electrode formulation may require the presence of buffer, additional electrolyte (e.g., Cl^- for the anode reaction), and antimicrobial, antioxidant, and cosolvent species [106]. In the case of reverse iontophoresis, the analyte of interest is just one component in the very complex mixture formed by the endogenous electrolytes. When multiple coions compete for transporting the charge, the relative concentration and mobility of the drug as compared to that of all others are the key parameters that determine drug flux. Briefly, the presence of competing ions in the electrode formulation usually results in (a) reduction of the drug transport number and (b) fluxes which increase proportionally with drug concentration [66,97,107]. It appears that the drug molar fraction may be a better predictor than concentration, as the former includes information about its concentration relative to that of the competing species [59,32]. For example, the iontophoretic flux of lidocaine increased linearly with the drug molar fraction in the electrode formulation until a maximum value was achieved at a molar fraction of 1 (i.e., the single-ion situation) (Figure 14.5) [32].

That iontophoretic flux that depends on concentration is obviously essential for reverse iontophoretic applications. In this case, the drug flux is used as a predictor of (and must be proportional to, therefore) the interstitial (plasma) concentration of the analyte of interest. The validity of this assumption has been demonstrated for glucose [108,109,28,29], valproate [110], phenylalanine [111], phenytoin [112], and lithium [17,82], i.e., for compounds transported by either electroosmosis or electromigration or both. The electroosmotic extraction of glucose via the skin constitutes the basis for the GlucoWatch Biographer [113]. Similarly, lithiemia in bipolar patients can be easily and noninvasively predicted from the lithium extraction flux [82]. Interestingly, for lithium (an exogenous ion transported by electromigration), the proportionality constant between iontophoretic flux and serum level was very similar for the different subjects examined, such that the mean value from a first group of patients ($n = 12$) could be used to accurately predict lithiemia in a second group ($n = 11$). A more challenging drug is phenytoin, which is ~90% bound to plasma proteins. In this case [112], the iontophoretic flux responded to changes in the free phenytoin concentration, and was correctly sensitive to protein binding.

It should be mentioned that certain anomalous behavior has sometimes been observed. For example, a lack, or even an inverse effect, of drug concentration on the iontophoretic flux [32,33,96] has been reported for the lipophilic, cationic drugs such as propranolol and quinine [32,33], and for peptides such as nafarelin, leuprolide, calcitonin, and octreotide [34,96,114]. This phenomenon has been associated with adsorption of the drug onto the skin and a progressive leads to a reversal of its permselective properties (see Figure 14.6) [32,33,96].

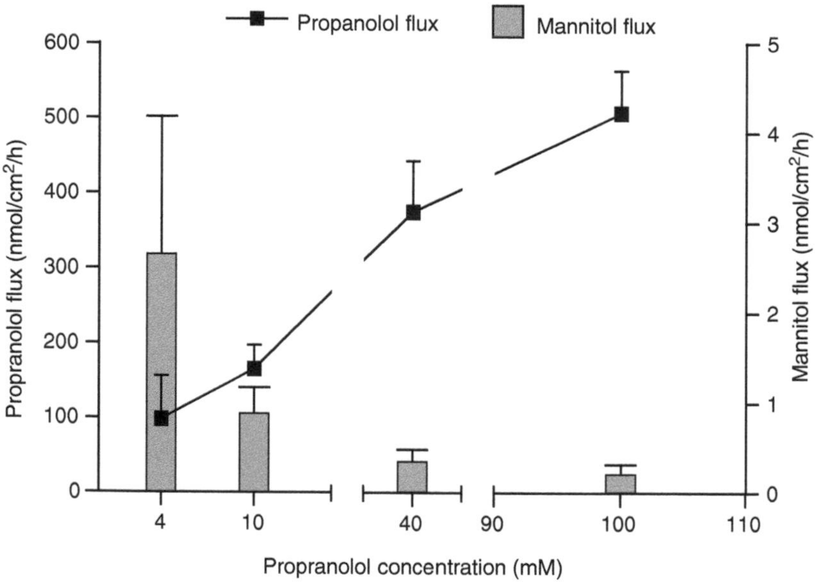

FIGURE 14.6 Iontophoretic fluxes (mean $\pm$ SD, $n \geq 4$) of the cationic drug propranolol, and of the electroosmotic marker mannitol, as a function of propranolol concentration in the anode solution. It is observed that (a) drug flux does not increase linearly with increasing concentration, and (b) electroosmosis is significantly impeded with increasing propranolol concentration. (Data from Marro, D. et al., *Pharm. Res.*, 18 (12), 1701, 2001.)

Finally, it is worth reiterating that transport numbers are relatively complex functions of the concentration and mobility of all the ions present in the system. Thus, while the relationship between lidocaine molar fraction in the binary system (lidocaine–sodium) and the drug flux has been well defined [32,78], the results cannot be directly extrapolated to a different anodal composition. That is, the drug flux depends not only on its molar fraction [59], but also on the mobilities of the competing ions [115].

14.5.5 Ionic Strength

From the above discussion, it is easily deduced (and has been often reported [18,116–120]) that a drug's transport number will decrease with increasing ionic strength in the electrode formulation. This is due, of course, to the increased number of coions available to compete with the drug to carry charge across the skin. The ionic strength of the vehicle also influences the electroosmotic contribution. A rapid and exponential decrease of the electroosmotic contribution with increasing ionic strength has been reported [24,26].

14.5.6 pH

The pH of the drug formulation must be carefully considered as it can critically influence iontophoretic flux. Whenever possible, the pH is adjusted to ensure maximal ionization of the drug, and maximal transport. However, the situation may be complicated, and each drug–vehicle combination must be carefully considered. For example, the tripeptide, tyrosine-releasing hormone (TRH), is better delivered in its unionized form at pH 8 by electroosmosis than at pH 4 at which it is 99% cationic [11]; this is because, under acidic conditions, the skin's permselectivity is reversed, electroosmosis proceeds from cathode to anode, and the transport number of the positively charged TRH is very small. In contrast, for the weak acid, 5-fluorouracil ($pK_a \sim 8.5$) at pH 8.0 (~50% ionization), electromigration was clearly more efficient than electroosmosis [121]. Finally, it is interesting to compare the efficiency of the electromigration and electroosmotic mechanisms at physiological pH and ionic strength (i.e., conditions that cannot be modified in reverse iontophoresis). For phenytoin [112], the iontophoretic extraction flux obtained at the cathode by electroosmosis was very similar to that found at the anode by electromigration. The proportions of unionized and anionic drug at physiological pH are 89% and 11%, respectively, confirming the general perception that electromigration is a significantly more efficient mechanism of transport than electroosmosis.

It should be recalled that a pH gradient exists across the skin, varying from ~5.5 at the outer epidermal surface to 7.4 in the viable tissue. Thus, the degree of ionization of a drug may change as it transports through the barrier [122]. It is not clear whether this pH gradient can be controlled *in vivo* by buffering the electrode formulation in contact with the skin. In any case, the range of pH values possible is limited because of the risk of skin irritation and because the high mobilities of OH^- and H_3O^+ reduce dramatically the transport number of the drug once the pH moves significantly away from physiological value.

Finally, the pH of the electrode solution also influences the skin's permselective properties, a subject which we now explore briefly.

14.5.7 Skin Permselectivity

Like many biological membranes [123,124], the skin has a net negative charge and is therefore cation-permselective [10,24,25,76,77,80,93,125]. The negative charge is thought to originate from the anionic amino acid residues of proteins in the skin, and from other charged species, such as fatty acids. However, very few specifics are actually known about the source of the skin's charge, nor how it varies along the different transport pathways. The so-called

TABLE 14.3
Isoelectric Points of Various Skin Models

Skin Model	Permeant	p*I*	Current Density (mA/cm^2)	Ref.
Human	Mannitol	4.8[a]	0.5	76
Human	Na	2.0–3.0[b]	0.5	125
Porcine	5-FU	4[a]	0.5	121
Porcine	Mannitol	4.4[a]	0.5	76
Neonatal porcine	Mannitol	3.5–3.75[c]	0.64	77
Hairless mouse	Mannitol	4.5–4.6[c]	0.51	77
Rabbit	Na	2.0–3.0[c]	0.5	125

[a]Dermatomed skin.
[b]Heat-separated epidermis.
[c]Full thickness.

isoelectric point (p*I*) corresponds to the pH value at which the skin has no net charge (Table 14.3) and has been estimated by indirect methods (measurements of membrane potential and inorganic ion transport numbers), and by the determination of anodal and cathodal fluxes of neutral compounds as a function of pH [25,76,77,80,93,125]. The p*I* of human and other mammalian skin models falls between 3 and 5 confirming that the membrane has a negative charge under physiological conditions.

In summary, there is evidence that the skin presents a weak cation permselectivity [25,76,77,80,93,125], which can be reversed by acidifying the pH of the solutions bathing the skin [10,23,76,77]. At $pH > pI$, the skin is negatively charged and electroosmotic flow proceeds in the anode-to-cathode direction. At $pH < pI$, the skin becomes positively charged and electroosmotic flow reverses to the cathode-to-anode direction. Under the application of an electric field, counterions (cations at physiological pH) are preferentially admitted into the skin. As a consequence, the sodium and chloride transport numbers are 0.6 and 0.4, respectively, during transdermal iontophoresis (in contrast to their values in a neutral membrane: $t_{Na} = 0.45$; $t_{Cl} = 0.55$) [126].

14.6 CONCLUSION

Transdermal and topical drug delivery can be effectively enhanced by iontophoresis, a technique that presents a number of clear advantages: (a) iontophoresis acts on the drug and not on the membrane, (b) control of drug fluxes and individual dose regimens are easily achievable by manipulating the current applied, (c) iontophoresis is relatively noninvasive and safe, and (d) ion transport proceeds in both directions across the skin, and iontophoresis can therefore be used for both drug delivery and noninvasive monitoring. The development of theoretical models, together with the experimental research performed over the last 25 years, has allowed the mechanisms of iontophoretic transport, the pathways of penetration, and the key factors determining drug flux (as well as the limitations of the technique) to be elucidated. Iontophoresis is now established as a mature technique for drug delivery and clinical monitoring, as evidenced by the recent commercialization of practical devices. It is anticipated that others will soon follow to consolidate the method as a safe, versatile, and efficient technology for the future.

ACKNOWLEDGMENT

The financial support of Vyteris, Inc. (Fair Lawn, NJ, USA), the U.S. National Institutes of Health (EB-001420), and the Parkinson's Disease Society is gratefully acknowledged.

REFERENCES

1. Gangarosa, L.P., and J.M. Hill. 1995. Modern iontophoresis for local drug delivery. *Int J Pharm* 123:159.
2. Zempsky, W.T., and M.A. Ashburn. 1998. Iontophoresis: Non-invasive drug delivery. *Am J Anaesthesiol* 25:158.
3. Behar-Cohen, F.-F., et al. 2002. Trans-scleral coulomb-controlled iontophoresis of methylprednisolone into the rabbit eye: Influence of duration of treatment, current intensity and drug concentration on ocular tissue and fluid levels. *Exp Eye Res* 74:51.
4. Halhal, M. 2004. Iontophoresis: From the lab to the bed side. *Exp Eye Res* 78:751.
5. Yoshizuma, M.O., et al. 1991. Experimental trans-scleral iontophoresis of ciprofloxacin. *J Ocul Opthalmol* 7:163.
6. Kalia, Y.N., et al. 2004. Iontophoretic drug delivery. *Adv Drug Deliv Rev* 56 (5):619.
7. Leboulanger, B., R.H. Guy, and M.B. Delgado-Charro. 2004. Reverse iontophoresis for non-invasive transdermal monitoring. *Physiol Meas* 25 (3):R35.
8. Delgado-Charro, M.B., and R.H. Guy. 1998. Percutaneous penetration and transdermal drug delivery. *Prog Dermatol* 32:1.
9. Delgado-Charro, M.B., and R.H. Guy. 2001. Transdermal drug delivery. In *Drug delivery and targeting for pharmacists and pharmaceuticals scientists*, ed. A.M. Hillery, A.W. Lloyd, and J. Swarbrick, 207. London: Taylor & Francis.
10. Pikal, M.J., and S. Shah. 1990. Transport mechanisms in iontophoresis. II. Electroosmotic flow and transference number measurement for hairless mouse skin. *Pharm Res* 7 (3):213.
11 Burnette, R.R., and D. Marrero. 1986. Comparison between the iontophoretic and passive. Transport of thyrotropin releasing hormone across excised nude mouse skin. *J Pharm Sci* 75 (8):738.
12. Sekkat, N., Y.N. Kalia, and R.H. Guy. 2004. Porcine ear skin as a model for the assessment of transdermal drug delivery to premature neonates. *Pharm Res* 21:1290.
13. Phipps, J.B., R.V. Padmanabhan, and G.A. Lattin. 1989. Iontophoretic delivery of model inorganic and drug ions. *J Pharm Sci* 78 (5):365.
14. Denet, A.-R., R. Vanbever, and V. Preat. 2004. Skin electroporation for transdermal and topical delivery. *Adv Drug Deliv Rev* 56:5.
15. Brett, C.M.A., and A.M.O. Brett. 1998. *Electrochemistry: Principles, methods, and applications*, 1st ed., New York: Oxford University Press, chap. 2.
16. Green, P.G. 1996. Iontophoretic delivery of peptide drugs. *J Control Release* 41:33.
17. Leboulanger, B., et al. 2004. Reverse iontophoresis as a non-invasive tool for lithium monitoring and pharmacokinetic profiling. *Pharm Res* 21 (7):1214.
18. Bellantone, N.H., et al. 1986. Enhanced percutaneous absorption via iontophoresis. I. Evaluation of an *in vitro* system and transport of model compounds. *Int J Pharm* 30 (1):63.
19. Del Terzo, S., C.R. Behl, and R.A. Nash. 1989. Iontophoretic transport of a homologous series of ionized and nonionized model compounds: Influence of hydrophobicity and mechanistic interpretation. *Pharm Res* 6 (1):85.
20. Padmanabhan, R.V., et al. 1990. *In vitro* and *in vivo* evaluation of transdermal iontophoretic delivery of hydromorphone. *J Control Release* 11 (3):123.
21. Meyer, B.R., et al. 1990. Transdermal versus subcutaneous leuprolide: A comparison of acute phamacodynamic effect. *Clin Pharmacol Ther* 48:340.
22. Gangarosa, L.P., et al. 1980. Increased penetration of non-electrolytes into mouse skin during iontophoretic water transport (iontohydrokinesis). *J Pharmacol Exp Ther* 212:377.
23. Burnette, R.R. 1987. Characterization of the permselective properties of excised human skin during iontophoresis. *J Pharm Sci* 76 (10):765.

24. Pikal, M.J. 1992. The role of the electroosmotic flow in transdermal iontophoresis. *Adv Drug Deliv Rev* 9:201.
25. Kim, A., et al. 1993. Convective solvent flow across the skin during iontophoresis. *Pharm Res* 10 (9):1315.
26. Santi, P., and R.H. Guy. 1996. Reverse iontophoresis—Parameters determining electroosmotic flow: I. pH and ionic strength. *J Control Release* 38:159.
27. Santi, P., and R.H. Guy. 1996. Reverse iontophoresis—Parameters determining electroosmotic flow: II. Electrode chamber formulation. *J Control Release* 42:29.
28. Sieg, A., R.H. Guy, and M.B. Delgado-Charro. 2003. Reverse iontophoresis for noninvasive glucose monitoring: The internal standard concept. *J Pharm Sci* 92 (11):2295.
29. Sieg, A., R.H. Guy, and M.B. Delgado-Charro. 2004. Non-invasive glucose monitoring by reverse iontophoresis *in vivo*: Application of the internal standard concept. *Clin Chem* 50 (8):1383.
30. Brand, R.M., and R.H. Guy. 1995. Iontophoresis of nicotine *in vitro*: Pulsatile drug delivery across the skin? *J Control Release* 33:285.
31. Sieg, A., R.H. Guy, and M.B. Delgado-Charro. 2004. Electroosmosis in transdermal iontophoresis: Implications for non-invasive and calibration-free glucose monitoring. *Biophys J* 87:3344.
32. Marro, D., et al. 2001. Contribution of electromigration and electroosmosis to iontophoretic drug delivery. *Pharm Res* 18 (12):1701.
33. Hirvonen, J., and R.H. Guy. 1997. Iontophoretic delivery across the skin: Electroosmosis and its modulation by drug substances. *Pharm Res* 14 (9):1258.
34. Lau, D.T.W., et al. 1994. Effect of current magnitude and drug concentration on iontophoretic delivery of octreotide acetate (Sandostatin®) in the rabbit. *Pharm Res* 11:1742.
35. Meyer, B.R., et al. 1988. Successful transdermal administration of therapeutic doses of a polypeptide to normal human volunteers. *Clin Pharmacol Ther* 44:607.
36. Cullander, C. 1992. What are the pathways of iontophoretic current flow through mammalian skin? *Adv Drug Deliv Rev* 9:119.
37. Burnette, R.R. 1989. Iontophoresis. In *Transdermal drug delivery*, eds. J. Hadrgraft and R.H. Guy. New York and Basel: Marcel Dekker, chap. 11.
38. Cullander, C., and R.H. Guy. 1991. Sites of iontophoretic current flow into the skin: Identification and characterization with the vibrating probe electrode. *J Invest Dermatol* 97:55.
39. Gibson, L.E., and R.E. Cooke. 1959. A test for concentration of electrolytes in sweat in cystic fibrosis of the pancreas utilizing pilocarpine by iontophoresis. *Pediatrics (Evanston)* 23:545.
40. Sato, K., et al. 1993. Generation and transit pathway of H^+ is critical for inhibition of palmar sweating by iontophoresis in water. *J Appl Physiol* 75:2258.
41. Turner, N.G., and R.H. Guy. 1998. Visualization and quantitation of iontophoretic effect pathways using confocal microscopy. *J Invest Dermatol Symp Proc* 3:136.
42. Turner, N.G., and R.H. Guy. 1997. Iontophoretic transport pathways: Dependence on penetrant physicochemical pathways. *J Pharm Sci* 86 (12):1385.
43. Bath, D.B., et al. 2000. Scanning electrochemical microscopy of iontophoretic transport in hairless mouse skin. Analysis of the relative contribution of diffusion, migration, and electroosmosis to transport in hair follicles. *J Pharm Sci* 89:1537.
44. Scott, E.R., J.B. Phipps, and H.S. White. 1995. Direct imaging of molecular transport through skin. *J Invest Dermatol* 104:142.
45. Bath, D.B., H.S. White, and E.R. Scott. 2000. Visualization and analysis of electroosmotic flow in hairless mouse skin. *Pharm Res* 17 (4):471.
46. Scott, E.R., A.I. Laplaza, H.S. White, and J.B. Phipps. 1993. Transport of ionic species in skin: Contribution of pores to the overall skin conductance. *Pharm Res* 10:1699.
47. Uitto, O.D., and H.S. White. 2003. Electroosmotic pore transport in human skin. *Pharm Res* 20:646.
48. Pechtold, L.A.R.M., et al. 2001. X-ray microanalysis of cryopreserved human skin to study the effect of iontophoresis on percutaneous ion transport. *Pharm Res* 18:1012.
49. Monteiro-Riviere, N.A., A.O. Inman, and J.E. Riviere. 1994. Identification of the pathway of iontophoretic drug delivery: Light and ultrastructural studies using mercuric chloride in pigs. *Pharm Res* 11:251.

50. Finkelstein, A., and A. Mauro. 1977. Physical principles and formalisms of electrical excitability. In *Handbook of Physiology*, sec. 1, The nervous system, vol. I, Cellular biology of neurones, Part 1, ed. S.R. Geiger. Bethesda, MA: American Physiological Society, chap. 6.
51. Laksminarayanaiah, N. 1969. *Transport phenomena in membranes*. London: Academic Press.
52. Oh, S.Y., L. Leung, D. Bommannan, R.H. Guy, and R.O. Potts. Effect of current, ionic strength and temperature on the electrical properties of skin. *J Control Release* 27:115.
53. Sims, S.M., W.I. Higuchi, and V. Srinivasan. 1992. Skin alteration and convective solvent flow effects during iontophoresis II. Monovalent anion and cation transport across human skin. *Pharm Res* 9:1402.
54. Li, S.K., et al. 2003. *In vitro* and *in vivo* comparisons of constant resistance AC iontophoresis and DC iontophoresis. *J Control Release* 91:327.
55. Keister, J.C., and G.B. Kasting. 1986. Ionic mass transport through a homogenous membrane in the presence of a uniform electric field. *J Membr Sci* 29:15.
56. Kasting, G.B. 1992. Theoretical models for iontophoretic delivery. *Adv Drug Deliv Rev* 9:177.
57. Kasting G.B., and J.C. Keistyer. 1989. Application of electrodiffusion theory for a homogeneous membrane to iontophoretic transport through skin. *J Control Release* 8:195.
58. Srinivasan, V., and W.I. Higuchi. 1990. A model for iontophoresis incorporating the effect of convective solvent flow. *Int J Pharm* 60:133.
59. Phipps, J.B., and R. Gyory. 1992. Transdermal ion migration. *Adv Drug Deliv Rev* 9:137.
60. Lai, P.M., and M.S. Roberts. 1999. An analysis of solute structure–human epidermal transport relationships in epidermal iontophoresis using the ionic mobility: Pore model. *J Control Release* 58:323.
61. Roberts, M.S., P.M. Lai, and Y.G. Anissimov. 1998. Epidermal iontophoresis: I. Development of the ionic mobility–pore model. *Pharm Res* 15 (10):1569.
62. Lai, M.P., and M.S. Roberts. 1998. Epidermal iontophoresis: II. Application of the ionic mobility–pore model to the transport of local anaesthetics. *Pharm Res* 15 (10):1579.
63. Masada, T., et al. 1989. Examination of iontophoretic transport of ionic drugs across skin: Baseline studies with the four-electrode system. *Int J Pharm* 49:57.
64. Srinivasan, V., et al. 1989. Transdermal iontophoretic drug delivery: Mechanistic analysis and application to polypeptide delivery. *J Pharm Sci* 78:370.
65. Srinivasan, V., W.I. Higuchi, and M.H. Su. 1989. Baseline studies with four electrode system: The effect of skin permeability increase and water transport on the flux of a model uncharged solute during iontophoresis. *J Control Release* 10:157.
66. Luzardo-Alvarez, A., M.B. Delgado-Charro, J. Blanco-Méndez. 2001. Iontophoretic delivery of ropinirole hydrochloride: Effect of current density and vehicle formulation. *Pharm Res* 18 (12): 1714.
67. Luzardo-Alvarez, A., M.B. Delgado-Charro, and J. Blanco-Méndez. 2003. *In vivo* iontophoretic administration of ropinirole hydrochloride. *J Pharm Sci* 92:2450.
68. Sims, S.M., W.I. Higuchi, and V. Srinivasan. 1991. Skin alteration and convective solvent flow effects during iontophoresis: I. Neutral solute transport across human skin. *Int J Pharm* 69:109.
69. Li, S.K., A.-H. Ghanem, K.D. Peck, and W.I. Higuchi. 1997. Characterization of the transport pathways induced during low to moderate voltage iontophoresis in human epidermal membrane. *J Pharm Sci* 87:40.
70. Peck, K.D., et al. 1996. Quantitative description of the effect of molecular size upon electroosmotic flux enhancement during iontophoresis for a synthetic membrane and human epidermal membrane. *J Pharm Sci* 85 (7):781.
71. Peck, K.D., A.-H. Ghanem, and W.I. Higuchi. 1994. Hindered diffusion of polar molecules through and effective pore radii estimates of intact and ethanol treated human epidermal membrane. *Pharm Res* 11 (9):1306.
72. Peck, K.D., J. Hsu, S.K. Li, A.H. Ghanem, and W. Higuchi. 1998. Flux enhancement effects of ionic surfactants upon passive and electroosmotic transdermal transport. *J Pharm Sci* 87:1161.
73. Li, S.K., A.H. Ghanem, K.D. Peck, and W.I. Higuchi. 1997. Iontophoretic transport across a synthetic membrane and human epidermal membrane: A study of the effects of the permeant charge. *J Pharm Sci* 86:681.

74. Pikal, M.J. 1990. Transport mechanisms in iontophoresis. I. A theoretical model for the effect of electroosmotic flow on flux enhancement in transdermal iontophoresis. *Pharm Res* 7 (2):118.
75. Pikal, M.J., and S. Shah. 1990. Transport mechanisms in iontophoresis. III. An experimental study of the contribution of electroosmotic flow and permeability change in transport of low and high molecular weight solutes. *Pharm Res* 7 (3):222.
76. Marro, D., R.H. Guy, and M.B. Delgado-Charro. 2001. Characterization of the iontophoretic permselectivity properties of human skin and pig skin. *J Control Release* 70:213.
77. Luzardo-Alvarez, A., et al. 1998. Iontophoretic permselectivity of mammalian skin: Characterization of hairless mouse and porcine membrane models. *Pharm Res* 15 (7):984.
78. Marro, D., Y.N. Kalia, M.B. Delgado-Charro, and R.H. Guy. 2001. Optimizing iontophoretic drug delivery: Identification and distribution of the charge-carrying species. *Pharm Res* 18:1709.
79. DeNuzzio, J.D., and B. Berner. 1990. Electrochemical and iontophoretic studies of human skin. *J Control Release* 11:105.
80. Phipps, J.B., R.V. Padmanabhan, and G.A. Lattin. 1988. Transport of ionic species through skin. *Solid State Ionics* 28–30:1778.
81. Leboulanger, B., J.M. Aubry, G. Gondolfi, R.H. Guy, and M.B. Delgado-Charro. 2004. Non-invasive lithium monitoring by reverse iontophoresis *in vivo*. *Clin Chem* 50:2091.
82. Turner, N.G., et al. 1997. Iontophoresis of poly-L-lysines: The role of molecular weight? *Pharm Res* 14:1322.
83. Guy, R.H., M.B. Delgado-Charro, and Y.N. Kalia. 2001. Iontophoretic transport across the skin. *Skin Pharmacol Appl Skin Physiol* 14 (Suppl. 1):35.
84. Green, P.G., et al. 1991. Iontophoretic delivery of a series of tripeptides across the skin *in vitro*. *Pharm Res* 8 (9):1121.
85. Roberts, M.S., et al. 1997. Solute structure as a determinant of iontophoretic transport. In *Mechanisms of transdermal drug delivery*, eds. R.H.G. Potts and R.H. Guy. New York: Marcel Dekker, chap. 9.
86. Ruddy, S.B., and B.W. Hadzija. 1992. Iontophoretic permeability of polyethylene glycols through hairless rat skin: Application of hydrodynamic theory for hindered transport through liquid-filled pores. *Drug Des Discov* 8:207.
87. Yoshida, N.H., and M.S. Roberts. 1993. Solute molecular size and transdermal iontophoresis across excised human skin. *J Control Release* 25:177.
88. Tashiro, Y., et al. 2001. Effect of lipophilicity on *in vivo* iontophoretic delivery—I. NSAIDs. *Biol Pharm Bull* 24 (3):278.
89. Tashiro, Y., et al. 2001. Effect of lipophilicity on *in vivo* iontophoretic delivery—II. β-Blockers. *Biol Pharm Bull* 24 (6):671.
90. VanOrman Huff, B., G.G. Liversidge, and G.L. McIntire. 1995. The electrophoretic mobility of tripeptides as a function of pH and ionic strength: Comparison with iontophoretic flux data. *Pharm Res* 12:751.
91. Yoshida, N.H., and M.S. Roberts. 1995. Prediction of cathodal iontophoretic transport of various anions across excised skin from different vehicles using conductivity measurements. *J Pharm Pharmacol* 47:883.
92. van der Geest, R., et al. 1996. Iontophoresis of bases, nucleosides and nucleotides. *Pharm Res* 13:553.
93. Kasting, G.B., and L.A. Bowman. 1990. DC electrical properties of frozen, excised human skin. *Pharm Res* 7 (2):134.
94. Green, P.G., et al. 1992. Transdermal iontophoresis of amino acids and peptides *in vitro*. *J Control Release* 21:187.
95. Delgado-Charro, M.B., and R.H. Guy. 1994. Characterization of convective solvent flow during iontophoresis. *Pharm Res* 11 (7):929.
96. Delgado-Charro, M.B., and R.H. Guy. 1998. Iontophoresis of peptides. In *Electronically controlled drug Delivery*, eds. B. Berner and S.M. Dinh, 129. London: CRC Press, chap. 7.
97. Nugroho, A.K., et al. 2004. Transdermal iontophoresis of rotigotine: Influence of concentration, temperature and current density in human skin *in vitro*. *J Control Release* 96:159.

98. Thysman S., and V. Préat. 1993. *In vivo* iontophoresis of fentanyl and sufentanil in rats: Pharmacokinetics and acute antinociceptive effects. *Anesth Analges* 77 (1):61.
99. Santi, P., et al. 1997. Transdermal iontophoresis of salmon calcitonin can reproduce the hypocalcemic effect of intravenous administration. *Farmaco* 52 (6):445.
100. Suzuki, Y., et al. 2001. Iontophoretic pulsatile transdermal delivery of human parathyroid hormone (1–34). *J Pharm Pharmacol* 53:1227.
101. Suzuki, Y., et al. 2002. Prevention of bone loss in ovariectomized rats by pulsatile transdermal iontophoretic administration of human PTH (1–34). *J Pharm Sci* 91:350.
102. Nakamura, K., et al. 2003. Transdermal administration of salmon calcitonin by pulse depolarization-iontophoresis in rats. *Int J Pharm* 218:93.
103. Hirvonen, J., F. Hueber, and R.H. Guy. 1995. Current profile regulates iontophoretic delivery of amino acids across the skin. *J Control Release* 37:239.
104. Lopez-Castellano, A., R.H. Guy, and M.B. Delgado-Charro. 1999. Effect of area on the iontophoretic transport of phenylalanine and propranolol across the skin. In *Proceedings of the international symposium on controlled release of bioactive materials*, 26, 405.
105. Ledger, P.W. 1992. Skin biological issues in electrically enhanced transdermal delivery. *Adv Drug Deliv Rev* 9:289.
106. Scott, E.R., et al. 2000. Electrotransport systems for transdermal delivery: A practical implementation of iontophoresis. In *Handbook of pharmaceutical controlled release technology*, ed. D.L. Wise. New York: Marcel Dekker, chap. 31.
107. Lopez, R.F.V., et al. 2001. Iontophoretic delivery of 5-aminolevulinic acid (ALA): Effect of pH. *Pharm Res* 18 (3):311.
108. Glikfeld, P., R.S. Hinz, and R.H. Guy. 1989. Noninvasive sampling of biological fluids by iontophoresis. *Pharm Res* 6:988.
109. Tamada, J.A., et al. 1999. Noninvasive glucosa monitoring. Comprehensive clinical results. *JAMA* 282:1839.
110. Delgado-Charro, M.B., and R.H. Guy. 2003. Transdermal reverse iontophoresis of valproate: A non-invasive tool for therapeutic drug monitoring. *Pharm Res* 20:1508.
111. Merino, V., et al. 1999. Noninvasive sampling of phenylalanine by reverse iontophoresis. *J Control Release* 61:65.
112. Leboulanger, B., R.H. Guy, and M.B. Delgado-Charro. 2004. Non-invasive monitoring of phenytoin by reverse iontophoresis. *Euro J Pharm Sci* 22 (5):427.
113. Pitzer, K.R., et al. 2001. Detection of hypoglycaemia with the GlucoWatch Biographer. *Diabet Care* 24:881.
114. Thysman, S., C. Hanchard, and V. Préat. 1994. Human calcitonin delivery in rats by iontophoresis. *J Pharm Pharmacol* 46:725.
115. Mudry, B., R.H. Guy, and M.B. Delgado-Charro. 2005. Rational optimization of iontophoretic transport across the skin. In *32th Annual meeting & exposition of the controlled release of bioactive materials*. Miami, USA.
116. Clemessy, M., et al. 1995. Mechanisms involved in iontophoretic transport of angiotensin. *Pharm Res* 12:998.
117. Thysman, S., V. Preat, and M. Roland. 1992. Factors affecting iontophoretic mobility of metoprolol. *J Pharm Sci* 81:7.
118. Lopez, R.F.V., et al. 2003. Optimization of aminolevulinic acid delivery by iontophoresis. *J Control Release* 88 (1):65.
119. Craane-van Hinsberg, W.H.M., et al. 1994. Iontophoresis of a model peptide across human skin *in vitro*: Effects of iontophoresis protocol, pH and ionic strength on peptide flux and skin impedance. *Pharm Res* 11:1296.
120. Brand, R.M., and P.L. Iversen. 1996. Iontophoretic delivery of a telomeric oligonucleotide. *Pharm Res* 13:851.
121. Merino, V., et al. 1999. Electrorepulsion versus electroosmosis: Effect of pH on the iontophoretic flux of 5-fluorouracil. *Pharm Res* 16 (5):758.
122. Sage, B.H., and R.A. Hoke. 1996. US Patent 5,494,679, February 27.

123. Rojanasakul, Y., and J.R. Robinson. 1989. Transport mechanism of the cornea: Characterization of barrier permselectivity. *Int J Pharm* 55:237.
124. Gandhi R.B., and J.R. Robinson. 1991. Permselective characteristics of rabbit buccal mucosa. *Pharm Res* 8:1199.
125. Nicoli, S., et al. 2003. Characterization of the permselective properties of rabbit skin during transdermal iontophoresis. *J Pharm Sci* 92 (7):1482.
126. Laksminarayanaiah, N.N. 1969. Properties of monolayers and bilayers. In *Transport phenomena in membranes*, 444. London: Academic Press, chap. 9.

15 Electroporation as a Mode of Skin Penetration Enhancement

Michael C. Bonner and Brian W. Barry

CONTENTS

15.1 INTRODUCTION

As the human skin is the largest single organ of the body, it may at first sight be attractive to formulators as an accessible means of drug input. This route of delivery holds many advantages that include avoidance of GI and liver first pass effects, controlled and continuous drug delivery, easy removal of the dosage form, and good patient compliance.

However, the skin's function as a barrier to xenobiotics ensures a difficult passage for most drugs both into and through the skin [1]. The main reasons for the good barrier properties of the organ lie within the highly organized lipid matrix within the stratum corneum, the outermost layer of the epidermis. The lipids in the stratum corneum arrange themselves in the lamellae between the corneocytes. These lipids may have a crystalline, gel, or liquid crystalline character and their arrangement provides great resistance to molecular penetration of the membrane [2,3].

It is therefore desirable to devise strategies both to enhance the penetration of molecules, which can already breach the skin barricade passively to some extent, and also to widen the spectrum of drug molecules that can penetrate the skin at therapeutically beneficial doses. Many tactics have been utilized to help overcome the barrier function. These include chemical means (e.g., chemical penetration enhancers or entrapment of molecules within lipid vesicles) or physical methods (such as ultrasound, microneedles, or electrical methods). Two important electrical methods are iontophoresis and electroporation.

15.2 IONTOPHORESIS

Iontophoresis can be described as an electrically facilitated movement of molecules or ions across a membrane, typically utilizing a constant, low, physiologically acceptable electric current. Less frequently a uniform, low, electrical potential difference may be used. Constant voltage iontophoresis is less often used because the electrical resistance of skin varies and so an identical potential difference applied to two subjects may produce different—and potentially hazardous—current densities in their skin. The normal construction of an iontophoretic system comprises an ionic drug in a formulation placed under an electrode of similar polarity and applied to the skin. A grounding or return electrode completes the circuit and when placed a short distance away, it takes care to ensure that there is no electrical contact between the two electrodes [4–7]. The general principle of iontophoresis is that of electrorepulsion, i.e., like charges repel each other; and so, for example, a negatively charged drug will be delivered under a cathode where it will be discharged into the skin and toward the anode. Clearly, a positively charged drug would be transported under an anode. A constant current mode of iontophoretic delivery is normally used with a maximum charge density of 0.5 mA/cm^2 of skin surface as this is generally regarded as clinically tolerable.

In addition to the electrorepulsive effect, iontophoresis may also enhance drug delivery by electroosmosis. Small ions like Na^+, if included, for example, as part of a buffer system in a formulation, may be iontophoresed into the skin and carry with them solvating water molecules. These molecules may in turn carry other penetrants with them into the membrane, and in this way iontophoresis may also enhance the delivery of uncharged drug species. Iontophoresis, as well as electroosmosis, may also generally disorganize the intercellular lipids and produce a membrane more permeable to penetrants.

15.3 ROUTE OF IONTOPHORETIC PENETRATION

Under iontophoresis, drug transport is largely thought to occur down a so-called "shunt route" [8,9]. This route comprises the skin appendages—sweat glands and hair follicles. Contributions to iontophoretic transport have also been suggested to come from an intercellular route [10] and also from aqueous pores induced by the application of current [11]. Authors have suggested that the iontophoretic pathway taken depends on the physicochemical properties of the penetrant with the more hydrophilic species favoring the shunt route whereas the more lipophilic permeants tend to select an intercellular pathway [12].

15.4 ELECTROPORATION

15.4.1 Mechanism of Skin Disorganization by Electroporation

Electroporation (sometimes referred to as electropermeabilization) involves the application of high-voltage pulses (normally from 100 to 1000 V) for a very short duration (from micro- to milliseconds). The pulses are thought to produce transient aqueous pores in the lipid bilayers of the stratum corneum. These apertures provide pathways for drug penetration through the-horny layer [13–17]. The technique of electroporation is more usually applied to the unilamellar phospholipid bilayers of cell membranes. Though electroporation has been in use since the early 1970s to permeabilize the bilayer membranes of cells, allowing foreign DNA or other agents to enter them [17–18], its feasibility for transdermal drug delivery was first demonstrated by Prausnitz et al. [19]. The electrical behavior of the human epidermal membrane (HEM) as a function of the magnitude and duration of the applied voltage mimics closely the break-downand recovery of bilayer membranes seen during electroporation [20]. In contrast to iontophoresis, electroporation acts mainly on the skin with less contribution of electromigration due to the short pulse "on" time. The approximately 100 multilamellar bilayers of the stratum corneum require around 100 V pulses for electroporation, or about 1 V per bilayer [17]. As with iontophoretic transport, some studies have indicated that high-voltage pulse-induced transdermal delivery of charged or even neutral drugs could be controlled by an appropriate use of the electric parameters, i.e., pulse voltage, width, and number [21–23].

Although it is generally believed that electroporation involves the creation of aqueous pathways (pores) in the stratum corneum [24], this mechanism remains somewhat controversial. These proposed channels have not yet been identified in any microscopic study due to their small size (about 10 nm), sparse distribution (<0.1% of the total skin area), and short-lived nature (millisecond to second) [16,25].

Notably, during electroporation highly localized pockets of molecular transport are observed, revealed by real-time video imaging for fluorescent molecules [13]. Studies with snake skin, which has no appendages, revealed a localized area for transport of fluorescent molecules after high-voltage pulsing, whereas low-voltage iontophoresis induced little molecular penetration through the membrane [26].

A possible mechanism proposed to explain the perturbation of the lipid barrier of the horny layer during electroporation is a heat dissipation phenomenon [27]. Although it was stated to be a nonthermal process, the temperature rose after electric pulses of 100 V (transmembrane voltage) applied for 1 ms [28]. A "pore" is supposed to form very rapidly, preceding any potential temperature rise [25,28]. However, localized heating may also occur at sites of large current density, especially with long pulses [29]. Even though convection propagates the heat front across the skin, the Joule heating could be sufficient for the melting of the skin lipids with phase transitions around 70°C, suggesting that the temperature rise has an important role to play [29,30]. The area involved in drug transport during electroporation was subdivided into two distinct domains: a number of local transport regions (LTRs) with radii ranging from 10 to 100 μm depending on pulse voltage, and surrounding local dissipation regions (LDRs) larger than LTRs with their size depending on the length of the applied pulse [31]. Experimental and theoretical investigations of localized heating by skin electroporation predict that LTR formation could be partly explained by Joule heating during pulsing and that temperature rise is relatively small in the LDR that surrounds the LTR [24,30,32].

The mechanisms of molecular transport during electroporation are expected to involve passive diffusion and electrically driven transport during the brief pulsing time, but unlike iontophoresis, electroosmosis is believed to be of much less significance [22,23,33]. The increase in horny layer water content, and consequently disorganization of stratum corneum, was also observed but seemed to be less important than during iontophoresis [16]. The barrier effect of the horny layer also reduces for an additional period after pulsing, further enhancing molecular penetration.

15.4.2 Safety Concerns with Electroporation

It is noteworthy that some therapeutic applications, such as transcutaneous electrical neural stimulation, involve application to the skin of electric pulses of up to hundreds of volts [5]. However, a safety limitation is the major concern associated with the use of electroporation, even though several reports indicated that the damage to the skin was mild and reversible [16,23]. The only skin alteration seen with electroporation was slight erythema that decreased within a few hours [34]. Patients submitted to electrochemotherapy seemed to tolerate well the application of 10,000 V/cm for 100 μs square-wave pulses [35]. However, to avoid pain during electroporation, milder conditions such as lower voltage, shorter pulses, or improved electrode design could be used [36].

15.4.3 Applications of Electroporation

As electroporation is still largely in the experimental stage and may possibly be never used clinically, many workers have employed model nontherapeutic molecules to demonstrate the feasibility of its use, such as sulforhodamine, calcein, and caffeine. However, more therapeutic and biotechnologically derived molecules have now been investigated for transdermal delivery, including, for example, fentanyl [22], metoprolol [21], timolol [36], flurbiprofen [37], cyclosporin [38], heparin [39], oligonucleotides [40], and genes [41].

15.4.4 Pulsing Protocols for Electroporation

Recent work by Rashid et al. [42] examined pulsing protocols for electroporation of estradiol and L-glutamic acid. Any changes occurring in the skin permeability barrier as a result of electrical application, and the extent of their reversibility, were also evaluated. The mechanisms of molecular transport through the skin under the influence of constant direct current or high-voltage pulsing were also considered.

Current was delivered to the membranes through silver/silver chloride (Ag/AgCl) electrodes for iontophoresis, whereas stainless steel electrodes were employed for electroporation studies. Of the two model compounds, L-glutamic acid carries a net negative charge of −1 at pH 7.4, whereas estradiol is nonionized. Hence, they were delivered under the cathode and anode, respectively.

The experimental protocol involved three consecutive stages of treatment to the same HEM: a first passive permeation stage, which lasted for 3 h, followed by a 2 h electrical treatment period during which electroporation or iontophoresis or both protocols were applied to the skin; and finally a second passive stage (2 h) evaluated possible reversibility of skin barrier function following electrical treatment.

Various pulsing protocols have been utilized for *in vitro* electroporation studies in different laboratories, and Vanbever et al. [21] suggested that many short duration pulses may be less efficient for transdermal drug delivery enhancement than a few long duration pulses. Therefore, Rashid et al. [42] studied the effect of 5 square-wave electrical pulses of 100 V magnitude and 100 ms pulse width with a pulse spacing of 1 min on epidermal penetration of estradiol and L-glutamic acid and compared it to that of 50 pulses (100 V for 10 ms) with a spacing of 10 s. These regimens had the same total "on" time (500 ms). Pulses were applied to the membranes twice; at the start, and halfway through, a 2 h electrical period. Square-wave pulses were used as they can be set at a constant predetermined magnitude and width, thus easily producing reproducible constant voltage, and are thought to be more tolerable *in vivo* than exponentially decaying pulses [43].

To delineate the effects of electroporation on estradiol and L-glutamic acid penetration through HEM, instantaneous fluxes were calculated from cumulative penetration profiles, and are presented as a function of time in Figure 15.1 and Figure 15.2, respectively. Apparent flux values for both permeants during three stages of the experiment were also calculated from cumulative penetration plots, and are recorded in Table 15.1. Apparent fluxes during first and second passive stages were measured as the regression slopes of the linear part of each cumulative penetration plot during these stages. Fluxes for both permeants during the electrical treatment stage were determined by taking an average of instantaneous fluxes between the two time points during which electroporation pulses were applied to the HEM.

For both estradiol (Figure 15.1) and L-glutamic acid (Figure 15.2), there was a slow, gradual increase in epidermal transport during the first passive stage. Upon pulsing (either protocol), the penetration rates for both compounds increased significantly in relation to their respective first passive fluxes ($P < 0.05$). However, fluxes for these permeants would have

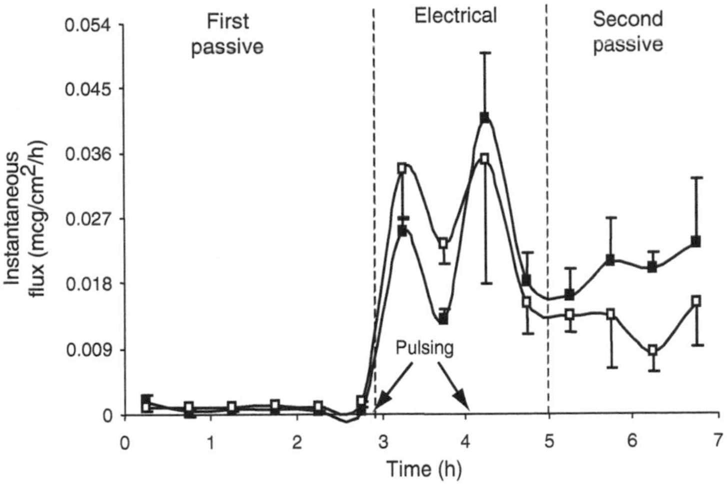

FIGURE 15.1 Instantaneous flux of estradiol through human epidermal membrane. Membranes were electroporated by either 5 × 100 V pulses with a pulse width of 100 ms (closed symbols) or 50 × 100 V pulses with a pulse width of 10 ms (open symbols) ($n = 3$).

been in their pre-steady states during the first passive period, so to measure a more realistic enhancement in penetration, it was more relevant to compare all the fluxes under steady-state conditions. The authors had previously determined steady-state passive fluxes of estradiol. These values were used to calculate the respective enhancement ratios (ER; the ratio of the apparent flux during electroporation to the apparent steady-state flux) and damage ratios (DR; the ratio of the apparent postpulsing flux to the apparent steady-state passive flux), which are also presented in Table 15.1

As shown by ERs (Table 15.1), both pulsing protocols significantly increased the penetration of estradiol and L-glutamic acid in relation to their respective steady-state passive fluxes

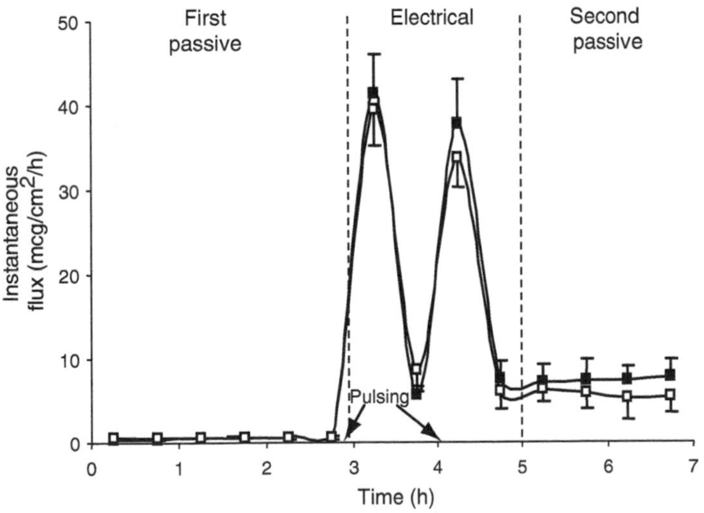

FIGURE 15.2 Instantaneous flux of L-glutamic acid through human epidermal membrane. Membranes were electroporated by either 5 × 100 V pulses with a pulse width of 100 ms (closed symbols) or 50 × 100 V pulses with a pulse width of 10 ms (open symbols) ($n = 3$).

TABLE 15.1
Apparent Fluxes of Estradiol and L-Glutamic Acid through Human Epidermis During a Three Stage Experiment for Two Pulsing Protocols. Enhancement Ratios and Damage Ratios for Both Compounds are also Given ($n = 3$)

Stage	Apparent Flux ($\mu g/cm^2/h$)			
	Estradiol		L-Glutamic Acid	
	5 × 100 V, 100 ms	50 × 10 V, 10 ms	5 × 100 V, 100 ms	50 × 10 V, 10 ms
First passive	0.0013 ± 0.0004	0.0017 ± 0.0002	0.33 ± 0.06	0.45 ± 0.03
Electrical	0.033 ± 0.004	0.034 ± 0.012	39.5 ± 4.48	35.4 ± 3.44
Second passive	0.02 ± 0.005	0.013 ± 0.004	7.56 ± 1.59	5.55 ± 0.40
Enhancement ratio[a]	11.2	11.5	55.6	49.9
Damage ratio[b]	6.80	4.41	10.6	7.81

[a]The ratio of the apparent flux during electrical period to the apparent steady-state passive flux.
[b]The ratio of the apparent flux during second passive stage to the apparent steady-state passive flux.

($P < 0.05$). For each permeant, ER with few pulses was not significantly different from that with many pulses ($P > 0.05$). In addition, although the DR during the second passive stage for either permeant was generally higher after few pulses, it was not significantly different from that with many pulses ($P > 0.05$). These results suggest that both pulsing protocols are equally efficient at enhancing epidermal penetration of these permeants.

Figure 15.1 and Figure 15.2 also show that postelectroporation fluxes for both permeants always remained elevated throughout the second passive period, and were considerably higher than respective steady-state passive diffusion values. This may have been due to alterations in membrane barrier properties during the course of pulsing, which would have increased its passive permeability. If no damage occurred to the skin permeability barrier during the course of pulsing, then DR would simply have been equal to 1, i.e., penetration after termination of the electrical protocol would not have been different from the apparent steady-state passive diffusion. However, DR values were always considerably greater than 1 (Table 15.1), which may indicate membrane damage. Alternatively, higher fluxes during the second passive stage may have arisen from efflux from the membranes, which became loaded during electroporation; or both mechanisms may have operated.

Instantaneous fluxes profiles (Figure 15.1 and Figure 15.2) also show that on the first pulsing round, penetration rates of both compounds increased considerably with either electroporation protocol. Fluxes then decreased during the "rest" period between two rounds of pulsing. A second run of pulsing further raised the fluxes, which fell again during the recovery period. This demonstrated the ability of electroporation to enhance repeatedly the transdermal transport of a molecule in a pulsatile manner. Various hormone peptides such as the leutinizing hormone-releasing hormone and the growth hormone may be delivered in such a manner, and electroporation may be a useful way of achieving this goal.

15.4.5 Electroporation and Iontophoresis Combination

In contrast to low-current iontophoresis, which utilizes an electrical driving force to push permeants into and across the skin, electroporation increases transdermal transport primarily by acting on the skin itself [44]. Therefore, combining the two techniques may enhance the penetration of drug molecules even further.

Electroporation has been combined with iontophoresis to enhance the penetration of peptides such as vasopressin, neurotensin, calcitonin, and LHRH, as well as other compounds such as defibrase and dextran sulfate [44–49]. However, failures have also been reported [50,51].

Rashid et al. [42] compared the effects of combined electroporation and iontophoresis on transepidermal transport of estradiol and L-glutamic acid with that of either method alone. A three-stage experimental protocol was used comprising first passive, then electrical, and finally second passive stages. Although the duration of the electrical period was 2 h, current was only applied for a total of 110 min in order to make a direct comparison between transport under iontophoresis alone and combined electroporation and iontophoresis. The electroporation protocol involved 5 square-wave electrical pulses (100 V for 100 ms) with a spacing of 1 min, applied at the start, and at 1 h, a 2 h electrical period. For the combined treatment, 5 square-wave pulses (100 V for 100 ms) spaced at 1 min intervals were selected at the beginning, and halfway through, a 2 h electrical period, using stainless steel electrodes. Immediately after the last pulse, these electrodes were replaced by Ag/AgCl electrodes, and a relatively high constant direct current of 0.8 mA/cm^2 was applied.

Results are presented as instantaneous fluxes versus time plots for estradiol and L-glutamic acid in Figure 15.3 and Figure 15.4, respectively. Apparent fluxes for the two permeants during three stages of the experiment are also shown in Table 15.2, along with respective enhancement and DRs.

During the first passive stage, both permeants' penetration slowly built up. Upon application of all three electrical protocols, the apparent fluxes for both molecules were always significantly enhanced in relation to their respective steady-state passive values ($P < 0.05$). During the second passive period, penetration rates considerably reduced, though the apparent fluxes were always significantly greater after electroporation or combined treatment than with iontophoresis alone ($P < 0.05$).

Table 15.2 shows that for estradiol, the ER during electroporation (8.7) was greater than during iontophoresis (2.5, $P < 0.05$). In contrast, the apparent L-glutamic acid fluxes were

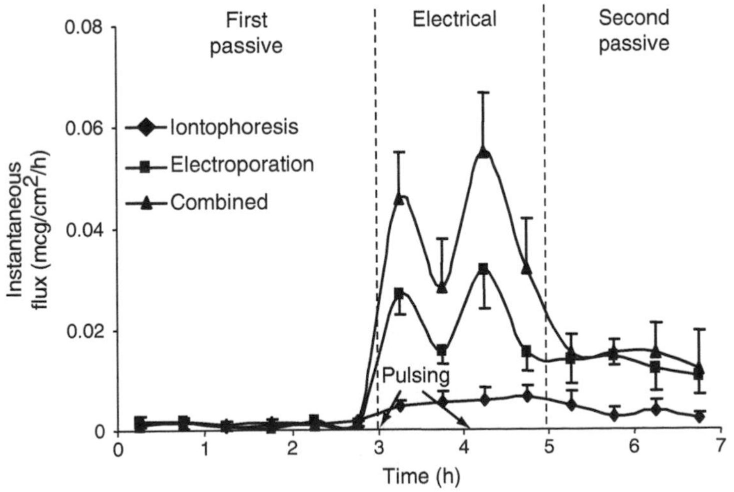

FIGURE 15.3 Instantaneous flux profiles of estradiol through human epidermal membrane during a three stage experiment. Electrical procedures were iontophoresis (0.8 mA/cm^2), electroporation (5 ×100 V pulses with a pulse width of 100 ms), or electroporation followed by iontophoresis ($n = 5$–9).

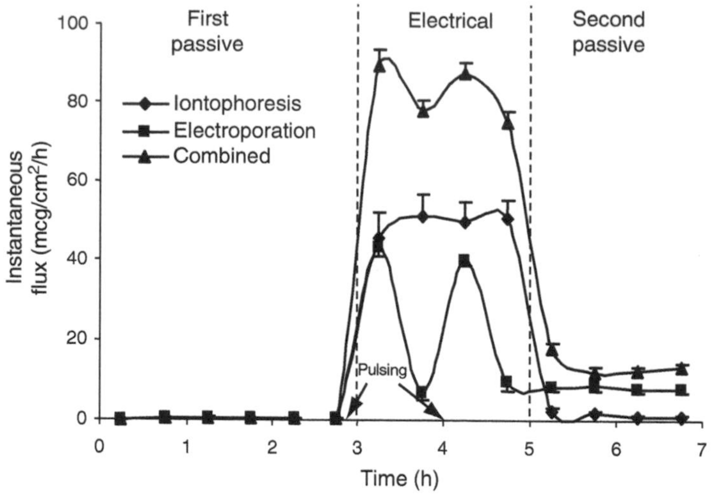

FIGURE 15.4 Instantaneous flux profiles of L-glutamic acid through human epidermal membrane during a three stage experiment. Electrical procedures were iontophoresis (0.8 mA/cm^2), electroporation (5×100 V pulses with a pulse width of 100 ms), or electroporation followed by iontophoresis ($n = 5$–9).

promoted to the same extent following both these protocols (ER of 58.7 with electroporation and 67.9 with iontophoresis, $P > 0.05$). It should also be noted that the overall enhancement in penetration with either electrical protocol was always appreciably higher for L-glutamic acid than for estradiol.

Iontophoresis by definition is the process of transport of ions into or through a tissue by the use of an applied potential difference across the tissue [52]. Depending on the physicochemical characteristics of a molecular species, electrorepulsion is usually the primary mechanism of transdermal transport for ions, whereas electroosmosis and increased passive diffusion (as a result of the reduced barrier properties) are more prominent for neutral species [53]. In contrast, enhancement in flux for neutral or weakly charged species during electroporation arises predominantly from the reduced barrier properties of the membrane, whereas direct electrorepulsion is usually of secondary importance [25].

For estradiol, a lipophilic and unionized molecule delivered from the anode, flux enhancement during iontophoresis was most likely due to electroosmosis with some contribution from increased passive diffusion. In contrast, L-glutamic acid is highly ionized at pH 7.4, and increase of its penetration under the influence of constant direct current was more likely to arise from direct interaction between the ions and the imposed electric field; passage would be against electroosmotic flow since the anode was in the receiver chamber of the diffusion cell. During electroporation, the contribution of electroosmotic flow to overall estradiol transport was unlikely to be of any significance due to the short pulsing time (1 s in this study) and, therefore, the enhancement in its epidermal penetration likely arose from increased passive diffusion. In contrast, both electrorepulsion and reduced permeability of the skin were probably responsible for increase in L-glutamic acid flux through the electroporated skin. Therefore, under the influence of an electric field, while electrorepulsion would forcibly push L-glutamic acid ions into and across the skin, estradiol would only move passively and be carried across the membrane with electroosmotic flow; this may explain the differences in ER values for the two permeants.

As the mechanisms of penetration and acceleration during iontophoresis or electroporation are different, the application of a constant direct current after pulsing may raise penetration even further in comparison with either method alone. Table 15.2 shows that for L-glutamic

TABLE 15.2
Apparent Flux Values of Estradiol and L-Glutamic Acid through Human Epidermis during Three Stages of the Experiment for Three Electrical Protocols; Iontophoresis, Electroporation, or Combined Electroporation and Iontophoresis. Enhancement Ratios and Damage Ratios for Both Compounds are also Given $5 \leq n \leq 9$

	Apparent Flux ($\mu g/cm^2/h$)					
	Estradiol			L-Glutamic Acid		
Stage	Iontophoresis	Electroporation	Combined	Iontophoresis	Electroporation	Combined
First passive	0.0014 ± 0.0002	0.0018 ± 0.0005	0.0014 ± 0.0002	0.31 ± 0.06	0.37 ± 0.06	0.43 ± 0.05
Electrical	0.008 ± 0.001	0.029 ± 0.003	0.041 ± 0.005	48.2 ± 5.43	41.7 ± 1.86	81.6 ± 3.06
Second passive	0.003 ± 0.002	0.014 ± 0.002	0.016 ±0.003	1.37 ± 0.38	8.69 ± 1.36	13.3 ± 1.16
Enhancement ratio[a]	2.5	8.7	12.3	67.9	58.7	115
Damage ratio[b]	1.38	4.32	4.82	1.93	12.2	18.7

[a]The ratio of the apparent flux during electrical period to the apparent steady-state passive flux for estradiol and L-glutamic acid, respectively.
[b]The ratio of the apparent flux during second passive stage to the apparent steady state passive flux for estradiol and L-glutamic acid, respectively.

acid, ER during combined treatment (115) was approximately twofold higher than during either protocol on its own (58.7 with electroporation and 67.9 with iontophoresis), suggesting an additive effect. For estradiol, the ER during the combined protocol (12.3) was approximately fivefold greater than during iontophoresis (2.5), and higher than the ER during electroporation (8.7). However, as the sum of the apparent fluxes during individual protocols (0.008 $\mu g/cm^2/h$ with iontophoresis and 0.029 $\mu g/cm^2/h$ with electroporation) was approximately equal to the apparent flux during the combined protocol (0.041 $\mu g/cm^2/h$), the overall effect once again was additive.

Bommannan et al. [45] have shown synergism with iontophoresis and electroporation combined. In their studies, application of a single exponentially decaying pulse with an initial amplitude of 1000 V and a time constant of 5 ms prior to the initiation of 0.5 mA/cm^2 iontophoresis increased five- to tenfold the flux of luteinizing hormone-releasing hormone across excised human epidermis. Riviere et al. [46] confirmed these results with porcine skin *in vitro*. Lack of synergism in the work by Rashid et al. [42] was attributed to the time elapsed between the final pulse and the switching-on of iontophoresis. Skin resistance is largely attributed to the stratum corneum lipids, and any changes during electroporation are likely to arise from alterations in lipid organization within the stratum corneum [25]. High-voltage pulsing of the skin reduces the electrical resistance by up to three orders of magnitude within microseconds after initiation of the first pulse; the tissue starts to recover immediately after cessation of the pulse. The extent of this drop and recovery is primarily determined by the amplitude and duration of the pulses [28,33,54,55]. Following the termination of pulsing, there are normally four distinguishable phases of resistance recovery. The very fast recovery (within 30 ms), where the resistance achieves between 10% and 80% of the pre-pulse value, is largely attributed to the closing kinetics of electrically created pathways. Other phases exist on the time order of 1 s, 10 s, and minutes, up to several hours. The recovery process on the order of 1 s is thought to be due to the reduction in the pathway size, whereas the others are likely to arise from the slower rearrangement of the lipid system [29]. Since it took approximately 1–2 min between the final pulse and switching-on of iontophoresis in Rashid's study [42], three phases of barrier recovery would have been completed and skin would most likely have been less permeable than immediately after pulsing, which in turn may have hindered the expected synergistic effect.

Jadoul and Préat [56] have also proposed a similar explanation for the lack of synergistic effects on transdermal delivery of domperidone with combined electroporation (1 pulse of 1000 V with a time constant of 4 ms) and iontophoresis (0.4 mA/cm^2) despite the fact that iontophoresis was switched on within a "few seconds" after electroporation. Combined pulsing and iontophoresis also did not improve penetration of sodium nonivamide acetate through nude mouse skin [51]. Therefore, when combing the two protocols, it should be more efficient to use a system that delivers current during or immediately after pulsing without delay.

Table 15.2 also illustrates that for either compound, DRs after a combined protocol or electroporation alone were always significantly higher than respective postiontophoresis DRs ($P < 0.05$). Prausnitz et al. [19], Bommannan et al. [45], Jadoul and Préat [56], and Hu et al. [57] have all reported similar elevations in postpulsing fluxes. It should also be noted that while postiontophoresis DR values for both estradiol (1.38) and L-glutamic acid (1.93) were close to the ideal value of 1, i.e., when permeation after termination of the electrical protocol was not different from the apparent steady-state passive diffusion, values were markedly higher for the other two protocols (postelectroporation DR values for estradiol and L-glutamic acid were 4.32 and 12.2, respectively, whereas postcombined protocol DR measurements were 4.82 and 18.7). These observations may indicate less damage to the skin permeability barrier and greater recovery of the barrier properties with iontophoresis (albeit a high current density of 0.8 mA/cm^2 was used) than for pulsing of the skin.

15.4.6 Chemical Enhancement of Electroporation-Mediated Delivery

Several reports in the literature have indicated that transdermal delivery may be further increased by combining chemical excipients with electroporation. These investigations included macromolecules like dextrans [58], cyclodextrins [59] and even simple salts such as calcium chloride [60]. Other workers have, however, looked at encapsulation of compounds within lipid vesicles (liposomes) as potential candidates for electroporation-mediated delivery [61,62]. A fuller treatment of this combination is described in Chapter 17.

15.4.7 Future Prospects for Electroporation-Mediated Delivery

The technique of using high-voltage pulses to facilitate drug delivery through the skin is still at an early stage. Notably, many of the compounds inserted in this way are model fluorescent materials. Agreement is still to be reached on the optimal values for electrical parameters such as pulse length, magnitude, and duration. It is likely that regulators will want to be satisfied that any changes to the stratum corneum barrier function are readily reversible. Additionally, the development of instrumentation for use in a domiciliary setting will prove challenging and this technique, if finally approved, is likely to be used only in the clinic—assuming further progress is made. However, its potential for use as a delivery mechanism for biotechnology-derived molecules ensures that the technique will continue to be studied in the hope of swift, controlled, pain-free transport of peptides, proteins, and nucleotides.

REFERENCES

1. Barry, B.W. 2001. Novel mechanisms and devices to enable successful transdermal delivery. *Eur J Pharm Sci* 14:101.
2. White, S.H., D. Mirejovsky, and G.I. King. 1988. Structure of lamellar lipid domains and corneocytes envelops of murine stratum corneum. An x-ray diffraction study. *Biochemistry* 27:3725.
3. Bouwstra, J.A., et al. 1994. The lipid and protein structure of mouse stratum corneum: A wide and small angle diffraction study. *Biochim Biophys Acta* 1212:183.
4. Green, P.G., et al. 1993. Iontophoretic drug delivery. In *Pharmaceutical skin penetration enhancement*, eds. K.A. Walters, and J. Hadgraft, 311. New York: Marcel Dekker Inc.
5. Banga, A.K. 1998. *Electrically assisted transdermal and topical drug delivery*. London: Taylor and Francis.
6. Guy, R.H. 1998. Iontophoresis—recent developments. *J Pharm Pharmacol* 50:371.
7. Pikal, M.J., and S. Shah. 1990. Transport mechanisms in iontophoresis. III. An experimental study of the contributions of electroosmotic flow and permeability change in transport of low and high molecular weight solutes. *Pharm Res* 7:222.
8. Cullander, C., and R.H. Guy. 1991. Sites of iontophoretic current flow into the skin: Identification and characterization with the vibrating probe electrode. *J Invest Dermtol* 97:55.
9. Cullander, C. 1992. What are the pathways of iontophoretic current flow through mammalian skin? *Adv Drug Deliv Rev* 9:119.
10. Monteiro-Riviere, N.A., A.O. Inman, and J.E. Riviere. 1994. Identification of the pathway of iontophoretic drug delivery: Light and ultrastructural studies using mercuric chloride in pigs. *Pharm Res* 11:251.
11. Graaff, A.M., et al. 2003. Combined chemical and electrical enhancement modulates stratum corneum structure. *J Control Release* 90:49.
12. Turner, N.G., and R.H. Guy. 1997. Iontophoretic transdermal pathways: Dependence on penetrant physicochemical properties. *J Pharm Sci* 86:1385.
13. Pliquett, U.F., et al. 1996. Imaging of fluorescent molecules and small ion transport through human stratum corneum during high voltage pulsing: Localized transport regions are involved. *Biophys Chem* 58:185.

14. Prausnitz, M.R., et al. 1996. Transdermal transport efficiency during skin electroporation and iontophoresis. *J Control Release* 38:205.
15. Higuchi, W.I., et al. 1999. Mechanistic aspects of iontophoresis in human epidermal membrane. *J Control Release* 62:13.
16. Jadoul, A., J. Bouwstra, and V. Préat. 1999. Effect of iontophoresis and electroporation on the stratum corneum. *Adv Drug Deliv Rev* 35:89–105.
17. Weaver, J.C., and Y. Chizmadzhev. 1996. Theory of electroporation: A review. *Bioelectrochem Bioenerg* 41:135.
18. Teissié, J., et al. 1999. Electropermeabilization of cell membranes. *Adv Drug Deliv Rev* 35:3.
19. Prausnitz, M.R., et al. 1993. Electroporation of mammalian skin: A mechanism to enhance transdermal drug delivery. *Proc Natl Acad Sci U S A* 90:10504.
20. Inada, H., W.I. Higuchi, and A. Ghanem. 1994. Studies on the effects of applied voltage and duration on human epidermal membrane alteration/recovery and the resultant effects upon iontophoresis. *Pharm Res* 11:687.
21. Vanbever, R., N. Lecoutturier, and V. Préat. 1994. Transdermal delivery of metoprolol by electroporation. *Pharm Res* 11:1657.
22. Vanbever, R., E. Le Boulengé, and V. Préat. 1996. Transdermal delivery of fentanyl by electroporation. I. Influence of electrical factors. *Pharm Res* 13:559.
23. Vanbever, R., M. Leroy, and V. Préat. 1998. Transdermal penetration of neutral molecules by skin electroporation. *J Control Release* 54:243.
24. Weaver, J.C., T.E. Vaughan, and Y. Chizmadzhev. 1999. Theory of electrical creation of pathways across skin transport barriers. *Adv Drug Deliv Rev* 35:21.
25. Prausnitz, M.R. 1999. A practical assessment of transdermal drug delivery by skin electroporation. *Adv Drug Deliv Rev* 35:61.
26. Chen, T., R. Langer, and J.C. Weaver. 1998. Skin electroporation causes molecular transport across the stratum corneum through localised regions. *J Investig Dermatol Symp Proc* 3:159.
27. Chizmadzhev, Y., et al. 1995. Mechanism of electroinduced ionic species transport through a multilamellar lipid system. *Biophys J* 68:749.
28. Prausnitz, M.R. 1996. The effects of electric current applied to the skin: A review for transdermal drug delivery. *Adv Drug Deliv Rev* 18:395.
29. Pliquett, U. 1999. Mechanistic studies of molecular transdermal transport due to skin electroporation. *Adv Drug Deliv Rev* 35:41.
30. Pliquett, U.F., and C.A. Gusbeth. 2000. Perturbation of human skin due to application of high voltages. *Bioelectrochemistry* 51:41.
31. Pliquett, U., et al. 1998. Local transport regions in human stratum corneum due to long and short "high voltage" pulses. *Bioelectrochem Bioenerg* 47:151.
32. Pliquett, U.F., G.T. Martin, and J.C. Weaver. 2002. Kinetics of the temperature rise within human stratum corneum during electroporation and pulsed high voltage iontophoresis. *Bioelectrochemistry* 57:65.
33. Pliquett, U.F., and J.C. Weaver. 1996. Electroporation of human skin: Simultaneous measurement of changes in the transport of two fluorescent molecules and in the passive electrical properties. *Bioelectrochem Bioenerg* 39:1.
34. Préat, V., and R. Vanbever. 1999. *In vivo* efficacy and safety of skin electroporation. *Adv Drug Deliv Rev* 35:77.
35. Heller, R., R. Gilbert, and M.J. Jaroszeski. 1999. Clinical applications of electrochemotherapy. *Adv Drug Deliv Rev* 35:119.
36. Denet, A., and V. Préat. 2003. Transdermal delivery of timolol by electroporation through human skin. *J Control Release* 88:253.
37. Cruz, M.P., et al. 1997. Transdermal delivery of fluriprofen in the rat by iontophoresis and electroporation. *Pharm Res* 14:309.
38. Wang, S., M. Kara, and T.R. Krishnan. 1997. Topical delivery of cyclosporin A co-evaporate using electroporation technique. *Drug Dev Ind Pharm* 23:657.
39. Prausnitz, M.R., et al. 1995. Transdermal delivery of heparin by skin electroporation, *Biotechnology* 13:1205.

40. Brand, R.M., A. Wahl, and P.L. Iverson. 1998. Effects of size and sequence on the iontophoretic delivery of oligonucleotides. *J Pharm Sci* 87:49.
41. Zhang, L., and G.A. Hoffman. 1997. Electric pulse mediated transdermal drug delivery. *Proc Int Symp Control Rel Bioact Mater* 24:27.
42. Rashid, M.H. 2005. Electrically assisted enhancement of human skin penetration by model molecules. PhD. thesis, University of Bradford, U.K.
43. Dujardin, N., et al. 2002. *In vivo* assessment of skin electroporation using square wave pulses. *J Control Release* 79:219.
44. Banga, A.K., S. Bose, and T.K. Ghosh. 1999. Iontophoresis and electroporation: Comparisons and contrasts. *Int J Pharm* 179:1.
45. Bommannan, D.B., et al. 1994. Effect of electroporation on transdermal iontophoretic delivery of luteinizing hormone releasing hormone (LHRH) *in vitro*. Pharm Res 11:1809.
46. Riviere, J.E., et al. 1995. Pulsatile transdermal delivery of LHRH using electroporation: Drug delivery and skin toxicology. *J Control Release* 40:229.
47. Chang, S.L., et al. 2000. The effect of electroporation on iontophoretic transdermal delivery of calcium regulating hormones. *J Control Release* 66:127.
48. Zhao, H.Y., et al. 2002. Effect of electroporation and iontophoresis on skin permeation of Defibrase—a purified thrombin-like enzyme from the venom of *Agkistrodon halys ussuriensis* Emelianov. *Pharmazie* 57:482.
49. Badkar, A.V., and A.J. Banga. 2002. Electrically enhanced transepidermal delivery of a macromolecule. *J Pharm Pharmacol* 54:907.
50. Cheng, T., R. Langer, and J.C. Weaver. 1999. Charged microbeads are not transported across the human stratum corneum *in vitro* by short high-voltage pulses. *Bioelectrochem Bioenerg* 48:181.
51. Fang, J.Y., et al. 2002. Transepidermal iontophoresis of sodium nonivamide acetate V. Combined effect of physical enhancement methods. *Int J Pharm* 235:95.
52. Masada, T., et al. 1989. Examination of iontophoretic transport of ionic drugs across skin: Baseline studies with the four-electrode system. *Int J Pharm* 49:57.
53. Li, K.S., et al. 1997. Iontophoretic transport across a synthetic membrane and human epidermal membrane: A study of the effects of permeant charge. *J Pharm Sci* 86:680.
54. Pliquett, U., R. Langer, and J.C. Weaver. 1995. Changes in the passive electrical properties of human stratum corneum due to electroporation. *Biochim Biophys Acta* 1239:111.
55. Prausnitz, M.R. 1996. Do high voltages cause changes in skin structure? *J Control Release* 40:321.
56. Jadoul, A., and V. Préat. 1997. Electrically enhanced transdermal delivery of domperidone. *Int J Pharm* 154:229.
57. Hu, Q., et al. 2000. Enhanced transdermal delivery of tetracaine by electroporation. *Int J Pharm* 202:121.
58. Vanbever, R., M.R. Prausnitz, and V. Preat. 1997. Macromolecules as novel transdermal transport enhancers for skin electroporation. *Pharm Res* 14:638.
59. Murthy, S.N., et al. 2004. Cyclodextrin enhanced transdermal delivery of piroxicam and carboxyfluorescein by electroporation. *J Control Release* 99:393.
60. Tokudome, Y., and K. Sugibayashi. 2003. The effects of calcium chloride and sodium chloride on the electroporation-mediated skin permeation of fluorescein isothiocyanate (FITC)-dextrans *in vitro*. *Biol Pharm Bull* 26:1508.
61. Badkar, A.V., et al. 1999. Enhancement of transdermal iontophoretic delivery of a liposomal formulation of colchicine by electroporation. *Drug Deliv* 6:111.
62. Essa, E.A., M.C. Bonner, and B.W. Barry. 2003. Electroporation and ultradeformable liposomes; human skin repair by phospholipid. *J Control Release* 92:163.

16 Ultrasound in Percutaneous Absorption

Joseph Kost and Lior Wolloch

CONTENTS

16.1 INTRODUCTION

Sonophoresis is defined as the transport of drugs through intact skin under the influence of an ultrasound. Ultrasound at various frequencies in the range of 20 kHz to 16 MHz has been used to enhance skin permeability [1–3]. This chapter attempts to present sonophoresis, experimental variables, possible mechanisms of action, and clinical applications.

16.2 PHYSICAL CHARACTERISTICS OF ULTRASOUND

Ultrasound is defined as a sound having a frequency above 18 kHz. Most modern ultrasound devices are based on the piezoelectric effect. This is achieved by applying pressure to quartz crystals and some polycrystalline materials, such as lead–zirconate–titanium or barium titanate, causing electric charge to develop on the outer surface of the material. Application of a rapidly alternating potential across the opposite faces of a piezoelectric crystal will therefore induce corresponding alternating dimensional changes, thereby converting electrical energy into vibrational (sound) energy [4].

The ultrasound wave is longitudinal in nature (i.e., the direction of propagation is the same as the direction of oscillation). Longitudinal sound waves cause compression and expansion of the medium at a distance of half the wavelength, leading to pressure variations in the medium. The resistance of the medium to the propagation of sound waves is dependent on the acoustic impedance (Z), which is related to the mass density of the medium (ρ) and the speed of propagation (C) according to the following equation:

$$Z = \rho \times C \tag{16.1}$$

The specific acoustic impedances for skin, bone, and air are 1.6×10^6, 6.3×10^6, and 400.0 kg/(m^{2}s), respectively [4].

As ultrasound energy penetrates the body tissues, biological effects can be expected to occur if the tissues absorb the energy. The absorption coefficient (a) is used as a measure of the absorption in various tissues. For an ultrasound consisting of longitudinal waves with perpendicular incidence on homogeneous tissues, the following equation applies:

$$I(x) = I_0 e^{-ax} \tag{16.2}$$

where $I(x)$ is the intensity at depth x, I_0 is the intensity at the surface, and a is the absorption coefficient.

To transfer ultrasound energy to the body, a coupling medium is required to overcome the high impedance of air. The many types of coupling medium currently available for ultrasound transmission can be broadly classified as oils, water–oil emulsions, aqueous gels, and ointments.

There are three distinct sets of ultrasound conditions based on frequency range and applications [5]:

High-frequency or diagnostic ultrasound in clinical imaging (3–10 MHz)

Medium-frequency or therapeutic ultrasound in physical therapy (0.7–3.0 MHz)

Low-frequency or power ultrasound for lithotripsy, cataract emulsification, liposuction, tissue ablation, cancer therapy, dental descaling, and ultrasonic scalpels (18–100 kHz)

16.3 BIOLOGICAL EFFECTS OF ULTRASOUND

Ultrasound over a wide frequency range has been used in medicine for the past century. For example, therapeutic ultrasound has been used for physical therapy, low-frequency ultrasound has been used in dentistry, and high-frequency ultrasound has been used for diagnostic purposes [6]. The utility of ultrasound is continuously expanding and new clinical applications are constantly being developed, including the use of high-intensity focused ultrasound for tumor therapy [7], lithotripsy [8], ultrasound-assisted lipoplasty [9], and ultrasonic surgical instruments [10,11].

Significant attention has thus been given to investigating the effects of ultrasound on biological tissues. Ultrasound affects biological tissues via three main effects: thermal, cavitational, and acoustic streaming.

16.3.1 Thermal Effects

The absorption of ultrasound increases the temperature of the medium. Materials that possess higher ultrasound absorption coefficients, such as bone, experience severe thermal effects as compared to muscle tissue, which has a lower absorption coefficient [5]. The increase in the temperature of the medium upon ultrasound exposure at a given frequency varies directly with the ultrasound intensity and exposure time. The absorption coefficient of a medium increases directly with ultrasound frequency resulting in temperature increase.

A recent study [12] suggested the use of a new safety parameter, time to threshold (TT). TT indicates the time a threshold temperature rise is exceeded and how long a tissue can be safely exposed to ultrasound, provided the safe threshold is known.

16.3.2 Cavitational Effects

Cavitation is the formation of gaseous cavities in a medium upon ultrasound exposure. The primary cause of cavitation is ultrasound-induced pressure variation in the medium. Cavitation involves either the rapid growth and collapse of a bubble (inertial cavitation) or the slow oscillatory motion of a bubble in an ultrasound field (stable cavitation). Collapse of cavitation bubbles releases a shock wave that can cause structural alteration in the surrounding tissue [13]. Tissues contain air pockets trapped in the fibrous structures that act as nuclei for cavitation upon ultrasound exposure. The cavitational effects vary inversely with ultrasound frequency and directly with ultrasound intensity. Cavitation might be important when low-frequency ultrasound is used, when gassy fluids are exposed, or when small gas-filled spaces are exposed.

16.3.3 Acoustic Streaming Effects

Acoustic streaming is the development of unidirectional flow currents in fluids generated by sound waves. The primary causes of acoustic streaming are ultrasound reflections and other distortions that occur during wave propagation [14]. Oscillations of cavitation bubbles might also contribute to acoustic streaming. The shear stresses developed by streaming velocities might affect the neighboring tissue structures. Acoustic streaming might be important when the medium has an acoustic impedance that is different from that of its surroundings, the fluid in the biological medium is free to move, or when continuous wave application is used. The potential clinical value of acoustic streaming has only been minimally explored to date. Nightingale et al. [15] used acoustic streaming to help distinguish cystic from solid breast lesions. The study concentrated on detecting the presence or absence of acoustic streaming as an indicator of whether a lesion was cystic or solid. Shi et al. [16] used acoustic streaming detection as a tool for distinguishing between liquid blood and clots or soft tissue in hematoma diagnosis.

16.3.4 Effect on Skin

Various investigators have reported histological studies of animal skin exposed to ultrasound under various conditions to assess the effect of ultrasound on living skin cells. Levy et al. [17] performed histological studies of hairless rat skin exposed to therapeutic ultrasound and reported that the application of ultrasound (1 MHz, 2 W/cm^2) induced no damage. Tachibana [18] performed similar studies on rabbit skin exposed to low-frequency ultrasound (105 kHz, 5000 Pa pressure amplitude) and also reported no damage to the skin upon ultrasound application. Mitragotri et al. [19,20] performed histological studies of hairless rat skin exposed to low-frequency ultrasound (20 kHz, 12.5–225 mW/cm^2) and found no damage to the epidermis and underlying living tissues. Using scanning electron microscopy, Yamashita et al. [21] investigated the effects of ultrasound with a frequency of 48 kHz (0.5 W/cm^2) on the surface of hairless mice and human skin. They found that the effect on mice skin was much more significant than on human skin; following ultrasound exposure, the outer layer in mice stratum corneum (SC) was totally removed and pores were observed, whereas in human skin some removal of keratinocytes around hair follicles was observed. This effect was attributed mostly to cavitation. Boucaud et al. [22] evaluated the effect of low-frequency ultrasound (20 kHz) on hairless mice skin and human skin. Human skin samples that were exposed to low-intensity ultrasound (<2.5 W/cm^2) showed no histological change. Further microscopic examination using transmission electron microscopy confirmed a lack of structural modification.

Based on the above, the effect of ultrasound on skin is derived directly from the application parameters, which include application duration, frequency, and intensity.

16.4 TRANSDERMAL DRUG DELIVERY

Although the first published report on sonophoresis dates back to the 1950s with Fellinger and Schmid's [23] report presenting the successful treatment of polyarthritis of the hand's digital joints using hydrocortisone ointment with sonophoresis, only in 2004 did the Food and Drug Administration (FDA) approve the first ultrasound device for transdermal application [24]. In the past two decades, with the development of transdermal delivery as an important means of systemic drug administration, researchers have been investigating the possible application of ultrasound in transdermal drug delivery systems [17,25]. Ultrasound has been evaluated and its effect well demonstrated at various frequencies using different molecules [26–31].

Simultaneous application of therapeutic ultrasound and drug was the most commonly used approach in early trials [29,32,33]. The ultrasound conditions corresponded to frequencies in the range of 0.75–3.00 MHz and an intensity range of 0.0–2.4 W/cm^2, chosen to avoid potential safety issues. Typical enhancements induced by therapeutic ultrasound have been approximately 10-fold [34]. This enhancement might be sufficient for local delivery of certain drugs, such as hydrocortisone, but not for the systemic delivery of most drugs. The effect of low-frequency ultrasound (frequencies below 100 kHz) on transdermal transport has been found to be significantly stronger. Tachibana and Tachibana [18,35] reported that the use of low-frequency ultrasound (48 kHz) enhanced transdermal transport of insulin across diabetic rat skin. Merino et al. [36] compared the enhancing transdermal effect of low (20 kHz) and high (10 MHz) ultrasound and observed significantly increased permeation only for low-frequency ultrasound.

Low-frequency ultrasound has also been used by Mitragotri et al. [19,20] to enhance the transport of various low-molecular-weight drugs, as well as high-molecular-weight proteins (including insulin, γ-interferon, and erythropoietin), across human cadaver skin *in vitro*. The experimental findings suggest that, among all the ultrasound-related phenomena evaluated (cavitation, thermal effects, generation of convective velocities, and mechanical effects), cavitation plays the dominant role in low-frequency sonophoresis, suggesting that application of low-frequency ultrasound should enhance transdermal transport more effectively. Mitragotri et al. [20] found that the enhancement induced by low-frequency ultrasound is up to 1000-fold higher than that induced by therapeutic ultrasound. For example, application of ultrasound (20 kHz, 225 mW/cm^2, 100 ms pulses applied every second) to a chamber glued onto the back of the rat and filled with insulin solution (100 U/ml) reduced the blood glucose level of diabetic hairless rats from approximately 400 to 200 mg/dl in 30 min [19].

Boucaud et al. [37] also demonstrated a dose-dependent hypoglycemia in hairless rats exposed to ultrasound and insulin. For an energy dose of 900 J/cm^2, approximately 75% reduction in glucose levels was reported.

Although the simultaneous application may be used to achieve a temporal control over skin permeability, it requires the patient to use a wearable ultrasound device. In an attempt to make this convenient to the user, Smith et al. [38] proposed the use of a low-profile light cymbal array ultrasound that is a flextensional transducer. The design of the cymbal transducer integrates two metal caps exposed onto a lead–zirconate–titanate (PZT) ceramic. The fundamental mode of vibration is the flexing of the end cap caused by the radial motion of the ceramic and the axial motion of the piezoelectric disk. Lee et al. [39] studied the use of the cymbal array (2×2 transducers) ultrasound ($37 \times 37 \times 7$ mm^3, 22g, 20 kHz, and 100 mW/cm^2) to deliver insulin to hyperglycemic rats and rabbits [39,40]. For the 5 and 10 min ultrasound exposure groups, the glucose concentration in rats decreased from the baseline (blood glucose levels prior to the beginning of the experiment) to −200 and −174 mg/dl, respectively, measured after 1 h. Similar results were observed in rabbits.

The inconvenience of a wearable ultrasound device was also addressed by Mitragotri et al. [41,42], who proposed the use of ultrasound as a brief pretreatment process followed by passive diffusion, therefore not requiring a wearable device. Several studies have been reported on the use of the pretreatment-type sonophoresis. One of the challenges in pretreatment-type sonophoresis is that the degree of skin permeabilization needs to be determined prior to drug placement. This challenge is addressed by quantifying the effect of ultrasound on skin by conductance measurements. Formal relationships between skin conductance (or resistance) and skin permeability have been developed [43]. During *in vitro*, *in vivo*, or clinical experiments, skin conductance is continuously measured and ultrasound application is terminated when skin conductance reaches a predetermined value, which varies depending on the goal of the experiment.

Pretreatment of skin by low-frequency ultrasound (20 kHz, approximately 7 W/cm^2) was found to enhance skin permeability to insulin. In these experiments, ultrasound was applied to increase rat skin conductivity by about 60-fold. Insulin (500 U/ml) was placed on ultrasound-treated skin. The blood glucose level of rats decreased by about 80% within 2 h [2]. No change in blood glucose level was found when insulin was placed on untreated skin. The effect of ultrasound pretreatment on macromolecule transport was also assessed *in vivo* using rats for low-molecular-weight heparin (LMWH) [44]. Transdermal LMWH delivery was measured by monitoring anti-Xa activity in blood (Xa—a plasma factor in the coagulation procedure). No significant anti-Xa activity was observed when LMWH was placed on non-treated skin. However, a significant amount of LMWH was transported transdermally after ultrasound pretreatment. Anti-Xa activity in the blood increased slowly for about 2 h, after which it increased rapidly before achieving a steady state after 4 h at a value of about 2 U/ml [44]. The effect of transdermally delivered LMWH was observed well beyond 6 h in contrast to intravenous or subcutaneous injections, which resulted only in transient biological activity.

Katz et al. [45] used pretreatment with low-frequency ultrasound (55 kHz) to shorten the lag time for an analgesic agent (EMLA cream (AstraZeneca International, Wilmslow, U.K. [46])) to be effective. EMLA cream is a mixture of two local anesthetics (lignocaine and prilocaine). It is indicated for use on normal intact skin to induce local analgesia about 60 min after application. The study was conducted on 42 human subjects and pain score and patient preference were measured. After ultrasound pretreatment (4–14 s), and then 5, 10, and 15 min after EMLA cream application, pain scores and overall preference were statistically indistinguishable from EMLA cream application for 60 min. Becker et al. [47] demonstrated the practical clinical utility and effectiveness of sonophoresis. Eighty-seven patients were enrolled in the study. The intervention was a brief (15 s) ultrasound pretreatment (similar to Katz et al. [45]) using the SonoPrep (Sontra Medical Corp., Franklin, MA) followed by 5 min of 4% liposomal lidocaine cream and standard care intravenous cannulation. The ultrasound device significantly reduced patient perception of the pain of intravenous cannulation. There were no adverse side effects noted in any participant during the follow-up period. Based on these and additional studies, Sontra Medical in August 2004 received 510(k) marketing clearance from the U.S. Food and Drug Administration to market the SonoPrep ultrasonic skin permeation device and procedure tray for use with topical lidocaine. (See Figure 16.1.)

16.5 IMMUNIZATION

Transcutaneous immunization is a new technique in which vaccine antigens in a solution are applied on the skin to induce an antibody response without systemic or local toxicity [48]. The primary advantage of transcutaneous immunization is the presentation of immunogens to antigen presenting cells (APCs) within the skin, specifically Langerhans cells that are highly

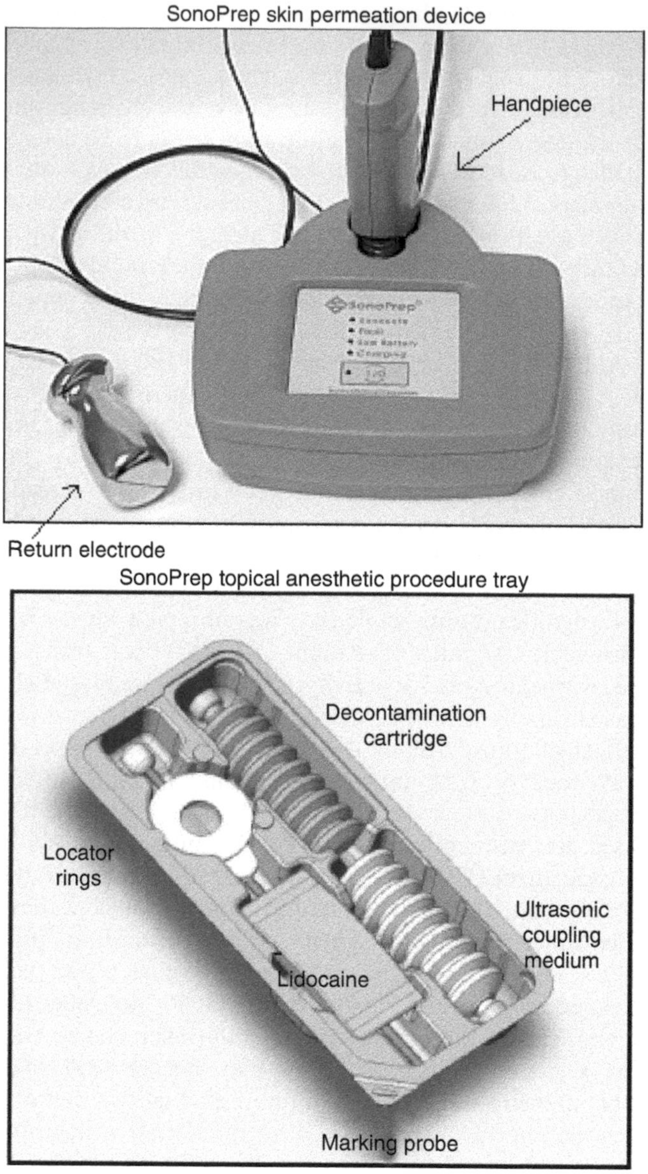

FIGURE 16.1 Sontra's SonoPrep skin permeation device + the topical anesthetic procedure tray. (From http://www.sontra.com. With permission.)

potent immune cells replete within the epidermis [49]. Langerhans cells are in close proximity to the outermost layer of the skin, the stratum corneum, and represent a network of immune cells that occupies about 20% of the skin's total surface area despite composing only 1% of the epidermis cell population [49]. Langerhans cells initiate immune responses by acting as professional APCs, taking up and processing antigens, and subsequently presenting antigenic peptides to naive T-cells in the lymph nodes [50,51]. Transcutaneous immunization has been shown to generate both systemic (IgG/IgM response) and mucosal immunities (IgA response), whereas conventional needle based injections often only generate systemic immunity [52,53].

Tezel et al. [54] have demonstrated a significant IgG response and activation of Langerhans cells in the epidermis to a model vaccine, tetanus toxoid, after pretreatment with a 20 kHz, 100 J/cm^2 ultrasound in BALB/c mice.

The authors also demonstrated that low-frequency ultrasound acts as a transcutaneous immunization adjuvant that eliminates the requirement of toxins to elicit an immune response. Delivery of as little as 1.3 μg of tetanus toxoid by low-frequency ultrasound generated an immune response comparable to that induced by 10 μg subcutaneous injection.

16.6 GENE THERAPY

A possible additional application of ultrasound as a topical enhancer, which seems to show promise, lies in the field of topical gene therapy [55,56]. Gene therapy is a technique for correcting defective genes that are responsible for disease development, most commonly by replacing an "abnormal" disease-causing gene with the "normal" gene. A carrier molecule (vector) is usually used to deliver the therapeutic gene to the target cell. Topical delivery of the vector–gene complex can be used to target cells within the skin, as well as for the systemic circulation. The identification of genes responsible for almost 100 diseases affecting the skin has raised the option of using cutaneous gene therapy as a therapeutic method [57]. The most obvious candidate diseases for cutaneous gene therapy are the severe forms of particular genodermatoses (monogenic skin disorders), such as epidermolysis bullosa and ichthyosis. Other applications might be healing of cutaneous wounds such as severe burns and skin wounds of diabetic origin [58].

Topical gene therapy acquires the penetration of a large complex to or through the skin. Ultrasound pretreatment of the skin will increase its permeability and permit the delivery of the vector–gene complex.

16.7 TRANSDERMAL MONITORING

Considerable efforts have been directed toward developing painless and convenient methods to measure blood analytes, particularly glucose, including implantable sensors, minimally invasive skin microporation, approaches involving laser or miniaturized lancets and non-invasive technologies such as near-infrared spectroscopy, transdermal permeation enhancers, and reverse iontophoresis [59]. One of the fundamental problems in noninvasive transdermal diagnostics is obtaining sufficient quantities of analyte for detection. Ultrasound, particularly at low frequencies, has been shown to increase skin permeability, hence allowing sufficient amounts of clinically relevant analytes, including glucose, to be collected for the purpose of noninvasive monitoring [42,60–62].

The technique was assessed on type 1 diabetic volunteers to determine whether a single short application of ultrasound (<2 min) was sufficient to extract glucose noninvasively across human skin for several hours, and to determine whether transdermal glucose flux varied in response to variations in blood glucose concentrations. Additional experiments to further assess the duration of ultrasound-induced permeability found that the skin permeability remained high for about 15 h and decreased to its normal value by 24 h [42]. A comparison of venous blood glucose levels and noninvasively extracted glucose fluxes after ultrasound pretreatment showed close correlation. Site-to-site variability of skin permeability after ultrasound application was also evaluated within the same patient and between patients. The site-to-site variability was about the same as patient-to-patient variability. This indicates the necessity of one-point calibration between transdermal glucose flux and one blood sample, which can then be used to predict subsequent blood glucose values.

Based on such a calibration, the correlation was assessed between transdermal glucose flux and blood glucose values (mean relative error of 17%). Sontra Medical [42] has been developing this technology for noninvasive continuous sensing of glucose. A minimally invasive system that continuously measures glucose flux through ultrasonically permeated skin was reported [61]. In this study, the glucose level of ten diabetes patients was monitored over a period of 8 h. A good correlation was observed between sensor output reading and blood glucose measurements. The investigators recognized several areas where improvements should be made to increase sensitivity, reduce variations, and increase accuracy.

Lee et al. [62] have demonstrated lately the possibility to quantify glucose concentrations in interstitial fluid (ISF) by an electrochemical biosensor after permeability enhancement with a 20 min light cymbal ultrasound array ($37 \times 37 \times 7$ mm^3, 22 g, 20 kHz, and 100 mW/cm^2). The authors present a good correlation to blood glucose concentrations in hyperglycemic rats.

16.8 MECHANISM

The mechanism of improved transdermal transport by ultrasound has been studied for the past 20 years. In spite of the large number of studies that have been published, the mechanism is still not well understood and characterized. A possible mechanism of improved percutaneous transport by ultrasound suggested by several groups [17,32,63] is that ultrasound might interact with the structural lipids located in the intercellular channels of the stratum corneum. This is similar to the postulated effects of some chemical transdermal enhancers that act by disordering lipids. Tachibana [18] and Simonin [63] postulated that the energy of ultrasonic vibration enhanced transdermal permeability through the transfollicular and transepidermal routes, suggesting that microscopic bubbles (cavitation) produced at the surface of the skin by ultrasonic vibration might generate a rapid liquid flow, thereby increasing skin permeability.

Mitragotri et al. [20] evaluated the role played by various ultrasound-related phenomena, including cavitation, thermal effects, generation of convective velocities, and mechanical effects. The authors hypothesized that transdermal transport during low-frequency ultrasound application occurs across the keratinocytes rather than the hair follicles. They suggested that cavitation causes disorder of the stratum corneum lipids, resulting in significant water penetration into the disordered lipid region. This might cause the formation of aqueous channels through the intercellular lipids of the stratum corneum through which permeants could move. Tang et al. [64] studied the relative roles of enhanced diffusion due to ultrasound-induced skin alteration and enhancement due to ultrasound-forced convection. The findings (theoretical and experimental) suggest that, for low-frequency ultrasound, the relative contribution depends on the *in vitro* skin model studied. Specifically, convection plays an important role when heat-stripped stratum corneum is exposed to ultrasound, whereas its effect is minimal when full-thickness skin is utilized. In addition, the effective pore radius of the skin estimated using heat-stripped stratum corneum during ultrasound exposure is much larger than that within full-thickness skin.

All recent studies indicate that cavitation plays an important role in the enhancing mechanism. Several attempts have been made to establish a suitable mathematical model that will describe the enhancement phenomenon and predict the enhancement ratio for different drugs in various conditions [64,65].

Tezel and Mitragotri [66] describe a theoretical analysis of the interaction of cavitation bubbles with the stratum corneum lipid bilayers. Three modes were evaluated—shock-wave emission, microjet penetration into the stratum corneum, and impact of microjet on the

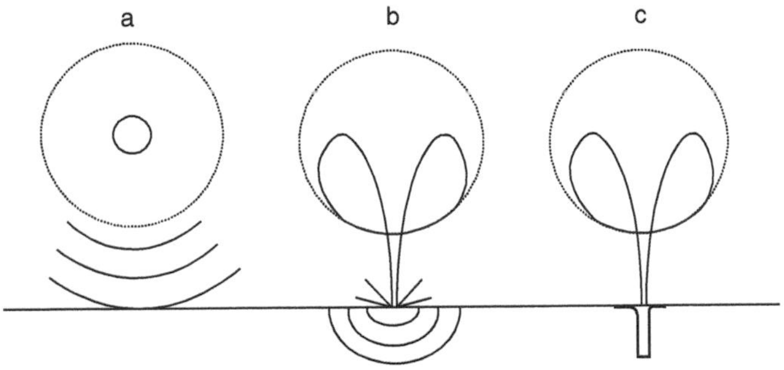

FIGURE 16.2 Three possible modes through which inertial cavitation may enhance SC permeability. (a) Spherical collapse near the SC surface emits shock waves, which can potentially disrupt the SC lipid bilayers. (b) Impact of an acoustic microjet on the SC surface. The microjet possessing a radius about one tenth of the maximum bubble diameter impacts the SC surface without penetrating into it. The impact pressure of the microjet may enhance SC permeability by disrupting SC lipid bilayers. (c) Microjets may physically penetrate into the SC and enhance the SC permeability. (From Mitragotri, S., and Kost J., *Adv. Drug Deliv. Rev.*, 56, 589, 2004. With permission.)

stratum corneum. Their suggested model predicts that both microjets and spherical collapses might be responsible for the enhancement effect. Figure 16.2 demonstrates these three possible modes [2].

Kushner et al. [67] have demonstrated experimentally that all these possible effects do not appear as a homogenous pattern on the skin surface, but rather there are highly permeable localized transport regions where mass transfer occurs through the stratum corneum, as shown in Figure 16.3. Hence, a possible improvement in homogeneity of the ultrasound

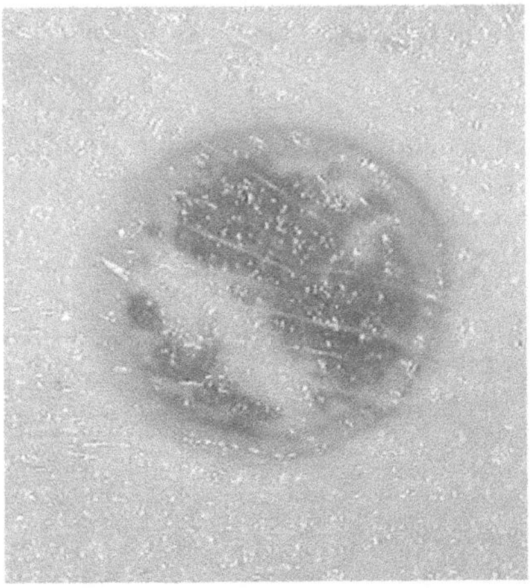

FIGURE 16.3 Demonstration of the highly permeable localized transport regions (LTRs), which occurred after ultrasound application, on pig full-thickness skin sample. The darker regions are the LTRs, visualized in presence of Potassium Indigo. (From *Wolloch L. and Kost J.; Unpublished data.*)

effect, resulting in increased skin's surface fraction of these permeable transport regions, will lead to an increase in the sonophoretic transdermal transport.

16.9 SYNERGISTIC EFFECTS

Ultrasound might enhance transdermal transport by inducing skin alteration, as well as by inducing active transport (forced convection) in the skin. Various other means of transport enhancement, including chemicals [33,68–70], iontophoresis [71], and electroporation [72], might enhance transport synergistically with ultrasound. Mitragotri et al. [73] evaluated the synergistic effect of low-frequency ultrasound with chemical enhancers and surfactants, including sodium lauryl sulfate (SLS) and a model permeant, mannitol. Application of ultrasound alone as well as SLS alone, both for 90 min, increased skin permeability about threefold for SLS and eightfold for ultrasound. However, combined application of ultrasound and 1% SLS solution induced an increase in skin permeability to mannitol by about 200-fold. The application of ultrasound simultaneously with SLS results in enhanced mass transfer and improved penetration and dispersion of the surfactant.

A recent work by Lavon et al. [74] suggests that the synergistic effect of SLS and ultrasound when applied simultaneously can also be attributed to the modification of the stratum corneum pH profile when exposed to ultrasound. The altered pH profile that results in improved SLS lipophylic solubility, together with improved SLS penetration and dispersion, can explain the synergistic enhancing effect on transdermal transport (see Figure 16.4).

Ultrasound also exhibited a synergistic effect with electroporation [72]. Ultrasound reduced the threshold voltage for electroporation as well as increased transdermal transport at a given electroporation voltage. The enhancement of transdermal transport induced by the combination of ultrasound and electroporation was higher than the sum of the enhancement induced by each enhancer alone.

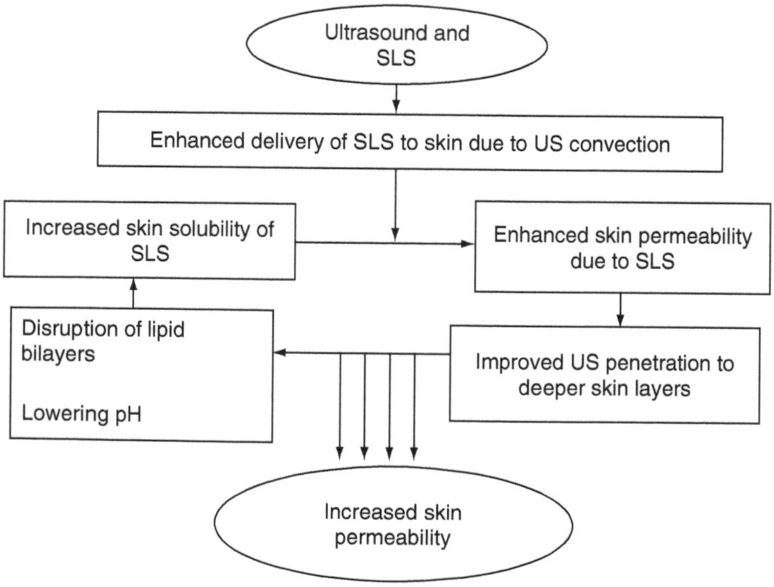

FIGURE 16.4 Schematic presentation of ultrasound/SLS synergistic effect on skin permeability. (From Lavon, I., Grossman, N., and Kost, J., *J. Control. Release*, 107, 484, 2005. With permission.)

Combined application of ultrasound and iontophoresis also has practical implications. The combination of ultrasound and electric current offers a twofold higher enhancement than that offered by each of them individually under the same conditions. Since ultrasonic pretreatment reduces skin resistivity, a lower voltage is required to deliver a given current during iontophoresis compared to that in controls. This should result in lower power requirements as well as possibly less skin irritation [71].

16.10 SAFETY

The utility of ultrasound in medicine as a technical tool, as well as a therapeutic agent, is constantly increasing. In view of this, much concern is directed to the issues of ultrasound bioeffects and safety. The World Federation for Ultrasound in Medicine and Biology (WFUMB [75]) has issued several publications related to safety of ultrasound bioeffects, addressing specifically thermal bioeffects [76] and nonthermal bioeffects [77], in an attempt to reach an international consensus and adopt a policy on safety guidelines. The use of ultrasound as an aid to increasing skin permeability is based on its nonthermal bioeffects, mostly cavitation. In view of this, much attention should be paid to the issue of ultrasound safety, short and long effects on skin structure, reversibility, and the effect of free radicals generated during the cavitation process within the skin.

REFERENCES

1. Lavon, I., and J. Kost. 2004. Ultrasound and transdermal drug delivery. *Drug Discov Today* 9:670.
2. Mitragotri, S., and J. Kost. 2004. Low-frequency sonophoresis: A review. *Adv Drug Deliv Rev* 56, 589.
3. Prausnitz, M.R., S. Mitragotri, and R. Langer. 2004. Current status and future potential of transdermal drug delivery. *Nat Rev Drug Discov* 3:115.
4. Wells, P.N.T. 1993. Physics of ultrasound. In *Ultrasonic exposimetry*, ed. P.A. Lewin, 35. Boca Raton, FL: CRC Press.
5. Suslick, K.S. 1988. *Ultrasound: its chemical, physical and biological effects*, New York: VCH.
6. Mitragotri, S. 2005. Healing sound: The use of ultrasound in drug delivery and other therapeutic applications. *Nat Rev Drug Discov* 4:255.
7. van Wamel, A., et al. 2004. Radionuclide tumour therapy with ultrasound contrast microbubbles. *Ultrasonics* 42:903.
8. Leveillee, R.J., and L. Lobik, 2003. Intracorporeal lithotripsy: Which modality is best? *Curr Opin Urol* 13:249.
9. Hong, J.P., et al. 2004. Ultrasound-assisted lipoplasty treatment for axillary bromidrosis: Clinical experience of 375 cases. *Plast Reconstr Surg* 113:1264.
10. Takahashi, S., et al. 2003. Exposure of the coronary artery using an ultrasonic scalpel. *J Thorac Cardiovasc Surg* 125:1533.
11. Whaley, D.H., D.L. Adamczyk, and E.A. Jensen. 2003. Sonographically guided needle localization after stereotactic breast biopsy. *Am J Roentgenol* 180:352.
12. Lubbers, J., R.T. Hekkenberg, and R.A. Bezemer. 2003. Time to threshold (TT), a safety parameter for heating by diagnostic ultrasound. *Ultrasound Med Biol* 29:755.
13. Williams, A.R. 1986. *Ultrasound: biological effects and potential hazards*, New York: Academic Press.
14. Clarke, L., A. Edwards, and E. Graham. 2004. Acoustic streaming: An *in vitro* study. *Ultrasound Med Biol* 30:559.
15. Nightingale, K.R., P.J. Kornguth, and G.E. Trahey. 1999. The use of acoustic streaming in breast lesion diagnosis: A clinical study. *Ultrasound Med Biol* 25:75.
16. Shi, X., et al. 2001. Color doppler detection of acoustic streaming in a hematoma model. *Ultrasound Med Biol* 27:1255.

17. Levy, D., et al. 1989. Effect of ultrasound on transdermal drug delivery to rats and guinea pigs. *J Clin Invest* 83:2074.
18. Tachibana, K. 1992. Transdermal delivery of insulin to alloxan-diabetic rabbits by ultrasound exposure. *Pharm Res* 9:952.
19. Mitragotri, S., D. Blankschtein, and R. Langer. 1995. Ultrasound-mediated transdermal protein delivery. *Science* 269:850.
20. Mitragotri, S., D. Blanckschtein, and R. Langer. 1996. Transdermal drug delivery using low-frequency sonophoresis. *Pharm Res* 13:411.
21. Yamashita, N., et al. 1997. Scanning electron microscopic evaluation of the skin surface after ultrasound exposure. *Anat Rec* 247:455.
22. Boucaud, A., et al. 2001. Clinical, histologic, and electron microscopy study of skin exposed to low-frequency ultrasound. *Anat Rec* 264:114.
23. Fellinger, K., and J. Schmidt. 1954. *Klinik and therapies des chromischen gelenkreumatismus*, 549. Austria: Maudrich Vienna.
24. http://www.sontra.com.
25. Kost, et al. 1988. Ultrasound enhancement of transdermal drug delivery. Cambridge, MA: Massachusetts Institute of Technology. US Patent 4,767,402.
26. Boucaud, A., et al. 2001. *In vitro* study of low-frequency ultrasound-enhanced transdermal transport of fentanyl and caffeine across human and hairless rat skin. *Int J Pharm* 228:69.
27. Kost, J. 1989. Ultrasound as a Transdermal Enhancer. In *Percutaneous absorption*, ed. Maibach, H. I, 595. New York and Basel: Marcel Dekker.
28. Mitragotri, S., and J. Kost. 2001. Transdermal delivery of heparin and low-molecular weight heparin using low-frequency ultrasound. *Pharm Res* 18:1151.
29. Bommannan, D., et al. 1992. Sonophoresis. I. The use of high-frequency ultrasound to enhance transdermal drug delivery. *Pharm Res* 9:559.
30. Fang, J.Y., et al. 1999. Effect of low frequency ultrasound on the *in vitro* percutaneous absorption of clobetasol 17-propionate. *Int J Pharm* 191:33.
31. Mutoh, M., et al. 2003. Characterization of transdermal solute transport induced by low-frequency ultrasound in the hairless rat skin. *J Control Release* 92:137.
32. Mitragotri, S., et al. 1995. A mechanistic study of ultrasonically-enhanced transdermal drug delivery. *J Pharm Sci* 84:697.
33. Johnson, M.E., et al. 1996. Synergistic effects of chemical enhancers and therapeutic ultrasound on transdermal drug delivery. *J Pharm Sci* 85:670.
34. Cagnie, B. 2003. Phonophoresis versus topical application of ketoprofen: Comparison between tissue and plasma levels. *Phys Ther* 83:707.
35. Tachibana, K., and S. Tachibana. 1991. Transdermal delivery of insulin by ultrasonic vibration. *J Pharm Pharmacol* 43:270.
36. Merino, G., et al. 2003. Frequency and thermal effects on the enhancement of transdermal transport by sonophoresis. *J Control Release* 88:85.
37. Boucaud, A., et al. 2002. Effect of sonication parameters on transdermal delivery of insulin to hairless rats. *J Control Release* 81:113.
38. Smith, N.B., S. Lee, and K.K. Shung. 2003. Ultrasound-mediated transdermal *in vivo* transport of insulin with low-profile cymbal arrays. *Ultrasound Med Biol* 29:1205.
39. Lee, S., R.E. Newnham, and N.B. Smith. 2004. Short ultrasound exposure times for noninvasive insulin delivery in rats using the lightweight cymbal array. *IEEE Trans Ultrason Ferroelectr Freq Control.* 51:176.
40. Lee, S. 2004. Noninvasive ultrasonic transdermal insulin delivery in rabbits using the light-weight cymbal array, *Diabetes Technol Ther* 6:808.
41. Mitragotri, S., and J. Kost 2000. Low-frequency sonophoresis: A noninvasive method of drug delivery and diagnostics. *Biotechnol Prog* 16:488.
42. Kost, J., et al. 2000. Transdermal monitoring of glucose and other analytes using ultrasound. *Nat Med* 6:347.
43. Tezel, A., A. Sens, and S. Mitragotri 2003. A theoretical description of transdermal transport of hydrophilic solutes induced by low-frequency sonophoresis. *J Pharm Sci* 92:381.

44. Kost, J., and S. Mitragotri. 2000. Transdermal delivery of heparin and low-molecular weight heparin using low-frequency ultrasound. *Pharm Res* 18:1151.
45. Katz, N.P., et al. 2004. Rapid onset of cutaneous anesthesia with EMLA cream after pretreatment with a new ultrasound-emitting device. *Anesth Analg* 98:371.
46. http://www.astrazeneca.com.
47. Becker, B.M., et al. 2005. Ultrasound with topical anesthetic rapidly decreases pain of intravenous cannulation. *Acad Emerg Med* 12:289.
48. Glenn, G.M., et al. 1998. Skin immunization made possible by cholera toxin. *Nature* 391:851.
49. Babiuk, S., et al. 2000. Cutaneous vaccination: The skin as an immunologically active tissue and the challenge of antigen delivery. *J Control Release* 66:199.
50. Roitt, I.M., J. Brostoff, and D. Male. 1998. *Immunology*, 5th ed. London: Mosby.
51. Stoitzner, P., et al. 2003. Visualization and characterization of migratory Langerhans cells in murine skin and lymph nodes by antibodies against Langerin/CD207. *J Invest Dermatol* 120:266.
52. Gockel, C.M., S. Bao, and K.W. Beagley. 2000. Transcutaneous immunization induces mucosal and systemic immunity: A potent method for targeting immunity to the female reproductive tract. *Mol Immunol* 37:537.
53. Glenn, G.M., et al. 1998. Transcutaneous immunization with cholera toxin protects mice against lethal mucosal toxin challenge. *J Immunol* 161:3211.
54. Tezel, A., et al. 2005. Low-frequency ultrasound as a transcutaneous immunization adjuvant. *Vaccine* 23:3800.
55. Cao, T., X.J. Wang, and D.R. Roop. 2000. Regulated cutaneous gene delivery: The skin as a bioreactor. *Hum Gene Ther* 11:2297.
56. Vogel, J.C. 2000. Nonviral skin gene therapy. *Hum Gene Ther* 11:2253.
57. Uitto, J., and L. Pulkkinen. 2000. The genodermatoses: Candidate diseases for gene therapy. *Hum Gene Ther* 11:2267.
58. Khavari, P.A., O. Rollman, and A. Vahlquist. 2002. Cutaneous gene transfer for skin and systemic diseases. *J Intern Med* 252:1.
59. Sieg, A., R.H. Guy, and M.B. Delgado-Charro. 2005. Noninvasive and minimally invasive methods for transdermal glucose monitoring. *Diabetes Technol Ther* 7:174.
60. Mitragotri, S., et al. 2000. Transdermal extraction of analytes using low-frequency ultrasound. *Pharm Res* 17:466.
61. Chuang, H., E. Taylor, and T.W. Davison. 2004. Clinical evaluation of a continuous minimally invasive glucose flux sensor placed over ultrasonically permeated skin. *Diabetes Technol Ther* 6:21.
62. Lee, S., et al. 2005. Glucose measurements with sensors and ultrasound. *Ultrasound Med Biol* 31:971.
63. Simonin, J.P. 1995. On the mechanisms of *in vitro* and *in vivo* phonophoresis. *J Control Release* 33:125.
64. Tang, H., et al. 2001. Theoretical description of transdermal transport of hydrophilic permeants: Application to low-frequency sonophoresis. *J Pharm Sci* 90:545.
65. Tezel, A., A. Sens, and S. Mitragotri. 2003. Description of transdermal transport of hydrophilic solutes during low-frequency sonophoresis based on a modified porous pathway model. *J Pharm Sci* 92:381.
66. Tezel, A., and S. Mitragotri. 2003. Interactions of inertial cavitation bubbles with stratum corneum lipid bilayers during low-frequency sonophoresis. *Biophys J* 85:3502.
67. Kushner, J. IV, D. Blankschtein, and R. Langer. 2004. Experimental demonstration of the existence of highly permeable localized transport regions in low-frequency sonophoresis. *J Pharm Sci* 93:2733.
68. Tezel, A., et al. 2002. Synergistic effect of low-frequency ultrasound and surfactants on skin permeability. *J Pharm Sci* 91:91.
69. Mitragotri, S. 2000. Synergistic effect of enhancers for transdermal drug delivery. *Pharm Res* 17:1354.
70. Meidan, V.M., et al. 1998. Phonophoresis of hydrocortisone with enhancers: An acoustically defined model. *Int J Pharm* 170:157.
71. Le, L., J. Kost, and S. Mitragotri. 2000. Combined effect of low-frequency ultrasound and iontophoresis: Applications for transdermal heparin delivery. *Pharm Res* 17:1151.

72. Kost, J., et al. 1996. Synergistic effect of electric field and ultrasound on transdermal transport. *Pharm Res* 13:633.
73. Mitragotri, S., et al. 2000. Synergistic effect of low-frequency ultrasound and sodium lauryl sulfate on transdermal transport. *J Pharm Sci* 89:892.
74. Lavon, I., N. Grossman, and J. Kost. 2005. The nature of ultrasound-SLS synergism during enhanced transdermal transport. *J Control Release* 107:484.
75. http://www.wfumb.org.
76. Anonymous. 1998. Update on thermal bioeffects issues. *Ultrasound Med Biol* 24:S1.
77. Anonymous. 2000. Constructing e-business. *Proceedings of the Institution of Civil Engineers, Civil Engineering* 138:149.

17 Combined Chemical and Electroporation Methods of Skin Penetration Enhancement

Michael C. Bonner and Brian W. Barry

CONTENTS

17.1 INTRODUCTION

Electroporation can be defined as an electrical enhancement strategy whereby relatively high voltages are applied to the skin for very brief periods in order to create temporarily aqueous pores within the membrane. These apertures provide pathways for molecular penetration through the skin. Its potential for increasing the transdermal delivery of molecules was first demonstrated in the 1990s [1] and since then the technique has been combined with other enhancement strategies to examine if penetration of dermally absorbed drugs can be further increased, e.g., estradiol [2] or to broaden the range of candidate drugs, which may be delivered by this route, e.g., insulin [3]. This chapter will examine a number of recent reports, which combined electroporation and chemical penetration enhancement strategies, with a particular emphasis on the use of lipids and lipid vesicles as suitable chemical adjuvants.

17.2 CHEMICAL PENETRATION ENHANCERS OF ELECTROPORATION-MEDIATED DELIVERY

17.2.1 MACROMOLECULES

Vanbever et al. [4] hypothesized that administration of macromolecules during electroporation would, in some fashion, stabilize aqueous pores produced by high-voltage pulsing. This group studied mannitol transdermal transport under electroporation with coadministered macromolecules (heparin, dextran sulfate, neutral dextran, and polylysine). They found that while skin electroporation increased mannitol delivery, the use of macromolecules further raised transport up to fivefold. The enhancement was not observed during passive diffusion or under low-voltage iontophoresis when the macromolecules were coapplied. The authors discussed that while all the macromolecules selected did enhance transport, those of the greatest size and charge were most effective. Weaver et al. [5], using heparin chemically to enhance transdermal transport by electroporation, observed both an increase in postpulse skin permeability and a persistent lower skin resistance. The authors proposed that as heparin molecules were long enough to span five to six lipid bilayers within the stratum corneum, their results could be explained by entrapment of heparin molecules within the membrane, thus maintaining pathways induced by the pulsing for extended times.

17.2.2 Cyclodextrins

Cyclodextrins are cyclic oligosaccharides, which form inclusion complexes with drug molecules and are frequently used to enhance either solubility or stability of therapeutic entities. Although they have been infrequently used in dermal delivery systems, one group examined their ability to enhance delivery of model molecules under electroporation [6]. Transport of piroxicam from a suspension across porcine skin was promoted by approximately fourfold in the presence of beta-cyclodextrin. When a pulsing protocol was applied (60 pulses of 100 V, duration 1 ms) a further sixfold penetration enhancement was produced. For carboxyfluorescein, a more hydrophilic species than piroxicam, addition of hydroxypropylcyclodextrin improved penetration by 1.5-fold under passive diffusion and the pulsing protocol further improved input by a further 50-fold. Electroporation provided greater enhancement of hydrophilic permeant penetration, whereas cyclodextrins improved delivery of more lipophilic species. It was also noted that the cyclodextrins retarded resistance recovery of the membrane postpulsing, which was attributed to their ability to extract skin lipids. The authors also demonstrated that carboxyfluorescein could be delivered *in vivo* by electroporation to the mouse, and noted that coadministration of hydroxypropylcyclodextrin approximately doubled the bioavailability of the molecule.

17.2.3 Keratolytic Agents

Based on observed penetration of test molecules under electroporation, Zewert et al. [7] suggested that transport was diminished with increasing molecular size. They proposed that although pulsing produced straight-through aqueous pathways, leading to easy passage of ions and charged molecules through the stratum corneum, the penetration of macromolecules would be hindered by the keratin matrix within corneocytes. Thus the delivery of a keratolytic should enhance macromolecule penetration. To this end, they delivered sodium thiosulfate under electroporation, then measured molecular fluxes of an oligonucleotide (approximately 7 kDa), lactalbumin (15 kDa), and IgG (150 kDa). The use of 1 M sodium thiosulfate during pulsing improved the nucleotide flux by 10,000-fold and the lactalbumin flux by 1000-fold. IgG flux increased 1–2 h after pulsing, but this could not be compared to a flux during the electroporation as the thiosulfate could disrupt bonds linking the IgG subunits. The investigators proposed two stages of aqueous pathway development. The high-voltage pulsing induced routes that perforated the multilamellar bilayer membranes between the corneocytes large enough to permit passage of small molecules and ions; then addition of the keratolytic enlarged these pathways for passage of the macromolecules. This hypothesis would indicate two molecular barriers to transport, due to lipid bilayers and keratin.

17.2.4 Electrolytes

Various electrolytes have also been shown to enhance delivery of drugs through electroporation. In a study on penetration of rat skin by calcein [8], a wide range of cations augmented the effects of electroporation. Monovalent chlorides increased calcein delivery under electroporation by 10- to 15-fold with potassium chloride exerting the greatest effect. A range of sodium halides also enhanced delivery but the greatest increase in penetration arose when calcium chloride was coadministered with calcein. This salt enhanced penetration of the model compound by 83-fold compared to electroporation of the calcein alone. Anodal electroporation provided a 10-fold improvement of delivery over cathodal electroporation. Interestingly, the authors found no correlation of enhancement of delivery and conductivity of the electrolyte solution. They stated that calcium ions delayed the build-up recovery of the stratum corneum and in some way the sizes of the pores induced by the pulsing differed when calcium ions were introduced. To investigate further the effect of calcium chloride on electroporation-mediated delivery the same authors examined lipid mobility in the mouse stratum corneum [9]. Using ATR-FTIR spectroscopy both electroporation and electrolyte treatment separately (and jointly) increased C–H stretching frequencies *in vivo*, indicating increased lipid fluidity in the stratum corneum. This effect was observed up to 6 h after treatment. In addition the investigators produced liposomes made from stratum corneum lipids and entrapped within these calcein, with or without electrolytes ($CaCl_2$ and NaCl). These vesicles were electroporated (10×100 V pulses, 10 ms) and release of the calcein was measured. This was highest with calcium chloride, which caused around 10 times the liberation of calcein compared to NaCl or distilled water. Calcium chloride also caused electroporated skin to increase transepidermal water loss by around 1.4-fold over a membrane electroporated under distilled water. A general conclusion provided by the authors was that calcium chloride provided a "high destroying" effect on the stratum corneum or delayed the recovery of the membrane's barrier properties.

17.2.5 Lipids and Lipid Vesicles

Recent investigations have indicated a role for lipids in enhancing dermal delivery of challenging molecules under electroporation, as workers anticipated synergy between electrical pulsing and disruption of stratum corneum bilayers through addition of extraneous lipids. Studies by Sen et al. [3,10,11] produced some interesting findings on lipid selection. Under their experimental conditions, neutral or cationic lipids alone did not enhance dermal penetration of model compounds. A dispersion of two lipids, dioleylphosphatidylglycerol and dioleylphosphatidylcholine (DOPG and DOPC), enhanced delivery of small fluorescent probes (up to 590 molecular weight) through porcine skin and also the insertion of a 4 kDa dextran (giving a 15-fold enhancement in the presence of the DOPG/DOPC dispersion). However, penetration of larger dextrans was not significantly increased. Skin resistance measurements indicated that lipids delayed recovery of the membrane's barrier properties. Interestingly, the authors asserted that the lipids were not encapsulating the target molecules and their method of penetration enhancement was more economic compared to liposomal enclosure. Evidence showing that phospholipids with saturated acyl chains were superior to those with unsaturated acyl chains at enhancing electroporation-mediated delivery was provided from a study using dimyristoylphosphatidylserine (DMPS). This phospholipid produced greater transport of fluorescent probes compared to unsaturated alternatives [11]. By fluorescence microscopy, the investigators revealed that the probes penetrated in local transport regions (LTRs) produced by the pulsing. They found DMPS at the center of the LTRs and extended throughout the stratum corneum. Their deduction was that, after cathodal pulsing, the anionic phospholipids were well retained in the stratum corneum and arranged as loose layers or vesicles rather than in multilamellar forms. This retention led to prolongation of subcutaneous recovery. In an extension of this work, DMPS has been used to assist the electroporation-mediated delivery of fluorescent-labeled insulin through porcine skin [3]. Pulsing of the epidermis for 10 min (100 V, 1 Hz, and 1 ms pulse width) in the presence of the lipid enhanced the protein transport by approximately 20-fold, compared to pulsing without DMPS. Fluorescence imaging of the insulin within the membrane indicated that its transport route was primarily within the lipid regions of the skin. The workers were especially encouraged by the insulin delivery obtained as they surmised that the molecules delivered were in a hexameric state and thus had a molecular weight of around 36 kDa.

Other groups have used lipids as an adjuvant to electroporation-mediated delivery but arranged in the form of vesicles (liposomes). Liposomes may be defined as arrangements of lipids in one or more bilayers that can entrap hydrophilic species either in an aqueous core, or for more lipophilic materials, within the bilayers themselves. The first of these reports [12] discussed the delivery of liposomally entrapped colchicine under both iontophoresis (continuous low-constant current or voltage applied across the membrane) and electroporation. It was observed that electroporation (250 V, 20 ms duration, 10 pulses/min for 5 min) delivered significantly less drug than did iontophoresis (0.5 mA/cm^2), although, interestingly, when the electrical techniques were combined, colchicine delivery exceeded that by either electrical protocol alone.

Another recent study considered the effect of electroporation on vesicular estradiol delivery through human epidermal membrane [11]. In this study, specially formulated liposomes, having a claimed feature of ultradeformability, were used to entrap estradiol. A pulsing protocol (5 pulses of 100 V, duration 100 ms, and 1 min spacing) promoted penetration of estradiol from a saturated aqueous solution by 16-fold over that from passive drug diffusion from the solution. Skin deposition of the drug (measured by assay of the drug in a solubilized epidermal membrane) was also improved 14-fold. Somewhat surprisingly, from a liposome formulation, the electroporation protocol did not markedly improve estradiol penetration over passive delivery from the vesicles (Figure 17.1). Relatively small enhancements were obtained of 1.3-fold for drug penetration and 1.6-fold for deposition (Figure 17.2).

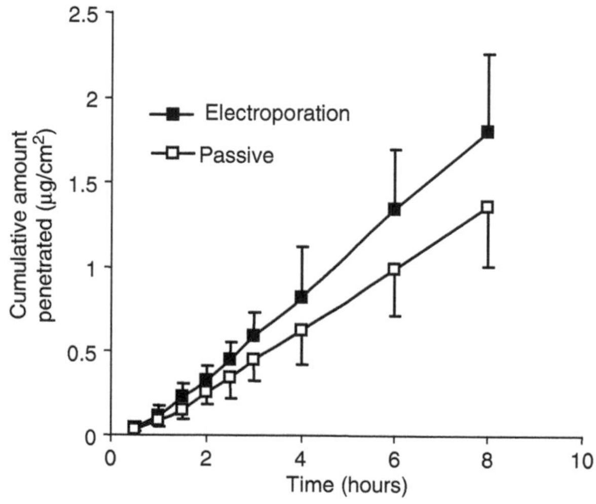

FIGURE 17.1 Human epidermal penetration profiles of estradiol ($n = 6$) from ultradeformable liposomes from passive delivery and electroporation (5 pulses of 100 V, 100 ms, and 1 min spacing).

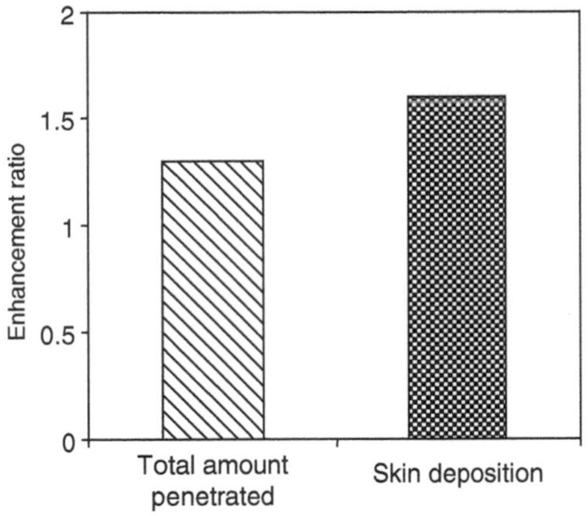

FIGURE 17.2 Enhancement ratios of total amount of estradiol penetrated and skin deposition from ultradeformable liposomes after an electroporation protocol (5 pulses of 100 V, 100 ms, and 1 min spacing) relative to passive diffusion from the vesicles.

Both total amount of drug penetrated and skin accumulation were less than those from the electroporation-mediated delivery from the saturated aqueous solution. This was a surprising finding, as it was expected that the phospholipid vesicles would augment the physical enhancement from the electroporation. The authors concluded that a large voltage had dissipated due to interaction between the applied voltage and the liposomes. This interaction lessened the effect of the pulsing protocol on the skin, probably reducing the number of the electrically created aqueous pores. The lipid vesicles were, however, suspended in saturated drug solution and so the amount of drug penetrated from the vesicle formulation would have been expected to be at least similar to that from the saturated solution alone. It thus appeared that skin that had been permeabilized by electroporation had its barrier function in some way repaired by the phospholipids.

In a further experiment to examine this protective effect, the group electroporated epidermal membranes, then treated half of these with empty (drug-free) vesicles and the remainder with deionized water as a control. After washing the membranes, penetration of estradiol from a saturated solution was monitored for 2.5 h (stage I) followed by a second pulsing stage with drug penetration measured for a further 2 h (stage II). When skin was pretreated with phospholipid vesicles, drug penetration was approximately halved compared to water-treated donors during stage I. For stage II the reduction was 15-fold in the lipid-treated membranes, providing further evidence of barrier repair from the phospholipid.

17.3 CONCLUDING REMARKS AND FUTURE PROSPECTS

Combination of electroporation and chemical enhancement may be attractive to formulators attempting to deliver challenging molecules dermally. Although there exist relatively few literature reports of the combination, it is interesting to note the range of chemical adjuvants used. These range from large hydrophilic molecules, to simple electrolytes and keratolytic agents. Lipids and lipid vesicles have also been employed with somewhat mixed results. Although certain dispersions promoted delivery of molecules as large as insulin, other workers indicated a potential penetration retarding effect of phospholipids when combined with pulsing protocols.

It is, however, likely that any future combinations of the strategies will face challenges, not least of which are potential safety and economic considerations, particularly as the techniques involve prolongation of an electrically compromised skin barrier with all the safety issues that involve.

REFERENCES

1. Prausnitz, M.R., et al. 1993. Electroporation of mammalian skin: A mechanism to enhance transdermal drug delivery. *Proc Natl Acad Sci USA* 90:10504.
2. Essa, E.A., M.C. Bonner, and B.W. Barry. 2003. Electroporation and ultradeformable liposomes; human skin barrier repair by phospholipid. *J Control Release* 92:163.
3. Sen, A., M.E. Daly, and S.W. Hui. 2002. Transdermal insulin delivery using lipid enhanced electroporation. *Biochim Biophys Acta* 1564 (1):5.
4. Vanbever, R., M.R. Prausnitz, and V. Preat. 1997. Macromolecules as novel transdermal transport enhancers for skin electroporation. *Pharm Res* 14:638.
5. Weaver, J.C., et al. 1997. Heparin alters transdermal transport associated with electroporation. *Biochem Biophys Res Commun* 234 (3):637.
6. Narasimha Murthy, S., et al. 2004. Cyclodextrin enhanced transdermal delivery of piroxicam and carboxyfluorescein by electroporation. *J Control Release* 99 (3):393.
7. Zewert, T.E., et al. 1999. Creation of transdermal pathways for macromolecule transport by skin electroporation and a low toxicity, pathway-enlarging molecule. *Bioelectrochem Bioenergetics* 49 (1):11.
8. Tokudome, Y., and K. Sugibayashi. 2003. The synergic effects of various electrolytes and electroporation on the *in vitro* skin permeation of calcein. *J Control Release* 92 (1–2):93.
9. Tokudome, Y., and K. Sugibayashi. 2004. Mechanism of the synergic effects of calcium chloride and electroporation on the *in vitro* enhanced skin permeation of drugs. *J Control Release* 95 (2):267.
10. Sen, A., et al. 2002. Enhanced transdermal transport by electroporation using anionic lipids. *J Control Release* 82 (2–3):399.
11. Sen, A., Y. Zhao, and S.W. Hui. 2002. Saturated anionic phospholipids enhance transdermal transport by electroporation. *Biophys J* 83 (4):2064.
12. Badkar, A.V., et al. 1999. Enhancement of transdermal iontophoretic delivery of a liposomal formulation of colchicine by electroporation. *Drug Deliv* 6:111.

18 Stratum Corneum Bypassed or Removed

James C. Birchall

CONTENTS

18.1 INTRODUCTION

The skin represents an attractive gateway for the localized and systemic delivery of therapeutically active molecules due to its ready accessibility, avoidance of gastrointestinal degradation and liver inactivation, monitoring capability and potential for improved patient compliance. The ability to deliver therapeutic quantities of medicaments to and through the skin, however, is dependent on the physicochemical properties of the candidate drug and the significant barrier properties of the target tissue. After all, a primary function of the skin is to restrict the ingress of external matter through physical blockade and immune surveillance. Whereas traditional transdermal delivery techniques use formulation strategies to promote the transport of small molecules through the stratum corneum, a growing number of delivery techniques that aim to bypass or disrupt the skin barrier have been developed. Such strategies, including the use of chemical enhancers [1], iontophoresis [2], electroporation [3], and sonophoresis [4] may supplement traditional transdermal delivery strategies or provide a means of delivery for new drug candidates, including macromolecules. Despite these advances, and a few exceptions, it could be argued that effective transdermal delivery is still generally restricted to a small number of low molecular weight, weakly lipophilic, and potent therapeutic molecules. Increasing emphasis on the administration of biotechnology-derived macromolecular, particulate- and DNA-based medicines requires the development of further delivery strategies and devices that circumnavigate the stratum corneum barrier to promote delivery of a wide range of therapeutics to the underlying epidermis, dermis, and possibly the systemic circulation. Clearly, as the requirement for the efficient delivery of larger molecules and nanoparticles increases more radical methods of disrupting skin barrier function are required.

18.2 THE SKIN AS A BARRIER TO GENE DELIVERY

The primary role of the skin is to serve as a physical and immunological barrier to the invasion of foreign material. Simply, the skin can be considered as a structure composing three distinct layers, the epidermis, dermis, and hypodermis, and hosting a number of adnexal features such as hair follicles, sebaceous glands, and sweat glands. In humans the uppermost layer of the skin, the epidermis, ranges from 50 to 150 μm in thickness. The external surface of the epidermis is comprised of flattened nonviable cells that have lost their nuclei following differentiation from the inner to the outer layer of the epidermis. This layer is termed the stratum corneum, approximately 15–20 μm in thickness, and its physical properties make it the principal barrier to the penetration and permeation of substances through the skin. The remainder of the epidermis is a progressively differentiated stratified epithelium. The range of cell types found in the epidermis includes keratinocytes, melanocytes, Langerhans cells, and Merkel cells. The differentiated keratinocytes arise from a pool of transient amplifying cells located at the basal layer of the epidermis, which in turn are derived from epidermal stem cells. Although stem cells can be isolated from the epidermis *in vitro* and typically express certain cell surface markers, these cells are presently impractical to selectively target *in vivo* [5]. The dermis is a connective medium underlying the epidermis, and acts as a protective layer against injuries and deformation and also maintains a role in thermal regulation. The dermis contains collagen, elastin, blood and lymphatic vessels, nervous elements, and scattered cells including fibroblasts, mast cells, macrophages, and lymphocytes. The hypodermis, the layer beneath the dermis, is composed of subcutaneous fat and blood vessels. Its main role is to maintain skin mobility, and to supply energy and insulate the body.

It has been firmly established that the structural basis of the skin permeability barrier in mammals is the stratum corneum [6], as once a compound crosses this barrier it can diffuse rapidly through deeper tissue and be taken up by the underlying capillaries. Therefore to deliver therapeutic compounds to the epidermis, dermis, or systemic circulation, delivery strategies must overcome the physical barrier afforded by the nature of the tightly packed dead cells of the stratum corneum. Conventional transdermal formulation strategies aim to enhance the delivery of small therapeutic molecules, less than 500 molecular weight in size, across the stratum corneum through the paracellular, transcellular, or intracellular routes. However, to deliver macromolecular products such as genes and proteins, more innovative and drastic methods of drug delivery are required.

18.3 METHODS FOR ENHANCING DRUG DELIVERY TO SKIN

The simplest, although crudest, method for bypassing the stratum corneum barrier is to administer medicines through direct injection. For over 150 years this technique of drug delivery, initially devised as a method for delivering opiates [7], has been used to administer medicaments to compartments of skin, muscle, and the systemic circulation. Whereas hypodermic injection will remain a routine drug delivery method due to its significant drug-loading capability and well-defined pharmacokinetics, alternative technologies are being developed to bypass the stratum corneum in a more sophisticated manner with an associated reduction in pain and diminished risk of phlebitis, hematoma, and thrombosis. Although modern fabrication methods are now used to manufacture individual needles, this chapter will focus on the innovative new technologies at the forefront of intra- and transdermal drug delivery.

18.3.1 Microfabricated Microneedle Arrays

In recent years, a particularly exciting alternative to conventional needle and syringe injection has been developed. Microneedle approaches are designed to circumnavigate the primary skin barrier without impinging on the underlying pain receptors and blood vessels (Figure 18.1). Microneedle manufacturers employ microelectromechanical systems (MEMS) technology, regularly utilized in the semiconductor and microelectronics industry, to create arrays of needles from a base material. Commonly the microneedles are fabricated from silicon and in these cases the microneedles are prepared using well-defined etching techniques [8]. Although the practical application of this technique has only been demonstrated for the first time within the past 10 years, the original concept for these delivery systems was described nearly 30 years ago [9].

Microneedles, so termed as they commonly range from 100 to 1000 μm in length, are designed to perforate the stratum corneum thus providing a direct and controlled route of access to the underlying tissue layers. When inserted into the skin, microneedles create microscopic punctures through the stratum corneum and into the viable epidermis. The length of the microneedle is controlled to ensure that the depth of penetration does not

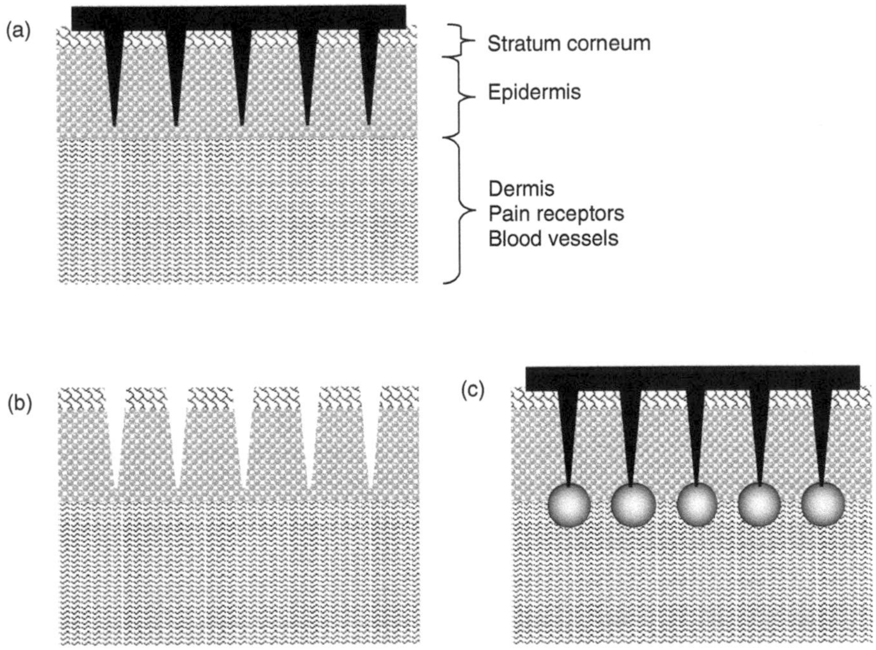

FIGURE 18.1 Schematic representation of the concept for microneedle-assisted delivery. (a) The microneedles penetrate the stratum corneum, to facilitate access of molecules to the viable epidermis, without impacting on the underlying nerve endings and blood vessels; (b) when removed the microneedles have created conduits for drug delivery; (c) hollow microneedles allow direct injection of the formulation.

impinge on the nerve fibers and blood vessels that reside primarily in the dermal layer. The micron-sized channels can therefore facilitate the delivery of both small and large molecular weight therapeutics into skin without causing pain or bleeding at the site of application [10].

The advantages of using microneedles include: (a) direct and controlled delivery of the medicament, (b) rapid exposure of large surface areas of epidermis to the delivery agents (microneedle arrays can contain over 1000 microneedles), (c) effortless, convenient, and painless delivery for the patient, (d) ability to manipulate the drug formulation, e.g., solution, suspension, emulsion, dry powder, and gel, (e) enhancing the impact of concomitant delivery methods such as transdermal patches, and (f) a minimally invasive methodology suited to patient self-administration without the need for medical supervision. A particularly important advantage lies in the ability to adapt the materials, types of structure, and dimensions of the needle to facilitate the delivery of macromolecules, nanoparticles, and vaccines [11,12].

18.3.1.1 Silicon Microneedle Structures

In collaboration with The Cardiff School of Engineering and Tyndall National Institute, Cork, our laboratories have characterized and exploited microneedles prepared using dry- and wet-etching methodologies.

At the start of the dry-etch microfabrication process a silicon wafer is coated with a positive photosensitive material. A standard high-resolution chromium-plated lithographic mask bearing the appropriate dot array pattern is used during the UV light exposure step to produce a photoresist etch mask. The surface is subsequently etched using a reactive blend of fluorinated and oxygen gases, with those regions protected by the photoresist mask resisting the etching process and leading to the formation of microneedles. The result, following removal of the photoresist, is an array of microneedles of approximately 150 μm in length. Figure 18.2a shows scanning electron micrographs of an array of cylindrical microneedles prepared by using this technique [12].

In an alternative preparation method, microneedles have also been prepared through wet-etching technologies using potassium hydroxide, KOH, at elevated temperatures. Solid microneedles are formed through a number of process steps involving a mask layout lithography step, crystal alignment, low-pressure chemical vapor deposition (LPCVD), plasma etching to create the mask on the wafer, wet etching using KOH, and mask release by cleaning procedures [13]. Figure 18.2b shows scanning electron micrographs of an array of pyramidal microneedles prepared by using this technique.

If skin is placed in a water bath under controlled conditions [14] the primary barrier to transdermal delivery, the epidermal membrane comprising the stratum corneum and viable epidermis, can be readily removed and used to analyze the penetration and diffusion of materials. Figure 18.3a and Figure 18.3b show the appearance of human breast epidermal membrane, with epidermis facing uppermost, following application of the cylindrical dry-etch and pyramidal wet-etch silicon microneedles, respectively. In each case the microneedles are clearly shown to pierce the stratum corneum and viable epidermis to facilitate controlled access of molecules to the target region of skin.

18.3.1.2 Microneedle Structures Prepared from Other Materials

Professor Mark Prausnitz and his research colleagues based at the Georgia Institute of Technology have been at the forefront of developments in the microfabrication processes used to create microneedle arrays from a range of base materials. The methods employed generally involve one or two fabrication steps or a single molding stage and use technologies that are readily scalable for industrial mass production [15]. As a result, in addition to using

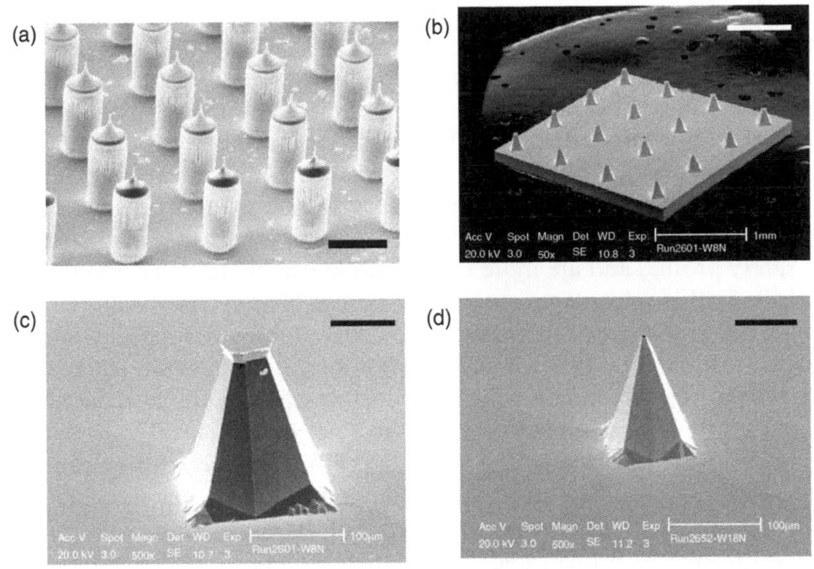

FIGURE 18.2 Scanning electron micrographs of silicon microneedles. (a) Silicon microneedles microfabricated using a modified form of the BOSCH deep reactive ion etching process. The microfabrication process was accomplished at CCLRC Rutherford Appleton Laboratory (Chilton, Didcot, Oxon, UK). The wafer was prepared at the Cardiff School of Engineering, Cardiff University, UK. Bar = 100 μm; (b–d) platinum-coated silicon microneedles prepared using a wet-etch microfabrication process performed at the Tyndall National Institute, Cork, Ireland. Bar = 1 mm (b), 100 μm (c,d).

silicon, a brittle material that regularly fractures on microneedle application, as the primary structural component, microneedles have been fashioned from metal, polymer, and glass. Metal microneedles were manufactured by electrodeposition of metal onto a polymer or silicon micromold. Glass microneedles were created by conventional drawn-glass micropipette techniques. Polymer microneedles were fabricated by melting polyglycolic acid, polylactic acid, or polylactic-*co*-glycolic acid into polydimethylsiloxane (PDMS) micromolds [16] with these structures demonstrating up to three orders of magnitude increase in the

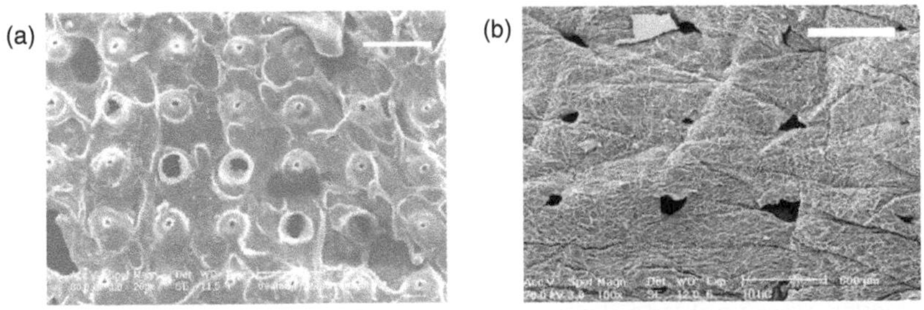

FIGURE 18.3 Scanning electron micrographs of epidermal membrane treated with dry-etch and wet-etch silicon microneedles. The epidermal membrane, consisting of stratum corneum and viable epidermis, was obtained by heat separation of full-thickness human breast skin. The tissue was immersed in distilled water preheated to 60°C for 60 s and the upper layers carefully peeled off from the dermal layer using tweezers. Epidermal membranes were treated with microneedles for 30 s at an approximate pressure of 2 kg/cm^2. (a) Dry-etch microneedle-treated epidermal membrane. Bar = 200 μm; (b) wet-etch microneedle-treated epidermal membrane. Bar = 500 μm.

permeability of human cadaver skin to a low molecular weight trace molecule, calcein, and a macromolecular protein, bovine serum albumin.

The ability to prepare microneedles from alternative material substrates confers important advantages to the industrial exploitation of the technique. Whereas silicon microneedles are generally fabricated in dedicated, and costly, clean room facilities, polymer and metal microdevices would be less expensive to mass produce as the raw materials are more available at a reduced cost and the single-step molding techniques do not require specialist amenities. From a clinical perspective many metals and polymers, as opposed to silicon, have established biomaterial safety profiles and are more robust and less likely to shear upon skin application and removal. A further advantage, specific to the use of polymeric material, relates to the ability to manipulate the recipe to formulate biodegradable microneedles capable of either *in situ* biological degradation following application or environmental degradation following termination of use. Recently, potentially multifunctional polymeric microneedles comprising a multilayer structure have also been prepared by Kuo and Chou [17].

A recent materials innovation in this field describes a method for coating microporous calcium phosphate onto stainless steel acupuncture needles [18]. The incorporation of trehalose into the porous coating acted as a model fast-dissolving reservoir for the potential loading and stabilizing of protein and DNA vaccines. The coated needles were shown to remain intact following contravention of the stratum corneum and allow access of the coated material to the viable epidermis. Importantly, no evidence of skin reaction, bleeding, or infection was observed.

18.3.1.3 Hollow Microneedles

Whereas solid microneedle arrays present the opportunity to create conduits through the restrictive skin barrier layer, the application of the formulation into the channel through dry-coating the microneedle array or coadministration of a solution, suspension, emulsion, or gel containing the medicament generally relies on passive delivery mechanisms. The capacity to microfabricate hollow microneedles, however, allows a controlled quantity of the medicament to be actively delivered from the tip of the inserted microneedle at a defined rate. In addition, hollow microneedles provide the opportunity to not only deliver substances but also to withdraw material from the skin for analysis, monitoring, and responsive purposes.

McAllister et al. [19] reported the fabrication of hollow tapered microneedles and microtubes using a combination of dry-etching processes, micromolding and selective electroplating or by direct micromolding. Another reported method for preparing hollow microneedles utilizes a combination of silicon microfabrication and copper electroplating technologies [20]. As the layer of copper on the square-pyramidal microneedles is designed to be thicker at the base than at the tip, structural stability is provided to the structure. One significant drawback, however, lies in the acknowledgment that copper can cause skin sensitization reactions and is prone to oxidation. As an alternative, tapered hollow nickel microneedles have also been prepared by a rapid, simple, and relatively inexpensive method [21].

A recent study used hollow microneedles with the following dimensions: 20–100 μm in diameter and 100–150 μm in length to deliver insulin through skin [22]. *In vivo* tests in diabetic animals, however, were unable to demonstrate any functional delivery of insulin through the hollow microneedles. It is therefore essential that the engineering processes evolve to ensure that both microneedle length and tip sharpness are optimized for systemic drug delivery to be seen *in vivo*.

The preparation of hollow microneedles affords the ability to combine the microneedle array with microfluidic channels to facilitate active insertion of the medicament into various layers of skin. This serves to increase the volume of medicament entering the skin, as opposed to passive diffusion of liquid into a microchannel created using a solid microneedle, and provides the opportunity to pulse or continuously replenish the medicament over a period of time. The extensive work of Zahn et al. [23] has shown that continuous pumping of fluid through microneedles can be achieved for more than 6 h.

18.3.1.4 Microneedle Application Forces

Although modern engineering techniques have fashioned a range of microneedle morphologies with potential utility for delivering drugs through the skin, it has become apparent that certain designs are more capable of easy insertion into, and removal from, the skin without causing damage to the microneedle structure. For instance, if the microneedles are not sharp enough, or are insufficiently spaced, then the skin will resist penetration due to excessive distribution of the imposed application force; i.e. a bed of nails effect. Conversely if the needles are too long or too widely distributed on the array the tip of the needle may not have sufficient rigidity and may break off on insertion. Wang et al. [24] specifically fabricated single hollow glass microneedles at a range of tip diameters and bevel angles to determine their ability to puncture human cadaver and *in vivo* rat skin. The authors found that the volume of liquid injection and interstitial fluid extraction depend on the microneedle geometry, with the larger needle tips facilitating more liquid injection and extraction. The same research group has also theoretically modeled and experimentally quantified two critical determinants, i.e., the force required to insert microneedles into skin and the force needles can withstand before fracturing [25]. The authors found that microneedle insertion force increases as a linear function of needle tip cross-sectional area with a measured insertion force of 0.1–3 N, making microneedle penetration into skin readily achievable by hand. The force required to fracture the microneedles increased with increasing wall thickness, microneedle angle, and tip radius and encouraging this force was always in excess of that required for adequate insertion of the microneedle. A further study proposed that a microneedle length of 600 μm was optimal when a combination of factors such as microneedle strength and robustness, minimal insertion pain, and skin damage are taken into consideration [26].

An interesting approach to assist microneedle insertion involves the use of vibrations to drive the needles into skin. Clearly, this has firm theoretical basis in nature as mosquitoes are able to pierce human skin using a vibratory cutting action at a frequency of 200–400 Hz. Coupling a vibratory actuator to hollow microhypodermic needles leads to more than 70% reduction in the force required to insert the needles into excised animal skin [27]. Ultrasound can be used to reduce the deformation of the skin on penetration and potentially provides a biological lubricating layer to assist microneedle insertion. The penetration force of microneedles into silicon rubber and vegetable skin simulates has shown to be reduced significantly when bonded to piezoelectric actuators [28].

From a skin application perspective it appears logical that rolling the microneedles onto the skin surface, in a manner analogous to printing presses, would result in greater penetration than direct application from above. Although the DermaRoller was developed for maximizing the penetration of cosmetically active materials, the cylindrical rolling design, comprising an array of metallic microneedles of varying lengths, should be further investigated for the delivery of other therapeutics [29].

18.3.1.5 Medical Applications for the Use of Microneedles

Microneedle arrays were originally designed to mediate the transdermal delivery of low molecular weight drugs or reporter molecules to the systemic circulation, resulting in the enhanced penetration of calcein [8], trypan blue [30], and methyl nicotinate [31]. In our

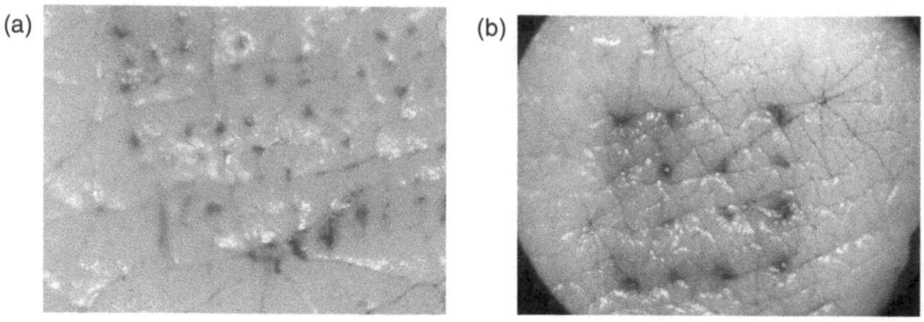

FIGURE 18.4 Methylene blue staining of dry-etch (a) and wet-etch (b) microneedle-treated human skin. Disruptions within the stratum corneum indicate microneedle penetration efficiency.

laboratories we routinely validate the creation of functional microchannels using methylene blue as a visual reporter molecule (Figure 18.4). Whereas microneedles are ideally suited for this purpose, the dimensions of the microchannels created within skin expand the applicability of this approach to the delivery of macromolecular therapeutics, such as proteins, nanoparticles, and nucleic acids.

The *in vivo* pharmacodynamic response to a therapeutic protein delivered through microneedles was recently demonstrated in a diabetic hairless rat model [32]. Insulin was selected as a nonpermeable clinically relevant protein with current drug administration issues, i.e., the requirement for regular painful and inconvenient injections. Laser ablation of a stainless steel sheet and subsequent bending of the planar microneedle shape was used to produce solid metallic microprojections of 1000 μm length. Coadministration of insulin solution and microneedles, followed by microneedle removal, enhanced the transdermal delivery of insulin and mediated an 80% reduction in blood glucose. The authors concluded that microneedles are capable of delivering physiologically relevant amounts of insulin with rapid pharmacodynamic action and appear to have broad applicability for a range of therapeutic macromolecules.

Several companies are currently developing microneedles for sustained drug delivery. Nanopass Technologies Ltd. (Haifa, Israel) has developed the Nanopump for the prolonged delivery of insulin through robust-engineered micropyramids [11]. Despite the clear advantages of this approach, as opposed to conventional invasive insulin administration devices, further studies are required to optimize the rate and quantity of delivery. BD Medical Systems (Franklin Lakes, NJ, USA) have developed a prefilled, disposable device, MicroInfusor, that is activated by the patient to enable relatively large volume administration without dermal back pressure.

Currently microneedle technology is exploited to deliver plasmid DNA (pDNA) into, and study the subsequent gene expression within, the viable epidermis [12,33]. The specific and efficient immune processing properties of skin have resulted in significant interest in the development of genetic vaccines [34–36] that can capitalize on the innate ability of Langerhans cells, powerful antigen-presenting cells (APCs) residing within the viable epidermis, to proficiently present antigen to stimulate an antigen-specific T-cell immune response. Clearly, microneedles represent a practicable method for targeting these cells in skin. In addition, the ability of microneedle arrays to penetrate cells for the intracellular delivery of DNA has also been shown to be possible in an *in vitro* environment in plant cells [37] and nematodes [38], with about 8% of the total progeny tested expressing the foreign gene in the latter case.

In our laboratories we have investigated microneedles for their potential to facilitate gene delivery to the viable epidermis of skin. Our studies have confirmed that the delivery of naked pDNA, i.e., pDNA formulated without additional complexing or targeting elements, through microneedle-facilitated microchannels results in measurable levels of reporter gene expression in excised human skin. Figure 18.5 shows typical results from skin transfection experiments. *En face* imaging, following delivery of pCMVβ reporter gene and staining with X-Gal, shows microchannels stained positive for reporter gene expression (Figure 18.5a and Figure 18.5b). Microchannel counterstaining with nuclear fast red shows that only a minority of microchannels were shown to be positive for gene expression (Figure 18.5a). These studies therefore validate the skin explant organ culture conditions in maintaining the cellular viability of excised skin and provide a realistic assessment of the current efficiency of the microneedle technique for facilitating gene transfer. Photomicrographs showing the high level of reporter gene expression in viable epidermal cells are presented in Figure 18.5c and Figure 18.5d. Further studies are currently utilizing optimized microneedle devices to facilitate more reproducible cutaneous gene delivery and explore additional factors that influence gene expression in epidermal cells.

BD Technologies (Research Triangle Park, NC, USA) have used microfabricated silicon microneedles, in their case termed microenhancer arrays (MEAs), to deliver genetic vaccines to skin [39]. The application protocol involved applying a solution of DNA to the surface of mouse skin and laterally scraping the microneedle array across the skin. The authors report a 2800-fold increase in reporter gene activity in comparison with conventional topical application of controls and a more proficient and reproducible immune response in comparison with needle injection. Encouragingly, lateral application of the MEA to human subjects confirmed that the devices can breach the skin barrier with negligible to minimal skin irritation, no damage to the microneedle array, and no reported incidence of infection.

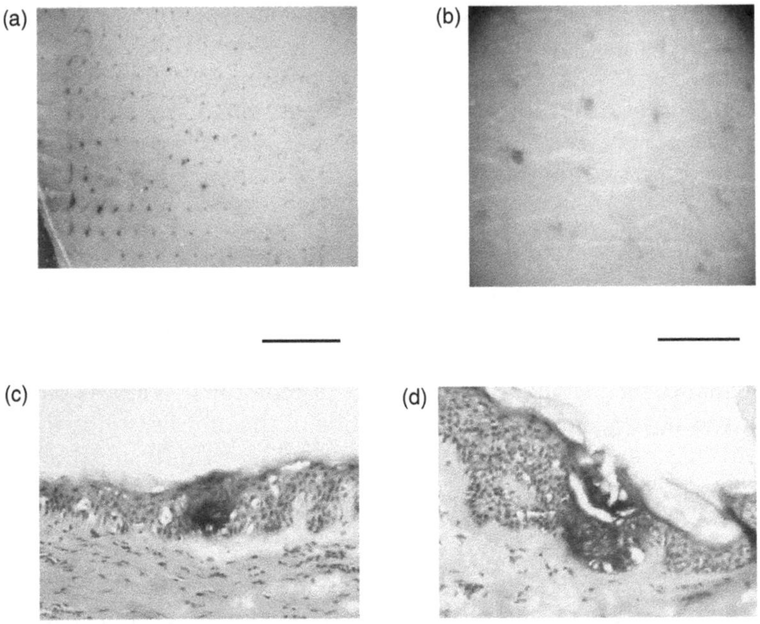

FIGURE 18.5 Light photomicrographs of microneedle-treated human skin stained for β-galactosidase expression. (a) *En face* stereomicroscopy of dry-etch microneedle-treated skin with nuclear fast red counterstaining; (b) *en face* stereomicroscopy of wet-etch microneedle-treated skin; (c) hematoxylin and eosin stained 12 μm cryosection of dry-etch microneedle-treated skin. Bar = 100 μm; (d) hematoxylin and eosin stained 12 μm cryosection of wet-etch microneedle-treated skin. Bar = 100 μm.

A further advantage in utilizing the microneedle approach for vaccination and for the delivery of proteins, peptides, and nucleic acids lies in the possibility of adjusting formulation to meet stability and cost requirements. 3M Drug Delivery Systems (St Paul, MN, USA) and ALZA Corporation (Mountain View, CA, USA) have developed methods to dry coat microneedles with vaccine or drug to eliminate common pharmaceutical and resource concerns over poor drug aqueous stability and cold-chain vaccine storage [11]. The future application of these technologies is likely to have a significant future role to play in mass immunization programs. ALZA have developed a microprojection patch (Macroflux) for the controlled delivery of medicaments. Macroflux comprises a stainless steel or titanium microprojection array, fabricated from metallic foil, which is used to create superficial pathways in skin and facilitate delivery of medicaments. This drug delivery system has been used to deliver the peptide hormone desmopressin [40] and protein antigens [41] dry coated onto the projections, or in conjunction with iontophoresis for the administration of therapeutically relevant quantities of oligonucleotides [42].

18.3.2 Other Methods for Disrupting Stratum Corneum

Whereas microneedles are designed to pierce through the outer layers of skin, the controlled destruction or adhesive removal of the stratum corneum represent alternative methods for overcoming the penetrative barrier to cutaneous drug delivery.

18.3.2.1 Tape Stripping

Since the early 1960s it has been recognized that the principal skin barrier can be removed using adhesive tape [43]. Tape stripping is now a routine analytical method for determining epidermal physiology and the permeation and dermatopharmacokinetics of substances within skin. The variability in the area and depth of stratum corneum removed by tape stripping is not only dependent on the type of tape used and the skin characteristics of the individual subject but also on the anatomical site, application pressure, duration of time in contact with the skin, and method of removal [44,45]. These factors, combined with the associated safety concerns of removing the skin barrier, confirm that tape stripping, although useful as a scientific tool, will never be appropriate for clinical exploitation.

18.3.2.2 Laser Methods

More controlled ablation and removal of the stratum corneum has been achieved using laser methods. A variety of lasers including the ruby laser [46], carbon dioxide laser [46], argon–fluoride laser [47], and erbium:YAG (yttrium–aluminum–garnet) laser [46,48–50] have been employed to enhance the percutaneous transport of molecules. Whereas physical disruption of skin barrier was not evident following application of the ruby laser [46], for the other lasers partial removal of stratum corneum (approximately 12% of diffusional area [49]), the appearance of granular structures on the stratum corneum surface and localized thermal effects led to the significant enhancement in skin permeation of 5-fluorouracil [46], 5-aminolaevulinic acid [48], dextran [47], hydrocortisone, gamma-interferon [49], indomethacin, and nalbuphine [50]. Furthermore the recovery of barrier function and epidermal thickness was demonstrable within 5 days following application of the most exploited laser (erbium:YAG; [50]).

18.3.2.3 Abrasive Methods

The method of microdermabrasion was developed as a technique to cosmetically treat photoaging, hyperpigmentation, acne, scars, and stretch marks [51,52]. The procedure involves trajecting inert sharp particles, e.g., aluminum oxide crystals or other abrasive substances, onto the skin and subsequent removal of the crystals and abraded material. Using this technique skin barrier function is transiently compromised with restoration within 24 h [53]. This method has been employed to enhance the topical permeation of vitamin C, 20-fold increases in flux and skin deposition are reported when compared with intact skin [54], and 5-aminolaevulinic acid, up to 15-fold increase in skin permeation [48]. Interestingly in this latter example, the permeation enhancement effect was significantly augmented by synergistic use of electroporation or iontophoresis. O Herndon et al. [55] described the use of a gas-entrained stream of 10–70 μm aluminum oxide particulates for the formation of microconduits for transdermal delivery and sample acquisition [55]. The 50–200 μm depth channels formed by microscission permitted rapid functional anesthesia following topical application of lidocaine and allowed for the removal of blood glucose for monitoring purposes. An alternative method for enhancing skin permeability uses standardized skin minierosion whereby an epidermal bleb is created by suction to facilitate diffusion of analgesia for postoperative pain relief [56].

18.3.2.4 Puncturing Methods

Ciernik et al. [57] reported the installation and expression of a solution of naked DNA in mice using high-frequency puncturing of skin with oscillating needles (1 cm length and 250 μm diameter). The authors demonstrated that this tattooing technique led to the enhanced expression of reporter gene compared with both direct subepidermal injection and topical application. This method was also used to demonstrate the induction of cytotoxic T-lymphocytes using a peptide oligonucleotide. Eriksson et al. [58] used a similar technique, termed microseeding, to deliver expression plasmids to intact porcine skin and partial thickness wounds and confirmed that the procedure proved more efficient than direct injection and particle-mediated gene transfer.

TransPharma Medical has developed an innovative technology (RF-Microchannels™) to create transient microchannel conduits in skin for direct and controlled access of molecules across the stratum corneum for diffusion to the underlying viable epidermis and dermis [59,60]. The technology comprises an intimately spaced array of microelectrodes, which are placed against the surface of skin to individually conduct an applied alternating electrical current at radio frequency (RF). During the application of RF energy, a frequency alternating current moves from the tip of the electrode into the surrounding tissue, creating localized heating and subsequent cell ablation. Microscopic studies have shown that the RF microchannels generated reside in the outer layer of skin and do not impact on the underlying blood vessels and nerve endings, thereby resulting in minimal skin trauma, bleeding, and neural sensations [59]. This technology has shown utility in the transdermal delivery of polar hydrophilic molecules including granisetron hydrochloride and diclofenac sodium [59] and has recently shown further applicability to high molecular weight medicaments [60].

The creation of physical openings through the stratum corneum can also be facilitated by applying an array of tiny-resistive elements to the skin surface and transmitting a short electric current. Localized resistive heating effects vaporize the cells of the stratum corneum leaving a microscopic hole, and are termed as micropore [61]. As natural desquamation processes repair the micropore, a transient window is provided for cutaneous drug delivery. Reporter gene expression using an adenoviral vector has been increased 100-fold, when compared with intact skin, using this method with similar increases in cellular and humoral immune responses following topical administration of adenovirus vaccine.

Further advances in skin resurfacing technologies, such as the use of gas jets and accelerated microdroplets (JetPeel [62]), may offer additional mechanisms for transporting medicaments through the stratum corneum barrier.

18.3.2.5 Ballistic Methods

The use of ballistic devices to propel materials through the stratum corneum and into the underlying tissue has been widely reported as a method for enhancing the delivery of anesthetics and biotechnology-derived drugs, such as proteins, peptides, vaccines, and nucleic acids, through the skin. The Powderject system accelerates pharmaceuticals in particle form and has shown utility for the delivery of conventional and genetic vaccines. Recent studies have identified optimum particle parameters and acceleration velocity to target specific skin layers and established sources of variability [63]. The commercially available Helios Gene Gun© (Bio-Rad Laboratories, Hercules, CA, USA) utilizes a helium cylinder to accelerate DNA-coated gold particles into target cells or tissues. The gold particles are typically around 1 μm in diameter and can penetrate through cell membranes, carrying the bound DNA, typically 0.5–5 μg/mg gold, into the cell cytoplasm. Subsequent dissociation of the DNA from the carrier particles allows the gene to be expressed [64]. Numerous studies have shown the successful delivery and expression of genes using this method, with both reporter genes and therapeutic plasmids having been transported to mammalian cells in culture [65], oral mucosa [66], cornea [67], and animal skin [68–70]. Whereas expression of the transgene is usually transient, lasting from a few days up to 4 weeks [71], Cheng et al. [68] reported sustained luciferase activity in rat dermis one and a half years after *in vivo* particle bombardment.

18.4 CONCLUSION

New technologies at the interface of engineering and biological sciences have provided the drug delivery specialist with fresh opportunities for administering a range of therapeutics to and through skin. The clinical exploitation of these technologies does not seem to be too remote; however, there remain some limitations in these approaches, which necessitate further investigation. It could be debated that creating channels, of any dimensions, in skin is inherently dangerous and creates the opportunity for lasting skin damage and infection. Do researchers in this area truly understand the kinetics and mechanisms of skin healing in response to these assaults on skin integrity? Where a physical structure, such as a microneedle, is used to penetrate the skin are any hyperproliferative, stress, or immune responses triggered? As these devices are often designed to be used by patients in the absence of clinical intervention is the hardware sufficiently simple to use in a safe and reproducible manner to create a consistent number of channels at a reliable depth? As the stratum corneum acts as our security blanket to infiltration by foreign bodies, are we wise in compromising it?

In my opinion these questions will be addressed, although it is clearly important that the biological scientists work with the engineering scientists to ensure the development of these technologies is treatment, disease, and patient focused rather than motivated by the technologies themselves.

ACKNOWLEDGMENTS

The significant contributions of Feriel Chabri, Marc Pearton, Sion Coulman, Ben Taunton, Dr. Chris Allender, and Dr. Keith Brain are gratefully acknowledged. I am indebted to Professor David Barrow, Tyrone Jones, and Kostas Bouris, Cardiff School of Engineering, Cardiff University, and Dr. Anthony Morrissey and Nicolle Wilke, Tyndall National Institute, Cork for their continued microfabrication support. I also thank Dr. Anthony Hann, Cardiff School of Biosciences, for assistance with electron microscopy and Drs. Alexander Anstey, Chris Gateley. and Helen Sweetland for clinical support.

REFERENCES

1. Barry, B.W. 1987. Mode of action of penetration enhancers in human skin. *J Control Release* 6:85.
2. Badkar, A.V., and A.K. Banga. 2002. Electrically enhanced transdermal delivery of a macromolecule. *J Pharm Pharmacol* 54:907.
3. Zhang, L., et al. 2002. Enhanced delivery of naked DNA to the skin by non-invasive *in vivo* electroporation. *Biochim Biophys Acta* 1572:1.
4. Smith, N., et al. 2003. Ultrasound-mediated transdermal transport of insulin *in vitro* through human skin using novel transducer designs. *Ultrasound Med Biol* 29:311.
5. Khavari, P.A., O. Rollman, and A. Vahlquist. 2002. Cutaneous gene transfer for skin and systemic diseases. *J Int Med* 252:1.
6. Menon, G.K., and P.M. Elias. 2001. The epidermal barrier and strategies for surmounting it: An overview. In *The skin and gene therapy*, eds. U. Hengge and B. Volc-Platzer. Heidelberg: Springer-Verlag, chap. 1.
7. McGrew, R., and M. McGrew. 1985. *Encyclopedia of medical history*. New York: McGraw-Hill.
8. Henry, S., et al. 1998. Microfabricated microneedles: A novel approach to transdermal drug delivery. *J Pharm Sci* 87:922.
9. Gerstel, M.S., and V.A. Place. 1976. Drug delivery device, US Patent 3,964,482.
10. Kaushik, S., et al. 2001. Lack of pain associated with microfabricated microneedles. *Anesth Analg* 92:502.
11. Chiarello, K. 2004. Breaking the barrier. *Pharm Tech* 28:46.
12. Chabri, F., et al. 2004. Microfabricated silicon microneedles for nonviral cutaneous gene delivery. *Br J Dermatol* 150:869.
13. Wilke, N., et al. 2004. Fabrication and characterization of microneedle electrode arrays using wet etch technologies. Abstract presented at EMN04, Paris.
14. Christophers, E., and A. Kligman. 1963. Preparation of isolated sheets of human stratum corneum. *Arch Dermatol* 88:702.
15. McAllister D.V., et al. 2003. Microfabricated needles for transdermal delivery of macromolecules and nanoparticles: Fabrication methods and transport studies. *Proc Natl Acad Sci USA* 100:13755.
16. Park, J.-H., M.G. Allen, and M.R. Prausnitz. 2005. Biodegradable polymer microneedles: Fabrication, mechanics and transdermal drug delivery. *J Control Release* 104:51.
17. Kuo, S.-C., and Y.A. Chou. 2004. A novel polymer microneedle arrays and PDMS micromolding technique. *Tamkang J Sci Eng* 7:95.
18. Shirkhanzadeh, M. 2005. Microneedles coated with porous calcium phosphate ceramics: Effective vehicles for transdermal delivery of solid trehalose. *J Mater Sci* 16:37.
19. McAllister, D.V., et al. 1999. Three-dimensional hollow microneedle and microtube arrays. Abstract presented at Transducers'99, Sendai, Japan.
20. Tan, P.Y.J., et al. 2002. Novel low cost fabrication of microneedle arrays for drug delivery applications. *Proc SPIE* 4936:113.
21. Kim, K., et al. 2004. A tapered hollow metallic microneedle array using backside exposure of SU-8. *J Micromech Microeng* 14:597.
22. Teo, M.A.L., et al. 2005. *In vitro* and *in vivo* characterization of MEMS microneedles. *Biomed Microdevices* 7:47.
23. Zahn, J.D., et al. 2004. Continuous on-chip micropumping for microneedle enhanced drug delivery. *Biomed Microdevices* 6:183.
24. Wang, P.M., M.G. Cornwell, and M.R. Prausnitz. 2002. Effects of microneedle tip geometry on injection and extraction in the skin. *Proceedings of the Second Joint EMBS/BMES Conference*, Houston.
25. Davis, S.P., et al. 2004. Insertion of microneedles into skin: Measurement and prediction of insertion force and needle fracture force. *J Biomech* 37:1155.
26. Aggarwal, P., and C.R. Johnston. 2004. Geometrical effects in mechanical characterizing of microneedle for biomedical applications. *Sens. Actuators B* 102:226.
27. Yang, M., and J.D. Zahn. 2004. Microneedle insertion force reduction using vibratory actuation. *Biomed Microdevices* 6:177.

28. Newton, A.M., A. Lal, and X. Chen. 2003. Ultrasonically driven microneedle arrays. National Nanofabrication Users Network, Cornell.
29. http://www.dermaroller.de
30. McAllister, D.V., et al. 2003. Microfabricated needles for transdermal delivery of macromolecules and nanoparticles: Fabrication methods and transport studies. *Proc Natl Acad Sci USA* 100:13755.
31. Sivamani, R.K., et al. 2005. Clinical microneedle injection of methyl nicotinate: Stratum corneum penetration. *Skin Res Technol* 11:152.
32. Martanto, W., et al. 2004. Transdermal delivery of insulin using microneedles *in vivo*. *Pharm Res* 21:947.
33. Coulman, S.A., et al. 2006. Minimally invasive delivery of macromolecules and plasmid DNA via microneedles. *Current Drug Delivery* 1:65.
34. Larregina, A.T. and L.D. Falo. 2000. Generating and regulating immune responses through cutaneous gene delivery. *Hum Gene Ther* 11:2301.
35. Peachman, K.K., M. Rao, and C.R. Alving. 2003. Immunization with DNA through the skin. *Methods* 31:232.
36. Partidos, C.D., et al. 2003. Immunity under the skin: Potential application for topical delivery of vaccines. *Vaccine* 21:776.
37. Trimmer, W., et al. 1995. Injection of DNA into plant and animal tissues with micromechanical piercing structures. *Proceedings of the IEEE Microelectro Mechanical Systems Workshop in Amsterdam*, 111.
38. Hashmi, S., et al. 1995. Genetic transformation of nematodes using arrays of micromechanical piercing structures. *Biotechniques* 19:766.
39. Mikszta, J.A., et al. 2002. Improved genetic immunization via micromechanical disruption of skin-barrier function and targeted epidermal delivery. *Nat Med* 8:415.
40. Cormier, M., et al. 2004. Transdermal delivery of desmopressin using a coated microneedle array patch system. *J Control Release* 97:503.
41. Matriano, J.A., et al. 2002. Macroflux® microprojection array patch technology: A new and efficient approach for intracutaneous immunization. *Pharm Res* 19:63.
42. Lin, W., et al. 2001. Transdermal delivery of antisense oligonucleotides with microprojection patch (Macroflux®) technology. *Pharm Res* 18:1789.
43. Tregear, R.T., and P. Dirnhuber. 1962. The mass of keratin removed from the stratum corneum by stripping with adhesive tape. *J Invest Dermatol* 38:375.
44. Jacobi, U., et al. 2005. Estimation of the relative stratum corneum amount removed by tape stripping. *Skin Res Technol* 11:91.
45. Loffler, H., F. Dreher, and H.I. Maibach. 2004. Stratum corneum adhesive tape stripping: Influence of anatomical site, application pressure, duration and removal. *Br J Dermatol* 151:746.
46. Lee, W.R., et al. 2002. The effect of laser treatment on skin to enhance and control transdermal delivery of 5-fluoruracil. *J Pharm Sci* 91:1613.
47. Fujiwara, A., et al. 2005. Partial ablation of porcine stratum corneum by argon-fluoride excimer laser to enhance transdermal drug permeability. *Lasers Med Sci* 19:210.
48. Fang, J.Y., et al. 2004. Enhancement of topical 5-aminolaevulinic acid delivery by erbium:YAG laser and microdermabrasion: A comparison with iontophoresis and electroporation. *Br J Dermatol* 151:132.
49. Nelson, J.S., et al. 1991. Mid-infrared laser ablation of stratum corneum enhances *in vitro* percutaneous transport of drugs. *J Invest Dermatol* 97:874.
50. Lee, W.R., et al. 2001. Transdermal drug delivery enhanced and controlled by erbium:YAG laser: A comparative study of lipophilic and hydrophilic drugs. *J Control Release* 75:155.
51. Tsai, R.Y., C.N. Wang, and H.L. Chan. 1995. Aluminium oxide crystal microdermabrasion. A new technique for treating facial scarring. *Dermatol Surg* 21:539.
52. Spencer, J.M. 2005. Microdermabrasion. *Am J Clin Dermatol* 6:89.
53. Song, J.Y., et al. 2004. Damage and recovery of skin barrier function after glycolic acid chemical peeling and crystal microdermabrasion. *Dermatol Surg* 30:390.
54. Lee, W.R., et al. 2003. Lasers and microdermabrasion enhance and control topical delivery of vitamin C. *J Invest Dermatol* 121:1118.

55. O Herndon, T., et al. 2004. Transdermal microconduits by microscission for drug delivery and sample acquisition. *BMC Med* 2:12.
56. Svedman, P., et al. 1996. Passive drug diffusion via standardized skin mini-erosion; methodological aspects and clinical findings with new device. *Pharm Res* 13:1354.
57. Ciernik, I.F., B.H. Krayenbühl, and D.P. Carbone. 1996. Puncture-mediated gene transfer to the skin. *Hum Gene Ther* 7:893.
58. Eriksson, E., et al. 1998. *In vivo* gene transfer to skin and wound by microseeding. *J Surg Res* 78:85.
59. Sintov, A.C., et al. 2003. Radiofrequency-driven skin microchanneling as a new way for electrically assisted transdermal delivery of hydrophilic drugs. *J Control Release* 89:311.
60. Levin, G., et al. 2005. Transdermal delivery of human growth hormone through RF-microchannels. *Pharm Res* 22:550.
61. Bramson, J., et al. 2003. Enabling topical immunization via microporation: A novel method for pain-free and needle-free delivery of adenovirus-based vaccines. *Gene Ther* 10:251.
62. Golan, J., and N. Hai. 2005. JetPeel: A new technology for facial rejuvenation. *Ann Plast Surg* 54:369.
63. Kendall, M., T. Mitchell, and P. Wrighton-Smith. 2004. Intradermal ballistic delivery of micro-particles into excised human skin for pharmaceutical applications. *J Biomech* 37:1733.
64. Lin, M.T.S., et al. 2000. The gene gun: Current applications in cutaneous gene therapy. *Int J Derm* 39:161.
65. Heiser, W.C. 1994. Gene transfer into mammalian cells by particle bombardment. *Anal Biochem* 217:185.
66. Wang, J., et al. 2001. Gene gun-mediated oral mucosal transfer of interleukin 12 cDNA coupled with an irradiated melanoma vaccine in a hamster model: Successful treatment of oral melanoma and distant skin lesion. *Cancer Gene Ther* 8:705.
67. Tanelian, D.L., et al. 1997. Controlled gene gun delivery and expression of DNA within the cornea. *Biotechniques* 23:484.
68. Cheng, L., P.R. Ziegelhoffer, and N.-S. Yang. 1993. *In vivo* promoter activity and transgene expression in mammalian somatic tissues evaluated by using particle bombardment. *Proc Natl Acad Sci USA* 90:4455.
69. Fuller, D.H., et al. 1997. Enhancement of immunodeficiency virus-specific immune responses in DNA-immunized rhesus macaques. *Vaccine* 15:924.
70. Yang, N.-S., et al. 1990. *In vivo* and *in vitro* gene transfer to mammalian somatic cells by particle bombardment. *Proc Natl Acad Sci USA* 87:9568.
71. Udvardi, A., et al. 1999. Uptake of exogenous DNA via the skin. *J Mol Med* 77:744.

Part V

Nasal Absorption Optimization

19 Physiological Factors Affecting Nasal Drug Delivery

Sian Tiong Lim, Ben Forbes, Marc B. Brown, and Gary P. Martin

CONTENTS

19.1 INTRODUCTION

The nasal route of drug delivery is used for the direct administration of medicines to the nose for treatment of local conditions or the systemic delivery of compounds that are not easily delivered by the oral route. It is also suggested that there may be a direct route for drug absorption to the central nervous system (CNS) from the olfactory region of the nose.

Nasal delivery has long been used to administer topically acting drugs to treat localized ailments such as nasal symptoms of common cold and allergy. Advantages of using the nasal route to treat local symptoms include the immediate targeting of relatively high drug concentrations, the simplicity of administration, reduced systemic exposure, and good patient acceptability. More recently, drugs with systemic actions have been marketed for nasal delivery in preference to oral delivery or injection. This takes advantage of the potentially rapid and high systemic availability of nasally administered compounds. Diseases or conditions for which nasal administration has been used to achieve such delivery include hormone replacement therapy (estradiol), osteoporosis (calcitonin), pain management (butorphanol, sumatriptan and zomitriptan), smoking cessation (nicotine), enuresis (desmopressin), endometriosis (nafarelin) and motion sickness (metoclopromide). The possibilities for CNS delivery via nasal administration are currently being investigated for the delivery of polar drugs to treat chronic CNS conditions such as Parkinson's disease or Alzheimer's disease.

19.2 ANATOMY AND PHYSIOLOGY OF THE NOSE

19.2.1 Nasal Passages

In humans, the nose is divided into two symmetrical left and right channels (passages) that are separated by the cartilaginous, bony, nasal septum. These parallel passages have distinct regions beginning with the vestibular area (nostrils), leading via the internal ostium (nasal valve) to the atrium (narrow transitional area), then to the main respiratory region (comprised of the nasal turbinates and olfactory region) and finally the nasopharynx where the channels converge (Figure 19.1) [1,2].

The lining of the vestibule, which contains a zone of short stiff hairs (vibrissae), acts as a baffle system for filtering out large (>10 μm) airborne particles. The stratified squamous epithelium of the vestibule is resistant to dehydration and inhaled irritants and poorly permeable. Between the vestibule and atrium, there is a constriction known as the internal

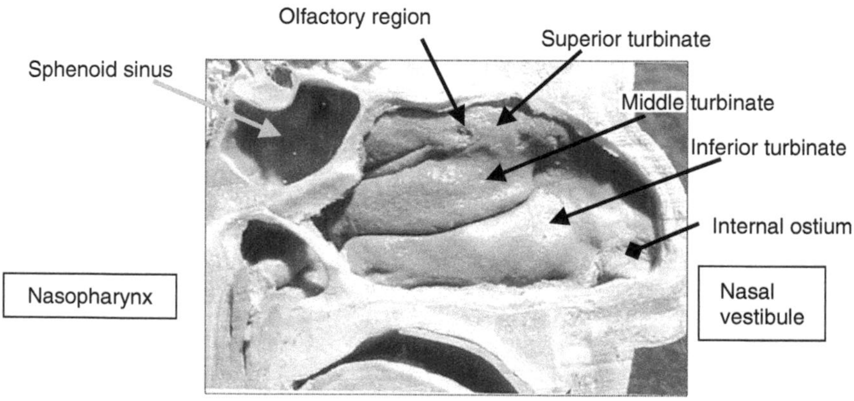

FIGURE 19.1 Lateral section of the nasal cavity, showing the extensive respiratory region (the turbinates) and the olfactory region. (From Jones, N., *Adv. Drug Deliv. Rev.*, 51, 5, 2001.)

ostium or nasal valve, which is the narrowest passageway of the entire airway (Figure 19.1). The incoming airstream is squeezed through this narrow passage, through the atrium, then curves into the main nasal passage—the respiratory region.

The respiratory region is formed of highly convoluted chambers (turbinates), which stretch from the atrium to the beginning of the nasopharynx. This region forms the middle and posterior thirds of the nasal cavity, is approximately 5–8 cm long, and is divided laterally into the superior, middle, and inferior turbinates. The olfactory region lies in this region and is situated on the roof of the nasal passage, accounting for 3%–5% of the surface area of the entire nasal cavity in humans. At a distance approximately 12–14 cm from the nostrils, the septum ends and the airways merge into the nasopharynx (Figure 19.1).

A primary function of the nasal cavity is olfaction. The location of the olfactory mucosa, recessed on the roof of the respiratory passage, means that it is accessed only by diffusion of volatile or airborne substances in inhaled air. The specialized epithelium features sensitive olfactory neurons that provide a direct link to the CNS. The structural features of the epithelium and potential for drug delivery are considered in detail in Section 19.6.1.

As normal breathing occurs via the nose, the nose also conditions air for inhalation for the lung. The anatomy of the nose permits intimate contact between inspired air and the mucosal surface, enabling air to be warmed and humidified by the rich blood flow and secretions of the epithelium. Air is also filtered in the nasal cavity, which is one of the most important physiological defensive mechanisms of the respiratory tract. It protects the body against any inhaled noxious materials or airborne contaminants such as dust, microorganisms, and allergens. The efficiency of particle removal is dependent upon a number of factors including the aerodynamic diameter of the inhaled particles. For example, as previously mentioned, particles greater than 10 μm are generally filtered out by the vibrissae at the nostrils, smaller particles (approximately 5–10 μm) are deposited in the nasal passages and subsequently cleared by the process of mucociliary clearance. Particles less that 2 μm are not normally filtered out and may enter the lungs.

19.2.2 Blood Supply

The nasal septum is supplied posteriorly by the sphenopalatine artery, superiorly by the anterior ethmoid artery, and antero-inferiorly by the superior labial branch of the facial artery. In addition, the inferior and lateral walls of the nasal cavity are supplied by the palatine and ethmoid artery, respectively, whereas the remainder of the blood supply to the nasal cavity comes from the sphenopalatine artery.

The direction of arterial blood flow in the nose runs anteriorly against inspiration. Blood vessels are arranged in such a manner as to provide an erectile capacity to the mucosa enabling the airway to widen and narrow. Blood flow through the autonomically controlled vasculature of the nasal tissue is of importance in the conditioning of inspired air.

19.2.3 Respiratory Epithelium

The nasal epithelium undergoes a transition in the atrium from the squamous epithelium of the vestibule to the more permeable epithelium of the respiratory region. The nonolfactory respiratory region is covered by typical airway epithelium, similar to that of the lungs. Basal, goblet, ciliated, and nonciliated cells form a pseudostratified, columnar, ciliated, highly vascular mucosa (Figure 19.2) [3].

The airway surface is formed by columnar ciliated and nonciliated cells interspersed with mucus-secreting goblet cells. The surface area of the epithelium is increased by the presence of numerous microvilli. The epithelial cells are held together at their apical surface by tight

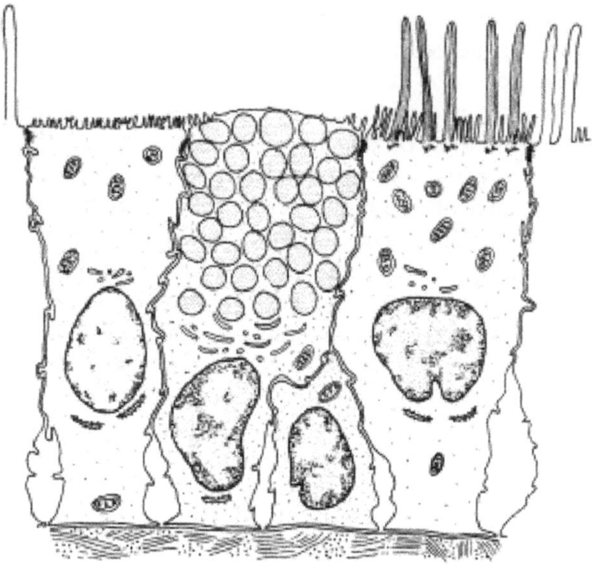

FIGURE 19.2 Diagrammatic representation of the cell types of the nasal mucosa as seen by transmission electron microscopy showing (I) nonciliated columnar cell, (II) goblet cell with mucus granules, (III) basal cell, and (IV) ciliated columnar cell. The epithelium is covered by a mucus layer 5–10 μm thick (not shown). (From Mygind, N. and Dahl, R., *Adv. Drug Deliv. Rev.*, 29, 3, 1998.)

junctions, which act as a physical barrier against the ingress of foreign material. The dimensions of these tight junctions provide a size-dependent barrier to the passage of larger compounds between adjacent cells (paracellular transport). Basal cells are located at the base of the epithelium and are precursors of columnar cells (ciliated and nonciliated) and goblet cells. The goblet cells are most dense in the posterior area of the nasal cavity and contain secretory granules in which dehydrated mucus glycoprotein is stored. The epithelium rests upon a layer of connective tissue fibrils called the basement membrane, underneath which is the submucosa (lamina propria), which contains a dense network of fenestrated capillaries that supply the tissue with a large volume of blood [4]. Glands lined by mucus and serous secretory cells are located in the submucosa with ducts leading to openings at the epithelial surface.

19.2.4 Mucus and Cilia

Nasal secretions consist mainly of heterogeneous secretory products originating from the goblet cells and submucosal glands, with additional contributions from lacrimal fluid and the vascular system. Together these secretions form a distinct two layers of airway lining fluid consisting of (i) a low-viscosity periciliary fluid layer bathing the epithelial cell surface and the cilia and (ii) a continuous mucous gel blanket overlying the periciliary fluid at the air interface. The periciliary fluid is a watery, ionic solution, maintained by transepithelial ion transport, which provides an environment within which the cilia are able to beat and perform their mucus clearing function [5]. The pH of periciliary fluid is 5.5–6.5 and any disruption to this can affect drug absorption, metabolism, toxicity, and physiological defense mechanisms.

The viscous mucus layer provides a physical protective barrier for the mucosa; trapping, binding, and clearing inspired particles. Approximately 1.5–2 L of mucus is secreted daily and the resultant gel layer varies between 5 and 10 μm in thickness. Mucus is composed of 95% water plus mucin glycoproteins (0.5%–5%), lipids in low proportions, mineral salts (1%), and free proteins (0.5%–1%). The protein content of mucus includes immunoglobulins, lysozyme, lactoferrin, and other enzymes with the exact composition varying [6].

The secretory mucins are the major functional constituents of mucus [7]. Respiratory mucins are high molecular weight glycoproteins ($2–40 \times 10^6$ Da) that comprise long, linear, apparently flexible threads, which vary in length from 0.5 to 10 μm. They are composed of subunits (monomers) each of about 500 nm in length, joined end to end via disulfide bonds. A characteristic of mucin is the high carbohydrate content of the molecule (70%–90% by weight), occurring as neutral and acidic oligosaccharide units. These units are assembled from varying numbers of monosaccharides arranged in linear as well as branched sequences. Sugars typically present in mucins are *N*-acetylglucosamine, *N*-acetylgalactosamine, galactose, fucose, and sialic acids. Characteristically, mucins lack uronic acids (which are found in proteoglycans) and mannose (which is a principal sugar in asparagine-linked oligosaccharides of serum). The sugar chains are attached to the central protein core through *N*-acetylgalactosamine residues via "O-glycosidic" linkage to serine or threonine.

Mucus is cleared continuously by cilia, which extend through the periciliary fluid to hook the mucus layer. The claw-like tips of the cilia engage with the mucus layer and transport the latter toward the nasopharynx [8,9]. Cilia are complex motile structures that extend from the surface of columnar ciliated cells; the number of cilia per cell is approximately 200 with a cell density of 6–8 cilia per μm^2. These hair-like protrusions range in length between 5 and 10 μm and width from 0.1 to 0.3 μm.

A typical cross section of a cilium shows a ring formed by nine pairs of microtubules and two central tubules, i.e., the so-called nine + two pattern. Each doublet contains an A and a B subfibril with an inner and an outer dynein arm (a complex protein with ATPase activity) located on the A subfibril with radial spokes extending toward the central doublet. The ciliary membrane, which is an extension from the cell membrane of the epithelial cell, encloses the microtubules. The motion of the cilia is dependent on the sliding of the outer doublets past one another with the energy provided by adenosine triphosphate (ATP) through dynein ATPase activity.

The coordination of cilia motility is controlled by neural innervation, chemical pacemaking, and hormonal stimulation, and the effects of ions such as calcium and potassium. Cilia beat in a coordinated fashion to achieve the unidirectional propulsion of mucus with the frequency of the ciliary beat, which is dependent upon the environment. Human nasal cilia have been reported to beat, *in vitro*, with an average frequency of 10 Hz. Any rheological abnormalities in the mucus gel, particularly those that alter the elastic properties of mucus, can greatly affect clearance and undermine ciliary activity in the nasal cavity. In such circumstances, the adhesiveness of the mucus layer may allow it to remain in contact with the underlying cell layers and retain any substances introduced into the nasal cavity.

19.3 PHYSIOLOGICAL BARRIERS TO NASAL DRUG DELIVERY

The physiological barriers to drug absorption from the nasal cavity include losses during drug delivery, mucus permeability, epithelial permeability, drug clearance, degradation, health, and the environment. The relative importance of these barriers is different for small and large molecular weight molecules (Table 19.1). These barriers are considered below.

19.3.1 Delivery to the Nasal Cavity

Administration of drug to the therapeutic target site or absorption site in the nasal cavity is the first step in nasal drug delivery. Uniform distribution over the affected area is desirable for the treatment of nasal symptoms, whereas targeting of the favorable absorption sites

TABLE 19.1
Relative Contribution of Physiological Barriers to the Absorption of Drugs via the Nasal Cavity

Barrier	Small Molecules (% loss)	Large Molecules (% loss)
Degradation	0–15	0–5
Clearance[a]	0–30	20–50
Deposition (anterior loss)	10–20	10–20
Health and environment	10–20	10–40
Epithelial permeability[a,b]	0–30	20–50
Mucus layer	<1	<1

Source: Gizurarson, S., *Adv. Drug Deliv. Rev.*, 11, 329, 1993.
[a]Depends on excipients.
[b]Depends on physicochemical characteristics of the drug.

within the nasal passages is the aim for systemic delivery. Targeting the comparatively small inaccessible olfactory region provides a challenge if nose-to-brain delivery is the objective.

Constraints on nasal delivery include a restricted capacity (i.e., the volume of formulation that can be delivered is limited), attainment of dose accuracy, and the reproducibility of delivery. The distribution of drug in the nasal cavity is highly dependent on the delivery device, formulation, and administration technique. In turn, the distribution of drug will affect permeability, residence time, and metabolism in the nasal cavity.

19.3.2 Permeability

Systemic drug delivery by the nasal route offers many potential advantages over some other routes, including rapid absorption and high systemic availability. As with most sites of drug absorption, the availability of a drug is affected by the area available for absorption, blood flow, and the intrinsic permeability of the epithelium. Other important considerations include contact time between the drug and the absorption site (Section 19.3.3), metabolism of the drug prior to and during absorption (Section 19.3.4), and pathology of the absorbing tissue (Section 19.6.2).

The total surface area of the nasal cavity is about 150 cm^2, with the area available for absorption enhanced by the convolutions of the turbinates and the presence of microvilli on the surface of the ciliated and unciliated cells of the respiratory epithelium. The arterial supply of the nose is particularly rich in the respiratory epithelium where the Kiesselbach's plexus lies, an area that is rich in numerous capillary loops. The nasal blood flow has been shown to be sensitive to the action of a variety of inhaled compounds, both locally or systemically acting. Clonidine has been shown to decrease the blood flow whereas histamine and phenylephrine have been shown to induce the converse effect. Such direct changes to blood flow are important in determining the rate and extent of drug absorption from the nasal cavity.

The permeability barrier in the respiratory region of the nose includes the mucus layer and the airway cells. Mucus potentially affects drug delivery by acting as a barrier to diffusion. *In vitro* studies have indicated that the presence of mucus has the potential to retard the transport of many compounds, although mechanisms are complex and the impact of mucus is difficult to predict.

In an evaluation of the diffusion of drugs across the gastrointestinal mucus, the most important physicochemical characteristic influencing the diffusion of most species was lipophilicity, whereas molecular size had more influence for larger peptide drugs [10]. Binding of drugs to the mucin is important for positively charged drugs such as amikacin, gentamicin, and β-lactam antibiotics, which bind electrostatically to the negatively charged components in mucus. When the diffusion of a range of β-lactam or aminoglycoside antibiotics through rat intestinal mucus was evaluated, cephaloridine and gentamicin were found to be significantly bound and the degree of binding was found to be dependent upon the pH and ionic strength of the mucus [11].

Interestingly, the use of mucolytic drugs, which alter the viscoelasticity of mucus, has been shown to increase the absorption of intranasally administered human growth hormone (22 kDa). In contrast, other studies have shown that antibodies (150–970 kDa) are able to diffuse through cervical mucus relatively unimpeded, a finding that suggests that the diffusion barrier to antibodies presented by mucus in the nasal cavity might be relatively minor [12].

Overall, it appears that poorly understood interactions (ionic, osmotic, hydrophobic) between the mucus components and drug compounds are likely to be more important in determining the magnitude of the mucus permeability barrier than the gel structure of the mucus alone. As nasally administered drugs are retained in contact with the nasal mucosa for a relatively short period of time, any retardation of diffusion by mucus might be expected to lead to marked changes in bioavailability, although the barrier presented by mucus is suggested to be proportionately low (see Table 19.1).

The nasal epithelium possesses selective absorption characteristics similar to those of a semipermeable membrane, i.e., it allows a rapid passage of some compounds while preventing the passage of others. The process of transportation across the nasal mucosa involves either passive diffusion, via paracellular or transcellular mechanisms, or occurs via active processes mediated by membrane-bound carriers or membrane-derived vesicles involving endo- or transcytosis.

Hydrophilic drug molecules are predominantly absorbed by paracellular absorption through the tight junctions between adjacent epithelial cells. Diffusion is driven by a concentration gradient along the aqueous pathway of the intercellular space of the cells and restricted by tight junctions at the apical surface of the cells such that this route is limited mainly to small hydrophilic molecules. This route is dependent upon the molecular weight and, although slower than transcellular absorption, can be fast enough to give a high systemic availability for low molecular weight polar compounds (<1000 Da).

The effect of molecular size has been demonstrated in a systematic study of a wide range of drugs with molecular weights varying from 160 to 34,000 Da [13]. The results indicated that nasal absorption decreases exponentially as a function of increasing molecular weight. The rate-limiting molecular weight cut off for absorption from the nasal cavity was found to be 1000 Da compared to 300 Da for the oral route. Similarly, a range of different-sized polyethylene glycols were well absorbed from the nasal route up to molecular weights of 2000 Da, after which permeability enhancers were required to improve bioavailability [14]. A comparison of the nasal absorption of uncharged and cationic dextrans in rabbits found that, as expected, the plasma level of the uncharged dextran decreased as its molecular weight increased. In contrast, the plasma level of the positively charged dextran increased up to a molecular weight of 9000 Da, only decreasing when the molecular weight rose further to 17,200 Da [15]. It was speculated that the increasing molecular weight of the FITC DEAE dextran facilitated an interaction with the negatively charged components of mucus and nasal epithelial cells (e.g., sialic acid residues), which enhanced permeability. This encouraged the view that relatively large molecules could be absorbed.

Lipophilic drug molecules are absorbed across the nasal epithelium by passive transcellular diffusion. For small, unionized molecules, this provides a rapid efficient transport mechanism, often resulting in plasma concentration profiles resembling that of intravenous injection and bioavailabilities of up to 100%.

Certain ionized and hydrophilic molecules that are unable to partition into the hydrophobic environment of the membrane lipids may utilize the facilitated transport mechanisms that are present for ions and small molecules, such as glucose. For example, carrier-mediated transport can be an absorption mechanism for peptides and amino acids. The amino acid L-tyrosine is absorbed through the nasal cavity via a carrier-mediated process, which can be increased by esterification of its carboxyl group [16].

In summary, nasal epithelial intercellular junctions are less restrictive compared to the gastrointestinal tract. Such polar pathways will mainly be responsible for the transport of water-soluble compounds, providing a relatively slow, but significant route which is dependent on the molecular weight of the diffusing species. Secondly, transcellular (lipoidal) pathways permit extremely rapid absorption of lipophilic drugs with a rate dependency based on cell membrane partitioning.

19.3.3 Mucociliary Clearance

The cilia, through coordinated beating, transport the overlying mucus layer and upper surface of the periciliary fluid in a unidirectional manner, whereas the periciliary fluid closest to the cell surface is moved back and forth [8,9]. The rate of mucociliary clearance from the nasal cavity to the nasopharynx is highly variable in different regions and under different environmental conditions. Mucociliary activity is regulated by several factors such as temperature, intracellular Ca^{2+} and cAMP levels, and by extracellular ATP. Particles are transported within the viscoelastic mucus blanket at uniform rates irrespective of their viscosity, size, density, or composition. Even large particles, with diameters up to 500 μm, are expelled from the nasal cavity within 10 to 20 min.

Major determinants of the efficiency of mucociliary clearance are cilia density, periciliary fluid, and composition of mucus. Some drugs and excipients, such as preservatives in drug formulations, may diminish the ciliary movement in the nasal cavity and trachea. A suggested adverse effect of ciliostasis (permanently or momentarily arrest or impairment of ciliary activity) is lower respiratory tract infection as a result of impaired nasal microbiological defense.

19.3.4 Metabolism

The nasal epithelium possesses a defensive enzymatic barrier including the presence of phase I and phase II enzymes [17], plus proteolytic enzymes that provide a formidable barrier to the nasal delivery of drugs. Therefore, although hepatic first-pass metabolism that occurs after oral absorption is avoided, the suggestion made over the years that the nasal mucosa might be largely deficient in enzyme activity is now recognized as misguided.

The influence of enzymatic degradation is obviously dependent on the susceptibility of the drug itself. For example, progesterone and propanolol are absorbed rapidly from the nose into the blood without undergoing degradation, but nafarelin acetate has poor systemic availability, perhaps as a result of enzymatic degradation [18–20]. Drug degradation in the nasal mucosa is important not only when considering the nasal delivery of drugs, but also due to the toxicological implications as a consequence of enzymatic transformation of inhaled environmental pollutants or other volatile chemicals. The phase I cytochrome P-450 enzymes can convert some airborne chemicals to reactive metabolites, which may be involved in forming deoxyribonucleic acid (DNA) adducts, increasing the risk of carcinogenesis in the nasopharynx. The cytochrome P-450 activity in the olfactory region of the nasal epithelium is higher even than in the liver, mainly because of a three- to fourfold higher NADPH-cytochrome P-450 reductase content [21].

The relatively low systemic availability of peptides from the nasal cavity led to the study of proteolytic activity in both *in vivo* and *in situ* systems [22,23]. The presence of both endo- and exopeptidases was found including aminopeptidases A, B, and N, leucine aminopeptidase, and microsomal aminopeptidase with up to 85% of the enkephalin hydrolysis in nasal mucosa homogenate being due to aminopeptidases. The presence of diaminopeptidase, postprolyl cleaving enzyme, angiotensin-converting enzyme, and endopeptidases has also been indicated [24]. This high enzyme activity may explain why enhancing the lipophilicity of peptidase-labile peptides does not have any great effect on bioavailability from the nasal cavity [18]. The increase in lipophilicity could simply increase the partitioning of the peptides into areas of higher enzyme activity where they are metabolized.

The antibacterial enzyme lysozyme is also found in nasal secretions. Lysozyme is produced by the epithelium and mucus glands where it can attack the cell walls of susceptible micro-organisms, its action being optimal at the slightly acidic microclimate pH. The pH of nasal mucus varies with age, sleep, rest, emotion, infection, and diet. When it is cold, or during rhinitis or sinusitis, the pH tends to be alkaline, which deactivates the lysozyme in mucus and therefore increases the risk of microbial infection. Under normal conditions, the nasal secretions, as indicated earlier, have a pH of 5.5 to 6.5, which is the optimum pH for the activity of lysozyme.

19.3.5 Toxicity

In chronic administration to the nasal cavity, or in the treatment of conditions where normal resistance to injury is impaired, concerns about biocompatibility are pertinent. This is particularly important if the use of absorption enhancers is being considered.

Direct toxic effects such as irritation, inflammation, or increased permeability will bring about symptoms such as the sneeze reflex, nasal discomfort, and hypersecretion with the possibility of underlying pathological changes such as squamous metaplasia, cilia erosion, plasma exudation, epithelial necrosis, inflammatory remodeling, or neutrophil accumulation. Indirect adverse effects can also occur and any alteration to normal nasal homeostasis should be avoided. For example, a reduction in mucociliary clearance can cause rhinitis, sinusitis, and an increased susceptibility to airway infections, and consequently ciliary movement should not be altered by any nasal medication. In the context of absorption enhancers, the rate and extent of recovery of normal nasal epithelial function after nasal administration is a prime consideration.

19.4 STRATEGIES FOR OVERCOMING NASAL DELIVERY BARRIERS

19.4.1 Delivery Devices

Delivery devices have a profound impact on drug deposition, for example the tendency for anterior versus more uniform distribution achieved by nasal sprays and solutions, respectively. The simple presentation as nasal drops is simple, economic, and convenient, but it is likely that more sophisticated presentations will be required for many compounds in development. At present typical delivery devices include solutions, nasal sprays (solutions and suspensions), gels, and powders. While the turbinate region is often regarded as the optimal target site for deposition of drugs for systemic absorption on account of the surface area and permeability of the epithelium, it is also a region of rapid mucociliary clearance. Hence the greater anterior deposition resulting from nasal sprays actually results in greater bioavailability on account of the extended residence time.

Current innovations in delivery devices are aimed at improving dose precision, avoiding the requirement for potentially toxic preservatives, enhancing ergonomy and patient usability, and incorporating specialized microsphere/polymer formulations [25].

19.4.2 Molecular Modification

Molecular modification has been advocated as a method for overcoming the barriers posed by the nasal mucosal lining and the enzymatic activity of the nasal epithelium [26]. Two main strategies have been outlined: (i) minimizing degradation through chemical derivitization or covalent attachment to a polymeric carrier and (ii) absorption enhancement through prodrug design or modification of the active molecule.

Classical prodrug strategies such as readily hydrolysable ester compounds have been tested experimentally with some success. For example, the enhanced absorption of ester prodrugs of acyclovir has been reported following modifications that conferred both resistance to degradation and an increased lipophilicity to the parent compound [27]. A limitation of this approach was the level of carboxyesterase activity present in the nasal mucosa (higher than that of the lungs), which resulted in rapid presystemic cleavage of the prodrug ester linkage before the enhanced absorption effect could occur.

In the modification of peptide such as enkephalins for nasal delivery, it is not surprising that the protection of the N-terminal group from the action of aminopeptidases is of prime importance considering the abundance of these peptidases in the nasal epithelium. Protection from C-terminal degradation and an increased lipophilicity are also valuable strategies if these can be achieved without extensively sacrificing aqueous solubility or unduly increasing molecular size or conformational rigidity.

19.4.3 Formulation Approaches

As a result of the physiological constraints outlined, a number of formulation strategies have been devised to optimize or improve nasal delivery of drugs (Figure 19.3).

Bioadhesive formulations and microsphere delivery systems in particular have attracted much attention. As drug formulations are usually rapidly removed from the site of deposition by the mucociliary clearance, increasing the retention time of drug in the nasal cavity via bioadhesion can increase bioavailability [28]. Bioadhesion may be defined as the ability of a material (synthetic or biological) to adhere to a biological tissue for an extended period of time. When applied to a mucous membrane, a bioadhesive polymer may adhere primarily to the mucus layer or epithelial cell surface in a phenomenon known as mucoadhesion [29,30]. The bioadhesive properties of a wide range of materials have been evaluated over the last decade.

Theoretically, bioadhesive dosage forms can provide a platform for sustained/optimized drug delivery that can: (i) localize drug at a specific site for absorption, thereby enhancing bioavailability, (ii) produce intimate contact between the dosage form and the mucosa, improving absorption and allowing for localized modification of tissue permeability, and (iii) prolong residence time, resulting in a delivery system that improves bioavailability or achieves sustained release dosing.

Powdered bioadhesive formulations have attracted much interest for their ability to overcome mucociliary clearance in the nasal cavity. For example, powder formulations of freeze-dried mixtures of insulin and excipients such as crystalline cellulose, hydroxypropyl cellulose, or carbopol 934 have been shown to enhance the absorption of insulin in dogs from the nasal cavity compared to freeze-dried mixtures of insulin and lactose as an excipient [28,31]. The improved delivery was attributed to both the improved dispersion of insulin and the high

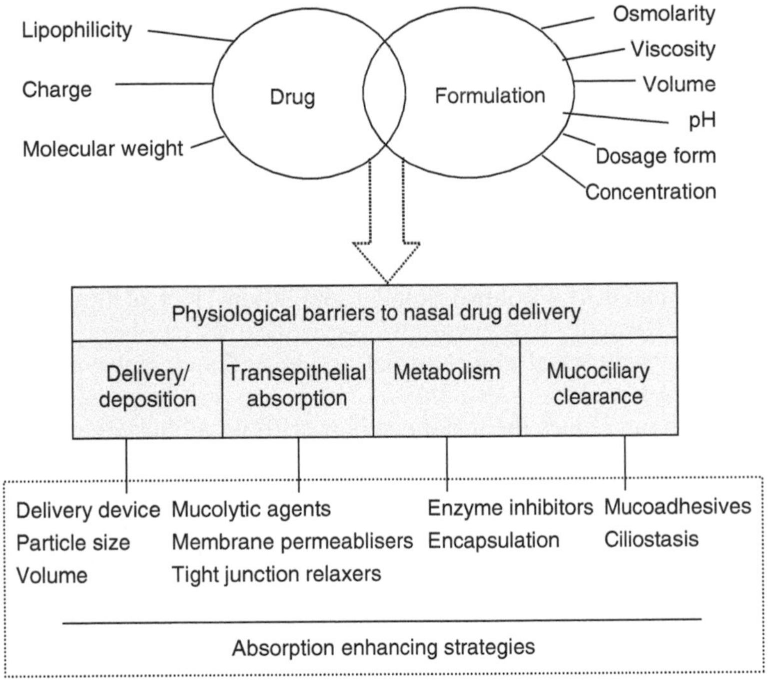

FIGURE 19.3 Drugs for administration via the nasal route have specific formulation requirements which, depending on their ability to overcome physiological barriers, may require an absorption-enhancing strategy.

viscosity of the formulation impairing mucociliary clearance. Similarly, the addition of methyl cellulose to nasal formulations also increases retention time [32]. A bioadhesive liposome formulation has also been shown to provide a therapeutic concentration of nifedipine for prolonged periods of time when delivered via the nose [33]. The administration of the liposomal formulations produced a rapid increase in plasma levels of the drug for an initial 1 to 2 h period followed by a constant plasma level for another 8 to 10 h. In comparison, a simple nifedipine solution achieved a high plasma level followed by a rapid decline.

Mucoadhesive microspheres of hyaluronic acid and chitosan for nasal delivery have been prepared and tested for the delivery of a model drug, gentamicin [34–36]. Hyaluronic acid and mixed hyaluronic acid–chitosan microspheres prolonged the absorption and increased the bioavailability of gentamicin *in vivo* compared to a control solution of the drug. This enhancement was attributed to the bioadhesiveness of hyaluronic acid and the combined bioadhesive and penetration enhancing properties of chitosan. When microsphere formulations were prepared using a combination of the polymers, gentamicin bioavailability increased by twofold compared to administration of the hyaluronic acid formulation alone and 40-fold compared to administration of a solution of the drug.

19.4.4 Absorption Enhancers

Penetration enhancers have been used to facilitate the absorption of higher molecular weight molecules. The mode of action of the surfactant enhancers is often attributed to membrane damage [37]. However, studies in epithelial cell monolayers suggest that some surfactant-based absorption enhancers act primarily by increasing the permeability of tight junctions [38]. Nevertheless, except for the chelators and nonsurfactants, which exert their influence almost entirely by making the tight junctions between cells more accessible, most penetration enhancers act by increasing membrane fluidity and disturbing the integrity of the membrane to some extent. Thus a concern is that long-term use of enhancers will cause permanent damage to cell membranes and cilia, especially if used in a formulation to treat chronic diseases.

Another type of absorption enhancer, which has been shown to have a better safety profile, is cyclodextrin (CD) [39]. CDs have been shown to form inclusion complexes with lipophilic drugs, thereby improving their aqueous solubility and stability. A powdered insulin formulation containing dimethyl-β-cyclodextrin improved the absolute bioavailability of insulin by 13% in rabbits compared to a control liquid formulation (1%) of insulin with dimethyl-β-cyclodextrin [40]. Recently, hydroxypropyl β-cyclodextrin has been shown to be more effective for enhancing the nasal absorption of acyclovir than a range of other absorption enhancers *in vivo* [41].

Enzyme inhibitors can reduce the metabolic barrier to nasal delivery. The selection of an inhibitor is made on the basis of its ability to inhibit effectively the enzyme primarily responsible for the degradation of a particular compound. The coadministration of peptidase and protease inhibitors such as bacitracin, bestatin, amastatin, and aminoboronic acid derivatives has been found to promote the absorption of LHRH and growth hormone [42,43]. Aminopeptidase inhibitors in particular are effective in improving the bioavailability of enkephalins [44].

19.5 EXPERIMENTAL MODELS OF PHYSIOLOGICAL BARRIERS TO NASAL DRUG ABSORPTION

19.5.1 Nasal Epithelial Models

Drug permeability, metabolism, and toxicity can be evaluated *in vitro* using models of the nasal epithelial in preparation for *in vivo* experiments. Human nasal epithelial cell cultures and animal nasal mucosa mounted in Ussing chambers provide convenient, simple systems in which drug targeting and absorption mechanisms can be investigated under defined, controlled conditions [45].

Primary culture of human nasal epithelial cells can provide suitable epithelial models for the study of drug permeability and metabolism. Healthy human cells from the nonolfactory respiratory region are used to establish the cultures with cells obtained from surgical samples after medical procedures or postmortem biopsies. Nontraumatic procedures such as nasal brushings/scrapings can also be used to harvest cells and have the advantage of being repeatable, although limited numbers of cells are obtained. After harvesting it is necessary to culture the cells to form confluent layers of nasal epithelial cells, using optimized culture conditions to ensure maximal differentiation and the formation of appropriate barrier properties to model the nasal respiratory epithelium. Limitations of primary cultures include the restricted source of cells, interculture variation, and the necessity to establish and verify optimal culture conditions according to experimental requirements.

There are a number of nasal and lung airway cell lines that can be cultured as epithelial cell layers *in vitro*. The RPMI 2650 cell line is of human nasal septal origin and has been advocated as a suitable cell line for use in permeability and metabolism studies [45]. However, limited formation of barrier function by cell layers limits the usefulness of this cell line for drug transport applications. The 16HBE14o- and Calu-3 cell lines of human tracheobronchial origin can also be used to model the nasal epithelium on the basis of the similarity of the airway epithelium in lung and the nose in terms of cell type and morphology. Both these cell lines form suitable cell layers for the investigation of drug delivery applications and are finding increased use for these purposes [46]. It must be remembered, however, that cell lines do not provide a physiological representation of the mixed cell population of the nasal epithelium and, although there are advantages for studying the permeability in cell layers with tight intercellular barrier properties, the nasal epithelium is less restrictive.

Nasal tissue from animals can be mounted in Ussing chambers, permitting experiments similar to those performed in cell cultures to be performed in intact tissues. The nasal tissues of rabbits, dog, sheep, and cattle have been used and such experiments provide the reassurance that the *ex vivo* system is representative of the nasal mucosa *in vivo*. The limitations of this technique are the requirement for the use of fresh tissue, the limited duration of tissue viability, and interspecies variation in tissue permeability and metabolic capacity.

19.5.2 Mucociliary Clearance Measurement

It is possible to study cilia beat frequency and drug interaction with mucus independently using models such as tissue explants or cultures of ciliated cells or purified mucus preparations. However, these are not able to provide information about mucociliary transport rates. For such studies, a model incorporating the integrated cilia and mucus components is required such as the frog palate. Drugs, preservatives and absorption enhancers, and bioadhesive formulations have been extensively studied in this model [9].

The excised frog palate model is a simple preparation and the most commonly used model to investigate the role of mucus rheology on mucociliary transport. The mucosa is similar to mammalian mucosa with regard to histology, morphology, and function (Figure 19.4). Nasal drug formulations and their components can be evaluated by direct application to the mucus layer of this model (mucus nondepleted). Alternatively, the effect of nasal gels and rheologically modified mucus can be studied by the application to the mucosa after removal of the native mucus (mucus-depleted model). Extrapolation of mucociliary transport rates to clearance *in vivo* is complicated by large intra- and interindividual variations in clearance rates, methodological differences, and higher sensitivity of the *in vitro* technique. Thus, the major

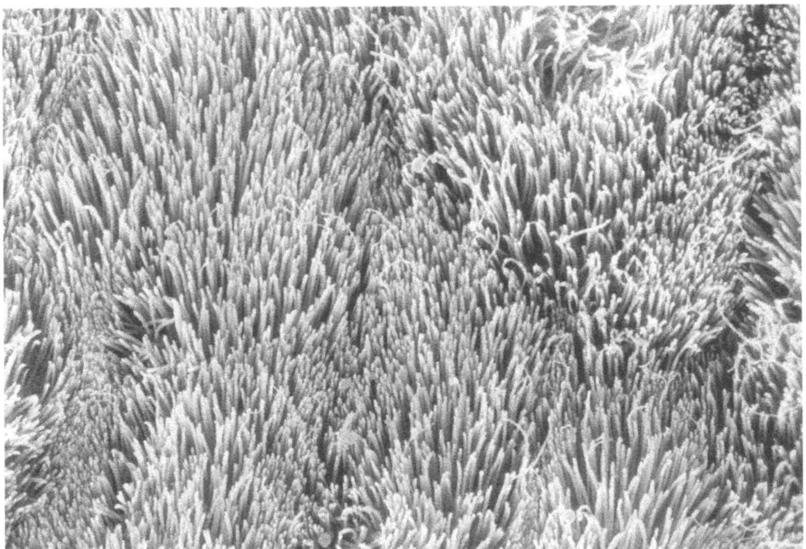

FIGURE 19.4 Scanning electron micrograph of the mucus-depleted ciliated frog palate. The cilia appear in clumps corresponding to the shape of the epithelial cells. (Micrograph provided by Kelly Pritchard.)

application of the technique is screening compounds and formulations for effects on mucociliary mechanisms, rather than prediction of effects *in vivo*.

Mucociliary clearance can be studied *in vivo* in humans using gamma scintigraphy to follow the clearance of radiolabeled solutions from the nasal cavity or by measuring the transport rate of radiolabeled markers administered to the nasal epithelium. Simpler methods are used to monitor the appearance of a strongly colored dye (by visual inspection) or sweet tasting substance such as saccharin (by taste) at the pharyngeal cavity.

19.5.3 *In Situ* Perfused Nasal Cavity

The *in situ* perfused nasal cavity model allows the study of absorption from the intact nasal cavity [47]. In this preparation, drug-containing fluid is introduced into the nasal cavity at the nasopharynx and exits through the nostrils. The fluid may then be recirculated or the model can be operated as a single pass system depending on experimental requirements and drug disappearance is monitored to establish losses to absorption. The benefit of this model in providing absorption data from the intact nasal cavity must be weighed against the limitations, i.e., that perfusion provides a nonphysiological presentation of the drug, which can falsely prolong residence times and provide enhanced absorption or toxicity. In addition, it is not possible to evaluate the effects of formulations such as powders in this model.

19.5.4 *In Vivo* Models

Ultimately, new nasal delivery systems must be tested using *in vivo* models in which it is possible to perform full pharmacokinetic and pharmacodynamic studies. The selection of an appropriate animal species is important and variation in nasal cavity anatomy, mucosal permeability, and metabolic capacity must be considered. Although rats can be used, rabbits are generally accepted as being the best small animal model as they have larger nasal cavities, facilitating drug administration and have sufficient blood volume to permit repeat sampling. If possible, *in vivo* protocols that require anesthesia should be avoided as many anesthetic agents inhibit mucociliary clearance.

19.6 SPECIAL CONSIDERATIONS IN NASAL DELIVERY

19.6.1 Nose–Brain Delivery

The olfactory region located in the poorly accessible recessed roof of the nasal passages offers the potential for certain compounds to circumvent the blood–brain barrier and enter into the brain [48]. The olfactory sensory cells are in contact with both the nasal cavity and the CNS and this neuronal connection constitutes a direct pathway to the brain. By utilizing this pathway drugs would not only circumvent the blood–brain barrier, but also avoid any hepatic first-pass effect and degradation in the blood compartment, a particularly important issue in the case of peptide drugs.

It has been suggested, with the support of experiments conducted in animals, that the rapid euphoria experienced after nasal administration of cocaine is the result of direct absorption to the CNS. Studies in man have also demonstrated a centrally mediated effect of nasally administered drug that was not seen with intravenous injection [48].

The specialized olfactory mucosa features olfactory sensory neurons, which span the nose–brain barrier. At the epithelial surface nonmotile cilia extend from swellings, which lead via a dendritic extension to the neuronal cell body, from which an extended axon penetrates the cribriform plate of the skull to synapse in the olfactory bulb. The sensory neurons are supported by sustentacular epithelial cells.

Theoretically, substances can be absorbed via the olfactory route by two different mechanisms. Firstly, endocytotic or pinocytotic uptake followed by transport along the axon of the olfactory sensory cells could lead to drug delivery to the olfactory bulb, followed by diffusion to the rest of the brain. This route of transport is relatively slow (hours) and probably not relevant in terms of drug administration. The second route of transport involves paracellular or transcellular transport across the sustentacular cells to the basolateral side of the olfactory epithelium. Diffusion through the perineural space around the sensory nerve axons as they penetrate the cribriform plate into the subarachnoid space completes the nose to brain absorption. This route enables relatively quick (minutes) absorption to the cerebrospinal fluid of hydrophilic and semilipophilic substances and is considered the most likely route of olfactory drug absorption.

Lipophilic drugs are not candidates for this route on account of their rapid absorption to the systemic circulation from the nasal cavity and favorable properties for penetration of the blood–brain barrier. Rather, it is large molecular weight or polar drugs for diseases such as Parkinson's disease and Alzheimer's disease that are relevant for this route. While the present discussion may suggest a clear possibility of nasal administration and subsequent absorption of drug substance directly to the CSF, clinically relevant drug delivery to the central brain via olfactory absorption remains a challenge. Many questions can be raised regarding the feasibility of "nose to brain" absorption given the transport mechanisms involved. The obstacles in targeting absorption via the olfactory pathway and distribution to the central brain may be too great to achieve therapeutic drug concentrations, but this area obviously warrants further investigation in the future.

19.6.2 Disease

Acute or chronic disease can compound the interindividual variations already discussed that contribute to unpredictable nasal drug delivery. For example, respiratory infection may alter respiratory secretion and a number of conditions can lead to mucociliary dysfunction, both of which can greatly affect the clearance rate of nasally administered drugs. Predicting the effects of disease is not straightforward, for example an excessive production of mucus coupled with a reduced capacity for mucus clearance would concomitantly enhance the mucus barrier to drug absorption but increase residence time. Similarly, inflammation may increase epithelial permeability, while nasal hypersecretion would result in dilution and possible changes in the pH. Such effects could result in either enhanced or reduced efficiency of nasal drug delivery.

Chronic diseases include cystic fibrosis in which nasal mucus is thick and viscous as a result of abnormal chloride transport across the membrane of the epithelial cells, leading to reduced water secretion. Similarly, chronic sinusitis also reduces nasal mucociliary clearance due to an increase in the rheological properties of mucus.

Infection and allergy are acute conditions that can affect the nasal cavity. Bacterial infection stimulates secretory cells to synthesize and discharge mucus at a faster rate and can induce an increase in the number of goblet cells. Certain bacteria such as *Streptococcus pneumoniae* and *Streptococcus aureus* produce toxins that disrupt epithelial cells leading to an increase in permeability and damage to the mucociliary clearance system. Rhinovirus infection promotes plasma exudation and increases interleukin (IL)-8 and gel phase mucin secretion, which may contribute to a progression from watery rhinorrhea to mucoid discharge, with mild neutrophilic infiltration during the common cold. Alterations in mucociliary clearance have been extensively studied in asthma, but less frequently in allergic rhinitis.

19.7 CONCLUSION AND CHALLENGES FOR THE FUTURE

The nasal route is generating increasing interest as a route for the administration of local treatments and a cost-effective and patient-friendly alternative to injection for systemic delivery [49]. The special advantages of nasal delivery make it attractive for (i) crisis treatment where rapid onset of action is desirable (e.g., pain, migraine, panic attacks), (ii) systemic delivery of compounds that at present can only be delivered by injection (peptides/pro-proteins/vaccination), and (iii) direct targeting of the CNS (polar drugs for the treatment of CNS disorders).

To take full advantage of these opportunities offered by nasal delivery, innovative approaches to overcome the biological barriers to delivery are being developed. Advances in formulation design, nasal delivery systems, penetration enhancers, enzyme inhibitors, and bioadhesive polymers require an understanding of the biological barriers that they seek to overcome. Furthermore, appropriate models in which to evaluate new delivery strategies are required. These should be capable of identifying any toxic effects of formulations or excipients, while avoiding misleading results brought about by poor experimental design or model selection. *In vitro* optimization should be undertaken to explore fully fundamental concepts and optimize formulations prior to *in vivo* testing, thereby improving the chances of success and complying with the ethical principles of replacement, refinement, and reduction of animal experimentation.

The challenge for the future is not to let physiological barriers prevent the realization of the unique opportunities offered by nasal drug delivery for safe and effective drug therapy.

REFERENCES

1. Jones, N. 2001. The nose and paranasal sinuses physiology and anatomy. *Adv Drug Deliv Rev* 51:5.
2. Chien, Y.W., and S.F. Chang. 1985. Transnasal systemic medications, in *Fundamentals, development concepts and biomedical assessment*, ed. Y.W. Chien, 1. Amsterdam: Elsevier.
3. Mygind, N., and R. Dahl. 1998. Anatomy, physiology and function of the nasal cavities in health and disease. *Adv Drug Deliv Rev* 29:3.
4. Theate, L.G., S.S. Spicer, and A. Spock. 1981. Histology, ultrastructure and carbohydrate cytochemistry of surface and glandular epithelium of human nasal mucosa. *Am J Anat* 162:243.
5. Marom, Z., J. Shelhamer, and M. Kaliner. 1984. Nasal mucus secretion. *Ear Nose Throat J* 63 (2):85.
6. Carlstedt, I., et al. 1985. Mucus glycoproteins—A gel of a problem. *Essays Biochem* 20:40.
7. Lethem, M.I. 1993. The role of tracheobronchial mucus in drug administration to the airways. *Adv Drug Deliv Rev* 11:271.
8. Batts, A.H. 1993. Mucociliary clearance and drug delivery via the respiratory tract. *Adv Drug Deliv Rev* 11 (3):299.
9. Martin, E., et al. 1998. Nasal mucociliary clearance as a factor in nasal drug delivery. *Adv Drug Deliv Rev* 29:13.
10. Lahred, A.W., et al. 1997. Diffusion of drugs in native and purified gastrointestinal mucus. *J Pharm Sci* 86 (6):660.
11. Niibuchi, J.J., Y. Aramaki, and S. Tsuchiya. 1986. Binding of antibiotics to rat intestinal mucin. *Int J Pharm* 30:181.
12. Saltzman, W.M., et al. 1994. Antibody diffusion in human cervical mucus. *Biophys J* 66 (2):508.
13. McMartin, C., et al. 1987. Analysis of structural requirements for the absorption of drugs and macromolecules from the nasal cavity. *J Pharm Sci* 76:535.
14. Donovan, M.D., G.L. Flynn, and G.L. Amidon. 1990. Absorption of polyethylene glycols 600 through 2000: The molecular weight dependence of gastrointestinal and nasal absorption. *Pharm Res* 7:863.
15. Maitani, Y., Y. Machida, and T. Nagai. 1989. Influence of molecular weight and charge on nasal absorption of dextran and DEAE-dextran in rabbits. *Int J Pharm* 49:23.

16. Huang, C.H., et al. 1985. Mechanism of nasal absorption of drugs. I: Physico-chemical parameters influencing the rate of *in situ* nasal absorption of drugs in rats. *J Pharm Sci* 74:608.
17. Sarkar, M.A. 1992. Drug metabolism in the nasal mucosa. *Pharm Res* 9:1.
18. Hussain, A.A., S. Hirain, and R. Bawarshi. 1981. Nasal absorption of natural contraceptive steroids in rats—Progesterone absorption. *J Pharm Sci* 70 (4):466.
19. Hussain, A., et al. 1980. Nasal absorption of propanolol in humans. *J Pharm Sci* 69:1411.
20. Anik, S.T., et al. 1984. Nasal absorption of nafarelin acetate, the decapeptide [D-NAL(2)6] LHRH, in rhesus monkeys. 1. *J Pharm Sci* 73 (5):684.
21. Dahl, A.R. 1986. Possible consequences of cytochrome P-450 dependent monooxygenases in nasal tissues, in *Toxicology of the nasal passages*, ed.G.S. Barrow, 263. Washington D.C.: Hemisphere.
22. Hussain, A., et al. 1985. Hydrolysis of leucine enkephalin in the nasal cavity of the rat—A possible factor in the low bioavailability of nasally administered peptides. *Biochem Biophys Res Commun* 133 (3):923.
23. Hirai, S., et al. 1981. Absorption of drugs from the nasal mucosa of rat. *Int J Pharm* 7:317–325.
24. Hayakawa, E., et al. 1989. Effect of sodium glycocholate and polyoxyethylene-9-lauryl ether on the hydrolysis of varying concentrations of insulin in the nasal homogenates of the albino rabbit. *Life Sci* 45 (2):167.
25. Devillers, G. 2003. Exploring a pharmaceutical market niche and trends: Nasal spray drug delivery. *Drug Deliv Technol* 3:38.
26. Krishnamoorthy, R., and A.K. Mitra. 1998. Prodrugs for nasal drug delivery. *Adv Drug Deliv Rev* 29:135–146.
27. Shao, Z., and A.K. Mitra. 1994. Bile salt-fatty acid mixed micelles as nasal absorption promoters. III. Effects of nasal transport and enzymatic degradation of acyclovir prodrugs. *Pharm Res* 11:243.
28. Nagai, T., and Y. Machida. 1990. Bioadhesive dosage forms for nasal administration, in *Bioadhesive drug delivery systems*, eds. V. Lenearts, and R. Gurny, 169. Florida: CRC Press, chap. 9.
29. Park, H., and J.R. Robinson. 1987. Mechanisms of mucoadhesion of poly(acrylic)acid hydrogels. *Pharm Res* 4:457.
30. Duchene, D., F. Touchard, and N.A. Peppas. 1988. Pharmaceutical and medical aspects of bioadhesive systems for drug administration. *Drug Dev Ind Pharm* 14:283.
31. Nagai, T., et al. 1984. Powder dosage form of insulin for nasal administration. *J Control Release* 1:15.
32. Harris, A.S., et al. 1988. Effect of viscosity on particle size, deposition and clearance of nasal delivery systems containing desmopressin. *J Pharm Sci* 77:405.
33. Vyas, S.P., S.K. Goswami, and R. Singh. 1995. Liposomes based nasal delivery system of nifedipine: Development and characterisation. *Int J Pharm* 118:23.
34. Lim, S.T., et al. 2000. Preparation and evaluation of the *in vitro* drug release properties and mucoadhesion of novel microspheres of hyaluronic acid and chitosan. *J Control Release* 66:281.
35. Lim, S.T., et al. 2001. *In vivo* and *in vitro* characterization of novel microparticulates based on hyaluronan and chitosan hydroglutamate. *AAPS Pharmsci Tech* 2 article 20.
36. Lim, S.T., et al. 2002. *In vivo* evaluation of novel hyaluronan/chitosan microparticulate delivery systems for the nasal delivery of gentamicin in rabbits. *Int J Pharm* 231:73.
37. Merkus, F.W.H.M., et al. 1993. Absorption enhancers in nasal drug delivery—Efficacy and safety. *J Control Release* 24:201.
38. Hochmann, J.H., J.A. Fix, and E.L. Lecluyse. 1994. *In vitro* and *in vivo* analysis of the mechanism of absorption enhancement by palmitoylcarnitine. *J Pharmcol Exp Ther* 269 (2):813–822.
39. Merkus, F.W.H.M., et al. 1999. Cyclodextrins in nasal drug delivery. *Adv Drug Deliv Rev* 36 (1):41.
40. Schipper, N.G.M., et al. 1993. Nasal insulin delivery with dimethyl-β-cyclodextrin as an absorption enhancer in rabbits-powder more effective than liquid formulations. *Pharm Res* 10:682.
41. Chanvanpatil, M.D., and P.R. Vavia. 2004. The influence of absorption enhancers on nasal absorption of acyclovir. *Eur J Pharm Biopharm* 57 (3):483.
42. Raehs, S.C., et al. 1988. The adjuvant effect of bacitracin on nasal absorption of gonadorelin and buserelin in rats. *Pharm Res* 5 (11):689.
43. O'Hagan, D.T., and L. Illum. 1990. Absorption of peptides and proteins from the respiratory tract and the potential for development of locally administered vaccine. *Crit Rev Ther Drug* 7 (1):35.

44. Hussain, M.A., and B.J. Aungst. 1992. Nasal absorption of leucine enkephaln in rats and the effects of aminopeptidase inhibition, as determined from the percentage of the dose unabsorbed. *Pharm Res* 9 (10):1362.
45. Schmidt, M.C., et al. 1998. *In vitro* cell models to study nasal mucosal permeability and metabolism. *Adv Drug Deliv Rev* 29:51.
46. Forbes, B., and C. Ehrhardt. 2005. Human respiratory epithelial cell culture for drug delivery applications. *Eur J Pharm Biopharm* 60:193.
47. Hirai, S., et al. 1981. Absorption of drugs from the nasal mucosa of rats. *Int J Pharm* 7:317.
48. Illum, L. 2004. Is nose-to-brain transport of drugs in man a reality. *J Pharm Pharmacol* 56:3.
49. Illum, L. 2003. Nasal drug delivery—Possibilities, problems and solutions. *J Control Release* 87:187.
50. Gizurarson, S. 1993. The relevance of nasal physiology to the design of drug absorption studies. *Adv Drug Deliv Rev* 11:329.

20 Nasal Delivery of Peptide Drugs

Dennis J. Pillion, John J. Arnold, and Elias Meezan

CONTENTS

20.1 INTRODUCTION

By virtue of their size and charge, peptide molecules are not the ideal candidates for transfer into the systemic circulation following instillation in the nose. Among the many barriers to absorption that must be overcome are mucociliary clearance, extracellular enzymatic destruction, the lipophilic bilayer membrane of nasal epithelial cells, the potential for nasal epithelial cells to degrade any peptide molecules that cross the lipid bilayer, and the potential to establish futile cycles of endocytosis and exocytosis on the apical surface of polarized epithelial cells. Indeed, in the face of these multiple barriers, it seems all the more remarkable that any substantial absorption of peptide drugs from the nose has ever been observed. Despite these barriers, recent reports have confirmed that many peptide drugs, ranging in sizes up to 30,000 Da, have been successfully delivered through the nasal route when they were formulated with an absorption-enhancing agent [1–3]. This chapter will focus on the unique barriers to be overcome and the therapeutic opportunities that are now becoming available as a result of successful nasal delivery of peptide drugs. Currently, the vast majority of nonpeptide therapeutic drugs administered nasally are administered to patients for local effects, rather than systemic effects. However, some small nonpeptide drugs, such as cocaine, are delivered nasally to produce effects elsewhere in the body. Characteristically, the absorption of peptide and nonpeptide drugs administered by the nasal route displays a very rapid pharmacokinetic profile and avoids first-pass metabolism of the drug by the liver [1,4–6]. Numerous different types of absorption-enhancing agents that improve the bioavailability of peptide drugs have now been described [1–14]. This chapter will discuss these agents in some detail, in so far as they have been shown to increase peptide drug absorption from the nasal cavity. Several review articles have been published on the topic of peptide drug delivery through the nasal route [2,9,12,15–17]. This chapter will build on this information and focus on recent developments in the field of peptide drug delivery. Current nasal peptide products and novel peptide drugs that are likely candidates for future development into nasal therapeutic agents will be discussed.

20.2 NASAL DRUG DELIVERY

Simple spray devices can deliver nasal formulations to the anterior portion of the nasal cavity. More sophisticated spray devices have been developed to deliver nasal formulations to the medial and posterior portions of the nasal cavity.

Nasal delivery of peptide drugs does not require sophisticated devices. Peptide drugs, like nonpeptide drugs, can be absorbed quickly and noninvasively following nasal instillation. Furthermore, many people already use nasally administered drugs, including over the counter and prescription drugs. Nasal drug delivery obviates the need for needles and reduces the production of hazardous medical waste. Hence, delivery of peptide drugs through the nasal route, rather than by subcutaneous injection, is expected to provide convenience, good patient acceptance, and improved compliance.

Despite these advantages, the development of nasal peptide drug formulations has been slow. Several disadvantages to the use of the nasal cavity as a practical site for peptide drug administration must be overcome [10,11]. The chief disadvantages of the nasal route include the low intrinsic permeability of the absorptive surface in the nasal cavity to larger peptide drugs, the potential for enzymatic degradation, and the rapid removal of drug formulations from the nasal cavity by mucociliary clearance. Other considerations that can constrain the utilization of nasal peptide drugs include the potential for local toxicity, and the realization that some peptide drugs, such as insulin, must be delivered to, and absorbed from, the nasal cavity with meticulous accuracy and reproducibility.

20.2.1 Anatomy and Physiology of the Nasal Cavity

The nose is equipped with a unique cellular architecture to perform several functions, including filtration of inspired particles, humidification of inspired air, olfaction, and some immunological functions [18,19]. The nose is not specifically designed for nutrient or peptide drug absorption. However, the large absorptive capacity of the nasal epithelium has now been fully appreciated because of the extremely high bioavailability of nasally applied peptide drugs observed under certain experimental conditions (described below).

Anatomically, the nasal cavity is divided into two halves by the nasal septum. Each half possesses an approximate total surface area of 75 cm^2 and an approximate total volume of 7.5 mL [20]. The two nasal cavities are separated by the septal wall and possess inferior, middle, and superior turbinates.

Several distinct anatomical areas compose the nasal passageways, including the vestibular, the atrial, the respiratory, and the olfactory regions. Anteriorly, the nares open into the vestibule, a tough, keratinized region of the nasal cavity that possesses hairs or vibrissae and stratified squamous epithelial cells. The atrium has both stratified squamous and pseudostratified columnar epithelial cells. The respiratory region of the nasal cavity contains pseudostratified columnar epithelial cells that have microvilli and can be ciliated or nonciliated. Adjacent to these epithelial cells are mucus-producing goblet cells that cover the nasal cavity with a thin film of mucus. The inferior, middle, and superior turbinates are folds in the nasal cavity that greatly increase the surface area of this region. The respiratory cavity is extremely well vascularized. The olfactory region is a relatively small area of the nasal cavity that is located at the top of the respiratory region and consists of a collection of specialized olfactory nerve cells that are critical in the ability to sense odors. Delivery of peptide and nonpeptide drugs directly to the central nervous system (CNS) via the olfactory region is currently under study [21–26].

Inspired air, which is humidified and warmed in the nose, moves rapidly from the nasal cavity to the nasopharynx and into the trachea. Most airborne particles that are inspired are entrapped in the mucus layer of the respiratory region. Subsequently, these particles are removed from the nasal cavity to the nasopharynx and esophagus and eliminated via the gastrointestinal tract through a process known as mucociliary clearance.

20.2.2 Mucociliary Clearance

Mucociliary clearance ordinarily serves as a defense mechanism of the nasal cavity that is critical for the entrapment and removal of inhaled noxious chemicals [27], but in this chapter we will describe the role it plays in regulating nasal peptide drug absorption. The mucus layer that covers the nasal cavity surface is released from goblet cells that are localized between and among the nasal epithelial cells that comprise the bulk of the nasal cavity surface, as well as from mucous glands that are located below the uppermost layer of cells. Mucus functions to protect the respiratory tract from noxious insults, control water balance, and regulate ion transport. Mucus is mainly composed of water (~95%) but also contains glycoproteins, lipids, and other cellular debris [28]. Serous glands are located beneath the surface of the nasal cavity and produce a watery solution that influences the viscosity of the nasal mucus. Consequently, the precise composition of the mucus layer may vary due to environmental factors such as temperature or cigarette smoke, as well as disease states.

The impact of the mucus layer of the nasal cavity on peptide drug absorption is poorly understood, but in the absence of absorption enhancers it could play a significant role in limiting drug absorption. Several agents that enhance nasal peptide drug absorption have demonstrable effects on the mucus layer (described below).

The mucus layer also bathes the cilia of ciliated epithelial surface cells and provides a stimulus for ciliary motility (i.e., ciliary beating). The cilia consist of microtubules with a $9+2$ configuration (nine pairs of peripheral microtubules and two central microtubules) that beat rhythmically to rapidly move mucus from the anterior to the posterior portion of the nasal cavity. To successfully cross the nasal permeability barrier, peptide drugs must penetrate the mucus layer and cross the epithelial cell layer, and do so in a limited time, because mucociliary clearance will limit the time of exposure of the peptide to the absorptive surface [19–21,27–29]. Typically, drugs or inspired particles that are delivered nasally are removed via mucociliary clearance, with a clearance time of approximately 15 min in humans; however, this transit time can vary from person to person and can be impacted by the addition of mucoadhesive agents to the formulation [30–37].

20.2.3 Enzymatic Degradation

Enzymatic degradation and metabolism is another potential barrier to peptide drug absorption from the nasal cavity. One role of the nasal cavity is to defend against inhaled pollutants. Cytochrome P-450 enzymes, which are capable of metabolizing xenobiotics, have been found in the nasal epithelium [38]. Furthermore, the nasal cavity is replete with peptidases (i.e., aminopeptidases, carboxylesterases, and glutathione transferases) that could potentially degrade peptides such as insulin [18,28,39]. Proteolytic enzyme inhibitors and chemical modifications by the attachment of polyethylene glycol (PEG) moieties have been shown to increase the stability and nasal absorption of calcitonin [40–42]. The combined actions of mucociliary clearance (described above) and enzymatic degradation would be expected to remove or degrade biologically active peptide drugs in the nasal cavity within 15–30 min after administration of the drug. Additional enzymatic degradation of peptide drugs would ensue following internalization of any peptide drug in the process of vesicular transcytotic movement (see below). Internalized peptide drugs, along with plasma membrane constituents that undergo endocytosis, could be shuttled vectorially across the cell for exocytosis at the basolateral surface. This process would result in the absorption of the intact peptide drug into the circulation. However, two other pathways are also available for endocytosed vesicles, i.e., recycling to the apical surface of the cell and exocytosis back into the nasal cavity, or merger with lysosomal vesicles intracellularly and degradation within the cell. Both of these processes would not result in peptide drug absorption. A fourth possible fate of nasal peptide insulin is binding to specific receptors on either the apical or the basolateral surface of the epithelial cells that line the nasal cavity. Peptide binding to its receptor would trigger a cellular response and would also initiate the internalization of the peptide drug. As noted below, the very high rates of bioavailability achieved in nasal drug delivery studies are remarkable when these possible degradative processes are considered.

20.2.4 Epithelial Cell Layer

Perhaps the chief barrier to peptide drug absorption from the nasal cavity is the nasal epithelial cell layer. Passive, transcellular movement of small, uncharged, and highly lipophilic drugs, such as cocaine can occur, as these drugs can diffuse across the phospholipid bilayer of the cell membrane. Vesicular transcytotic movement of drugs involves the endocytosis and internalization of drugs presented at the apical surface and the subsequent release of the drugs at the basolateral surface. This type of vectorial vesicular transcytotic movement has been demonstrated for some peptides such as calcitonin [43–45]. An alternative route of drug permeation through the nasal epithelial cell layer is through the tight junctions between cells, known as the paracellular route. Small hydrophilic drugs with molecular weights less than 1 kDa are able to traverse the epithelial cell layer by paracellular movement. Larger peptide drugs, such as insulin (5.7 kDa), are unable to gain access to the systemic circulation through the paracellular route in the absence of an absorption enhancer.

20.3 ABSORPTION ENHANCERS

One strategy employed for overcoming the nasal absorption barrier to peptide drugs is coadministration of absorption enhancers. Such reagents have been shown to increase the absorption of peptide drugs ranging in size from 1000 to 31,000 Da. Several distinct mechanisms of action have been proposed for absorption enhancers: alteration of the rheological (fluidity) properties of the mucus, alteration of the cilia (i.e., paralysis of ciliary beating or removal of cilia from the epithelial cells that line the nasal cavity), enhanced paracellular transport through loosening of tight junctions between cells, enhanced vesicular transcytotic transport, damage to the epithelial cell layer, and inhibition of proteolytic enzymes [1–4,11,12,20,29,46–48]. Several absorption enhancers that have been shown to increase peptide drug bioavailability from the nose are described in Table 20.1 and discussed in detail below.

20.3.1 Cyclodextrins

Cyclodextrins (CDs) are cyclic oligosaccharides composed of six or more monosaccharide units with a central cavity that can form inclusion complexes with hydrophobic molecules. The most extensively studied CDs are alpha-, beta-, and gamma-CDs, which possess six, seven, and eight glucopyranose units, respectively. Presently, CDs are used in pharmaceutical formulations to increase drug solubility and dissolution and to enhance low molecular weight drug absorption through molecular encapsulation [9,49]. Merkus et al. [50] reported that, among the CD derivatives studied as potential nasal insulin absorption promoters in rats, dimethyl-beta-cyclodextrin (DMBCD) was found to be the most effective, whereas alpha-CD was less effective and beta- and gamma-CD had negligible effects on insulin absorption. The most common explanation of CD action is interaction with cholesterol found on the cellular membrane, which can be selectively accommodated in the central cavity of the CD ring and thereby removed from the membrane [51]. The interaction of cyclodextrins with the cell membrane also transiently opens tight junctions, which may explain their ability to facilitate peptide absorption across the nasal mucosa [52].

Since only some CDs are effective at increasing peptide bioavailability, the architecture of the central cavity of CDs appears to be critical for nasal peptide drug absorption. Previous work has demonstrated that DMBCD is effective in promoting the absorption of insulin [53,54], calcitonin [55], and low molecular weight heparin [56]. Since DMBCD stimulated the nasal absorption of two distinct peptides and a third drug that was not a peptide, it seemed unlikely that the only mechanism of DMBCD action was direct binding to the drug or inhibition of proteolytic enzymes. When DMBCD was administered concomitantly with another absorption enhancer, dodecylmaltoside (DDM), the mixture was ineffective at promoting either insulin or calcitonin absorption [54]. This result was unexpected because both reagents were used at concentrations that stimulated insulin and calcitonin absorption from the rat nasal cavity when used alone. Through a series of dose-escalation studies, it was discovered that the alkyl chain of DDM was able to insert into the cavity of DMBCD. Formation of the DMBCD–DDM complex eliminated the ability of either DMBCD or DDM to increase insulin transport. Hence, the capacity of DMBCD to increase peptide drug absorption required that the central cavity of the cyclodextrin be empty. Interestingly, administration of other CDs with empty central cavities, including beta- and gamma-CD was totally ineffective at increasing peptide drug absorption.

20.3.2 Chitosan

Chitosan is a cationic polysaccharide produced from the deacetylation of chitin, a component of crab and shrimp shells [7,57,58]. Chitin is composed of units of 2-deoxy-2-(acetylamino) glucose joined by glycosidic bonds that form a linear polymer. Illum et al. [7,57,58] demonstrated the ability of chitosan to increase the bioavailability of insulin and other small peptides and polar macromolecules in different animal models. In both the sheep and rat models, the addition of chitosan at concentrations of 0.2%–0.5% to nasal formulations of insulin resulted in significant increases in plasma insulin and reductions in blood glucose. Reversibility studies indicated that the effect of chitosan on the nasal absorption of insulin

TABLE 20.1
Proposed Mechanisms of Action of Selected Absorption Enhancers

Absorption Enhancer	Concentrations	Experimental System	Peptides	Proposed Mechanisms of Action[a]	Ref.
Cyclodextrins	2%–30%	Rat, rabbit	Calcitonin, insulin, buserelin	A, B, C, D, E,	9, 49–56, 110–112
Chitosans	0.1%–1%	Rat, sheep	Calcitonin, insulin, goserelin growth hormone	C, F	57–68
Bile salts	0.2%–2%	Rat, rabbit	Calcitonin, insulin	A, B, C	4,69–74
Saponins	0.025%–0.5%	Rat	Insulin	G	75–78
Soybean steryl glucosides	0.1%–3.5%	Liposomes, rabbit	Insulin	C, G	79–81, 113, 114
Phosphatidylcholines	0.1%–0.5%	Rabbit, rabbit nasal mucosa	Insulin, growth hormone	C, G, H	82, 83
Alkylglycosides and sucrose esters of fatty acids	0.03%–0.5%	Rat, 16HBE14o$^-$ cells	Calcitonin, glucagon, insulin, leptin, growth hormone erythropoietin	C, G, H, I	1, 6, 10, 84–90

[a]A, Inhibition of proteolytic enzymes; B, dissociation of hexameric to monomeric form of insulin; C, loosening of tight junctions; D, increase in membrane fluidity due to cholesterol removal; E, reversible ciliostasis; F, mucoadhesion and prolonged residence time; G, incorporation into lipid bilayer and membrane perturbation; H, insulin internalization, increased transcellular transport; and I, correlation with CMC.

was transient, and histological studies suggested that the enhancer had little effect on nasal epithelial cell morphology. Studies in human volunteers have confirmed that nasal administration of chitosan formulations resulted in significantly longer nasal clearance times [59]. Hence, chitosan may decrease mucociliary clearance of, and prolong the residence time of, insulin and other peptides within the nasal cavity [60–67]. Microsphere formulations of chitosan were cleared at a slower rate than solution formulations from the sheep nasal cavity, which may give such preparations a greater potential for promoting peptide absorption [65]. Furthermore, chitosan also produces a transient loosening of the tight junctions of confluent Caco-2 cells derived from a human colon cancer [68]. The net result of chitosan action is cytoskeletal contraction, opening of tight junctions, and increasing paracellular drug absorption [7,57,58,65].

20.3.3 Bile Salts and Derivatives

Nearly two decades ago, bile salts and their derivatives such as sodium glycocholate (NaG), sodium taurocholate, and sodium taurodihydrofusidate (NaTDHF) were demonstrated to effectively promote nasal insulin absorption [3]. Subsequently, bile salts and their derivatives were extensively studied for their ability to promote the absorption of a variety of peptide drugs from several alternative delivery sites [69–71]. The mechanisms of action by which the bile salts and their derivatives promote the increased nasal absorption of peptide drugs are still not defined at the molecular level but may involve the alteration and fluidization of the nasal epithelial cell membranes, increase in transcytotic movement of peptides via endocytotic vesicles, or the inhibition of certain proteolytic enzymes capable of degrading peptides before they can successfully cross the nasal epithelium. Inclusion of NaG in an insulin formulation resulted in a significant reduction in enzymatic degradation of insulin [70]. Through circular dichroism and α-chymotrypic degradation studies, a dose–response relationship between increasing concentrations of NaG and the presence of monomeric insulin has been shown [70]. More recently, sodium taurocholate was found to increase the disaggregation of hexameric insulin in a dose-dependent manner [72]. Unfortunately, *in vivo*, bile salts and derivatives produce nasal irritation, stinging, and lacrimation, which may be a consequence of cellular swelling and mucus discharge from goblet cells [4,73,74].

20.3.4 Saponins

Saponins and saponin derivatives belong to a family of structurally related compounds that can be isolated from plants. A particular purified saponin, designated QS-21, is an amphipathic quillaic acid 3, 28-*O*-bisglycoside purified from the bark of the *Quillaja saponaria* tree. QS-21 has been shown to be a highly immunogenic compound. This compound is an excellent excipient in vaccines and can be used to generate a strong immune response [75]. Derivatives of QS-21, such as DS-1 and DS-2, have been shown to increase the nasal absorption of gentamicin in mice and rats [76]. These derivatives differ from QS-21 in that they do not induce a similarly strong immune response. Pillion et al. [77,78] demonstrated that DS-1 and DS-2 can increase the nasal and ocular absorption of insulin in anesthetized rats. In most cases, the efficacy of these compounds for nasal absorption correlated with surfactant strength indicators such as critical micellar concentration (CMC) and hemolysis of sheep erythrocytes. However, despite the fact that DS-1 and DS-2 share similar CMC and hemolytic titers, DS-1 was significantly more potent in stimulating insulin uptake than DS-2. Hence, the molecular architecture of the saponin derivatives must be an important determinant of their potency on nasal peptide drug absorption.

20.3.5 Soybean-Derived Steryl Glucosides

Soybean-derived sterol mixture (SS), soybean-derived steryl glucosides (SG), and their individual components have been extensively studied for their ability to promote the nasal absorption of drugs, particularly insulin [79,80]. Maitani et al. [79] demonstrated that the nasal administration of SG plus insulin to rabbits resulted in significant reductions in blood glucose. The effect of SG was dose dependent to 1%, with a plateau being reached thereafter. Muramatsu et al. [81] have demonstrated that SG perturbs the phospholipids in artificial membranes (i.e., liposomes). Furthermore, circular dichroism studies with insulin in the presence or absence of SG have indicated that the enhancer had little effect on the dissociation of insulin hexamers to monomers. These results suggest that the action of SS and SG involves interaction with the nasal membrane rather than interaction with insulin molecules.

An alternate mechanism suggested to explain the effects of SG and Sit-G on peptide drug absorption involves chelating tight junction-associated calcium (Ca^{++}) or increasing intracellular Ca^{++}. Ca^{++} is critical for the maintenance of intact tight junctions. These agents may increase insulin permeation by two independent processes: increasing vesicular transcytotic transport, as well as increasing paracellular transport.

20.3.6 Phosphatidylcholines

Didecanoyl-L-alpha-phosphatidylcholine (DPPC) is a medium-chain phospholipid that has been demonstrated to increase the absorption of insulin from the nasal cavity [82,83]. DPPC has also been shown to rapidly and significantly increase the absorption of growth hormone (GH) following nasal administration to rabbits [48]. To elucidate the mechanism of action of DPPC, investigators have conducted *in vitro* studies with excised rabbit nasal mucosa [82]. DPPC lowered the transepithelial electrical resistance (R_{TE}) and increased the transmucosal transport of several marker compounds (sucrose and PEG 4000), as well as insulin. These changes were found to be reversible. Carstens et al. [82] concluded that DPPC increased the transport of insulin in part by the reversible loosening of tight junctions. Nevertheless, transmission electron microscopic experiments conducted in rabbits 15 min after treatment with DPPC demonstrated that tight junctions were unchanged morphologically [48]. Furthermore, when gold-labeled growth hormone was visualized in the tissue, growth hormone was found in the nuclei and cytoplasm of both ciliated epithelial cells and goblet cells, localized in endocytotic vesicles, not exclusively in paracellular channels. Thus, DPPC stimulated growth hormone absorption by increasing vesicular transcytotic pathways and by increasing paracellular involvement.

20.3.7 Alkylglycosides and Sucrose Esters of Fatty Acids

The Alkylglycosides (AGs) and Sucrose esters of fatty acids (SEFAs) are families of nonionic glycosurfactants that have been used for their ability to gently extract membrane proteins with a minimal loss of functionality. These compounds can be synthesized and purified economically, with a worldwide production of thousands of tons per year. Chemically, AGs and SEFAs are a group of uncharged amphipathic compounds that consist of an aliphatic hydrocarbon chain attached to a sugar moiety. Certain AGs and SEFAs such as dodecanoyl sucrose have enjoyed widespread use as food-grade emulsifiers and in cosmetic preparations.

Individual members of the AG and SEFA family have been studied for their ability to promote the nasal and ocular absorption of peptide drugs in rats, mice, cats, dogs, and monkeys [1,6,10,53,54,84–90]. Dose-escalation studies were conducted in rats to determine the potencies of each of the AGs and SEFAs as enhancers for the ocular and nasal absorption of insulin and to determine the contribution of the alkyl chain and the sugar moiety. Insulin absorption kinetics were measured directly by radioimmunoassay. Following nasal administration of insulin in the absence or presence of increasing amounts of each respective reagent, insulin bioavailability was determined by measuring the reduction in blood glucose concentrations. Parallel studies were conducted in rats receiving the same dose of insulin by subcutaneous injection. Xylazine and ketamine anesthesia was used to block endogenous insulin secretion by the rats. In this experimental system, exquisite sensitivity to exogenously applied insulin could be achieved. In terms of potency, studies with both nasal and ocular insulin administration indicated that shorter chained AGs and SEFAs, coupled to glucose such as hexylglucose (C_6), heptylglucose (C_7), octylglucose (C_8), and nonylglucose (C_9), were ineffective or minimally effective at promoting insulin absorption from the eye or nose [85,88]. Intermediate-length AGs and SEFAs linked to disaccharides such as decanoyl sucrose (C_{10}), decylmaltoside (C_{10}), or octylmaltoside (OM) (C_8) were more effective in the promotion of nasal and ocular insulin administration. However, longer chain AGs and SEFAs linked to disaccharides such as DDM (C_{12}), tridecylmaltoside (C_{13}), tetradecylmaltoside (TDM) (C_{14}), and dodecanoyl sucrose (C_{12}) were very potent and could promote nasal insulin administration even at concentrations as low as 0.03%–0.06%. None of the other absorption-enhancing agents tested previously were effective at this low dose. One possible explanation for this result was that the ability of the AG and SEFA to enhance insulin absorption was directly correlated to the amphipathic nature of the molecule, and the fact that the surfactants were uncharged. As alkyl chain length increased, lipophilicity increased and potency increased. A finding that was consistent with data from other investigators was that the agents that promoted the absorption of insulin most potently were well above their CMCs. The CMC of the AGs and SEFAs is inversely proportional to chain length. Initially, TDM (C_{14}) and dodecanoyl sucrose (C_{12}) were the largest commercially available AG and SEFA species, as well as the most lipophilic and the most potent. Subsequently, longer chained AGs and SEFAs (pentadecylmaltoside [C_{15}], hexadecylmaltoside [HDM] [C_{16}], tridecanoyl sucrose [C_{13}], and tetradecanoyl sucrose [C_{14}]) were synthesized and tested in formulations containing insulin to determine if these agents would result in even greater increases in nasal insulin absorption in anesthetized rats [89]. Results from these experiments indicated that, among the family of SEFAs, tridecanoyl sucrose (C_{13}) and tetradecanoyl sucrose (C_{14}) were the most effective at promoting nasal insulin absorption. An interesting finding was that, among the family of alkylmaltosides, increasing alkyl chain length correlated with increasing potency of nasal insulin absorption only up to a length of C_{14}. Pentadecylmaltoside (C_{15}) and hexadecylmaltoside (C_{16}) were less potent than TDM (C_{14}) at stimulating nasal insulin uptake. Thus, these results indicate that simple surfactant chain length and increasing hydrophobicity do not completely explain the promotion of nasal insulin absorption and that a carbon chain length of C_{14} is optimal. This result may be due to the ability of the TDM (C_{14}) and tetradecanoyl sucrose (C_{14}) to interact with the nasal membrane in a way that optimizes transcytotic insulin movement. Absorption studies conducted in anesthetized rats with monomeric insulin (i.e., lyspro) and hexameric insulin (i.e., regular) indicated that DDM (C_{12}) promoted the absorption of both forms of insulin in a similar manner [6]. A similar result has recently been reported for absorption of insulin with tetradecylmaltoside via the pulmonary route in rats [16]. Hence, the dissociation of hexameric insulin was not the underlying mechanism that accounted for the absorption-promoting action of the AGs and SEFAs. Cross-comparison of the results obtained utilizing the nasal and ocular routes of insulin administration demonstrated that nasal administration of insulin plus various AGs and SEFAs produced a more potent hypoglycemic response than ocular administration in rats [6,84,85]. Furthermore, previous studies have shown that the corneal surface is relatively impermeable to insulin, even in the presence of potent absorption enhancers such as saponins [77,78,91,92]. The maximal volume that can be applied to the eye without substantial spillage is only 20–30 μL per eye, whereas the nasal cavity can accommodate 0.1–0.5 mL more readily. Consequently, these data indicate that the nasal cavity has advantages over the ocular route for peptide drug delivery.

20.4 INTRANASAL INSULIN DELIVERY

Type 1 diabetes mellitus is a disease that results from the autoimmune destruction of the pancreatic β-cells and the subsequent loss of insulin production. Therapy for type 1 diabetes mellitus requires lifelong replacement of insulin in a time-sensitive manner that must be matched with food, exercise, stress, and other hormonal and therapeutic events that impact serum glucose levels. Type 2 diabetes mellitus is a progressive disease that starts with a state of insulin resistance that initially causes overproduction of insulin in a futile attempt to maintain normoglycemia. Subsequently, insulin production wanes and hyperglycemia worsens. Initial therapy for type 2 diabetes mellitus can include a variety of oral agents that promote insulin secretion, insulin sensitivity, and glucose uptake into muscle and fat cells, or that dampen glucose release from the liver. Eventually, insulin replacement therapy is needed to maintain glycemic control. More than 18 million people in the United States suffer from type 2 diabetes mellitus and another 40 million are thought to be prediabetic. Nearly 1 million people in the United States suffer from type 1 diabetes mellitus. Results from the Diabetes Control and Complications Trial [93] and United Kingdom study [94] have demonstrated that intensive insulin therapy and glucose monitoring significantly delay the progression of diabetes-related complications. For this reason, the market for insulin is very large and continues to grow as the obesity and diabetes mellitus pandemic worsens worldwide.

Since its discovery, isolation, and purification in the early twentieth century, insulin has been administered to diabetic patients exclusively by injection until the recent introduction of inhaled insulin. Insulin possesses certain physiochemical properties that contribute to its limited absorption from the gastrointestinal tract, and requires subcutaneous injection to achieve clinically relevant bioavailability. With a molecular size of 5.7 kDa, insulin is a moderately sized polypeptide composed of two distinct peptide chains designated the A chain (21 amino acid residues) and the B chain (30 amino acid residues) and joined by two disulfide bonds. Like all polypeptides, insulin is a charged molecule that cannot easily penetrate the phospholipid membrane of the epithelial cells that line the nasal cavity. Furthermore, insulin monomers self-associate into hexameric units with a molecular mass greater than 30 kDa, which can further limit its passive absorption. Despite these constraints, successful delivery of insulin via the nasal route has been reported in humans and animals when an absorption enhancer was added to the formulation.

Nasal delivery of insulin is only one of many routes of insulin delivery currently under study. Oral, ocular, pulmonary, transdermal, vaginal, rectal, and buccal delivery have been or still are being developed. Pulmonary delivery of insulin in humans is now approved in the United States. Pulmonary administration yields rapid insulin absorption, not greatly different than subcutaneous injection of rapid-acting insulin. The success of inhaled insulin marks the beginning of a new era in peptide drug delivery. First, it provides strong evidence in favor of the hypothesis that diabetic patients can be managed using a rapid-acting insulin formulation that is not injected, when coupled with an injection of a longer-acting insulin. Second, it provides additional evidence that other peptide drugs, at least those in the same molecular mass range as insulin (5700 Da), can be delivered by a route other than subcutaneous injections. The future may include nasal drug delivery of growth hormone, interferon, glucagon, and a number of other therapeutic agents currently administered by injection (see below).

20.4.1 Mechanisms of Action of Alkylglycosides

Several potential pathways exist for drug absorption across the nasal mucosa. Absorption enhancers have been shown to modulate the epithelial membrane permeability barrier by increasing paracellular movement between cells or by increasing vesicular transcytotic movement through permeabilized cell membranes [95–97]. In the case of AG stimulation of nasal insulin delivery, more than one transport pathway could be increased. The absorption-enhancing properties of AGs and SEFAs appear to be due to a direct interaction with the epithelial cell barrier rather than with insulin because DDM promoted the absorption of both regular and lyspro insulin in a similar fashion [6]. Furthermore, AGs and SEFAs increased nasal and ocular absorption of other peptide and nonpeptide drugs, such as glucagon, calcitonin, leptin, growth hormone, and low-molecular-weight heparin [1,5,84,86].

Immortalized human bronchial epithelial cells ($16HBE14o^-$ cells) were also used to conduct insulin transport studies in the presence and absence of TDM [53]. The $16HBE14o^-$ cell line served as a surrogate for human nasal epithelial cells [98]. Unfortunately, it has been difficult to grow and propagate nasal epithelial cells in culture and even more difficult to mimic the mixture of ciliated and nonciliated epithelial cells, as well as goblet cells that are colocalized on the surface of the nasal cavity. As a result, a completely authentic cultured nasal epithelial cell system is not available. Of the immortalized nasal cell lines (RPMI 2650, BT, NAS 2BL) available, none have the properties of $16HBE14o^-$ cells, which grow to confluency, express well-formed tight junctional belts and mimic the ability of nasal tissue to restrict the passage of many compounds. In this experimental system, it was possible to determine the permeation and degradation of radiolabeled [^{125}I]-insulin applied to the apical surface of cells grown on transwells, in the presence and absence of two distinct absorption enhancers: tetradecylmaltoside (C_{14}) (0.125%) and DMBCD (1%). The transport of mannitol was also measured, as an indication of paracellular transport from the apical to the basolateral side of the chamber. Interestingly, results indicated that either TDM or DMBCD, when used alone, could increase paracellular trafficking, as evidenced by decreased R_{TE} and increased movement of mannitol from the apical to the basolateral chamber. However, only TDM had the additional effect of promoting [^{125}I]-insulin movement from the apical to the basolateral chamber.

Fluorescein isothiocyanate (FITC)-labeled insulin was used to visualize the pathway insulin took when traversing the nasal epithelial layer *in vivo* following the administration of TDM. FITC-insulin was visualized in the cytoplasm of nasal epithelial cells, but only when added in the presence of TDM. FITC-insulin was not found exclusively in the paracellular spaces, either in the presence or absence of TDM. These results were consistent with animal studies in which transmission electron microscopy of nasal septal tissue exposed to formulations containing 0.125% TDM showed increased endocytotic internalizations compared to septal tissue from rats treated with saline. Taken together, these results suggest that the AGs increase both the transcytotic and paracellular pathways.

20.4.2 Reversibility of Alkylglycoside Effects on Insulin Absorption and Nasal Morphology

Nasal administration of formulations containing insulin plus 0.125% TDM concurrently at time 0 caused a rapid and significant increase in plasma insulin levels and a corresponding decrease in blood glucose levels (described above). When an interval of 2 h elapsed between TDM addition and insulin administration, a significant attenuation was noted in the maximal increase in plasma insulin, as well as in the maximal reduction in blood glucose levels [10]. The experimental protocol described above was then used to assess the amount of insulin absorbed when the interval between TDM administration and insulin administration was changed to 1, 4, or 8 h. The changes in the AUC_{0-120} of plasma insulin levels and blood glucose levels were plotted as a function of time after TDM treatment. Essentially complete reversal of the permeation-enhancing action of tetradecylmaltoside (C_{14}) on the nasal mucosa was noted at 8 h after TDM administration [10]. When other peptides and nonpeptide drugs were utilized in this protocol in place of insulin, similar but distinct patterns of reversibility were observed (see below). Hence, TDM effects on nasal absorption of drugs are reversible.

Following exposure to TDM, the apical membrane of epithelial cells exhibited a more fluidized appearance, as evidenced by the increased number of endocytotic vesicles and wisp-like cilia [10]. Samples obtained 2 h after TDM administration displayed a distinct appearance, with widened intercellular spaces and occasionally ejected, apoptotic nuclei. Four hours after TDM administration, the appearance of the nasal epithelial cell cilia had essentially returned to normal, although they were less dense than in controls, and several goblet cells appeared to be actively producing mucus. In nasal septa removed from untreated animals, the tight junctional complexes viewed at higher magnification were intact, and cell–cell junctions appeared closed. However, tight junctions and cell–cell junctions were often difficult to identify 15 min after the addition of 0.125% TDM. Two hours postaddition of TDM, cell–cell junctions appeared to be widened both apically and basolaterally, in comparison to those seen in untreated samples. However, 4 h post-TDM administration, tight junctions and cilia appeared fully intact and basolateral junctions appeared less widened. Administration of TDM resulted in rapid early changes (~15 min) in nasal cell morphology that included a reduction in the density of cilia, an increase in apical cell membrane endocytosis, and a perturbation of tight junctions. Later changes (~2–4 h) in nasal cell morphology may represent recovery of normal cellular architecture, including an apparent recovery of tight junction integrity and the appearance of normal cilia.

20.5 INTRANASAL DELIVERY OF OTHER PEPTIDES

Since a comprehensive listing of every potential peptide and macromolecule that could be delivered nasally is not possible, this section will focus on: (1) the drugs currently in use as nasal products in humans, (2) drugs currently in use as injectables in humans that will gain increased utilization if and when they can be formulated for nasal delivery, and (3) drugs that are currently in development for utilization in humans (Table 20.2).

20.5.1 Peptides Currently Used in Nasal Formulations

Arginine vasopressin is a nonapeptide synthesized in the posterior pituitary. Its physiological roles are to serve as a vasoconstrictor to conserve fluid volume at times of severe hemorrhage, and to increase water reabsorption in the collecting ducts of the nephron. These actions are achieved by binding two distinct vasopressin receptors, V_1 and V_2 found in vascular tissues and the kidney, respectively. Patients with inadequate vasopressin production can receive intravenous, intramuscular, subcutaneous, or intranasal replacement with synthetic vasopressin. As vasopressin has a short half-life in the systemic circulation, a D-amino acid analog, desmopressin, was developed to provide a longer-acting congener of vasopressin. Desmopressin can be administered intranasally, intravenously, subcutaneously, or orally. Desmopressin is administered nasally at bedtime to increase water reabsorption in children who display excessive nocturia and enuresis. The bioavailability of nasal desmopressin is approximately 20-fold greater than oral desmopressin tablets and approximately 3.2% compared to intravenous delivery. This modest bioavailability is achieved in the absence of any absorption-enhancing agent in the formulation.

TABLE 20.2
Nasal Peptide Drug Candidates

Status	Peptide	Use
Currently used in nasal formulations	Arginine vasopressin, desmopressin	*Endocrine disorders*: Nocturnal enuresis; bleeding, hemophilia A, diabetes insipidus
	Nafarelin acetate	Endometriosis
	Calcitonin	Osteoporosis
Currently used as injectables	Amylin (Pramlintide)	*Diabetes mellitus*: Diabetes mellitus
	GLP-1-related peptides (Exendin-4, exenatide, byetta)	Diabetes mellitus
	Insulin	Diabetes mellitus
	Glucagon	Severe hypoglycemia
	FSH, LH, HCG	*Endocrine disorders*: Infertility
	Ganirelix acetate	Infertility
	Triptorelin	Prostate cancer
	GnRH and analogs (gonadorelin, leuprolide, goserelin)	Endometriosis, prostate and breast cancer, control of ovulation
	Cetrorelix	Premature ovulation
	Octreotide, somatostatin	Acromegaly, carcinoid, bleeding
	Teriparatide (PTH 1-34)	Osteoporosis
	Human growth hormone	Dwarfism, AIDS wasting
	Vasoactive intestinal peptide	Vasodilation
	Enfuvirtide	*Antiviral*: HIV fusion inhibitor
	Interferon-alpha, human granulocyte-stimulating factor	Chronic hepatitis C; malignant melanoma
	Erythropoietin	Miscellaneous: Anemia
	G-CSF (filgrastim)	Neutropenia
	Glial-derived neurotrophic factor	Parkinson's
	Interferon-beta	Multiple sclerosis
Currently in development	Leptin	*Obesity*: Obesity, hypothalamic amenorrhea
	PYY	Obesity, diabetes mellitus
	Ghrelin	Obesity, wasting
	IGF-1	*Endocrine disorders*: Short stature, AIDS wasting, dwarfism, burns

Nafarelin is a decapeptide analog of gonadotropin-releasing hormone (GnRH), a peptide hormone produced in the hypothalamus. GnRH stimulates follicle-stimulating hormone (FSH) and luteinizing hormone (LH) release from the pituitary gland. Nafarelin has a much longer half-life than GnRH. Nafarelin can be administered nasally to induce hypogonadism and to treat endometriosis and central precocious puberty.

Calcitonin is a peptide hormone produced in the thyroid gland that serves to lower serum calcium and phosphate levels by inhibiting bone resorption. Calcitonin has been used in the treatment of a variety of diseases, such as primary hyperparathyroidism, Paget's disease, and postmenopausal osteoporosis [99,100]. Salmon calcitonin has a longer half-life than human calcitonin. Salmon calcitonin, 3.6 kDa, is available as a nasal formulation that contains only benzalkonium chloride as a preservative, without an absorption enhancer, and as a parenteral product for injection. The direct effect of benzalkonium chloride on the nasal mucosa is under investigation [74] and this preservative may actually increase the absorption of calcitonin from the nose. Unfortunately, the bioavailability of the marketed nasal preparation is low and variable (0.3%–30%) [101]. In the rat, addition of TDM to a nasal calcitonin formulation caused a significant increase in nasal calcitonin bioavailability, from less than 5% to greater than 75% [1,90]. Development of a safe effective nasal calcitonin formulation containing an absorption enhancer could increase the bioavailability of calcitonin and improve the efficacy of this product in the treatment of osteoporosis, Paget's disease, and hyperparathyroidism.

20.5.2 Peptides Currently Used as Injectables

Several peptide products used in the treatment of diabetes mellitus, in addition to insulin, are currently administered by subcutaneous injection and these drugs are candidates for development of nasal formulations. Glucagon-like peptide-1 (GLP-1)-related peptides stimulate the insulin response to glucose and diminish the release of glucagon after a meal. These effects diminish the excessive postprandial increase in glucose observed after a meal in persons with type 2 diabetes mellitus. GLP-1-related peptides must be administered by subcutaneous injection before meals in order to be effective. This requirement for injection before each meal is likely to impact the utilization of these products by persons with type 2 diabetes. Exendin-4 is a GLP-1-related peptide with a molecular mass of 4.2 kDa. The development of a GLP-1-related peptide nasal formulation containing an absorption enhancer would allow patients to self-administer one of these drugs just before a meal without the need for a subcutaneous injection.

Amylin is a peptide produced by pancreatic β-cells along with insulin. Amylin reduces postprandial hyperglycemia by slowing absorption of carbohydrates from the intestine and reducing hepatic glucose release. Amylin levels are diminished in patients with type 1 diabetes. Pramlintide (Symlin) is a synthetic analog of amylin used to reduce postprandial hyperglycemia. Pramlintide must be injected before meals in order to be effective. Patients with type 1 diabetes, and patients with type 2 diabetes who use insulin, have been shown to achieve better glycemic control when pramlintide therapy was initiated. Development of a nasal pramlintide formulation containing an absorption enhancer would allow patients to self-administer this drug before meals without the need for a subcutaneous injection. Like the GLP-1 products described above, amylin is likely to be underutilized by persons with type 2 diabetes as long as subcutaneous injections are required multiple times per day.

Insulin administration by inhalation is now approved in the United States. Inhaled insulin is rapidly absorbed following administration with, or just before, meals. Nasal insulin is likely to fill a similar niche in the treatment of diabetes mellitus. The development of a safe, economical nasal insulin formulation containing an absorption enhancer will change the paradigm of current therapeutic management of type 1 diabetes and is likely to also change the treatment of patients with type 2 diabetes. As with inhaled insulin, utilization of nasal insulin will require precise dosing capabilities and an exquisite margin of safety since the product will be utilized 3–6 times daily for the entire life of the patient.

Glucagon is currently used to rescue patients with diabetes mellitus from severe hypoglycemic episodes. When glucose levels fall precipitously, patients become unresponsive and sometimes present with convulsions. An intramuscular injection of glucagon stimulates hepatic release of glucose. Unfortunately, administration of glucagon, which is packaged as a dry powder with a diluent, is often difficult for caregivers confronted with an emergency situation. A nasal glucagon formulation would provide improved portability, convenience, and safety. Animal studies have confirmed that nasal administration of glucagon in a formulation containing an absorption enhancer is effective [86].

In addition to the drugs described above, which would enhance the treatment of patients with diabetes mellitus, additional drugs described in Table 20.2 are candidates for delivery in nasal formulations containing absorption enhancers. For example, somatropin, the recombinant form of growth hormone, has been utilized as replacement therapy in children with growth hormone deficiencies [102,103]. Currently somatropin is administered by injection every evening. This approach to GH replacement therapy fails to mimic the natural pattern of hourly GH secretion from the pituitary. Introduction of a safe effective GH nasal formulation would allow smaller, more frequent GH dosing, which could prove to be more efficacious. Similarly, several peptide drugs used to treat endocrine disorders are currently administered by injection and they fail to achieve an adequate mimicry of natural pituitary hormone secretion. The development of a nasal formulation containing an absorption enhancer would improve the dosing regimen of these hormone analogs and would likely improve their efficacy.

Epoetin alfa, recombinant erythropoietin, is a glycoprotein that simulates erythrocyte production. Epoetin alfa is administered three times weekly subcutaneously or intravenously. Epoetin is used to treat anemia in patients with chronic renal failure, HIV infection, and patients receiving chemotherapy [104]. Development of a safe, effective nasal formulation of epoetin alfa, containing an absorption enhancer could once again improve the efficacy of epoetin alfa therapy, and reduce the number of injections required in these sensitive patient populations.

20.5.3 Peptides Currently in Development

Leptin is a peptide hormone secreted by adipocytes. Recombinant human leptin has been investigated for its potential as an antiobesity agent [105,106]. Women with hypothalamic amenorrhea display reduced levels of leptin. Leptin administration to these women improves reproductive and neuroendocrine function [107]. Nasal administration of leptin to rats in the presence of either TDM (1) or LPC [108] caused a significant increase in serum leptin levels. Increased serum leptin levels were associated with reduced food consumption [108]. The development of an effective nasal formulation of leptin containing an absorption enhancer may allow more frequent dosing with leptin and thereby overcome the limited efficacy observed following subcutaneous injections of large doses of this hormone.

Peptide YY (PYY) is a peptide hormone produced by L-cells in the intestinal tract after eating. It appears to act centrally to diminish appetite. It may serve normally to produce a sense of satiety. Diminished PYY production would be expected to cause increase in appetite and increased food intake. Hence, a biologically active PYY derivative, PYY 3–36, is under investigation as a potential agent to treat obesity and type 2 diabetes. A nasal PYY 3–36 product, if developed, would be an attractive alternative to injected PYY 3–36 because of the rapid and direct nose-to-brain drug transport that occurs following nasal drug delivery.

Ghrelin is a peptide secreted from the gastric fundus, upstream from the site of PYY secretion. Ghrelin has opposite effects from PYY, i.e., ghrelin stimulates hunger and food intake just before a meal begins. Ghrelin could potentially be used clinically in patients with abnormally low appetites. Development of novel ghrelin antagonists would likely produce a reduction in hunger and food intake. Hence, a ghrelin antagonist, taken alone or in combination with a PYY agonist, could curb appetite and be a useful antiobesity, weight-reduction agent.

Insulin-like growth factor-1 (IGF-1) is a peptide, 7649 Da, with structural homology to proinsulin. It binds preferentially to IGF-1 receptors on the surface of target cells. It has a lower affinity for insulin receptors. The receptors for insulin and IGF-1 share some structural and functional homology. IGF-1 is produced in the liver and in other organs in response to growth hormone action. IGF-1 that reenters the circulation is bound to

IGF-binding proteins. Protein binding of IGF-1 serves to increase its half-life. Thorne et al. [109] demonstrated that in rats, nasally administered IGF-1 was rapidly transported into the brain. IGF-1 uptake into the brain appeared to follow two routes, through olfactory nerve tracts and through trigeminal associated extracellular pathways. The net effect of nasal IGF-1 administration in rats was to rapidly elicit biological effects at several sites in the brain.

20.6 THE FUTURE OF NASAL PEPTIDE DRUG DELIVERY

Nasal drug delivery is likely to achieve a prominent role in a large number of therapeutic regimens that currently utilize peptide drugs delivered by injection. The normal physiology of the hypothalamic–pituitary system features hourly pulsatile release of peptide hormones that rapidly bind to cell surface hormone receptors and subsequently are cleared from the systemic circulation through proteolysis, glomerular filtration, and endocytosis at the target cell. This pattern is not reproduced by subcutaneous or intramuscular injections of large doses of exogenous peptide drugs. This pharmocokinetic and pharmacodynamic problem may be largely responsible for the decades of suboptimal therapy for numerous endocrine disorders, including diabetes mellitus, dwarfism, acromegaly, infertility, diabetes insipidus, and anemia. The powerful allure of nasal peptide drug delivery is the realization that patients may be able to self-administer smaller more frequent doses of peptide drugs. Patient compliance is expected to increase and the generation of contaminated needles is expected to decrease. The efficacy of drugs such as erythropoietin and growth hormone is expected to improve as a result of multiple daily administrations.

The introduction of safe, effective absorption-enhancing agents represents the critical element limiting the widespread use of nasal peptide formulations. Progress in this area has been slow to date, but several agents appear to have the potential to enable consistent and efficient absorption of peptide drugs from the nose.

ACKNOWLEDGMENT

The technical assistance of Libby Wilson in the preparation of this chapter is sincerely appreciated. The research studies described herein were supported in part by a grant from the NIEHS, NIH, and STTR program.

REFERENCES

1. Arnold, J.J., et al. 2004. Correlation of tetradecylmaltoside induced increases in nasal peptide drug delivery with morphological changes in nasal epithelial cells. *J Pharm Sci* 93:2205.
2. Illum, L. 2002. Nasal drug delivery: New developments and strategies. *Drug Discov Today* 7:1184.
3. Aungst, B.J., N.J. Rogers, and E. Shefter. 1988. Comparison of nasal, rectal, buccal, sublingual, and intramuscular insulin in efficacy and the effects of a bile salt absorption enhancer. *J Pharm Exp Ther* 244:23.
4. Marttin, E., et al. 1997. Confocal laser scanning microscopic visualization of the transport of dextrans after nasal administration to rats: Effects of absorption enhancers. *Pharm Res* 14:631.
5. Arnold, J.J., et al. 2002. Nasal administration of low molecular weight heparin. *J Pharm Sci* 91:1707.
6. Pillion, D.J., S. Hosmer, and E. Meezan. 1998. Dodecylmaltoside-mediated nasal and ocular absorption of lyspro-insulin: Independence of surfactant action from peptide multimer dissociation. *Pharm Res* 15:1641.
7. Illum, L., V. Dodane, and K. Iqbal. 2002. Chitosan technology to enhance the effectiveness of nasal drug delivery. *Drug Deliv Tech* 2:40.

8. Maitani, Y., et al. 2000. The enhancing effect of soybean-derived sterylglucoside and beta-sitosterol beta-D-glucoside on nasal absorption in rabbits. *Int J Pharm* 200:17.
9. Merkus, F.W., et al. 1999. Cyclodextrins in nasal drug delivery. *Adv Drug Deliv Rev* 36:41.
10. Maggio, E.T. 2005. Recent developments in intranasal drug delivery technology are creating new vistas for peptide and protein therapeutics. *Drug Deliv Comp Report, Spring/Summer* 34:37.
11. Donovan, M.D., and Y. Huang. 1998. Large molecule particulate uptake in the nasal cavity: The effect of size on nasal absorption. *Adv Drug Deliv Rev* 29:147.
12. Lee, V.H., A. Yamamoto, and U.B. Kompella. 1991. Mucosal penetration enhancers for facilitation of peptide and protein drug absorption. *Crit Rev Ther Drug Carrier Syst* 8:91.
13. Mitra, R., et al. 2000. Lipid emulsions as vehicles for enhanced nasal delivery of insulin. *Int J Pharm* 15:127.
14. Varshosaz, J., H. Sasrai, and R. Alinagari. 2004. Nasal delivery of insulin using chitosan microspheres. *J Microencapsul* 21:761.
15. Sayani, A.P., and Y.W. Chien. 1996. Systemic delivery of peptides and proteins across absorptive mucosae. *Crit Rev Ther Drug Carrier Syst* 13:85.
16. Hussain, A., and F. Ahsan. 2005. State of insulin self-association does not affect its absorption from the pulmonary route. *Eur J Pharm Sci* 25:289.
17. Davis, S.A., and L. Illum. 2003. Absorption enhancers for nasal drug delivery. *Clin Pharmacokinet* 42:1107.
18. Mygind, N., and R. Dahl. 1998. Anatomy, physiology and function of the nasal cavities in health and disease. *Adv Drug Deliv Rev* 29:3.
19. Gizuarson, S. 1993. The relevance of nasal physiology to the design of drug absorption studies. *Adv Drug Deliv Rev* 11:329.
20. Wermeling, D.P., J.L. Miller, and A.C. Rudy. 2002. Systemic intranasal drug delivery: Concepts and applications. *Drug Deliv Tech* 2:56.
21. Frey, W.H. 2002. Bypassing the blood–brain barrier to deliver therapeutic agents to the brain and spinal cord. *Drug Deliv Tech* 2:46.
22. Sakane, T., et al. 1995. Direct drug transport from the rat nasal cavity to the cerebrospinal-fluid. The relation to the molecular-weight of drugs. *J Pharm Pharmacol* 47:379.
23. Chou, K.J., and M.D. Donovan. 1997. Distribution of antihistamines into the CSF following intranasal delivery. *Biopharm Drug Disp* 18:335.
24. Illum, L. 2000. Transport of drugs from the nasal cavity to the central nervous system. *Eur J Pharm Sci* 11:1.
25. Chow, H.H.S., Z. Chen, and G.T. Matsura. 1999. Direct transport of cocaine from the nasal cavity to the brain following intranasal cocaine administration in rats. *J Pharm Sci* 88:754.
26. Dufes, C., et al. 2003. Brain delivery of vasoactive intestinal peptide (VIP) following nasal administration to rats. *Int J Pharm* 255:87.
27. Chien, Y.W. 1995. Biopharmaceutics basis for transmucosal delivery. *STP Pharma Sci* 5:718.
28. Druce, H.M. 1986. Nasal physiology. *Ear Nose Throat J* 65:201.
29. Khanvilkar, K., M.D. Donovan, and D.R. Flanagan. 2001. Drug transfer through mucus. *Adv Drug Deliv Rev* 48:173.
30. Dondeti, P., H. Zia, and T.E. Needham. 1996. Bioadhesive and formulation parameters affecting nasal absorption. *Int J Pharm* 27:115.
31. Marttin, E., et al. 1998. Nasal mucociliary clearance as a factor in nasal drug delivery. *Adv Drug Deliv Rev* 29:13.
32. Witschi, C., and R.J. Mrsny. 1999. *In vitro* evaluation of microparticles and polymer gels for use as nasal platforms for protein delivery. *Pharm Res* 16:382.
33. Schipper, N.G.M., et al. 1999. Chitosans as absorption enhancers of poorly absorbable drugs 3: Influence of mucus on absorption enhancement. *Eur J Pharm Sci* 8:335.
34. Lim, S.T., et al. 2000. Preparation and evaluation of the *in vitro* drug release properties and mucoadhesion of novel microspheres of hyaluronic acid and chitosan. *J Control Release* 66:281.
35. Illum, L., et al. 2001. Bioadhesive starch microspheres and absorption enhancing agents act synergistically to enhance the nasal absorption of polypeptides. *Int J Pharm* 222:109.

36. Ugwoke, M.I., N. Verbeke, and R. Kinget. 2001. The biopharmaceutical aspects of nasal mucoadhesive drug delivery. *J Pharm Pharmacol* 53:3.
37. Edsman, K., and H. Hagerstrom. 2005. Pharmaceutical applications of mucoadhesion for the nonoral routes. *J Pharm Pharmacol* 57:3.
38. Dahl, A.R., and J.L. Lewis. 1993. Respiratory tract uptake of inhalants and metabolism of xenobiotics. *Ann Rev Pharmacol Toxicol* 32:383.
39. Aceto, A., et al. 1989. Glutathione transferase in human nasal mucosa. *Arch Toxicol* 63:427.
40. Morimoto, K., et al. 1991. Effects of viscous hyaluronate–sodium solutions on the nasal absorption of vasopressin and an analogue. *Pharm Res* 8:471.
41. Na, D.H., et al. 2004. Stability of PEGylated salmon calcitonin in nasal mucosa. *J Pharm Sci* 93:256.
42. Shin, B.S., et al. 2004. Nasal absorption and pharmacokinetic disposition of salmon calcitonin modified with low molecular weight polyethylene glycol. *Chem Pharm Bull* 52:957.
43. Schmidt, M.C., et al. 1998. Translocation of human calcitonin in respiratory nasal epithelium is associated with self-assembly in lipid membrane. *Biochemistry* 37:16582.
44. Cremaschi, D., et al. 1996. Endocytosis inhibitors abolish the active transport of polypeptides in the mucosa of the nasal upper concha of the rabbit. *Biochim Biophys Acta Biomemb* 1280:27.
45. Lang, S., et al. 1998. Permeation and pathways of human calcitonin (hCT) across excised bovine nasal mucosa. *Peptides* 19:599.
46. Lutz, K.L., and T.J. Siahaan, 1997. Molecular structure of the apical junction complex and its contribution to the paracellular barrier. *J Pharm Sci* 86:977.
47. Marttin, E., et al. 1995. Effects of absorption enhancers on rat nasal epithelium *in vivo*. Release of marker compounds in the nasal cavity. *Pharm Res* 12:1151.
48. Agerholm, C., et al. 1994. Epithelial transport and bioavailability of intranasally administered human growth hormone formulated with the absorption enhancers didecanoyl-L-alpha-phosphatidylcholine and alpha-cyclodextrins in rabbits. *J Pharm Sci* 83:1706.
49. Rajewski, R.A., and V.J. Stella. 1996. Pharmaceutical applications of cyclodextrins 2. *In vivo* drug delivery. *J Pharm Sci* 85:1142.
50. Merkus, F.W., et al. 1991. Absorption enhancing effect of cyclodextrins on intranasally administered insulin in rats. *Pharm Res* 8:588.
51. Francis, S.A., et al. 1999. Rapid reduction of MDCK cell cholesterol by methyl-beta-cyclodextrin alters steady state transepithelial electrical resistance. *Eur J Cell Biol* 78:473.
52. Marttin, E., J.C. Verhoef, and F.W.H.M. Merkus. 1998. Efficacy, safety and mechanism of cyclodextrins as absorption enhancers in nasal delivery of peptide and protein drugs. *J Drug Target* 6:17.
53. Ahsan, F., et al. 2003. Effects of permeability enhancers, tetradecylmaltoside and dimethyl-beta-cyclodextrin, on insulin movement across human bronchial epithelial cells ($16HBE14o^-$). *Eur J Pharm Sci* 20:27.
54. Ahsan, F., et al. 2001. Mutual inhibition of the insulin absorption-enhancing properties of dodecylmaltoside and dimethyl-beta-cyclodextrin following nasal administration. *Pharm Res* 18:608.
55. Yetkin, G., et al. 1999. The effect of dimethyl-beta-cyclodextrin and sodium taurocholate on the nasal bioavailability of salmon calcitonin in rabbits. *STP Pharma Sci* 3:249.
56. Yang, T., et al. 2004. Cyclodextrins in nasal delivery of low-molecular-weight heparins: *In vivo* and *in vitro* studies. *Pharm Res* 21:1127.
57. Illum, L. 1998. Chitosan and its use as a pharmaceutical excipient. *Pharm Res* 15:1326.
58. Illum, L., N.F. Farraj, and S.S. Davis. 1994. Chitosan as a novel delivery system for peptide drugs. *Pharm Res* 11:1186.
59. Soane, R.J., et al. 1999. Evaluation of the clearance characteristics of bioadhesive systems in humans. *Int J Pharm* 178:55.
60. Aspden, T.J., L. Illum, and O. Skaugrud. 1996. Chitosan as a nasal delivery system: Evaluation of insulin absorption enhancement and effect on nasal membrane integrity using rat models. *Eur J Pharm Sci* 4:23.
61. Dodane, V., and V.D. Vilivalam. 1998. Pharmaceutical applications of chitosan. *Pharm Sci Technol Today* 1:246.

62. Fernandez-Urrusuno, R., et al. 1999. Enhancement of nasal absorption of insulin using chitosan nanoparticles. *Pharm Res* 16:1576.
63. Illum, L., et al. 2000. Novel chitosan based delivery systems for the nasal administration of a LHRH-analogue. *STP Pharma Sci* 10:89.
64. Janes, K.A., P. Calvo, and M.J. Alonso. 2001. Polysaccharide colloidal particles as delivery systems for macromolecules. *Adv Drug Deliv Rev* 47:83.
65. Van der Lubben, I.M., et al. 2001. Chitosan and its derivatives in mucosal drug and vaccine delivery. *Eur J Pharm Sci* 14:201.
66. Dyer, A.M., et al. 2002. Nasal delivery of insulin using novel chitosan based formulations: A comparative study in two animal models between simple chitosan formulations and chitosan nanoparticles. *Pharm Res* 19:998.
67. Hinchcliffe, M., I. Jabbal-Gill, and A. Smith. 2005. Effect of chitosan on the intranasal absorption of salmon calcitonin in sheep. *J Pharm Pharmacol* 57:681.
68. Artursson, P., et al. 1994. Effect of chitosan on the permeability of monolayers of intestinal epithelial cells (Caco-2). *Pharm Res* 11:1358.
69. Lee, W.A., et al. 1991. Intranasal bioavailability of insulin powder formulations: Effect of permeation enhancers-to-protein ratio. *J Pharm Sci* 80:725.
70. Li, Y., Z. Shao, and A.K. Mitra. 1992. Dissociation of insulin oligomers by bile salt micelles and its effect on alpha-chymotrypsin-mediated proteolytic degradation. *Pharm Res* 7:864.
71. Deurloo, M.J., et al. 1989. Absorption enhancement of intranasally administered insulin by sodium taurodihydrofusidate (STDHF) in rabbits and rats. *Pharm Res* 6:853.
72. Johansson, F., et al. 2002. Mechanisms for absorption enhancement of inhaled insulin by sodium taurocholate. *Eur J Pharm Sci* 17:63.
73. Merkus, F.W.H.M., N.G.M. Schipper, and J.C. Verhoef. 1996. The influence of absorption enhancers on intranasal insulin absorption in normal and diabetic subjects. *J Control Release* 41:69.
74. Marttin, E., et al. 1996. Acute histopathological effects of benzalkonium chloride and absorption enhancers on rat nasal epithelium *in vivo*. *Int J Pharm* 141:151.
75. Kensil, C.R., and R. Kammer. 1998. QS-21: A water soluble triterpene glycoside adjuvant. *Expert Opin Investig Drugs* 9:1475.
76. Recchia, J., et al. 1995. A semisynthetic *Quillaja saponin* as a drug delivery agent for aminoglycoside antibiotics. *Pharm Res* 12:1917.
77. Pillion, D.J., et al. 1996. Structure–function relationship among *Quillaja saponins* serving as excipients for nasal and ocular delivery of insulin. *J Pharm Sci* 84:518.
78. Pillion, D.J., et al. 1995. DS-1, a modified *Quillaja saponin*, enhances ocular and nasal absorption of insulin. *J Pharm Sci* 84:1276.
79. Maitani, Y., et al. 1995. The effect of soybean-derived sterol and its glucoside as an enhancer of nasal absorption of insulin in rabbits *in vitro* and *in vivo*. *Int J Pharm* 117:129.
80. Ando, T., et al. 1998. Nasal insulin delivery in rabbits using soybean-derived sterylglucoside and sterol mixtures as novel enhancers in suspension dosage forms. *Biol Pharm Bull* 21:862.
81. Muramatsu, K., et al. 1994. Effect of soybean-derived sterol and its glucoside mixtures on the stability of dipalmitoyl-phosphatidylcholine and dipalmitoylphosphatidylcholine/cholesterol liposomes. *Int J Pharm* 107:1.
82. Carstens, S., et al. 1993. Transport of insulin across rabbit nasal mucosa *in vitro* induced by didecanoyl-L-alpha-phosphatidylcholine. *Diabetes* 42:1032.
83. Jacobs, M.A., et al. 1993. The pharmacodynamics and activity of intranasally administered insulin in healthy male volunteers. *Diabetes* 42:1649.
84. Ahsan, F., et al. 2003. Sucrose cocoate, a component of cosmetic preparation, enhances nasal and ocular peptide absorption. *Int J Pharm* 251:195.
85. Pillion, D.J., et al. 1994. Insulin delivery in nosedrops: New formulations containing alkylglycosides. *Endocrinology* 135:1386.
86. Pillion, D.J., et al. 1995. Systemic absorption of insulin and glucagon applied to the eyes of rats and a diabetic dog. *J Ocul Pharmacol* 11:283.
87. Morgan, R.V. 1995. Delivery of systemic regular insulin via the ocular route in cats. *J Ocul Pharmacol Ther* 4:565.

88. Pillion, D.J., J.A. Atchison, and E. Meezan. 1994. Alkylglycosides enhance systemic absorption of insulin delivered topically to the rat eye. *J Pharmacol Exp Ther* 271:1274.
89. Pillion, D.J., et al. 2002. Synthetic long-chain alkylglycosides as enhancers of nasal insulin absorption. *J Pharm Sci* 91:1458.
90. Ahsan, F., et al. 2001. Enhanced bioavailability of calcitonin formulated with alkylglycosides following nasal and ocular administration in rats. *Pharm Res* 18:1742.
91. Chiou, G.C., C. Chuang, and M.S. Chang. 1988. Reduction of blood glucose concentrations with insulin eye drops. *Diabetes Care* 11:750.
92. Chiou, G.C., and C.Y. Chuang. 1989. Improvement of systemic absorption of insulin through eyes with absorption enhancers. *J Pharm Sci* 78:815.
93. The DCCT Research Group. 1993. The effect of intensive treatment of diabetes on the development and progression of long-term complications in insulin-dependent diabetes mellitus. *N Engl J Med* 329:977.
94. Turner, R.C., et al. 1998. Risk factors for coronary artery disease in non-insulin dependent diabetes mellitus: United Kingdom Prospective Diabetes Study (UKPDS: 23). *Br Med J* 316:823.
95. Dimitrijevic, D., A. Shaw, and A. Florence. 2000. Effects of some non-ionic surfactants on transepithelial permeability in Caco-2 cells. *J Pharm Pharmacol* 52:157.
96. Duizer, E., et al. 1998. Absorption enhancement, structural changes in tight junctions and cytotoxicity caused by palmitoyl carnitine in Caco-2 and IEC-18 cells. *J Pharmacol Exp Ther* 287:395.
97. Liu, D.-Z., E.L. Lecluyse, and D.R. Thakker. 1999. Dodecylphosphocholine-mediated enhancement of paracellular permeability and cytotoxicity in Caco-2 cell monolayers. *J Pharm Sci* 88:1161.
98. Wan, J., et al. 2000. Tight junction properties of the immortalized human bronchial epithelial cell lines Calu-3 and (16HBE14o$^-$). *Eur Respir J* 15:1058.
99. Marcus, R. 2000. *Melmon and Morrelli's Clinical Pharmacology*, 4th ed., 703. New York: Carruthers, Hoffman, Melmon, & Nierenberg.
100. Torring, O., et al. 1991. Salmon calcitonin treatment by nasal spray in primary hyperparathyroidism. *Bone* 12:311.
101. Miacalcin® Nasal Spray. Novartis Pharmaceutical Corp. East Hanover, NJ (1998). Package Insert.
102. Guyda, H.J. 1999. Four decades of growth hormone therapy for short children: What have we achieved? *J Clin Endocrinol Metab* 84:4307.
103. Ranke, M.B., et al. 1999. Dosing of growth hormone in growth hormone deficiency. *Horm Res* 51(Suppl. 3):70.
104. Hillman, R.S. 1996. *The pharmacological basis of therapeutics*, eds. J.G. Hardman and L.E. Limbird, 1312. New York: McGraw-Hill.
105. Crowley, V.E., G.S. Yeo, and S. O'Rahilly. 2002. Obesity therapy: Altering the energy intake-and-expenditure balance sheet. *Nat Rev Drug Disc* 1:276.
106. Heymsfield, S.B., et al. 1999. Recombinant leptin for weight loss in obese and lean adults: A randomized, controlled, dose-escalation trial. *J Am Med Assoc* 282:1568.
107. Welt, C.K., et al. 2004. Recombinant human leptin in women with hypothalamic amenorrhea. *N Engl J Med* 351:987.
108. Shimizu, H., et al. 2005. Inhibition of appetite by nasal leptin administration in rats. *Int J Obes* 29:858.
109. Thorne, R.G., et al. 2004. Delivery of insulin-like growth factor-1 to the rat brain and spinal cord along olfactory and trigeminal pathways following intranasal administration. *Neuroscience* 127:481.
110. Krishnamoorthy, R., et al. 1995. Cyclodextrins as mucosal absorption promoters 4. Evaluation of nasal mucotoxicity. *Eur J Pharm Biopharm* 41:296.
111. Matsubara, K., et al. 1995. Improvement of nasal bioavailability of luteinizing hormone-releasing hormone agonist, buserelin, by cyclodextrin derivatives in rats. *J Pharm Sci* 84:1295.
112. Irie, T., and K. Uekama. 1997. Pharmaceutical applications of cyclodextrins 3. Toxicological issues and safety evaluation. *J Pharm Sci* 86:147.
113. Yamamoto, T., et al. 1998. High absorbency and subchronic morphologic effects on the nasal epithelium of a nasal insulin powder dosage form with a soybean-derived sterylglucoside mixture in rabbits. *Biol Pharm Bull* 21:866.
114. Maitani, K., K. Nakamura, and K. Kawano. 2005. Application of sterylglucoside-containing particles for drug delivery. *Curr Pharm Biotechnol* 6:81.

Part VI

Drug Absorption from Vagina and Uterus

21 Vagina and Uterus as Drug-Absorbing Organs

R. Karl Malcolm, Stephen D. McCullagh, Ryan J. Morrow, and A. David Woolfson

CONTENTS

21.1 INTRODUCTION

Recent years have witnessed an increased interest in delivering drugs to the vagina and uterus. Much of this interest has stemmed from new perspectives on a range of female health applications, including contraception, hormone replacement therapy (HRT), cervicovaginal cancer, and the prevention and treatment of sexually transmitted diseases (STDs). Typically, these clinical applications take advantage of vaginal administration for provision of a direct local effect within the vaginal fluid or vaginal tissue. Consequently, systemic side effects are minimized and the magnitude of the dose may be much reduced compared to more conventional routes of drug delivery. However, employing the vagina as a portal for the systemic absorption of drugs, such as is required for steroidal contraception, has also been shown to be useful, particularly for drug molecules that undergo extensive first-pass hepatic metabolism. The renewed interest in vaginal drug delivery has also prompted reexamination of the anatomy, physiology, histology, and immunology of the vagina and uterus, leading to a much better understanding of the key parameters and mechanisms influencing vaginal drug absorption. This is best exemplified by the developments in the fields of uterine targeting, microbicides for HIV prevention, and vaginal vaccinations. It is, therefore, the purpose of this chapter to describe the various concepts and review recent advances in the use (and potential use) of the vagina and the uterus for the absorption of drug substances. The term absorption is interpreted in its widest sense so as to cover both local tissue and systemic absorption.

21.2 VAGINAL ANATOMY AND PHYSIOLOGY

A number of excellent articles have been published detailing the anatomy and physiology of the human vagina [1–4], and therefore only a brief overview is provided here.

21.2.1 Vaginal Anatomy

The vagina is a highly expandable, slightly s-shaped, fibromuscular, collapsible tube situated between the rectum, which lies posterior to it, and the urethra and bladder, which lie anterior to it (Figure 21.1). It extends from the lower part of the uterine cervix to the external part of the vulva known as the labia minor. The vault of the vagina is divided into four areas relative to the cervix. These are the posterior fornix, which is capacious, the anterior fornix, which is shallow, and two lateral fornices. The anterior wall of the vagina averages 6 to 7 cm in length, whereas the posterior wall is slightly longer (approximately 7.5 to 8.5 cm) due to the intrusion of the cervix below the vault. Recent advances in three-dimensional imaging techniques, such as magnetic resonance imaging (MRI), have also provided new insights into vaginal anatomy [5–8].

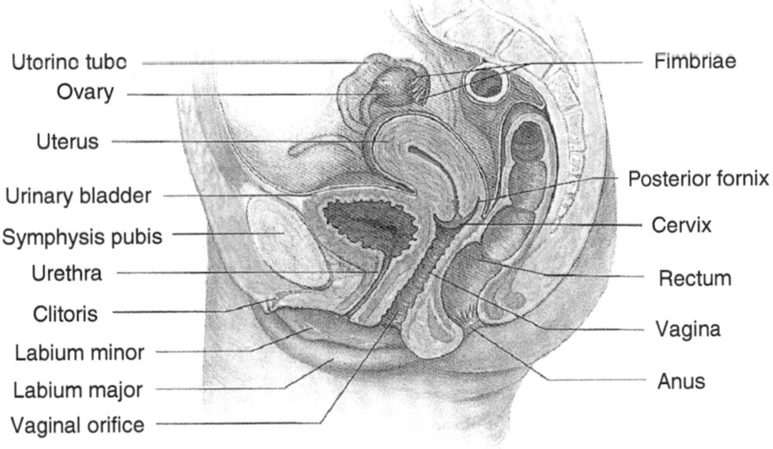

FIGURE 21.1 Schematic representation of the female reproductive system.

21.2.2 Vaginal Histology

The vagina consists of three distinctive tissue layers: the mucosa, the muscularis, and the tunica adventitia (Figure 21.2). A more detailed description of the histological and immunological characteristics of these layers is provided in Table 21.1. Several studies have demonstrated that the thickness of the human vaginal epithelial layer remains relatively constant throughout the menstrual cycle [9–11] (Table 21.2), being neither influenced by anovulatory states, nor by the normal cyclic fluctuations in circulating hormones, which so effect the endometrial lining in the uterus.

However, changes in hormonal levels of estrogen and progesterone do affect the character of the cells within the vaginal epithelium. During the early follicular phase (also known as the proliferative stage), exfoliated vaginal epithelial cells have vesicular nuclei and are basophilic. During the late follicular phase, the vaginal epithelial cells display pyknotic nuclei and are acidophilic due to the influence of rising estrogen levels. As progesterone rises during the luteal phase, the acidophilic cells decrease in number and are replaced by an increasing number of leukocytes.

The blood supply to the vagina is through the uterine arteries and the internal iliac artery. Blood returns to the venous system through veins that empty into the internal iliac vein. Lymphatic drainage is through the external and internal iliac lymph nodes and superficial inguinal lymph nodes.

21.2.3 Vaginal Fluid

For a vaginally administered drug to exert a local or systemic effect, it must possess at least some degree of solubility in vaginal fluid. It is therefore important to consider the nature of vaginal fluid and, in particular, the characteristics of vaginal fluid that may effect vaginal absorption.

Vaginal fluid consists primarily of transudate that passes through the vaginal wall from the blood vessels. It is mixed with vulval secretions from sebaceous and sweat glands, with minor contributions from Bartholin's and Skene's glands [1,11]. The fluid then becomes contaminated with cervical mucus, produced by glandular units within the cervical canal, and sloughed cells from the vaginal epithelia. The amount, composition, and physical characteristics of

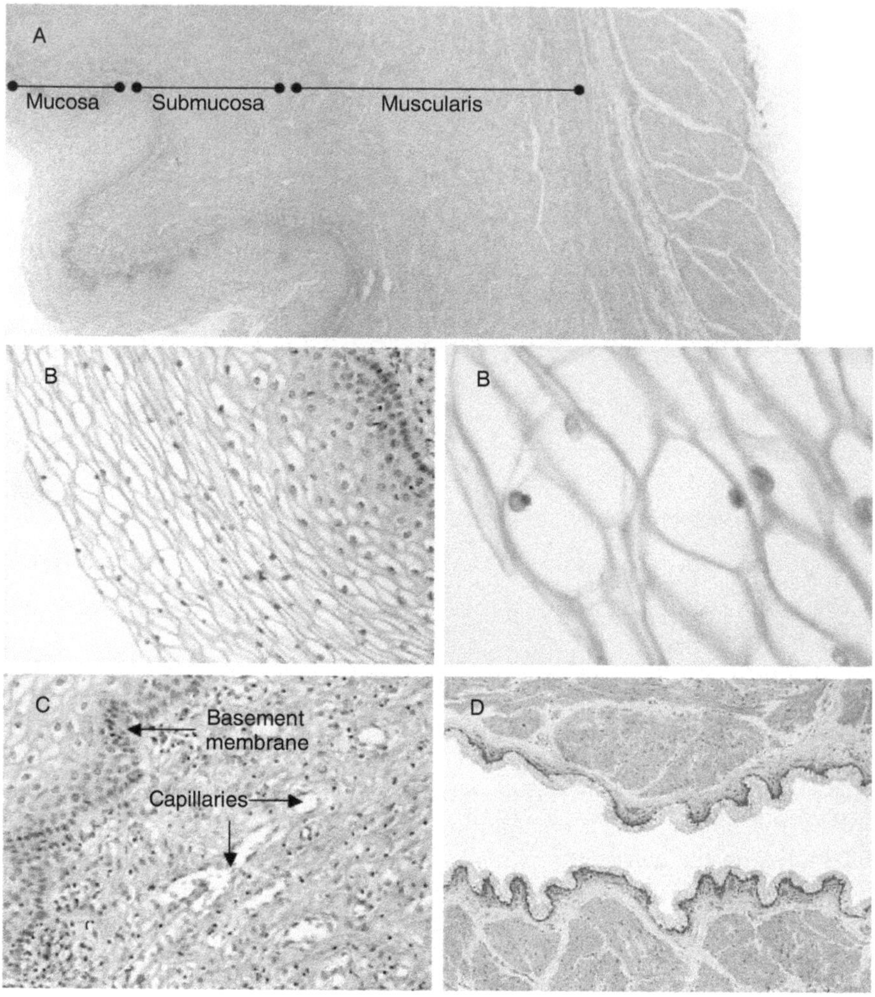

FIGURE 21.2 A, Histology of the vagina showing the mucosa, submucosa, and muscularis; B, non-keratinized stratified squamous epithelium of the vagina; C, submucosa of the vagina containing abundance of connective tissue and capillaries; D, the cervix lined by simple highly columnar epithelium containing mucous cells. Images used and modified with kind permission from Rose M Chute, Biology Department, North Harris College, Houston, Texas, US (http://science.nhmccd.edu/biol/reproductive/vagina.htm)

cervical mucus change with the menstrual cycle, making its production estrogen-dependent. The amount of mucus is minimal immediately after menstruation and becomes more transparent, viscous, and elastic before ovulation. At the time of ovulation, the amount of cervical secretions further increases, resulting in an increase in the overall volume of vaginal fluid. Consequently, there is an increase in fibrosity, pH, and mucin content and a decrease in the viscosity, cellularity, and albumin concentration. Endometrial and oviductal fluids may also contribute to the chemical composition of vaginal fluid. Enzymes, enzyme inhibitors, protein, carbohydrates, amino acids, alcohols, hydroxyketones, and aromatic compounds are also components of vaginal fluid. Previous studies suggest that approximately 6 g of vaginal fluid are produced daily with approximately 0.5 to 0.75 g present at any one time in the vagina [12–21], depending upon the menstrual cycle and the presence of exogenous hormones.

TABLE 21.1
Human Vaginal Epithelial Cell-Layer Thickness

Ref.	**Epithelial Cell-Layer Thickness on Different Days of Menstrual Cycle**																							
	1	**2**	**3**	**4**	**5**	**6**	**7**	**8**	**9**	**10**	**11**	**12**	**13**	**14**	**15**	**16**	**17**	**18**	**19**	**20**	**21**	**22**	**23**	**24**
Zondek and Friedmann (1936)				13–30 Cells (regular menstruation);										15–32 Cells (primary amenorrhea);						12–29 Cells (secondary amenorrhea)				
Burgos and de Vargas-Linares (1978)										22				45					33					23
Patton et al. (2000)			←28→							←28→											←26→			

TABLE 21.2
Histological Description of the Human Vagina

Layers	Sublayers	Histological Description	Immune Cells Present in Normal Vaginal Mucosa
Mucosa	Epithelium 15–200 mm	Approximately 25 layers (menstrual cycle dependent) of nonkeratinized, stratified, squamous cells; five cell types identified:	T8 lymphocytes
		(i) *Basal*: Small, typically columnar or squamous in shape with small nuclei, very little cytoplasm, and microvilli present on cell surface	T4 lymphocytes
		(ii) *Parabasal*: Polygonal in shape, similar in size and structure to basal cells	Langerhans cells
		(iii) Transitional	
		(iv) Intermediate: Largest cells, exhibit microvilli, abundant in cytoplasm and glycogen	Macrophages
		(v) Superficial: Outermost layer during follicular phase of cycle	Neutrophils
	Lamina propria	Loose, highly vascularized connective tissue (extremely rich in blood vessels and lymphatics)	Plasma cells
	Submucosa	Connective tissues made of collagen and elastic fibers, and blood vessels	B lymphocytes
Muscularis		Smooth muscle fibers arranged in an outer longitudinal layer and an inner circular layer	
Adventitia		Thin, fibrous layer composed of dense sheath of collagen and elastic fibers; contains blood vessels, lymph vessels, and nerves	

↑ Increasing number of cells in mucosa

The vaginal fluid in healthy mature women is maintained at a pH of between 3.5 and 5 by the commensal microorganism *Lactobacillus acidophilus*, which produces lactic acid from glycogen contained in the sloughed mature cells of the vaginal mucosa. The acidic nature of the vaginal fluid is of great practical importance as it offers natural resistance to the colonization of various microorganisms [22–27]. The pH of vaginal fluid rises during menstruation, but it may also increase after periods of frequent acts of coitus as both vaginal transudate, formed during coitus, and ejaculate are alkaline. Elevated pH has been associated with increased transmissibility of HIV infection [28–30]. Physiologically, the anterior fornix of vagina has the lowest pH, which gradually rises toward the vestitube. Intravaginal pH may also be affected by the presence of cervical mucus, which has a pH in the range 6.5 to 9, and by the amount of lubricating vaginal secretions. These changes could influence the solubility, uptake, and release profile of pH-sensitive substances within the vagina. It is also worth noting that vaginal fluid has limited buffering capacity [31,32]. Also, changes in the volume

TABLE 21.3
Recipe for Vaginal Fluid Simulant

Compound	Concentration (g/L)
Sodium chloride	3.5
Potassium hydroxide	1.4
Calcium hydroxide	0.22
Bovine serum albumin	0.018
Lactic acid	2.0
Acetic acid	1.0
Glycerol	0.16
Urea	0.40
Glucose	5.0
HCl (aq)	To pH 4.2
Water	To 1000 mL

Source: From Owen, H.O. and Katz, D.F., *Contraception* 59, 91, 1999. With permission.

and physical makeup of the fluid, particularly in response to sexual excitement, may further influence drug release, solubility, and permeation characteristics [33].

Synthetic fluid mediums, for use in *in vitro* studies, have been developed to simulate certain properties of vaginal and cervical fluids [20,34], and their compositions are detailed in Table 21.3 and Table 21.4, respectively.

21.2.4 Vaginal Flora

A wide range of microorganisms may be present within the human vagina [35–38]. A summary of the major species observed is presented in Table 21.5. The vagina becomes colonized soon after birth with corynebacteria, staphylococci, nonpyogenic streptococci, *Escherichia coli*, and *L. acidophilus*. During reproductive life, from puberty to menopause, the vaginal epithelium contains glycogen due to the actions of circulating estrogens. The predominate lactobacillus species metabolize glycogen to lactic acid, which is responsible for the acidic pH of the vaginal fluid. Also, lactic acid and other products of metabolism inhibit colonization by

TABLE 21.4
Recipe for Cervical Fluid Simulant

Compound	Concentration (g/L)
Guar gum (cross-linked with sodium borate)	10.0
Dried porcine gastric mucin	5.0
Imidurea	3.0
Methylparaben	1.5
Propylparaben	0.2
Dibasic potassium phosphate	2.6
Monobasic potassium phosphate	15.7
Water	(To 1000 mL)

Source: From Burruano, B.T., Schnaare, R.L., and Malamud, D., *Contraception* 66, 137, 2002. With permission.

TABLE 21.5
Major Species of Microorganisms Present in the Human Vagina

Species		Comment	Species		Comment
Bacteroides sp.	–		*Neisseria* sp.	+	Gram-negative cocci, *Neisseria gonorrhoeae* associated with gonorrhoea
Candida albicans	+	Causative microorganism for yeast infections, such as thrush	*Neisseria meningitidis*	+	
Corynebacteria	+		*Proteus* sp.	+	
Enterococcus faecalis	+		*Staphylococcus epidermis*	++	
Escherichia coli	+		*Staphylococcus aureus*	+	Present in approximately 5% or women; overgrowth may cause vaginitis or toxic shock syndrome
Garnerella vaginalis	++	Facultatively anaerobic gram-variable rods, one of the most frequent causes of vulvovaginitis, diagnosed by the presence of masses of *Gardnerella bacilli*	*Streptococcus mitis*	+	
Lactobacilli sp.	++	Major constituent of normal vaginal flora, facultative or anaerobic gram-positive rods; increased numbers during pregnancy, during ovulation, at menopause, and during steroid treatment	*Streptococcus pneumoniae*	–	
Mycoplasmas	+	*Hominis* species has been linked to premature labor and birth; also associated with bacterial vaginosis	*Streptococcus pyogenes*	–	

++, Very common; +, common; –, rare.
Source: Adapted from *Todar's Online Textbook of Bacteriology*; http://textbookofbacteriology.net/normalflora.html)

exogenous microbes. The resulting low pH of the vaginal epithelium prevents establishment of most bacteria as well as the potentially pathogenic yeast, *Candida albicans*.

21.3 UTERINE ANATOMY AND PHYSIOLOGY

The human uterus is a thick, muscular, pear-shaped organ composed of three major anatomical divisions—corpus, isthmus, and cervix (Figure 21.3). The main uterine cavity is known as the corpus and comprises the upper two-thirds of the uterus. It is several centimeters in

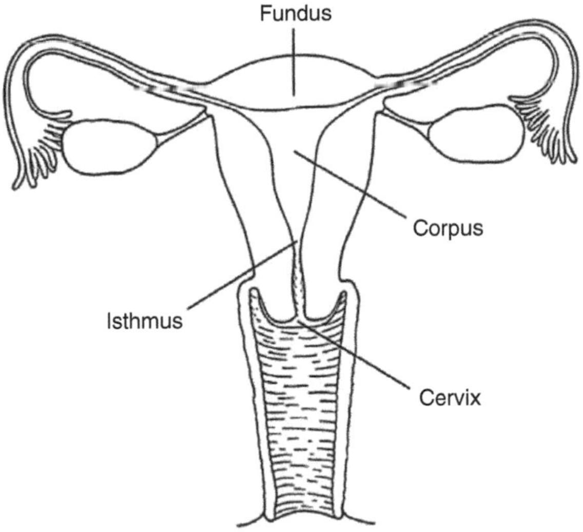

FIGURE 21.3 Anatomical features of the human uterus.

diameter and connected on both sides to the fallopian tubes. Below the corpus, the uterus narrows to form the isthmus, which terminates, through the cervical canal, in a constricted opening into the vagina known as the cervix. The main functions of the uterus include: transportation of sperm from vagina to uterine tubes for fertilization; site for implantation, nourishment and protection of embryo and fetus; expulsion of mature fetus at the end of pregnancy.

From a histological perspective, the uterine tissue consists of three well-defined layers—the *perimetrium* (tunica serosa), the *myometrium* (tunica muscularis), and the *endometrium* (tunica mucosa and submucosa) (Figure 21.4). The endometrium is the mucosal lining of the uterus consisting of epithelial tissue, tubular glands, and connective stroma tissue. Hormonal changes associated with the menstrual cycle cause the superficial two-thirds surface of the endometrium (also known as the functionalis or stratum functionale) to be sloughed in a process known as menstruation. However, the basalis (or stratum basale) component of the endometrial layer remains intact and permits regeneration of sloughed funtionalis layer. The highly vascularized myometrium consists of a thick inner circular layer and a thinner outer longitudinal layer of smooth muscle. The region in between the two layers of smooth muscle contains large blood vessels. The perimetrium consists of loose connective tissue containing a large number of lymphatic vessels.

21.4 VAGINAL DRUG ABSORPTION

According to historical records, women have been administering a wide range of substances to the vagina for thousands of years. One of the earliest such records is the Kahun Papyrus, an Egyptian treatise on gynecology dating back to 1825 BC. In addition to describing various methods of diagnosing pregnancy, determining the sex of the fetus, and treating toothache during pregnancy, the manuscript also describes the formulation and use of vaginal pastes and inserts to prevent pregnancy or to treat local infection. For example, a common contraceptive formulation involved the use of a vaginal suppository containing crocodile dung mixed with honey and sodium carbonate. In the Papyrus of Ebers, another ancient Egyptian document on medicine, women were advised to grind together dates, acacia (tree bark), and a

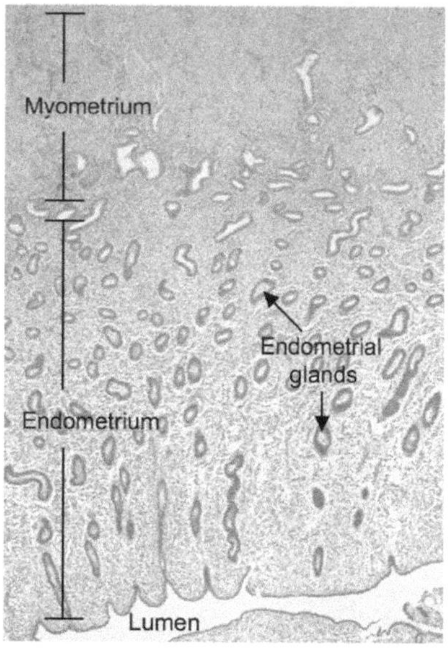

FIGURE 21.4 Histology of the uterus showing the endometrium and the myometrium. The perimetrium is not shown.

touch of honey into a moist paste, which was then administered to the vagina using dip seed wool. The contraceptive efficacy of these formulations was likely to be reasonably good. The viscous nature of the mixture would have prohibited sperm from passing through the cervical canal, whereas the acacia component would have fermented in aqueous environment of the vagina to produce lactic acid, which not only restores and maintains the natural, slightly acidic pH of the vagina but is also a potent spermicidal agent (sperm are immobilized at normal acidic vaginal pH and semen is slightly alkaline at approximately pH 7.4). In fact, lactic acid pessaries are still used today to restore normal vaginal ecology.

More alarmingly, during the nineteenth century, abortion, suicide, and homicide attempts using vaginally administered arsenic and other poisons were not uncommon. Thus, it has long been realized that exogenous chemicals, when administered intravaginally, could find their way into the systemic circulation. However, the modern concept of vaginal drug absorption was not formalized until 1918 with the publication of David Macht's seminal paper "The absorption of drugs and poisons through the vagina" in which the absorption of alkaloids, inorganic salts, esters, and antiseptics through the vagina was described, thus demonstrating for the first time the potential for systemic drug delivery through this route [39].

Much of the fundamental research investigating the mechanistic aspects of vaginal absorption was conducted during the 1970s. For example, between 1975 and 1977, Higuchi and colleagues published a series of four papers describing methods for determining the factors influencing the vaginal absorption and permeability in a rabbit model [40–43]. The studies demonstrated that (i) drug solutions of *n*-butanol administered to the vagina permeated across the vaginal membrane according to first-order kinetics, and were influenced by pH and buffer type, (ii) permeability coefficients for a homologous series of unbranched aliphatic alcohols increased with increasing number of methylene groups, and were in good agreement with a model in which that absorption barrier consists of an aqueous diffusion layer in series with a membrane having two parallel pathways (a lipoidal and an aqueous pore pathway) for solute transport, (iii) the absorption characteristics of progesterone and hydrocortisone from silicone matrices were determined by two mechanisms, namely silicone matrix diffusion-control and aqueous layer diffusion-control, and (iv) membrane pH was critical in order to rationalize vaginal absorption of weakly ionized drugs.

The administration of drug substances through the vagina may be effected to induce either a local or a systemic effect. Local therapies have primarily focused on the treatment of vaginal infections and postmenopausal vaginal atrophy, although, more recently, there has been considerable interest in the possibility of administering vaginal microbicides for the prevention of HIV infection. Vaginal administration for systemic therapies takes advantage of the highly vascularized vaginal tissue to provide circulating blood levels of drug molecules that might otherwise be extensively metabolized by first-pass hepatic metabolism if administered orally. For example, intravaginal delivery of estrogens, which are subject to extensive first-pass hepatic metabolism, requires a significantly lower dose than the oral route in order to achieve the required serum concentration of circulating steroid. As for all mucosal tissue, the absorption of substances from the vagina depends on a number of factors, including molecular size and volume, lipophilic and hydrophilic character, degree of ionization, local pharmacological action of the substance, thickness of the vaginal wall, and transport efficiency in the blood or lymph. Systemic drug absorption across the vaginal epithelium membrane involves a number of well-defined steps: drug release from the delivery system, drug dissolution in vaginal fluid, and absorption across vaginal epithelium. The local pharmacological action of the drug, thickness of the vaginal wall, presence of cervical mucous, and the presence of specific cytoplasmic receptors may also be important. Drug absorption is also modified by changes in the thickness of the vaginal wall influenced by the ovarian cycle or by pregnancy and by postmenopausal changes in vaginal epithelium and intravaginal pH.

Local intravaginal drug treatment follows drug release, dissolution, and delivery throughout the vaginal space. The significant absorption capability of the vagina for exogenous substances, including drugs, is now well recognized. Originally regarded as relatively passive and impermeable to foreign agents, the intravaginal absorption of numerous compounds has now been noted. The emphasis in human clinical trials has been primarily on the systemic delivery of contraceptives and, more recently, on estrogenic and progestogenic compounds for HRT. Currently, there is substantial interest in the intravaginal delivery of therapeutic peptides and proteins [44,45].

The pathways for drug diffusion across vaginal epithelium are essentially similar to other epithelial tissues [44] and are well represented by the fluid mosaic model as a lipid continuum interspersed with aqueous pores, the latter forming an aqueous shunt route [46]. The lipid continuum predominates in vaginal drug absorption. The permeability coefficient of a drug across a vaginal epithelial barrier membrane may be considered as the product of the amount of drug penetrating the membrane per unit time per unit area drug (flux) and the membrane thickness, divided by the drug concentration in the delivery vehicle. For drugs with a high vaginal membrane permeability coefficient, absorption is mainly controlled by permeability across the hydrodynamic diffusion layer formed by vaginal fluid sandwiched between the vaginal epithelial membrane and the delivery device [47]. For drugs with a low vaginal membrane permeability, vaginal absorption is mainly controlled by permeability across the vaginal epithelium [47]. Consequently, in vaginal drug delivery, for systemic drug absorption to occur, the penetrant substance must have sufficient lipophilicity to diffuse through the lipid continuum of the membrane, but also require some degree of aqueous solubility to ensure dissolution in vaginal fluid. This is sometimes a difficult compromise to achieve.

The sequence of events leading to the absorption of a vaginally administered substance depends, in part, on the nature of the delivery system that is employed, i.e., whether it is solid or semisolid, swellable or erodible, soluble or insoluble, immediate or controlled release.

The bioavailability of an intravaginally administered drug can be modified by the use of chemical penetration enhancers, typically acting on epithelial tight junctions to provide an alternative intercellular penetration route that may be particularly significant in the vaginal absorption of higher molecular weight species, such as therapeutic peptides and proteins. The overall permeability of vaginal epithelium to penetrant species is greater than the rectal, buccal, or transdermal routes, but less than the nasal and pulmonary routes [45].

21.5 UTERINE DRUG ABSORPTION

A number of drug substances are known to act directly upon the uterus, including uterine relaxants (e.g., β-agonists) and stimulants (e.g., prostanoids, oxytocin). The administration of drugs to the uterus is achieved by the application of a formulated product to the vagina or the cervix. However, it has been demonstrated that the mechanism by which the drug is transported from the cervicovagina to the uterus is not limited to passive diffusion, but is facilitated by a preferential transport mechanism termed as the first uterine pass effect.

21.5.1 First Uterine Pass Effect

During the course of the early 1990s, two studies reported that vaginal administration of progesterone produced uterine effects that could not be reasonably accounted for on the basis of circulating progesterone levels [48,49]. Similar findings had previously been reported for terbutaline (a uterine relaxant) [50] and were published subsequently for danazol (used to treat endometriosis) [51]. The results suggested a mechanism whereby drugs administered vaginally were preferentially transported to the uterus through a so-called first uterine pass effect. Several theoretical mechanisms have been proposed to account for the phenomenon: (i) direct (passive) diffusion through the tissues; (ii) passage from the vagina to the uterus through the cervical lumen; (iii) transport through the venous or lymphatic circulatory systems; and (iv) countercurrent vascular exchange involving diffusion between adjacent uterovaginal veins and arteries [52]. Confirmation of the existence of the direct mechanism first uterine pass effect has been established in a human *ex-vivo* uterine perfusion model wherein [^{3}H]-progesterone applied to the vaginal tissue associated with freshly hysterectomized uteri diffused into and reached steady state within the entire uterus within 5 h of application. The nature of the perfusion model employed demonstrated that vaginal to uterus transport of the steroid could not be attributed to conventional distribution by the circulatory system [53].

Evidence for the alternative direct transport mechanism involving aspiration through the cervical canal into the uterus and facilitated by uterine contractions has also been provided using vaginally administered sperm-sized ^{99m}Tc-labeled macroaggregates of human serum albumin [54–56]. However, ^{99m}Tc activity was observed in the uterus within minutes of administration, compared to the 5 to 6 h for the perfusion model, concurring with the general observation that transport rates through the cervical lumen can be extremely rapid. However, it is widely considered that the countercurrent mechanism, whereby substances are transported from the vagina (through the vaginal vein) to the uterus (through the counterflowing uterine artery), is of primary importance in understanding the first uterine pass effect [52,57–60]. The vein-to-artery transport is facilitated by the fact that in women the utero-ovarian veins are known to form a plexus on top of and in intimate contact with the ovarian artery, thus providing a large surface area favoring direct partitioning of substances between

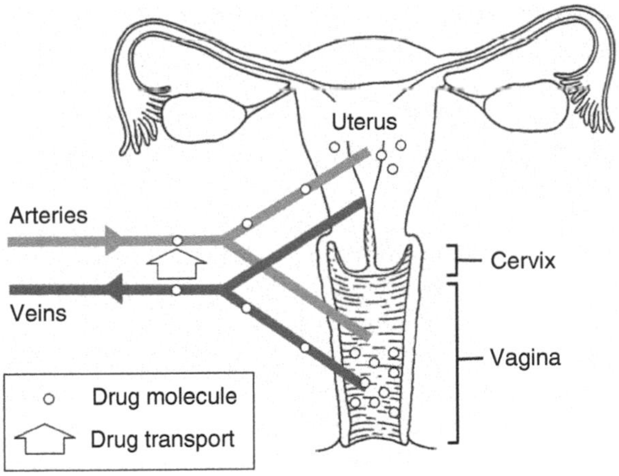

FIGURE 21.5 Illustration of the first uterine pass effect.

vessels and in accordance with the concentration gradient established by vaginal application. The mechanism is described pictorially in Figure 21.5.

It has recently been demonstrated that the extent of the first uterine pass effect might be dependent on the exact location within the vagina of the administered formulation [61]. The study showed preferential vaginal to uterine distribution of estradiol when a single estradiol tablet was placed in the upper third of the vagina, whereas no effect was observed for placement in the lower third. This has significant practical implications for optimizing clinical outcomes relating to vaginal estradiol administration. In particular, the results suggest that improved endometrial maturation in premenopausal women may be achieved by placing vagina estradiol tablets in the upper third of the vagina (making use of the first-pass effect), whereas improved treatment of postmenopausal vaginal atrophy might be achieved by placing the vaginal estradiol tablet in the lower third of the vagina in order to minimize the risk of endometrial hyperplasia. Further studies are warranted to establish if this effect of vaginal placement is critical to the clinical outcomes of other therapeutic agents acting upon the vagina or the uterus.

21.6 VAGINAL AND UTERINE DRUG DELIVERY PLATFORMS

There is a wide range of drug delivery formulations available for use in administration of drugs to the vagina and uterus. Often, these systems are adaptations of conventional oral and topical formulations (gels, creams, ointments, tablets, capsules, etc.), although a number of systems specifically designed for optimized vaginal and uterine delivery are also known (vaginal rings, intrauterine devices [IUDs], vaginal films, etc.). Selection of the most appropriate form of delivery system will ultimately depend on a consideration of the following questions:

- Is a local or systemic effect required? This will influence, for example, the choice between a traditional dosage form, such as a semisolid cream or gel, and a system that promotes increased intravaginal residence, with a concomitant increased possibility of absorption across vaginal epithelium, for example, from intravaginal ring (IVR) systems or systems utilizing a bioadhesive polymer component.
- Is a site-specific application required or is it preferable for the drug to be distributed rapidly throughout the vaginal space, as might be the case for an anti-infective treatment?

Site-specific delivery will require the use of a self-locating system, typically a bioadhesive formulation, although an IVR, due to its elastomeric nature, will remain located high in the vaginal space. Conversely, for rapid distribution throughout the space, semisolid or fast-dissolving solid systems will be required. For semisolids, flow properties and viscoelastic character will be the critical determinants of their ability to spread rapidly from their point of application.

- Is drug release to be immediate or sustained and, if the latter, what is the duration of release that is required? Consideration of this question is a good example of the intimate relationship that must exist between delivery system design and clinical requirements. Drug release by the intravaginal route is most commonly immediate but a viscoelastic semisolid can be designed to offer some increase in duration of delivery, as can solid hydrogels or intravaginal tablets. For controlled, zero-order release sustained over prolonged periods (days, extending to months), IVRs and IUDs are most suitable, provided it is compatible with the physicochemical nature of the drug to be delivered.
- Is a controlled-release system required? This question will primarily relate to systemic drug delivery applications, typically for potent drugs such as steroid sex hormones and, perhaps, peptides or peptidomimetic agents.
- What is the physicochemical and pharmacological nature of the penetrant? The drugs to be delivered should be considered in relation to its polarity and partition characteristics, molecular weight and size, in respect of either water-based or more hydrophobic systems.
- What cultural and patient acceptability factors may be important, for example, a tendency for the preparation to leak following initial application? Vaginal delivery may not be universally acceptable in all cultures, and within cultures the preference for systems that can be self-inserted and removed, or considerations relating to leakage, will vary considerably.
- What are the economic implications associated with the use of specifically designed, controlled-release polymeric systems compared with, for example, simple semisolid instillations? A cost–benefit analysis is essential in deciding upon the choice of delivery system. Capital costs, for example, are considerably higher for specific intravaginal systems, such as rings and IUDs, than for those capable of manufacture on generic equipment, such as semisolids and intravaginal tablets. These costs must be considered in relation to the type of active to be delivered and the disease state.

Table 21.6 summarizes the available delivery platforms in respect of their relevant physicochemical considerations in relation to the type of drug that can be delivered, and the ability to sustain and control the delivery process. More substantive descriptions of each formulation type are now discussed.

21.6.1 Vaginal Tablets

Vaginal tablets represent a commonly used intravaginal delivery system, having the advantages of ease of manufacture, ease of insertion and economy. They can be formulated with one or more mucoadhesive polymers in an attempt to improve intravaginal residence time. Vaginal tablets have been used as a delivery system for cervical ripening and the induction of labor [62], spermicidal agents [63], antifungal agents [64], analgesia [65], and urge incontinence [66].

21.6.2 Vaginal Pessaries and Suppositories

Vaginal pessaries or suppositories (the terms often used interchangeably) containing such substances as natural gums, fatty acids, alum, and rock salts were originally used in ancient Egyptian times as contraceptives. One of the earliest technical papers describing a

TABLE 21.6
Description of Vaginal Formulations

Vaginal Formulation Type	Description	Indication	Examples	Advantages	Disadvantages
Capsule/ovule	Sometimes known as shell pessaries. They are generally similar to soft capsules and come in various shapes, usually ovoid to ease insertion to the vagina	Fungal infection Bacterial vaginosis	Gyno-Daktarin 200 mg vaginal capsule (Janssen-Cilag) miconazole nitrate Cleocin (Pharmacia) clindamycin phosphate	• Ease of manufacture • Inexpensive • Ease of insertion	• Frequent applications • Poor retention in vagina
Intravaginal cream	Intravaginal creams are semisolid preparations formulated to produce a product that is miscible with vaginal secretions. They are generally delivered to the vagina using an applicator device	Fungal infection Contraceptive HRT/vaginal atrophy	Nystan (Squibb) nystatin Ortho-Crème (Janssen-Cilag) nonoxynol 9 Ortho-Gynest (Janssen-Cilag) estriol 0.01% Ovestin (Organon) estriol 0.01% Premarin (Wyeth) conjugated estrogens Estrace (Warner Chilcott) estradiol 0.01%	• Ease of manufacture • Inexpensive	• Messy semisolid gel formulations make administration difficult • Frequent applications • Not possible to remove if adverse reaction • Poor retention in vagina
Foam	Medicated foams consist of a liquid with a large volume of gas dispersed within it. The product is usually supplied in the form of a liquid enclosed within a pressurized container. Generally an applicator is used during insertion	Spermicidal contraception	Delfen (Janssen-Cilag) nonoxynol 9 (Discontinued 1/11/2004)	• Ease of manufacture • Inexpensive	• Messy semisolid foam formulations make administration difficult • Frequent applications • Not possible to remove if adverse reaction poor retention in vagina

(*Continued*)

TABLE 21.6 (Continued)
Description of Vaginal Formulations

Vaginal Formulation Type	Description	Indication	Examples	Advantages	Disadvantages
Gel	Gels, depending on their application, contain either a hydrophilic or lipophilic liquids gelled by a suitable agent. They are often supplied in a single dose container or provided with an appropriate applicator to aid insertion	Maintenance of vaginal acidity Bacterial vaginosis Spermicidal contraception	Aci-jel (Janssen-Cilag) acetic acid 0.94% Zidoval (3 M) metronidazole 0.75% Metrogel (Sandoz) metronidazole 0.75% Duragel (SSL) nonoxynol 9	• Hydrogels offer prolonged residence time due to bioadhesive nature • Ease of manufacture • Inexpensive	• Messy semisolid gel formulations make administration difficult • Frequent applications • Not possible to remove if adverse reaction • Relatively short retention in vagina
Intrauterine device (IUD)	An IUD is a small medical device that is inserted through the cervix and placed within the uterus. The device may contain a drug-loaded polymeric component allowing the controlled delivery of an active	Contraceptive	Mirena (Schering Health) levonorgestrel Progestasert IUD (Alza corporation) progesterone	• Accurate controlled dosing • Easy to modify release profile through device design • Minimal patient intervention • Immediately effective • Long lasting	• Insertion and removal requires visit to clinician • Involuntary IUD expulsion • May cause more difficult menstrual periods • Possible risk to future fertility
Intravaginal ring (IVR)	The IVR is a polymeric ring-shaped device designed to deliver an active to the vagina. The device is inserted into the vagina where it positions itself near the cervix	HRT/vaginal atrophy contraceptive	Estring (Pharmacia & Upjohn) estradiol Femring (Warner Chilcott) estradiol-3-acetate NuvaRing (Organon) etonogestrol	• Accurate controlled dosing for up to 6 months • Easy to modify release profile through device design • Minimal patient intervention • Removal possible with quick reduction in plasma levels • Cost efficient	• Initially expensive to manufacture • Occasional involuntary ring expulsion • Drug wastage

Pessary	Pessaries are solid, single dose preparation formulated with a suitable volume and consistency to be inserted into the vagina. The active may either be dispersed or dissolved in a medium, which is soluble or dispersible in water or may melt at body temperature	Fungal infection Spermicidal Contraception Cervical ripening	Lomexin (Akita) fenticonazole nitrate Orthoforms (Janssen-Cilag) nonoxynol 9 Propess (Ferring) dinoprostone	• Ease of manufacture • Ease of insertion • Inexpensive	• Frequent applications • Poor retention in the vagina
Tablets	Tablets are single dose, solid preparations manufactured from compressing the active with a range of excipients into a specified shape	HRT/vaginal atrophy	Vagifem (Novo Nordisk) estradiol	• Ease of manufacture • Ease of insertion • Inexpensive	• Frequent applications
Medicated tampons	Medical tampons are solid, single dose preparations designed to be inserted into the vagina for a limited period of time. They can be designed to both absorb the menstrual fluid and simultaneously release an active into the vagina	Maintenance of vaginal acidity	Ela pH tampon (Rostram Ltd) citric acid/lactic acid	• Ease of manufacture ease of insertion • Inexpensive	• May be prone to bacterial infection

suppository-based vaginal device was published in 1947 by Rock et al. [67]. These suppository or pessary systems are now most commonly used to administer drugs to promote cervical ripening before childbirth [68,69], although other applications, primarily for local drug delivery to the vagina, have also been reported [70–73].

21.6.3 Intravaginal Gels

Mucoadhesive hydrogels are weakly cross-linked polymers, which are able to swell in contact with water and to spread onto the surface of mucus [74]. Their ability to achieve an intimate contact with an absorbing membrane, to localize drug delivery systems at a certain place, and to extend residence time has led to their application as vaginal drug delivery systems [75]. It should be understood that vaginal epithelium is not strictly a mucosal epithelium because it possesses no mucus-producing goblet cells; it is nevertheless bathed in vaginal fluid containing cervical mucus, thus mucoadhesion is an appropriate process to consider in this context. Overall, mucoadhesion can be understood as a two-step process, in which the first adsorptive contact is governed by surface energy effects and a spreading process. In the latter phase, the diffusion of polymer chains across the polymer–mucus interface may enhance the final bond [76]. Water plays an important role in mucoadhesion, the invading water molecules liberating polymer chains from their twisted and entangled state and, thus, exposing reactive sites that can bond to tissue macromolecules. The adhesion of dried hydrogels to moist tissue can be quite substantial, with water uptake from the tissue surface facilitating surface dehydration and exposing surface depressions that may act as anchoring locations [77]. Thus, the majority of hydrogels applied for intravaginal are based on mucoadhesive polymers, including poly (acrylic acid) and polycationic materials such as chitosan. However, other hydrogels systems are also investigated.

21.6.4 Vaginal Creams

Vaginal creams are semisolid preparations, and are typically semisolid emulsions of either the oil-in-water or water-in-oil type. They may be formulated from synthetic or natural ingredients to produce a product that is miscible with vaginal secretions. Vaginal creams have been used in a wide variety of proprietary products for the local vaginal administration of antifungal, antibacterial, and antiatrophic agents.

21.6.5 Intravaginal Rings

IVRs are torus-shaped polymeric devices designed to release one or more incorporated drugs in a controlled fashion [78,79]. Compared to other vaginal delivery systems, they offer certain advantages including accurate and sustained dosing, extending over a period of months if required, and minimal need for patient intervention. The simplest IVR device contains drug evenly dispersed throughout the ring volume, such that drug release rates are proportional to both the drug loading and the surface area of the device. A plot of cumulative drug release versus the square root of time yields a linear plot for this IVR system. Sandwich and reservoir-type IVR systems were developed to provide a near constant rate of release of the drug throughout the delivery period, according to zero-order release kinetics. The sandwich, or shell, ring design consists of a narrow drug-containing layer located below the surface of the ring and positioned between a nonmedicated central core and a nonmedicated outer band. The position of the drug core close to the surface ensures sandwich IVR devices are best suited to the delivery of drugs having poor polymer diffusion characteristics. The small, constant release of drug from this IVR system can help to reduce side effects compared with matrix devices, and also minimizes the cost of the device owing to the relatively low drug

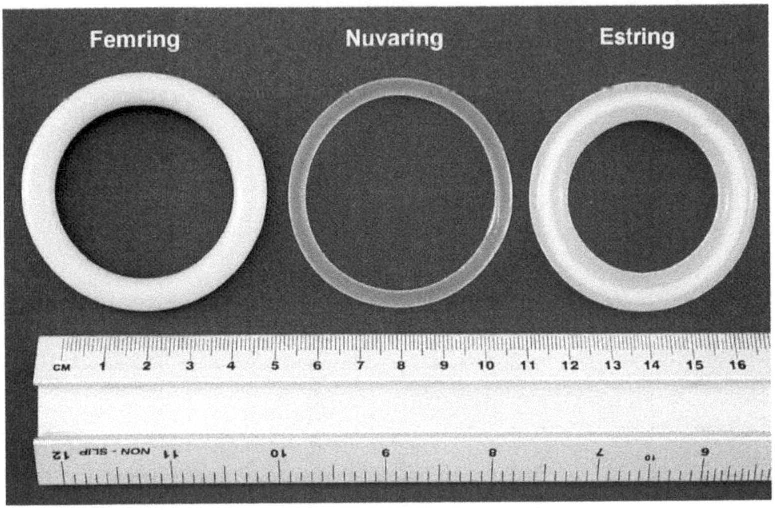

FIGURE 21.6 Commercial intravaginal rings.

loading. However, the multilayer design of these systems and the relatively complex manufacturing process has so far prevented them from being commercialized. In contrast, the core, or reservoir, design of IVR, which has been commercialized successfully in Estring, Femring, and NuvaRing, has the drugs present within a centralized core covered by a drug-free outer polymer sheath (Figure 21.6). With injection-molded silicone reservoir rings (as exemplified by Estring and Femring), two or more drug-loaded core sections having independent release rates may be incorporated into the same ring, thereby allowing multiple drug administration. The rate of drug release can be easily modified in these systems by changing the length of the drug-loaded core and the cross-sectional diameter of the sheath layer. Longer core lengths and narrow sheath diameters increase the rate of drug release. Although the option of including several distinct cores in vaginal ring systems manufactured by extrusion processes is not so easy to achieve, multiple drug administration may still be achieved through incorporation of several drug components into the central extruded core, as exemplified by NuvaRing, which contains a mixture of etonogestrel and ethinyl estradiol within the central poly(ethylene-*co*-vinyl acetate) core. Reservoir devices, irrespective of design type and method of manufacture, produce near zero-order release profiles making these devices particularly useful in clinical settings were constant, reproducible drug levels are required.

Details of all IVRs either commercialized or in development are provided in Table 21.7.

21.6.6 Intrauterine Devices

An IUD is a small T-shaped device, made from a metal or flexible polymer, which is fitted within the uterine cavity for the purpose of preventing conception. The most widely used IUDs are copper-bearing IUDs, although inert (unmedicated) and progestin-releasing IUDs (levonorgestrel or progesterone) are also available. Their contraceptive mode of action involves production of a local sterile inflammatory reaction causing lysis of the ovum and the sperm. Since the 1970s, WHO has conducted 10 large trials to compare and evaluate the safety and efficacy of six different IUDs [80]. The newest generations of copper IUDs combine high continuation rates with very low pregnancy rates [81]. Since little can be done to increase the efficacy of these devices, recent research has focused on developing devices to

TABLE 21.7
Intravaginal Rings Currently being Marketed or Developed

IVR Product	Company	Indication	Elastomer Description	Active(s) [loading]	Release Rate
Estring	Pharmacia	HRT	Silicone	Estradiol (2 mg)	7.5 μg/d
NuvaRing	Organon	Contraception	Ethylene vinylacetate	Etonogestrel/Ethinyl estradiol (11.7/2.7 mg)	120/15 μg/d
Femring	Warner Chilcott	HRT	Silicone	Estradiol acetate (12.4/24.8 mg)	50 and 100 μg/d as estradiol acetate
—	Population Council/QPharma	Contraceptive	Silicone	Nestorone/Ethinylestradiol	Approx. 150/15 μg/d
—	Population Council/QPharma	Contraceptive	Silicone	Progesterone	In trial
—	Population Council/QPharma	Contraceptive	Silicone	Ethinyl estradiol/Norethindrone acetate	In trial
—	Population Council/QPharma	Contraceptive	Silicone	Nesterone	In trial
—	FEI Technologies/Barr	Incontinence	—	Oxybutynin	In trial
—	World Health Organization	Contraceptive	—	Levonorgestrel	Approx. 20 μg/d
—	FEI Technologies	Anesthetic	—	Lidocaine	In trial

address side effects, particularly bleeding and pain, which account for a significant number of removals. The levonorgestrel-releasing IUD, a device with high effectiveness and acceptability, reduces menstrual blood loss compared to preinsertion levels [82]. The levonorgestrel-releasing IUD, Mirena, has been available in Europe for 10 years and has been used by 2 million women; it was approved for sale in the United States in December 2000. Frameless IUDs, such as the Gynefix [83–85], have been specifically designed to reduce cramping and pain. This device consists of a surgical nylon thread that holds copper sleeves and is anchored to the uterine fundus during insertion. It recently became available in Europe, and is licensed for 5-year use. Studies suggest that the Gynefix is as effective as the Copper T380A, and expulsion rates are less than 1 per 100 women years.

Very recently, clinical assessments of an IUD releasing danazol for the treatment of endometrial hyperplasia and endometriosis-related pelvic pain have been reported [86,87].

21.7 CLINICAL APPLICATIONS

21.7.1 Vaginal Microbicides

STDs are a major global health problem. There are now more than 25 recognized STDs and the past decade has witnessed an unprecedented rise in the number of cases in both developed and developing countries. It has been estimated that more than 333 million new cases of syphilis, gonorrhea, Chlamydia, and trichomoniasis occur each year [88–90]. The health implications associated with untreated STDs can be extremely serious, and include problems and diseases such as ectopic pregnancy, meningitis, septicemia, pelvic inflammatory disease, infertility, and nerve and heart damage. Also, it is well established that the presence of certain STDs increase the risk of contracting HIV, particularly because of disruption of the vaginal epithelium [91,92]. Of all the STDs, HIV is the most notorious and is often categorized separately due to its severity. HIV has killed more people worldwide than any other infectious disease with approximately 22 million deaths since 1981 [93]. Furthermore, it is estimated that 39.4 million people are carrying the AIDS virus, of which 65% are in sub-Saharan Africa [94], and with approximately 15,000 new infections taking place each day the pandemic is not likely to have peaked [95].

As STDs are spread by sexual contact, their transmission could be significantly reduced or prevented by people adhering to monogamous relationships or abstaining from sex. Worldwide prevention activities, and in particular those associated with HIV prevention, have therefore focused on encouraging people to abstain from sex and from risky sexual behavior. However, in an increasingly sexually liberal society, with many teenagers and young adults having sex with numerous different partners, it seems that these options are unlikely to occur. There is the added problem of a significant number of monogamous women becoming infected with STDs by their husbands or partners who do not remain monogamous or who are abusing drugs [96]. For the past two decades, condom use had been widely promoted as they provide a highly effective measure for preventing HIV and other STDs when used properly and systematically. Although cooperation of men to use condoms has increased in a few countries the proper and regular use of condoms is unsatisfactory [97]. Furthermore, their correct use is often constrained by a number of factors including [93,95,98]:

- The compromise of sexual intimacy
- Religious teachings and beliefs
- Concerns about raising suspicions of promiscuity or infidelity
- Women often lack the power to negotiate the use of condoms
- Women may wish to become pregnant

- Poor awareness of risks of STD
- Lack of availability in developing countries
- Affordability in developing countries

Subsequently, condom use and behavioral intervention have only been partially successful in reducing the transmission of HIV and other STDs, and so there is an urgent need for additional interventions.

The ideal solution for preventing transmission of STDs, and HIV in particular, would be the development of an effective vaccine. However, rapid mutation of the HIV genome, in part exacerbated by the genetic pressure exerted by the use of antiretrovirals for treatment, has so far frustrated attempts by scientists to develop an effective vaccine. Also, an additional hurdle with respect to vaccine development is that various strains of HIV exist, and as such multiple vaccines may be required for complete HIV vaccination. It has been estimated that a suitable vaccine will not be realized for at least another decade [98]. Meanwhile, highly active antiretroviral treatment (HAART) is routinely used to delay onset of full-blown AIDS and to dramatically prolong life expectancy for those who are HIV positive. Unfortunately the treatments are highly expensive and therefore not widely available and affordable in developing countries where HIV is most prevalent. Thus, there is an obvious need to develop alternative, and inexpensive, HIV prevention strategies. As the major mode of HIV transmission is now through heterosexual vaginal intercourse (approximately 90% in sub-Saharan Africa [99,100]), there is a strong case for developing vaginally administered substances, so called vaginal microbicides, to prevent HIV transmission. These are chemical substances when applied to the vagina before intercourse have the ability to reduce or prevent the sexual transmission of HIV and other STDs. Although there are no microbicides currently available on the market, numerous products are under development and a few are undergoing phase III clinical trials (Table 21.7). A potential microbicide product could be administered vaginally as any number of formulation types, including gel, cream, suppository, film, slow-releasing sponges, or IVRs. The mechanism of action of a microbicide could also vary. Microbicides currently under development can prevent viral or bacterial infections by a number of different modes (Table 21.8) [4,93,95,101,102] including: killing or inactivating the microorganism; increasing the natural defenses of the vagina; inhibiting pathogen fusion or cell entry; preventing the pathogen from replicating once inside the cell; and creating a barrier between the pathogen and the vaginal wall.

21.7.1.1 Killing or Inactivating the Pathogens

There are a great number of different microbicides that act by killing or inactivating invading pathogens. They include a large number of surfactants and detergents that act by disrupting and emulsifying the lipid cell membranes of the pathogens and have the advantage of possessing a broad killing ability. Nonoxynol-9 (N-9) was one of the first compounds to undergo developmental research as a possible microbicide, particularly for the prevention of HIV. N-9 is widely employed as the active ingredient in various spermicides due to its surfactant properties, which cause rapid disaggregation of the lipid membranes of spermatozoa. A number of initial studies examining the effect of N-9 on a variety of sexually transmitted bacteria and viruses showed promising results for STD prevention [103–105]. However, several phase III clinical trials in Africa, where different N-9 formulations were tested, revealed that it did not provide significant protection against HIV and several other STDs [106–108]. Furthermore, the cell-destroying properties of N-9 have also been found to damage the epithelium of the vaginal wall and, subsequently, cause vaginal ulceration thus rendering the woman more susceptible to the transmission of HIV. The results of such clinical

TABLE 21.8
HIV Microbicides Currently in Development

Microbicide Name	Microbicidal Agent	Mechanism of Action	Phase I	Phase II	Phase III	Formulation
Acidform/Amphora	Acid buffer system	Acid buffer	×			Gel
Benzalkonium chloride	Benzalkonium chloride	Surfactant	×			Suppository
Buffergel	Polyacrylic acid (Carbopol 974)	Acid buffer		×		Carbopol gel
Human monoclonal antibodies	C2F5, C2G12, C4E10	Entry/fusion inhibitor		×		Gel
Calanolide A	Calanolide A	NNRTI		×		Gel
Carraguard	Carrageenan	Adsorption inhibitor			×	Gel
Cellulose acetate phthalate	Cellulose acetate phthalate	Entry/fusion inhibitor	×			Gel
Emmelle	Dextrin-2-sulfate	Adsorption inhibitor		×		Gel
Invisible Condom	Sodium lauryl sulfate	Entry/fusion inhibitor	×			Gel
LactinVaginal Capsule	*Lactobacillus crispatus*	Vaginal buffer	×			Capsule
Polystyrene sulfonate	Polystyrene sulfonate	Adsorption inhibitor	×			Gel
Praneem	Polyherbal extracts	Uncharacterized	×			Tablet
PRO 2000	Naphthalene sulfonate polymer	Entry/fusion inhibitor		×		Gel
Savvy	C31G	Surfactant			×	Cream
VivaGel	SPL7013 polyanionic dendrimer	Entry/fusion inhibitor	×			Gel
Tenofovir	PMPA	NRTI		×		Gel
UC781	UC781	NNRTI	×			Lipophilic gel
Ushercell	Cellulose sulfate	Adsorption inhibitor			×	Gel

trials has raised cautions regarding the use of surfactants as microbicides, and it is unlikely that N-9 will undergo further investigation as a vaginal microbicide [103]. Furthermore, those microbicides with surfactant and detergent properties that are currently under development will have to provide sufficient protection against invading pathogens while maintaining the epithelium of the vaginal wall, in order to be considered as potentially useful products.

Additional compounds investigated as potential microbicides, which also kill or inactivate pathogens by destroying their cell membranes, are lipids. It has been observed that several free fatty acids and 1-monoglycerides kill a number of pathogens [109–113]. In a recent study Thormar et al. [114] manufactured pharmaceutical hydrogels containing the lipid monocaprin and found that the hydrogels were extremely potent inactivators of sexually transmitted bacteria and viruses. Likewise, Lampe et al. [115] have discovered that synthetic derivatives of breast milk, which are related to the antimicrobial monoglyceride esters found in human breast milk [111] are able to kill *Chlamydia trachomatis* directly. These synthetic analogs of breast milk are currently under investigation to determine their action against other STDs.

Oxidizing agents (peroxides and peroxidases), antimicrobial peptides, and monoclonal antibodies also kill or inactivate sexually transmitted pathogens. However, these compounds are found naturally in the vagina and are often introduced to enhance the natural defense mechanisms, and are described below.

21.7.1.2 Enhancers of Natural Vaginal Defense Mechanisms

The natural defense systems of the vagina are capable of inhibiting the transmission of microorganisms by the production of a number of chemicals (lactic acid, hydrogen peroxide, lactin, and acidolin) that help to maintain a low vaginal pH and make it inhospitable to invading pathogens. Semen, however, is alkaline and so lowers the pH of the vagina during sexual intercourse. This reduces the natural acidic defense mechanism, rendering the vagina susceptible to infection. Both Buffergel and Acidform are acidic vaginal gels designed to maintain the acidity of the vagina in the presence of semen, thus maintaining an environment hostile to microorganisms. The vaginal defenses can also be enhanced by the use of formulations that contain monoclonal antibodies. Such antibodies occur naturally in the vagina and their levels can be increased to artificially boost the human immune response to STD pathogens. For example, it has been demonstrated that a microbicide containing the human monoclonal antibody b12 (an antibody that binds to the gp120 protein of HIV—refer to Section 21.7.1.3) effectively protected monkeys against the simian–human immunodeficiency virus (SHIV)—a hybrid of the human and simian immunodeficiency virus [116].

Antibiotic peptides and peroxides or peroxidases can also be introduced to enhance the vaginal defense mechanisms. FemCap (manufacturers of the cervical cap) are currently developing a gel that contains N-9 and hydrogen peroxide. The peroxide component minimizes the adhesion of microbicides to the vaginal wall, and kills a number of pathogens and enhances the effects of N-9 [117].

21.7.1.3 Inhibitors of HIV Binding and Entry

There are a number of chemical compounds under development that can inhibit the attachment (fusion) of infecting pathogens to host cells. These agents include peptides and proteins that block the surface proteins or receptors of a specific virus. For example, HIV possesses two specific proteins on its surface, gp41 and gp120, that allow it to attach itself to target cells. By blocking these proteins, fusion inhibitors prevent the virus from attaching to host cells and thus the virus cannot reproduce. Cyanovirin-N is one such protein that inhibits the binding of HIV to target cells by interacting with N-linked mannose oligosaccharides of the gp120 protein of the viral surface thus preventing its duplication [118–120]. Other agents in this class are less specific and produce a general mode of action against all pathogens. They include polyanionic sulfated polysaccharides (dextran sulfate and carrageenan—Carraguard) and sulfonated anionic polymers (PRO 2000). Their mode of action is related to the negative anionic charge they possess on their surface that inhibits the pathogens from binding to their host cells, thus preventing microorganism replication.

21.7.1.4 Inhibitors of Viral Replication

This group of microbicides inhibits the replication of the pathogen after it has been fused with healthy cells, i.e., postfusion inhibitors. These drugs, now investigated as potential microbicides, were originally developed and administered to patients who had contracted HIV. They include reverse transcriptase inhibitors (RTIs) and protease inhibitors (PIs). RTIs slow down or inhibit the enzyme reverse transcriptase, which is present in HIV. This enzyme controls the replication of HIV in healthy cells and prevents HIV from duplicating itself, resulting in low viral levels in the blood. Some RTIs have such a high affinity for reverse transcriptase (e.g., UC781) that they can enter HIV cells and bind to its reverse transcriptase even before it has been fused with target cells, thereby rendering it harmless [103]. These high affinity RTIs are ideal candidates for vaginal microbicides.

PIs inhibit the digestive protease enzyme of HIV, which it uses to break the proteins of healthy cells into smaller pieces. These infected smaller pieces of protein then carry on infecting new host cells. PIs therefore slow down HIV proteases and subsequently inhibit the infection of new cells [94].

21.7.1.5 Physical Barriers

As barrier methods of protection are extremely effective at reducing the spread of STDs, researchers have been examining the possibility of new barrier products that women can use and thereby take control. To date the female condom has proved to be unpopular and has been described as both noisy and uncomfortable [121]. An alternative female condom is under development at the Universite Laval in Quebec, Canada, where researchers are currently in the process of developing a physical barrier in the form of an invisible condom. The invisible condom is a thermoreversible gel that hardens when inserted into the vagina or the rectum due to the increased temperature of the human body. The gel, which breaks down several hours after application, has been shown in laboratory studies to provide sufficient protection against herpes simplex virus (HSV) and HIV. Although the invisible condom will provide women with the autonomy to protect themselves, and it has the major disadvantage that is unlikely to go unnoticed by the user's partner.

Another form of physical barrier is the Protectaid sponge—a contraceptive sponge that is currently available as an alternative to hormonal contraception. The sponge provides two modes of protection. Firstly, it acts as a physical barrier by blocking sperm from entering the uterus and thus will also potentially block the entry of sexually transmitted pathogens. Secondly, the sponge contains a number of spermicides, which act as a chemical barrier and thus kill sperm. However, the spermicides present also possess antiviral properties and thus should also inactivate and kill sexually transmitted pathogens. The distinct advantage of the sponge, although currently used specifically as a contraceptive, is that additional or alternative drugs could be incorporated into the sponge. Thus, it can be used to deliver various microbicides as well as the spermicides for the contraceptive properties. The Protectaid sponge has the major benefit that it does not interfere with sexual intercourse. It can go unnoticed by both the woman and their partner and it can be inserted several hours before intercourse.

Finally, many of the gels mentioned earlier such as carrageenan, dextran sulfate, and PRO 2000 as well as acting as fusion inhibitors also can act as physical barriers. Such gels and creams provide a physical barrier between pathogens and vulnerable cells in the epithelium (cell wall) of the vagina thus providing a dual mode of protection.

Combined formulations: As there is such a wide variety of compounds with a diverse range of actions for preventing STDs, so there is an obvious approach for the development of formulations combining two or more of the above-listed microbicides to offer a broader range of protection. As some of the sulfated polysaccharides and the sulfonated polymers will form gels when mixed with water, they could be used as drug carriers for other types of microbicide while also offering protection against STDs. There are numerous combinations and possibilities that could be studied as potential dual action (or more) microbicides.

The above summary of the numerous possible microbicides in development provides great optimism for the future production of first-generation vaginal microbicides for the prevention of heterosexual transmission of HIV and other STDs. Considerable progress has been made in understanding the mechanisms of the transmission STDs and the necessary procedures to reduce transmission. Although there have been some setbacks, it is clear that vaginal microbicides could be a fast, relatively inexpensive, and effective means of reducing the spread of STDs, at least in the short term.

21.7.2 Vaginal Vaccination

Recently, there has been considerable interest in utilizing mucosal vaccination approaches to induce an immune response in the female reproductive tract (Table 21.9). These techniques are developed primarily as a method to protect the vagina and uterus against invading pathogens [122–124] and as a potential strategy for contraception [125,126]. Although the vaginal mucosal epithelium is considered part of the common mucosal immune system, it displays a number of unique characteristics that distinguishes it from other components of the system. For example, it is known that the vagina is lacking in organized lymphoepithelial structures, such as the Peyer's patches in the intestine or the bronchus-associated lymphoid tissues in the respiratory tract, which are necessary for the production of the precursors of mucosal immunoglobulin A (IgA) and disseminating the mucosal immune response to remote effector sites [127–130]. In contrast to other mucosal secretions, where secretory IgA (S-IgA) is the prevalent type [131,132], higher levels of immunoglobulin G (IgG) rather than IgA are found. Studies have shown that the IgG production is primarily dependent on the systemic immune system with any IgA antibodies found in the cervicovaginal fluid predominantly secreted by the plasma cells around the cervix [133].

In the vagina, another important difference is that the concentration levels of the most common immunoglobulin isotypes present, as well as the activity of antigen-presenting cells, have been shown to display a greater hormone dependency than in the other mucosal systems [124,134,135]. Thus, it is important to recognize the effect of the estrus cycle and the effect of other therapies such as contraception and HRT have on the vaccination schedule and the subsequent immune response [123,136,137]. A combination of these factors and the consequence of the pathological effects on reproductive tissues of the different immune responses have led a number of researchers to suggest the intravaginal route of immunization may not be successfully exploited for the development of vaccines [129,138]. Nevertheless, it has been shown that mucosal immunity is most efficiently induced by local presentation of antigens [123] and that the vagina is capable of mounting an immune response to locally administered antigens [139–141]. Thus, research has continued on generating humoral immunity within the female reproductive tract. Indeed, it has been suggested that vaccine research for diseases that

TABLE 21.9
Vaccine Vectors being Developed for Intravaginal Immunization

Vaccine Vector Type	Description/Mode of Action	Advantages	Disadvantages
Live attenuated organism	These types of vaccines induce a protective response by replicating within the host. They are engineered with one or more of the disease-promoting genes having been removed, so as to reduce its pathogenicity, but still retain some of the antigens of the virulent form	• Immunological vaccine of choice • Customized by host producing a diverse and persistent immune response to the pathogen	• May revert during replication or mutation to a more pathogenic form within the host • May induce a disease state in a host with a compromised immune system • Temperature sensitive
Whole inactivated/nonlive organism	An organism that has been inactivated by chemicals, irradiation or other means so it is not infectious	• Inherently safer than the live attenuated organism as replication or mutation to a more pathogenic form cannot occur • Easier to manufacture than other vaccine vector types	• The immunity induced may diminish over time requiring repeat boosting • Noncustomization by host • Localized at the site of administration • If the process of inactivation fails during manufacture the recipient may be infected with the pathogen • Difficulty in distinguishing between antibodies produced by the vaccine or from the wild-type organism leading to problems with diagnosis

(*Continued*)

TABLE 21.9 (Continued)
Vaccine Vectors being Developed for Intravaginal Immunization

Vaccine Vector Type	Description/Mode of Action	Advantages	Disadvantages
Recombinant bacterial vector	Commensal bacteria, capable of colonizing mucosal surfaces, which are genetically engineered to express viral, bacterial, or eukaryotic antigens to produce an immune response	• A single inoculum is sufficient to establish a colony to induce an antigen-specific immune response	• Noncustomization by host • Difficult to prepare • Temperature sensitive • May induce an antivector response
DNA/plasmid vector	Gene and genes of interest are removed from the pathogen and inserted into a nonreplicating DNA or plasmid vector. These genes are then expressed *in vivo* producing an immune response	• Induce minimal antivector response, which allows the same vector to be used more than once • The method allows genes that confer pathogenicity or virulence to be excluded • Inherently safer than the live attenuated organism as replication or mutation to a more pathogenic form cannot occur • Can be prepared and purified easily • More thermally stable than live preparations	• Noncustomization by host • The immunity induced may diminish over time requiring repeat boosting
Protein subunit/peptides/virus-like particle vector	Chemically synthesized sections of viral proteins which are noninfectious which are designed to stimulate specific immunity against a virus	• Inherently safer than the live attenuated organism as replication or mutation to a more pathogenic form cannot occur • More thermally stable than live preparations	• Localized at the site of administration • Limited immunity • More difficult to prepare than DNA/plasmid vectors as it requires additional steps for the expression and purification of the protein

are transmitted heterosexually should focus on the stimulation of protective immunity at the most common site of exposure in the female—the vagina [124].

21.7.2.1 Uterovaginal Vaccine Vectors

A range of different vaccine vectors has been developed over time to provoke an immune response within the body [127,142]. However, it has only been comparatively recently that they have been applied to inducing mucosal immunity within the uterovaginal tract. The general vector platforms that have been used include attenuated viruses, live viruses, commensal bacteria, DNA vectors, and protein subunit/peptide or virus-like particles (Table 21.9). The choice of vector is dependent on a number of factors such as the pathogenic virus and bacterial type and the length of duration of immunity required.

21.7.2.2 Attenuated Virus

Attenuated viruses are engineered with one or more of the disease-promoting genes removed such that the ability of the virus to infect or produce disease is much reduced while still retaining the ability to stimulate a strong immune response. By replicating within the host a long-lasting immune response can be induced. Attenuated virus vaccination techniques have been investigated by a number of research groups with varying success rates. In guinea pigs challenged by intradermal, intranasal, or intravaginal routes with vaccine recombinants expressing HSV-2 glycoprotein D, vaginal HSV shedding, and incidence of acute genital disease was reduced compared to control experiments [143]. However, the intravaginal vaccination was the most effective method for reducing primary and recurrent HSV disease. In a similar study, the vaccine potential of a genetically disabled HSV-1, administered intravaginally, intranasally, and orally, was investigated [144]. The results demonstrated that the intranasal route gave the most effective protection against primary vaginal disease. However, intravaginal immunization gave the biggest reduction in challenge virus titers in the vagina. In all cases, vaccination resulted in substantially more protection from the disease than an inactivated virus platform.

Parr and Parr [145], using a murine model, compared nasal and vaginal immunizations using a strain of attenuated HSV-2 for protection against HSV-2 wild-type vaginal infection. Vaginal immunization was performed both after progestin treatment with Depo-Provera (DP), and scarification after estradiol treatment. Both types of vaginal immunizations increased the number of IgG plasma cells in the vagina and the secretion/serum titer ratio of IgG antiviral antibody, indicating local production of virus-specific IgG in these groups. Intranasal and both intravaginal treatments protected all mice from neurological disease after challenge, but vaginal DP immunization induced the greatest immunity against reinfection of the vaginal epithelium.

21.7.2.3 Nonlive Vaccination

In a nonlive vaccination, immunization is achieved by using an inactivated form of the microorganism. Inactivation may be accomplished by chemical, irradiation, or other means. A large number of studies have been reported investigating such killed vaccines as a mucosal vaccination strategy, and a number of recent publications are summarized here; poliovirus administered through intravaginal and intrauterine routes [146], *C. albicans* administered vaginally for protection against vaginal candidiasis [147], multistrain bacterial vaccine for protection against recurrent urinary tract infections [148], gp120-depleted HIV-1 vaccine [149].

21.7.2.4 Commensal Bacteria

Utilizing commensal bacteria as a vaccine vector is a promising technique, whereby bacteria capable of colonizing mucosal surfaces are genetically engineered to express viral, bacterial, or eukaryotic antigens within the female reproductive tract, thus inducing the required immune response. Medaglini et al. [141] investigated the use of recombinant commensal bacteria as live vehicles for their mucosal vaccines. They developed a genetic system whereby *Streptococcus gordonii*, a bacterium present in the normal microbial flora of the human oral cavity, was engineered to express the E7 protein of the human papillomavirus (HPV) type 16. Their results have shown that a single inoculum of recombinant bacteria was sufficient to establish colonization of the murine vagina and consequently induce a papillomavirus-specific vaginal IgA and serum IgG response.

21.7.2.5 DNA/Plasmid Vectors

DNA/plasmid vector immunization is accomplished by genetically engineering gene and genes of interest from the required pathogen into a nonreplicating DNA or plasmid vector. The vectors are then used to transfect mucosal tissue thus inducing a specific immune response to the antigen. Schreckenberger et al. [142] investigated whether DNA vaccination would induce an intravaginal mucosal antibody response against HPV 6bL1 in rabbits by the muscular, vaginal, or rectal route [142]. The results showed that only the intravaginal route induced a specific IgA response and this was detectable until at least 14 weeks after immunization.

21.7.2.6 Protein Subunit/Peptides or Virus-Like Particle Vectors

The effects of chemically synthesized, noninfectious sections of viral proteins on stimulating an immune response were investigated by a number of research teams with mixed results. Pyles et al. [150] investigated the intravaginal mucosal administration of two immunostimulatory sequence (ISS)—containing phosphorothioate-stabilized oligonucleotides for antiherpetic efficacy in animal models [151]. The results indicated that vaginal epithelial application of ISS with mice lethally challenged with HSV-2 delayed disease onset and reduced the number of animals that developed signs of disease. The ISS treatment was found to significantly increase the survival rates over those of controls.

Conversely, O'Hagan et al. [151] investigated an enzymatically cleaved glycoprotein fragment from influenza virus hemagglutinin (TOPS). This was used to assess an intravaginal antigen delivery system comprising lysophosphatidylcholine and degradable starch microspheres [152]. These were compared with intramuscular immunization with TOPS absorbed to an aluminum hydroxide gel. The results showed that the highest level of antibodies in serum and vaginal wash samples were induced by intramuscular immunization with the intravaginal immunization producing no enhancement in the levels of antibodies in either serum or vaginal wash samples.

21.7.2.7 Adjuvants

Although it has been shown that delivering an antigen to the mucosal tissues of the female reproduction tract produces an effective immune response a number of researchers have explored the possibility of enhancing the response by using a nonspecific stimulatory adjuvant.

Intravaginal vaccination with whole cell and cholera toxin B subunit (CTB) oral cholera vaccine provided a greater success rate in providing a mucosal immune response in the female genital tract than an oral vaccination [152]. This study demonstrated that in a single individual, systemic immunity did not directly reflect the local antibody response in the mucosal tissue. Their data suggest that a strong local immune response is most effectively achieved by local intravaginal vaccination whereas systemic IgG responses require peroral rather than genital vaccine administration.

Johansson et al. [153], using the same cell and B subunit (CTB) oral cholera vaccine, suggest that vaginal vaccination given on days 10 and 24 in the menstrual cycle induced strong specific antibody responses in the cervix with 58-fold IgA and 16-fold IgG increases. In contrast, only modest responses were seen after nasal vaccination and in the group vaccinated intravaginally during different times of their menstrual cycle. However, they concluded that a combination of nasal and vaginal vaccination strategy for inducing protective antibody responses in both cervical and vaginal secretions provided that the vaginal vaccination is given on optimal time points in the cycle.

Vaginal immunization experiments with a cholera vaccine containing killed vibrios and CTB have been conducted in both the follicular (V-FPimm) and luteal (V-LPimm) menstrual cycle phase. With both producing comparable cervical CTB-specific IgA responses, however, only the V-FPimm induced cervical IgA2-restricted Ab to the bacterial lipopolysaccharide (LPS) vaccine component and induced CTB-specific IgA in rectal secretions.

21.7.3 Vaginal Contraception

There has been a resurgence of interest in vaginal contraception methods as a result of recent concerns about the prolonged use of oral combined contraception. The issue of compliance is also a major stumbling block for the contraceptive pill. Although the theoretical effectiveness of the pill is close to 100%, various studies have demonstrated that the real life pregnancy rate is almost 8 per 100 women during their first year of use [154,155].

Vaginal contraception by vaginal diaphragm or by cervical cap is totally reversible and never causes complications. There may be short-term reversible side effects with the diaphragm, such as cystitis, uretritis, and hemorroids. No side effects are associated with the use of cervical caps. Vaginal contraception has the added advantage of exercising notable prophylactic actions on the diffusion of venereal diseases and of other vaginal infections such as trichomoniasis and candidosis. It is also possible that vaginal contraception offers protection against cervical neoplasia. Failure rate of diaphragm use is an average 10/100 women years and for the cervical cap is about 7.6/100 women years, when both devices are properly used. Vaginal contraception needs to be used in conjunction with spermicidal agents. Spermicidal agents can be used alone and can be very effective. They are, however, not well accepted by most couples, who resent the interruption of the sexual act. Two experimental models of vaginal sponge are now under study. These vaginal sponges can be left in place for some time and insertion is very easy.

21.7.3.1 Vaginal Contraceptive Film

Vaginal contraceptive film (VCF) is a 5×5 cm translucent film containing the spermicide and surfactant nonoxynol-9. In use, the product is placed at the cervix at least 15 min before intercourse, where it quickly dissolves in the aqueous environment of the vagina. The application is considered to be effective for approximately 3 h, although a new insert does need to be applied before every act of intercourse. Advantages of the VCF include: simple to use, no messiness or discharge, no foreign body sensation, purchased without prescription, can be used without partner's knowledge, can be used in conjunction with condoms, may decrease the risk of contracting some STDs. Disadvantages include: insertion of film may interrupt intercourse, sensitivity to film and irritation, not as effective as other contraceptive methods, do not provide adequate protection against HIV and other STDs.

21.7.3.2 Contraceptive Vaginal Rings

Nonhormonal methods of contraception are generally recommended to lactating (breast-feeding) women, as they are generally safer for mother and child, and do not interfere with lactation. However, a number of hormonal contraceptive methods, specifically delivering progestogens, have also been developed in a bid to increase contraceptive choice for these women. Orally inactive progestogens are the preferred choice since, although they are transferred from mother to child by breast milk, they undergo rapid first-pass metabolism in the infant, thereby minimizing side effects. However, it is clear that these progestogens need to be administered to the mother by a nonoral route in order to be effective in fertility control. The possibility of using the natural hormone progesterone for contraception, in the form of implantable subdermal pellets, was first reported in 1982 [156,157]. However, despite providing excellent fertility control, the unacceptably high incidence of pellet expulsion prevented further development [158].

21.7.3.3 Progesterone Vaginal Ring

Vaginal rings as a method for providing sustained delivery of steroids to the body was first reported in a 1970 patent [159]. Unlike subdermal implants, these had the advantage of self-insertion and removal. Presently, the Population Council, New York, is actively pursuing the concept of a sustained release, progestogen-only vaginal ring for contraceptive control in lactating women. Their work has focused on two specific progestogens: progesterone itself and 16-methylene-17α-acetoxy-19- (Nesterone). The progesterone vaginal ring consists of silicone elastomer loaded with 22.5% w/w progesterone and is designed to maintain effective systemic levels of the steroid release over a 3-month period of continuous use. The matrix design releases according to $t^{1/2}$ kinetics, such that serum progesterone levels are not constant over the period. Typical serum progesterone levels reported range from approximately 30 nmol/L in the first week after insertion to 10 nmol/L after 16 weeks use [160–163]. Clinical studies have demonstrated a high contraceptive efficacy with the progesterone-only ring, with cumulative pregnancy rates similar to those observed with the copper T 380A IUD [161,162–167]. The progesterone ring has been approved for use in Chile and Peru.

21.7.3.4 Nesterone Vaginal Ring

The Population Council is also developing Nesterone for potential application in nonoral, sustained-release, contraception, and hormone therapy delivery systems. This new, orally inactive progestogen offers several advantages over more conventional progestogens. It binds almost exclusively to the progesterone receptor without interfering with the receptors of other steroids, i.e., it does not exhibit androgenic or estrogenic activity [168,169]. As a consequence of its enhanced receptor binding, Nesterone is exceptionally potent, having progestational potency 100 times greater than progesterone and 10 times greater than levonorgestrel [170–172]. The ring may also be used for contraception after weaning. Clinical studies to evaluate the contraceptive potential of reservoir-type, silicone vaginal rings containing Nesterone (50% w/w Nesterone-loaded silicone elastomer core surrounded by a nonmedicated silicone elastomer sheath) have demonstrated high contraceptive efficacy [169]. A recent review of the clinical applications of Nesterone has been published [170].

21.7.3.5 NuvaRing

Despite the head start afforded to progestogen-only vaginal rings, the first contraceptive ring to reach market was the combined contraceptive ring, NuvaRing. It is a nonbiodegradable, flexible, transparent, colorless vaginal ring loaded with 11.7 mg etonogestrel and 2.7 mg ethinyl estradiol, and providing sustained daily release rates of 120 and 15 mug daily, respectively. Similar to the combined oral contraception regime, the ring is worn continuously for 3 weeks, and then removed for 1 week to permit menstruation before application of a new ring for the subsequent cycle. During use, the steroids are released from the ring and absorbed systemically by the vaginal tissue. The near zero-order release profile is achieved because of the reservoir-type design, manufactured by the coextrusion of a steroid-loaded poly(ethylene-*co*-vinyl acetate) core and a nonmedicated poly(ethylene-*co*-vinyl acetate) sheath layer. The low, continuous dose of steroids provided by the vaginal ring offers several advantages compared with the combined contraceptive pill, most notably reduced maximum serum levels [173], a lower incidence of side effects [174], and better cycle control [175–177]. Furthermore, user compliance, satisfaction, and acceptability with the vaginal ring are also high [177–179]. Maximum serum concentrations of etonogestrel and ethinyl estradiol are achieved approximately 1 week after ring insertion, and are typically 30%–40% of those obtained with a combined oral contraceptive. Thereafter, steroid concentrations decrease in a gradual linear manner [180]. Absolute bioavailability has also been shown to be higher for NuvaRing compared with combined oral contraception [181].

21.7.4 Hormone Replacement Therapy

HRT provides women with the female hormones estrogen and progestogen, either individually or as a combination of both. HRT is required to replace the depleting hormone levels women experience with menopause, or after a hysterectomy. The decreased hormone levels can produce a variety of symptoms including hot flushes, sleep disturbances, depression, anxiety, loss of libido, vaginal dryness, and a number of urogenital complaints. HRT has been proven to be extremely effective at reducing many of the symptoms associated with menopause and the patient has a variety of dosage forms available for treatment including systematic dosage forms such as injectable, oral, transdermal, and implant preparations, as well as local dosage forms such as vaginal creams, pessaries, and vaginal rings [181]. Systemic dosage forms have a general mode of action and thus will reduce the majority of symptoms associated with the menopause. Most local dosage forms however, are only designed to provide localized absorption of estrogen to reduce urogenital complaints and are intended to avoid systemic uptake.

Oral dosage forms are the most common means of administering hormone replacement therapies, as they are easy to take and can be easily controlled or stopped by the patient. However, oral dosage forms have the distinct disadvantage of being subject to first-pass hepatic metabolism. Such hepatic metabolism can lead to complications such as cholelithiasis, hypertension, and intravascular clotting. Other disadvantages of oral doses include a low bioavailability (3%–5%), stable serum levels are not achieved and poor patient compliance often poses a problem. Subsequently, HRT has focused on the development of alternative means of delivering female hormones for women suffering from menopausal complaints.

Transdermal patches such as Climara, Vivelle, Estraderm, Alora, and FemPatch are a convenient means of delivering female hormones into systemic circulation. Patches provide constant hormone levels, aid with patient compliance and allow for lower dosage levels due to avoidance of first-pass metabolism. The only noticeable problem with transdermal patches is that they may cause skin irritation. For those women who suffer from skin irritation, an alternative to transdermal patches are HRT gels such as Oestrogel and Sandrena. These gels can be simply applied to the skin on the arms and shoulders or inner thighs on a daily basis. The gel will dry in a few minutes and provides a daily dose of estrogen. The disadvantage of such gels is the variation in the dose, which is dependent on the amount of gel applied to the skin.

HRT implants are another means of achieving constant hormone levels while removing the responsibility of the patient to take or apply a daily dose of HRT. Implants are inserted in the fat under the skin (usually in the lower abdomen) by a simple surgical procedure involving a local anesthetic and a few stitches. However, this type of HRT has the major disadvantage that if the therapy does not suit then it is almost impossible to remove the implants.

Several formulations exist for HRT to be delivered by the vaginal route. The majority of these formulations are designed for only the localized delivery of female hormones to counteract uncomfortable vaginal complaints associated with the menopause such as vaginal dryness, dyspareunia, dysuria, and urinary urgency. There are several vaginal creams available (e.g., Ortho-Gynest, Ovestin, Premarin, Estrace), which contain low doses of female hormone and are intended to provide localized absorption of drug with minimal systemic uptake. Such low doses are unlikely to cause any of the complications associated with systemic estrogen treatment such as blood clots, increased risk to breast cancer, etc. However, low localized doses will not reduce more general menopausal symptoms such as hot flushes, sleep disturbances, depression, anxiety, etc. Other women may require localized estradiol delivery to the vagina if their systemic HRT does not produce a satisfactory urogenital response and thus the localized delivery of estradiol can supplement their systemic HRT [182].

IVRs are alternative formulations that offer a controlled and sustained delivery of female hormones for up to 3 months. The benefits of rings are that they are convenient, they avoid the messiness of creams and gels, they go unnoticed to both patient and partner during sexual intercourse and they aid with patient compliance. However, unlike implants the ring can be easily removed if the HRT is to be stopped for any reason. Estring is a vaginal ring made of silicone elastomer containing a reservoir core of estradiol hemihydrate (2 mg) that is designed for localized delivery of estradiol to reduce postmenopausal urogenital complaints. Estradiol is released in low amounts (approximately 7.5 μg/d) for 90 days before the ring is removed and replaced. Femring on the other hand, is a silicone vaginal ring with a reservoir core of estradiol-3-acetate (12.4 mg or 24.8 mg). Femring is designed to release higher concentrations of estradiol-3-acetate (50 or 100 μg/d) for systemic uptake, thereby providing a more centralized role in reducing postmenopausal complaints.

It should be noted that those HRT formulations, which only provide replacement of estradiol, need to be supplemented with progesterone tablets if the patient still has an intact womb. Estrogen stimulates the growth of the womb endometrium, which can lead to endometrial cancer if the growth is unopposed. Progestogen is required to oppose estrogen's effect on the womb lining and reduce the risk of cancer. Obviously, if a woman has had her womb surgically removed (a hysterectomy) then endometrial cancer is not a risk, and only estradiol is required for the HRT.

Recent concerns about the safety of systemically administered sex hormones have lead to increased interest in delivering the lowest possible dose of hormones directly to key target tissues, thereby minimizing systemic and local side effects. In the context of uterine drug delivery, the clinical applications might include contraception and HRT. Although estrogen replacement therapy is very effective in treating the symptoms of menopause, it is also associated with an increased risk of endometrial hyperplasia, a precursor for endometrial cancer. This risk may be significantly reduced by the combined use of a progestogen and an estrogen, known as HRT. The progestogen may be administered continuously (every day along with the estrogen) or cyclically (10–12 d each month). Several options for delivery of the progestogen are available, including oral and transdermal routes. Targeted uterine delivery of levonorgestrel, using a novel frameless IUD known as FibroPlant, has recently been evaluated to oppose estrogen stimulation of the endometrium [183–186]. The device, which releases 14 mcg daily levonorgestrel, was shown to significantly reduce endometrial proliferation and bleeding.

21.7.5 Urinary Incontinence

Urinary incontinence is a socially incapacitating problem affecting up to 55% of women over 65 years of age [187]. Normal urinary continence and bladder control requires a complex interaction between the brain, central nervous system, and pelvic organs. When any component of this system loses normal function, urinary control can be affected, resulting in urinary incontinence. This condition is usually categorized into four types, i.e., stress, urge, functional, and overflow incontinence [188]. The drug of choice in the treatment of urinary incontinence is oxybutynin, an anticholinergic tertiary amine with additional muscle relaxant and local anesthetic properties [189]. Oxybutynin blocks acetylcholine released from parasympathetic nerves in the urinary bladder, thus preventing contractions of the detrusor muscle and exerting a direct spasmolytic effect on the organ [190–193].

Oxybutynin has a relatively low oral bioavailability (6%) due to an extensive first-past presystemic metabolism after administration [194]. Indeed, the absence of intact oxybutynin in urine suggests that the major elimination pathway of this drug is hepatic metabolism [195]. The compound is readily converted to its stable toxic metabolite, *N*-desethyloxybutynin by cytochrome P450-mediated oxidation [196]. Following oral administration of oxybutynin, peak plasma concentrations are reached within 1 h. The short half-life of this drug (less than 2 h), when administered as a conventional oral formulation, necessitates multiple 5 mg daily dosing [197].

Although oxybutynin has a well-documented therapeutic effect in patients with bladder instability [198], the drug has a high incidence of adverse effects, observed in 30% to 80% of recipients [199]. The occurrence of these untoward effects necessitates treatment discontinuation in up to 25% of patients, depending on the dosage [200]. In order to reduce the incidence of adverse reactions, a number of oral controlled-release systems have been reported [191,201,202].

The possibility of delivering oxybutynin by a nonoral route may be a more effective method of reducing the drug dosage and incidence of adverse reactions associated with oral oxybutynin. The vaginal route offers potential in this regard, because it is located adjacent to the bladder. Also, continuous delivery of oxybutynin to the bladder would further enhance clinical outcomes by maintaining blood levels and improving patient compliance. In a bid to optimize delivery of oxybutynin, a number of vaginal controlled-release formulations have been reported. Cylindrical silicone elastomer inserts releasing 0.5, 1.0, and 5.0 mg/d oxybutynin over a 7-d period have been investigated in a rabbit model [203]. The results demonstrated consistent dose-dependent vaginal absorption, resulting in stable plasma levels by day 3. The amounts of oxybutynin released were sufficient to have a clinical effect on the bladder, and levels of *N*-desethyloxybutynin, the active metabolite considered to be responsible for side effects, were minimal. A similar silicone elastomer formulation, this time in the form of a reservoir-type IVR, has also been reported for the controlled delivery of oxybutynin [204]. A vaginal ring for the treatment of urinary incontinence is developed by Barr Laboratories [205].

21.7.6 Pain Relief

Morphine is an opioid analgesic used routinely to treat moderate-to-severe pain. It is commonly provided in a range of formulations for oral administration, including immediate, extended, and sustained-release tablets and capsules. It also comes as a suppository for rectal application. Other routes of administration have also been investigated, including buccal [206,207], sublingual [208–210], and intramuscular [210]. In a unique case report, morphine has also been administered by the intravaginal route, using suppositories and immediate and sustained-release tablets, as an alternative to patient-controlled analgesia [65]. Unusually, the patient was unable to have morphine administered orally or rectally owing to previous surgery, and preferred not to have to be tied to equipment and lines for parenteral administration. The vaginal dose was titrated from the oral dose according to the patient's response. Vaginal administration of morphine was continued successfully for approximately 6 weeks. Although vaginal absorption is expected to be less efficient compared with gastrointestinal absorption, bioavailability will be significantly increased owing to the avoidance of first-pass metabolism, which is significant in the case of morphine.

21.7.7 Antifungals

A range of antifungal agents, including butoconazole, clotrimazole, econazole, miconazole, terconazole, and tioconazole, in the form of vaginal creams, ointments, and suppositories, are presently used to treat vaginal candidiasis (yeast infections). These substances are generally poorly absorbed from the gut, skin, and buccal mucosa in humans [211]. For vaginal application, the antifungal is often administered in relatively low doses, which are insufficient to produce systemic levels [70,212,213]. However, systemic absorption of the antifungal agent miconazole has been reported using high-dose pessary formulations (1200 mg) [71].

21.7.8 Pelvic Endometriosis

Endometriosis is a benign condition in which the tissue that lines the uterus grows in locations outside the uterus, usually in various regions of the pelvis and occasionally on the ovaries. It is commonly treated with oral danazol (400 to 800 mg), a synthetic androgen, which acts on the pituitary to prevent secretions of luteinizing hormone (LH) and follicle-stimulating hormone (FSH), thereby inducing a state of amenorrhea during which the endometrial tissue shrinks away. However, the androgenic character of danazol often induces side effects, such as liver dysfunction, excessive hair growth, weight gain, and acne. There is considerable evidence that danazol exhibits a direct inhibitory action on endometriotic tissue [214–218], suggesting that vaginal administration might be effective. Matrix-type vaginal rings providing controlled release of danazol have been evaluated in women and have demonstrated significantly reduced endometrial mass and reduced side effects [219,220]. Also, compared with oral therapy, danazol administered vaginally by a ring device does not produce detectable systemic levels, reflecting the very low levels released (<3 mg) from the ring, and resulting in no inhibition of ovulation. More recently, a danazol-loaded gel [221] and IUD that reduce chronic pelvic pain in women with endometriosis have been reported [86].

21.8 CONCLUSIONS

Drug delivery to the uterus and, particularly, to the vagina is of increasing importance in women's health issues. Delivery systems for these routes can broadly be subdivided into those that are adaptations of technologies used for other applications, such as semisolids or vaginal tablets that may incorporate mucoadhesive polymers or particulate carriers, and those systems specifically designed for vaginal or uterine application. More recent developments in the latter categories involve solid polymeric carriers, including vaginal rings and intrauterine delivery devices, which can offer controlled delivery of a range of actives for both local and systemic administration. Perhaps the great challenge in drug delivery remains the ability to deliver biomolecules by a nonparenteral route. The delivery of biomolecular actives that have implications for female health, particularly with respect to ameliorating the HIV/AIDS pandemic, is therefore likely to provide one of the main drives for the further development of novel intravaginal and intrauterine drug delivery technologies.

REFERENCES

1. Deshpande, A.A., C.T. Rhodes, and M. Danish. 1992. Intravaginal drug delivery. *Drug Dev Ind Pharm* 18:1225.
2. Tindall, V.R. 1987. *Jeffcoate's principles of gynaecology*. London: Butterworths.
3. Jaszczak, S., and E.S.E. Hafez. 1980. The vagina and reproductive processes. In *Human reproduction—Conception and contraception*, chap. 11. Hagerstown: Harper and Row.
4. Malcolm, R.K., et al. 2004. Vaginal microbicides for the prevention of HIV transmission. In *Biotechnology and genetic engineering reviews*, chap. 3, 21 vols, ed. S.E. Harding. Andover: Intercept Ltd.
5. Suh, D.D., et al. 2004. MRI of female genital and pelvic organs during sexual arousal. *J Psychosom Obstet Gynecol* 25:153.
6. Barnhart, K.T., E.S. Pretorius, and D. Malamud. 2004. Lesson learned and dispelled myths: Three-dimensional imaging of the human vagina. *Fertil Steril* 81:1383.
7. Barnhart, K.T., et al. 2001. Distribution of a spermicide containing nonoxynol-9 in the vaginal canal and the upper female reproductive tract. *Hum Reprod* 16:1151.
8. Barnhart, K.T., et al. 2001. Distribution of topical medication in the human vagina as imaged by magnetic resonance imaging. *Fert Steril* 76:189.
9. Zondek, B., and M. Friedmann. 1936. Are there cyclic changes in the human vaginal mucosa? *JAMA* 106:1051.
10. Patton, D.L., et al. 2000. Epithelial cell layer thickness and immune cell populations in the normal human vagina at different stages of the menstrual cycle. *Am J Obstet Gynecol* 183:967.
11. Burgos, M.H., and R. de Vargas-Linares. 1978. Ultrastructure of the vaginal mucosa. In *The human vagina*, eds. E.S.E. Hafez and T.N. Evans, 63–93. Amsterdam: Elsevier/North Holland Biomedical Press.
12. Godley, M.J. 1985. Quantitation of vaginal discharge in healthy volunteers. *Br J Obstet Gynecol* 92:739.
13. Perl, J.I., G. Milles, and Y. Shimozato. 1959. Vaginal fluid subsequent to panhysterectomy. *Am J Obstet Gynecol* 78:285.
14. Lissimore, N., and D.W. Currie. 1939. Studies in vaginal fluid. Paper 1. Pathological considerations. *Am or Eur J Obstet Gynecol* 46:673.
15. Odeblad, E. 1964. Intracavity circulation of aqueous material in the human vagina. *Acta Obstet Gynecol Scand* 43:360.
16. Lapan, B., and M.M. Friedman. 1950. Glycogen and reducing substances in vaginal mucus: Gestational and cyclical variations. *Am J Obstet Gynecol* 59:361.
17. Preti, G., G.R. Huggins, and G.D. Silveberg. 1979. Alterations in the organic compounds of vaginal secretions caused by sexual arousal. *Fertil Steril* 32:47.
18. Wagner, G., and R.J. Levin. 1980. Electrolytes in vaginal fluid during the menstrual cycle of coitally active and inactive women. *J Reprod Fert* 60:17.
19. Wagner G. 1979. Vaginal transduction. In *The biology of the fluids of the female genital tract*, eds. F.K. Beller and G.F.B. Shumacher, 25–34. New York: Elsevier/North Holland.
20. Owen, H.O., and D.F. Katz. 1999. A vaginal fluid simulant. *Contraception* 59:91.
21. Dusitsin, N., et al. 1967. Histidine in human vaginal fluid. *Obstet Gynecol* 29:125.
22. Croughan, W.S., and Behbehani, A.M. 1988. Comparative study of inactivation of herpes simplex virus types 1 and 2 by commonly used antiseptic agents. *J Clin Microb* 26:213.
23. Sparling, P.F. 1999. Biology of *Neisseria gonorrhoeae*. In *Sexually transmitted diseases*, eds. K.K. Holmes, et al., 433–449. New York: McGraw-Hill.
24. Pettit, R.K., S.C. McAllister, and T.A. Hamer. 1999. Response of gonococcal clinical isolates to acidic conditions. *J Med Microbiol* 48:149.
25. Graves, S., T. Gotv, and F. Trewartha. 1980. Effect of pH on the motility and virulence of *Treponema pallidum* (Nichols) and *Treponema paraluis* cuniculi *in vitro* under anaerobic conditions. *Br J Vener Dis* 56:269.
26. Sturn, A.W., and H.C. Zanen. 1984. Characteristics of *Haemophilus ducreyi* in culture. *J Clin Microbiol* 19:672.

27. Nagy, E., M. Petterson, and P.-A. Mardh. 1991. Antibiosis between bacteria isolated from the vagina of women with and without signs of bacterial vaginosis. *Acta Pathol Microbiol Immunol Scand* 99:739.
28. Martin, L.S., J.S. McDougal, and S.L. Loskoski. 1985. Disinfection and inactivation of the human T lymphotropic virus type III/lymphadenopathy-associated virus. *J Infect Dis* 152:400.
29. Ongradi, J., et al. 1990. Acid sensitivity of cell-free and cell-associated HIV-1: Clinical implications. *AIDS Res Human Retrovirol* 6:1433.
30. O'Connor, T.J., et al. 1995. The activity of candidate virucidal agents, low pH and genital secretions against HIV-1 *in vitro*. *Int J Sex Trans Dis AIDS* 6:267.
31. Wagner, G., and R. Levin. 1984. Human vaginal pH and sexual arousal. *Fertil Steril* 41:389.
32. Tevi-Benissan, C., et al. 1997. *In vivo* semen-associated pH neutralization of cervicovaginal secretions. *Clin Diagn Lab Immunol* 4:367.
33. Masters, W.H., and V.E. Johnson. 1966. *Human sexual response*. Boston: Little, Brown.
34. Burruano, B.T., R.L. Schnaare, and D. Malamud. 2002. Synthetic cervical mucus formulation. *Contraception* 66:137.
35. Hillier, S.L., et al. 1993. The normal vaginal flora, H_2O_2-producing lactobacilli, and bacterial vaginosis in pregnant women. *Clin Infect Dis* 16 (Suppl. 4):S273.
36. Priestley, C.J.F., B.M. Jones, and L. Goodwin. 1997. What is normal vaginal flora? *Genitourin Med* 73:23.
37. Fontaine, E.A., E. Claydon, and D. Taylor-Robinson. 1996. Lactobacilli from women with or without bacterial vaginosis and observations on the significance of hydrogen peroxide. *Microb Ecol Health Dis* 9:135.
38. Valore, E.V., et al. 2002. Antimicrobial components of vaginal fluid. *Am J Obstet Gynecol* 187:561.
39. Macht, D. 1918. The absorption of drugs and poisons through the vagina. *J Pharmacol Pathol* 10:509.
40. Hwang, S., et al. 1977. Systems approach to vaginal delivery of drugs IV: Methodology for determination of membrane surface pH. *J Pharm Sci* 66:778.
41. Ho, N.F.H., et al. 1976. Systems approach to vaginal delivery of drugs III: Simulation studies interfacing steroid release from silicone matrix and vaginal absorption in rabbits. *J Pharm Sci* 65:1576.
42. Hwang, S., et al. 1976. Systems approach to vaginal delivery of drugs II: *In situ* vaginal absorption of unbranched aliphatic alcohols. *J Pharm Sci* 65:1578.
43. Yotsuyanagi, T., et al. 1975. Systems approach to vaginal delivery of drugs I: Development of *in situ* vaginal drug absorption procedure. *J Pharm Sci* 64:71.
44. Richardson, J.L., and L. Illum. 1992. Routes of delivery—Case studies. The vaginal route of peptide and protein drug delivery. *Adv Drug Deliv Rev* 8:341.
45. Sayani, A.P., and Y.W. Chien. 1996. Systemic delivery of peptides and proteins across absorptive mucosae. *Crit Rev Ther Drug Carrier Syst* 13:85.
46. Singer, S.J., and G.L. Nicolson. 1972. The fluid mosaic model of the structure of cell membranes. *Science* 175:720.
47. Chien, Y.W. 1992. *Novel drug delivery systems*, 2nd ed. New York: Marcel Dekker.
48. De Ziegler D., et al. 1992. Effects of luteal estradiol on the secretory transformation of human endometrium and plasma gonadotropins. *J Clin Endocrinol Metab* 74:322.
49. Miles, R.A., et al. 1994. Pharmacokinetics and endometrial tissue levels of progesterone after administration by intramuscular and vaginal routes: A comparative study. *Fertil Steril* 62:485.
50. Kullander, S., and L. Svanberg. 1985. On resorption and the effects of vaginally administered terbutaline in women with premature labour. *Acta Obstet Hynaecol Scand* 64:613.
51. Mizutani, T., et al. 1995. Danazol concentrations in ovary, uterus, and serum and their effect on the hypothalamic-pituitary-ovarian axis during vaginal administration of danazol suppository. *Fertil Steril* 63:1184.
52. Cicinelli, E., and de Ziegler D. 1999. Transvaginal progesterone: Evidence for a new functional 'portal system' flowing from the vagina to the uterus. *Human Reprod Update* 5:365.
53. Bulletti, C., et al. 1997. Targeted drug delivery in gynaecology: The first uterine pass effect. *Hum Reprod* 12:1073.

54. Kunz, G., et al. 1996. The dynamics of rapid sperm transport through the female genital tract: Evidence from vaginal sonography of uterine peristalsis and hysterosalpingscintigraphy. *Hum Reprod* 11:629.
55. Kunz, G., et al. 1998. Sonographic evidence for the involvement of the utero-ovarian counter-current system in the ovarian control of directed uterine sperm transport. *Hum Reprod Update* 4:667.
56. Wildt, L., et al. 1998. Sperm transport in the female genital tract and its modulation by oxytocin as assessed by hypersalpingoscintigraphy, hysterotonography, electrohysterography and Doppler sonography. *Hum Reprod Update* 4:655.
57. Einer-Jensen, N. 1992. Counter-current transfer between blood vessels in the ovarian adnex. *Acta Obstet Gynecol Scand* 71:566.
58. Einer-Jensen, N. 1993. Rapid adsorption and local redistribution after vaginal application in gilts. *Acta Vet Scand* 34:1.
59. Einer-Jensen, N., et al. 2002. Uterine first pass effect in postmenopausal women. *Hum Reprod* 17:3060.
60. Cicinelli, E., et al. 2000. Mechanisms of uterine specificity of vaginal progesterone. *Hum Reprod* 15:159.
61. Cicinelli, E., et al. 2004. "First uterine pass effect" is observed when estradiol is placed in the upper but not the lower third of the vagina. *Fertil Steril* 81:1414.
62. Lim, J.M.H., E.B.S. Soh, and S. Raman. 1995. Intravaginal misoprostol for termination of midtrimester pregnancy. *Aust NZ J Obstet Gynaecol* 35:54.
63. Digenis, G.A., et al. 1999. Novel vaginal controlled-delivery systems incorporating coprecipitates of nonoxynol-9. *Pharmaceut Develop Tech* 4:421.
64. Stein, G.E., and N. Mummaw. 1993. Placebo-controlled trial of itraconazole for treatment of acute vaginal candidiasis. *Antimicrob Agents Chemother* 37:89.
65. Ostrop, N.J., J. Lamb, and G. Reid. 1998. Intravaginal morphine: An alternative route of administration. *Pharmacotherapy* 18:863.
66. Enzelsberger, H., et al. 1991. Effects of intravaginal application of estriol tablets in women with urge incontinence. *Geburtshilfe und Frauenheilkunde* 51:834.
67. Rock, J., R.H. Barker, and W.B. Bacon. 1947. Vaginal absorption of penicillin. *Science* 105:13.
68. Yamashita, A., S. Oshima, K. Matsuo, K. Ito, A. Ito, and Y. Mori. 1991. Pharmacological studies of intravaginally applied dehydroepiandrosterone sulfate. *Folia Pharmacologica Japonica* 98:31.
69. Tan, L.K., and S.K. Tay. 1999. Two dosing regimens for preinduction cervical priming with intravaginal dinoprostone pessary: A randomised clinical trial, *Br J Obstet Gynaecol* 106:907.
70. Abrams, L.S., and H.S. Weintraub. 1983. Disposition of radioactivity following intravaginal administration of 3H-miconazole nitrate. *Am J Obstet Gynecol* 147:970.
71. Daneshmend, T.K. 1986. Systemic absorption of miconazole from the vagina. *J Antimicrob Chemother* 18:507.
72. Borgida, A.F., J.F. Rodis, W. Hanlon, A. Craffey, L. Ciarleglio, and W.A. Campbell. 1995. Second trimester abortion by intramuscular 15-methyl-prostaglandin F2-alpha or intravaginal prostaglandin E(2) suppositories—A randomized trial. *Obstet Gynecol* 85:697.
73. Mircioiu, C., et al. 1998. Pharmacokinetics of progesterone in postmenopausal women: 1. Pharmacokinetics following intravaginal administration. *Eur J Drug Metabol Pharmacokin* 23:391.
74. Junginger, H.E. 1991. Mucoadhesive hydrogels. *Pharmazeutische Industrie* 53:1056.
75. Brannonpeppas, L. 1993. Novel vaginal drug-release applications. *Adv Drug Deliv Rev* 11:169.
76. Lehr, C.M., et al. 1992. A surface energy analysis of mucoadhesion: Contact angle measurements on polycarbophil and pig intestinal mucosa in physiologically relevant fluids. *Pharm Res* 9:70.
77. Coury, A.J., et al. 1984. Recent developments in hydrophilic polymers. *Med Dev Dia Ind* 6:28.
78. Woolfson, A.D., R.K. Malcolm, and R. Gallagher. 2000. Drug delivery by the intravaginal route. *Crit Rev Ther Drug Carrier Syst* 17:509.
79. Malcolm, R.K. 2003. The intravaginal ring. In *Modified-release drug delivery technology*, eds. M.J. Rathbone, J. Hadgraft, and M.S. Roberts. New York: Marcel Dekker, chap. 64.
80. World Health Organization (WHO). 2002. The IUD—worth singing about. *WHO Progress in Reproductive Health Research* 60:1. Available at: www.who.int/reproductive-health/hrp/progress/60/Progress60.pdf

81. Johns Hopkins Center for Communication Programs, Population Information Program. IUDs. Population Reports, 23, 1995. Available at: www.jhuccp.org/pr/b6edsum.stm.
82. Luukkainen, T., and J. Toivonen. 1995. Levonorgestrel-releasing IUD as a method of contraception with therapeutic properties. *Contraception* 52:269.
83. Kishen, M. 1998. Gynefix. *IPPF Medical Bulletin*, 32.
84. Van Os, W., and D. Edelman. 1998. New Directions in IUD. *Dev Adv Contracept* 14:41.
85. Wildemeersch, D., et al. 1999. GyneFIX. The frameless intrauterine contraceptive implant—An update for interval, emergency and postabortal contraception. *Br J Fam Plan* 24.
86. Corbellis, L., et al. 2004. A danazol-loaded intrauterine device decreases dysmenorrheal, pelvic pain, and dyspareunia associated with endometriosis. *Fertil Steril* 82:239.
87. Tamaoka, Y., et al. 2004. Treatment of endometrial hyperplasia with a danazol-releasing intrauterine device: A prospective study. *Gynecol Obstet Invest* 58:42.
88. UNAIDS (2002). Report on the global HIV/AIDS epidemic, July 2002. Joint United Nations Programme on HIV/AIDS, Geneva. Available at http://www.unaids.org/html/pub/Global-Reports/Barcelona/BRGlobal_AIDS_Report_en_pdf.pdf.
89. Gerbase, A.C., et al. 1998. Global prevalence and incidence estimates of selected curable STDs. *Sex Trans Infect Suppl* 1:S12.
90. Fauci A.S. 1999. The AIDS epidemic. *N Engl J Med* 341:1046.
91. Wasserheit, N.J. 1992. Epidemiological synergy: Interrelationships between HIV and infection and other STDs. *Sex Trans Dis* 19:61.
92. Stratton P., and N.J. Alexander. 1993. Prevention of sexually transmitted infections: Physical and chemical barrier methods. *Infect Dis Clin North Am* 7:841.
93. Van de Wijgert, J., and C. Coogins. 2002. Microbicides to prevent heterosexual transmission of HIV: Ten years down the road. *AIDScience* 2:1. Available at: www.aidscience.org/Articles/aidscience015.asp
94. Avert—A guide to HIV and AIDS treatment. Available at: www.avert.org
95. amfAR Aids Research—Microbicides: A New Weapon Against HIV. Available at: www.amfar.org
96. AIDS Action—Policy Facts. Microbicides. March 2002. Available at: www.thebody. com/aac/brochures/microbicides.html.
97. Knodel, J., and M. VanLandingham. 1996. Thai views of sexuality and sexual behaviour. *Health Trans Rev* 6:179.
98. The Terrence Higgins Trust. Briefing on Microbicides: Prevention for Women. Available at: www.iapac.org/clinmgt/women/microb.html.
99. AIDS Epidemic Update 2004, Joint United Nations Programme of HIV/AIDS, Available at: www.unaids.org/wad2004/report.html.
100. Royce, R.A., A. Sena, W. Cates, Jr., and M.S. Cohen. 1997. Sexual transmission of HIV. *N Engl J Med* 336:1072.
101. AIDS Epidemic update (2002). UNAIDS/WHO, December 2002. Available at: www.who.int/hiv/pub/epidemiology/epi2002/en/.
102. Stone, A. 2002. Microbicides: A new approach to preventing HIV and other sexually transmitted infections. *Nat Rev Drug Discov* 1:997.
103. Singh, B., J.C. Cutler, and H.M.D. Utidjian. 1972. Studies of the development of a vaginal preparation providing both prophylaxis against venereal disease and other genital infections and contraception: II. Effect *in vitro* of vaginal contraception and non-contraceptive preparations on *Treponema pallidum* and *Neisseria gonorrhoeae*. *Br J Ven Dis* 48:57.
104. Hicks, D., L. Martin, and J. Getchell. 1985. Inactivation of HTLV-III/LAV-infected cultures of normal human lymphocytes by nonoxynol-9 *in vitro*. *Lancet* 2:1422.
105. Judson. F.N., et al. 1989. *In vitro* evaluation of condoms with and without nonoxynol-9 as physical and chemical barriers against *Chlamydia trachomatis*, herpes simplex virus type 2, and human immunodeficiency virus. *Sex Trans Dis* 16:51.
106. Van Damme, L., et al. 2002. Effectiveness of COL-1492, a nonoxynol-9 vaginal gel, on HIV-1 transmission in female sex workers: A randomised controlled trial. *Lancet* 360:971.
107. Kreiss J., et al. 1992. Efficacy of nonoxynol 9 contraceptive sponge use in preventing heterosexual acquisition of HIV in Nairobi prostitutes. *JAMA* 268:477.

108. Roddy, R.E., et al. 1998. A controlled trial of nonoxynol-9 film to reduce male-to-female transmission of sexually transmitted diseases. *N Engl J Med* 339:504.
109. Isaacs, C.E., R.E. Litov, and H. Thormar. 1995. Antimicrobial activity of lipids added to human milk, infant formula, and bovine milk. *Nutr Biochem* 6:362.
110. Kabara, J.J. 1978. Fatty acids and derivatives as antimicrobial agents. In *The pharmacological effect of lipids*, ed. J.J. Kabara, 1–14. St. Louis: American Oil Chemists Society.
111. Shibasaki, I., and N. Kato. 1978. Combined effects on antibacterial activity of fatty acids and their esters against gram-negative bacteria. In *The pharmacological effects of lipids*, ed. J.J. Kabara, 15–24. St. Louis: American Oil Chemists Society.
112. Thormar H., C.E. Isaacs, H.R. Brown, M.R. Barshatzky, and T. Pessolano. 1987. Inactivation of enveloped viruses and killing of cells by fatty acids and monoglycerides. *Antimicrob Agents Chemother* 31:27.
113. Welsh, J.K., M. Arsenakis, R.J. Coelen, and J.T. May. 1979. Effect of antiviral lipids, heat, and freezing on the activity of viruses in human milk. *J Infect Dis* 140:322.
114. Thormar, et al. 1999. Hydrogels containing monocaprin have potent microbicidal activities against sexually transmitted viruses and bacteria *in vitro*. *Sex Trans Infect* 75:181.
115. Lampe, M.F., et al. 1998. Killing of *Chlamydia trachomatis* by novel antimicrobial lipids adapted from compounds in human breast milk. *Antimicrob Agents Chemother* 42:1239.
116. Fienberg, M.B., and J.P. Moore. 2002. AIDS vaccine models: Challenging challenge viruses. *Nat Med* 8:207.
117. Microbicide Research and Development—Progress Report. January 2002. Alliance for Microbicide Development.
118. Esser, M.T., et al. 1999. Cyanovirin-N binds to gp120 to interfere with CD4-dependent human immunodeficiency virus type 1 virion binding, fusion, and infectivity but does not affect the CD4 binding site on gp120 or soluble CD4- induced conformational changes in gp120. *J Virol* 73:4360.
119. O'Keefe, B.R., et al. 2000. Analysis of the interaction between the HIV-inactivating protein cyanovirin-N and soluble forms of the envelope glycoproteins gp120 and gp41. *Mol Pharmacol* 58:982.
120. Bolmstedt, A.J., et al. 2001. Cyanovirin-N defines a new class of antiviral agent targeting N-linked, high-mannose glycans in an oligosaccharide-specific manner. *Mol Pharmacol* 59:949.
121. Cariati, S. 2002. Invisible condoms for women on the way, Society for Womens Health Research. Available at: http://www.womens-health.org/press/newsservice/011702.htm.
122. Parr, M.B., and E.L. Parr. 1985. Immunohistochemical localization of immunoglobins A, G and M in female genital tract. *J Reprod Fert* 74:361.
123. Forrest, B.D. 1991. Women, HIV, and mucosal immunity. *Lancet* 337:835.
124. Kozlowski, P.A., et al. 2002. Differential induction of mucosal and systemic antibody responses in women after nasal, rectal, or vaginal immunization: Influence of the menstrual cycle. *J Immunol* 169:566.
125. Upadhyay, S. 1995. Reproductive tract immunity for the control of fertility: New strategies for immunocontraception. *Mucos Immun Update* 3:7.
126. Alexander, N.J., D.L. Fulgham, and E. Goldberg, E. 1992. Contraceptive vaccine development: Secretory immune response in mice and monkeys. *Vaccine Res* 1:331.
127. Mestecky, J., et al. 1997. Current options for vaccine delivery systems by mucosal routes. *J Control Release* 48:243.
128. Russel, M.W. 2002. Immunization for protection of the reproductive tract: A review. *Am J Reprod Immunol* 47:265.
129. O'Hagan, D.T. 1998. Microparticles and polymers for the mucosal delivery of vaccines. *Adv Drug Deliv Rev* 34:305.
130. Parr, E.L., M.B. Parr, and M. Thapar. 1988. A comparison of specific antibody responses in mouse vaginal fluid after immunization by various routes. *J Reprod Immunol* 14:165.
131. Russell, M.W., and J. Mestecky. 1998. Humoral immune responses to microbial infections in the genital tract. *Microb Infect* 34:305.
132. Parr, M.B., and E.L. Parr. 1994. Mucosal immunity in the female and male reproductive tracts. In *Handbook of mucosal immunology*, eds. P.L. Ogra, W. Strober, J. Mestecky, J.R. McGhee, M.E. Lamm, and J. Bienenstock, 667. San Diego: Academic Press.

133. Bernkop-Schnurch, A., and M. Hornof. 2003. Intravaginal drug delivery systems. *Am J Drug Deliv* 1:241.
134. Whalen, R.G. 1996. DNA vaccines for emerging infectious diseases: What if? *Emerg Infect Dis* 2:168.
135. Kozwolski, P.A., M.R. Neutra, T.P. Flanigan, and S. Cu-Uvin. 2000. Induction of antibodies in women after vaginal immunization during the mid-proliferative or mid-secretory phase of the menstrual cycle. *FASEB J* 14:A1248.
136. Gillgrass, A.E., A.A. Ashkar, K.L. Rosenthal, and C. Kaushic. 2003. Prolonged exposure to progesterone prevents induction of protective mucosal responses following intravaginal immunization with attenuated herpes simplex virus type 2. *J Virol* 77:9845.
137. Kaushic, C., A.A. Ashkar, L.A. Reid, and K.L. Rosenthal. 2003. Progesterone increases susceptibility and decreases immune responses to genital herpes infection. *J Virol* 77:4558.
138. Li Wan Po, A., E. Rogers, M. Shepphard, and E.M. Scott. 1995. Delivery systems for non-parenteral vaccines. *Adv Drug Deliv Rev* 18:101.
139. Kozlowski, P.A., et al. 1999. Mucosal vaccination strategies for women. *J Inf Dis* 179:S493.
140. Medaglini, D., et al. 1997. Commensal bacteria as vectors for mucosal vaccines against sexually transmitted diseases: Vaginal colonization with recombinant streptococci induces local and systemic antibodies in mice. *Vaccine* 15:1330.
141. Schreckenberger, C., et al. 2001. Induction of an HPV 6bL1-specific mucosal IgA response by DNA immunization. *Vaccine* 19:227.
142. Chattergoon, M., J. Boyer, and D.B. Weiner. 1997. Genetic immunization: A new era in vaccines and immune therapeutics. *FASEB J* 11:753.
143. Bernstein, D.I. 2000. Effect of route of vaccination with vaccinia virus expressing HSV-2 glycoprotein D on protection from genital HSV-2 infection. *Vaccine* 18:1351.
144. McLean, C.S., D. NiChallanain, I. Duncan, M.E.G. Boursnell, R. Jennings, and S.C. Inglis. 1996. Induction of a protective immune response by mucosal vaccination with a DISC HSV-1 vaccine. *Vaccine* 14:987.
145. Parr, E.L., and M.B. Parr. 1999. Immune responses and protection against vaginal infection after nasal or vaginal immunization with attenuated herpes simplex type-2. *Immunology* 98:639.
146. Ogra, P.L., and S.S. Ogra. 1973. Local antibody response to poliovaccine in the human female genital tract. *J Immun* 110:1307.
147. Cardenas-Freytag, L., et al. 2002. Partial protection against experimental vaginal candidiasis after mucosal vaccination with heat-killed *Candida albicans* and the mucosal adjuvant LT(R192G). *Med Mycol* 40:291.
148. Uehling, D.T., et al. 2003. Phase 2 clinical trial of a vaginal mucosal vaccine for urinary tract infections. *J Urol* 170:867.
149. Dumais, N., et al. 2002. Mucosal immunization with inactivated human immunodeficiency virus plus CpG oligodeoxynucleotides induces genital immune responses and protection against intravaginal challenge. *J Infect Dis* 15:1098.
150. Pyles, R.B., D. Higgins, C. Chalk, A. Zalar, J. Eiden, C. Brown, G. VanNest, and L.R. Stanberry. 2002. Use of immunostimulatory sequence-containing oligonucleotides as topical therapy for genital herpes simplex virus type 2 infection. *J Virol* 76:11387.
151. OHagan, D.T., D. Rafferty, S. Wharton, and L. Illum. 1993. Intravaginal immunization in sheep using a bioadhesive microsphere antigen delivery system. *Vaccine* 11:660.
152. Wassen, L., K. Schon, J. Holmgren, M. Jertborn, and N. Lycke. 1996. Local intravaginal vaccination of the female genital tract. *Scand J Immunol* 44:408.
153. Johansson, E.L., L. Wassen, J. Holmgren, M. Jertborn, and A. Rudin. 2001. Nasal and vaginal vaccinations have differential effects on antibody responses in vaginal and cervical secretions in humans. *Infect Immun* 69:7481.
154. Moreno, L., and N. Goldman. 1991. Contraceptive failure rates in developing countries: Evidence from the demographic and health surveys. *Int Fam Plan Perspect* 17:44.
155. Jones, E.F., and J.D. Forrest. 1992. Contraceptive failure rates based on the 1988 NSFG. *Fam Plan Perspect* 24:12.

156. Croxatto, H.B., et al. 1982. Plasma progesterone levels following subdermal implantation of progesterone pellets in lactating women. *Acta Endocrinol* 100:630.
157. Croxatta, H.B., et al. 1982. Fertility regulation in nursing women. II. Comparative performance of progesterone implants versus placebo and copper T. *Am J Obstet Gynecol* 144:201.
158. Diaz, S., et al. 1984. Fertility regulation in nursing women. VI. Contraceptive effectiveness of a subdermal progesterone implant. *Contraception* 30:311.
159. Duncan, G.W. 1970. Medicated devices and methods. US Patent 3,545,439.
160. Massai, R., et al. 1999. Pre-registration study on the safety and contraceptive efficacy of a progesterone-releasing vaginal ring on Chilean nursing women. *Contraception* 60:9.
161. Massai, R., et al. 2000. Vaginal rings for contraception in lactating women. *Steroids* 65:703.
162. Diaz, S., et al. 1985. Fertility regulation in nursing women. VIII. Progesterone plasma levels and contraceptive efficacy of a progesterone-releasing vaginal ring. *Contraception* 32:603.
163. Shaaban, N.N. 1991. Contraception with progestogens and progesterone during lactation. *J Steroid Biochem Molec Biol* 40:705.
164. Croxatto, H.B., and S. Diaz, S. 1991. Progesterone vaginal rings for contraception during breast-feeding. In *Female contraception and male fertility regulation*, vol. 2, eds. B. Runnebaum, T. Rabe, and L. Kiessel, 135–142. New Jersey: The Parthenon Publishing Group.
165. Sivin, I., et al. 1997. Contraceptives for lactating women: A comparative trial of a progesterone-releasing vaginal ring and the copper T 380A IUD. *Contraception* 55:225.
166. Diaz, S., et al. 1997. Fertility regulation in nursing women. IX. Contraceptive performance, duration of lactation, infant growth, and bleeding patterns during use of progesterone vaginal rings, progestin-only pills, Norplant® implants and copper T 380-A intrauterine devices. *Contraception* 56:223.
167. Chen, J.H., et al. 1998. The comparative trial of TCu 380A IUD and progesterone-only vaginal ring used by lactating women. *Contraception* 57:371.
168. Massai, R., et al. 2000. Vaginal rings for contraception in lactating women. *Steroids* 65:703.
169. Sitruk-Ware, R., et al. 2003. Nestorone®: Clinical implications for contraception and HRT. *Steroids* 68:907.
170. Kumar, N., et al. 2000. Nesterone®: A progestin with a unique pharmacological profile. *Steroids* 65:629.
171. Shapiro, E.L., et al. 1962. 16-Alkylated progestogens. *J Med Pharm Chem* 5:975.
172. Bottela, J., et al. 1986. Interaction of new 19-norprogesterone derivative with progestagen, mineralocortoid and glucocorticoid receptors. *J Pharmacol* 17:699.
173. Killick, S. 2002. Complete and robust ovulation inhibition with NuvaRing. *Eur J Contracept Reprod Health Care* 7:13.
174. Veres, S., L. Miller, and B. Burington. 2004. A comparison between the vaginal ring and oral contraceptives. *Obstet Gynecol* 104:555.
175. Oddson, K., et al. 2005. Superior cycle control with a contraceptive vaginal ring compared with an oral contraceptive containing 30 mu g ethinylestradiol and 150 mug levonorgestrel: A randomized trial. *Hum Reprod* 20:557.
176. Vree, M. 2002. Lower hormone dosage with improved cycle control. *Eur J Contracept Reprod Health Care* 7:25.
177. Dieben, T.O.M., F.J.M.E., Roumen, and D. Apter. 2002. Efficacy, cycle control, and user acceptability of a novel combined contraceptive vaginal ring. *Obstet Gynecol* 100:585.
178. Szarewski, A. 2002. High acceptability and satisfaction with NuvaRing use. *Eur J Contracept Reprod Health Care* 7:31.
179. Novak, A., et al. 2003. The combined contraceptive vaginal ring, NuvaRing: An international study of user acceptability. *Contraception* 67:187.
180. Timmer, C.J., and T.M.T. Mulders. 2000. Pharmacokinetics of etonogestrel and ethinylestradiol released from a combined contraceptive vaginal ring. *Clin Pharmacokin* 39:233.
181. Gabrielsson, J., et al. 1995. New kinetic data on estradiol in light of the vaginal ring concept. *Matutitas* S35–S39.
182. Bachmann, G. 1995. The estradiol vaginal ring—A study of existing clinical data. *Matutitas* S21–S39.

183. Wildemeersch, D., et al. 2004. Endometrial safety with a low-dose intrauterine levonorgestrel-releasing system after 3 years of estrogen substitution therapy. *Maturitas* 48:65.
184. Wildemeersch, D., and E. Schacht. 2001. Treatment of menorrhagia with a novel 'frameless' intrauterine levonorgestrel-releasing drug delivery system: A pilot study. *Eur J Contracept Reprod Health Care* 6:93.
185. Wildemeersch, D., and E. Schacht. 2000. Contraception with a novel 'frameless' intrauterine levonorgestrel-releasing drug delivery system: A pilot study. *Eur J Contracept Reprod Health Care* 5:234.
186. Wildemeersch, D., et al. 2003. Miniature, low-dose, intrauterine drug-delivery systems. *Annals NY Acad Sci* 997:174.
187. Couture, J.A., and L. Valiquette. 2000. Urinary incontinence. *Ann Pharmacother* 34:646.
188. Hattori, T. 1998. Drug treatment for urinary incontinence. *Drugs Today* 34:125.
189. Pietzko, A., et al. 1994. Influences of trospium chloride and oxybutynin on quantitative EEG in healthy-volunteers. *Eur J Clin Pharmacol* 47:337.
190. Goldenberg, M.M. 1999. An extended-release formulation of oxybutynin chloride for the treatment of overactive urinary bladder. *Clin Ther* 21:634.
191. Anderson, R.U., et al. 1999. Once daily controlled versus immediate release oxybutynin chloride for urge urinary incontinence. *J Urol* 161:1809.
192. Gupta, S.K., and G. Sathyan. 1999. Phamacokinetics of an oral once-a-day controlled-release oxybutynin formulation compared with immediate-release oxybutynin. *J Clin Pharmacol* 39:289.
193. Gupta, S.K., et al. 1999. Quantitative characterisation of therapeutic index: Application of mixed-effects modelling to evaluate oxybutynin dose-efficacy and dose-side effect relationships. *Clin Pharmacol Ther* 65:672.
194. Lukkari, E., et al. 1998. Cytochrome P450 specificity of metabolism and interaction of oxybutynin in human liver microsomes. *Pharmacol Toxicol* 82:161.
195. Douchamps, J., et al. 1988. The pharmacokinetics of oxybutynin. *Eur J Clin Pharmacol* 35:515.
196. Yaich, M., et al. 1998. *In vitro* cytochrome p450 dependant metabolism of oxybutynin to N-deethyloxybutynin in humans. *Pharmacogenetics* 8:4.
197. Nilsson, C.G., et al. 1997. Comparison of a 10mg controlled release oxybutynin tablet with a 5 mg oxybutynin tablet in urge continence patients. *Neurourol Urodyn* 16:533.
198. Castleden, C.M., et al. 1990. Oxybutynin and urinary incontinence. *Age Ageing* 19:72.
199. Brendler, C.B., et al. 1989. Topical oxybutynin chloride for relaxation of dysfunctional bladders. *J Urol* 141:1350.
200. Yarker, Y.E., K.L. Goa, and A. Fitton. 1995. Oxybutynin—A review of its pharmacodynamic and pharmacokinetic properties and its therapeutic use in detrusor instability. *Drug Ageing* 6:243.
201. Birns, J., E, Lukkari, and J.G.A. Malonelee. 2000. Randomised controlled trial comparing the efficacy of controlled-release oxybutynin tablets (10 mg once daily) with conventional oxybutynin tablets (5 mg twice daily) in patients whose symptoms were stabilized on 5 mg twice daily of oxybutynin. *BJU Int* 85:793.
202. Versi, E., et al. 2000. Dry mouth with conventional and controlled-release oxybutynin in urinary incontinence. *Obstet Gynecol* 95:718.
203. Schroder, A. et al. 2000. Absorption of oxybutynin from vaginal inserts: Drug blood levels and the response of the rabbit bladder. *Urology* 56:1063.
204. Woolfson, A.D., R.K. Malcolm, and R.J. Gallagher. 2003. Design of a silicone reservoir intravaginal ring for the delivery of oxybutynin. *J Control Release* 91:465.
205. www.barrlabs.com
206. Manara, A.R., et al. 1989. Pharmacokinetics of morphine following the buccal route. *Br J Anaesth* 62:498.
207. Bell, M.D.D., et al. 1985. Buccal morphine—A new route for analgesia? *Lancet* 1:71.
208. Hirsch, J.D. 1984. Sublingual morphine sulfate in chronic pain management. *Clin Pharm* 3:585.
209. Pannuti, F., et al. 1982. Control of chronic pain in very advanced cancer patients with morphine hydrochloride administered by oral, rectal and sublingual routes. Clinical report and preliminary results on the morphine pharmacokinetics. *Pharmacol Res Commun* 14:369.

210. Davis, T., et al. 1993. Comparative morphine pharmacokinetics following sublingual, intramuscular and oral administration in patients with cancer. *Hospice J* 9:85.
211. Daneshmend, T.K., and D.W. Warnock. 1983. Clinical pharmacokinetics of systemic antifungal drugs. *Clin Pharmacokin* 8:17.
212. Brugmans, J., et al. 1972. Systemic antifungal potential, safety, biotransport and transformation of miconazole nitrate. *Eur J Clin Parmacol* 5:93.
213. Vukovich, R.A., A. Heald, and A. Darragh. 1977. Vaginal absorption of two imidazole antifungal agents, econazole and miconazole. *Clin Pharmacol Therap* 21:121.
214. Chamness, G.C., R.H. Asch, and C.J. Pauerstein. 1980. Danazol binding and translocation of steroid receptors. *Am J Obstet Gynecol* 136:426.
215. Taketani, Y., and M. Mizuno. 1985. Direct action of danazol on endometrial cells *in vitro*. *Sanka to Fujinka* (in Japanese) 8:89.
216. Surrey, E.S., and J. Halme. 1992. Direct effects of medroxyprogesterone acetate, danazol, and leuprolide acetate on endometrial stromal cell proliferation *in vitro*. *Fertil Steril* 58:273.
217. Braun, D.P., H. Gebel, and W.P. Dmowski. 1994. Effect of danazol *in vitro* and *in vivo* on monocyte-mediated enhancement of endometrial cell proliferation in women with endometriosis. *Fertil Steril* 62:89.
218. Chin, K., et al. *In vitro* augmentative effect of danazol on apoptosis of endometrial cells. *Abstract of 18th Annual Meeting of Japan Endometriosis Society*, p. 17, 1997.
219. Igarashi, M., et al. 1998. Novel vaginal danazol ring therapy for pelvic endometriosis, in particular deeply infiltrating endometriosis. *Hum Reprod* 13:1952.
220. Igarashi, M. 1990. A new therapy for pelvic endometriosis and uterine adenomyosis: Local effect of vaginal and intrauterine danazol application, Asia-Oceania. *J Obstet Gynaecol* 16:1.
221. Dmowski, W.P., and T.I. Janicki. 2004. Intravaginal danazol is efficacious with limited side-effect profile in the treatment of endometriosis. *Obstet Gynecol* 103:60S.

22 Strategies to Improve Systemic and Local Availability of Drugs Administered via Vaginal Route

Giuseppina Sandri, Silvia Rossi, Franca Ferrari, Maria Cristina Bonferoni, and Carla Caramella

CONTENTS

22.1 INTRODUCTION

The vagina, as a site for drug delivery, is characterized by several features, which can be exploited in order to achieve desirable therapeutic effects. Considerable progress has been made in the last decade in this research area and the high number of patents and scientific papers published in pharmaceutical journals indicates an increasing interest in this field (Figure 22.1).

Although anatomy and physiology, microflora, and secretions of the vagina are well understood, the scientific knowledge of drug delivery via the vaginal route is still limited and relatively unexplored despite its potentiality as a noninvasive route of drug administration. The presence of a thick network of blood vessels renders the vagina an excellent route for drug delivery to achieve both systemic and local effects. The main advantages of drug delivery via the vagina are related to the bypass of first-pass metabolism, the ease of administration, and the high permeability of low-molecular-weight drugs. On the other hand, drug delivery systems intended for vaginal application must overcome several drawbacks such as leakage and limited residence time of the formulations, menstrual cycle and sexual relations, local irritations, and personal care habits. Moreover, the possible sensitivity of normal bacterial microflora, in particular *Lactobacillus acidophilus*, to dosage form ingredients, should be carefully evaluated. Finally, the considerable variability in absorption rate and extent due to changes in epithelium thickness must be considered.

This chapter deals with recent strategies designed to improve the systemic and local bioavailability of drugs administered via the vaginal route; after a brief description of the vaginal anatomy and physiology, the paper will focus on the various drug delivery systems currently in development. Special attention will be paid to the difficulties associated with systemic vaginal delivery and attempts to overcome them. Section 22.9 will be devoted to the possibility to achieve immunization via vaginal route. Finally experimental models to study vaginal tolerability and absorption will be described.

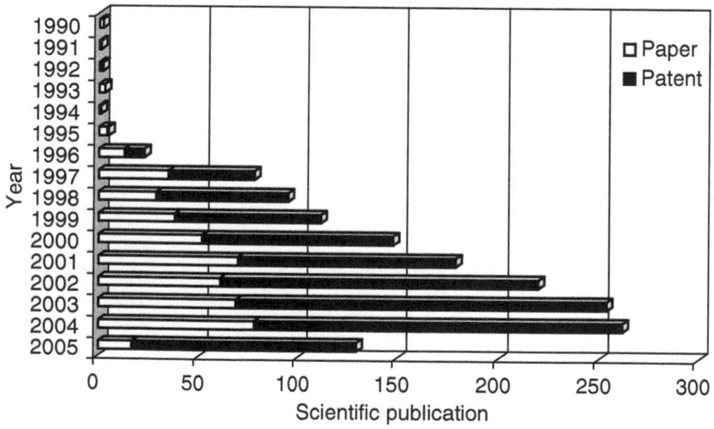

FIGURE 22.1 Number of scientific papers and patents published on the topic as a function of publication year (2005: until June). (From SciFinder Scholar, 2004; keywords entered: vaginal drug delivery systems.)

22.2 ANATOMY AND PHYSIOLOGY OF THE VAGINA RELEVANT TO DRUG DELIVERY

Human vagina is an S-shaped fibromuscular collapsible tube, 6–10 cm long, extending from cervix to uterus and located between the bladder and the rectum [1]. It presents unique features in terms of secretions (pH and composition) and microflora, which must be considered during the development and the evaluation of vaginal drug delivery systems [2].

The histology and physiology of the vagina may vary both with menstrual cycle and age. Until puberty, the thickness of the vaginal epithelium remains relatively thin and the pH of the vaginal fluids is close to neutrality. Between the menarche and the menopause, the pH of the vaginal environment is between 4 and 5 and the thickness of the epithelium increases up to approximately 200 μm with changes during the menstrual cycle linked to the estrogenic secretions. In menopause or after an ovariectomy, with the end of estrogenic secretions, the vaginal epithelium becomes thin and atrophic, the glycogen content of the exfoliating cells decreases, and consequently the pH increases. The administration of estrogens can restore the previous situations [2,3].

The vaginal epithelium is squamous, pluristratified, nonkeratinized, and lies over smooth elastic fibers of the muscular coat and on the loose connective tissue which, together, confer an excellent elasticity to the vagina. The extracellular matrix is composed of lipophilic substances (polar and neutral lipids, poor in glycosaminoglycans and in cholesterol). The blood and lymphatic vessels are well distributed and abundant in the vagina walls. The blood leaving the vagina enters the peripheral circulation via a rich venous plexus, which empties primarily into the internal iliac veins and secondarily into the inferior hemorrhoidal veins [4]. This enables drugs absorbed from the vagina to escape first-pass metabolism [2,4,5].

The epithelial cells are joined by desmosomes and gap junctions; tight junctions are absent or not completely formed [6]. Although the vaginal epithelium is usually considered as a mucosal surface, the vagina does not possess glands and goblet cells and does not produce mucin. The large amount of fluids secreted by the vagina is composed of cervical secretions (in particular cervical mucus and endometrial and tubal fluids) that transudate through the epithelium from extracellular matrix and blood vessels, exfoliating epithelial cells, and leukocytes [7,8]. The human vaginal fluid is hydrophilic and contains mucus, enzymes, enzyme inhibitors, carbohydrates, amino acids, alcohols, hydroxyketones, and aromatic compounds [9]. The lactic acid produced from glycogen by *L. acidophilus*, a bacterium dwelling in the vagina, is responsible for the acidic pH of the vaginal secretions; in particular, lactic acid acts as a buffer maintaining the pH from 3.8 to 4.2 [2,9]. The menstrual, cervical, uterine secretions, and the ejaculate may cause an increase in the pH of vaginal fluid. The presence of glycogen and carbohydrates in the vaginal fluid influences the microflora of the vagina; besides *L. acidophilus*, the prevalent microorganisms in the vaginal environment are bacteria belonging to Bacterioids and *Staphylococcus* species [4].

The vaginal fluid is also characterized by a certain enzymatic activity due to bacterial flora and enzymes present in the secretions and in the epithelium. The external and basal cell layers of the vagina express the majority of the enzymatic activity [2,5,9].

The outer cell layers possess β-glucuronidase, acid phosphatase, some α-naphthylesterase, and diaphorase with small amounts of phosphoamidase and succinic dehydrogenase. The basal cell layers present high activity of the enzymes found in the citric acid cycle, fatty acid metabolism, and 17-ketosteroidogenesis; moreover, they contain β-glucuronidase, succinic dehydrogenase, and diaphorase with small amounts of acid phosphatase and α-naphthylesterase. Levels of alkaline phosphatase (ALP), lactate dehydrogenase (LDH), aminopeptidase, and esterase are all high in the follicular phase of the menstrual cycle but fall down immediately before ovulation [2,5,6,10].

Between the cited enzymes, proteases represent the prominent barrier for the absorption of peptides and proteins into the systemic circulation [5,6,10].

Variation in hormonal levels (especially in estrogens) during the menstrual cycle determines the alteration of epithelial cell thickness, causing the widening of the intercellular channels and consequently variation in pH, composition, and secretion amount. Moreover, hormonal changes influence enzymatic activity of endopeptidases and aminopeptidases, causing further complications for drug delivery and therapeutic schedule [2,3,5,6,10].

22.3 VAGINAL MUCOSA AS A ROUTE FOR DRUG ADMINISTRATION

Traditionally the vaginal cavity has been employed for the delivery of drugs specifically related with female pathologies such as vaginitis caused by bacteria, fungi, protozoa, or virus. Nowadays, a particular interest has been devoted to microbicides used to prevent sexually transmitted diseases, including AIDS [9]. Moreover, labor-inducing drugs, spermicidal agents, and steroids have been administered via vaginal route. In recent years, vaginal delivery systems have been exploited as an alternative to parenteral formulations for drugs that cannot be profitably administered *per os* due to hepatic or gastrointestinal degradation or inducing side effects in the gastrointestinal tract. Modern technologies have highlighted vaginal drug delivery systems capable of maintaining proper pharmacological profiles; these make the vagina an excellent route for drug delivery [1,4].

Vaginal administration of drugs presents different advantages. Besides the avoidance of first-pass metabolism, it can reduce the occurrence and severity of gastrointestinal diseases and of hepatic side effects following hormone administration. The main limitations to vaginal administration of drugs are due to gender specificity and to menstrual cycle variations.

After administration, either drug penetration into the deep cell layers to obtain a local effect or drug permeation across the epithelium to reach the systemic circulation can occur.

The pathways of drug absorption are transcellular and paracellular. Moreover, absorption can be due to the vesicular- or receptor-mediated transport or it can occur through the aqueous pores on the mucosa surface.

The vaginal epithelium can be considered as a diffusion layer with aqueous pores and lipophilic pathways. Drug permeation or penetration is related to the hydrophilic–lipophilic balance of the drug. Small lipophilic molecules are thought to be absorbed by transcellular diffusion at therapeutic doses, whereas hydrophilic molecules are believed to cross the epithelium through aqueous, paracellular pore channels. Drug diffusion depends not only on drug molecular weight, partition coefficient, and ionization, but also on its surface charge and chemical nature [2,6].

Drug penetration (local effect) or drug permeation (systemic effect) occurs in two main steps: drug dissolution in the vaginal fluids and formation of a hydrodynamic layer on the vaginal walls, followed by penetration of the epithelium according to a concentration gradient. As a consequence, the penetrating or permeating substance must possess not only sufficient lipophilicity to diffuse through the lipidic components of the membrane, but also a certain degree of aqueous solubility to ensure dissolution into the vaginal fluids [2].

Drug stability and absorption are also affected by its compatibility towards vaginal fluids (especially for pH, enzymes, and microflora) and vaginal membrane passage [5].

The increase in vaginal fluids due to estrogens and sexual stimulations improve the absorption of poorly water-soluble drugs; on the contrary, a greater presence of mucus on the mucosal membrane slows down the contact between the drug and the absorptive epithelium [3]. Moreover, the vaginal fluids could remove and wash away the formulations from application and action site or cause an erratic drug distribution. Furthermore pH variation, following the menstrual cycle, affects drug ionization (by varying the degree of ionization of weak electrolytes) and consequently drug-release profiles and drug absorption [2,6].

The variations in thickness of the vaginal epithelium due to hormone variation during menstrual cycle also influence drug absorption [3]. Besides drugs, vagina is involved in the uptake of natural compounds present in seminal fluids, enzymatic and microbial metabolites, and electrolytes. The absorption of such molecules elicits physiological responses in other parts of the genital apparatus (i.e., prostaglandins in seminal fluids cause oviductal contraction within minutes after their absorption) [6].

Steroids such as estradiol, estrone, progesterone, medroxyprogesterone acetate, norgestrel, norethisterone, and testosterone can be successfully absorbed through the vaginal walls, as determined by the measurement of plasmatic levels of the drugs after applications in humans. In particular, lipophilic steroids (such as progesterone and estrone) are much more readily absorbed than more polar steroids (hydrocortisone and testosterone) [11,12]. Also prostaglandins are readily absorbed even though *in vitro* studies have shown that drug release of PGE_2 from vaginal preparations may vary depending on the pH of the vaginal environment [12]. The vaginal permeability of a series of linear chain aliphatic alcohols increases in rabbits and monkeys as the chain length and lipophilicity increase [13]. Studies carried out *in vitro* on rabbit vaginal epithelium have demonstrated a linear relationship between the epithelium permeability of a series of progestins and their lipophilicity [12]. Antimicrobials show a much more variable absorption from the vagina than steroids. Sulfanilamide, econazole, metronidazole, triclosan, and hexachlorophene demonstrated consistent absorption in therapeutic levels when administered in rats whereas penicillin, sulfatiazole, azalomycin, miconazole, and clotrimazole showed a little but not therapeutically useful absorption [14,15]. Insulin and thyroid-stimulating hormone (TSH) are characterized by an absorption rate highly dependent on the stage of estrous cycle if administered in rats and rabbits; this is probably due to the influence of the hormonal levels on the thickness and porosity of the vaginal epithelium, thus modifying the hydrophilic pathways [16].

The molecular weight cutoff for absorption may be higher for vagina than for other mucosal areas. Concerning this, even though the overall permeability of the vagina towards penetrating species is greater than that of rectum, skin, and buccal mucosa (but less than that of nasal and pulmonary mucosa), the bioavailability of a vaginally administered drug can be modified by the use of chemical penetration enhancers and by an appropriate dosage form. These become particularly important for vaginal absorption of high-molecular-weight molecules such as peptides and proteins [5].

The vaginal route can be used for uterine targeting of active ingredients. It has been observed that the uterine effects of vaginally administered progesterone were higher than those expected on the basis of drug plasma levels, suggesting a preferential distribution of the drug from the vagina to the uterus before being transported in the systemic circulation [2,17]. In particular, it was found that after i.m. administration of progesterone plasma levels were about sevenfold higher than those obtained after vaginal administration, but uterus levels were about 10-fold lower [18]. Similar findings were also obtained with other drugs such as terbutaline and danazol. This peculiarity, called "first uterine pass effect" can be exploited for drugs intended to act on the uterus itself such as utero-relaxants (terbutaline) or utero-contractants (misoprostol) [17].

22.4 CONVENTIONAL DRUG DELIVERY SYSTEMS

Conventional formulations intended for vaginal application include solutions, foams, creams, gels, tablets, and suppositories, in particular pessaries.

22.4.1 Solutions and Foams

Solutions (lavages) and foams are liquid preparations. The foams differ from the solutions in the presence of a suitable propellant, in the formulation, and the type of container, a pressurized delivery device. Plasma concentration profiles obtained after solution administration are characterized by a burst effect followed by a rapid decrease below therapeutic levels, due to the low residence time of the formulation in the vaginal cavity. Such preparations are designed to achieve a local effect particularly in case of inflammations or infections caused by bacteria or yeasts (anaerobic bacteria or *Candida* species). Nonoxynol-9 (N-9) foam is used as a contraceptive and against sexually transmitted diseases [19].

22.4.2 Creams and Gels

The greatest number of vaginal medications is represented by creams or gels. In the aqueous vaginal environment, hydrophilic formulations, in particular hydrogels, swell and release the drug mainly by diffusion [20]. Swelling increases the spreading of formulation on the mucosal surface and consequently the contact between the drug and the vaginal epithelium. Such formulations are commonly used for local delivery of antibacterial drugs and contraceptives or spermicides.

Vaginal infections, which represent the most common pathology of this organ, are usually caused by anaerobic bacteria and *Gardnerella* species and by yeasts like *Candida albicans* or parasites like *Tricomonas vaginalis* [2].

The comparison of oral and vaginal administration of metronidazole for the treatment of vaginitis due to *Tricomonas* has demonstrated that the local administration is not as effective as the systemic one; only in 46% of the women, the treatment with a 0.75% vaginal gel administered twice daily for 7 days determined recovery whereas 250 mg tablet administered orally 3 times daily healed 100% of the subjects examined [21]. Moreover, the vaginal treatment determined a higher relapse percentage (30.4%) than the oral one (12.5%). In another study, similar therapeutic effects were achieved after 2 times administration of 500 mg metronidazole tablets *per os* daily (85% recovery) and a vaginal gel (75% recovery); the latter formulation presented, however, the advantage of fewer systemic adverse effects. Metronidazole and clindamycin vaginal creams were found to be as effective as orally administered drugs in the treatment of bacterial vaginosis. Moreover, these formulations were characterized by good vaginal tolerability following administration [22].

Various studies led to the development of antiviral, antibacterial semisolid formulations to reduce mucosal and perinatal virus transmissions in absence of an effective prophylactic anti-HIV vaccine and for the treatment of other sexually transmitted pathologies. Concerning this, the vaginal delivery of N-9, which is also active against *Neisseria gonorrhoeae* and *Chlamydia trachomatis*, has been studied. An N-9 gel based on polyoxypropylene–polyoxyethylene polymer has been evaluated to reduce the side effects of N-9 [23]. N-9 solutions administered to rabbits showed dose-dependent toxic effects, evidenced by bleeding, irritation, epithelial disruption, and necrosis accompanied by loss of epithelial cells integrity and accumulation of leukocytes. These impressive side effects are often associated with vaginal ulcers that may even increase HIV transmission by disrupting the cervical and the vaginal mucosa. On the contrary, an N-9 gel demonstrated *in vitro* in human cervical and colon epithelial cells to prevent local mucosal inflammation and *in vivo* after vaginal application in rabbits to markedly reduce N-9 toxicity [23].

A gelly microemulsion loaded with a phenyl phosphate derivative of zidovudine has been developed to prevent the HIV transmission [24]. This new drug also presents a spermicidal activity. The gel formulation is characterized by low cytotoxicity towards human vaginal, ectocervical and endocervical cells *in vitro* and good tolerability *in vivo*. Unlike N-9 solution, the spermicidal activity of the microemulsion is not associated with cytotoxicity on the genital apparatus and repeated vaginal administrations do not cause local inflammations in a rabbit model.

Other vaginal gels with virucidal properties or capable of interfering with sexually transmitted diseases (in particular with HIV virus) have been reported [25]. In particular, gel formulations based on a naphthalene sulfonate polymer named PRO 2000 and dextrin sulfate have been developed as topical microbicides to protect against HIV and other sexually transmitted pathogens.

Vaginal gels are also employed to deliver drugs used in the cervical ripening and labor induction. PGE_2 (dinoprostone), PGE_1 (misoprostol), and oxytocin are frequently loaded in vaginal or intracervical formulations [6]. The safety and the efficacy of gel formulations based on the above-cited drugs are fully described in the literature [2].

22.4.3 Tablets

This dosage form presents the advantages of ease of manufacturing, ease of insertion, and economy. Conventional tablets for vaginal application may contain binders, disintegrants, and other excipients commonly used in tablets intended for oral administration. Absorption across the vaginal mucosa is impaired if the drug is hydrophobic or release-retarding materials are present in the formulation.

Vaginal tablets based on *Lactobacilli* were proposed to maintain or restore a vaginal environment, capable of limiting the growth of pathogenetic microorganisms [26]. Freeze-dried tablets loaded with different strains of *Lactobacilli* were prepared using skim milk powder and malt extract as excipients. A vaginal effervescent tablet formulation containing three different *Lactobacilli* strains ensured a rapid and complete distribution of viable microorganisms on the vaginal mucosa [27].

22.4.4 Suppositories (Pessaries)

The suppositories (pessaries) are very common vaginal drug delivery systems, quite similar to those intended for rectal administration, which can ensure a sustained drug release by dissolving or melting, depending on the choice of excipients. Aqueous vaginal fluids can determine dissolution of hydrophilic preparations, whereas the presence of lipophilic components enables melting at physiological temperature. Actually suppositories represent the systems more commonly used for vaginal ripening before childbirth and local treatment of vaginal infections.

The ripening action of dehydroepiandrosterone sulfate after vaginal administration of a pessary and intravenous injections was found to be comparable [28].

Pessaries based on metronidazole were effective against vaginal candidosis even though metronidazole is poorly absorbable across the vaginal epithelium [29]. The prolonged residence of the formulation at the application site together with the rapid drug release favored the contact of the drug with the vaginal mucosa and the adjacent tissues. This resulted in an initial fast absorption phase, followed by a slow elimination phase mainly due to the leakage of the formulation.

The pharmacological effects of vaginal pessaries based on PGE_2 were compared with the more effective drug 15 M $PGF_{2\alpha}$ administered intramuscularly to determine second trimester abortion [30]. The mean times required for rupture of membrane, delivery of fetus and placenta were significantly shorter for women receiving PGE_2 intravaginally in comparison with those receiving the parenteral formulation. The cumulative abortion rate after 24 h was 96% for the pessary and 69% for the intramuscular injection.

A suppository formulation based on progesterone was proposed to prevent cystic hyperplasia of the endometrium and possible cancer formation during postmenopausal long-term estrogen therapy [31].

A vaginal suppository based on bromocriptine was employed for the therapy of hyperprolactinemia [32]. The rationale of the local vaginal delivery of bromocriptine lies in the noteworthy side effects consequent to oral therapy: gastrointestinal disorders, extensive hepatic degradation, and hypotension. The pessary based on bromocriptine proved to be effective in lowering serum prolactin to normal levels after 20 days of local therapy; the treatment was well tolerated by the majority of the patients and a minimal vaginal irritation was observed.

A vaginal suppository based on ornidazole, a drug characterized by the activity against bacteria and protozoa, demonstrated the same safety and effectiveness in the treatment of vaginitis as that of a tablet formulation; in addition, the slippery and smooth surface of the suppository-facilitated application and the irritation in the vulva–vaginal region was reduced in comparison with that induced by the tablet formulation [33].

22.5 MODIFIED RELEASE FORMULATIONS

These formulations are characterized by excipients and technology capable of prolonging and controlling drug release. Modified release systems intended for vaginal application include films and rings. Other drug delivery devices enabling site-specific release will be discussed in Section 22.7.

22.5.1 Films

Films are polymeric formulations produced by lamination or casting. Films containing N-9 and based on polyvinyl alcohol (PVA) [34] and hydroxypropyl methylcellulose (HPMC) [35] were developed for contraception. HPMC was finally preferred due to its better compatibility for vaginal mucosa and lower cost. In addition, the lower hygroscopicity of HPMC film rendered it more stable to humid climates and less likely to stick to fingers during application. De-epithelization occurred in few cases without symptoms: the damage of the vaginal epithelium was associated to N-9 dose and not to the polymer.

22.5.2 Rings

Vaginal rings are circular, torus-shaped, drug delivery devices designed to prolong drug release after application. They measure approximately 5.5 cm in diameter with a circular cross section of 4–9 mm [6,12].

In the simplest system, the drug is homogenously dispersed within the polymeric ring (matrix-type devices). Drug-release rate is proportional to drug loading and surface area of the system. Drug release from such a device occurs in different steps. As soon as the device contacts dissolution medium, the drug located at the ring surface dissolves and is released fast. For this reason a burst effect is observed. Then drug present in the inner part of the device is slowly released. At first, the drug molecules undergo diffusion through the polymer closer to the ring surface. As the dissolution of the drug continues, a boundary layer depleted of drug forms, which separates the inner drug loaded region of the polymer matrix from the dissolution medium. Water can penetrate into the channels created by drug depletion and can determine drug dissolution at the depletion boundary. Release rate decreases as boundary thickness increases until complete dissolution of drug loaded into the device.

Sandwich or reservoir-type intravaginal ring (IVR) devices have been developed to obtain zero-order drug release.

The sandwich-type device is characterized by an inner layer containing the drug, which is between the nonmedicated inner and outer regions. In the reservoir-type ring, the drug is dispersed in a central core, which is encapsulated in an outer polymeric layer, free of drug. This system provides a constant release rate throughout the delivery period. The release is achieved by the initial partition of the drug from the core into the polymer sheath and by the subsequent diffusion through the polymer up to the device surface. The partition coefficient of the drug in the polymer and its diffusion coefficient through the polymer layer represent the key parameters governing drug-release rate. Finally, a continuous diffusion of the drug from the device surface towards the hydrodynamic layer constituted by the vaginal fluids occurs. This type of ring can present multiple cores, each containing a different drug to enable the administration of several drugs in the same device.

Vaginal rings are usually based on elastomeric polymers like poly(dimethylsiloxane) or silicone, even though in recent years devices based on ethylene–vinyl acetate and styrene–butandiene block copolymers have been examined. The rate of drug release from rings can be modified by increasing the polymer layer depth or by changing the polymer type. The polymer properties to be considered are flexibility, resistance towards mechanical stresses, and eventually adhesion properties to the mucosal surface. Vaginal rings can remain in place for prolonged time periods (from 21 days up to months) according to therapeutic indications. They are designed for hormonal replacement therapy and contraception.

The use of contraceptive devices containing various steroids such as medroxyprogesterone acetate, levonorgestrel, and norethindrone have been studied since the 1970s [12]. At the beginning, these systems were designed to release doses capable of inhibiting ovulation; subsequently, these devices were set up to release lower doses using a contraceptive mechanism similar to that of the progesterone minipill.

For most contraceptive applications, rings are placed in the vaginal cavity for 21 days followed by a ring-free period lasting a week.

Since 1997, three comparative studies have been reported on progesterone IVR device and copper intrauterine device (IUD). In a one-year multicenter study on 1536 breast-feeding women, the effectiveness of continuous regimen of four sequential mg/day rings did not differ from that obtained by IUD application: pregnancy rate was 1.5%, lactation performance and health and weight increase of the infants were similar. Ring users complained of more vaginal problems but presented fewer vaginal disorders upon clinical examination [36].

NuvaRing (NV Organon, Berlex Inc., Montville, NJ, USA) is a flexible, transparent, contraceptive vaginal ring containing two active ingredients, etonogestrel and ethinyl estradiol. The ring releases 120 μm/day of etonogestrel and 15 mg/day of ethinyl estradiol over a 3-week period. The clinical trials of NuvaRing demonstrated that this device is an effective contraceptive with efficacious cycle control and good patience compliance [37]. The possibility to coadminister NuvaRing with a cream or suppository formulation based on different antifungal drugs to prevent the concomitant vaginitis that could arise during ring estrogen therapy has also been investigated [38]. The concomitant administration of miconazole nitrate slightly increases the systemic exposure to the estrogens contained in the ring; the coadministration of antifungal-loaded suppository determined a greater effectiveness in comparison with the cream. The release rate of etonogestrel and ethinyl estradiol is higher in the presence of the antimycotic agent. This is probably due to the lipophilicity of miconazole that can facilitate the hormone release. Such a behavior is not expected to compromise the ring contraceptive efficacy and safety.

Femring (Galen Pharmaceutical Ltd., Rockaway, NJ, USA) and Estring (Pfizer, New York, NY, USA) are estrogen-releasing rings used for estrogen therapy. Both devices are based on silicone elastomer. Femring contains an acetate derivative of estradiol, a prodrug, which is active for 3 months. Estradiol acetate is hydrolyzed to estradiol before its release from the device. Estring releases 7.5 μg/day of drug [6].

The controlled delivery of estrogen is also useful in the treatment of urogenital atrophy affected by plasmatic estrogen level. After the menopause, the vaginal and urethral epithelia become progressively deprived of estrogen and the tissues lose thickness, rugation, moisture, vascularization, and elasticity [39]. Furthermore, the increase in vaginal pH determined by the deficiency of estrogen provokes vaginal and urinary infections. The estrogen replacement therapy restores the hormone level and the vaginal acidic pH. The use of a ring device for such a therapy offers several advantages: enhanced patience compliance due to the reduced dosing, greater convenience, avoidance of irritations, self-administration, and removal.

Rings based on norethindrone acetate at two dosages, 15 and 20 μg/day, have been developed. Both dosages present good performance in terms of lack of pregnancy and luteal activities: 0.9% of cycles presented ovulation with the higher dosage system and 1.2% with the lower dosage device [40]. Although there was significantly more luteal activity in women using the lower dosage device, pregnancy did not occur in any of these cycles probably due to other action mechanisms of the steroid such as effects on endometrium, corpus luteum, and cervical mucus. Even though drug serum levels were consistently high during the application of both rings, slightly higher incidence of unscheduled bleeding was evident among the users of the lower dosage device. The experimental evidence suggests that the higher dosage ring had marginally better performance with a lower incidence of luteal activity and slightly better bleeding patterns. The lower dosage device released a dose closer to the lower margin of contraceptive efficacy, although the effect on lipid metabolism was marginally better. The incidence of nausea and other side effects did not occur by the dosage.

Rings containing danazol were designed to control and treat pelvic endometriosis. Danazol acts directly on endometriotic tissue, inhibits DNA synthesis, and induces cell apoptosis. Danazol locally administered did not show inhibition of ovulation; on the contrary, oral administration produced a lack of menstrual cycle. Consequently, the main mechanism of action after local administration was reported to be the direct action on endometriotic cells. Danazol probably diffuses through vaginal mucosa and reaches the deep layers, where infiltrating endometriosis takes place. This behavior allows for the avoidance of the side effects, which characterize oral therapy [41].

In conclusion, contraception therapies with vaginal rings present excellent efficacy with little risk of side effects and appear to be as effective as oral contraception and IUD, in terms of pregnancy prevention [42]. Moreover, rings are characterized by important benefits like ease of use, long-term schedule, and user-controlled application. These versatile drug delivery systems appear to be safe, effective, and acceptable.

22.6 COMPARISON OF THE FUNCTIONALITY OF CONVENTIONAL AND MODIFIED DELIVERY SYSTEMS

The vaginal retention and antiviral (anti-HIV) effect of different N-9 formulations were compared [43]: Advantage 24 (cream) (Lake Pharmaceutics, Columbia Laboratories Inc., Livingston, NJ, USA), Conceptrol (cream) (Ortho Pharmaceuticals, Family Health International, Research Triangle Park, NC, USA), Deflon Foam (foam) (Ortho Pharmaceuticals, Janssen-Ortho Inc., Toronto, Ontario, Canada), Semicid (suppository) (Whitehall Pharmaceuticals, Whitehall-Robbins Healthcare, Madison, NJ, USA), and VCF (film) (Apothecus Pharmaceutical Corporation, Oyster Bay, NY, USA). Even though there was a great variability in retention caused by the loss of formulation from the vagina, Deflon Foam (foam) was characterized by the best retention, although not statistically different from Conceptrol (cream). The protection against HIV transmission theoretically lasted for 8 h for Deflon Foam and 4 h for Conceptrol. The authors did not report a hypothesis for formulation behavior but the longer permanence of Deflon Foam is probably related to the higher area covered by the foam in comparison with the semisolid formulation, which is more subjected to leakage.

A comparative study of a low-dose estradiol vaginal ring and estrogen cream for post-menopausal urogenital atrophy demonstrated the similar efficacy and safety of the two formulations [39]. In both cases, endometrial biopsy revealed a low incidence of endometrial proliferation and in both groups of women treated with one of the two products the physiological pH was restored to normality for the majority of subjects. However, the vaginal ring, due to product comfort, ease of use, and dosing reduction, was characterized by superior patient compliance in comparison with the cream.

Conventional formulations present several drawbacks related to their difficult retention into the vaginal cavity, poor spreadability (nonuniformity in formulation distribution), and leakage into the undergarments, which all result in poor patience compliance. Residence time of the formulations, drug distribution in the vagina, and the dose administered can vary. Spreading on the vaginal mucosa is considerably affected by the type of preparation and is greater for solutions, suspensions, and foams in comparison with semisolid or solid dosage forms, like tablets.

Tablets are flat with an oval-, pear-, or bullet-shape to facilitate the insertion into the vagina but often present a hard surface and dried ingredients that may cause undesirable irritations; on the other hand, pessaries are characterized by a smooth surface and by excipients dissolving or melting in physiological conditions, but they are messy to apply and undergo leakage.

Conventional systems do not offer sufficient flexibility in controlling drug-release rate and sustaining the release over time periods extending from days to months. Therefore specific modified release vaginal delivery systems are continuously under development and are based on mucoadhesive systems. Penetration enhancement may represent a necessary feature for certain delivery systems, particularly when the absorption regards a macromolecule (such as a peptide or a protein).

22.7 BIOADHESIVE DELIVERY SYSTEMS

Both for local and systemic therapy to prevent poor retention, messiness, and leakage associated with conventional vaginal dosage forms, it is useful and sometimes necessary to prolong and improve the contact between the drug and the mucosa. The presence of mucoadhesive agents in the formulation could enable the attainment of such a goal.

It must be noted that mucoadhesion is an appropriate process to be considered in the case of vaginal administration even though vaginal epithelium is not strictly a mucosal epithelium, due to the absence of mucus-producing globlet cells. Nevertheless, such an epithelium is bathed in vaginal fluids containing cervical mucus, which represents a potential site of interaction of mucoadhesive agents [44]. Vaginal bioadhesive drug delivery systems are largely based on nonspecific mucoadhesion and include gels, tablets, films, and microparticulate systems.

22.7.1 Bioadhesion: Definition and Mechanisms

The term bioadhesion refers to any bond formed between two biological surfaces or to a bond between a biological and a synthetic surface [44]. In the case of bioadhesive drug delivery systems, the term bioadhesion is typically used to describe the adhesion between polymers, either synthetic or natural, and a soft tissue (i.e., vaginal mucosa). The target of these systems may be either the mucus layer lining the mucosa, the epithelial cells, or a combination of the two.

Two steps have been described in adhesive bond formation: (1) intimate contact of the mucoadhesive agent and of mucus or mucosa consequent to wetting and (2) formation of physical or chemical bonds between the biological substrate and the mucoadhesive agent, preceded, in the case of polymeric materials, by interpenetration and diffusion between the bioadhesive and the mucin glycoprotein.

The different theories that have been proposed to explain the bioadhesion phenomenon are hereafter summarized. The design of bioadhesive drug delivery systems should take into account the mechanisms on which the bioadhesion phenomenon is based. An insight into bioadhesion theories could help formulators to design bioadhesive drug delivery systems with optimal performances.

22.7.1.1 Bioadhesion Theories

The same theories relevant to adhesion, developed to explain and predict the performance of glues, adhesives, and paints, have also been applied to bioadhesive systems [44]. These include the electronic, absorption, wetting, diffusion, and fracture theories.

The electronic theory is based on the assumption that the bioadhesive material and the target biological substrate have different electronic structures, and when they come in contact, electronic transfer occurs, causing the formation of a double layer of electrical charges at the bioadhesive interface [44]. The bioadhesive force is believed to be due to the attractive forces of the electrical double layer. There are discordant opinions on this item—are the electrostatic forces the causes or rather the results of the contact between the bioadhesive material and the biological substrate [44,45]?

A confirmation of the soundness of electronic theory was derived from a recent study, performed by Bogotaj et al. [46]. They measured the zeta potential of different polymer dispersions and mucosal homogenates and found a correlation between such a parameter and the force necessary to detach a polymer dispersion from the biological substrate. The adsorption theory states that the bioadhesive bond is due to van der Waals interactions, hydrogen bonds, and other related weak interactions [44].

The wetting theory is applicable to liquid bioadhesive systems. According to this theory, the ability of a bioadhesive material to spread and determine an intimate contact with the biological substrate plays a major role in bond formation [44]. This theory uses interfacial tensions to predict spreading and, in turn, bioadhesion. In the past, the surface energy of both bioadhesive materials and tissues or mucus have been extensively studied to predict the bioadhesive performance [47–49].

The diffusion theory states that the interpenetration and entanglement of bioadhesive polymer chains and mucin glycoprotein chains (mucus main component) produce semipermanent adhesive bonds (Figure 22.2). It is believed that bond strength increases with the degree of penetration of the polymer chains into the mucus layer [44]. In the past, attenuated total reflection infrared spectroscopy (ATR-FTIR) was used to assess the extent of chain interpenetration at the polymer–mucin interface [50]. The use of rheological analysis to assess the mucoadhesive properties of different materials finds its rationale in this theory [51–56]. Chain interlocking, conformational changes, and chemical interactions (like hydrogen and van der Waals bonds), which occur between polymer and mucin during the bioadhesion phenomenon, produce changes in the rheological behavior of the two macromolecular species. The changes in viscous and viscoelastic behavior of polymeric solutions when mixed with mucin have been used as indicators of mucoadhesive joint strength.

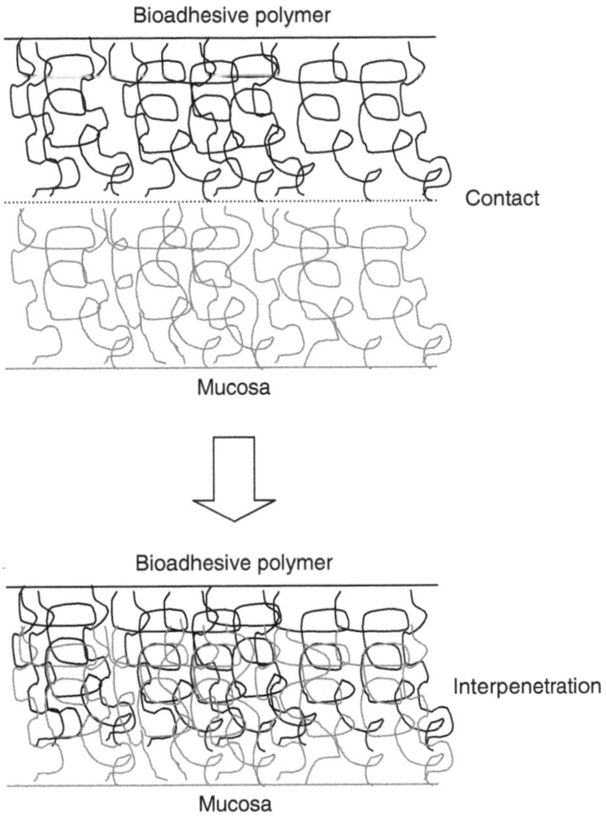

FIGURE 22.2 Schematic representation of interpenetration of polymer chains and mucin glycoproteins.

22.7.2 Bioadhesive Drug Delivery Systems

22.7.2.1 Bioadhesive Gels

Various bioadhesive gels are already available on the marketplace with the target of lubricating the vaginal mucosa and to prolong the contact of the drug (usually hormones) with the mucosa. The advantages of gel in comparison with tablets and films are its greater contact area and, generally, less irritation of the mucosa.

A nonhormonal bioadhesive gel, Replens (Lil' Drug Store Products Inc., Cedar Rapidis, IA, USA), based on polycarbophil and intended to retain moisture and lubricate the vagina is commercially available. The formulation remains in the vagina for 2–3 days and maintains the vaginal mucosa at physiologic acidic pH [6]. A study designed to evaluate the efficacy of Replens on symptoms of vaginal atrophy in postmenopausal women, in comparison with Ortho-Dienestrol (Janssen-Ortho Inc., Toronto, Ontario, Canada), an estrogenic vaginal cream, showed that Replens enables a full therapy for all symptoms of vaginal atrophy and represents an alternative treatment to local estrogens [57].

Miphil (Columbia Laboratories Inc., Livingston, NJ, USA), a marketed bioadhesive gel based on polycarbophil and capable of reducing vaginal pH, has been used in combination with oral tinidazol for the treatment of bacterial vaginosis. In a multicenter, randomized, investigator-blinded, controlled clinical trial, the combination of the antibiotic with the vaginal gel produced a more rapid normalization of the vaginal microflora and higher cure rate in comparison with a vaginal antibiotic therapy alone [58].

With the aim of maintaining and restoring the physiological acidic conditions that can be compromised for example in postmenopausal age, with irritation of the mucosa, itching, and unpleasant smell, a mucoadhesive formulation capable of sustained release of lactic acid, based on a chitosan lactate gel has also been proposed [59,60].

Crinone (Serono International S.A., Geneva, Switzerland) is a vaginal progesterone gel designed to provide prolonged release of progesterone over 48 to 72 h after a single vaginal application. The gel delivery system uses the bioadhesive properties of polycarbophil, a polymer known to adhere to mucosa for extended periods [61]. It appears to have a substantial local uterine effect (first uterine pass effect) before it reaches the systemic circulation. Crinone provided a prolonged release of progesterone in postmenopausal women [62]. Some authors suggested that the use of Crinone after *in vitro* fertilization–embryo transfer is associated with higher biochemical pregnancy rate in comparison with the administration of intramuscular progesterone-in-oil [63].

Recently, a new bioadhesive gel, Acidform (Conrad, Arlington, VA, USA), was designed with acid-buffering and viscosity-retaining properties to maintain the acidic vaginal milieu (the low pH inactivates many pathogens and spermatozoa), to form a protective layer over the vaginal epithelium (minimizing the contact with pathogenic organism), and to provide long-term vaginal retention [64]. Acid-buffering, bioadhesive, viscosity retaining, and spermicidal properties of Acidform were compared *in vitro* with those of marketed formulations. The results obtained suggest that such a formulation presents advantages over the presently marketed formulation and might be used itself as an antimicrobial contraceptive or as a vehicle for an active ingredient with antimicrobial and contraceptive properties. Phase I clinical studies provided initial information about the safety of Acidform and showed the formation of a layer over the vaginal epithelium [65].

Mucoadhesive gels intended for vaginal administration of N-9, a spermicidal agent, have been reported in the literature [6,66]. They consist of varying levels of drug and ethylenediaminetetraacetate (EDTA), a chelating agent, loaded in hydrated Carbopol 934P [67,68].

A mucoadhesive thermosensitive gel (MTG) has been proposed by Chang et al. [69] as a more convenient dosage form for drug mucosal administration. The liquid applied to the mucosa can gellify as a result of physical changes induced by the physiological environment. The transition is induced by an increase in temperature (from ambient temperature to physiological one). The formulation was composed by a mixture of two different grades of poloxamer and polycarbophil and was intended for vaginal administration of clotrimazole. *In vivo* antifungal activity of clotrimazole, tested against *C. albicans* vaginitis in female rats, was significantly prolonged after vaginal delivery of MTG. Moreover, the vaginal delivery of clotrimazole in MTG enhanced the viability of epithelial cells without affecting the morphology of vaginal mucosa [70].

Wang and Lee [71] investigated the feasibility of a Carbopol 934P–HPMC-based gel formulation for vaginal drug delivery. Differential scanning calorimetry was used to investigate the effect of gel on the conformational changes of rat vaginal membrane. Carbopol in combination with HPMC showed good biocompatibility and did not cause any conformational changes in the rat vaginal membrane.

The bioadhesive properties of hydrophobic polybasic gels containing *N*,*N*-dimethylaminoethyl methacrylate-*co*-methyl methacrylate were investigated in view of their use as vaginal drug delivery systems [72]. The bioadhesive properties of such gels make them suitable for site-specific and pH-controlled drug delivery.

Bioadhesive gels were also reported in the literature as vehicles for microparticulate formulations. Such formulations will be described in Section 22.7.2.5.

22.7.2.2 Bioadhesive Tablets

Bioadhesive vaginal tablets are relatively easy and inexpensive to manufacture. They are characterized by a greater irritation effect and a smaller contact area with mucosa in comparison with mucoadhesive gels, but on the other hand, they should show a more prolonged residence time at the application site. Different tablets have been investigated for their capability to adhere to cow vaginal mucosa and their swelling properties [73]. They consisted of Carbopol 934, polyvinylpyrrolidone, pectin, and ethylene maleic anhydride resins and their mixtures. The most favorable formulation was that based on a mixture of Carbopol 934 and pectin in a 2:1 weight ratio, which showed the highest bioadhesive strength and swelling volume and the lowest vaginal pH reduction.

Polycarbophil-based bioadhesive tablets of metronidazole were tested for adhesion on bovine submaxillary mucin [74]. In a more recent study, metronidazole tablets based on a mixture of modified starch–polyacrylic acid showed an increased potential for the treatment of bacterial vaginosis [75].

Recently, a bioadhesive vaginal tablet has been designed for the release of clotrimazole. The tablet was based on chitosan derivatized with thioglycolic acid [76]. By the introduction of thiol groups, the bioadhesive properties of chitosan have been significantly improved. The immobilization of thiol groups also produced an improvement in cohesive properties of the polymeric carrier matrix, thus avoiding an irritating vaginal secretion due to eroded polymer fragments. A dependence of the controlled release of clotrimazole on the amount of covalently attached thiol group was also described.

To achieve more effective treatment of vaginal candidosis, ketoconazole was formulated in bioadhesive tablets that increase the time of contact of drug with the vaginal mucosa and thus its therapeutic effect [77]. The tablets were based on sodium carboxymethylcellulose (NaCMC), polyvinylpyrrolidone, or HPMC.

22.7.2.3 Bioadhesive Pessaries

New bioadhesive pessaries, capable of prolonging the permanence of the drug at the application site, have been recently developed [78]. They were prepared by adding increasing amounts of bioadhesive polymers to melted semisynthetic triglycerides. The bioadhesive polymers include polycarbophil, HPMC, and hyaluronic acid sodium salt. In particular, the formulation that guaranteed the greatest permanence of the drug (clotrimazole) at the target area was the one containing polycarbophil at the highest concentration (0.3% w/w).

22.7.2.4 Bioadhesive Films

Bioadhesive films have also been investigated for vaginal delivery of drugs. These systems present the advantage of remaining more flexible and thinner and covering a wider area of the mucosa in comparison with tablets. Moreover, in many parts of the world, vaginal films are preferred over gels due to their esthetic appeal and ease of application (without applicator) [79]. Recently, a rapid dissolving bioadhesive vaginal film based on sodium polystyrene sulfonate (PSS), a novel contraceptive antimicrobial agent, was developed [80]. Such a film was designed to dissolve rapidly (in less than 3 min) upon contact with vaginal fluids to form a smooth, viscous, and bioadhesive gel. PSS film has demonstrated various sperm function inhibition and antimicrobial properties without affecting the normal vaginal microflora. It is safe for vaginal administration and it appears as a suitable vaginal microbicide for the prevention of sexually transmitted diseases.

In another recent study, a novel bioadhesive patch containing 5-aminolevulinic acid (a photosensitizing drug) was developed [81]. It could be profitably used in the management of intraepithelial neoplasias by means of photodynamic therapy, which combines the action of visible light with appropriate wavelength on a photosensitizing drug to cause the destruction of selected cells.

The patch was based on poly(methyl vinylether/maleic anhydride) and has been shown to remain located firmly to mucous surface for up to 4 h. The drug concentration within mucosa was shown to exceed by a considerable margin that needed to photoinduce cell death in a model cell line. The patch matched the correspondent conventional creams with respect to the amount of drug penetrated into the tissue, but offered the advantage of remaining in place for extended time periods without the need of occlusion.

Bioadhesive films based on mixtures of chitosan and polyacrylic acid and intended for vaginal delivery of acyclovir were recently developed in our laboratory [82]. Such films showed good mucoadhesive properties and the capability, depending on the polymer-mixing ratio, of enhancing drug penetration into pig vaginal mucosa. Moreover, they were capable of prolonging drug release and were characterized by elastic properties, which make them resistant towards mechanical stress.

22.7.2.5 Microparticulate Systems

Microparticulate systems present the advantage, in comparison with single-unit solid systems, to guarantee a wider contact area between the drug and the mucosa. Different microparticulate systems intended for vaginal administration were developed; they include liposomes, microcapsules, and microspheres. Such systems can possess intrinsic bioadhesive properties or can be loaded in a vehicle with bioadhesive properties.

Recently a liposome carrier system, capable of providing controlled and sustained release of drugs intended for the local treatment of gynecological diseases, was developed [83]. In order to optimize liposome preparation, liposomes containing calcein as a hydrophilic model substance and egg phosphatidylcoline and egg phosphatidylglycerol sodium as lipid membrane components were prepared by several methods: proliposome, polyol dilution, film, detergent dialysis, and high-pressure homogenization. Proliposome and polyol dilution have been demonstrated to be the optimal preparation methods because they are characterized by a high trapping efficiency of the model drug. Such liposomes were incorporated in Carbopol gel in order to prolong for extended time periods the contact of the systems with the mucosa.

More recently, vaginal tablets were prepared by mixing NaCMC microcapsule containing ketoconazole with effervescent granules [84]. Since NaCMC is a good bioadhesive polymer, it was used as a coating material. A convenient sustained drug release was obtained with microcapsules in comparison with two different commercial pessaries. Moreover microcapsules demonstrated good flowability properties.

Different microsphere formulations have been proposed for the delivery of peptides and protein via vaginal mucosa. This will be considered in Section 22.8.

22.7.3 Bioadhesion Testing

Various mechanical testing methods have been used to assess the bioadhesive properties of materials and formulations. Review of the literature reveals that the technique most commonly used is the tensile test [82,85]. This test provides the measure of the force needed to detach a layer of the tested material or formulation from a mucosal substrate as a function of the displacement occurring at the bioadhesive interface. Besides maximum force of detachment, another parameter provided by tensile test is the work of adhesion calculated as the area under the force versus displacement curve. Such a parameter gives more complete

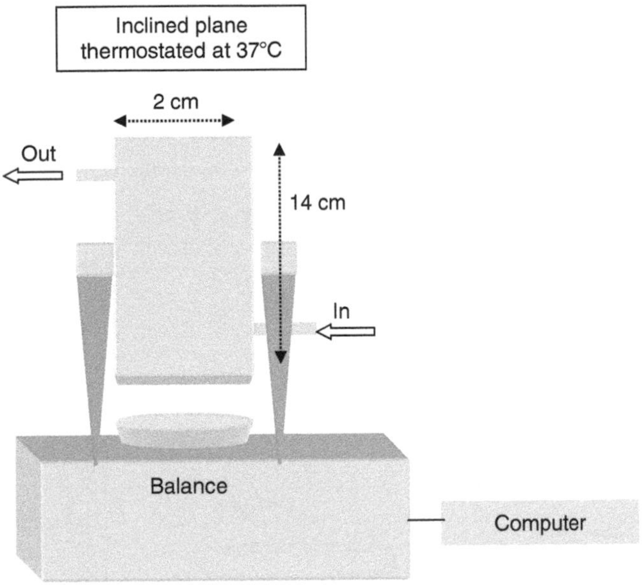

FIGURE 22.3 Schematic representation of the inclined plane apparatus.

information on bioadhesive potential of the tested sample, since it considers the sample behavior during the whole detachment process. Sustained load and cyclic stress testing have also been reported in the literature to evaluate the durability of the bioadhesive joint [82,85].

Artificial membranes soaked in animal mucin dispersions or animal model mucosae are used as biological substrates. Another apparatus proposed for *in vitro* measurements of bioadhesive properties of liquid formulations (polymer solutions or pessaries upon melting) consists of a thermostated inclined plane over which a mucosal membrane or a mucin film is layered. This test measures, as a function of time, the amount of formulation that after contact with the biological substrate, drops on a microbalance placed under the inclined plane [86] (Figure 22.3).

Recently a new apparatus resulting from a modification of the Setnikar and Fantelli method for the liquefaction time of rectal suppositories has been proposed to evaluate the permanency of drug loaded in suppositories on the vaginal mucosa [78] (Figure 22.4) The vaginal physiology is recreated by means of a cellophane tube in which a solution of mucin and lactic acid is placed. The tube is tied to both ends of a condenser and each end of the tube is open. Water circulates through the condenser at such a rate that the lower half of the cellophane tube collapses and the upper half gapes. The formulation is dropped into the cellophane tube and a plate is placed under the condenser to collect the discharged liquid (fraction A). At the end of the measurement, the cellophane tube is cut horizontally into two parts (fraction B and fraction C). Drug amounts extracted from the two parts of the cellophane tube and present in the discharged liquid are quantified. The drug amounts recovered in the three different fractions are indicative of the amount of drug washed from the vagina by the vaginal fluid (A), amount of drug that remained in the simulated application site (B), and the amount of drug that remained in the vagina after melting of the pessary (C).

A modified Franz diffusion cell has been developed in our laboratory in order to simultaneously measure the amount of drug released by diffusion from a semisolid formulation and the amount of drug washed away by a fluid stream simulating the removal action exerted by

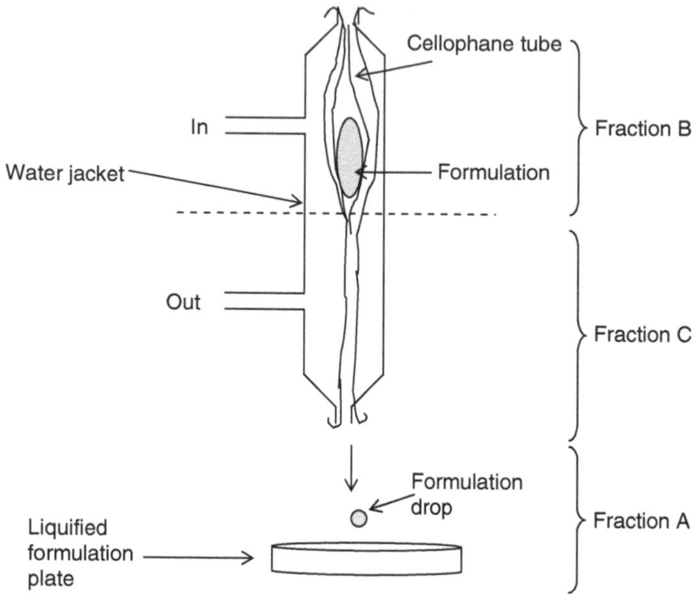

FIGURE 22.4 Schematic representation of the modified Setnikar and Fantelli apparatus.

physiological fluids at the application site [87,88]. Such a stream is obtained by fluxing a buffer tangential to the sample at a constant rate. The modified diffusion cell differs from the standard Franz cell for the presence of two side arms in the donor chamber which enable the buffer to stream over the formulation (Figure 22.5).

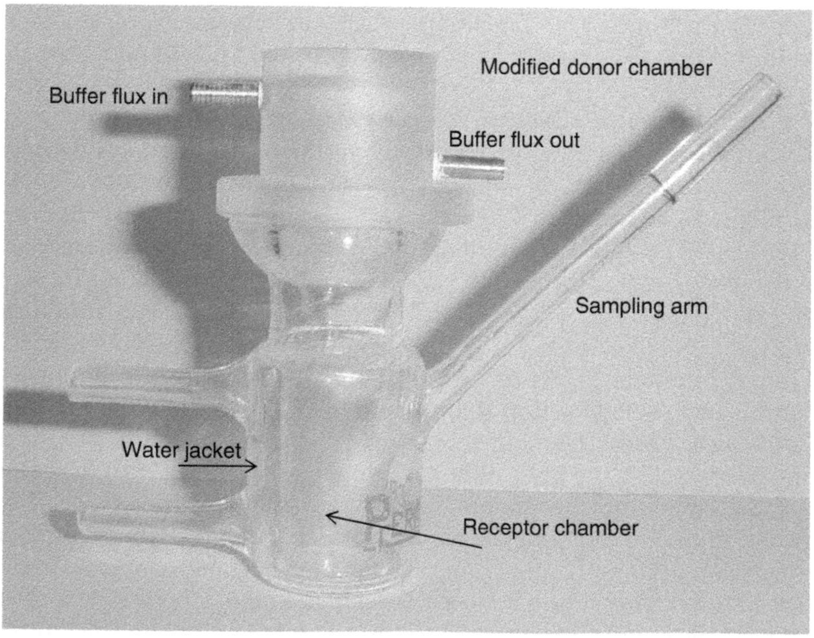

FIGURE 22.5 Photograph of modified Franz cell.

22.8 VAGINAL DELIVERY OF PEPTIDES AND PROTEINS

The vaginal route has considerable potential for the systemic delivery of peptides and proteins, particularly of those specifically used for female-related diseases. The rapid hydrolytic and enzymatic degradation occurring to peptides and proteins in the gastrointestinal tract and the "first-pass" metabolism renders oral administration impractical for these molecules. The preferred administration route for these drugs is the parenteral one, which is however invasive and usually encounters poor patient compliance since peptides and proteins, often characterized by very short half-lives, require frequent injections to achieve an effective therapy. In the last decade, as an alternative to repeated injections, the absorption of peptides and proteins via transmucosal routes has been investigated and, among these, the vaginal route has gained an increasing interest.

This route, however, is characterized by some intrinsic limitations, principally due to the barrier properties of the mucosa (related to anatomical characteristics and enzymatic activity) and the short residence time of the formulation, caused by mechanical stress at the application site. In Figure 22.6, the physicochemical barrier properties of vaginal and buccal epithelia in rabbit are compared [89]. Such mucosae present similar anatomical characteristics (squamous pluristratified epithelium, poor in tight junctions) that differentiate them from other mucosal epithelia, like the nasal and rectal ones. Vaginal mucosa is characterized by a lower resistance, a comparable permselectivity towards positively charged molecules, and a higher permeability towards a hydrophilic model drug than buccal mucosa. In Figure 22.7, the half-lives of insulin and calcitonin in homogenates or extracts of freshly isolated rabbit mucosae are compared [89]. The short half-lives are indicative of the rapid enzymatic degradation of these macromolecules. It can be observed that calcitonin undergoes a slower degradation in vaginal than in nasal extract whereas insulin is characterized by a higher degradation rate in vaginal than in nasal homogenate, but lower than that observed in buccal homogenate.

In recent years, efforts have been made to find methods to increase peptide and protein absorption via the vaginal mucosa by the use of penetration enhancers, enzyme inhibitors, and bioadhesive drug delivery systems.

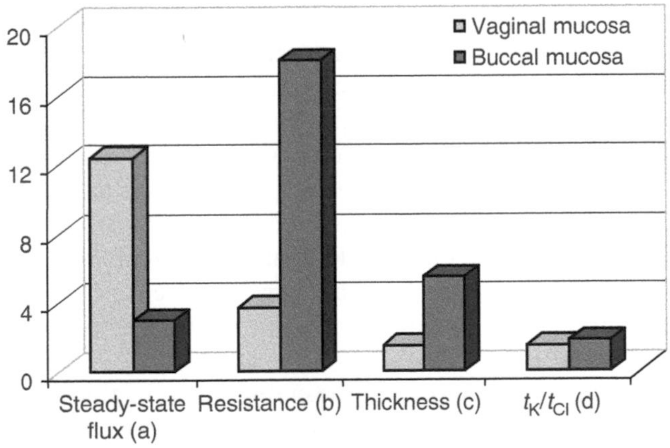

FIGURE 22.6 Permeability and permselectivity of vaginal and buccal epithelia in the rabbit. (a) Flux of 6-carboxyfluoroscein, a hydrophilic molecule, by *in vitro* perfusion studies steady-state flux ($\mu g/cm^2/h \times 10^6$), (b) resistance ($\Omega\ cm^2 \times 10^{-2}$), (c) thickness ($\mu m \times 10^{-2}$), and (d) ratio of potassium transport number to chloride transport number, which is calculated from electrical measurements, used as indicative of the epithelium selectivity for positively charged molecules. (Modified from Sayani, A.P. and Chien, Y.W., *Crit. Rev. Ther. Drug Carrier Syst.* 13, 85, 1996.)

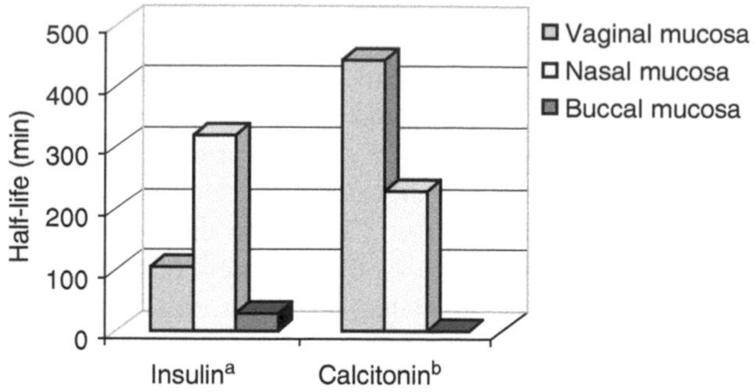

FIGURE 22.7 Half-life values of insulin and calcitonin in homogenates or extracts of freshly isolated rabbit mucosae (in the absence of any inhibitors). (a) In homogenates, and (b) in extracts. (Modified from Sayani, A.P. and Chien, Y.W., *Crit. Rev. Ther. Drug Carrier Syst.* 13, 85, 1996.)

It is clear from the literature that the choice of a permeation enhancer must be based on the following features: effectiveness, safety, chemical inertness, lack of biological activity, and rapid reversibility of effect. A variety of enhancers having different physicochemical properties have been studied: citric acid, lysophosphatidylglycerol, palmitoylcarnitine chloride, lysophosphatidylcholine, polyoxyethylene-*p*-lauryl ether, and sodium taurodihydrofusidate [1,2,89,90]. No decrease in blood glucose levels was found after vaginal administration of insulin in rats in the absence of any enhancer [90]. The coadministration of all the above-cited enhancers, except citric acid, produced a significant increase in hypoglycemia. The histological changes in the vaginal epithelium after treatment with enhancers were variable and often severe. Among the enhancers tested, lysophosphatidylglycerol emerged as a promising absorption enhancer with minimal effects on mucosal histology. Recently benzalkonium chloride has demonstrated to possess good penetration enhancement properties towards human vaginal mucosa, even though no proteic drug was considered [91]. Hyaluronic acid, chitosan derivatives, and mixtures of chitosan and polyacrylic acid have been demonstrated to improve the penetration into vaginal mucosa of a low-molecular-weight drug, characterized by poor aqueous solubility and absorption property, belonging to Class IV of BCS [82,86,88,92]. The effect of such polymers on vaginal mucosa permeability towards high-molecular-weight and hydrophilic compounds as peptides and proteins has not yet been proven.

To improve the absorption of peptides and proteins, the employment of different enzymatic inhibitors has been suggested. For example, some authors have demonstrated the usefulness of inhibitors of both exo- and endopeptidases to promote calcitonin absorption across rat vaginal mucosa [93]. A direct relationship was found between the effect of peptidase inhibitors of *in vitro* degradation constant of calcitonin in vaginal mucosa homogenates and *in vivo* promotion of drug absorption, indicated by the decrease in plasma calcium levels after calcitonin administration in rats. The peptidase inhibitors tested were pepstatin, bestatin, and leupeptin.

In a recent study, some authors proposed the employment of thiolated polymers obtained by the immobilization of thiol groups on well-established multifunctional polymers such as poly(acrylates) and chitosans as functional agents having mucoadhesive, penetration enhancement, and enzyme-inhibiting properties [94,95].

Different drug delivery systems have been proposed for vaginal delivery of peptides and proteins. The first one was a mucoadhesive gel based on polyacrylic acid intended for vaginal administration of insulin [96]. More recently, microparticulate systems such as starch and hyaluronan ester (HYAFF) microspheres have been proposed for vaginal delivery of insulin and calcitonin, respectively [2,97,98]. Richardson and coworkers demonstrated that the hypoglycemic response to vaginal-administered insulin in rats was improved using starch microspheres in comparison with an insulin solution, and that it was further enhanced by the addition of lysophosphatidylcholine [2]. HYAFF microspheres were administered in rats and sheep as pessary formulation [97,98]. Vaginal administration of calcitonin in HYAFF microspheres resulted in a more pronounced decrease in plasma calcium levels than the calcitonin solution. The enhanced absorption of calcitonin is thought to be due to the close contact of the microspheres with the vaginal epithelium, resulting in high local drug concentration and thus in increased gradient for absorption. A variety of pessary vehicles were assessed, but of those tested, Suppocire BS_2X was the only one that did not reduce calcitonin absorption. It melts at physiological temperature and forms a fine emulsion upon contact with the aqueous environment of the vagina, thus favoring the dispersion of microspheres and the release and absorption of the drug. In a subsequent study, the tolerability and pharmacokinetics of the vaginal HYAFF formulation were compared with that of a commercially available nasal spray, both administered at the same dose [98]. Both treatments were well tolerated with few side effects and resulted in a gradual absorption of drug, avoiding the initial high peak associated with subcutaneous injections.

22.9 IMMUNIZATION VIA THE VAGINAL ROUTE

The female genital tract is a component of the common mucosal immune system (CMIS) [99]. Antigen-presenting cells (dendritic and Langherans cells), T and B cells, and macrophages populate cervix and vagina of the human female genital tract, indicating the potential for the production of mucosal immunity in the genital tract by means of local application [100]. These features combined with the ease of access of vaginal surface make local immunization feasible, presenting some difficulties. It is well established that the mucosal and systemic compartments of the immune system display a significant degree of mutual independence. Several studies have shown that there is also a compartmentalization within the mucosal immune system itself, so that mucosal immunization induces stronger immune responses at or adjacent to the site of induction than at distant areas [101]. The genital apparatus participates in the mucosal immune system with some unique features. The CMIS is not uniform in all mucosal areas; whereas in mucosal areas the mucosal predominant immunoglobulin (Ig) is the immunoglobulin A (IgA); the dominant Ig present in the genital tract is the IgG derived from the circulation [102]. In fact the female genital tract lacks organized lymphoid structures resembling Peyer's patches of the intestines, where mucosal immune responses are initiated and subsequently disseminated to the other effector sites. The activity and the number of antigen-presenting cells and the permeability of the epithelium to proteins change during the course of the estrous cycle [100]. The mucus layer lining the vaginal epithelium also prevents the entry of foreign matter. The elimination of macromolecules that penetrate the mucus depends on both the absorption of macromolecules by the epithelium and the shedding of mucus. Without adjuvants and penetration enhancers, the absorption of macromolecules by the epithelium is slow ($<5\%$ per hour); therefore, the washing action of vaginal fluids represents the limiting step to contact between a vaccine and mucosa [103]. However the female reproductive tract possesses the ability to initiate an immune response against infections caused by various sexually transmitted pathogens such as *Herpes Simplex* virus type 2 (HSV-2) [104] and *N. gonorrhoeae* [105]. Several studies regarding genital tract immunization have shown that administration of nonreplicating antigens into vagina can result in the generation of specific immunity [106]. However, these responses have been shown to confer only a modest immunity and are not disseminated either to the other mucosal sites or the systemic compartment. In conclusion, based on these findings, it is generally believed that the genital tract is unable to respond well to nonreplicating antigens following intravaginal immunization [107]. Despite these limits, there is a considerable interest in the induction of immune response at mucosal surface, since this may prove a key element in successful vaccination strategies against a number of pathogens entering the body via this route. Moreover the local vaginal vaccination is proposed to halt the unintended pregnancy with the local production of sperm-specific antibodies in the female genital tract.

HSV-2 is responsible for a sexually transmitted disease and its primary site of access is the mucosa of the genital tract. Induction of mucosal immune response against HSV-2 specifically in the genital tract may be fundamental in preventing infections. The majority of vaccines is administered via parenteral route and although such route is able to induce a strong systemic immune response, this is not always a guarantee of significant immune response at mucosal site [107]. By contrast, mucosal immunization can induce strong, long-lasting mucosal and systemic immune responses and consequently high levels of protection against subsequent antigen contact [108]. The major obstacle to the development of mucosal vaccines for humans is the lack of adjuvants that combine efficacy with safety [107].

In recent years, several studies demonstrated successful DNA immunization via the vaginal route [6]. Moreover, there is increasing evidence that prolonged induction of immunoglobulin IgA secretion provides protection against sexually transmitted pathogens and spermatozoa [108].

A wide variety of novel approaches for the development of mucosal immune responses by administration of antigens in different delivery systems have been investigated, in many cases, with considerable success. These approaches include: (1) the coadministration of immunogens with adjuvants active at the mucosal surface, (2) the coupling of immunogens to carrier molecules that promote their uptake at the mucosal inductive site, and (3) the expression of antigens in live attenuated bacterial or viral vectors, which can promote the colonization of mucosal tissue and the incorporation of antigens into a variety of microparticulate and adhesive vehicles, which are taken up in mucosal inductive sites.

Shen et al. [103] investigated the potential use of a poly(ethylene-*co*-vinyl acetate) (EVAc) matrix to control the exposure of DNA to the vaginal tract and induce local immunity [103]. EVAc matrix was loaded with a plasmid pcDNA3/LDH-C4 inducing LDH-C4 gene. LDH-C4 encodes a peptide derived from the sperm-specific isoenzyme of LDH; such a peptide was shown to suppress fertility in female mice, rabbit, and baboons. The contraceptive effect, following such a matrix administration, was reversed 1 year after the last immunization. The partial contraceptive effect has been attributed to antibody-induced agglutination of spermatozoa. The matrix prepared by means of solvent evaporation technique was characterized by the controlled release of DNA, providing a prolonged DNA contact with the mucosal surface. The initial burst determined sensitization of immune cells and the initialization of immune response; subsequently, the continuous release induced a long-term antigen activation and maintained the pool of antibodies and antibody-producing cells and the memory cells. Moreover, this system demonstrated the protection of DNA from the vaginal environment. The DNA released was able to transinfect cells within the reproductive apparatus, in particular, the vaginal tract and cervix and uterine horns. The effect persisted for 28 days in the vagina and cervix.

McLean et al. [109] proposed to achieve a mucosal immunization against *Herpes Simplex* virus type 1 (HSV-1) using a suspension of disabled virus, disabled infectious single cycle (DISC) vaccine. The disabled virus maintains the capability of infecting the cells for a single cycle replication. For this reason, the genetically disabled DISC HSV-1 vaccine can be delivered effectively via a mucosal surface. Mice vaccinated intravaginally with DISC HSV-1 vaccine showed a delayed-type hypersensitivity and cytotoxic T-cell responses against HSV-1. Such a vaccine may represent a safe and effective way of immunization against a variety of HSV-1 [109].

Kwant and Rosenthal. [107] described the induction of strong immune response in the genital tract by means of the intravaginal administration of recombinant glycoprotein B (rgB) of HSV-2 associated with a synthetic oligodeoxynucleotide (ODN), containing immunostimulatory CpG motifs, as mucosal adjuvant (CpG ODN). The CpG ODN can act as an effective mucosal adjuvant for nonreplicating antigens when intravaginally administered. Intravaginal immunization with rgB plus adjuvants induced high levels of glycoprotein B (gB)-specific IgA and IgG in genital secretions and serum and also conferred a protection from genital HSV-2 challenge and maintained low levels of viral shedding. The vaccine activates the innate immune responses. CpG ODN directly activates antigen-presenting cells including dendritic cells to upregulate the major histocompatibility complex (MHC) class II. For these reasons, the authors concluded that the vaccine is able to induce local and systemic antibody production and provide a protection against the vaginal infection of HSV-2 [107].

Saltzman et al. [110] illustrated the possibility to administer various monoclonal and polyclonal antibodies via the vaginal route by means of EVAc matrices. The monoclonal and polyclonal antibodies tested were mouse monoclonal IgG against human chorionic gonadotropin (hCG), mouse monoclonal IgM against mouse sperm, polyclonal mouse IgG antibodies, and murine monoclonal IgG1. Polymeric matrices can provide a long-term high-dose local (vaginal) delivery of antibodies. The experiments described the biodistribution of the antibodies after prolonged vaginal administration. The polymer matrices were characterized by prolonged active antibody release by means of a diffusion mechanism through the porous matrix. The antibody presence following vaginal administration was found in vaginal lavages, blood, and other tissues, in particular in the gastrointestinal apparatus. The antibody elimination following administration was mainly determined by the mucus shedding (probably increased by the device presence). These studies suggested that the employment of polymeric matrices can represent a new approach to maintain suitable serum antibody levels for prolonged periods via a route of administration alternative to the parenteral one. Moreover, this investigation demonstrated that the vaginal epithelium is not a barrier to antibody diffusion [110].

Yao et al. [111] proposed to induce immunocontraception by means of vaginal inoculation of a recombinant lactobacillus (*L. casei*) excreting human chorionic gonadotropin β antigen (hCGβ) [111]. The antigen-specific antibody was produced in the reproductive tract after the inoculation of the microorganisms. *Lactobacillus* was chosen as vector because it is a part of nonpathogenic normal vaginal flora of the female reproductive tract and it is capable of spontaneously adhering to the vaginal mucosa. The microorganism *L. casei* expressed the hCGβ antigen in a stable manner; hCGβ antigen protein was detectable even 2 weeks after mouse inoculation via vaginal route in the vaginal lavage. Moreover, after treatment, INF-γ and IL-4 were secreted at high levels indicating that the specific immune response was evoked through a common mucosal immune mechanism (CMIM) [111].

22.10 EXPERIMENTAL MODELS TO STUDY VAGINAL TOLERABILITY AND ABSORPTION

The use of vertebrates to evaluate tolerability and absorption of drug administered via the vaginal route has been widely criticized on the basis of scientific and ethical considerations. Studies on animals can be substituted by validated *in vitro* tests as described in the guideline issued by Committee for Proprietary Medicinal Products (CPMP), now Committee for Medicinal Products for Human Use (CHMP) [112]. Before an *in vitro test* can be considered valid, this test must undergo a procedure aimed at establishing its relevance and reliability. The relevance of the alternative test has to be compared with accepted *in vivo* standard methods.

A great number of animal models have been employed in the study of vaginal absorption of drugs. It is necessary to consider the correlation between the anatomy and physiology of the vaginal tissues in the animal model with respect to the human ones and if there is a correlation between the modification of the vaginal mucosa following the menstrual or estrous cycle and the vaginal absorption in animal models and women.

Various animals have been employed as models: rats, rabbits, sheep, and monkeys.

The rat vaginal epithelium is constituted by pluristratified squamous cell layers that are subjected to pronounced changes during the estrous cycle. In rats, the estrous cycle lasts 4–5 days. At estrous cycle (ovulation), the vaginal epithelium is thick with superficial flat layers of cornified cells exfoliating into the vaginal lumen; immediately after ovulation the epithelium becomes thinner again. The cyclic changes in thickness and consequently in histology of rat vaginal mucosa impair hydrophilic drug absorption during ovulation [113,114]. This is probably due to the formation of cornified epithelial cells immediately after ovulation. Such a feature is considered the major difference between the human and rat mucosa; the changes in histology of the human vaginal epithelium are not as pronounced as in rats [5].

To overcome the differences between the human and the vaginal mucosa and the differences in histology dependent on the estrous cycle, an ovariectomized rat model, with or without hormone treatments, was proposed [5]. After ovariectomy, the rat vaginal epithelium becomes thin and atrophic (two layers in thickness); this determines that the susceptibility to drug-induced damage is greater than that of the human epithelium [115]. Alternatively, ovariectomized rats were given a low estradiol dose by subcutaneous injection 24 h before the absorption study. In this model, the rat vaginal epithelium becomes thicker (approximately 40 μm) and consists of a basal cuboidal cell layer covered by several flat cell layers and by an outer cuboidal cell layer. In this model, the vaginal epithelium is closer to physiologic conditions and the epithelium is more resistant to epithelial damage [5]. Another advantage is that the rat is a quadruped, and this allows for the exclusion of the confounding effect of gravity on pelvic floor, which plays a role in bipedal and semibipedal animals [115–117].

Rabbits have been extensively used by Yotsuyanagi et al. [118]. In rabbits, ovulation is consequent to coitus and in the absence of male contact the vaginal epithelium histology is thought to stay relatively unchanged [5]. This should result in reduced variability of vaginal absorption in comparison with animal models subjected to changes in histology due to estrous cycle. A drawback is represented by the fact that the anatomy and histology of vagina is markedly different in rabbits and women. In particular, the epithelium is characterized by the presence of both ciliated and nonciliated cells and only the lower third of the vaginal tract is lined by a stratified squamous epithelial cells similar to the human vaginal epithelium. This advises against the employment of rabbits for *in vivo* drug absorption study [5]. Nevertheless, rabbits are frequently used for vaginal irritation tests, because they present a simple cuboidal or columnar epithelium more sensitive to mucosal irritants than the stratified squamous epithelium of human vagina [119].

Monkeys, in particular Rhesus macaque and squirrel monkey, have been considered as models for vaginal drug absorption because of the similarities between nonhuman primate and human anatomy. The vagina of rhesus macaques is characterized by connective tissue attachments, localization of steroid hormone receptors, duration, and cyclic changes in vaginal physiology similar to that of humans [119].

Sheep have also been investigated as animal models to study drug vaginal absorption. The sheep ovarian cycle lasts for approximately 16 days [120] and the consequent changes in histology of vaginal epithelium are not pronounced and are similar to those of other species [121]. The stage of estrous cycle may have a great effect on the vaginal absorption of less readily absorbable drug [5].

To overcome the ethical problems and the costs related to the employment of models based on vertebrate animals, alternative methods based on *in vitro* methods (primary cell culture and immortalized cell lines) and also phylogenetically lower organisms (invertebrates, plants, and microorganisms) have been considered. *In vitro* assays that employ primary human cell cultures or immortalized cell lines derived from cervical and vaginal epithelial cells have been developed to evaluate the cytotoxicity of formulations designed to be administered via the vaginal route [81,122,123]. The HeLa cell line (derived from intraepithelial neoplasia) [79], well-characterized human papillomavirus (HPV-16/E6E7), immortalized vaginal (Vk2/E6E7), ectocervical (Ect1/E6E7), and endocervical (End1/E6E7) epithelial cell lines were employed to evaluate toxicity [122,123].

Recently, Dhondt et al. [124] developed an alternative test based on an invertebrate, for the evaluation of irritation potential of drug delivery systems administered via the vaginal route [124]. The test considers a terrestrial slug named *Arion lusitanicus* as a model organism. The body wall of slug consists of a single-layer epithelium composed of cells and mucous gland cells overlying the connective tissue. Slugs exposed to irritating substances produce mucus to protect the body wall; the amount of mucus produced by the slugs during a repeated contact period is a measure for irritation. The release of proteins, LDH and ALP, from the body walls enables the estimation of membrane damage [125].

The slug mucosal irritation test was found to be useful to estimate the irritation potential of bioadhesive formulations intended for vaginal administration. The results obtained using the slug mucosal irritation test are comparable to those of vaginal irritation test in rabbits. Additionally, a previous study showed that the results obtained with slug mucosal irritation test were comparable to human vaginal tolerance data [124]. The slug mucosal irritation test seems to be a promising alternative to screen new vaginal semisolid formulations for local tolerance in early stages of development. The use of this test can possibly contribute to reduce the use of vertebrates in preclinical studies [125].

22.11 FINAL REMARKS

The increasing knowledge of physiology, histology, and immunology of vagina suggests new and challenging uses of this route for local and also systemic administration of drugs. Vaginal administration allows lower doses, avoidance of first-pass metabolism, less frequent administration, ease of access and self-medication and, depending on drug physicochemical properties, good absorption potential. On the other hand, the main limitations to vaginal administration of drugs are due to gender specificity, menstrual cycle, and hormone level variations. Cultural aspects and personal care habits can lead to poor patient compliance. All these features still limit the current use of vaginal route to the administration of sex hormones, anti-infectives, and drugs used for induction of labor, cervical ripening, or pregnancy termination. The systemic delivery especially of peptidic drugs such as insulin and calcitonin has been extensively studied but it has not yet reached consolidated clinical application. In any case, when the risk–benefit balance is evaluated, it must be taken into consideration that each vaginal formulation is introduced into a delicate microflora and environmental equilibrium that should be respected and preserved.

REFERENCES

1. Bernkop-Schnürch, A., and M. Hornof. 2003. Intravaginal drug delivery systems: Design, challenges, and solutions. *Am J Drug Deliv* 1:241.
2. Woolfson, A.D., R.K. Malcom, and R. Gallagher. 2000. Drug delivery by the intravaginal route. *Crit Rev Ther Drug Carrier Syst* 17:509.

3. Fosberg, J.-G. 1995. A morphologist's approach to the vagina—age-related changes and estrogen sensitivity. *Maturitas* 22:S7.
4. Alexander, N.J., et al. 2004. Why consider vaginal drug administration? *Fertil Steril* 82:1.
5. Richardson, J.L., and L. Illum. 1992. Routes of drug delivery: Case studies. The vaginal route of peptide and protein drug delivery. *Adv Drug Deliv Rev* 8:341.
6. Hussain, K., and S. Ahsan. 2005. The vagina as a route for systemic drug delivery. *J Control Release* 103:301.
7. Thompson, I.O.C., et al. 2001. A comparative light-microscopic, electron-microscopic and chemical study of human vagina and buccal epithelium. *Arch Oral Biol* 46:1091.
8. Burgos, M.H., and C. Roig de Vargas-Linares. 1978. Ultrastructure of the vaginal mucosa. In *Human reproductive medicine: The human vagina*, eds. E.S.E. Hafez and T.E. Evans, 63–93. New York: North Holland Publishing.
9. Vermani, K., and S. Garg. 2000. The scope and potential of vaginal drug delivery. *PTTS* 10:359.
10. Lee, V.H.L. 1988. Enzymatic barriers to peptide and protein absorption. *Crit Rev Drug Carrier Syst* 5:69.
11. Song, Y., et al. 2004. Mucosal drug delivery: Mechanisms, methodologies, and applications. *Crit Rev Drug Carrier Syst* 21:195.
12. Brannon-Peppas, L. 1993. Novel vaginal drug release applications. *Adv Drug Deliv Rev* 11:169.
13. Hwang, S., et al. 1977. System approach to vaginal drug delivery of drugs. II. *In situ* absorption of unbranched aliphatic alcohols. *J Pharm Sci* 65:1574.
14. Owada, E., et al. 1977. Vaginal drug absorption in rhesus monkeys. I. Development of methodology. *J Pharm Sci* 66:216.
15. Corbo, D.C., J.-C. Liu, and Y.W. Chien. 1990. Characterization and barrier properties of mucosal membranes. *J Pharm Sci* 79:202.
16. Knuth, K., M. Amiji, and J.R. Robinson. 1993. Hydrogel delivery systems for vaginal and oral applications. *Adv Drug Deliv Rev* 11:137.
17. Bulletti, C., et al. 1997. Targeted drug delivery in gynaecology: The first uterine pass effect. *Hum Reprod* 112:1073.
18. Miles, R.A., et al. 1994. Pharmacokinetics and endometrial tissue levels of progesterone after administration by intra-muscular and vaginal routes: A comparative study. *Fertil Steril* 62:485.
19. Braditch-Crovo, P., et al. 1996. Quantitation of vaginally administered nonoxynol-9 in premenopausal women. *Contraception* 55:261.
20. Mandal, T.K. 2000. Swelling-controlled release system for the vaginal delivery of miconazole. *Eur J Pharm Biopharm* 50:337.
21. DuBouchet, L., et al. 1998. A pilot study of metronidazole vaginal gel versus oral metronidazole for the treatment of *Trichomonas vaginalis* vaginitis. *Sex Transm Dis* 25:176.
22. Lamont, R.F., et al. 2003. The efficacy of vaginal clindamycin for the treatment of abnormal genital tract flora pregnancy. *Infect Dis Obstet Gynecol* 11:181.
23. Gagne, N., et al. 1999. Protective effect of thermoreversible gel against the toxicity of nonoxynol-9. *Sex Transm Dis* 26:177.
24. D'cruz, O.J., and F.M. Uckun. 2001. Gel microemulsion as vaginal spermicides and intravaginal drug delivery vehicles. *Contraception* 64:113.
25. Rusconi, S., et al. 1996. Naphthalene sulphonate polymers with CD-4 blocking an anti-immunodeficiency virus type 1 activities. *Antimicrob Agents Chemother* 40:234.
26. Mastromarino, P., et al. 2002. Characterization and selection of vaginal *Lactobacillus* strains for the preparation of vaginal tablets. *J Appl Microbiol* 93:884.
27. Catalanotti, P., et al. 1994. Effects of cetylmethylammonium naproxenate on the adherence of *Gardnerella vaginalis*, *Mobiculunus curtisii* and *Lactobacillus acidophilus* to vaginal epithelial cells. *Sex Transm Dis* 21:338.
28. Yamashida, A., et al. 1991. Pharmacological studies on intravaginally applied dehydroepiandrosterone sulphate. *Folia Pharmacol Jpn* 98:31.
29. Vukovich, R.A., A. Heald, and A. Darragh. 1977. Vaginal absorption of two imidazole antifungal agents, econazole and miconazole. *Clin Pharmacol Ther* 21:121.

30. Jain, J.K., and D.R. Mishell. 1994. A comparison of intravaginal misoprostol with prostaglandin E_2 for termination of second trimester pregnancy. *N Engl J Med* 331:290.
31. Fachin, R., et al. 1997. Transvaginal administration of progesteron. *Obstet Gynecol* 90:396.
32. Darwish, A.M., et al. 2005. Evaluation of novel vaginal bromocriptine mesylate formulation: A pilot study. *Fertil Steril* 83:1055.
33. Baloglu, E., et al. 2003. A randomized trial of a new ovule formulation of ornidazole for the treatment of bacterial vaginosis. *J Clin Pharm Ther* 28:131.
34. Mauck, C.K., et al. 1997. A phase I comparative study of three contraceptive vaginal films containing nonoxynol-9. *Contraception* 56:97.
35. Mauck, C.K., et al. 1997. An evaluation of the amount of nonoxynol-9 remaining in the vagina up to 4 h after insertion of a vaginal contraceptive film (VCF®) containing 70 mg of nonoxynol-9. *Contraception* 56:103.
36. Novak, A., et al. 2003. The combined contraceptive vaginal ring, NuvaRing: An international study of user acceptability. *Contraception* 67:187.
37. Chen, J.H., et al. 1998. The comparative trial of Tcu 380A IUD and progesterone-releasing vaginal ring used by lactating women. *Contraception* 57:371.
38. Verhorven, C.H.J., et al. 2004. The contraceptive vaginal ring NuvaRing and antimycotic co-medication. *Contraception* 69:129.
39. Bachmann, G. 1995. Urogenital aging: An old problem newly recognized. *Maturitas* 16:145.
40. Weisberg, E., et al. 1997. Effect of different insertion regimens on side effects with a combination contraceptive vaginal ring. *Contraception* 56:233.
41. Igarashi, M., et al. 1998. Novel vaginal danazol ring therapy for pelvic endometriosis, in particular deeply infiltrating endometriosis. *Hum Reprod* 13:1952.
42. Johansson, E.D.B., and R. Sitruk-Ware. 2004. New delivery systems in contraception: Vaginal rings. *Am J Obstet Gynecol* 190:S54.
43. Witter, F.R., et al. 1999. Duration of vaginal retention and potential duration of antiviral activity of five nonoxynol-9 containing intravaginal contraceptives. *Int J Gynecol Obstet* 65:165.
44. Chickering, D.E., and E. Mathiowitz. 1999. Definitions, mechanisms, and theories of bioadhesion. In *Bioadhesive drug delivery systems: Fundamentals, novel approaches, and development*, eds. E. Mathiowitz, D.E. Chickering, III, and C.M. Lehr. Basel: Marcel Dekker.
45. Derjaguin, B.V., et al. 1977. On the relationship between the molecular component of the adhesion of elastic particles to a solid surface. *J Colloid Interface Sci* 58:528.
46. Bogotaj, M., et al. 2003. The correlation between zeta potential and mucoadhesion strength on pig vesical mucosa. *Biol Pharm Bull* 26:743.
47. Lehr, C.M., et al. 1993. A surface energy analysis of mucoadhesion: II. Prediction of mucoadhesive performance by spreading coefficients. *Eur J Pharm Sci* 1:19.
48. Esposito, P., I. Colombo, and M. Lovrecich. 1994. Investigation of surface properties of some polymers by a thermodynamic and mechanical approach: Possibility of predicting mucoadhesion and biocompatibility. *Biomaterials* 15:177.
49. Shojaei, A.H., and X. Li. 1997. Mechanisms of buccal mucoadhesion of novel copolymers of acrylic acid and polyethylene glycol monomethyl ether monomethacrylate. *J Control Release* 47:151.
50. Jabbari, E., N. Wisnieswski, and N. Peppas. 1993. Evidence of mucoadhesion by chain interpenetrating at a poly(acrylic acid)/mucin interface using ATR-FTIR spectropscopy. *J Control Release* 26:99.
51. Hassan, E.E., and J.M. Gallo. 1990. A simple rheological method for the *in vitro* assessment of mucin–polymer bioadhesive bond strength. *Pharm Res* 7:491.
52. Mortazavi, S.A., B.G. Carpenter, and J.D. Smart. 1992. An investigation of the rheological behaviour of the mucoadhesive/mucosal interface. *Int J Pharm* 83:221.
53. Rossi, S., et al. 1995. Influence of mucin type on polymer–mucin rheological interactions. *Biomaterials* 16:1073.
54. Ferrari, F., et al. 1997. Technological induction of mucoadhesive properties on waxy starches by grinding. *Eur J Pharm Sci* 5:277.
55. Rossi, S., et al. 1999. Model-based interpretation of creep profiles for the assessment of polymer–mucin interaction. *Pharm Res* 16:1456.

56. Rossi, S., et al. 2001. Characterization of chitosan hydrochloride–mucin rheological interaction: Influence of polymer concentration and polymer:mucin weight ratio. *Eur J Pharm Sci* 12:479.
57. Bygderman, M., and M.L. Swahn. 1996. Replens versus dienoestrol cream in the symptomatic treatment of vaginal atrophy in postmenopausal women. *Maturitas* 23:259.
58. Milani, M., E. Barcellona, and A. Agnello. 2003. Efficacy of the combination of 2 g oral tinidazole and acidic buffering vaginal gel in comparison with vaginal clindamycin alone in bacterial vaginosis: A randomized, investigator-blinded, controlled trial. *Eur J Obstet Gynecol Reprod Biol* 109:67.
59. Caramella, C.M., M.C. Bonferoni, and P. Giunchedi. 2003. Compositions with controlled release of lactic acid at vaginal level. Patent WO 03/000224.
60. Bonferoni, M.C., et al. 2001. Characterization of vaginal chitosan gel intended for lactic acid delivery. In *AAPS annual meeting*, Denver, 21–25.
61. Franchin, R., et al. 1997. Transvaginal administration of progesterone. *Obstet Gynecol* 90:396.
62. Roux, F.C., et al. 1996. Morphometric, immunohistological and three-dimensional evaluation of the endometrium of menopausal women treated by estrogen and Crinone, a new slow release vaginal progesterone. *Hum Reprod* 11:357.
63. Damario, M.A., et al. 1999. Crinone 8% vaginal progesterone gel results in a lower embryonic implantation efficiency after *in vitro* fertilization–embryo transfer. *Fertil Steril* 72:830.
64. Garg, S., et al. 2001. Properties of a new acid-buffering bioadhesive vaginal formulation (ACID-FORM). *Contraception* 64:67.
65. Amaral, E., et al. 1999. Study of the vaginal tolerance to Acidform, acid-buffering bioadhesive gel. *Contraception* 60:361.
66. Lee, C.H., and Y.W. Chien. 1996. *In vitro* permeation study of a mucoadhesive drug delivery for controlled delivery of nonoxynol-9. *Pharm Dev Technol* 1:135.
67. Lee, C.H., M. Anderson, and Y.W. Chien. 1996. The characterization of *in vitro* spermicidal activity of chelating agent on human sperm. *J Pharm Sci* 85:649.
68. Lee, C.H., R. Bagdon, and Y.W. Chien. 1996. Comparative *in vitro* spermicidal activity and synergistic effect of chelating agents with nonoxynol-9 on human sperm functionality. *J Pharm Sci* 85:91.
69. Chang, J.Y., et al. 2002. Rheological evaluation of thermosensitive and mucoadhesive vaginal gels in physiological conditions. *Int J Pharm* 241:155.
70. Chang, J.Y., et al. 2002. Prolonged antifungal effects of lotrimazole-containing mucoadhesive thermosensitive gels on vaginitis. *J Control Release* 82:39.
71. Wang, Y., and C.H. Lee. 2002. Characterization of a female controlled drug delivery system for microbicides. *Contraception* 66:281.
72. Quintanar-Guerrero, D., et al. 2001. *In vitro* evaluation of the bioadhesive properties of hydrophobic polybasic gels containing *N,N*-dimethylaminoethyl methacrylate-*co*-methyl methacrylate. *Biomaterials* 22:957.
73. Baloglu, E., et al. 2003. An *in vitro* investigation for vaginal bioadhesive formulations: Bioadhesive properties and swelling states of polymer mixtures. *Farmaco* 58:391.
74. Lejoyeaux, F., et al. 1989. Bioadhesive tablets: Influence of the testing medium composition on bioadhesion. *Drug Dev Ind Pharm* 15:2037.
75. Bouckaert, S., et al. 1995. Preliminary efficacy study of bioadhesive vaginal metronidazole tablet in the treatment of bacterial vaginosis. *J Pharm Pharmacol* 47:970.
76. Kast, E.K., et al. 2002. Design and *in vitro* evaluation of a novel bioadhesive vaginal drug delivery system for clotrimazole. *J Control Release* 81:347.
77. Karasulu, H.Y., et al. 2004. Efficacy of a new ketoconazole bioadhesive vaginal tablet on *Candida albicans*. *Farmaco* 59:163.
78. Ceschel, G.C., et al. 2001. Development of a mucoadhesive dosage form for vaginal administration. *Drug Dev Ind Pharm* 27:541.
79. Coggins, C., et al. 1998. Women's preferences regarding the formulation of over-the-counter vaginal spermicides. *AIDS* 12:1389.
80. Garg, S., et al. 2005. Development and characterization of bioadhesive vaginal films of sodium polystyrene sulfonate (PSS), a novel contraceptive antimicrobial agent. *Pharm Res* 22:584.

81. McCarron, P.A., et al. 2004. Phototoxicity of 5-aminolevulinic acid in the HeLa cell line as an indicative measure of photodynamic effect after topical administration to gynaecological lesions of intraepithelial form. *Pharm Res* 21:1871.
82. Rossi, S., et al. 2003. Development of films and matrices based on chitosan and polyacrylic acid for vaginal delivery of acyclovir. *STP Pharma Sci* 13:181.
83. Pavelic, Z., N.S. Skalko-Basnet, and R. Schubert. 2001. Liposomal gels for vaginal drug delivery. *Int J Pharm* 219:139.
84. Karasulu, H.Y., et al. 2002. Sustained release bioadhesive effervescent ketoconazole microcapsules tabletted for vaginal delivery. *J Microencap* 19:357.
85. Vermani, K., S. Garg, and L.J.D. Zaneveld. 2002. Assemblies for *in vitro* measurement of bioadhesive strength and retention characteristics in simulated vaginal environment. *Drug Dev Ind Pharm* 28:1133.
86. Sandri, G., et al. 2004. Mucoadhesive and penetration enhancement properties of three grades of hyaluronic acid using porcine buccal and vaginal tissue, Caco-2 cell lines, and rat jejunum. *J Pharm Pharmacol* 56:1083.
87. Bonferoni, M.C., et al. 1999. A modified Franz diffusion cell for simultaneous assessment of drug release and washability of mucoadhesive gels. *Pharm Dev Technol* 4:45.
88. Rossi, S., et al. 1999. Drug release and washability of mucoadhesive gels based on sodium carboxymethylcellulose and polyacrylic acid. *Pharm Dev Technol* 4:55.
89. Sayani, A.P., and Y.W. Chien. 1996. Systemic delivery of peptides and proteins across absorptive mucosae. *Crit Rev Ther Drug Carrier Syst* 13:85.
90. Richardson, J.L., L. Illum, and N.W. Thomas. 1992. Vaginal absorption of insulin in the rat: Effect of penetration enhancers on insulin uptake and mucosal histology. *Pharm Res* 9:878.
91. Van der Bijl, P., et al. 2002. Enhancement of transmucosal permeation of cyclosporine by benzalkonium chloride. *Oral Dis* 8:168.
92. Sandri, G., et al. 2004. Assessment of chitosan derivatives as buccal and vaginal penetration enhancers. *Eur J Pharm Sci* 21:35.
93. Nakada, Y., M. Miyake, and N. Awata. 1993. Some factors affecting the vaginal absorption of human calcitonin in rats. *Int J Pharm* 89:169.
94. Valenta, M.K., et al. 2001. Evaluation of the inhibitory effect of thiolated poly(acrylates) on vaginal membrane bound aminopeptidase N. *J Pharm Pharmacol* 54:603.
95. Bernkop-Schnurch, A., et al. 2004. Thiomers: Potential excipients for non-invasive peptide delivery systems. *Eur J Pharm Biopharm* 58:253.
96. Morimoto, K., et al. 1982. Effective vaginal absorption of insulin in diabetic rats and rabbits using polyacrylic acid aqueous gel bases. *Int J Pharm* 12:107.
97. Rochira, M., et al. 1996. Novel vaginal delivery systems for calcitonin. II. Preparation and characterization of HYAFF microspheres containing calcitonin. *Int J Pharm* 144:19.
98. Richardson, J.L., and T.I. Armstrong. 1999. Vaginal delivery of calcitonin by hyaluronic acid formulations. In *Bioadhesive drug delivery systems: Fundamentals, novel approaches, and development*, eds. E. Mathiowitz, D.E. Chickering, III, and C.M. Lehr. Basel: Marcel Dekker.
99. McDermott, M.R., and J. Bienenstock. 1979. Evidence for a common mucosal immunologic system. I. Migration of B immunoblast into intestinal, respiratory, and genital tissues. *J Immunol* 122:1892.
100. Parr, M.B., et al. 1991. Langherans cells phagocytose vaginal epithelial cells undergoing apoptosis during the murine estrus cycle. *Biol Reprod* 45:252.
101. Johansson, E.L., et al. 1998. Antibodies and antibody secreting cells in the female genital tract after vaginal or intranasal immunization with cholera toxin B subunits or conjugate. *Infect Immunol* 66:514.
102. Russel, M.W. 2002. Immunization for protection of the reproductive tract: A review. *Am J Repr Immunol* 47:265.
103. Shen, H., E. Goldber, and W.M. Saktzman. 2003. Gene expression and mucosal immune response after vaginal DANN immunization in mice using a controlled delivery matrix. *J Control Release* 86:339.
104. Merriman, H., et al. 1984. Secretory IgA antibody in cervicovaginal secretions from women with genital infection due to herpes simplex virus. *J Infect Dis* 149:505.

105. Hedges, S.R., et al. 1998. Cytokine and antibody responses in women infected with *Neisseria gonorrhoeae*: Effect of concomitant infections. *J Infect Dis* 178:742.
106. Wassen, L., et al. 1996. Local intravaginal vaccination of the female genital tract. *Scand J Immunol* 44:408.
107. Kwant, A., and K.L. Rosenthal. 2004. Intravaginal immunization with viral subunit protein plus CpG oligodeoxynucleotides induces protective immunity against HSV-2. *Vaccine* 22:3098.
108. McGhee, J.R., et al. 1992. The mucosal immune system: From fundamental concepts to vaccine development. *Vaccine* 10:75.
109. McLean, C.S., et al. 1996. Induction of a protective immune response by mucosal vaccination with a DISC HSV-1 vaccine. *Vaccine* 14:987.
110. Saltzman, W.M., et al. 2000. Long-term vaginal antibody delivery: Delivery systems and biodistributions. *Biotechnol Bioeng* 67:253.
111. Yao, X.Y., et al. 2004. Inoculation of *Lactobacillus* expressing hCGβ in the vagina induces an anti-hCGβ antibody response in murine vaginal mucosa. *J Reprod Immunol* 63:111.
112. Committee for Proprietary Medicinal Product. 1997. Replacement of animal studies by *in vitro* model, CMPC/SWP/728/95. Available at http://www.emea.eu.int/pdfs/human/swp/072895en.pdf, accessed on January 7, 2005.
113. Richardson, J.L., and L. Illum. 1992. Routes of delivery: Case of studies. The vaginal route of peptide and protein drug delivery. *Adv Drug Deliv Rev* 8:341.
114. Okada, H., et al. 1983. Vaginal absorption of a potent luteinizing hormone-releasing hormone analogue (leuprolide) in rats. IV. Evaluation of vaginal absorption and gonadotropin responses by radioimmunoassay. *J Pharm Sci* 73:289.
115. Hsu, C.C., et al. 1984. Topical vaginal drug delivery. I. Effects of the oestrous cycle on vaginal membrane permeability and diffusivity of vidabirine in mice. *J Pharm Sci* 72:674.
116. Richardson, J.L., et al. 1989. Vaginal administration of gentamicin to rats. Pharmaceutical and morphological study using absorption enhancer. *Int J Pharm* 56:29.
117. Moalli, P.A., et al. 2005. A rat model to study the structural properties of the vagina and its supportive tissues. *Am J Obstet Gynecol* 192:80.
118. Yotsuyanagi, T., et al. 1975. Systems approaches to vaginal drug delivery of drugs. Development of *in situ* vaginal drug absorption procedure. *J Pharm Sci* 64:71.
119. Eckstein, P., et al. 1969. Comparison of tolerance tests of spermicidal preparations in rabbits and monkeys. *J Reprod Fertil* 20:85.
120. Legan, S.J., and F.J. Karsh. 1979. Neuroendocrine regulation of the oestrous cycle and seasonal breeding in the ewe. *Biol Reprod* 20:74.
121. Robinson, T.J. 1959. The estrous cycle of ewe and doe. In *Reproduction in domestic animals*, eds. H.H. Cole and P.T. Cupps, 291–333. New York: Academic Press.
122. Fichorova, R.N., L.D. Tucher, and D.J. Anderson. 2001. The molecular basis of nonoxynol-9-induced vaginal inflammation and its possible relevance to human immunodeficiency virus type I transmission. *J Infect Dis* 184:418.
123. Maguire, R.A., N. Bergman, and D.M. Phillips. 2001. Comparison of microbicides for efficacy in protecting mice against challenge with herpes simplex virus type 2, cytotoxicity, antibacterial properties, and sperm immobilization. *Sex Transm Dis* 28:259.
124. Dhondt, M.M.M., et al. 2005. The evaluation of the local tolerance of vaginal formulations containing dapivirine using the slug mucosal irritation test and the rabbit vaginal irritation test. *Eur J Pharm Biopharm* 60:419.
125. Dhondt, M.M.M., E. Adriaens, and J.P. Remon. 2004. The evaluation of the local tolerance of vaginal formulations with or without nonoxynol-9, using the slug mucosal irritation test. *Sex Transm Dis* 31:229.

Part VII

Systemic Absorption Through the Ocular Route

23 Eye Structure and Physiological Functions

Clive G. Wilson, Ekaterina M. Semenova, Patrick M. Hughes, and Orest Olejnik

CONTENTS

23.1 INTRODUCTION

Eyesight is unquestionably the most important modality of perception that humans possess. It is so central to our way of life that irrational fear of losing vision, termed Scoptophobia or Scotomaphobia, is totally incapacitating for some individuals. Creationists suggested that an organ as special as the human eye could not have come about by chance; it is therefore *prima facie* evidence of divine intervention. More rational people point out design imperfections: for example, the invertebrates seem to have acquired the services of a better architect. In the squid and the octopus, the photoreceptor cells point forward toward the incoming light; whereas in humans, the structural elements—rods and cones—are aimed backward, away from the light source. Moreover, the nerve fibers that must carry signals from the retina to the brain must pass in front of the receptor cells, partially impeding the penetration of light to the receptors [1]. Whatever the anatomical failings at inception, this wonderful structure is also subject to all the vicissitudes of time and disease. Preservation of sight as the mean age of the population increases, and thus as blindness becomes more prevalent will remain a key healthcare objective.

23.2 EXTERNAL FEATURES OF THE EYE

The human eye is a spherical structure that, in three dimensions, has the outside profile of a smaller sphere embedded within a larger sphere. The organ is the biological embodiment of a very familiar physical object—a camera, with the ability to control the focus, light intensity, and depth of field of an image through movement of the external and internal musculature under intentional and autonomic nervous control. The film of this camera is the brain itself, the stimuli being carried by the afferent pathways from the photoreceptors to the optic nerve that hands on the information to the visual cortex at the back of the brain through the optic tracts. As the information is relayed to the brain, a few fibers leave the main tracts to enter the nerve nuclei that control the visual reflexes. The purpose of this feed helps both eyes work in concert to improve the stereo image—the consensual response—and also to adjust light input and lens power via a separate pathway.

The case of the camera is the skull. The forehead and brow partially protect the eye that is enclosed in the bony orbit. To facilitate stereoscopic vision, the adult eyes are positioned such that the focus is able to converge to achieve a binocular view between 20 cm and infinity. The eyelids provide partial protection to sudden changes of light and also serve to compress and spread the tear film maintaining a clear optical surface. Fibrous tarsal plates provide the framework for the movable folds of modified skin (Figure 23.1a and Figure 23.1b).

23.2.1 Eyelids

The eyes are very vulnerable and must be protected from desiccation and the accumulation of particle debris to preserve visual acuity. The deliberate act of instillation of an ophthalmic formulation therefore triggers protective mechanisms that attempt to restore the normal tear film. The primary responses in humans are blinking and increased production of tears. The palpebral (literally meaning "of the eyelids") parts of the orbicularis oculi muscle, particularly of the upper lid, are used in gentle lid closure. Orbital parts of the muscle are recruited for a stronger forced-downward movement of the upper lid. This action is associated with a reflex upward movement of the eye, which is effective before the lids are fully closed (Bell's phenomenon). The orbicularis muscle aids tear drainage, and underneath this muscle is the tissue plane that contains the blood and nervous supply to the lids. The process of drainage is assisted actively by blinking and a negative pressure in the lacrimal sac.

The mucins of the eyelid margin trap particulate matter, which on squeezing of the lids move the material onto the base of the eyelashes. Eyelid closure results in compression of the film, temporarily increasing the thickness of the tear film—mucins in the tears must support changes in dimensions and anchor the hydrated film to the surface. The elasticity of the mucins, provided by uncoiling regions in the structure coupled with the binding of water, maintains the hydration of the cornea.

23.2.2 Eyelashes

The eyelashes are attached to the free edges of the eyelids. They are short, stout, curved hairs, arranged in a double or triple row and penetrate deep into the dermis. Those of the upper eyelid, more numerous and longer than those of the lower, curve upward; those of the lower eyelid curve downward, so that they do not interlace in closing the lids. The lifetime of a lash is around 150 days. The hair follicles are similar to those found elsewhere in the body but lack the arrector pili smooth muscle. Near the attachment of the eyelashes are the openings of a number of glands, the Meibomiam glands, arranged in several rows close to the free margin of the lid; these secrete a complex mixture of oils and waxes to lubricate the surface and provide an important component of tears. Administration of material directly onto the lid margin for

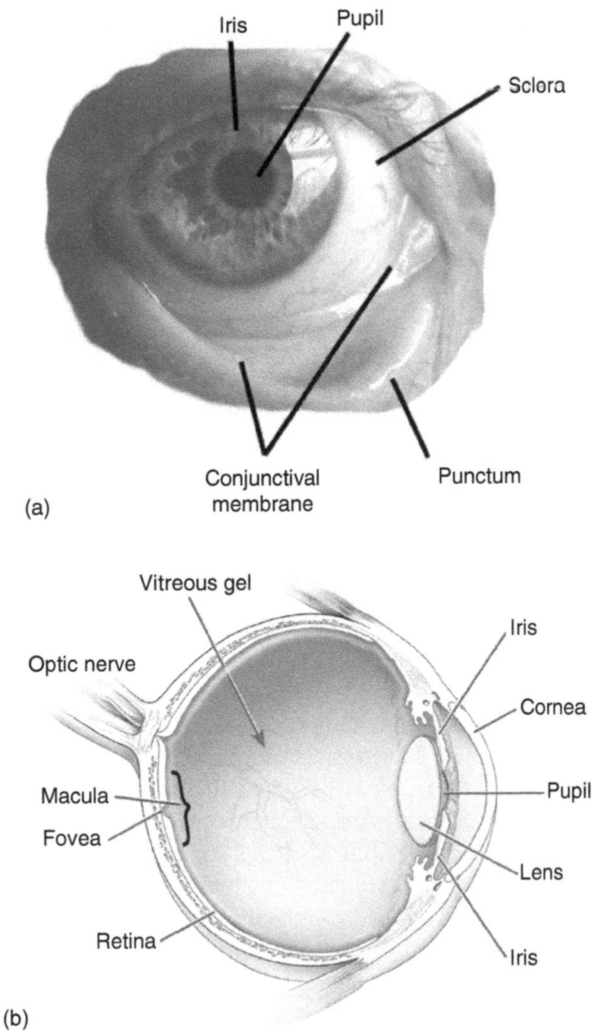

FIGURE 23.1 (a) The external eye. (b) The internal eye structures. (Courtesy The National Eye Institute Archive.)

religious purposes or cosmetic effect is practiced in eastern cultures, the material being soot (kohl) or toxic metal sulfides of lead and antimony (surma). The material lines the outer edge of the eyelid near the gland exits; in the past, these heavy metal poisons may have had an effect in reducing ocular parasite burdens.

A similar effect can be seen following ocular administration of micronized carbon suspensions in perfluorodecalin. This vehicle is a nonsolvent and because the powder is delivered dry, the particles attach during the first reflexive blink and disperse to the lash roots (Figure 23.2) [2].

23.2.3 Nasolacrimal Apparatus

At the inner corners of the eyelids, two small punctae or openings exteriorises to the globe and drain the tearfilm to the nasolacrimal duct. These openings are approximately 0.3 mm in diameter and sit on top of an elevated mound—the papilla lacrimalis—that are visible as pale

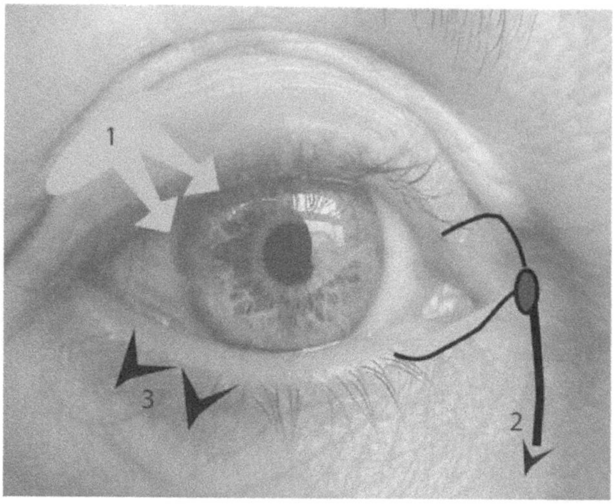

FIGURE 23.2 Lacrimal flow and drainage. (1) The lacrimal gland produces tears that are distributed by blinking. (2) At normal rates of flow, tears are cleared by nasolacrimal drainage. (3) Sudden increases produced by instillation or reflex tearing overspill the lower fornix.

blips on the eyelid margin close to the conjunctival margin. Tears enter these openings to drain down through the canaliculi that join at the lacrimal sac, lying inside a gap between maxillary and lacrimal bones. The sac consists of a double-layered epithelium that continues to the nasolacrimal duct and then joins the nasal turbinates. The sac is surrounded by a venous plexus that connects to the drainage territory of the face and flows directly to the heart. Ophthalmic β-blocker formulations of drugs such as timolol show marked effects on exercise tolerance from miniscule doses due to this highly efficient absorption pathway. This pulsed entry is probably responsible for bronchospasm or cardiovascular side effects associated with β-blockade from topical ophthalmic administration of these agents. One of the benefits of thickening ophthalmic solutions is the reduction in the rate of transfer from eye into the nose, reducing the rapid upswing in drug concentration [3]. Current eyedroppers attempt to minimize loss from the eye as instillation of more than 30 μL into the eye exceeds the temporary capacity of the *cul-de-sac* causing spillage down the cheek and rapid nasolacrimal drainage. Delivery of much smaller volumes has been shown to be more efficient [4] although devices have been generally unsuccessful in overcoming the problems caused by the surface tension of water.

The synchronized movement of the eyelids spreads the precorneal tearfilm across the cornea and pushes it toward the nasolacrimal duct. Precorneal drainage is quite efficient. An aqueous instilled dose leaves the precorneal area within 5 min of instillation in humans. Most of the drug absorbed by transcorneal penetration, without retention modification, is spread across the cornea by the eyelids in the first minutes postdosing. In the precorneal space transcorneal penetration is limited by solution drainage, lacrimation and tear dilution, tear turnover, conjunctival absorption, and the corneal epithelium. Slowing down tear film turnover has well-established benefits to topical ocular drug delivery.

Chitosan has been shown to increase precorneal drug residence times. The cationic chitosan slows tear drainage by increasing viscosity and by mucoadhesion with the negatively charged mucin. Up to a threefold increase of the corneal residence time has been achieved by the addition of chitosan to topical vehicles [5]. Carbomer gels at 0.3% have also been shown to be effective in prolonging the tear film break-up time [6]. Hyaluronic acid has been reported to stabilize the tear film, resulting in a delay of the break up time [7]. In the study, it was found that a concentration of at least 0.1% was necessary to delay the break up time. A clinical study of a 0.1% formulation demonstrated a clear benefit of hyaluronan over saline in both objective and subjective assessments [8]. Other polymers that have been successfully used to increase precorneal retention time of instilled fluids include hydroxypropyl methylcellulose and carboxymethylcellulose.

23.2.4 The Conjunctivae

A thin mucous membrane overlying a connective tissue layer with many fine blood vessels, the substantia propria, attached to the eyelid margin and extending to the corneoscleral junction is a predominant feature of the outer eye. The conjunctiva covers the frontal part of the eye, apart from the cornea, and is connected to the epidermis of the eyelids. It is important for tear film stability and for maintaining anti-inflammatory status. The tissue, visible as a pink flat mass covering the anterior sclera, is the bulbar conjunctiva and is loosely attached; the tissue that is lining the inside surface of the eyelids (palpebral conjunctiva) is more firmly anchored. The conjunctivae are composed of squamous nonkeratinized cells and the total area exceeds that of the cornea by approximately 400%. Together with the nasal mucosa, the conjunctivae provide a route for systemic access of the drug; the nasal tissue contributes more than 80% of the absorptive capacity. Although the conjunctiva is structurally very similar to the cornea, it shows greater permeability. Several workers have calculated that large polar solutes 20–40 kDa across the conjunctival epithelium by restricted diffusion through a pore-equivalent size of 5.5 nm [9].

The conjunctivae serve as a secondary reservoir for drugs, and instillation of one preparation within 5 min of the first usually results in a reduction of the concentration of drug from the primary instillation.

23.2.5 Cornea and Sclera

The sclera and the cornea are the toughest and outermost layers of the eye and resist the normal internal pressure of 13 to 19 mmHg. This intraocular pressure (IOP) gives the eye its shape and maintains its dimensions that are necessary for sharp vision. The sclera covers 5/6 of the eye's surface and the cornea the remaining 1/6. Although the principal structural element of both tissues comprises of type 1 collagen fibers, differences in size and orientation of the fibers, degree of hydration, and presence of mucopolysaccharides are responsible for differences in transparency. The avascular cornea receives nourishment from the tear film, the aqueous humor, and the limbal vessels. In contrast, the sclera is vascularized and is supplied by several blood vessels, particularly in the uppermost layers (episclera).

The cornea is the transparent window of the frontal globe and is formed from several layers: an external hydrophobic epithelium about 50 μm thick bounded by Bowman's layer, followed by the hydrophilic stromal layer, and finally the thin Descemet's membrane and the endothelium, which controls the transparency of the cornea when alive (Figure 23.3). The epithelium is tightly packed, containing approximately 100-fold more lipid than the stroma; the high lipid content of this tissue, coupled with the close proximity of the conjunctivae, provides a tissue reservoir for fat-soluble drugs, which must be then rendered more hydrophilic to complete transcorneal transfer. This sequential arrangement of barriers provides the rationale for ester-linked prodrugs, for example, prednisolone acetate.

The corneal epithelium is composed of 5–6 layers of cells, increasing to 8–10 at the periphery with a total thickness of 50–100 μm. The cells at the base are columnar, but as

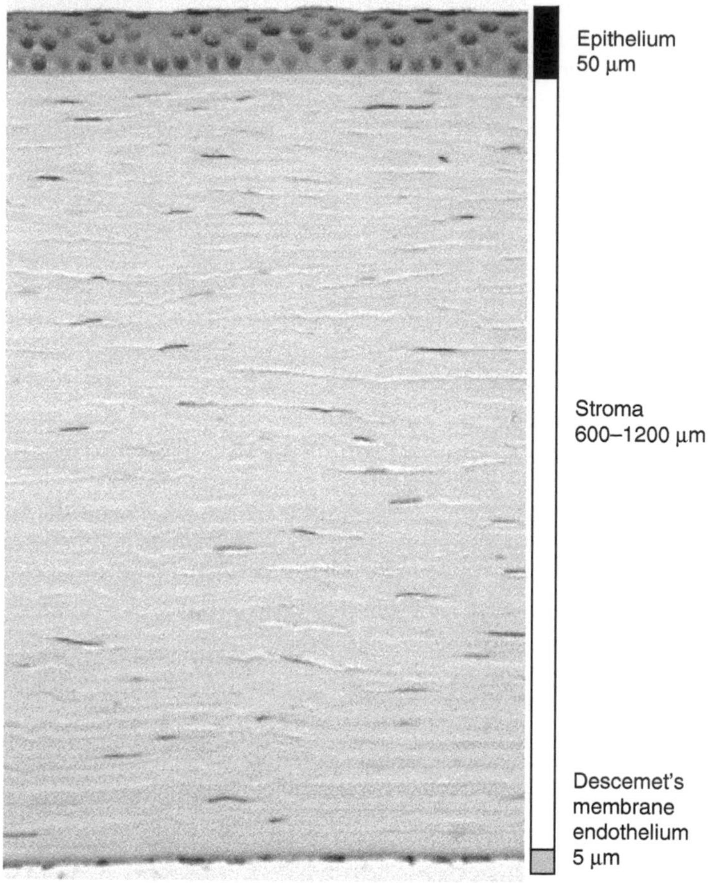

FIGURE 23.3 The corneal structure showing the sequential hydrophobic and hydrophilic barriers.

they are squeezed forward by new cells, they become flatter so that three groups of cells are usually identified: basal cells, an intermediate zone of 2–3 layers of polygonal cells (wing shaped), and squamous cells. This structure is illustrated in the diagram (Figure 23.3). The permeability of the intact corneal epithelium is low until the outermost layer is damaged, suggesting that tight junctions exist between the cells of the outer layer—the function of which is to exclude all solute movement, except that which occurs by partitioning through the apical and basal plasma membranes of the surface epithelial cells. The cells of the outer layer possess microvilli on their anterior surface which presumably help to anchor the precorneal tear film.

The stroma, or substantia propria, consists of collagenous lamellae running parallel to the surface and superimposed on each other. The stroma is fibrous with a high content of water, approximately 78% of the total weight at the normal hydration; 15% of collagen; 1% each for glycosaminoglycans and salts; and the remaining 5% is noncollagenous proteins [10]. The composition of the sclera is similar to the stroma [11] but the diameters of the fibers are markedly different: 48—113 nm for the cornea and 118–1268 nm for the sclera [12]. The cornea is continuous with the sclera and the transparency results from the regular dimensions and packing array of the fibers. On the interior surface of the stoma there is a 5–10 μm thick layer, Descemet's membrane, which is secreted by the endothelium. The endothelium consists of a single layer of flattened cells approximately 5 μm high and 20 μm wide. These cells form a regular mosaic with close contact between them. The endothelium is about 200 times more permeable than the epithelium and thus represents a weak barrier.

A passive flux of water continually flows across the endothelial layer toward the stroma, which has a tendency to swell. An active pump mechanism pulls an aqueous flux in the opposite direction which controls corneal turgescence [13]. Corneal deturgescence is an ATP-dependent process of the endothelial cells; and as such any disruption of the endothelium may result in corneal oedema, thereby affecting corneal transparency. The specific distribution of different proteoglycans across the cornea has recently been implicated in water gradients across the cornea. This water gradient serves to diminish dehydration of the front of the cornea, which is exposed to the atmosphere.

The sclera is the outer white tough part of the eye, which is an important structural element, with the site of insertion of extraocular muscles. It covers 80% of the exterior surface and is white and nontransparent. It borders the transparent cornea at the pars planar. The sclera is divided into three layers: episclera, stroma, and lamina fusca. Only a limited number of blood vessels, originating from arteriolar branches of the anterior ciliary vessels, are found and superficial vessels are mainly confined to the loose outer episclera. Scleral permeability approximates that of the corneal stroma and has been shown to be permeable to solutes up to 70 kDa in molecular weight [14].

The stroma is composed of dense bundles of collagen fibers that crisscross in layers as illustrated in the diagram (Figure 23.3). The outer fibers are thicker than the inner fibers. The main components of the stroma in addition to collagen fibers, which comprise 75% of the scleral dry weight, are fibroblasts and proteoglycans.

The lamina fusca is light brown in colour, attributable to the layer of melanocytes that coat the membrane in an irregular fashion.

23.2.6 The Corneal Tearfilm

The curved, bulging cornea provides about 60% of the refractive power of the eye but unlike the lens, this component is fixed. Since focus is maintained using the combination of the cornea and lens, the consequences of a sudden change in the dimensions of the cornea is an attempt by the ciliary muscles to change the shape of the lens, which requires time. Specifically, the presence of water on the surface changes the refraction of the incident beam: immediately after an eye drop is instilled, the ability of the eye to form a focused image is lost and the sensation of the liquid plus the distortion causes blinking to restore the normal tear thickness. The excess liquid is held by expansion of the lower tear meniscus; and at a rate according to viscosity, the contents will be mixed and diluted by additional tear secretions.

The anterior surfaces of the eye—the cornea, the conjunctiva and the anterior chamber—are in constant apposition to potentially detrimental stimuli. Fluid is lost throughout the day from the tear film through evaporation and nasolacrimal drainage. In general, this fluid is replaced. The tear film is uniquely engineered to minimize these losses, keeping the anterior surface bathed in a moist, nutrient-rich environment. The corneal surface is kept wet by the conjunctivae and lacrimal secretions that are secreted and drained by the lacrimal system outside the eye. The system consists of the secretory lacrimal gland along with accessory glands in the orbit above the eye and the nasal drainage system. The lacrimal gland releases tears through many excretory ducts but is spread evenly across the cornea by blinking, leading to a tear film with a thickness of approximately 7 μm [15]. Tears normally have pH values in a range between 7.3 and 7.7 [16] and are secreted at a rate of 0.5–2.2 μL/min. However, the rate of secretion depends strongly on environmental conditions (e.g., temperature, wind and humidity), age, disease, and psychological state.

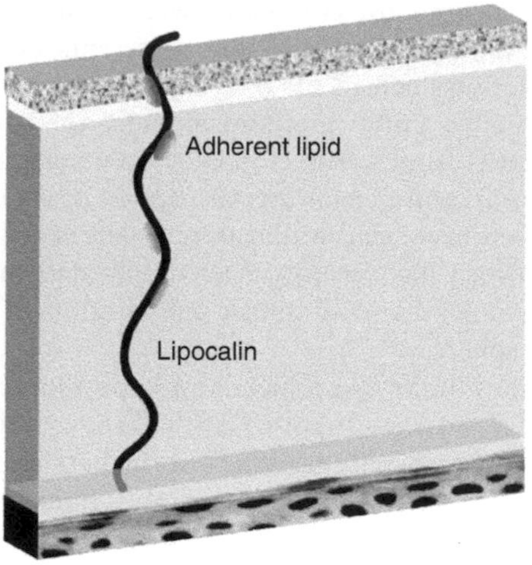

FIGURE 23.4 A model of tear film stabilization.

The traditional model of the tear film was of a three-layered structure (Figure 23.4): an adhered precorneal mucin layer (thickness 0.02–0.05 μm); a middle aqueous layer (4–7 μm), and an external lipid layer (0.01 μm). The mucin of the tear film aqueous layer is produced by conjunctival goblet cells and is necessary for corneal wetting. The function of the protein- and carbohydrate-containing mucin layer was to improve adherence and stability of the aqueous layer. Without adequate mucin the tear film is unstable and unable to wet the corneal epithelium, resulting in tear film breakup. The glycoproteins further serve to make the tear film thixotropic, allowing stability at rest and shear thinning during blinking. Loss of these glycoproteins decreases viscosity of the tear film and precorneal residence time. Pflugfelder's team [17] has also identified a sialomucin complex secreted into the tear film that participates with mucin in maintaining the stability of the tear film. The middle layer composed mainly of water, proteins, and inorganic ions such as Na^+, K^+, Cl^-, and HCO_3^- represent 98% of the total tear film volume.

The lipids of the external tear film namely, neutral oils, phospholipids, sterol esters, various waxes, and other lipids are requisite for regulation of evaporation. These lipids are produced by the Meibomiam and other glands. The ability of the glandular structures situated in the conjunctiva and the lid propia to produce the tear film may be impaired due to exposure to pollutants and irritants or as a result of age-related dysfunction.

This view has been challenged as the mixture of water, mucin, and oils must be anchored to the surface and kept clear: it is suggested therefore that an alternative model is needed. In a current model, this function is aided by tear lipocalin (TL), which is the principal lipid-binding protein in tears comprising 15%–35% of the protein content. The protein contains a conserved aromatic group, tryptophan, which appears to be essential in the maintenance of the secondary structure [18]. The two-phase model, suggested by McCulley and by Greiner and colleagues [19–21], proposes that the aqueous–mucin layer is covered by two thin layers of lipid. More polar phospholipids align to the aqueous–mucin layer whereas the air–tear surface layer is enriched with cholesterol. Glasgow and others [22] have suggested that the role of TL is to stretch from an anchoring point on the glycocalyx through the aqueous phase reaching through both lipid layers. This process increases the surface tension of the exterior film and returns the hydrophobic lipids back into the aqueous phase. The risk of corneal infection is reduced by immunoglobulins and antibacterial proteins such as lysozyme and lactoferrin in the tear fluid.

23.3 THE INTERNAL EYE STRUCTURES

This complex globe must support an internal lens, which by the movement of the ciliary muscles, focuses a sharp image upon the retina. The retina is a highly specialized neuronal tissue that enters the globe at the back of the eye and lines the wall. Color perception is mediated via the cones, whereas the more widely distributed rods are sensitive to lower levels of light and form a black-and-white image. Light regulation is provided by the iris, which under autonomic control adjusts the light received by the retina. The whole structure needs a tough but elastic external support, the sclera, which is sufficient to maintain dimensions but allows blood flow through the choroids to service the high metabolic demand of the retina. The structure is inflated from within by the secretion of aqueous humor from the ciliary body. This drains through the canal of Schlemm and to a lesser extent through the uveal tract.

23.3.1 Uveal Tract

The uveal tract of the eye is composed of the iris, the ciliary body, and the choroid. The iris muscle is located in front of the lens, which extends into the ciliary body. The iris regulates the amount of light entering the eye and the opening of the pupil depends on factors such as light intensity, emotional state, vigilance, and accommodation. Although the epithelium and stroma of the iris are both pigmented, differences in the amount of melanocytes in the stroma and thus light scattering determine a person's eye color. Blue, green, and gray irises are due to relatively few melanocytes in the stroma, and heavily pigmented brown irises are due to more melanocytes in the iris [23]. Many drugs are known to bind to melanin. It has been shown that the pigmentation of the iris can have a significant effect on the ocular residence time of some drugs.

The ciliary body, situated posterior to the iris, performs several functions. It connects the anterior part of the choroid to the circumference of the iris, and contains the ciliary muscles necessary for accommodation. The ciliary body secretes aqueous humor into the posterior together with nutrients to nourish the lens. In the ciliary body, melanin is located only in the outer pigmented epithelium [24]. Aqueous humor is actively secreted and passively filtered by the ciliary body. Although the rate of secretion is about 2 μL/min the same volume is drawn-off via Schlemm's canal from where it is conducted into veins [25].

23.3.2 The Lens

The lens is suspended in the proximal quarter of the eye just behind the iris. Like the cornea, it is remarkable in being optically transparent as the tissue is made up of a single cell type that is precisely aligned. The cells are encased in a capsular bag that has tendon attachments to the ciliary muscles allowing change in dimensions. The lens is composed of soluble protein fibers and is completely surrounded by the lens capsule [15]. Since there is no regeneration of the lens fiber, any damage may be retained throughout life. This makes the lens more susceptible to metabolic changes associated with disease states as this may lead to permanent opacity [26]. The human lens continues to grow throughout life and at all decades from 10 to 70 years; the lens is heavier in males than in females. As the lens hardens it becomes less elastic contributing to loss of accommodation and becomes yellow. Patients often comment after intraocular lens insertion on the vividness of colored objects, previously masked by the ageing lens.

23.3.3 Internal Humors

For modeling drug disposition in the eye following direct injection, the relative volumes of the aqueous spaces and the rates of turnover are important determinants of kinetics. Two spaces are clearly discernable as illustrated in Figure 23.5.

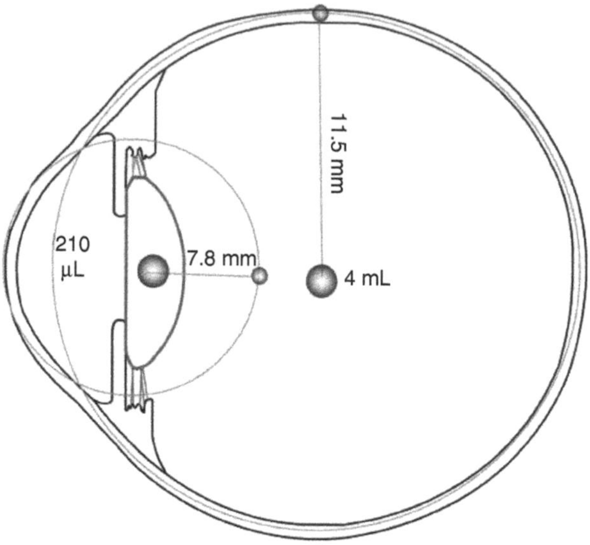

FIGURE 23.5 The eye is effectively a sphere within a sphere with radii of 7.8 and 11.5 mm as shown. The relative volumes of vitreous humor and anterior chamber fluid are indicated.

The aqueous humor can be aspirated during eye surgery and about 150 μL can be routinely obtained. Photographic measurements using a slit lamp showed that the volumes of the anterior chamber measured in both eyes of 39 normal subjects (mean age, 28 years; range, 19–56 years) were 209 ± 37 μL. Some fluid remains behind the iris in front of the lens, and researchers studying the conditions associated with abnormal iris pigmentation have proposed that blinking deforms the cornea creating a pressure wave that pushes the iris against the lens and empties the posterior chamber into the anterior chamber [27,28]. The aqueous humor is similar in composition to and isosmolar with plasma, but contains much less protein (30 mg/ml vs 80 mg/ml); has higher ascorbate, pyruvate, and lactate than plasma; and less glucose and urea.

The aqueous humor is formed by the ciliary epithelium and flows into the posterior chamber providing nourishment for the lens and the epithelia. It flows through the pupil and slowly circles in the anterior chamber entering the canal of Schlemm and also drains through the low-pressure pathway of the episcleral veins (Figure 23.6). This pathway, uveoscleral outflow, is thought to account for about 35% of drainage in monkeys and older estimates for humans are 5%–20% [29]. From here it enters the systemic circulation. Drugs administered to the cornea will be prevented from further inward diffusion by this outflow. The resistance to outflow generates a small intraocular pressure of about 16 mmHg; abnormally high pressures (>21 mmHg) are strongly associated with retinal anoxia (glaucomatous damage). The rate of production is probably around 2–3 μL/min [30], yielding estimates of turnover of 1.5% per min or an equivalent half-life of 47 min. The epithelium of the iris and of the ciliary body pump anionic drugs from the aqueous into the blood: this limits entry into the aqueous from the blood, a phenomenon which is termed the blood–aqueous barrier.

The vitreous humor volume is about 4 ml in an adult. Its viscosity is 2–4 times that of water and is dependent on the concentration of sodium hyaluronate. Although it has the outward appearance of a transparent, viscoelastic gel, it contains fine diameter type II collagen fibers (8–12 nm diameter) that entrap the large coiled hyaluronic acid molecules. These fibers give the gel a spherical structure with a dent in the anterior surface (the hyaloid fossa). The main constituent of the vitreous is water (98%) with a refractive index of 1.33, but the gel is

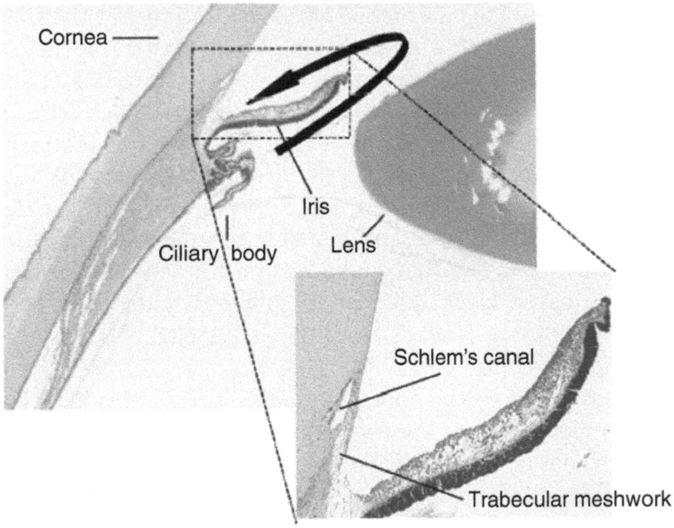

FIGURE 23.6 Formation and drainage of aqueous humor.

heterogeneous and two zones can be distinguished: a cortical zone, characterized by more densely arranged collagen fibrils, and a more liquid central vitreous region. The whole structure is anchored rather like a barrage balloon to various structures including peripheral retina and pars plana, the posterior capsule, and the retina along the margins of the optic disk. After adolescence, portions of the vitreous become more fluid with the appearance of fluid filled vacuoles being richer in soluble proteins and hyaluronic acid. Local variations in vitreal anatomy most likely reflect small variations in the density of vitreous tracts that are condensations of collagen.

Although head movements cause some stirring of the vitreous, the movement of molecules within the vitreous is predominantly through a process of diffusion. The vitreous is largely an unstirred fluid body and, due to its geometry, exerts a large diffusional barrier to drug elimination. The vitreous behaves isotropically toward diffusing molecules and the diffusion coefficient of molecules in the vitreous is only marginally less than the diffusion coefficient of the molecule in free solution. As the animal gets larger, convective effects become more important. Acid orange 8 diffusivity within bovine vitreous humor is about half that in free solution [31]. Convective flow has been suggested to be responsible for no more than 30% of the intravitreal transport in humans—in the mouse it is recognized as being unimportant. Net inflow of liquid moves from posterior anterior chamber to the vitreous, and drug elimination from an intravitreal injection is predominantly through the retina or diffusion to the posterior chamber. Compounds that can cross the blood–retinal barrier (BRB) will be cleared across the retina and have relatively short half-lives. Compounds that cannot traverse the BRB must diffuse to the retrozonluar space and are cleared through the anterior chamber. These compounds typically have longer vitreal half-lives dependent on molecular weight and steric factors.

23.3.4 The Choroid and Retinal Blood Supply

The retinal pigment epithelium and the photoreceptors are the most metabolically active tissues in the body and contain large numbers of mitochondria. These generate a lot of heat and waste products and must be supplied by a fast-moving, low resistance network. The retina is supplied by two separate systems that do not overlap; the inner retina is supplied by the retinal vascular system and the outer retina by the choroid below the sclera. The choroid is a highly pigmented, loose, connective tissue having an apparent thickness of approximately 220 μm at the posterior pole and 100 μm anteriorly. The choroid supplies nutrition to the retina. It also contains pigment cells known as melanocytes. These cells produce melanin, which absorbs excess light being reflected back through the retina. In many retinal diseases there is a significant change to melanin distribution and content. When this occurs and melanin's light-absorbing ability is reduced, light may be scattered in the retinal pigment epithelial cells (RPE) and the choroid, and contrast in image definition may be impaired. Histologically, there are five layers that make up the human choroid: Bruch's membrane (a modified connective tissue layer 2–4 μm in thickness), the choriocapillaris (provides nutrition to the photoreceptors), Haller's layer (large vessels), Sattler's layer (medium sized vessels), and the suprachoroid [32].

The choriocapillaris is worth special consideration as it forms a critical barrier between the sclera and retina. The inner vessels of the choroid (see Figure 23.7) form a diffuse monolayer, the choriocapillaris, which is fenestrated and therefore permeable. The fenestrations, approximately 60 nm in diameter, allow nutrient waste product exchange, but the adjacent RPE have tight junctions and form an outer BRB, which prevents access to the subretinal space above the RPE. The choroidal blood flow is anteriorly directed. Together with the fenestrated nature of the choriocapillaris these pose a large hurdle to retinal drug delivery. Systemically or periocularly administered drugs must establish a concentration gradient across the RPE for productive absorption while fighting anterior clearance through the choriocapillaris. The inner retina is supplied by retinal vessels from the branching of the ophthalmic artery; these vessels are blind ended and obstruction results in loss of supply to a complete area (capillary infarction). The vessels are not fenestrated and like the vessels of the brain, do not leak small

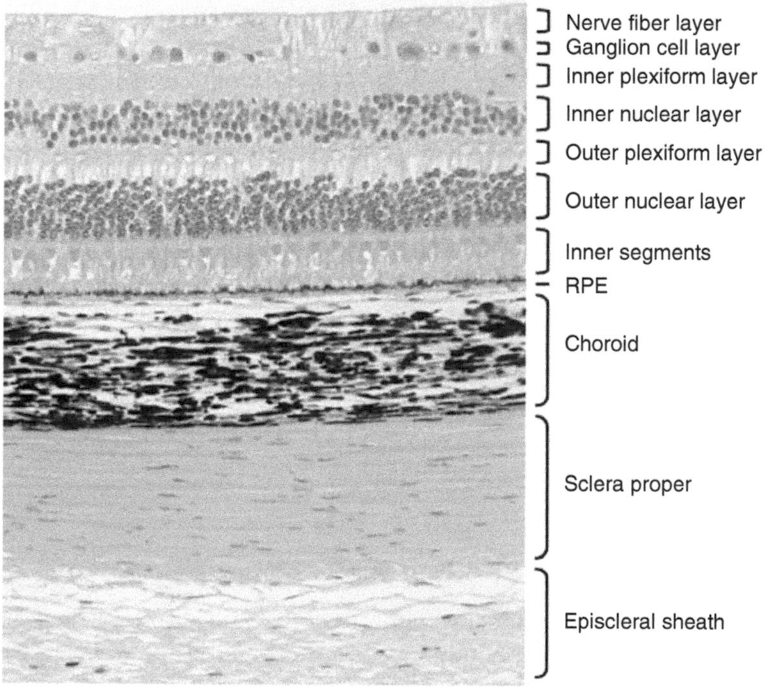

FIGURE 23.7 Retina and sclera.

molecules such as fluorescein unless they are ruptured. This is in contrast to capillary beds elsewhere in the body that are leaky. This structure is sometimes referred to as the retinal capillary endothelial barrier. There is outward pumping of anions across the retina by the RPE and the endothelial cells of the retinal vessels.

The suprachoroid is 30 μm in thickness and comprises thin interconnected lamellae of melanocytes, fibroblasts, and connective tissue fibers. These are separated by a thin space known as the suprachoroidal space. In this space hydrostatic pressure is a few mm of Hg lower than the IOP. The small gradient allows drainage of aqueous humor, through the tissue spaces of the ciliary muscles, into the suprachoroidal space. Venous drainage takes place through a series of large vortex veins; each of these drains a sector of the choroid into the superior and inferior ophthalmic veins of the orbit.

23.3.5 The Retina

The retina extends forward to the sclera as a globe-shaped wineglass almost external to the skull. That part of the sclera devoid of retina is the pars planar, which is used as an access point for injection or for close delivery to the iris and ciliary body (ICB). When stripped from its basement membrane and opened out, the collapsed retina is a circular disk approximately 42 mm in diameter and 0.5 mm in thickness. The organization of the retina is based on a three-neuron chain (photoreceptor cell–bipolar cell–ganglion cell) and accompanying cells (horizontal, amacrine, and Müller cells)

The retina consists of specialized photoreceptors for color (cones) and monochrome receptors active at lower light levels (rods) that connect back to the brain through the optic nerve at the back of the eye. There are different cone cell types (each with a different photopigment) for each of the three primary colors: red, blue, or green. Rods and cones are energy transducers; the photosensitive molecule rhodopsin absorbs photons of light and transduces it into chemical signals. Rods are widely dispersed; cones are mainly located in the central macula area particularly at the fovea, which is devoid of rods. The distribution of both receptors and supporting ganglions decreases to the rim of the structure. The retinal ganglion cells are stacked to a thickness of four to five around the fovea, thinning toward the retinal periphery to a single noncontiguous cell layer. The functions of RPE not only include phagocytosis of photoreceptor outer segments and active transport of materials from the blood supply, but also production of the interphotoreceptor matrix (IPM), which contributes to the adhesion of the retina to the RPE. RPE cells support the vitamin A metabolism in the visual cycle by moving retinoids between photoreceptor and pigment epithelium cells via the aqueous space of the IPM.

23.3.6 Retinal Pigmented Epithelium

The retinal pigmented epithelium (RPE) is in intimate, anatomic, and functional contact with the retina. Separated from the retina by Bruch's membrane, the RPE serves to regulate nutrients to the retina, phagocytize retinal debris, remove metabolic end products, and control the visual cycle. From a mass transport perspective, the RPE represents the outer BRB.

The BRB is anatomically separated into an inner and outer blood barrier. The RPE is a tight, ion-transporting barrier; and paracellular transport of polar solutes across the RPE from the choroid is restricted. This is reflected by the transepithelial electrical resistance (TEER) of the cell layers. It has been reported that the choroidal TEER (~ 9 ohm cm^2) is less than 10% the total resistance of isolated bovine RPE–choroid (100–150 ohm cm^2). Passive RPE diffusion has been shown to be a function of lipophilicity. The endothelium of the retinal vessels represents the inner BRB and offers considerable resistance to systemic penetration of drugs.

Carrier-mediated membrane transport proteins on the RPE selectively transport nutrients, metabolites, and xenobiotics between the choriocapillaris and the cells of the distal retina, and include amino acid [33–35], peptide [36], dicarboxylate, glucose [37], monocarboxylic acid [38,39], nucleoside[40], and organic anion and organic cation [41] transporters. Membrane barriers such as the efflux pumps, including multidrug resistance protein (P-gp), and multidrug resistance-associated protein (MRP) pumps have also been identified on the RPE. Exploitation of these transport systems may be the key to circumventing the outer BRB.

23.3.7 Posterior Pole: the Optic Nerve and Fovea

The nerves from the retina are collected at the back of the eye at the optic nerve, which connects via the optic chiasm to the brain. In the central field there is the most sensitive area of vision—the macula and the blind spot. The macula is in the center of the field of vision; visual field loss during old age and macular degeneration are significant causes of blindness in the Western world. The blind spot is the point at which the nerves enter and exit the eye and since there is no retina at this point, slightly off the midline axis, the brain adjusts the image to fill in the information. In severe macular degeneration caused by the accumulation of drusen (dry form) just underneath the retina or the more serious neovascularization associated with the wet form, the buckled and deficient spot in central vision is more immediately apparent. Directing therapy toward the macula is currently one of the biggest challenges facing ocular drug delivery companies because access through the systemic route is limited by the blood–ocular barrier, and topically by the large diffusional distances between convenient point of application and target tissue.

REFERENCES

1. Attributed to Professor F. Zindler. See http://www.2think.org/eye.shtml
2. Zhu, Y.P., et al. 1999. Dry powder dosing in liquid vehicles: ocular tolerance and scintigraphic evaluation of a perfluorocarbon suspension. *Int J Pharm* 191:79.
3. Wilson, C.G. 2004. Topical drug delivery in the eye. *Exp Eye Res* 78:737.
4. Martini, G.L., et al. 1997. The use of small volume ocular sprays to improve the bioavailability of topically applied ophthalmic drugs. *Eur J Pharm Biopharm* 44:121.
5. Felt, O., et al. 1999. Topical use of chitosan in ophthalmology: tolerance assessment and evaluation of precorneal retention. *Int J Pharm* 180:185.
6. Sullivan, L.J., et al. 1997. Efficacy and safety of 0.3% carbomer gel compared to placebo in patients with moderate-to-severe dry eye syndrome. *Ophthalmology* 104:1402.
7. Hamano, T., et al. 1996. Sodium hyaluronate eye drops enhance tear film stability. *Jpn J Ophthalmol* 40:62.
8. Condon, P.I., et.al. 1999. Double blind, randomised, placebo controlled, crossover, multicentre study to determine the efficacy of a 0.1% (w/v) sodium hyaluronate solution (Fermavisc) in the treatment of dry eye syndrome. *Br J Ophthalmol* 83:1121.
9. Ahmed, I. 2003. The noncorneal route in ocular drug delivery. In *Ophthalmic Drug Delivery Systems*, ed. A.K. Mitra. 335. Marcel Dekker, New York.
10. Fatt, I., B.A. Weissmann. 1992. Physiology of the eye. An introduction to the vegative functions, Butterworth-Heinemann, Boston.
11. Edwards, A., M.R. Prausnitz. 1998. Fiber matrix model of sclera and corneal stroma for drug delivery to the eye. *AIChE Journal* 41:214.
12. Meek, K.M., N.J. Fullwood. 2001. Corneal and scleral collagens—a microscopist's perspective. *Micron* 32:261.
13. Maurice, D.M., J. Polgar. 1977. Diffusion across the sclera. *Exp Eye Res* 25:577.
14. Olsen T.W., et al. 1998. Human sclera: thickness and surface area. *Am J Ophtahlmol* 83:237.
15. Berman, E.R. 1991. *Biochemistry of the Eye*. New York: Plenum Press.

16. Stjernschantz, J., M. Astin. 1993. Anatomy and physiology of the eye. Pathophysiological aspects of drug delivery. In *Biopharmaceutics of Ocular Drug delivery*, vol. 5, ed. P. Edman, 1. London: CRC Press.
17. Pflugfelder, S.C., et al. 2000. Detection of sialomucin complex (MUC4) in human ocular surface epithelium and tear fluid. *Invest Ophthalmol Vis Sci* 41:1316.
18. Gasymov, O.K., et al. 1999. Binding studies of tear lipocalin: the role of the conserved tryptophan in maintaining structure, stability and ligand affinity. *Biochem Biophys Acta* 1433:307.
19. McCulley, J.P., W.E. Shine. 1977. A compositional based model for the tear film lipid layer. *Trans Am Ophthalmol Soc* 95:79.
20. McCulley, J.P., W.E. Shine. 1998. Meibomiam secretions in chronic blepharitis. *Adv Exp Med Biol* 438:319.
21. Greiner, J.V., et al. 1996. Meibomiam gland phospholipids. *Curr Eye Res* 15:371.
22. Glasgow, B.J., et al. 1999. Tear lipocalins: potential lipid scavengers for the corneal surface. *Invest Ophthalmol Vis Sci* 40:3100.
23. Prota, G. 1992. Melanins and Melanogenesis. London: Academic Press.
24. Raviola, G. 1977. The structural basis of the blood-ocular barriers. *Exp Eye Res* 25:27.
25. Faller, A. 1988. Der korper des Menschen, Einfurhung in Bau und Funktion, Thieme Verlag, Stuttgart.
26. Forrester, J.V., A.D. Dick, P. McMenamin, W.R. Lee. 1996. The eye, basic science in practice. London: W.B. Saunders.
27. Campbell, D.G. 1993. Iridotomy, blinking and pigmentary glaucoma. *Invest Ophthalmol Vis Sci* 34:725 (Suppl).
28. Liebmann, J.M., et al. 1995. Prevention of blinking alters iris configuration in pigment dispersion syndrome and in normal eyes. *Ophthalmology* 102:446.
29. Bill, A., C.I. Phillips. 1971. Uveoscleral drainage of aqueous humour in human eyes. *Exp Eye Res* 12:275.
30. Brubaker, R. 1984. The physiology of the aqueous humour formation. In *Glaucoma Applied Pharmacology in Medical Treatment*, eds. S.M. Drance and A.H. Neufield, 35. Orlando, FL: Grune & Stratton.
31. Xu, J., et al. 2000. Permeability and diffusion in vitreous humor: implications for drug delivery. *Pharm Res* 17(6):664.
32. Klagsbrun, M., J. Folkman. 1990. Handbook of *Experimental Pharmacology*, vol. 95. Berlin, Heidelberg: Springer-Verlag.
33. Miyamoto, Y., et al. 1991. Taurine uptake in apical membrane vesicles from the bovine retinal pigment epithelium. *Invest Ophthalmol Vis Sci* 32:2542.
34. Leibach, J.W., et al. 1993. Properties of taurine transport in a human retinal pigment epithelial cell line. *Curr Eye Res* 12:29.
35. Pow, D.V. 2001. Amino acids and their transporters in the retina. *Neurochem Int* 121:89.
36. Duvvuri, S., M.D. Gandhi, and A.K. Mitra. 2003. Effect of P-glycoprotein on the ocular disposition of a model substrate, quinidine. *Curr Eye Res* 27:345.
37. Mantych, G.J., G.S. Hageman, and S.U. Devaskar. 1993. Characterization of glucose transporter isoforms in the adult and developing human eye. *Endocrinology* 133:600.
38. Knott R.M., et al. 1999. A model system for the study of human retinal angiogenesis: activation of monocytes and endothelial cells and the association with the expression of the monocarboxylate transporter type 1 (MCT-1). *Diabetologia* 42:870.
39. Philp, N.J., et al. 2003. Polarized expression of monocarboxylate transporters in human retinal pigment epithelium and ARPE-19 cells. *Invest Ophthalmol Vis Sci* 44:1716.
40. Williams, E.F., I. Ezeonu, and K. Dutt. 1994. Nucleoside transport sites in a cultured human retinal cell line established by SV-40 T antigen gene. *Curr Eye Res* 13:109.
41. Han, Y.H., et al. 2001. Characterization of a novel cationic drug transporter in human retinal pigment epithelial cells. *J Pharmacol Exp Ther* 296:450.

24 Drug Delivery Systems for Enhanced Ocular Absorption

Muhammad Abdulrazik, Francine Behar-Cohen, and Simon Benita

CONTENTS

24.1 INTRODUCTION

To date, nearly all ocular therapeutics have been administered to the eye as simple aqueous solution eyedrops by instillation to the lower conjunctival sac. The main drawbacks of aqueous eyedrops are their inability to deliver lipophilic and insoluble molecules, their low retention time, and their limited ability to resist the washout effect of blinking and tears turnover. The expected 10% to 20% fraction of the applied topical dose that escapes the immediate washout by blinking and tearing is then challenged by tear fluid proteins and enzymes binding and metabolism, ocular permeation barriers, phagocytic activity and partial diversion to adjacent tissues, and systemic circulation. Thus, more drug elimination occurs before it reaches the target tissue. Therefore, it is estimated that only 1% or less of the administered dose can penetrate the ocular surface [1,2]. Moreover, when targeting a remote target tissue like the retina, the fraction of the administered dose that reaches the action site will be much less due to further anatomical barriers, aqueous humor turnover, intraocular metabolic activity, binding to intraocular pigmented tissues, and phagocytosis by cell line other than the targeted tissue cells.

Maurice [3] has estimated that the maximum drug concentration that can be found in the vitreous following administration of a traditional eyedrop to the conjunctival sac is approximately a hundred-thousandth of the dose comprised in the drop. Thus, with growing indications for pharmacological intervention in diseases of the posterior segment of the eye, it is obvious that the traditional aqueous eyedrops cannot meet this need. Efforts to improve efficacy of the traditional topical ocular drug delivery systems (DDS) face formulation challenges and penetration obstacles. Hydrophilic molecules can be formulated easily as simple aqueous eyedrops but their ocular penetration is critically limited by the intercellular zonulae occludentes and the hydrophobic nature of the epithelium cell membranes of the ocular surface. In contrast, lipophilic molecules can penetrate through cell membranes of the ocular surface epithelium but their formulation for topical ocular administration is complex because they are insoluble in simple aqueous solutions. The formulation challenge opens new opportunities for the development of novel DDS for lipophilic molecules in the form of liquids with a semiaqueous nature to maintain the comfort and ease of use that make aqueous eyedrops so popular despite their limited efficacy. To overcome the penetration obstacles, the spreading on ocular surface as well as the adhesion properties to this surface must first be improved. This will lead to prolonged retention on ocular surface, enhancement of ocular surface absorption, less frequent applications, and fewer side effects.

24.2 INDICATION FOR PHARMACOLOGICAL INTERVENTION IN MODERN OPHTHALMOLOGY

Efforts in terms of workforce and funds devoted globally by the industry and academia to develop new DDS for ophthalmology are much greater than one might expect based on the relative prevalence of ocular diseases alone. This approach is supported by the recent marketing of new drugs developed for previously incurable ocular diseases (wet age-related macular degeneration [wet-AMD], severe dry eye syndrome), clinical studies reporting that pharmacological intervention efficiency is comparable to surgical intervention efficiency (glaucoma), and the growing awareness of patients and clinicians to the diagnosis and therapy of some ocular diseases (mild dry eye). These changes have raised great expectations for a significant expansion of the ocular therapeutics market soon. Some examples of ocular drug delivery challenges that have emerged in the last decade follow. New therapeutics for AMD and choroidal neovascularization raised the issue of targeting a remote part of the eye, the posterior segment. The introduction of topical cyclosporin A (CsA) for the treatment of severe dry eye pointed out the difficulties of incorporating a large insoluble lipophilic molecule in an efficient yet comfortable and well-tolerated formulation. The introduction of the powerful prostaglandin analogs (Bimatoprost, Travoprost, and Latanoprost) emphasized the need for developing improved formulations to avoid side effects like conjunctival hyperemia (drug uptake by the conjunctiva) while enhancing the effect on target tissue (ciliary body and uveoscleral outflow tract). Remedy of mild dry eye requires the introduction of formulations with improved wettability and stability on ocular surface. Finally, the growing role of gene therapy in modern medicine in general and in ophthalmology in particular necessitates the development of delivery systems that can protect and enhance the penetration of charged hydrophilic molecules (e.g., oligonucleotides), which are impermeable and sensitive to environmental degradation.

Nevertheless, very limited research has been carried out to develop DDS for the posterior segment of the eye in the last decades. Whereas new antiangiogenic compounds and neurotrophic factors can open new therapeutic avenues in the treatment of AMD, retinal dystrophies and ocular manifestations of diabetes, the frequent intraocular injections of these compounds remain a problem and are associated with complications and discomfort for the patient. Thus, the need for efficient noninvasive DDS for the administration of drugs to the back of the eye is awaiting a ground-breaking multidiscipline advancement.

Pharmacological intervention is envisioned in the following ophthalmic indications: glaucoma, AMD, ocular allergies, dry eye, ocular infections, and ocular inflammations. For a better comprehension of the problems that can be encountered and need to be solved when developing a DDS for a specific indication, the respective target ocular tissues should be identified. The target tissues in glaucoma include the ciliary body and the uveoscleral outflow tract; in AMD, the choroid–retina interface; in ocular allergies, the ocular surface and eyelids; in dry eye, the ocular surface and secretary glands (aqueous, mucin, and oil); in ocular infections, the eyelids, tear drainage system, ocular surface, anterior segment tissues, and posterior segment tissues; in ocular inflammations, the eyelids, ocular surface, anterior uvea (iris), intermediate uvea (ciliary body, pars plana), posterior uvea (choroid), and optic nerve.

24.3 DRUG DELIVERY RELATED OCULAR PARAMETERS

Knowledge of the anatomy and physiology of the eye should contribute to the identification of an appropriate drug delivery strategy that will enhance the penetration of active molecules through ocular tissues without eliciting any adverse effect and to the understanding of the mechanism of entry, which may lead to a better tuning of the formulation parameters of an ocular DDS.

This section will focus only on drug delivery related issues of ocular physiology and anatomy that are essential for the understanding of the following discussion on drug permeation barriers and routes of drug delivery to the posterior segment of the eye. A detailed review of ocular anatomy and physiology can be found in Chapter 27.

24.3.1 Ocular Surface Parameters

24.3.1.1 Capacity of the Lower Conjunctival Sac

The preferred site for the instillation of eyedrops is the lower conjunctival sac (cul-de-sac). The anterior wall of this sac is the conjunctiva of the lower eyelid (palpebral conjunctiva), whereas the conjunctiva that covers the eyeball (bulbar conjunctiva) forms the posterior border, and the forniceal conjunctiva (the conjunctival bridge between bulbar and palpebral parts) lines the crater and the lateral borders of this space (Figure 24.1).

When the lower eyelid skin is pulled down gently by fingers, the lower conjunctival sac forms a funnel-shaped reservoir with the capacity to contain a drug droplet, but the conjunctival surface cannot support a volume of more than 25 μL if this volume is added quickly [4]. When the lower eyelid returns to its normal position the capacity of the conjunctival sac will shrink by 70% to 80% to less than 10 μL. The capacity of the conjunctival sac is further limited for untrained, incompliant, or disabled patients. Conjunctival pathologies such as cicatricial, allergic, or inflammatory process will also limit the capacity of the conjunctival sac.

24.3.1.2 Blinking

Blinking is an important defense mechanism of the eye. The brisk blinking reflex is usually fast enough to precede high-speed foreign bodies approaching the eye. Blinking is also essential for the periodic reforming of the trilamellar tear film. In order to maintain a uniform tear film on the ocular surface a blink must come before the breakup phenomenon of the tear film occurs. Blinking also activates a pumping mechanism for the drainage of tears through the lacrimal drainage apparatus. The blink rate in humans is 15 to 20 per min (roughly one

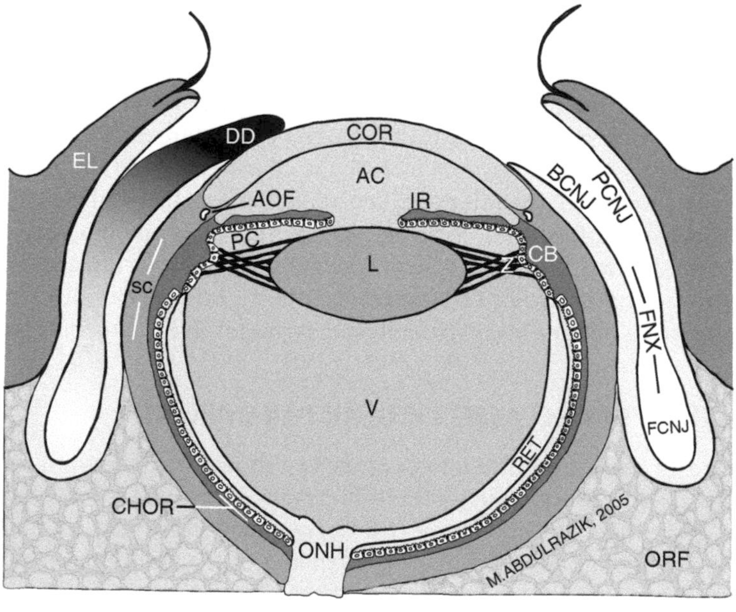

FIGURE 24.1 Ocular and periocular tissues. EL, eyelid; DD, drug droplet; COR, cornea; AC, anterior chamber; PC, posterior chamber; AOF, aqueous outflow tract (trabeculum meshwork and Schlemm canal); BCNJ, bulbar conjunctiva; PCNJ, palpebral conjunctiva; FCNJ, fornical conjunctiva; FNX, fornix (conjunctival sac/cul-de-sac); SC, sclera; IR, iris; CB, ciliary body; CHOR, choroid; RET, retina; ONH, optic nerve head; V, vitreous; L, lens; Z, lens zonulae; ORF, orbital fat.

blink every 5–7 s). In one study, a single drop of either fluorescein or fluorescein-labeled (FITC) dextran was applied to a rabbit eye. For both molecules, the tear film fluorescence remained constant after the first few minutes and dropped abruptly following the first blink. As the blink rate in rabbits is much slower than humans (roughly 5 per min), the author concluded that low blink rate can result in overestimation of the transcorneal penetration when data acquired from animal experiments is extrapolated to humans [5].

24.3.1.3 Tear Secretion

The fluid that is contained in the upper and lower conjunctival sacs, as well as the fluid that covers the exposed ocular surface, is derived from tear secretion and is drained to the nasal cavity and replaced continuously by tear secretion. The tear film is a trilamellar structure consisting of a thin mucin layer that is attached to the ocular surface epithelium, a monomolecular film of lipids that faces the air, and an aqueous layer in between that occupies most of the tear film thickness. The aqueous component is secreted by the main and the accessory lacrimal glands; the mucin is secreted by conjunctival goblet cells; whereas the lipid layer is secreted by meibomian glands in the eyelid margin. In unstimulated conditions, total tear film volume on ocular surface is approximately 6 to 8 μL, and the secretion rate of aqueous tears has been reported to be about 1.2 μL/min. Therefore, in normal human eyes, the basal tear turnover rate (per minute) is about 16% of the total tear film volume. Stimulated or reflex tearing occurs when the conjunctiva and cornea are stimulated by irritating stimulus. The tear film volume increases during reflex tearing to about 16 μL, with a range of 5 to 66 μL [6]. Thus stimulated tearing from any cause, ironically including some formulation parameters set to enhance solubility and solution stability, will cause accelerated wash out of an instilled drug droplet.

Usually drugs are instilled to the lower conjunctival sac in large volumes (50 μL), and in this range the unprovoked tear turnover was reportedly estimated to play only a minor role in the loss of instilled drug dose [7].

24.3.1.4 Tear Drainage

The tear film leaves the surface of the globe and eyelids, enters the upper and the lower punctum at the medial aspect of the lid margin, and enters the lacrimal sac before drainage to the nasolacrimal duct and the nasal cavity. However, much of the tear film is eliminated by direct evaporation or by absorption at the level of the lacrimal sac. The lacrimal outflow system is based on an active and dynamic pumping mechanism. Blinking cycle leads to changes in the drainage canaliculi that activate a pump mechanism that drains tears even with the head held in an inverted position. When the palpebral blink mechanism is impaired, tears accumulation leads to spillover to the skin of the lids and cheek [4].

Patton and Robinson [7] studied the influence of tear turnover, instilled solution drainage, and nonproductive absorption on ocular drug bioavailability. The authors concluded that instilled solution drainage was the main contributing factor to the loss of drug from the precorneal area.

Linden and Alm [8] studied the effect of reduced tear drainage on the intraocular penetration of topically applied fluorescein in healthy volunteers. In one eye of each participant both the lower and upper puncta were occluded by the insertion of punctal plugs. A single dose of 20 μL of a 2% solution of sodium fluorescein was instilled in the lower conjunctival sac of both eyes. There was a significant increase in aqueous fluorescein concentrations 1 to 8 h after application of fluorescein in the eye with the occluded puncta, to levels almost four times higher than in the other eye.

24.3.2 Barriers to Drug Permeation

24.3.2.1 Ocular Surface Barriers

The corneal and conjunctival superficial layers form the ocular surface that is in contact with the tear film (Figure 24.1). It is well accepted that one of the main tasks of the ocular surface is to create a defense barrier against penetration from undesired molecules. The ocular surface comprises the exposed interpalpebral surface and the surfaces of the conjunctival fornices. In humans the corneal surface is only 5% of the total ocular surface and the remaining 95% is occupied by the conjunctiva [9]. The cornea is made up of five layers (epithelium, Bowman's layer, stroma, Descemet's membrane, and endothelium), but only the outermost two to three layers of the corneal squamous epithelial cells form a barrier for intercellular drug penetration by intercellular zonulae occludentes (Figure 24.2) [10]. Corneal endothelium plays an important role in maintaining corneal transparency. Its active transport capability allows the maintenance of optimal hydration level of the corneal stroma [10–14]. Although historically termed corneal endothelium, this cell layer is not a true endothelium because it is derived from neuroectodermal origin rather than the mesoderm. Adjacent endothelial cells form a variety of tight junctions, but desmosomes are never seen between normal endothelial cells.

The superficial epithelial layer of the conjunctiva is continuous with the corneal epithelium and stretches over the conjunctival sacs toward the epidermis of the lids. The stratified epithelium varies in thickness from two to four layers at the junction with eyelid margin skin, to six to eight layers at the junction with corneal epithelium [15]. The substantia propria of the conjunctiva is divided into two layers, a superficial lymphoid layer and a deeper fibrous layer. The deeper fibrous layer consists of thick, collagenous, elastic tissue that is rich with vessels [15].

24.3.2.2 Ocular Wall Barriers

The skeleton of the eye globe consists of the rigid scleral collagenous shell that is lined internally by the uveal tract.

The sclera (Figure 24.1) covers the posterior 80% of the eye globe except for a small posterior opening occupied by the optic nerve head (ONH), whereas the rest of the globe is covered anteriorly by the cornea. The scleral stroma is composed of bundles of collagen, fibroblasts, and a moderate amount of ground substance. This tissue is essentially avascular but is lined superficially by the vascular episclera. A large number of channels penetrate the sclera to allow the passage of vessels and nerves to the choroid side. In humans, the scleral thickness is in the range of 0.3 to 1.0 mm with the posterior pole being the thickest [16].

The uveal tract consist of the iris anteriorly, the choroid posteriorly, and the ciliary body in between. The iris (Figure 24.1) is made up of connective tissue and blood vessels, with variable contents of melanocytes and pigment cells that are responsible for its color. The posterior surface of the iris is lined by a monolayer of iridial pigment epithelium (IPE) that is continuous peripherally with the nonpigmented epithelium of the ciliary body. The iris diaphragm subdivides the anterior segment of the eye into the anterior and posterior chambers. The pupil is a round opening in the center of the iris diaphragm that allows the passage of light toward the retina and aqueous humor from the posterior to the anterior chamber (Figure 24.1) and its diameter is changed by the action of the iris sphincter and dilator muscles [17].

The ciliary body is a triangular structure at the periphery of the border between the anterior and the posterior segments of the eye (Figure 24.1). Its uveal portion consists of comparatively large fenestrated capillaries, collagen fibrils, and fibroblasts. The ciliary body is lined by a double layer of epithelial cells, the nonpigmented and the pigmented epithelium with their apices facing each other. The basal side of the inner nonpigmented epithelium is embedded in the aqueous humor of the posterior chamber. The apical borders of the nonpigmented epithelium are sealed with zonulae occludentes that maintain the blood–aqueous barrier [18].

The choroid (Figure 24.1) is a highly vascularized tissue, with one of the highest blood flow rates among body tissues. Beside vessels and intercellular space collagen fibers, choroidal stroma consists of abundant melanocytes as well as variable contents of macrophages, lymphocytes, mast cells, and plasma cells. In the human eye this layer thickness averages 0.25 mm and consists of an innermost layer of fenestrated choriocapillaris, middle medium, and outer large vessels (both nonfenestrated). Bruch's membrane separates the choroid from the retinal pigment epithelium (RPE) and consists of a series of connective tissue sheets, including the basement membranes of the RPE and the choriocapillaris [19].

24.3.2.3 Retinal Barriers

The retina consists of 10 layers: the RPE, photoreceptor outer segments, external limiting membrane (ELM), outer nuclear layer, outer plexiform layer, inner nuclear layer, inner plexiform layer, ganglion cell layer, nerve fiber layer, and internal limiting membrane (ILM). The term neurosensory retina includes all the above except the most external layer, the RPE.

24.3.2.3.1 Retinal Pigment Epithelium

The RPE (Figure 24.1 and Figure 24.3) consists of a monolayer of hexagonal cells of neuroectoderm origin that starts from the ONH posteriorly and extends until it merges with the pigmented epithelium of the ciliary body anteriorly. Adjacent RPE cells are firmly attached by zonulae occludentes that play an important role in maintaining the outer blood–ocular (retinal) barrier (oBOB). The apical side of the RPE has numerous villous processes and faces the subretinal space. The basal side faces the choroid and is attached to a thin basement membrane that forms the inner layer of Bruch's membrane. The cytoplasm of the RPE contains melanosomes, lipofuscin granules, numerous mitochondria, and a large round nucleus. Beside the maintenance of the outer blood–retinal barrier, RPE is responsible for the phagocytosis of the photoreceptor outer segments, absorption of scattered light, heat exchange, metabolism, and transport [20].

24.3.2.3.2 Internal Limiting Membrane of the Retina

The ILM (Figure 24.1 and Figure 24.3) borders the neuroretina from the vitreous (internal) side. Its outer portion consists mostly of the basement membrane of Müller's cells, whereas the inner portion is formed by vitreous fibrils and mucopolysaccharides. This basement membrane covers the entire inner surface of the retina. Anteriorly it is around 50 nm in thickness, and posteriorly it thickens to around 2000 nm. ILM acts as a selective permeation barrier between the intercellular space of the retina and the vitreous [21].

24.3.2.3.3 External Limiting Membrane of the Retina

The ELM (Figure 24.1 and Figure 24.3) is a pseudomembrane formed by the attachment sites of adjacent photoreceptors and Müller's cells. The photoreceptors generally do not make any contact directly with each other. They are nearly always insulated from each other by Müller's cells. At the point of the ELM the plasma membrane of each photoreceptor and Müller's cell

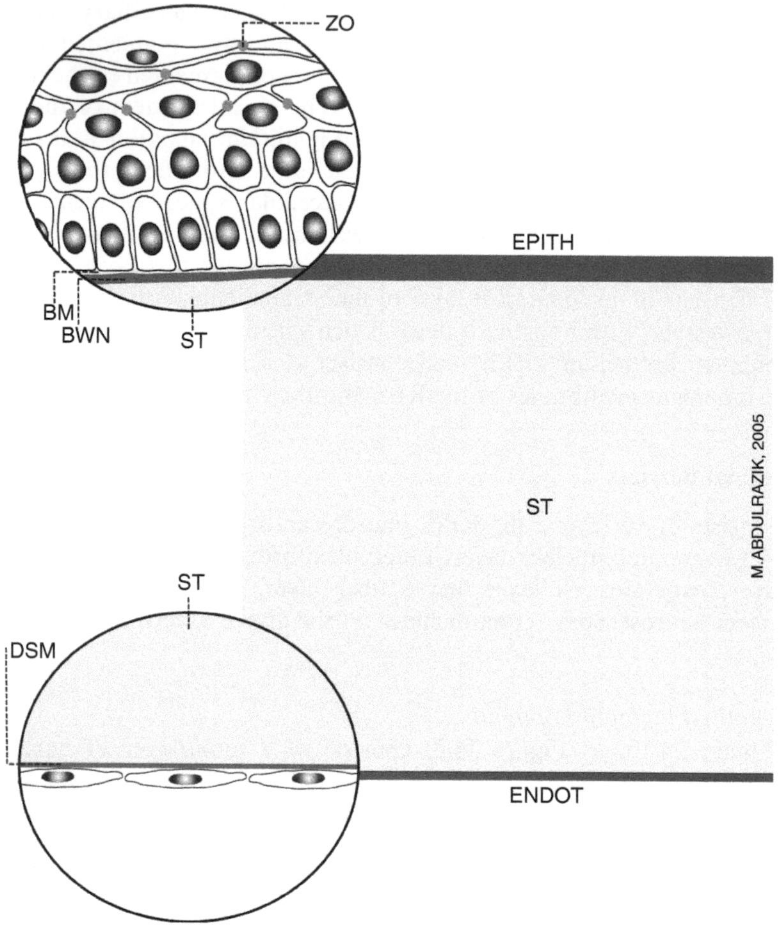

FIGURE 24.2 Cornea. EPITH, epithelium; ENDOT, endothelium; BM, basement membrane; BWN, Bowman's layer; ST, stroma; DSM, Descemet's membrane; ZO, zonulae occludentes.

is differentiated into an attachment band around its circumference. The dense line designated as ELM is a zone of intercellular junctions with a vertical extent of 1 μm. ELM acts as a selective permeation barrier between the intercellular space of the retina and the subretinal space [21].

24.3.2.4 Core of the Eye Parameters

The eye globe is a hollow structure filled anteriorly by the aqueous humor and posteriorly by the vitreous. The aqueous humor (Figure 24.1) is continuously formed by the ciliary body. It is secreted into the posterior chamber from which it passes through the pupil into the anterior chamber and is drained at the anterior chamber periphery. Due to the blood–aqueous barrier formed by zonulae occludentes of the nonpigmentary ciliary epithelium, macromolecules such as proteins can pass to the aqueous in very small quantities, regardless of their plasma concentrations. In humans, the aqueous humor protein level is around 20 mg/100 mL, less than 0.5% of the normal plasma total protein levels (Table 24.1) [22]. In the human eye, the rate of aqueous formation is around 2.5 μL/min, whereas in the case of rabbit eye it is

TABLE 24.1
Aqueous/Plasma Solutes Concentration Ratios in the Rabbit Eye

Substance	Aqueous/Plasma Concentration Ratio
Na^+	0.96
K^+	0.955
Ca^{2+}	0.58
Mg^{2+}	0.78
Cl^-	1.015
HCO_3^-	1.26
H_2CO_3	1.29
Glucose	0.86
Urea	0.87
Total protein	>0.05

Source: From Davson, H., in *Physiology of the eye*, 5th ed., Pergamon Press, New York, 1990, 3–95. Reprinted with permission from Elsevier.

approximately 3 to 4 μL/min [23]. Drugs available in the aqueous humor are subject to an elimination process that is proportional to the aqueous turnover rate [24].

The vitreous (Figure 24.1 and Figure 24.3) occupies 80% of the volume of the human eye, and its volume is approximately 4.0 mL. The vitreous consists of hyaluronic acid, collagen fibrils, and hyalocytes. Even though the water content of the vitreous is around 99%, this tissue has a gel-like structure and its viscosity is approximately twice that of water. The gel-like vitreous can be liquefied in a disease and aging processes. Drugs in the vitreous move by diffusion through the gel when it is formed or by convection when it is liquefied [25].

24.4 DRUG DELIVERY TO BACK OF THE EYE

Site-specific delivery to the back of the eye, including the vitreous, the choroid and particularly the retina, is the most challenging objective facing researchers in the field of therapeutic ophthalmology. There is a growing but unmet need for drug carriers that reach the retina at appropriate therapeutic levels following topical application.

24.4.1 Routes of Drug Permeation to the Posterior Segment

For decades, researchers adhered to a two-route model to describe the penetration of topically applied drugs to the intraocular tissues. In this model the corneal route comprises the penetration through the cornea to the aqueous humor, whereas the conjunctival route comprises the penetration through the conjunctiva to the sclera. Traditionally the importance of the corneal route was overestimated mainly because aqueous humor concentration of a drug was considered to represent the intraocular availability of that drug, and because of the misconception that drug found in the aqueous humor originated solely from drug permeation through the cornea. In fact, many researchers used the term nonproductive absorption to describe most of the conjunctival absorption [7,26,27]. The possibility that drug in the aqueous humor could originate not only from transcorneal penetration, and that drug absorbed by the noncorneal route could reach intraocular tissues by means of other than the aqueous humor of the anterior chamber, was raised two decades ago. Maurice [24] suggested that drugs, once absorbed into the conjunctiva, diffuse laterally into the scleral

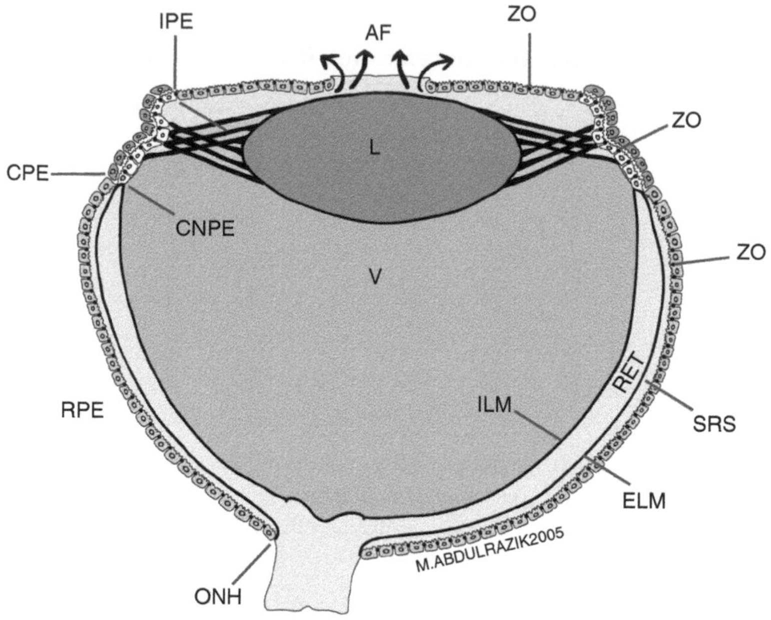

FIGURE 24.3 Barriers of the vitreoretinal compartment. AF, aqueous flow direction; IPE, iridial pigmented epithelium; CPE, ciliary pigmented epithelium; CNPE, ciliary nonpigmented epithelium; RPE, retinal pigment epithelium; ILM, internal limiting membrane of the retina; ELM, external limiting membrane of the retina; ZO, paracellular zonulae occludentes; RET, neurosensory retina; SRS, subretinal space; ONH, optic nerve head; V, vitreous; and L, lens.

border of the cornea and then directly enter the aqueous humor. Ahmed and Patton [28] reported that the noncorneal absorption route may be important for drugs that are poorly absorbed across the cornea due to their physicochemical properties. In this work aqueous humor timolol was found to originate from transcorneal penetration while most of the drug found in the iris and ciliary body was related to conjunctival and scleral penetration. Chien et al. [29] studied the ocular penetration pathways of three alpha 2-adrenergic agents. In contrast to the more lipophilic drugs where the corneal route was the predominant pathway for delivering drugs to the iris, the less lipophilic brimonidine underwent conjunctival and scleral penetration for its delivery into the ciliary body.

24.4.1.1 Routes of Drug Permeation to the Neurosensory Retina

The retina is the ultimate target tissue for pharmacological interventions in modern ophthalmology. The subretinal space and the vitreous border the retina from its outer and the inner sides, respectively (Figure 24.3). Therefore, drugs should pass through either of these two immune privileged and highly protected sites before reaching the retina, except in the instance of drugs administered to the systemic circulation that should cross the wall of retinal vessels to penetrate the retina.

The subretinal space (Figure 24.3) is the space between the RPE apical (internal) side and the ELM. Thus drugs available in the subretinal space should be able to penetrate through the ELM before reaching the neurosensory retina. Likewise, drugs available in the vitreous are challenged by the selective permeability of the ILM before reaching the neuroretina. To reach the subretinal space, topical drugs should penetrate the ocular surface, the corneoscleral collagenous shell, the choroid, and the encircling RPE monolayer (Figure 24.3). To reach the vitreous, drugs should penetrate the ocular surface and exhibit sufficient availability in the aqueous humor of the anterior chamber, or penetrate directly through the conjunctiva–sclera–uvea to the posterior chamber.

The permeation facility from the anterior to the posterior chamber through the narrow and apparently impermeable pupil-lens canal (Figure 24.1 and Figure 24.3) and against the current of the aqueous humor, or through the lens itself, is debatable [3]. An alternative is to permeate from the anterior chamber through the stroma of the ciliary body or iris. However, the latter scenario necessitates the ability to permeate through the ciliary nonpigmented epithelium (CNPE) or the iridial pigmented epithelium (IPE), both are monolayers sealed with zonulae occludentes, before reaching the posterior chamber (Figure 24.3).

24.4.2 Obstacles for the Permeation of Topically Applied Therapeutics

Cell membrane has a lipophilic character that functions as a selective barrier denying free diffusion from hydrophilic compounds. The encircling of cell contents by a cell membrane is an important evolutional milestone aimed at protecting cell contents from external environment and allowing the creation of an autonomic biological microsystems to carry out undisturbed precise biological reactions. The evolutional rationale of protecting biological microsystems by a lipophilic membrane might be that the main threat to the normal function of such systems comes from bioactive molecules like enzymes, peptides, and oligonucleotides rather than from lipophilic compounds. However, evidence has emerged in recent decades showing the existence of defense mechanisms against lipophilic molecules as well. Several studies have reported the expression of ATP-dependent drug extrusion pumps (multidrug-resistance protein 1; p-glycoprotein) in the surface epithelium of the cornea and conjunctiva, which are responsible for the extrusion of some lipophilic drugs from the surface epithelium back to the tear film [30–32].

24.4.2.1 Surface Epithelium

The superficial two to three cell layers of the corneal and conjunctival epithelium are the main barrier for the permeation of topically applied compounds. In this rate-limiting cell layer, the transcellular permeation is dictated by the lipophilicity of the cell membrane whereas the paracellular permeation is limited by the paracellular pore size and density. Vesicular penetration (e.g., receptor- or endocytosis-mediated) of macromolecules across surface epithelium is possible [33]. However, the proposed mechanism is energy consuming (e.g., incorporation into pinocytotic vesicles and phagosomes) and thus more feasible in cell lines with abundant intracellular energy sources like corneal endothelium and RPE [34–37].

Hamalainen et al. [38] reported that paracellular pores are fewer (1/16) and their size is smaller (1/2) in the cornea as compared to the conjunctiva. The total paracellular space in the conjunctiva was estimated to be 230 times greater than that in the cornea. Indeed, the conjunctiva was reported to be 15 to 25 times more permeable to polyethylene glycols (PEGs) than the cornea. In the cornea the reported paracellular pore density was $4.3 \times 10^6/cm^2$ with an estimated pore diameter of 2.0 nm, corresponding to a permeation facility for molecules with a Stokes' radius of 1.0 nm. The paracellular conjunctival epithelial pores measured approximately 3.0 nm in the bulbar conjunctiva and 4.9 nm in the palpebral conjunctiva, allowing the potential permeation of molecules with 1.5 and 2.5 nm Stokes' radius, respectively. Therefore, theoretically peptides and oligonucleotides with molecular weights (MW) of 5 kDa can permeate through the bulbar and palpebral conjunctiva whereas those with molecular weights of 10 kDa can permeate only through the palpebral conjunctiva. However, experimentally there was no significant difference in the permeability of palpebral and bulbar conjunctiva to PEGs [38]. Sasaki et al. [39] reported that *ex vivo* ocular surface permeability for topically applied hydrophilic macromolecules (4 and 10 kDa dextrans) was markedly low, although the permeability for the conjunctiva was much higher compared to the cornea.

24.4.2.2 The Corneoscleral Collagenous Shell

24.4.2.2.1 Obstacles for Permeation through the Cornea

After permeation through the corneal epithelium, molecules should then permeate through the hydrophilic corneal stroma and the corneal endothelium before reaching the aqueous humor. Thus, while the corneal epithelium favors the penetration of lipophilic compounds, the hydrophilic stroma favors the permeation of hydrophilic compounds. The corneal endothelium monolayer is not sealed by paracellular zonulae occludentes. Consequently this layer has only a minimal effect on the overall permeability of the cornea. Overall, the cornea as a whole is tailored to allow the permeation of moderately lipophilic compounds (close to log $P=2$) [40]. On the contrary, the reported optimal lipophilicity for compounds penetrating through an isolated epithelial layer of the cornea was around log $P=4$ [41].

24.4.2.2.2 Obstacles for Permeation through the Sclera

The permeability of the sclera is comparable to that of corneal stroma. Published reports have shown that bovine, human, and rabbit sclera are permeable even to high molecular weight compounds [42,43].

Ambati et al. [44] measured the permeability of rabbit sclera to a series of fluorescein-labeled hydrophilic compounds and found that scleral permeability decreased with increasing molecular weight and molecular radius but was quite permeable to large molecules, such as IgG (145 kDa). Thus when targeting the retina by the topical or subconjunctival routes the sclera is not a rate-limiting barrier and the main barriers are beyond this tissue at the choroid, RPE, and ELM.

24.4.2.3 Role of Ocular Tissue Vessels

The role of ocular tissue vessels in delivering or diverting topically applied drugs to or away from the target tissue is debatable. Chiang and Schoenwald [45] have shown that following topical instillation of clonidine, peak concentrations were reached in the iris and ciliary body at 2 min or at the first time interval that can be measured, whereas the peak concentration time for the aqueous humor was 15 min. These results imply that drug penetration to the ciliary body and iris is mediated by a fast delivery route that is probably a vascular one. Schoenwald et al. [46] studied the uptake and distribution of drugs by vessels in the rabbit eye following topical instillation. Following entry of drug into the conjunctival and scleral tissue a significant portion of the drug entered scleral vessels, as evidenced by the confocal laser scanning microscopy, and reached the ciliary body. However, another study showed that topically applied drugs, with assumable penetration into the conjunctival vasculature, were diverted from ocular tissues to the systemic circulation [47]. Flurbiprofen, a nonsteroidal anti-inflammatory agent, which is not ocularly metabolized, was used as a probe to show the proportion of topically applied drug lost to the systemic circulation following uptake by ocular and local vascular beds. The systemic bioavailability of flurbiprofen following the administration of topical ocular dose of 225 μg was 74% [48].

24.4.3 The Vitreoretinal Compartment

The neurosensory retina is an immune privileged tissue situated in a highly protected compartment with multiple defense lines (Figure 24.3). Like brain neurons, the retinal neurons need strictly controlled homeostasis for their proper function. Selective barriers separate, therefore, the retina from the adjacent tissues and the blood circulation. The vitreoretinal compartment occupies most of the intraocular volume and is encircled almost 360° by a series of epithelium monolayers with paracellular zonulae occludentes. This encircling barrier includes the RPE posteriorly, the iris posterior pigmented epithelium (IPE) anteriorly, and the nonpigmented epithelium of the ciliary nonpigmented epithelium (CNPE) body in between (Figure 24.3). The pupil and the ONH are the only gaps in this defensive encircling. However, drug diffusion through the pupil is very low due to the narrow and apparently impermeable iris-lens canal, the counterflow of the aqueous humor from the posterior to anterior chamber, the relatively impermeable lens, and the dense zonular diaphragm that connects the lens with the ciliary processes of the ciliary body (Figure 24.3). Furthermore, the bulk of the vitreous is a dilutional and to some extent metabolic obstacle that faces drugs on their way from the anterior vitreous border to the posterior retina.

At the inner border of ONH, the ILM becomes continuous with the basement membrane of fibrous astrocytes lining the internal surface of the ONH [21]. However, the lateral borders between the ONH and the adjacent choroid and retina are not well defined. Furthermore, it was reported [49] that microvessels in the prelaminar region of the ONH lack classical blood–brain barrier characteristics and display nonspecific permeability, possibly mediated by vesicular transport. Thus, there is a theoretical possibility that topically applied drugs can penetrate indirectly through the retrobulbar space and then, through the ONH, reach the posterior choroid and retina. It was reported that following retrobulbar administration of fluorescein, the dye rapidly accumulated in the ONH and penetrated later to the vitreous [50].

The encircling epithelial monolayer with paracellular zonulae occludentes provides a selective barrier separating the whole vitreoretinal compartment from the rest of the eye (Figure 24.1 and Figure 24.3). Therefore, only molecules that can permeate transcellularly, paracellularly, or by vesicular (i.e., receptor- or endocytosis-mediated) penetration through this encircling barrier line can reach the retina. However, the neurosensory retina is further protected by another two selective barriers, i.e., the ILM from the vitreous side and the ELM from the RPE side (Figure 24.1 and Figure 24.3).

24.4.3.1 Barriers of the Vitreoretinal Compartment

24.4.3.1.1 RPE, CNPE, and IPE

Pitkanen et al. [51] reported the isolated bovine RPE–choroid was up to 20 times more permeable to lipophilic than hydrophilic beta-blockers. Furthermore, the *in vitro* permeability of bovine RPE–choroid to hydrophilic compounds and macromolecules was 10 to 100 times less compared to sclera, whereas the permeability for lipophilic molecules was in the same range for both tissues. The isolated bovine RPE–choroid also exhibited differential permeation by molecular weight and Stokes radius. The permeation rate of 4, 10, and 20 kDa FITC dextrans was moderate compared to a good permeation rate for the 376 Da carboxyfluorescein and a poor penetration rate for 40 and 80 kDa FITC-dextrans. The permeability to carboxyfluorescein was 35 times more than to 80 kDa FITC-dextran [51]. In a study on the permeability of the human ciliary epithelium to a horseradish peroxidase, Tonjum and Pedersen [52] reported that ciliary and iridial epithelium contained a system of paracellular zonulae occludentes. Peroxidase was applied on the stromal side of ciliary body and iris specimens obtained from freshly enucleated eyes. The 40 kDa peroxidase was blocked apically in the lateral intercellular spaces of the CNPE whereas in the iris the progression of peroxidase was blocked apically in the lateral intercellular spaces of the IPE. Freddo [53] studied the intercellular junctions in the posterior IPE cells of the rhesus monkey by electron microscopy, freeze-fracture, and horseradish peroxidase. Intravenously injected horseradish peroxidase, which diffused from the ciliary body stroma, was prevented from reaching the posterior chamber by the presence of the zonulae occludentes between adjacent posterior IPE cells. The author concluded that these apicolateral zonulae occludentes are analogous in both location and degree of permeability to the zonulae occludentes present in the CNPE. In another study the functional barrier for macromolecules in IPE and RPE cells was reported to be similar in an *in vitro* setting [54].

24.4.3.1.2 ILM and ELM

According to published reports, linear molecules (e.g., dextran) larger than 40 kDa (corresponding to Stokes' radius of 45 A) and globular molecules (e.g., albumin) larger than 70 kDa (corresponding to Stokes' radius of 65 A) cannot diffuse across the ILM of the retina [55–58]. Bunt-Milam et al. [59] studied the pore size of zonulae adherentes in the ELM of the rabbit retina, by exposing the ELM side of isolated rabbit retinas to protein probes of known Stokes' radius. Proteins with a Stokes' radius of 30 A or less diffused freely through the neurosensory retina whereas those with a Stokes' radius greater than 36 A were blocked at the ELM. The authors concluded that the pore radius of the zonulae adherentes of the ELM ranges from 30 to 36 A, which is equivalent to 60% to 80% of the pore size in ILM.

24.4.3.1.3 Blood–Retina Barriers

Maurice [3] raised the possibility that drug levels detected in the vitreous and retina following topical application may represent indirect delivery through the systemic circulation. This hypothesis was based on doubts whether the vitreoretinal compartment could be reached by a drug dissolved in a droplet located in the conjunctival sac. The main arguments in support of this hypothesis are the large dilutional effect of tissues separating the drug source from the target tissues, and the possibility that a significant fraction of the applied dose could be absorbed by the systemic circulation through ocular and local vascular beds. However, blood-derived compounds approaching the retina from the fenestrated choroidal vasculature face the oBOB formed by the RPE monolayer, whereas blood-derived compounds reaching the retina through the retinal vessels face the obstacle of the inner blood–ocular barrier (iBOB) formed by the CNS-like retinal vasculature. The retinal blood vessels are analogous to the cerebral blood vessels, expressing blood–brain barrier-like properties. Stewart and Tuor [60] have reported the similarity of retinal and brain vasculature permeability in the rat. Compared to the brain vessels the more porous endothelium of the retinal vessels is compensated by a fourfold denser pericytes layer. This iBOB consists of a single layer of nonfenestrated endothelial cells with tight junctions. The endothelial layer is covered by a basal lamina with an interrupted layer of pericytes embedded in the surrounding basement membrane matrix. In humans, retinal vessels are impermeable to tracer substances such as fluorescein and horseradish peroxidase (40 kDa) [21].

24.5 ENHANCEMENT IN TOPICAL OCULAR DRUG DELIVERY

24.5.1 Polymeric Drug Delivery Systems

24.5.1.1 Introduction

24.5.1.1.1 Biodegradable Polymers as Drug Delivery Systems

Biodegradable polymers as DDS have been studied extensively over the last few decades. These polymers provide the advantage of being degraded and eliminated from the body thus avoiding the risk of toxic accumulation or the need for intervention to eliminate them.

Polylactic acid (PLA) and polyglycolic acid (PGA) and their copolymer polylactic-*co*-glycolic acid (PLGA) are among the most widely used biodegradable polymers. They are approved for human use by worldwide health authorities and are degraded into nontoxic compounds (lactic acid and glycolic acid, respectively) following hydrolysis and enzymatic cleavage. These polymers undergo bulk erosion and drug diffusion may change according to the erosion rate of the polymer matrix. Thus, drug burst phenomena are likely to occur depending on the MW and chemical structure of the polymers. PGA for example is not an appropriate candidate for prolonged controlled DDS as it is highly sensitive to hydrolysis. In contrast, other biodegradable polymers like polyorthoester and polyanhydride undergo surface erosion, and subsequently the drug release from such systems depends on the extent of the surface area, which is more controllable. Biodegradable polymeric systems were used as colloidal systems and inserts, and also as an ophthalmic DDS for subconjunctival, scleral, intracameral, or intravitreal application. However, this chapter will focus on noninvasive DDS.

24.5.1.1.2 Nonbiodegradable Polymeric DDS

Nonbiodegradable polymers allow the design of long-term sustained release DDS but carry the disadvantage of subsequent intervention to remove the DDS at the end of its functioning life. The virtually unlimited combinations that can be made from permeable (polyvinyl alcohol [PVA]) and impermeable polymers (ethylene vinyl acetate [EVA] and silicon laminate) allow the development of sophisticated DDS that can overcome the burst phenomena (mainly early burst), which disturb the planned zero-order release from microspheres and microcapsules made of biodegradable polymers. The PVA–EVA systems were used as implantable DDS for subconjunctival, scleral, or intraocular applications. A review of the use of contact lenses (CL) as noninvasive nonbiodegradable ocular DDS will be described below.

24.5.1.2 Nanoparticles

In general, particulate DDS can offer more possibilities for controlled drug release. Furthermore, particulate systems are basically more stable than other colloidal systems, and as they can be freeze-dried for long-term storage, extended shelf life is feasible. Particulate carriers meet the basic needs of advanced ocular drug carriers, being nontoxic, nonimmunogenic, biocompatible, uniform, and biodegradable in a predictable pace. In addition, they have the ability to provide protection for the delivered molecules while interacting with the ocular surface. Extensive research efforts are underway in the development of ophthalmic particulate delivery systems. The advantage of submicron particulate systems for ocular delivery was addressed by several studies. The uptake of a PLGA particulate carrier in Caco-2 cell culture line was shown to be time-, size-, and concentration-dependent [61]. Internalization of particulate carriers by corneal cells has been evidenced *in vitro* for nanoparticles but not for microparticles. In the *in vivo* setting conjunctival cells, but not corneal, have been shown to internalize nanoparticles after topical instillation [62–64].

Qaddoumi et al. [65] studied the uptake of PLGA nanoparticles in rabbit conjunctival epithelial cell culture. The highest uptake by cultured conjunctival cells was achieved for the smallest particles (100 nm), compared to larger 800 nm and 10 μm particles. A study of the fate of the tiny 100-nm particles following 2 h of cultured cells exposure to a 0.5 mg/mL dose showed that 6% was internalized by conjunctival epithelial cells, 1.5% was surface-bound, whereas the remainder of the dose was found in the donor medium. In an *in vivo* rabbit eye study [66] on the uptake of poly(hexyl cyanoacrylate) nanoparticles, 6 h postinstillation into the conjunctival sac, it was found that the fraction that was internalized by conjunctival epithelial cells was only 1% of the dose reflecting *in vivo* precorneal elimination factors neglected by the *ex vivo* setting of Qaddoumi et al. The adhesive properties of these polymeric systems contributed to the enhancement of corneal penetration of the drug as a result of the increased residence time of the drug in the precorneal environment [67]. Furthermore, surface modifications of nanoparticles with a sterically stabilizing layer can modulate their *in vivo* biodistribution and lower their tendency to aggregate with biomolecules. Longer residence times on the ocular surface can be achieved if the nanoparticles are coated with a mucoadhesive or charged polymer. In general, there is a drive for target-specific surface modifications based on the notion that both noncoated and pegylated particulate systems are nonspecific in their interaction with cells and macromolecules. Giannavola et al. [68] have made acyclovir-loaded nanospheres from poly-D,L-lactic acid (PLA) by a single step nanoprecipitation of the polymer. The effect on morphological changes was studied by the surface incorporation of pegylated 1,2-distearoyl-3-phosphatidylethanol amine. Study on rabbit eyes showed the ocular tolerability of PLA nanospheres. PLA nanosphere colloidal suspensions containing acyclovir provided a marked sustained drug release in the aqueous humor and significantly higher levels of acyclovir compared to the free drug formulation. When loaded in PLA nanospheres, acyclovir aqueous humor area under the curve (AUC) was seven times higher compared to the free drug formulation AUC, whereas the pegylation of the nanospheres almost doubled this augmentation [68].

De Campos et al. [69] have investigated the mechanism of interaction between nanoparticles of the cationic polysaccharide chitosan and the ocular surface based on the assumption that chitosan nanoparticles would increase the residence time of drugs in the precorneal area owing to their adhesive and electrostatic properties and enhance the penetration of drugs into the intraocular structures. Following instillation of fluorescein-loaded chitosan nanoparticles, fluorescein and chitosan solution or control solution of fluorescein, the ocular surface fluorescein concentrations were significantly higher for the nanoparticles compared to the chitosan solution at all times assayed with the exception of time 1 h for the cornea and time 1 to 2 h for the conjunctiva, suggesting that the interaction of chitosan with the ocular surface is more persistent when it is in a nanoparticulate form. Furthermore, the nanoparticulate form sustained the release of fluorescein for up to 24 h compared to shorter and gradual release for chitosan–fluorescein and control fluorescein solutions. Cell culture investigation results confirmed the nontoxicity of chitosan on ocular surface [69]. In another study [70], the nanoparticulate entrapment of CsA was achieved by ionic gelation of chitosan and comprised the premature dissolution of the CsA hydrophobic peptide in an acetonitrile–water mixture (1:1) before its incorporation into the chitosan solution, followed by the addition of the sodium tripolyphosphate solution. In an *in vivo* experiment on rabbit eyes, five applications at 10-min intervals of 10 μL of different formulations (chitosan nanoparticles, chitosan solution, and aqueous solution), each containing 320 μg/mL of CsA and [^{3}H]-CsA, were placed in the conjunctival sac of both eyes of animals. Significantly higher CsA levels were achieved and sustained up to 24 h in ocular surface tissues of the chitosan nanoparticulated CsA-treated group, with low levels in the iris and ciliary body samples. The authors related these results to the electrostatic attraction between the negatively charged ocular surface and the positively charged chitosan. Calvo et al. [71] reported that chitosan-coated poly(ε-caprolactone) (PECL) nanocapsules significantly increased the ocular bioavailability of indomethacin compared to uncoated and poly-L-lysine (PLL)-coated poly(ε-caprolactone) nanocapsules, even though the chitosan and the PLL coat exhibit a similar positive surface charge. It is apparent from these results that both the positive charge and the mucoadhesive properties of the chitosan play a role in the enhanced bioavailability of the drug incorporated into chitosan-coated nanocapsules. The possibility that chitosan can modify the barrier function of the ocular surface was not addressed by the aforementioned studies.

24.5.1.3 Nanocapsules

Nanocapsules are novel colloid systems with an oil core and polymeric coating. Polyalkylcyanoacrylate-based biodegradable submicroscopic colloidal nanocapsules prepared by polymerization of alkylcyanoacrylate have been investigated for the targeted delivery of drugs. Following the pioneer work of Couvreur et al. [72,73] these systems were further developed and shown to control the release of drugs [74–78], to be bioadhesive on the eye tissues, and to improve ocular bioavailability [66,79–82]. Indeed, nanocapsules present an advantage over nanoparticles as carriers for lipophilic drugs, as they can incorporate within their core oil phase marked amounts of lipophilic drugs [83,84]. Calvo et al. [85] have evaluated nanocapsules consisting of an oil core (Migliol 840) surrounded by a poly (ε-caprolactone) coat, as potential vehicles for the topical ocular administration of CsA. The investigated nanocapsules had a mean diameter size in the range of 210 to 270 nm, a negative zeta potential, and a maximum CsA loading capacity of 50%. The corneal levels of CsA achieved by the encapsulated CsA formulation were up to five times higher than for the control oil solution of CsA. Furthermore, the encapsulated CsA formulation sustained the release of CsA for up to 3 days, as reflected by the AUC values that were significantly higher than those of the control oil solution group. However, the corneal CsA levels in the encapsulated CsA group declined rapidly from C_{max} of 14 μg CsA per g of cornea at 2 h to less than 10 μg/g in the next time point. In certain circumstances, nanocapsules can be superior also as carriers of hydrophilic drugs. Ashton et al. [86] have shown that raising the pH of ophthalmic formulations from 7.4 to 8.4 favored a shift toward corneal rather than conjunctival penetration of some drugs, partly by increasing the fraction of the nonionized drug. This mechanism allows penetration enhancement of hydrophilic drugs through the hydrophobic corneal epithelium by introducing hydrophilic molecules in a nonionized form. The oil core of the nanocapsules offers an additional advantage in this regard since it can protect an incorporated nonionized hydrophilic drug from being transformed into an ionized form in the hostile precorneal environment. Marchal-Heussler et al. [87] incorporated carteolol in nanoparticles and nanocapsules in an attempt to improve the bioavailability and to reduce the systemic side effects compared to the commercial product. Carteolol hydrochloride is a hydrophilic nonselective beta-1 and beta-2 adrenergic antagonist, approved in the United States since 1992 as antiglaucoma agent. The commercial product Ocupress/Carteolol is available as 1% carteolol hydrochloride solution with 0.005% benzalkonium chloride. In this study, carteolol 0.2% to 1%, in its ionized and nonionized forms, with 0.005% benzalkonium chloride was incorporated into poly(ε-caprolactone) nanoparticles and nanocapsules. The hydrophilic carteolol hydrochloride was incorporated into the nanoparticles during their preparation process and achieved a 41% level of entrapped drug. When 0.2% of nonionized carteolol was incorporated in nanoparticles or nanocapsules, both formulations achieved more intraocular pressure (IOP) reduction than 1% carteolol hydrochloride aqueous solution, but the IOP reduction compared to the aqueous solution was statistically significant only with the nanocapsule formulation. The authors assumed that the release of carteolol from the oil core of the nanocapsules was easier than from the internal layers of the nanoparticles, and that the more lipophilic nonionized-carteolol can penetrate the hydrophobic corneal epithelium and could be transformed in the hydrophilic corneal stroma to an ionized form. Collectively, these studies have shown the prolongation of the precorneal elimination half-life that accompanied the evolution of the particulate colloidal systems, from around 1 to 3 min that was achieved by aqueous delivery systems to the range of 15 to 20 min for the particulate systems. Based on advances in other fields of drug delivery, it was hypothesized that further progress might be possible if these colloidal carrier systems could be incorporated into polymeric gel formulations. Le Bourlais et al. [88] evaluated the *ex vivo* absorption of CsA in bovine cornea.

Polyacrylic acid polymeric gels, polyisobutylcyanoacrylate (PIBCA) nanocapsules, and their combination, loaded with 1% CsA were compared with a CsA olive oil control formulation. After 24 h contact with the different formulations the highest CsA corneal concentrations were found for CsA loaded in a gel–nanocapsule combination. However, these study results must be viewed in light of the inherent shortcomings of a static *ex vivo* design. Desai and Blanchard [89] studied the additive effect of Pluronic F127-based gel delivery system (PF127MC) on the ability of a PIBCA nanocapsule system to prolong the release of pilocarpine using the miotic response in albino rabbit eyes as an *in vivo* model. A prolonged plateau in the miosis was seen with a PF127MC gel formulation containing a dispersion of the pilocarpine nanocapsules, compared with the 1% pilocarpine-loaded PIBCA dispersion alone, or a PF127MC gel formulation of 1% pilocarpine. The authors concluded that the combination of gel and nanocapsules provided an additive effect of both the viscous environment of the studied gel system (PF127MC) and the mucoadhesive ability of PIBCA nanocapsular system, allowing for longer contact time of pilocarpine-loaded nanocapsules with the ocular surface, thereby improving ocular bioavailability of pilocarpine. As in other colloidal carriers, liposomes and emulsions, surface modification could play a role in the biofate of particulate systems. Relying on this assumption, the interactions of three types of nanocapsules that differ in their surface properties, with the ocular surface of rabbit eyes were investigated [90]. PECL nanocapsules, chitosan-coated PECL nanocapsules, and poly(ethylene glycol) (PEG)-coated PECL nanocapsules were loaded with a fluorescent dye (rhodamine) to quantify and visualize their *ex vivo* and *in vivo* interaction with the ocular surface. All systems were able to enter the corneal epithelium by a transcellular pathway, as evidenced by the confocal laser scanning microscopy, but the depth of penetration differed according to the surface properties of the preparation. Whereas chitosan coating led to nanocapsule retention in the superficial epithelial layers, the PEG coating allowed deeper penetration of the nanocapsules along the whole epithelium.

24.5.1.4 Micelles

Amphiphilic molecules (surfactants) can assemble into nanoscopic supramolecular structures with a hydrophobic core and a hydrophilic shell micellar arrangement. As surfactant concentration is increased in aqueous solutions, the separated molecules aggregate into micelles upon reaching a concentration interval known as the critical micellar concentration (CMC).

In micellar systems, nonpolar molecules are solubilized within the internal micelle hydrophobic core, polar molecules are adsorbed on the micelle surface and substances with intermediate polarity are distributed along surfactant molecules in intermediate positions.

Micellar ocular DDS should be based on nontoxic and nonirritant materials and should be stable enough to achieve a reasonable shelf life. Attention should be paid to the CMC of detergents used for micellar assembly, since high CMC often renders them toxic and irritant to ocular tissues.

The nonionic triblock copolymer polyethylene oxide–polypropylene oxide–polyethylene oxide (PEO–PPO–PEO) has been widely used in medicine and has shown low toxicity. Wanka et al. [91] studied aggregation behavior of PEO–PPO–PEO polymeric micelles. In this study, the hydrophobic core of this micellar system consisted of dehydrated poly(oxypropylene) groups which were surrounded by an outer shell of hydrated poly(oxyethylene) groups. The feasibility of using PEO–PPO–PEO micelles as a topical ocular carrier for gene delivery was addressed in an *in vivo* study on nude mice and albino rabbits [92]. Each animal eye was treated with a topical application (10 μL for mouse and 50 μL for rabbit) of 0.08 mg/mL of plasmid and 0.3% (w/v) PEO–PPO–PEO polymeric micelles. After 2 days of three times per day topical delivery, the reporter expression was detected in the treated eyes around the iris, sclera, conjunctiva, and lateral rectus muscle of rabbit eyes and also in the intraocular tissues of nude mice. The potential of a micellar carrier for topical ocular delivery using pilocarpine as a model drug was evaluated in another study [93]. Micellar solution of pilocarpine for topical ocular delivery was prepared by a simple method of drug dissolution within an aqueous solution of a surface-active high molecular weight triblock copolymer, Pluronic F127. For this purpose, an aqueous solution of Pluronic F127 was prepared in concentrations above the CMC, where the copolymer is supposed to form micelles. The *in vivo* performance of this micellar system in the eye was examined by determining the miotic response in rabbits to a single instillation of 25 μL of micellar solution, with or without drug, to the lower conjunctival sac. The miotic effect of the micellar solution was enhanced by 20% and 50% in terms of t_{max} value and 10.2% and 64% in terms of AUC, for pilocarpine hydrochloride and pilocarpine base, respectively, when compared to the aqueous solution of the drug. These results suggest a better ocular bioavailability of the nonionized pilocarpine base incorporated in the hydrophobic core of the micelles. In another *in vivo* experiment in rabbits, the potential of polyoxyl stearate, a nonionic surfactant, as a micellar carrier for CsA delivery to the eye was studied [94]. CsA levels in the cornea following a single 50 μL topical administration were 60-fold higher for the 0.1% CsA suspension of polyoxyl 40 stearate micelles compared to 0.1% CsA castor oil solution.

24.5.1.5 Dendrimers

Dendrimers is a name used for polyamidoamine (PAMAM, Starburst (Dendritch, Inc., Midland, MI, USA)), polypropyleneimine–diaminobutane (DAB), polypropyleneimine–diaminoethane (DAE), polyethyleneoxide–carbosilane (CSi-PEO), polyether, and similar macromolecules with a highly branched globular nanometric structure that comprises branches radiating from a central core. Dendrimers vary in the flexibility of branches and type of peripheral functional groups. Branches can terminate at charged and uncharged amino, carboxyl, or hydroxyl groups. Most of the described dendrimers are liquid or semi-solid polymers that are not soluble in aqueous solutions and their *in vivo* toxicity remains a concern. However, innovative synthetic chemistry has allowed the formation of dendrimers tailored for drug delivery. Dendrimers are classified and designated according to the surface-terminating group quality and quantity. For amine and sodium carboxylate-terminating groups, the designation is PAMPAM; for amine and carboxylic acid groups, DAB; for amine groups, PEA; for polyethyleneoxide, CSi-PEO; and finally, for carboxylate and malonate groups, polyether.

According to the classification proposed by Malik et al. [95], full generations (G1, 2, 3, etc.) describe amine-terminating groups (except for polyether G0 or G2 where it refers to carboxylate and malonate groups, and for PAMAM G2(OH) and G4(OH) where it refers to hydroxyl group), whereas the half generations (G1.5, 2.5, 3.5, etc.) describe carboxylic acid or sodium carboxylate-terminating groups. The number of surface-terminating groups, and consequently the molecular weight of the macromolecule, rises with higher generations. Improved loading capacity of dendrimer was reported by the process of activation [96], a term used to describe a process comprising a heat treatment in a solvent that manipulates the structure of dendrimers where some branches are trimmed away, while keeping the general topology and size, resulting in more flexible molecules. The reduced structural density in the intramolecular core allows higher levels of drug loading and enhanced penetration capabilities.

Vandamme and Brobeck [97] studied the residence time as well as the miotic and mydriatic effect of different generations of fluorescein, pilocarpine nitrate, and tropicamide-loaded PAMAM, respectively. A single dose of 25 μL of drug-loaded dendrimer solution was applied each time on the conjunctival sac of albino rabbit eye. PAMAM dendrimeric solutions achieved pH, osmolality, and viscosity compatible with ocular dosage form formulations. All dendrimeric formulations showed prolonged residence time (>100 min) compared to phosphate buffer or hydroxypropylmethylcellulose solutions (<30 min). Among PAMAM dendrimers, residence time was longer for the solutions containing dendrimers with carboxylic (220–300 min) and hydroxyl (300 min) surface-terminating groups. Only the solution of polyacrylic acid (Carbopol 980 NF), a bioadhesive polymer, achieved residence time similar to the leading dendrimeric solutions (270 ± 30 min), but this solution was less tolerated on ocular surface. When sorted by the miotic and mydriatic effect, the best bioavailability was demonstrated by dendrimers with carboxylic (G1.5) and hydroxyl (G4.0-OH) surface-terminating groups. The authors assumed that the bioadhesive properties of dendrimers are related to their capacity to establish electrostatic and hydrophobic interactions and hydrogen bonds with the underlying ocular surface that will lead to a structure with a more rigid behavior that traps the instilled solution. Hudde et al. [98] studied the efficiency of activated PAMAM dendrimers to transfect rabbit and human corneas in *ex vivo* culture. The highest levels of marker gene expression were obtained with 36 μg of dendrimers mixed with 2 μg of plasmid DNA per quarter rabbit cornea on day 3 of incubation (18:1 w/w dendrimer/DNA ratio). Low efficiency was achieved with ratios below 14:1, whereas above 18:1 ratio, the efficiency remains high. Under optimized conditions 6% to 10% of the rabbit, and about 2% of the human, corneal endothelial cells expressed β-galactosidase following transfection with dendrimer–DNA complex. The authors concluded that the *ex vivo* transfection of endothelial cells of corneal donor tissue by DNA-loaded dendrimers is feasible and reproducible.

24.5.1.6 Solid Precorneal Inserts

24.5.1.6.1 Collagen Shields

Collagen shields were first introduced by Fyodorov in 1984 for use as a bandage contact lens following radial keratotomy and photorefractive surgery [99]. Collagen shields are manufactured from porcine scleral tissue and commercially available (bio-Cor, Bausch & Lomb) with three dissolution times of 12, 24, and 72 h, depending on the level of collagen cross-linking induced during the manufacture process. Hydrophilic drugs are entrapped within the collagen matrix when the dry shield is soaked in aqueous solution of the drug whereas water-insoluble drugs are incorporated into the shield during the manufacturing process.

The commercially available shields consist of a contact-lens-shaped dry product, which can absorb fluid up to 60% water content when soaked in aqueous solution, and is biodegradable in a time frame according to the programmed dissolution time. When compared with intensive topical treatment, collagen shields have been found superior with regard to the delivery of different antibiotics and antifungal agents in the rabbit model [100–102]. In experimental bacterial keratitis in animal models, the enhanced drug delivery ability of collagen shield was translated to enhanced bacterial eradication [103–107]. Improved results were reported also for the delivery of anti-inflammatory agents by collagen shields [108–111]. However, the complexity of the manufacturing process and the resulting blurred vision are serious drawbacks that have curbed the enthusiasm raised during the development of corneal shields. Furthermore, corneal shield self-administration is difficult for the average user and their positioning should be monitored since they can be easily dislocated.

24.5.1.6.2 Contact Lenses

Hydrogel contact lenses are better tolerated on ocular surface than collagen shields. Non-optical hydrogel contact lenses have been used for a few decades as bandage CL to promote ocular surface healing. This nonoptical indication was later extended when drug presoaked

CL began to be used as therapeutic CL. This approach was advocated as a way to overcome the washout effect of the tear film turnover and has resulted in prolonged drug retention time on ocular surface. Furthermore, therapeutic CL can act as a precorneal drug reservoir, allowing for higher drug concentrations to be in contact with ocular surface [112–115]. The main drawback of a contact lens as a therapeutic lens is the low drug-loading capacity, which is not sufficient to build up a therapeutic concentration in the eye for most drugs [116]. For hydrogel contact lenses, evidence has shown that the drug-loading capacity can deliver a drug amount, which is equivalent to only a small fraction of the dose that can be delivered by topical drug instillation [117]. Indeed, this DDS is barely represented in the modern array of ocular DDS [118].

Ongoing attempts are made to improve the drug-loading capacity of contact lenses. Manipulation of the monomer composition and the cross-linking pattern can alter the permeability and the drug loading capacity of the polymeric backbone. Molecular-imprinted polymers are prepared by the self-assembly of functional monomers and target molecules, followed by polymerization with a cross-linker in an inert solvent. The cavities that remain following target molecules removal can recognize the spatial features and bonding preferences of the target molecules, consequently improving the loading capacity of the polymer for such molecules. Hiratani et al. [119] used molecular imprinting techniques to produce soft contact lenses with a higher timolol loading capacity than simple hydrogel contact lenses presoaked with timolol. In an *in vivo* study on rabbits, the imprinted contact lens was able to sustain timolol release in the precorneal area and elicited greater tear fluid levels with a significantly lower dose.

24.5.1.7 Particle-Laden Solid Inserts

Grammer et al. [120] studied collagen shield with dissolution times of 12 h, that were presoaked with either hydrophilic or lipophilic fluorophore (4,5-carboxyfluorescein and *N*-[Lissamine rhodamine B sulfonyl]-diacyl-phosphatidylethanolamine, respectively), in a solution or unilamellar liposome suspension with different surface charges and bilayer fluidity. For the hydrophilic fluorophore, two to seven times higher concentrations were achieved in the collagen shield by immersion in aqueous solution than immersion in the liposome suspensions.

In the liposomal preparations, the negatively charged and cholesterol-containing preparation led to the highest concentration of the hydrophilic fluorophore, whereas the neutral and cholesterol-free preparation led to the highest concentration of hydrophobic fluorophore.

The release kinetics data results indicated that liposome surface charge and bilayer fluidity are of minor importance for the interaction of liposomes with collagen shields. Moreover, the release kinetics of hydrophilic or lipophilic substance from collagen shield was similar for both liposome-encapsulated and nonencapsulated drug. Thus, these results suggest that there is no added value for combined application of collagen shields and liposomes. In another study, the combination of collagen shield and liposomes enhanced CsA ocular penetration but failed to provide sustained release compared to CsA liposomal suspension [121].

Gulsen and Chauhan [122,123] introduced particle-laden hydrogel contact lenses as ocular DDS. Their studies have shown that poly(2-hydroxyethyl methacrylate) hydrogel lenses loaded with a lidocaine-encapsulated microemulsion stabilized with a silica shell are transparent and allow sustained release of lidocaine at therapeutic levels over a period of few days. The main drawback of drug delivery by the nanoparticle-laden gels is the decaying release rates. Furthermore, these drug-laden lenses need to be stored in a pouch saturated with the drug to prevent drug loss during storage. Moreover, the safety of this system is not clear since the components used in this study have not been used previously in eye care.

24.5.2 Liposomes

Liposomes are vesicles consisting of an aqueous compartment core surrounded by a lipid bilayer that mimics a cell membrane. The aqueous core can be surrounded by either a single lipidic bilayer (unilamellar liposomes) or concentric multiple bilayers (multilamellar liposomes). Liposomes with multiple bilayers are termed multilamellar vesicles (MLV), whereas the unilamellar liposomes are termed small unilamellar vesicles (SUV) or large unilamellar vesicles (LUV) according to their size (less or more than 100 nm accordingly) [124]. The bilayer of the liposomes is formed from a double array of phospholipids with their lipophilic tails facing each other and their hydrophilic heads embedded either in the aqueous core or in the aqueous solution that separate vesicles from each other. The typical liposome bilayer components are phosphatidylcholine (PC) as the main backbone and a small amount of cholesterol (Chol) to improve the stability of the bilayer array. However, other amphiphiles such as negatively charged phospholipids (phosphatidylethanolamine [PE], phosphatidylserine [PS], etc.) or cationic lipids (stearylamine, DOTAP, etc.) can be used to achieve the desired electrostatic bilayer properties. The incorporation of phospholipids with PEG moiety on the hydrophilic head increases the stability of the liposomal vesicle and decreases the rate of uptake by the reticuloendothelial system (RES), probably by steric interference with the foreign particle elimination mechanisms of the RES. These PEGylated liposomes have shown excellent prolonged circulation in the serum and are termed stealth liposomes [125]. One example of PEGylated liposome bilayer molecular composition consists of 10:1:0.5 of PC:Chol:PEG–PE. However, the nature and proportions of the bilayer components are usually tailored to meet the optimal conditions for a specific incorporated drug.

Liposomes are usually prepared by the reverse-phase evaporation method or its modification, consisting of vigorous mechanical mixing of the water phase and low boiling point organic solvents that contain the phospholipids and other bilayer components. Liposomes can incorporate a wide variety of molecules and peptides. Hydrophilic compounds should be incorporated in the aqueous phase whereas lipophilic compounds are added to the bilayer components. Because the lipid bilayer occupies only a small volume, the incorporating capacity of liposomes for lipophilic drugs is far more limited than for hydrophilic drugs. However, prodrug strategy can allow the incorporation of lipophilic drugs in the aqueous phase and vice versa [126]. The use of liposomes for topical ocular drug delivery was first reported in the early 1980s. Smolin et al. [127] reported that idoxuridine incorporated in liposomes was more effective in the treatment of acute and chronic herpetic keratitis in the rabbit eye than idoxuridine or the vehicle alone. Singh et al. [128] investigated the mechanism of liposome drug delivery to the eye by using colloidal gold as an electron-dense marker for MLV. Despite the evidenced presence of a gold-labeled MLV in the conjunctiva, it was not possible to confirm the absorption of vesicles in the surface epithelium of the cornea or conjunctiva. Ahmed and Patton [129] reported that the liposomal preparation of inulin favored the conjunctival route for penetration to the iris and ciliary body rather than the less efficient penetration through the corneal route to the aqueous humor. Schaeffer and Krohn [130] studied the ocular penetration of penicillin G and indoxole from liposomal preparations of different surface charge. A fourfold enhancement of transcorneal flux of penicillin G across isolated rabbit cornea was reported for positively charged unilamellar liposomes. This enhancement was observed only when the drug was entrapped in the liposomes but not when the liposomes were preformed in the absence of drug and mixed with penicillin G immediately before application to the cornea. The rank order of best corneal uptake for the different liposomal preparations was: positively charged > negatively charged > neutral liposomes. The authors suggested that the differences stem from the electrostatic interaction between the negatively charged corneal surface and the different preparations. McCalden and Levy [131] examined the ability of liposomes composed of different kinds of phospholipids to adhere to the surface of the cornea and reported that liposomes containing positively charged phospholipids are better retained than an albumin control preparation. The positively charged liposomal formulation of tropicamide was also found to be more effective than the neutral liposomal preparation in terms of pupil dilatory effect following topical instillation in the rabbit eye [132]. Law et al. [133] reported a discrepancy between *in vitro* and *in vivo* results for the ocular penetration of acyclovir loaded in liposomal formulations with positive, negative, or neutral charge. Whereas in terms of *in vitro* penetration the positively charged formulation was second to the negatively charged preparation, the *in vivo* experiment showed a better penetration of the positively charged preparation compared to the negatively charged formulation. The authors carried out morphological observations of the corneal surface and suggested that positively charged liposomes prolonged the residence time on corneal surface by forming a coating layer with more intimate contact with this surface in *in vivo* conditions. Milani et al. [134] studied the effects of different formulations of topical CsA on corneal allograft rejection in the rat eye. The mean survival time of the grafts was significantly higher for the treatment group receiving the liposomal formulation of CsA compared to groups treated with CsA formulated in olive oil, empty liposomes, or the untreated group. Bochot et al. [135] formulated teller gene in a simple solution, a 27% poloxamer 407 gel, a suspension of liposomes and in liposomes dispersed within a 27% poloxamer 407 gel. Following topical instillation to the rabbit eye, the concentration of teller gene in tissues of the anterior segment and the vitreous were measured. The rank order of best tissue availability was: solution > gel > liposomes (+gel). The authors concluded that topical liposomes may not be an effective delivery system for the administration of oligonucleotides to the superficial ocular tissues.

Despite the aforementioned reports demonstrating the superiority of some liposomal preparations in animal models, this drug delivery system still suffers from inherent limitations that restrict its potential as a topical ocular DDS. These limitations include instability and short shelf life, low drug-loading capacity (for lipophilic drugs in particular), limited drug retention on ocular surface, sensitivity to sterilization, and expensive large-scale manufacturing process.

24.5.3 Emulsions

Emulsions are two-phase systems formed from oil and water by the dispersion of one liquid (the internal phase) into the other (the external phase) and stabilized by at least one surfactant. Microemulsion, contrary to submicron emulsion (SME) or nanoemulsion, is a term used for a thermodynamically stable system characterized by a droplet size in the low nanorange (generally less than 30 nm). Microemulsions are also two-phase systems prepared from water, oil, and surfactant, but a cosurfactant is usually needed. These systems are prepared by a spontaneous process of self-emulsification with no input of external energy. Microemulsions are better described by the bicontinuous model consisting of a system in which water and oil are separated by an interfacial layer with significantly increased interface area. Consequently, more surfactant is needed for the preparation of microemulsion (around 10% compared with 0.1% for emulsions). Therefore, the nonionic-surfactants are preferred over the more toxic ionic surfactants. Cosurfactants in microemulsions are required to achieve very low interfacial tensions that allow self-emulsification and thermodynamic stability. Moreover, cosurfactants are essential for lowering the rigidity and the viscosity of the interfacial film and are responsible for the optical transparency of microemulsions [136].

In principle, in the preparation of medicated nanoemulsions, the drug is initially solubilized or dispersed together with an emulsifier in suitable single oil or oil mixtures by means of slight heating. The water phase containing the osmotic agent with or without an additional emulsifier is also heated and mixed with the oil phase by means of high-speed mixers. Further homogenization takes place to obtain the needed small droplet size range of the nanoemulsion. A terminal sterilization by filtration or steam then follows. The SME thus formed contains most of the drug molecules within its oil phase. This is a generally accepted and standard method to prepare lipophilic drug-loaded SME for ocular use. This process is normally carried out under aseptic conditions and nitrogen atmosphere to prevent both contamination and potential oxidation of sensitive excipients. The average oil droplet size of the SME or nanoemulsions is above 50 nm, the emulsion exhibits a milky appearance and the oil concentration is usually below 10%. Polar oils such as medium chain triglycerides (MCT) are commonly used as oil phase.

The rationale for introducing emulsions as a DDS for topical ocular application is their ability to incorporate within their oil inner phase lipophilic active molecules, which exhibit low water solubility and cannot be normally administered in an aqueous eyedrop formulation. Thus, the preferred emulsion for topical ocular application should be of the oil-in-water type formulation.

Muchtar et al. [137] have shown the merit of nanoemulsions for the formulation of lipophilic compounds in topical ocular therapy. Delta-8-tetrahydrocannabinol, a lipophilic cannabinoid, was formulated in a negatively charged SME. The ocular hypotensive effect of this topical formulation was studied on rabbits with ocular hypertension and on normotensive rabbits. The mean droplet size of the emulsion was 130 $\pm$ 41 nm and the zeta potential was -57.1 mV. This formulation showed stability for up to 9 months in terms of pH and droplet size. An intense and long-lasting IOP depressant effect was observed following topical application in the lower conjunctival sac of ocular hypertensive albino rabbits, but less effect was observed in the ocular normotensive group [137]. Similar results were obtained in a later study when HU-211, a nonpsychotropic synthetic cannabinoid, was formulated in a negatively charged SME and applied topically to the rabbit eye. The IOP reduction lasted 6 h with maximal magnitude of 24% of baseline in the treated eyes compared to 12.5% in the contralateral eye [138]. Naveh et al. [139] compared the IOP reduction effect of aqueous solution containing 2% pilocarpine hydrochloride and 1.7% pilocarpine base (equivalent to 2% of hydrochloride salt) incorporated in negatively charged SME. Maximum IOP reduction was 28.5% of baseline for the SME formulation compared to 18% for the aqueous solution. However the time to reach maximal effect was shorter for the aqueous solution compared to the SME formulation (2 and 5 h, respectively), but the IOP reduction effect was more sustained in the group receiving the SME formulation (29 compared to 11 h). The authors suggested that the sustained effect was related in part to the availability of pilocarpine in the oily phase. This advantage could be translated into less frequent applications for the SME formulation [137]. Zurowska-Pryczkowska et al. [140] incorporated pilocarpine in anionic SME and studied the interaction between the incorporated drug and the emulsion and the consequences of this interaction on both drug and emulsion stability. In a later study from the same group, it was found that the bioavailability of the drug was pH-dependent with the best miotic activity at pH 5.0 and 8.5. However, high pH cannot be considered for clinical use because of pilocarpine degradation in the emulsion, which occurs at a similar rate as in aqueous solutions. Thus, a pH of 5.0 was suggested for better stability and bioavailability [141]. In another report from the same research group, the partitioning of pilocarpine in the oil phase of anionic SME was increased by pilocarpine ion pairing. Surprisingly, the augmentation in pilocarpine content of the oil phase was not translated into a better ocular bioavailability. This finding was attributed to the components of the lipid emulsion (soybean oil and egg lecithin), which did not allow sufficient residence time on corneal surface [142]. In a study of topical indomethacin penetration to the rabbit eye, Calvo et al. [64] reported a

300% increased indomethacin ocular bioavailability following instillation of the indomethacin submicron SME compared to the performance of a commercial solution. The other studied colloidal systems (nanoparticles and nanocapsules) showed comparable performance. An endocytic mechanism of penetration into corneal epithelium cells was proposed based on confocal images obtained in this study. Melamed et al. [143] incorporated adaprolol, a novel soft beta-blocking agent, in anionic SME and showed the safety and the sustained IOP reduction effect of this formulation following topical ocular application to healthy volunteers. Garty and Lusky [144] compared the IOP reduction effect of twice a day topical pilocarpine anionic SME with four times a day commercial pilocarpine solution in ocular hypertensive patients. The two groups showed comparable results with 25% of baseline IOP reduction. This study has shown that the incorporation of pilocarpine in SME allows the achievement of the same clinical effect with less frequent administrations per day compared to pilocarpine solution. Ding [145] has formulated CsA, a highly lipophilic molecule, in an anionic castor oil-in-water SME. This formulation can incorporate up to 0.4% of CsA and was developed for the treatment of severe dry eye with inflammatory background. This indication requires only low concentrations of CsA in the ocular surface tissues for local immunomodulation without the need for deep intraocular penetration. Ding's formulation was stable up to 9 months but caused mild discomfort when administered to rabbit eyes [146]. Ocular pharmacokinetic data showed adequate penetration to ocular surface tissues and to the lacrimal gland whereas the penetration to the intraocular tissues was low and systemic absorption was minimal. This CsA formulation was challenged successfully in phase I and II clinical trials and formulations of 0.05% and 0.1% w/w of the drug were further evaluated in phase III trial [147,148]. Patients treated with both concentrations have shown statistically significant improvement in subjective and objective dry eye indexes compared to the placebo arm. However, the results for 0.1% w/w concentration were not superior to those obtained with the 0.05% w/w concentration. This anionic SME with CsA concentration of 0.05% w/w (Restasis, Allergan, Inc., Irvine, CA, USA) was approved in December 2002 by the United States Food and Drug Administration (FDA). Klang et al. [149] hypothesized that a positively charged SME would favor electrostatic interaction with the negatively charged ocular surface and prolong the local residence time. Indeed, piroxicam was formulated in a positively charged SME and was shown to be the most effective formulation for the delivery of the lipophilic piroxicam to the rabbit cornea following alkali burn. The effect of emulsion charge on the ocular penetration of indomethacin was studied [150]. Indomethacin was formulated in positively and negatively charged SMEs. The ocular penetration of both emulsions was evaluated and compared to a commercial ocular solution of indomethacin (Indocollyre hydro-PEG solution, Bausch & Lomb, Rochester, NY, USA). The positively charged SME achieved significantly higher drug levels than the negatively charged emulsion and the commercial solution in the aqueous humor and the sclera-retina. Ocular surface indomethacin levels were high and nondifferential to the tested formulation. The contact angle and the spreading coefficient of the different formulations on the isolated freshly excised rabbit cornea were studied [150] to elucidate the effect of the formulation charge on its interaction with the corneal surface. Lower contact angle and higher spreading coefficient were exhibited by the positively charged SME indicating better wettability properties on the cornea than either the saline or the negatively charged SME. Another study was conducted to investigate the effect of SME charge on the ocular penetration of the lipophilic and impermeable CsA. The drug was formulated in a positively charged SME and the optimization and characterization of this formulation was reported by Tamilvanan et al. [151]. Following one single instillation to the rabbit eye, CsA incorporated in positively charged SME achieved higher drug availability on ocular surface, particularly on the conjunctiva, compared to CsA in negatively charged SME. The penetration to the intraocular tissues was limited whereas blood levels were extremely low [152].

Following these encouraging results the CsA positively charged SME formulation was further developed and proceeded to phase I and II clinical trials.

In conclusion, the approval of Restasis by the FDA is an important milestone in lipid emulsion research for ophthalmic application. This approval reflects the achievements of the last decade in terms of the availability of better ingredients, improved manufacturing processes, feasibility of sterilization, and better understanding of the optimization process. In all of the comparative studies done so far, positively charged SME achieved better ocular bioavailability regardless of the studied drug. Research efforts are underway to further explore the mechanism of interaction of positively charged SMEs with ocular tissues and to translate the results of this research into enhanced clinical performance.

24.5.4 Iontophoresis

This section will focus on the evolution and latest advancements of electrically assisted drug delivery as a tool for enhanced ocular drug penetration. A detailed review of different aspects of ocular iontophoresis can be found in Chapter 26.

Iontophoresis is a method used to enhance the delivery of charged molecules across tissue barriers that are relatively impermeable for ionized compounds. This rapid and noninvasive technique utilizes low electrical current to drive ionized molecules across barriers. The main factors influencing iontophoretic drug delivery are the molecular weight, charge, and lipophilicity of the drug, current density, the duration of treatment, pH and the permeability of the tissue. One electrode is placed on the eye whereas the other is placed on a remote tissue. Therefore, ionic drugs are delivered from electrodes by current of the same polarity measured in milliamperes (mA). Thus, for negatively charged drugs, the negative pole (cathode) is placed on the eye and the positive pole (anode) on a remote location as the indifferent or ground electrode. The total delivered electric charge is proportional to the delivered drug. The optimal setting that the operator should seek must allow for maximum delivery while minimizing the current level. Thus, the efficiency of a iontophoretic device can be described in terms of the total amount of charge that must be passed to obtain the desired clinical effect (i.e., induction of local anesthesia upon delivery of lidocaine) and quantified in units of milliamperes minutes, calculated by multiplying the electrical current and the application time [153]:

$$\text{Current (mA)} \times \text{Application time (min)} = \text{Iontophoretic dose (mA min)}$$

The efficiency of the iontophoresis process is expressed by the ratio

$$[\text{Micrograms of drug per gram (or mL) of target tissue}]/[\text{Iontophoretic dose (mA min)}]$$

Iontophoresis has been used in medicine for decades to improve the penetration of various ionized compounds across tissue barriers. The introduction of this technique in ophthalmology dates to the early years of the twentieth century [154] and throughout the next seven decades iontophoresis was abandoned and rediscovered periodically [155–163]. In 1984, Hughes and Maurice [164] reported that the rate of penetration of fluorescein and gentamicin ions into the anterior chamber of the rabbit was increased more than 100-fold by the application of a 1.0 mA current through iontophoresis apparatus. This report was followed by a wave of enthusiasm for iontophoresis as an ocular drug delivery tool during the second half of the 1980s. This enthusiasm was translated into research leading to a considerable number of publications dedicated almost entirely to the enhancement of ocular penetration of antibiotic compounds. The results of these studies varied, but none of them resulted in clinical practice [165–182]. This era of experimental efforts was aborted by reports that showed the side effects of iontophoresis. Lam et al. [183] reported for 1.5 mA transscleral iontophoresis in rabbits that increments in application time beyond 1 min carried a growing risk of focal retinal necrosis. The size and severity of the lesions increased with the duration of application. This study and others did not come to clear conclusions on whether multiple iontophoresis sessions cause a cumulative damage. Side effects of ocular iontophoresis were related at that time to the high current used and to inconvenient equipment that could not be optimized to allow full control of the process parameters. In the first half of the 1990s, experimental data on iontophoresis-assisted ocular drug penetration became scarce as only a few research groups worked on iontophoresis [184–187]. These studies showed the efficacy of transscleral iontophoresis of antiviral agents for the treatment of cytomegalovirus retinopathy in the rabbit eye, using small localized probe with high current density. Yoshizumi et al. [188] reported, however, using high density iontophoresis, that focal burns in tissues at the area of probe placement were not more severe following 11 consecutive treatments over 21 days than after a single treatment. Sarraf and Lee [189] reviewed in 1994 one decade of experimental evidence that was published on ocular iontophoresis up to then. Most of the reviewed papers claimed superior results for iontophoresis-assisted ocular penetration compared to conventional drug application. Interest in ocular iontophoresis has been renewed in the last decade following the introduction of improved equipment. Compared to the earlier devices, modern iontophoresis devices are well standardized allowing for better control of the delivered current and can achieve reproducible drug delivery.

Behar-Cohen et al. [190] reported in 1997 the performance of a Coulomb-controlled iontophoresis (CCI) system. This system is based on a large surface of application, control of current delivery in function of resistance modifications, and low current density. Another original aspect of this device is its placement around the cornea, from the limbus to 3 mm posteriorly, at the border between the anterior and the posterior segments of the eye. The potential of iontophoretic delivery of dexamethasone using this system was shown against systemic and local treatment modalities by measuring the therapeutic effect in endotoxin-induced anterior and posterior uveitis in the rat eye. In another report from the same group [191], the safety of the CCI system was addressed along with an ocular pharmacokinetic study of hemisuccinate methylprednisolone, comparing ocular iontophoretic delivery in the rabbit eye with intravenous administration. The safety of the system was shown both clinically and by histological evaluation. Compared to intravenous administration and to no-current iontophoretic delivery, the CCI-assisted delivery achieved higher and more sustained tissue concentrations with negligible systemic absorption. This is the only device that has been actually used recently on patients in a clinical study [192].

Recently, Frucht-Pery et al. [193] have worked on corneal iontophoresis and shown the efficacy of iontophoresis to deliver gentamicin in the rabbit cornea compared to other modalities of gentamicin delivery. The general concept was the performance of the iontophoretic delivery on the corneal surface using a formulation of gentamicin in agar. Another study [194] from the same group showed that following the performance of iontophoretic delivery on the center of the cornea, gentamicin corneal concentrations in different concentric zones of the cornea were not homogenous and the highest concentration of the drug was found in the central cornea. Recently, similar experiments were reported from the same researchers [195]; however, better iontophoretic merit was claimed by using a novel hydrogel probe for the iontophoretic delivery of gentamicin to the rabbit eye.

Li et al. [196] studied the sites of ion delivery in the eye during iontophoresis. Using nuclear magnetic resonance imaging (MRI) and a probe ion (Mn^{2+}), the study compared transscleral with transcorneal iontophoresis. The results have shown that transscleral iontophoresis delivered the ion into the vitreous, whereas transcorneal iontophoresis delivered the ion into the anterior chamber. The delivery pathway was perpendicular to the electrode–eye interface beneath the electrode. In transscleral iontophoresis, the ion penetrated the sclera and traveled as far as 1.5 mm from the electrode–conjunctiva interface into the vitreous. For transcorneal iontophoresis, the ion penetrated the cornea and filled the entire anterior chamber.

In later reports, the added value of iontophoresis-assisted drug delivery to the rabbit eye of nonsteroidal anti-inflammatory drugs using CCI system was reported [197,198]. Vollmer et al. [199] studied iontophoretic ocular delivery of amikacin using 2, 3, and 4-mA current compared to no-current. Three identical studies were done. The potential of a higher iontophoretic current was demonstrated but the interstudy reproducibility was tissue-dependent, being low (8%) for the retina and choroid and satisfactory (51%) for the anterior segment tissues in the 4-mA group. Reported clinical experiments on humans are rare. Recently, the ocular iontophoretic delivery of methylprednisolone sodium succinate was shown to be effective in reversing corneal graft rejection in humans [192]. Using the CCI system, 88% of the 17 treated eyes showed complete reversal of the rejection process, without any side effects.

Horwath-Winter et al. [200] reported an advantageous therapeutic effect of iodide in the treatment of dry eye when the ion was delivered by 0.2 mA/7 min iontophoresis compared to mock iontophoresis.

One of the main motivations behind the renewed interest in ocular iontophoresis in recent years is the growing need for delivery systems that can enhance the ocular penetration of proteins, peptides, oligonucleotides, and plasmids, as these compounds are relatively impermeable charged macromolecules that fit in the methodology of electrically assisted penetration. The feasibility and the potential of using ocular iontophoresis to enhance the intraocular delivery of oligonucleotides and plasmids were shown in several studies [201–204].

Despite the aforementioned efforts, this technology has not yet reached the therapeutic arsenal in clinical ophthalmology.

24.5.4.1 Electroporation

Recently another electrically assisted drug delivery technology, electroporation, was proposed as an alternative or adjuvant to iontophoresis. Electroporation comprises the use of electric pulses to induce transient changes in the cell membrane architecture that turn it into more permeable barrier. Beside the permeabilization effect on cell membrane, it was postulated that this technique induces electrophoretic effect on charged macromolecules and drives them to move across the destabilized membrane [205].

Compared to other medical fields, the reported electroporation experience in ophthalmology is scarce and limited to gene therapy [206–212]. Although most studies claim the merit of this technique in ocular gene therapy, electroporation in ophthalmology lacks safety studies and standards for the optimal field strength at different ocular tissues. Recently, Bloquel et al. have developed a new concept for ocular electrotransfer, using the ciliary muscle as a target for the transfection of therapeutic plasmids. Using particular electrodes, specific and safe transfection of the ciliary muscle was achieved in the rat eye allowing for the production of proteins in the ocular media for at least 3 weeks, without systemic diffusion [213]. This new electrotransfer concept offers the advantage to use the ciliary muscle as a platform for the production of any therapeutic protein within the eye.

24.6 CONCLUSION

Until recently, the topical ophthalmic use of some interesting and promising either lipophilic or macromolecular therapeutic substances has been limited clinically because of their restrictive physicochemical properties, which has exhibited poor ocular bioavailability. It is possible that these substances may become more widely used in a clinical ocular setting if appropriate drug delivery techniques can be designed to overcome their intrinsic absorption drawbacks particularly in the posterior segment.

The options for ophthalmic pharmaceuticals continue to expand, as several recent approvals and novel medications are attracting the scientific interest of ophthalmologists and gaining popularity in the ocular market. Pharmaceutical investigators and pharmacologists have been trying to develop delivery systems that allow the fate of a drug to be controlled and the optimal drug dosage to arrive at the site of action in the eye by means of novel micro- or nanoparticulate or insert dosage forms. Ongoing efforts are being made to design systems or drug carriers capable of delivering the active molecules specifically to the intended ocular tissue in the eye, by increasing the therapeutic efficacy. Some of the promising results on how to improve the current arsenal of ophthalmic medications and delivery systems were presented in this chapter. Novel therapies and ocular DDS will continue to be revealed, providing invaluable aid in treatment of the entire spectrum of ophthalmic diseases soon.

REFERENCES

1. Burstein, N.L., and J.A. Anderson. 1985. Corneal penetration and ocular bioavailability of drugs. *J Ocul Pharmacol* 1:309.
2. Shell, J.W. 1984. Ophthalmic drug delivery systems. *Surv Ophthalmol* 29:117.
3. Maurice, D.M. 2002. Drug delivery to the posterior segment from drops. *Surv Ophthalmol* 47(Suppl 1):S41.
4. Records, R.E. 2002. The tear film. In *Duane's ophthalmology* on CD-ROM, Foundations of Clinical Ophthalmology, vol. 2, eds. W. Tasman and E.A. Jaeger. Philadelphia: Lippincott Williams & Wilkins, chap. 3.
5. Maurice, D. 1995. The effect of the low blink rate in rabbits on topical drug penetration. *J Ocul Pharmacol Ther* 11:297.
6. Norn, M.S. 1965. Tear secretion in normal eyes. Estimated by a new method: The lacrimal streak dilution test. *Acta Ophthalmol (Copenh)* 43:567.
7. Patton, T.F., and J.R. Robinson. 1976. Quantitative precorneal disposition of topically applied pilocarpine nitrate in rabbit eyes. *J Pharm Sci* 65:1295.
8. Linden, C., and A. Alm. 1990. The effect of reduced tear drainage on corneal and aqueous concentrations of topically applied fluorescein. *Acta Ophthalmol (Copenh)* 68:633.
9. Karesh, J.W. 2002. Topographic anatomy of the eye. In *Duane's ophthalmology* on CD-ROM, Foundations of Clinical Ophthalmology, vol. 1, eds. W. Tasman and E.A. Jaeger. Philadelphia: Lippincott Williams & Wilkins, chap. 1.
10. Smolek, M.K., and S.D. Klyce. 2002. Cornea. In *Duane's ophthalmology* on CD-ROM, Foundations of Clinical Ophthalmology, vol. 1, eds. W. Tasman and E.A. Jaeger. Philadelphia: Lippincott Williams & Wilkins, chap. 1.
11. Maurice, D.M. 1972. The location of the fluid pump in the cornea. *J Physiol* 221:43.
12. Hodson, S.A., and F. Miller. 1976. The bicarbonate ion pump in the endothelium which regulates the hydration of rabbit cornea. *J Physiol (Lond)* 263:563.
13. Lim, J.J. 1981. Na+ transport across the rabbit corneal endothelium. *Curr Eye Res* 1:255.
14. Riley, M.V. 1985. Pump and leak in regulation of fluid transport in rabbit cornea. *Curr Eye Res* 4:371.
15. Pepperl, J.E., et al. 2002. Conjunctiva. In *Duane's Ophthalmology* on CD-ROM, Foundations of Clinical Ophthalmology, vol. 1, eds. W. Tasman and E.A. Jaeger. Philadelphia: Lippincott Williams & Wilkins, chap. 1.
16. de la Maza, M.S., and C.S. Foster. 2002. Sclera. In *Duane's ophthalmology* on CD-ROM, Foundations of Clinical Ophthalmology, vol. 1, eds. W. Tasman and E.A. Jaeger. Philadelphia: Lippincott Williams & Wilkins, chap. 1.

17. Hutchinson, A.K., M.M. Rodrigues, and H.E. Grossniklaus. 2002. Iris. In *Duane's ophthalmology* on CD-ROM, Foundations of Clinical Ophthalmology, vol. 1, eds. W. Tasman and E.A. Jaeger. Philadelphia: Lippincott Williams & Wilkins, chap. 1.
18. Streeten, B.W. 2002. The ciliary body. In *Duane's Ophthalmology* on CD-ROM, Foundations of Clinical Ophthalmology, vol. 1, eds. W. Tasman and E.A. Jaeger. Philadelphia: Lippincott Williams & Wilkins, chap. 1.
19. Buggage, R.R., E. Torczynski, and H.E. Grossniklaus. 2002. Choroid and suprachoroid. In *Duane's Ophthalmology* on CD-ROM, Foundations of Clinical Ophthalmology, vol. 1, eds. W. Tasman and E.A. Jaeger, vol. 1. Philadelphia: Lippincott Williams & Wilkins, chap. 1.
20. Feeney-Burns, L., and M.L. Katz. 2002. Retinal pigment epithelium. In *Duane's ophthalmology* on CD-ROM, Foundations of Clinical Ophthalmology, vol. 1, eds. W. Tasman and E.A. Jaeger. Philadelphia: Lippincott Williams & Wilkins, chap. 1.
21. Park, S.S., J. Sigelman, and E.S. Gragoudas. 2002. The anatomy and cell biology of the retina. In *Duane's ophthalmology* on CD-ROM, Foundations of Clinical Ophthalmology, vol. 1, eds. W. Tasman and E.A. Jaeger. Philadelphia: Lippincott Williams & Wilkins, chap. 1.
22. Davson, H. 1990. The aqueous humor and the intraocular pressure. In *Physiology of the eye*, 5th ed., 3–95. New York: Pergamon Press.
23. Millar, C., and P.L. Kaufman. 2002. Aqueous humor: secretion and dynamics. In *Duane's ophthalmology* on CD-ROM, Foundations of Clinical Ophthalmology, vol. 2, eds. W. Tasman and E.A. Jaeger. Philadelphia: Lippincott Williams & Wilkins, chap. 6.
24. Maurice, D.M. 1984. Ocular pharmacokinetics. In *Pharmacology of the eye*, ed. M.L. Sears, 19–116. New York: Springer-Verlag.
25. Williams, G.A., and M.S. Blumenkranz. 2002. Vitreous humor. In *Duane's ophthalmology* on CD-ROM, Foundations of Clinical Ophthalmology, vol. 2, eds. W. Tasman and E.A. Jaeger. Philadelphia: Lippincott Williams & Wilkins, chap. 11.
26. Wilson, C.G. 2004. Topical drug delivery in the eye. *Exp Eye Res* 78:737.
27. Kaur, I.P., et al. 2004. Vesicular systems in ocular drug delivery: An overview. *Int J Pharm* 269:1.
28. Ahmed, I., and T.F. Patton. 1985. Importance of the noncorneal absorption route in topical ophthalmic drug delivery. *Invest Ophthalmol Vis Sci* 26:584.
29. Chien, D.S., et al. 1990. Corneal and conjunctival/scleral penetration of p-aminoclonidine, AGN 190342, and clonidine in rabbit eyes. *Curr Eye Res* 9:1051.
30. Saha, P., J.J. Yang, and V.H. Lee. 1998. Existence of a p-glycoprotein drug efflux pump in cultured rabbit conjunctival epithelial cells. *Invest Ophthalmol Vis Sci* 39:1221.
31. Kawazu, K., et al. 1999. Characterization of cyclosporin A transport in cultured rabbit corneal epithelial cells: P-glycoprotein transport activity and binding to cyclophilin. *Invest Ophthalmol Vis Sci* 40:1738.
32. Dey, S., et al. 2003. Molecular evidence and functional expression of P-glycoprotein (MDR1) in human and rabbit cornea and corneal epithelial cell lines. *Invest Ophthalmol Vis Sci* 44:2909.
33. Steuhl, P., and J.W. Rohen. 1983. Absorption of horse-radish peroxidase by the conjunctival epithelium of monkeys and rabbits. *Graefes Arch Clin Exp Ophthalmol* 220:13.
34. Tonjum, A.M. 1974. Vesicular transport of horseradish peroxidase across the rabbit corneal endothelium. *Acta Ophthalmol (Copenh)* 52:647.
35. van der Want, H.J., et al. 1983. Electron microscopy of cultured human corneas. Osmotic hydration and the use of a dextran fraction (dextran T 500) in organ culture. *Arch Ophthalmol* 101:1920.
36. Orzalesi, N., et al. 1982. Identification and distribution of coated vesicles in the retinal pigment epithelium of man and rabbit. *Invest Ophthalmol Vis Sci* 23:689.
37. Akeo, K., Y. Tanaka, and T. Fujiwara. 1988. Electron microscopic studies of the endocytotic process of cationized ferritin in cultured human retinal pigment epithelial cells. *In Vitro Cell Dev Biol* 24:705.
38. Hamalainen, K.M., et al. 1997. Characterization of paracellular and aqueous penetration routes in cornea, conjunctiva, and sclera. *Invest Ophthalmol Vis Sci* 38:627.
39. Sasaki, H., et al. 1995. Ocular permeability of FITC-dextran with absorption promoter for ocular delivery of peptide drug. *J Drug Target* 3:129.

40. Yoshida, F., and J.G. Topliss. 1996. Unified model for the corneal permeability of related and diverse compounds with respect to their physicochemical properties. *J Pharm Sci* 85:819.
41. Civiale, C., et al. 2004. Ocular permeability screening of dexamethasone esters through combined cellular and tissue systems. *J Ocul Pharmacol Ther* 20:75.
42. Maurice, D.M., and J. Polgar. 1977. Diffusion across the sclera. *Exp Eye Res* 25:577.
43. Olsen, T.W., et al. 1995. Human scleral permeability. Effects of age, cryotherapy, transscleral diode laser, and surgical thinning. *Invest Ophthalmol Vis Sci* 36:1893.
44. Ambati, J., et al. 2000. Diffusion of high molecular weight compounds through sclera. *Invest Ophthalmol Vis Sci* 41:1181.
45. Chiang, C.H., and R.D. Schoenwald. 1986. Ocular pharmacokinetic models of clonidine-3H hydrochloride. *J Pharmacokinet Biopharm* 14:175.
46. Schoenwald, R.D., et al. 1997. Penetration into the anterior chamber via the conjunctival/scleral pathway. *J Ocul Pharmacol Ther* 13:41.
47. Vuori, M.L., et al. 1993. Beta 1- and beta 2-antagonist activity of topically applied betaxolol and timolol in the systemic circulation. *Acta Ophthalmol (Copenh)* 71:682.
48. Tang-Liu, D.D., S.S. Liu, and R.J. Weinkam. 1984. Ocular and systemic bioavailability of ophthalmic flurbiprofen. *J Pharmacokinet Biopharm* 12:611.
49. Hofman, P., et al. 2001. Lack of blood–brain barrier properties in microvessels of the prelaminar optic nerve head. *Invest Ophthalmol Vis Sci* 42:895.
50. Miyake, K., and K. Ohtsuki. 1981. Fluorescence fundusphotography by retrobulbar administration of the dye, photochemical trans-illumination. *Jpn J Ophthalmol* 25:280.
51. Pitkanen, L., et al. 2005. Permeability of retinal pigment epithelium: Effects of permeant molecular weight and lipophilicity. *Invest Ophthalmol Vis Sci* 46:641.
52. Tonjum, A.M., and O.O. Pedersen. 1977. The permeability of the human ciliary and iridial epithelium to horseradish peroxidase: An *in vitro* study. *Acta Ophthalmol (Copenh)* 55:781.
53. Freddo, T.F. 1984. Intercellular junctions of the iris epithelia in *Macaca mulatta*. *Invest Ophthalmol Vis Sci* 25:1094.
54. Rezai, K.A., et al. 1997. Comparison of tight junction permeability for albumin in iris pigment epithelium and retinal pigment epithelium *in vitro*. *Graefes Arch Clin Exp Ophthalmol* 235:48.
55. Smelser, G.K., T. Ishikawa, and Y.F. Pei. 1965. In *Structure of the eye*, vol. II, ed. E.W. Rohen, 109–120. Stuttgart: Schattauer-Verlag.
56. Peyman, G.A., and D. Bok. 1972. Peroxidase diffusion in the normal and laser-coagulated primate retina. *Invest Ophthalmol* 11:35.
57. Marmor, M.F., A. Negi, and D.M. Maurice. 1985. Kinetics of macromolecules injected into the subretinal space. *Exp Eye Res* 40:687.
58. Kamei, M., K. Misono, and H. Lewis. 1999. A study of the ability of tissue plasminogen activator to diffuse into the subretinal space after intravitreal injection in rabbits. *Am J Ophthalmol* 128:739.
59. Bunt-Milam, A.H., et al. 1985. Zonulae adherentes pore size in the external limiting membrane of the rabbit retina. *Invest Ophthalmol Vis Sci* 26:1377.
60. Stewart, P.A., and U.I. Tuor. 1994. Blood-eye barriers in the rat: Correlation of ultrastructure with function. *J Comp Neurol* 22:566.
61. Speiser, P., and J. Kreuter. 1976. *In vitro* studies of poly(methylmethacrylate) adjuvants. *J Pharm Sci* 65:1624.
62. Desai, M.P., et al. 1997. The mechanism of uptake of biodegradable microparticles in Caco-2 cells is size dependent. *Pharm Res* 14:1568.
63. Zimmer, A., J. Kreuter, and J. Robinson. 1991. Studies on the transport pathway of PBCA nanoparticles in ocular tissues. *J Microencapsul* 8:497.
64. Calvo, P., et al. 1996. Improved ocular bioavailability of indomethacin by novel ocular drug carriers. *J Pharm Pharmacol* 48:1147.
65. Qaddoumi, M.G., et al. 2004. The characteristics and mechanisms of uptake of PLGA nanoparticles in rabbit conjunctival epithelial cell layers. *Pharm Res* 21:641.
66. Wood, R., et al. 1985. Ocular disposition of poly-hexyl-2 cyano 3-[14C] acrylate nanoparticles in the albino rabbit. *Int J Pharm* 23:175.

67. Marchal-Heussler, L., et al. 1990. Antiglaucomatous activity of betaxolol chlorhydrate sorbed onto different isobutylcyanoacrylate nanoparticle preparations. *Int J Pharm* 58:115.
68. Giannavola, C., et al. 2003. Influence of preparation conditions on acyclovir-loaded poly-D, L-lactic acid nanospheres and effect of PEG coating on ocular drug bioavailability. *Pharm Res* 20:584.
69. De Campos, A.M., et al. 2004. Chitosan nanoparticles as new ocular drug delivery systems: *In vitro* stability, *in vivo* fate, and cellular toxicity. *Pharm Res* 21:803.
70. De Campos, A.M., A. Sanchez, and M.J. Alonso. 2001. Chitosan nanoparticles: A new vehicle for the improvement of the delivery of drugs to the ocular surface. Application to cyclosporin A. *Int J Pharm* 224:159.
71. Calvo, P., J.L. Vila-Jato, and M.J. Alonso. 1997. Evaluation of cationic polymer-coated nanocapsules as ocular drug carriers. *Int J Pharm* 153:41.
72. Couvreur, P., et al. 1977. Nanocapsules: A new type of lysomotropic carrier. *FEBS Lett* 84:331.
73. Couvreur, P., et al. 1979. Polycyanoacrylate nanocapsules as potential lysosomotropic carriers: Preparation, morphological and sorptive properties. *J Pharm Pharmacol* 31:331.
74. AI-Khouri, N., et al. 1985. Development of a new process for the manufacture of polyisobutylcyanoacrylate nanocapsules. *Int J Pharmacol* 28:125.
75. El-Samaligy, M.S., P. Rohdevald, and H.A. Mahmoud. 1986. Polyalkyl cyanoacrylate nanocapsules. *J Pharm Pharmacol* 31:216.
76. Fessi, H., et al. 1989. Nanocapsule formation by interfacial polymer deposition following solvent displacement. *Int J Pharm* 55:R1.
77. Andrieu, V., et al. 1989. Pharmacokinetic evaluation of indomethacin nanocapsules. *Drug Des Del* 4:295.
78. Allemann, E., R. Gurny, and E. Doelker. 1993. Drug-loaded nanoparticles preparation methods and drug targeting issues. *Eur J Pharm Biopharm* 39:173.
79. Harima, T., et al. 1986. Enhancement of the myotic response of rabbits with pilocarpine-loaded polybutylcyanoacrylate nanoparticles. *Int J Pharm* 33:187.
80. Fitzgerald, P., et al. 1987. A gammascintigraphic evaluation of microparticulate ophthalmic delivery systems. Liposomes and nanoparticles. *Int J Pharm* 40:81.
81. Diepold, R., et al. 1989. Distribution of poly-hexyl-2-cyano [3-14C]acrylate nanoparticles in healthy and inflamed rabbit eyes. *Int J Pharm* 54:149.
82. Das, S.K., et al. 1995. Evaluation of poly(isobutylcyanoacrylate) nanoparticles for mucoadhesive ocular drug delivery. I. Effect of formulation variables on physicochemical characteristics of nanoparticles. *Pharm Res* 12:534.
83. Damge, C., et al. 1987. Polyalkylcyanoacrylate nanocapsules increase the intestinal absorption of a lipophilic drug. *Int J Pharm* 36:121.
84. Ammoury, N., et al. 1989. Physiochemical characterization of polymeric nanocapsules and *in vitro* release evaluation of indomethacin as drug model. *STP Pharma* 5:647.
85. Calvo, P., et al. 1996. Polyester nanocapsules as new topical ocular delivery systems for cyclosporin A. *Pharm Res* 13:311.
86. Ashton, P., S.K. Podder, and V.H. Lee. 1991. Formulation influence on conjunctival penetration of four beta blockers in the pigmented rabbit: a comparison with corneal penetration. *Pharm Res* 8:1166.
87. Marchal-Heussler, L., et al. 1993. Poly(ε-caprolactone) nanocapsules in carteolol ophthalmic delivery. *Pharm Res* 10:386.
88. Le Bourlais, C.A., et al. 1997. Effect of cyclosporine A formulations on bovine corneal absorption: *Ex-vivo* study. *J Microencapsul* 14:457.
89. Desai, S.D., and J. Blanchard. 2000. Pluronic F127-based ocular delivery system containing biodegradable polyisobutylcyanoacrylate nanocapsules of pilocarpine. *Drug Deliv* 7:201.
90. De Campos, A.M., et al. 2003. The effect of a PEG versus a chitosan coating on the interaction of drug colloidal carriers with the ocular mucosa. *Eur J Pharm Sci* 20:73.
91. Wanka, G., H. Hoffmann, and W. Ulbricht. 1994. Phase diagrams and aggregation behaviour of poly(oxythylene)–poly(oxypropylene)–poly(oxyethylene) triblock copolymers in aqueous solutions. *Macromolecules* 27:4145.

92. Liaw, J., S.F. Chang, and F.C. Hsiao. 2001. *In vivo* gene delivery into ocular tissues by eye drops of poly(ethylene oxide)–poly(propylene oxide)–poly(ethylene oxide) (PEO–PPO–PEO) polymeric micelles. *Gene Ther* 8:999.
93. Pepic, I., N. Jalsenjak, and I. Jalsenjak. 2004. Micellar solutions of triblock copolymer surfactants with pilocarpine. *Int J Pharm* 272:57.
94. Kuwano, M., et al. 2002. Cyclosporine A formulation affects its ocular distribution in rabbits. *Pharm Res* 19:108.
95. Malik, N., et al. 2000. Dendrimers: Relationship between structure and biocompatibility *in vitro*, and preliminary studies on the biodistribution of 125I-labelled polyamidoamine dendrimers *in vivo*. *J Control Release* 65:133.
96. Roberts, J.C., M.K. Bhalgat, and R.T. Zera. 1996. Preliminary biological evaluation of polyamidoamine (PAMAM) Starburst dendrimers. *J Biomed Mater Res* 30:53.
97. Vandamme, T.F., and L. Brobeck. 2005. Poly(amidoamine) dendrimers as ophthalmic vehicles for ocular delivery of pilocarpine nitrate and tropicamide. *J Control Release* 20:102.
98. Hudde, T., et al. 1999. Activated polyamidoamine dendrimers, a non-viral vector for gene transfer to the corneal endothelium. *Gene Ther* 6:939.
99. Fyodorov, S.N., et al. 1984. Efficiency of collagen covers: Application in cases in keratotomy. In *Eye microsurgery*, ed. S.N. Fydorov. Moscow: Research Institute of Eye Microsurgery.
100. Phinney, R.B., et al. 1988. Collagen-shield delivery of gentamicin and vancomycin. *Arch Ophthalmol* 106:1599.
101. O'Brien, T.P., et al. 1988. Use of collagen corneal shields versus soft contact lenses to enhance penetration of topical tobramycin. *J Cataract Refract Surg* 14:505.
102. Pleyer, U., et al. 1992. Use of collagen shields containing amphotericin **B** in the treatment of experimental *Candida albicans*-induced keratomycosis in rabbits. *Am J Ophthalmol* 113:303.
103. Sawusch, M.R., et al. 1988. Use of collagen corneal shields in the treatment of bacterial keratitis. *Am J Ophthalmol* 106:279.
104. Hobden, J.A., et al. 1988. Treatment of experimental *Pseudomonas keratitis* using collagen shields containing tobramycin. *Arch Ophthalmol* 106:1605.
105. Silbiger, J., and G.A. Stern. 1992. Evaluation of corneal collagen shields as a drug delivery device for the treatment of experimental *Pseudomonas keratitis*. *Ophthalmology* 99:889.
106. Clinch, T.E., et al. 1992. Collagen shields containing tobramycin for sustained therapy (24 hours) of experimental Pseudomonas keratitis. *CLAO J* 18:245.
107. Callegan, M.C., et al. 1994. Efficacy of tobramycin drops applied to collagen shields for experimental staphylococcal keratitis. *Curr Eye Res* 13:875.
108. Hwang, D.G., et al. 1989. Collagen shield enhancement of topical dexamethasone penetration. *Arch Ophthalmol* 107:1375.
109. Sawusch, M.R., T.P. O'Brien, and S.A. Updegraff. 1989. Collagen corneal shields enhance penetration of topical prednisolone acetate. *J Cataract Refract Surg* 15:625.
110. Reidy, J.J., B.M. Gebhardt, and H.E. Kaufman. 1990. The collagen shield. A new vehicle for delivery of cyclosporin A to the eye. *Cornea* 9:196.
111. Chen, Y.F., et al. 1990. Cyclosporine-containing collagen shields suppress corneal allograft rejection. *Am J Ophthalmol* 109:132.
112. Waltman, S.R., and H.E. Kaufman. 1970. Use of hydrophilic contact lenses to increase ocular penetration of topical drugs. *Invest Ophthalmol* 9:250.
113. Gasset, A.R., and H.E. Kaufman. 1970. Therapeutic uses of hydrophilic contact lenses. *Am J Ophthalmol* 69:252.
114. Hillman, J.S., J.B. Marsters, and A. Broad. 1975. Pilocarpine delivery by hydrophilic lens in the management of acute glaucoma. *Trans Ophthalmol Soc UK* 95:79.
115. Jain, M.R. 1988. Drug delivery through soft contact lenses. *Br J Ophthalmol* 72:150.
116. Wajs, G., and J.C. Meslard. 1986. Release of therapeutic agents from contact lenses. *Crit Rev Ther Drug Carrier Syst* 2:275.
117. Weiner, A.L. 1994. Polymeric drug delivery systems for the eye. In *Polymeric site-specific pharmacotherapy*, ed. A.J. Domb, 315–346. Chichester: Wiley.

118. McMahon, T.T., and K. Zadnik. 2000. Twenty-five years of contact lenses: The impact on the cornea and ophthalmic practice. *Cornea* 19:730.
119. Hiratani, H., et al. 2005. Ocular release of timolol from molecularly imprinted soft contact lenses. *Biomaterials* 26:1293.
120. Grammer, J.B., et al. 1996. Impregnation of collagen corneal shields with liposomes: Uptake and release of hydrophilic and lipophilic marker substances. *Curr Eye Res* 15:815.
121. Pleyer, U., et al. 1994. Ocular absorption of cyclosporine A from liposomes incorporated into collagen shields. *Curr Eye Res* 13:177.
122. Gulsen, D., and A. Chauhan. 2004. Ophthalmic drug delivery through contact lenses. *Invest Ophthalmol Vis Sci* 45:2342.
123. Gulsen, D., and A. Chauhan. 2005. Dispersion of microemulsion drops in HEMA hydrogel: A potential ophthalmic drug delivery vehicle. *Int J Pharm* 292:95.
124. Lee, V.H., et al. 1985. Ocular drug bioavailability from topically applied liposomes. *Surv Ophthalmol* 29:335.
125. Gabizon, A., H. Shmeeda, and Y. Barenholz. 2003. Pharmacokinetics of pegylated liposomal doxorubicin: Review of animal and human studies. *Clin Pharmacokinet* 42:419.
126. Cheng, L., et al. 2000. Intravitreal toxicology and duration of efficacy of a novel antiviral lipid prodrug of ganciclovir in liposome formulation. *Invest Ophthalmol Vis Sci* 41:1523.
127. Smolin, G., et al. 1981. Idoxuridine-liposome therapy for herpes simplex keratitis. *Am J Ophthalmol* 91:220.
128. Singh, M., et al. 1993. Liposomal drug delivery to the eye and lungs: A preliminary electron microscopy study. *J Microencapsul* 10:35.
129. Ahmed, I., and T.F. Patton. 1987. Disposition of timolol and inulin in the rabbit eye following corneal versus noncorneal absorption. *Int J Pharm* 38:9.
130. Schaeffer, H.E., and D.L. Krohn. 1982. Liposomes in topical drug delivery. *Invest Ophthalmol Vis Sci* 22:220.
131. McCalden, T.A., and M. Levy. 1990. Retention of topical liposomal formulations on the cornea. *Experientia* 46:713.
132. Nagarsenker, M.S., V.Y. Londhe, and G.D. Nadkarni. 1999. Preparation and evaluation of liposomal formulations of tropicamide for ocular delivery. *Int J Pharm* 190:63.
133. Law, S.L., K.J. Huang, and C.H. Chiang. 2000. Acyclovir-containing liposomes for potential ocular delivery. Corneal penetration and absorption. *J Control Release* 63:135.
134. Milani, J.K., et al. 1993. Prolongation of corneal allograft survival with liposome-encapsulated cyclosporine in the rat eye. *Ophthalmology* 100:890.
135. Bochot, A., et al. 1998. Comparison of the ocular distribution of a model oligonucleotide after topical instillation in rabbits of conventional and new dosage forms. *J Drug Target* 6:309.
136. Vandamme, T.F. 2002. Microemulsions as ocular drug delivery systems: Recent developments and future challenges. *Prog Retin Eye Res* 21:15.
137. Muchtar, S., et al. 1992. A submicron emulsion as ocular vehicle for delta-8-tetrahydrocannabinol: Effect on intraocular pressure in rabbits. *Ophthalmic Res* 24:142.
138. Naveh, N., et al. 2000. A submicron emulsion of HU-211, a synthetic cannabinoid, reduces intraocular pressure in rabbits. *Graefes Arch Clin Exp Ophthalmol* 238:334.
139. Naveh, N., S. Muchtar, and S. Benita. 1994. Pilocarpine incorporated into a submicron emulsion vehicle causes an unexpectedly prolonged ocular hypotensive effect in rabbits. *J Ocul Pharmacol* 10:509.
140. Zurowska-Pryczkowska, K., M. Sznitowska, and S. Janicki. 1999. Studies on the effect of pilocarpine incorporation into a submicron emulsion on the stability of the drug and the vehicle. *Eur J Pharm Biopharm* 47:255.
141. Sznitowska, M., et al. 2001. *In vivo* evaluation of submicron emulsions with pilocarpine: The effect of pH and chemical form of the drug. *J Microencapsul* 18:173.
142. Sznitowska, M., et al. 2000. Increased partitioning of pilocarpine to the oily phase of submicron emulsion does not result in improved ocular bioavailability. *Int J Pharm* 202:161.
143. Melamed, S., et al. 1994. Adaprolol maleate in submicron emulsion, a novel soft β-blocking agent, is safe and effective in human studies. *Invest Ophthalmol Vis Sci* 35:1387.

144. Garty, N., and M. Lusky. 1994. Pilocarpine in submicron emulsion formulation for treatment of ocular hypertension: A phase II clinical trial. *Invest Ophthalmol Vis Sci* 35:2175.
145. Ding, S. 1995. Nonirritating emulsions for sensitive tissue. US Patent 5,474,979.
146. Acheampong, A.A., et al. 1999. Distribution of cyclosporin A in ocular tissues after topical administration to albino rabbits and beagle dogs. *Curr Eye Res* 18:91.
147. Stevenson, D., J. Tauber, and B.L. Reis. 2000. Efficacy and safety of cyclosporin A ophthalmic emulsion in the treatment of moderate-to-severe dry eye disease: A dose-ranging, randomized trial. The Cyclosporin A Phase 2 Study Group. *Ophthalmology* 107:967.
148. Sall, K., et al. 2000. Two multicenter, randomized studies of the efficacy and safety of cyclosporine ophthalmic emulsion in moderate to severe dry eye disease. CsA Phase 3 Study Group. *Ophthalmology* 107:631.
149. Klang, S.H., et al. 1999. Evaluation of a positively charged submicron emulsion of piroxicam on the rabbit corneum healing process following alkali burn. *J Control Release* 57:19.
150. Klang, S., M. Abdulrazik, and S. Benita. 2000. Influence of emulsion droplet surface charge on indomethacin ocular tissue distribution. *Pharm Dev Technol* 5:521.
151. Tamilvanan, S., et al. 2001. Ocular delivery of cyclosporin A.I. Design and characterization of cyclosporin A-loaded positively charged submicron emulsion. *STP Pharm Sci* 11:421.
152. Abdulrazik, M., et al. 2001. Ocular delivery of cyclosporin A. II. Effect of submicron emulsion's surface charge on ocular distribution of topical cyclosporin A. *STP Pharm Sci* 11:427.
153. Kalia, Y.N., et al. 2004. Iontophoretic drug delivery. *Adv Drug Deliv Rev* 56:619.
154. Wirtz, R. 1908. Die iontophoresis in der augenheilkunde. *Klin Monatsbl Augenheilkd* 46:543.
155. Karbowski, M. 1939. Iontophoresis in ophthalmology. *Ophthalmologica* 97:166.
156. Smith, V.L. 1951. Iontophoresis in ophthalmology. *Am J Ophthalmol* 34:698.
157. Erlanger, G. 1954. Iontophoresis, a scientific and practical tool in ophthalmology. *Ophthalmologica* 128:232.
158. Erlanger, G. 1955. Iontophoresis, a scientific and practical tool in ophthalmology. *Ear Nose Throat J* 34:508.
159. Fielding, I.Z., S.H. Witzel, and H.L. Rmsby. 1956. Ocular penetration of antibiotics by iontophoresis. *Am J Ophthalmol* 42:89.
160. Tonjum, A.M., and K. Green. 1971. Quantitative study of fluorescein iontophoresis through the cornea. *Am J Ophthalmol* 71:1328.
161. Colasanti, B.K., and R.R. Trotter. 1977. Enhanced ocular penetration of the methyl ester of (+/–) alpha-methyl-para-tyrosine after Lontophoresis. *Arch Int Pharmacodyn Ther* 228:171.
162. Sisler, H.A. 1978. Iontophoretic local anesthesia for conjunctival surgery. *Ann Ophthalmol* 10:597.
163. Hill, J.M., et al. 1978. Iontophoresis of vidarabine monophosphate into rabbit eyes. *Invest Ophthalmol Vis Sci* 17:473.
164. Hughes, L., and D.M. Maurice. 1984. A fresh look at iontophoresis. *Arch Ophthalmol* 102:1825.
165. Fishman, P.H., et al. 1984. Iontophoresis of gentamicin into aphasic rabbit eyes. Sustained vitreal levels. *Invest Ophthalmol Vis Sci* 25:343.
166. Burstein, N.L., L.H. Leopold, and D.B. Bernacchi. 1985. Trans-scleral iontophoresis of gentamicin. *J Ocul Pharmacol* 1:363.
167. Barza, M., C. Peckman, and J. Baum. 1986. Transscleral iontophoresis of cefazolin, ticarcillin, and gentamicin in the rabbit. *Ophthalmology* 93:133.
168. Maurice, D.M. 1986. Iontophoresis of fluorescein into the posterior segment of the rabbit eye. *Ophthalmology* 93:128.
169. Barza, M., C. Peckman, and J. Baum. 1987. Transscleral iontophoresis of gentamicin in monkeys. *Invest Ophthalmol Vis Sci* 28:1033.
170. Barza, M., C. Peckman, and J. Baum. 1987. Transscleral iontophoresis as an adjunctive treatment for experimental endophthalmitis. *Arch Ophthalmol* 105:1418.
171. Choi, T.B., and D.A. Lee. 1988. Transscleral and transcorneal iontophoresis of vancomycin in rabbit eyes. *J Ocul Pharmacol* 4:153.
172. Rootman, D.S., et al. 1988. Iontophoresis of tobramycin for the treatment of experimental *Pseudomonas keratitis* in the rabbit. *Arch Ophthalmol* 106:262.

173. Hobden, J.A., et al. 1988. Iontophoretic application of tobramycin to uninfected and *Pseudomonas aeruginosa*-infected rabbit corneas. *Antimicrob Agents Chemother* 32:978.
174. Rootman, D.S., et al. 1988. Pharmacokinetics and safety of transcorneal iontophoresis of tobramycin in the rabbit. *Invest Ophthalmol Vis Sci* 29:1397.
175. Kondo, M., and M. Araie. 1989. Iontophoresis of 5-fluorouracil into the conjunctiva and sclera. *Invest Ophthalmol Vis Sci* 30:583.
176. Grossman, R., and D.A. Lee. 1989. Transscleral and transcorneal iontophoresis of ketoconazole in the rabbit eye. *Ophthalmology* 96:724.
177. Hobden, J.A., et al. 1989. Tobramycin iontophoresis into corneas infected with drug-resistant *Pseudomonas aeruginosa*. *Curr Eye Res* 8:1163.
178. Lam, T.T., et al. 1989. Transscleral iontophoresis of dexamethasone. *Arch Ophthalmol* 107:1368.
179. Maurice, D. 1989. Iontophoresis and transcorneal penetration of tobramycin. *Invest Ophthalmol Vis Sci* 30:1181.
180. Grossman, R.E., D.F. Chu, and D.A. Lee. 1990. Regional ocular gentamicin levels after transcorneal and transscleral iontophoresis. *Invest Ophthalmol Vis Sci* 31:909.
181. Hobden, J.A., et al. 1990. Ciprofloxacin iontophoresis for aminoglycoside-resistant pseudomonal keratitis. *Invest Ophthalmol Vis Sci* 31:1940.
182. Meyer, D.R., J.V. Linberg, and R.J. Vasquez. 1990. Iontophoresis for eyelid anesthesia. *Ophthalmic Surg* 21:845.
183. Lam, T.T., J. Fu, and M.O. Tso. 1991. A histopathologic study of retinal lesions inflicted by transscleral iontophoresis. *Graefes Arch Clin Exp Ophthalmol* 229:389.
184. Sarraf, D., et al. 1993. Transscleral iontophoresis of foscarnet. *Am J Ophthalmol* 115:748.
185. Lam, T.T., et al. 1994. Intravitreal delivery of ganciclovir in rabbits by transscleral iontophoresis. *J Ocul Pharmacol* 10:571.
186. Yoshizumi, M.O., et al. 1995. Ocular toxicity of iontophoretic foscarnet in rabbits. *J Ocul Pharmacol Ther* 11:183.
187. Yoshizumi, M.O., et al. 1996. Ocular iontophoretic supplementation of intravenous foscarnet therapy. *Am J Ophthalmol* 122:86.
188. Yoshizumi, M.O., et al. 1997. Determination of ocular toxicity in multiple applications of foscarnet iontophoresis. *J Ocul Pharmacol Ther* 13:529.
189. Sarraf, D., and D.A. Lee. 1994. The role of iontophoresis in ocular drug delivery. *J Ocul Pharmacol Ther* 10:69.
190. Behar-Cohen, F.F., et al. 1997. Iontophoresis of dexamethasone in the treatment of endotoxin-induced-uveitis in rats. *Exp Eye Res* 65:533.
191. Behar-Cohen, F.F., et al. 2002. Transscleral Coulomb-controlled iontophoresis of methylprednisolone into the rabbit eye: Influence of duration of treatment, current intensity and drug concentration on ocular tissue and fluid levels. *Exp Eye Res* 74:51.
192. Halhal, M., et al. 2004. Iontophoresis: From the lab to the bed side. *Exp Eye Res* 78:751.
193. Frucht-Pery, J., et al. 1996. Efficacy of iontophoresis in the rat cornea. *Graefes Arch Clin Exp Ophthalmol* 234:765.
194. Frucht-Pery, J., et al. 1999. The distribution of gentamicin in the rabbit cornea following iontophoresis to the central cornea. *J Ocul Pharmacol Ther* 15:251.
195. Frucht-Pery, J., et al. 2004. Iontophoresis-gentamicin delivery into the rabbit cornea, using a hydrogel delivery probe. *Exp Eye Res* 78:745.
196. Li, S.K., E.K. Jeong, and M.S. Hastings. 2004. Magnetic resonance imaging study of current and ion delivery into the eye during transscleral and transcorneal iontophoresis. *Invest Ophthalmol Vis Sci* 45:1224.
197. Voigt, M., et al. 2002. Ocular aspirin distribution: A comparison of intravenous, topical, and coulomb-controlled iontophoresis administration. *Invest Ophthalmol Vis Sci* 43:3299.
198. Kralinger, M.T., et al. 2003. Ocular delivery of acetylsalicylic acid by repetitive coulomb-controlled iontophoresis. *Ophthalmic Res* 35:102.
199. Vollmer, D.L., et al. 2002. *In vivo* transscleral iontophoresis of amikacin to rabbit eyes. *J Ocul Pharmacol Ther* 18:549.

200. Horwath-Winter, J., et al. 2005. Iodide iontophoresis as a treatment for dry eye syndrome. *Br J Ophthalmol* 89:40.
201. Asahara, T., et al. 2001. Induction of gene into the rabbit eye by iontophoresis: Preliminary report. *Jpn J Ophthalmol* 45:31.
202. Voigt, M., et al. 2002. Down-regulation of NOSII gene expression by iontophoresis of anti-sense oligonucleotide in endotoxin-induced uveitis, *Biochem Biophys Res Commun* 295:336.
203. Berdugo, M., et al. 2003. Delivery of antisense oligonucleotide to the cornea by iontophoresis. *Antisense Nucleic Acid Drug Dev* 13:107.
204. Davies, J.B., et al. 2003. Delivery of several forms of DNA, DNA–RNA hybrids, and dyes across human sclera by electrical fields. *Mol Vis* 9:569.
205. Mir, L.M., et al. 2005. Electric pulse-mediated gene delivery to various animal tissues. *Adv Genet* 54:83.
206. Oshima, Y., et al. 2002. Targeted gene transfer to corneal stroma *in vivo* by electric pulses. *Exp Eye Res* 74:191.
207. Blair-Parks, K., B.C. Weston, and D.A. Dean. 2002. High-level gene transfer to the cornea using electroporation. *J Gene Med* 4:92.
208. Dezawa, M., et al. 2002. Gene transfer into retinal ganglion cells by *in vivo* electroporation: A new approach. *Micron* 33:1.
209. Yu, W.Z., et al. 2003. Gene transfer of kringle 5 of plasminogen by electroporation inhibits corneal neovascularization. *Ophthalmic Res* 35:239.
210. Mamiya, K., et al. 2004. Effects of matrix metalloproteinase-3 gene transfer by electroporation in glaucoma filter surgery. *Exp Eye Res* 79:405.
211. Chen, Y.X., C.E. Krull, and L.W. Reneker. 2004. Targeted gene expression in the chicken eye by in ovo electroporation. *Mol Vis* 10:874.
212. Kachi, S., et al. 2005. Nonviral ocular gene transfer. *Gene Ther* 12:843.
213. Bloquel, C., et al. 2006. Plasmid electrotransfer of eye ciliary muscle: Principles and therapeutic efficacy using hTNF-α soluble receptor in uveitis. *FASEB J* 20:389.

25 Ocular Penetration Enhancers

Indu Pal Kaur and Anupam Batra

CONTENTS

25.1 INTRODUCTION

Successful delivery of drugs into the eye is extremely complicated because the eye is protected by a series of complex defense mechanisms that make it difficult to achieve an effective concentration of the drug within the target area of the eye. Drugs delivered in classical ophthalmic dosage forms (eye drops) into the lower cul-de-sac have a poor bioavailability due to these complex defense mechanisms. Furthermore, drugs administered systemically for their ocular action have poor access to the eye tissue because of the blood–aqueous barrier, which prevents drugs from entering the extravascular retinal space and the vitreous body [1,2].

After topical administration of an ophthalmic drug solution, the drug has to cross a succession of anatomical barriers before reaching the systemic circulation. These barriers (as shown in Figure 25.1) can be commonly classified as precorneal and corneal barriers [3].

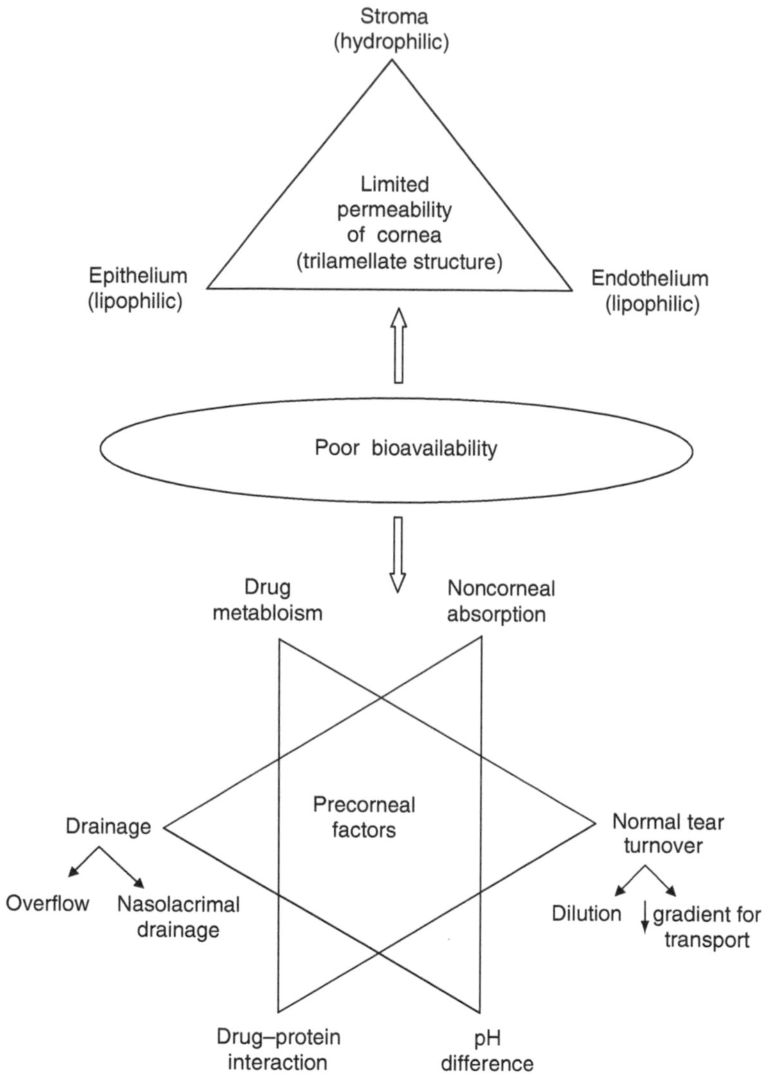

FIGURE 25.1 Factors affecting the poor bioavailability from ocular dosage forms.

25.1.1 Precorneal Barriers

The precorneal barriers are the first barriers that slow down the penetration of an active ingredient into the eye. They are comprised of the tear film and the conjunctiva [4].

1. *Tear film*: Precorneal tear film (as shown in Figure 25.2) can be considered as being made up of three layers [5]. These layers are
   a. Superficial lipid layer: The superficial lipid layer is 0.1 μm thick and derived from the meibomian glands and the accessory sebaceous glands of Zeiss [4]. This layer is composed of esters, triacylglycerols, free sterols, sterol esters, and free fatty acids [6].
   b. Aqueous layer: The aqueous layer is about 7 μm thick [5] and secreted by the lachrymal gland and the accessory glands of Krauss and Wolfring. This layer

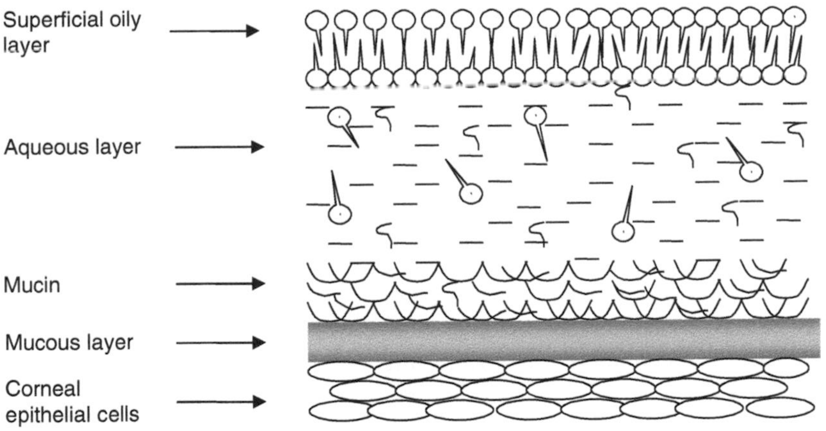

FIGURE 25.2 Schematic representation of the precorneal tear film.

contains inorganic salts, glucose, and urea as well as biopolymers, proteins, and glycoproteins.

c. Mucus layer: The mucus layer forms the bottom layer of the tear film lying adjacent to and wetting the corneal epithelium. It is elaborated by the goblet cells of the conjunctiva and forms a bridge between the hydrophobic corneal epithelial surface and the aqueous layer of the tear film lying immediately above it. The composition of mucus varies widely depending on the animal species, the anatomical location, and the normal or pathological state of the organism [7]. Its major constituents are the high molecular weight glycoproteins capable of forming slimy and viscoelastic gels containing more than 95% water [8]. These glycoproteins form disulfide as well as ionic bonds and physical entanglements, and consist of a peptide backbone, a major portion of which is covered with carbohydrates (grouped in various combinations) such as galactose, fructose, *N*-acetyl glucosamine, and sialic acid. At physiological pH, the mucus network usually carries a significant negative charge because of the presence of sialic acid and sulfate residues [9].

The tear film fulfills several important functions in the eye such as formation and maintenance of a smooth refracting surface over the cornea, lubrication of the eyelids, transportation of metabolic products (O_2 and CO_2) to and from the epithelial cells and cornea, and elimination of foreign substances and bactericidal action [10,11].

Tear film factors [4] that can influence the ocular bioavailability of a drug are as follows:

1. Solution drainage rate to the nasolachrymal duct
2. Lacrimation and tear turnover causing dilution
3. Drug binding to tear proteins resulting in a lower free drug concentration for ocular absorption
4. Enzyme metabolization of drugs
5. Electrolyte composition of tears
6. pH and buffer capacity of tears, which can also influence the therapeutic effect of drugs, their toxicity, and their stability

After ocular instillation, aqueous eye drop solutions and suspensions mix with the tear fluid and are dispersed over the eye surface. The greater part of the applied drug solution (25–50 μL) is rapidly drained from the eye surface so as to return to the normal resident tear volume of 7.5 μL. Thereafter, the preocular solution volume remains constant, but the drug concentration keeps on decreasing due to dilution by tear turnover and corneal and noncorneal absorptions (Figure 25.1). Furthermore, ocular administration of an irritating drug or vehicle can increase the drug loss from the precorneal area due to an increased tear flow rate (induced lacrimation). This not only decreases the concentration of the drug in contact with the cornea but also shortens the corneal contact time, resulting in a poor absorption of the applied drug.

Another important route of drug loss from the precorneal area is the noncorneal absorption that mostly occurs via the conjunctiva [12].

2. *Conjunctiva*: The conjunctiva is a thin mucous membrane lining inside the eyelids and the anterior sclera [12]. Its surface is an order of magnitude greater than that of the cornea [13] and two to three times (depending upon the drug) more permeable than the cornea [14,15]. Since the conjunctiva is highly vascularized, blood circulation removes most of the instilled drug before it can enter the inner ocular tissues.

   These precorneal factors collectively reduce the amount of drug that reaches the aqueous humor and eye tissues through the corneal epithelium to about 10% of the total drug instilled into the eye [4].

25.1.2 Corneal Barriers

Limited corneal area and poor corneal permeability (Figure 25.1) are the main corneal barriers that are responsible for the poor bioavailability of drugs applied topically as eye drops.

Although the surface of the cornea (1.3 cm^2) represents barely one-sixth of the total surface area of the eyeball, it is the cornea through which the drug reaches the inner tissues of the eye. The cornea, which is 1 mm thick at its edges and 0.5 mm thick at its center, is classically described as a heterogeneous tissue composed of the following [16]:

1. The epithelium and its basal layer in contact with Bowman's membrane
2. The corneal stroma, which alone represents nine-tenths of the total thickness of the cornea
3. Descemet's membrane and the lowermost corneal endothelium

The corneal epithelium in itself is also a layered structure, 50–100 μm thick, consisting of a deep layer of basal columnar cells, an intermediate layer of polyhedral cells, and a surface layer of squamous, polygonal-shaped cells. The outermost cells have skirting intercellular junctions, termed tight junctions (Figure 25.3). These form a strong barrier to nonlipophilic substances, allowing the preferential penetration of nonionized forms.

Bowman's membrane separating the stroma from the epithelial tissue is composed of a layer of collagen fibers, 8–14 μm thick, forming a tough and relatively impermeable barrier. The corneal stroma is composed essentially of collagen and represents approximately 90% of the thickness of the cornea. It is highly hydrophilic, porous, and an open-knit, thus allowing the free passage of hydrophilic substances but acting as a barrier to lipophilic molecules [4].

Descemet's membrane covers the posterior surface of the stroma. It is a single-cell layer, 5–10 μm thick, and also composed of collagen. The corneal endothelium, a single-cell layer lining the posterior surface of the stroma, is rich in phospholipids, permeable to lipid-soluble materials, and almost impermeable to ions.

Almost all of the ophthalmic drugs that have been studied so far appear to cross the cornea by simple diffusion involving the paracellular and transcellular pathways (Figure 25.3). The paracellular pathway anatomically involves the intercellular space and is the primary route of

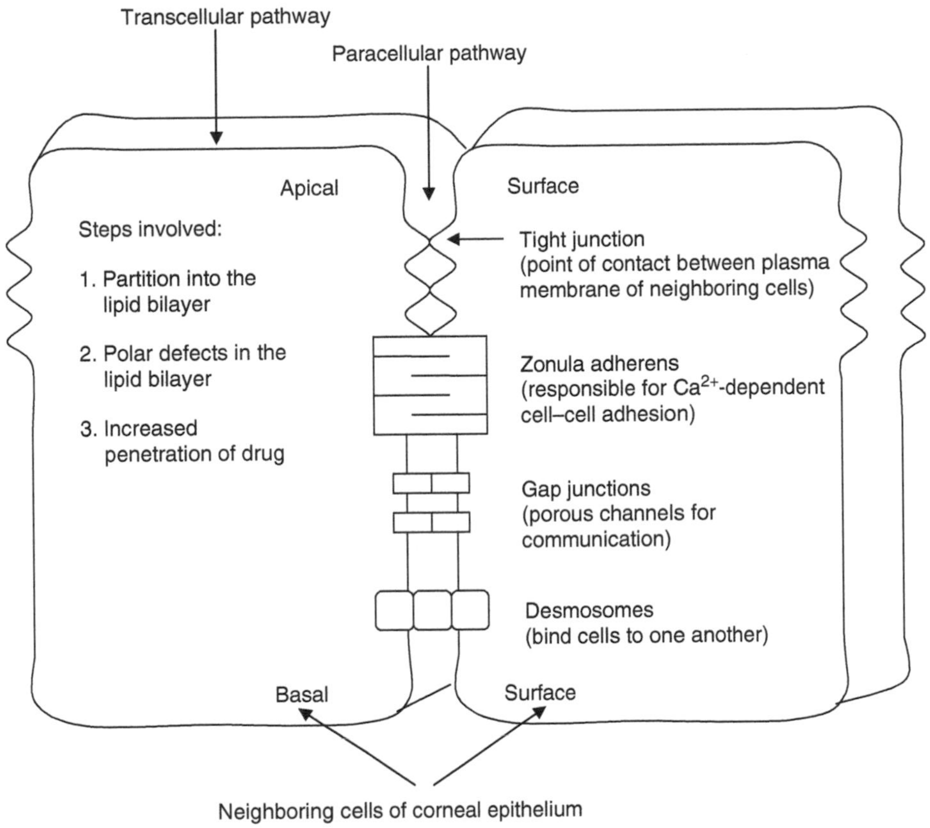

FIGURE 25.3 Schematic illustration of the transcellular and paracellular modes of action of penetration enhancers.

passive ion permeation [17]. Considering that the total surface area of the cornea exposed to the tears, attributable to the paracellular pathway, is rather small, it is supposed that most of the drugs would probably opt for the transcellular pathway in crossing the cornea. The principal drug properties governing drug absorption via this pathway are as follows:

1. The lipophilicity of the drug, as reflected by its *n*-octanol/buffer partition coefficient (PC), with the optimum PC being in the range 10–100.
2. pK_a, which determines the proportion of drug in its preferentially absorbable form at a given pH.
3. Solubility, drugs that are both lipid and water soluble pass through the cornea readily.

25.2 IMPROVING THE AVAILABILITY OF DRUGS USING VARIOUS FORMULATION APPROACHES

A major challenge in ocular therapeutics is to improve ocular bioavailability from less than 1%–3% to at least 15%–20%. The last three decades have witnessed continued efforts in this direction, and current investigations are being pursued along the following lines:

1. To increase the transcorneal passage of drugs by incorporating absorption promoters and penetration enhancers into the drug formulations.

2. To optimize the formulation vehicles for prolonged drug retention in the precorneal area and increase the contact time between the administered drug and the conjunctival and corneal epithelia by the following methods:

   - Addition of water-soluble, natural, synthetic, or semisynthetic viscolizers.
   - Use of drug carrier systems such as nanoparticles, microspheres, liposomes, etc., which would remain in the cul-de-sac for a long period of time, thus giving a sustained action.
   - Using a new approach that involves the utilization of the mucoadhesive property of polymers to improve the ocular absorption of poorly absorbed drugs.

Since much has already been published about the use of viscolizers and the drug carrier systems, along with their application to ophthalmics, the main focus of this chapter will be on penetration enhancers that can improve the ocular bioavailability so that an ocularly delivered drug can elicit its biological action. The aim of this chapter is to give an insight into the use of corneal penetration enhancers for innovative ophthalmic delivery approaches, to decrease the systemic side effects, and create a more pronounced effect. Ophthalmic formulations based on these penetration enhancers are simple to manufacture and their use can increase the rate and amount of drug transport.

25.3 USE OF PENETRATION ENHANCERS

This approach consists of transiently increasing the permeability characteristics of the cornea by appropriate substances, known as penetration enhancers or absorption promoters. It bears a strict analogy to techniques aimed at facilitating drug penetration through the skin and other epithelia such as the buccal, nasal, intestinal, or rectal.

An ideal penetration enhancer should have the following desirable characteristics [18]:

1. Its action should be immediate and unidirectional, and the duration of effect should be specific and predictable.
2. After the removal of the material from the applied surface, the tissue should immediately recover its normal barrier property.
3. It should not exhibit any systemic or toxic effects.
4. It should not irritate or damage the membrane surface.
5. It should be physically compatible with a wide range of drugs and pharmaceutical excipients.

Even though we are currently far from having penetration enhancers that fulfill all or most of the above requirements, several classes of ocular absorption enhancers, however, are being evaluated for their suitability under the conditions of use by various groups. This chapter aims to discuss various classes of ocular absorption enhancers, along with the basic modes of drug transport and the mechanism by which these penetration enhancers act. Further, new categories of penetration enhancers, which have been reported more recently, are also included.

25.3.1 Basic Mechanisms in Drug Transport through the Cornea

Drug transport across corneal epithelial cells can be categorized into two major groups: transcellular transport and paracellular transport

1. *Transcellular transport*: Basic mechanisms of transepithelial transport of drugs include passive transport of small molecules, active transport of ionic and polar compounds, and endocytosis and transcytosis of macromolecules.

The rate at which a drug molecule diffuses across the lipid bilayer of the cell membrane depends largely on the size of the molecule and its relative lipid solubility. In general the smaller and more hydrophobic the molecule, the more rapid will be its diffusion across the bilayer. However, cell membranes are also permeable to some small, water-soluble molecules such as ions, sugars, and amino acids [19].

2. *Paracellular transport*: Paracellular transport is the transport of molecules around or between cells. The paracellular pathway consists of the intercellular space with intercellular junctions. Further, the intercellular junction consists of the zonula occludens or tight junction at the most apical zone of contact. Adjacent to the tight junction is the adherens junction (Figure 25.3). Below these two junctions are the spotlike contacts of desmosomes and gap junctions [20,21]. Adherens junctions are formed by the transmembrane protein cadherin that is responsible for Ca^{2+}-dependent cell–cell adhesion, whereas the desmosomes are structures that bind cells to one another. Gap junctions include connexons and contain porous channels for communication between cells. These structures do not provide any barrier for drug penetration. The entry of molecules through the paracellular pathway is primarily restricted by tight junctions. The latter are band like structures in which the plasma membranes of neighboring cells are brought into close apposition and completely encircle the superficial epithelial cells. Both the assembly and barrier properties of tight junctions are influenced by the classical second messenger and signaling pathways, including tyrosine kinases, Ca^{2+}, protein kinase C, G proteins, calmodulin, cAMP, and phospholipase C [22,23]. Tight-junction permeability has been shown to depend on a number of factors [24], including (1) degree of maturation of epithelia, (2) response to physiological requirements, (3) change in environmental conditions such as osmolarity and ionic strength, (4) presence of drugs, vitamins, and hormones, and (5) concentration of calcium.

25.3.2 Mechanisms of Action of Penetration Enhancers

Penetration enhancers can show their action by acting either on the membrane components (transcellular mode of action) or on tight junctions (paracellular mode of action) as shown in Figure 25.3.

1. *Action on the membrane components*: Numerous studies have shown that the passive transcellular transport of hydrophilic compounds, including macromolecules such as peptides, can be enhanced by interaction of the penetration enhancers with both the phospholipid bilayer and the integrated proteins, thereby making the membrane more fluid and thus more permeable to both lipophilic and hydrophilic compounds.

   Fatty acids [25], monoglycerides [25], fatty acid–bile salt micelles [26], and surfactants [27] have all been shown to increase epithelial membrane permeability by affecting the membrane proteins or lipids [28].
2. *Action on the tight junctions*: As already mentioned, the intercellular tight junction is one of the major barriers to the paracellular transport of macromolecules and polar compounds. Calcium ions play an important role as cross-linking agents in mucus structure and the intercellular junctions [29]. Thus, the maintenance of the tight junction requires an unknown appropriate level of Ca^{2+} bound to or in the vicinity of the plasma membrane [30]. The chelation of Ca^{2+} can therefore lead to the disruption of these tight junctions, resulting in the increased paracellular transport of poorly absorbed drugs.

25.4 CLASSES OF PENETRATION ENHANCERS (TABLE 25.1)

25.4.1 Calcium Chelators

The calcium chelators such as EDTA have been reported to loosen the tight junctions between the superficial epithelial cells, thus facilitating paracellular transport [31,32]. The Ca^{2+} depletion does not act directly on the tight junctions; rather it induces global changes in the cells, including (1) the disruption of actin filaments, (2) the disruption of adherent junctions, (3) diminished cell adhesion of actin filaments, and (4) the activation of protein kinases [33].

The association of actin filaments inside the cell with the junctional complex also seems to be disrupted by the low extracellular Ca^{2+} [34,35]. Grass and Robinson [36] were among the

TABLE 25.1
Various Classes of Penetration Enhancers

Class	Mechanism	Examples of Drugs for which these Agents are Reported	Refs.
Calcium chelators			
• EDTA	Loosening of tight junctions between superficial epithelial cells	Glycerol	[36]
		Cromolyn	
		Timolol	[37,38]
		Atenolol	
Surfactants			
Nonionic surfactants			
• Polyoxyethylene-9-lauryl ether	Phospholipid acyl chain perturbation	Sulfadicramide	[45]
• Tween 80		Acetazolamide	[50]
• Span 60		Acetazolamide	[51]
Bile acids and salts			
• Na-deoxycholate • Na-taurocholate • Na-taurodeoxycholate	Change in rheologic properties of biological membrane due to mucolytic action	Glutathione 6-carboxy fluorescein FITC-dextran and Insulin	[52]
Preservatives			
• Benzalkonium chloride	Loosening of tight junctions and enlargement of intercellular spaces	Pilocarpine	[55]
		Prostaglandin	[57]
		Timolol	[58]
		Carbachol	[59]
• Cetylpyridinium chloride		Pilocarpine nitrate	[64]
		Penicillin	[65]
Miscellaneous			
Glycosides			
• Digitonin	Exfoliation of epithelium by solubilizing membrane cholesterol	Polyethylene glycols	[70]
• Saponin		Insulin	[67]
		β-Blockers	[39]
Fatty acids			
• Capric acid • Oleic acid • Short fatty acids	Loosening of tight junctions as well as membrane perturbation	β-Blockers	[39]

TABLE 25.1 (Continued)
Various Classes of Penetration Enhancers

Class	Mechanism	Examples of Drugs for which these Agents are Reported	Refs.
Azone	Loosening of epithelial cell junctions as well as change in membrane structure	Cyclosporine Cimetidine	[74]
Chitosan	Opening of tight junctions.	Ofloxacin	[79]
Tamarind seed polysaccharide	Accumulation of drug at membrane surface	Gentamycin Ofloxacin	[81]
		Rufloxacin	[82]
Polycarbophils and its derivatives			
• Polycarbophil	Loosening of tight junctions	Gentamycin	[93]
• Polycarbophil–cysteine conjugate		Sodium fluorescein	[94]
Cytochalasins	Alteration of cytoskeleton polymerization and thus loosening of tight junctions	Several peptides	[88]
Cyclodextrins	Solubilization of lipophilic drug and thus an increase in availability at the surface of biological barrier	Arachidonylethanolamide	[98]
		Ganciclovir dibutyrate	[99]
		Pilocarpine nitrate Cyclosporine A	[100]
		Acetazolamide	[50]

first to emphasize the positive effect of chelating agents on corneal drug absorption. They found that 0.5% EDTA doubled the ocular absorption of topically applied glycerol and cromolyn sodium. Similarly, 0.5% EDTA also showed an enhancing effect for timolol and atenolol, but reduced the transcorneal flux of levobunolol [37,38]. Sasaki et al. [39] reported small but significant permeability increases for timolol and the more lipophilic befunolol with 0.5% EDTA. EDTA acts on the tight junctions producing ultrastructural changes in the corneal epithelium, resulting in a water influx and a decrease of the overall lipophilic characteristics [31]. This effect may account for the permeability reduction observed in the case of more lipophilic drugs. The studies in our laboratory indicated that there was an increase in the intraocular pressure lowering effect of acetazolamide in the presence of 0.5% EDTA [40].

25.4.2 Surfactants

Unlike the chelators, surfactant enhancers have been suggested to increase drug and peptide permeability through the cell membranes or via the transcellular pathway [41]. Epithelial cells are surrounded by an outer cell membrane composed of a phospholipid bilayer with the protein molecules embedded in the lipid membrane. When present at low concentrations, these surfactants are incorporated into the lipid bilayer, forming polar defects that change the physical properties of the cell membranes. When the lipid bilayer is saturated, mixed micelles begin to form, resulting in the removal of phospholipids from the cell membranes and hence leading to membrane solubilization. It has therefore been suggested that a dose-dependent increase in the permeability of the cell membrane is responsible for surfactant-induced increases in permeability across different epithelia. Some surfactants have however been reported to increase the paracellular penetration of drugs by affecting the tight junctions [42,43]. Palmitoyl carnitine, a zwitterionic surfactant, has been shown to enhance drug absorption without causing significant changes to the morphology of intestinal cells both *in vivo* and *in vitro*, suggesting that this surfactant increases the paracellular penetration via the tight junction [43]. Disruption of tight-junction integrity of intestinal epithelial cells using surfactants, such as sodium caprate and sodium dodecyl sulfate, has also been reported [44].

Various classes of surfactants reported in the literature to act as corneal penetration enhancers include the following:

1. *Nonionic surfactants*: The corneal permeability of atenolol, timolol, and betaxolol was increased significantly by 0.05% Brij 35, Brij 78, and Brij 98(ICI Americas, Inc., USA), respectively [31]. The effect of sodium deoxycholate, polyoxyethylene-9-lauryl ether (nonionic surfactant), and L-α-lysophosphatidylocholine on sulfadicramide dialysis was studied through synthetic membranes and the animal cornea (*in vitro*). Polyoxyethylene-9-lauryl ether was found to be the most effective surfactant among these [45]. The promoting effects induced by polyoxyethylene ethers depend on the number of oxyethylene groups in a molecule and also on the HLB number. A significant effect is obtained if many such groups are present in the molecule. The use of this group to enhance the biological availability of various drugs applied to the eye [46,47] consists of a combination of two mechanisms: loosening the cell packing through local disturbances of Ca^{2+} metabolism and increasing the permeability of a given tissue by washing out proteins from the membranes [48,49]. The permeability of acetazolamide was found to be significant when 0.5% of the drug was dispersed with 1% Tween 80. The permeability compared well with a solution of acetazolamide in 10% hydroxypropyl-β-cyclodextrin (HP-β-CD) at the same concentration [50] and also when the same concentration of acetazolamide was entrapped in niosomes [51]. The increase in permeation of the latter system could further be explained in terms of the use of span 60 (in combination with cholesterol) for preparing the niosomal vesicles [51].
2. *Bile acids and salts*: Bile salts are amphipathic molecules that are surface active and self-associate to form micelles in aqueous solutions. These agents, e.g., deoxycholate, taurodeoxycholate, and glycocholate, act by changing the rheological properties of biological membranes [39]. Owing to their mucolytic properties, these agents can increase diffusion of drug molecules through the membrane by inducing a transient change in its structure and permeability [45]. The exact mechanism of the enhancement of corneal permeability by bile salts is still unclear, though it has been indicated that they promote the transport of hydrophilic and macromolecular compounds, mainly through a pore-like route, such as intercellular channels [52]. The effect of taurocholic acid and taurodeoxycholic acid on the *in vitro* rabbit corneal permeability of hydrophilic molecules such as 6-carboxy fluorescein and glutathione, and macromolecular compounds such as FITC-dextran and insulin was studied. Taurodeoxycholic acid markedly increased the corneal permeability of these penetrants at 2 and 10 mM concentrations [52]. Taurocholic acid has been reported to show a significant increase in the conjunctival absorption of β-blockers. However, it had a lower promoting activity on the corneal permeability [39]. Further, it has been reported that sodium taurocholate (Tc-Na) marginally increased the corneal permeabilities of these hydrophilic and macromolecular compounds, whereas sodium taurodeoxycholate (TDC-Na) produced a marked effect [52]. A difference in the physicochemical properties of these bile salts, e.g., their solubilizing activity, lipophilicity, and Ca^{2+} sequestration capacity, is probably related to their permeability-enhancing effects. The critical micellar concentrations (CMCs) of

dihydroxy bile salts are generally lower, and their aggregation numbers larger than those of trihydroxy salts. Thus, the solubilizing activity of sodium taurodeoxycholate is higher than that of sodium taurocholate. Moreover, sodium taurodeoxycholate possesses a higher lipophilicity and Ca^{2+} sequestration activity [52], leading to a greater loosening of the corneal epithelial barrier.

3. *Lysophosphatidilo lipids*: These are amphiphilic surfactants produced in a natural way from phospholipids by phospholipases. Their mechanism of action as a promoter is not fully understood. It is supposed that, like other surfactants, they can affect intracellular proteins and polar groups of phospholipids in intercellular spaces, which may favor the formation of channels permitting the penetration of water and substances dissolved therein [45].

25.4.3 Preservatives

Many researchers have demonstrated that some preservatives significantly increase the corneal permeability of ophthalmic drugs [53–56].

1. *Benzalkonium chloride* (*BAC*): Benzalkonium chloride shows the highest promoting effect on corneal drug penetration among the currently used preservatives. Camber and Edman [55] demonstrated that exposing the isolated porcine cornea to 0.017% BAC or 0.5% chlorbutanol for 4 h almost doubled the transcorneal flux of pilocarpine. The addition of BAC (0.005%) enhanced 2.5-fold the transport percentage and P_{app} of S-1033, a novel prostaglandin derivative used as an antiglaucoma medication [57]. Podder et al. [58] reported that BAC increased the ocular absorption of timolol in rabbits by approximately 80%, but that systemic timolol absorption was also increased and no change was observed in the absorption ratio between the eye and systemic circulation. Smolen et al. [59] demonstrated that BAC and another cationic surfactant, diethylaminoethyl dextran, enhanced the miotic response of topically applied carbachol. Instillation of 0.01% BAC shows no ocular irritation according to the Draize score, and at a concentration of 0.004%–0.01% BAC had no influence on the epithelial aerobic metabolism [60]. Corneal exposure to multiple drops of BAC leads to epithelial accumulation but no penetration into the anterior chamber. However, 0.01% BAC has been reported to cause cells of the corneal epithelium to peel at their borders [61]. Benzalkonium chloride has also been found to enlarge the intercellular spaces in the superficial cells of the cornea [62,63]. It has been established that rabbits are more sensitive to single doses of preservatives than humans, and it was found that 0.01% BAC increased the anterior chamber fluorescence level in rabbits but not in humans [53].
2. *Cetylpyridinium chloride*: Inclusion of 0.02% cetylpyridinium chloride with pilocarpine nitrate applied topically to the rabbit eye was found to cause a miotic effect 10 times that obtained in the absence of cetylpyridinium [64]. In another study, cetylpyridinium chloride was shown to enhance penicillin penetration across the isolated rabbit cornea [65] with an intact epithelial layer. The penetration was even more than that shown by the de-epithelialized corneas.

25.4.4 Glycosides

Some glycosides with surface activity have been used successfully as penetration enhancers. For example:

1. *Saponin*: Saponin is a type of glycoside widely distributed in plants. It is an amphiphilic compound that has surface activity. Purified quillaja saponin has been reported to

possess an ability to promote the nasal absorption of gentamicin antibiotics in mice and rats, and the nasal or ocular absorption of insulin in rats [66]. At 0.5% concentration, saponin was found to enhance the corneal penetration of β-blockers [39]. It increased the ocular permeability of hydrophilic β-blockers markedly, but the increase was slight in the case of lipophilic β-blockers. At 1.0% concentration, saponin has been reported to act as a promoter of systemic absorption of insulin administered via eye drops [67]. It is not known at present if the saponins act by simple detergent action to make the epithelial membrane more permeable to intermediate-sized peptides, or if there is a selective interaction between certain moieties of the excipient and certain sites on the epithelial surface that play a regulatory role in controlling peptide absorption [66].

2. *Digitonin*: Digitonin (a nonionic surfactant) possesses certain detergent characteristics and has the ability to permeabilize the membranes of a wide variety of cells [68]. In the cornea, digitonin has been found to selectively solubilize membrane cholesterol and cause exfoliation of the epithelium layer by layer [69]. Digitonin has been reported to greatly increase the corneal absorption, *in vitro*, of a series of polyethylene glycols with different molecular weights. However, its use also led to a significant alteration or sometimes a complete removal of the epithelial layer [70].

25.4.5 Fatty Acids

These agents act as absorption promoters by affecting both the cell membranes and the tight junctions, and by forming ion-pair complexes with cationic drugs [71]. Muranishi [72] demonstrated that fatty acids perturb the membrane structural integrity by their incorporation into the plasma membrane. The effect of capric acid on the paracellular pathway has been related to the Ca^{2+}-dependent contraction of the perijunctional actomyosin ring, as well as the chelation of Ca^{2+} around the tight junction. Caprylic acid was reported to interact mainly with proteins, whereas capric acid interacted with both proteins and lipids [39]. Capric acid has been reported to increase the ocular delivery of hydrophilic β-blockers, but causes only a slight improvement in the ocular delivery of lipophilic β-blockers [39]. Kato and Iwata [71,73] also reported that fatty acids significantly increased the corneal permeability of bunazosin, but this enhancing effect was attributed to ion-pair formation rather than membrane disturbance.

25.4.6 Miscellaneous

Several other agents or groups of substances have also been quoted in the literature to act as ocular promoters. Some of them are listed below:

1. *Azone*: Newton et al. [74] reported that azone, a transdermal absorption promoter, increased the ocular delivery of instilled cyclosporine and enhanced its immunosuppression activity. Four penetration enhancers (azone [laurocapram], hexamethylene lauramide, hexamethylene octanamide, and decylmethyl sulfoxide) were studied for cimetidine and all of them enhanced the corneal penetration of cimetidine. The effect of azone on the corneal permeability of a series of structurally unrelated drugs, ranging from hydrophilic to lipophilic, was also studied. Azone enhanced the transcorneal penetration of hydrophilic drugs but retarded the apparent drug permeation of lipophilic drugs across the cornea [75]. The mechanism of these penetration enhancers may have to do with the changes in the structure and fluidity of the barrier membrane because of its high lipophilicity and sequestration in the epithelium. It is also possible that the enhancer,

when highly concentrated in the epithelium, may loosen the epithelial cell junctions and facilitate the influx of water and hydrophilic compounds but retard the movement of lipophilic molecules by creating a more hydrated barrier [75].

2. *Chitosan*: Chitosan, a well-known cationic polymer of natural origin, was found to enhance the permeability of corneal epithelial cells by opening the tight junctions between them, thereby favoring paracellular drug transport [76,77]. More recently, it has been found that chitosan increases cell permeability by affecting both paracellular and intracellular pathways of epithelial cells in a reversible manner, without affecting cell viability or causing membrane wounds [78]. The permeability of the conjunctiva can also be enhanced by chitosan. Chitosan has an excellent ocular tolerance and favorable rheologic behavior, and it was initially used for its mucoadhesive property. Aqueous solution containing 1% w/v of chitosan in HCl significantly enhanced (190% increase of peak concentration in aqueous humor) intraocular penetration of ofloxacin [79].
3. *Tamarind seed polysaccharide* (*TSP; xyloglucan*): A mucoadhesive polymer extracted from tamarind seeds has been found to significantly increase the corneal accumulation and intraocular penetration of various antibiotics. Being similar to mucins, TSP is able to bind to the cell surface and intensify the contact between the drugs and the adsorbing biological membrane [80]. Ghelardi et al. [81] found that TSP significantly increased the corneal accumulation and intraocular penetration of gentamycin and ofloxacin when administered topically to healthy rabbits. Later Ghelardi et al. [82] also reported that TSP significantly increased the intraocular penetration of rufloxacin in the treatment of experimental *Pseudomonas aeruginosa* and *Staphylococcus aureus* keratitis in rabbits when administered topically in a drop regimen. It was also found to prolong the precorneal residence times of antibiotics and enhance drug accumulation in the cornea, probably by reducing the washout of topically administered drugs [82].
4. *Cremophore EL*: Cremophore EL, a polyethoxylated castor oil used as a vehicle in various intravenous injections, has also been found to enhance the intraocular penetration of the drugs. Bijl et al. [83] reported that for Cremophore EL (10% and 20%), flux values of cyclosporine A across the fresh cornea were significantly higher than for corneal tissue without the enhancer for the first 16 h of the experiment. Steady-state flux rates were reached for cyclosporine A across the corneal tissue after approximately 2 h in the presence of 10% and 20% Cremophore EL. This observation can probably be explained on the basis of increased aqueous solubility of cyclosporine A in the presence of this emulsifier. Cremophore EL also improved the penetration of cyclosporine A through the epithelial and stromal layers of the cornea [83].
5. *Cytochalasins*: Cytochalasins are a group of small, naturally occurring heterocyclic compounds that bind specifically to actin microfilaments, the major component of the cell cytoskeleton, and alter their polymerization [84,85]. In addition to its normal role in regulating cell contractibility, mobility, and cell-surface receptors, the cytoskeleton has been shown to participate in the regulation of epithelial tight-junction permeability [86]. The actin microfilament has been found to play a role in positioning junctional strands through its association with plasma membrane components and influencing the degree of opening of the occluding junctions [87]. Among the cytoskeleton-active agents, the specificity and efficacy of cytochalasin B on tight-junction permeability make it a very useful agent in peptide delivery. It has been shown to be effective in enhancing the corneal permeability with minimal membrane damage [88].
6. *Ionophores*: Mitra [89] showed that an ionophore such as lasalocid can also be used to enhance the corneal permeability of pilocarpine.
7. *Polycarbophils and its derivatives*: Polyacrylic acid derivatives such as Carbomer 934P and Polycarbophil (B.F. Goodrich Specialty Polymers of Cleveland, Ohio) that are

weakly cross-linked have been extensively studied for their possible use as penetration enhancers that exclusively trigger paracellular transport and show no toxic side effects.

These are mucoadhesive polymers and thought to penetrate the mucous layer, thus reaching the physical barrier and facilitating transport of hydrophilic molecules across the mucous layer [90,91]. These agents have also been found to remove Ca^{2+} from the intestinal epithelial cells by formation of poly(acrylic acid)–Ca^{2+} complexes and thus loosen the cellular barrier, especially by triggering the (reversible) opening of tight junctions [92]. Presently, these agents are under investigation for their use as ocular penetration enhancers. Lehr et al. [93] reported that polycarbophil formulations increased by two times the uptake of gentamycin by the bulbar conjunctiva.

More recently, polycarbophil–cysteine conjugates (thiolated polycarbophils) have been found to increase the permeation of some drugs. Hornof and Schnurch [94] found that the polycarbophil–cysteine conjugates increase the permeation of sodium fluorescein (2.2-fold) and dexamethasone phosphate (2.4-fold). The unmodified polymer however was ineffective.

8. *Cyclodextrins* (*CDs*): CDs are a group of cyclic oligosaccharides that have been shown to improve physiochemical properties of many drugs through the formation of inclusion complexes. CDs are expected to improve the ophthalmic delivery of drugs, either by increasing the solubility of poorly water-soluble drugs or altering the corneal permeability. However, they do not directly increase the corneal permeability. Whatever increase is observed, it is attributed to an increased fraction of drug present in an unionized state at the corneal surface [95,96]. Jarvinen et al. [97] and Freedman et al. [95] indicated the absence of any microscopic changes in the cornea treated with HP-β-CD. It may, however, be pointed out that CDs enhance penetration by increasing drug availability at the surface of the cornea.

   An area where CDs have a significant therapeutic benefit is in the solubilization of drugs intended for ophthalmic use. Corneal permeability favors moderately lipophilic drugs. These compounds often have a low aqueous solubility. Thus, the CDs are being extensively exploited for their usefulness in increasing the solubility of insoluble and poorly soluble drugs and keeping them at the corneal barrier.

   Loftssona et al. [98] found that with increases in the concentration of HP-β-CD, the flux of a hydrophobic drug arachidonylethanolamide increased. Later, Tirucherai and Mitra [99] found that HP-β-CD could also enhance the corneal permeation of the lipophilic drug ganciclovir dibutyrate (acyl ester prodrug of ganciclovir). Aktas [100] also found that addition of HP-β-CD to the simple aqueous solution of pilocarpine nitrate increased the permeation of pilocarpine nitrate by four times in an *in vitro* permeability study using isolated rabbit cornea.

9. *Bioadhesive polymers*: Bioadhesive polymeric systems, which have the capacity to adhere to the mucin coat covering the conjunctiva and corneal surfaces of the eye by noncovalent bonds, improve the bioavailability of the drug and hence can also be classified as penetration enhancers, in the sense that the total amount of drug penetrating the internal ocular tissue is increased by their use. These bioadhesive polymeric systems significantly prolong the drug residence time, because the clearance is controlled by a much slower rate of mucous turnover rather than tear turnover [101]. Mucoadhesives thus increase the residence time and also provide an intimate contact between the drug and the absorbing tissue, which may result in a high drug concentration in the local area and hence a high drug flux through the absorbing tissue [101].

25.5 EFFECT OF ABSORPTION PROMOTERS ON CONJUNCTIVAL MEMBRANES AND ITS SIGNIFICANCE

An absorption promoter can control the extent and pathway of the ocular and systemic absorption of instilled drug solution by altering not only the corneal but also the conjunctival drug penetration. Most of the agents discussed above show different effects on the corneal and conjunctival membranes. The different response of corneal and conjunctival barriers to absorption promoters can be exploited and used to control the extent and pathway of the ocular and systemic absorption of drugs instilled into the eye. The mechanism of action of absorption promoters and the barrier properties of membranes are the two factors that define drug permeability.

The conjunctiva, consisting of a thin mucous membrane and a highly vascularized tissue lining the inside of the eyelids and the anterior sclera, contributes to the noncorneal ocular absorption and to a large extent to the systemic absorption. It has a nonkeratinized, stratified, squamous epithelium, overlying a loose, highly vascular connective tissue. The superficial conjunctival epithelium has tight junctions that are the main barriers for drug penetration across this tissue. However, the intercellular spaces in the conjunctival epithelium are wider than those in the corneal epithelium. Therefore, the richness of the paracellular route makes the conjunctiva more permeable than the cornea. Taurocholic acid had the most potent effect on the conjunctival penetration of β-blockers. It has even been reported to increase the conjunctival penetration of macromolecules such as insulin [102]. Meanwhile, EDTA showed only a slight promoting activity on the conjunctival penetration of β-blockers [39].

In corollary to the above observations, conjunctival penetration enhancers can be employed to increase the ocular absorption of peptide drugs, e.g., insulin, for their systemic effects. The ocular route is a simple, noninvasive, more accurate, and less expensive alternative for the systemic delivery of peptides as compared to the buccal, nasal, rectal, vaginal, or dermal routes. It also avoids (1) pain in comparison to administration by parenteral injections and (2) gastrointestinal degradation since the drug enters the circulation directly [103]. Chiou and Chuang [67] and Yamamoto et al. [104] demonstrated that insulin was well absorbed and showed a significant hypoglycemic effect after its instillation (in combination with several absorption enhancers). Saponin was found to be the best enhancer of insulin absorption via this route [67], whereas EDTA, fusidic acid, bile salts, and some surfactants also showed an enhancing effect on insulin absorption [49,105,106]. Furthermore, insulin instilled with ophthalmic preservatives, especially BAC and paraben, showed a significant hypoglycemic response [107].

25.6 OTHER APPLICATIONS

In addition to the use of penetration enhancers in eye solutions, both for improving the corneal and the conjunctival (systemic) delivery of drugs, the use of these agents has also been extended to their incorporation into other ocular delivery systems. For example, solid ocular inserts made of polyvinylalcohol (PVA) containing sulfadicramide and some absorption promoters, e.g., polyoxyethylene-9-lauryl ether, L-(lysophosphatidylocholine), and deoxycholic acid sodium salt, have been reported. Such inserts showed an increase in the penetration of the drug through the animal cornea in *in vitro* studies [108].

Furthermore, addition of a penetration enhancer to the vehicle of an ophthalmic solution has been used to reduce the size of the drop instilled. Since this reduction in drop size results in a decreased washout and systemic drug loss, this would result in a decreased potential for systemic toxicity, at the same time improving the ocular absorption of poorly absorbed drugs [109]. Therefore, improved ocular bioavailability and therapeutic response could be obtained.

Moreover, the small drop instilled ensures less lacrimation and hence a decreased dilution and drainage that make the product more cost-effective.

25.7 LIMITATIONS TO PRACTICAL APPLICABILITY

In spite of the citation of a large number of virtual applications of penetration enhancers in the literature, the unique characteristics and high sensitivity of the corneal and conjunctival tissues require great caution to be applied in the selection of permeation enhancers for their capacity to affect the integrity of the epithelial surface. There is evidence that penetration enhancers themselves can penetrate the eye and may therefore lead to unknown toxicological implications; for example, BAC was found to accumulate in the cornea for days [110]. Similarly, EDTA was found to reach the iris–ciliary body in concentrations high enough to alter the permeability of the blood vessels in the uveal tract, indirectly accelerating drug removal from the aqueous humor [111]. A repeated application of 0.5% EDTA was observed to significantly alter the corneal epithelial architecture, although a single application was well tolerated [54]. Azone, at concentrations of 0.1% or higher, has also been reported to be irritating, discomforting, and toxic to the eyes [112]. Even saponin has been reported to cause eye irritation at a concentration of 0.5% [67]. Bile salts and surfactants were also shown to cause irritation of the eye and the nasal mucosa [56,113].

Ocular damaging and irritant agents can be identified and evaluated by the Draize rabbit test [114]. However, more recently this test has been criticized on the basis of ethical considerations and unreliable prognosis of human response. Alternative methods such as the evaluation of toxicity on ocular cell cultures have been recommended and are being indicated as promising prognostic tools [115–120]. Direct confocal microscopic analysis [121], hydration level of isolated corneas [122], and various other tests on isolated corneas or animal eyes have also been proposed for evaluation of ocular toxic effects.

In addition to the above-mentioned methods, measurement of the electrical properties of the cornea (electrical resistance and transcorneal potential) has proven to be a sensitive test for detecting damage of the corneal tissues [123]. Chetoni et al. [124] tested BAC, EDTA, polystearyl ether (PSE), polyethoxylated castor oil (PCO), cetylpyridinium chloride (CPC), and deoxycholic acid sodium salt (DC), for their potential to cause ocular damage, at different concentrations. All these agents, at all the tested dose levels, significantly altered the electrophysiological parameters, after a 5 h contact (which is indicative of ocular toxicity). These agents also significantly influenced the corneal hydration after a 1 h contact, which evidently was a consequence of alteration of the corneal epithelium. Burgalassi et al. [120] also tested these potential ocular permeation enhancers for cytotoxicity on cultures of rabbit and human corneal epithelial cells (HCE). The order of toxicity of these agents found in the WST-1 (cell proliferation reagent) test after an exposure of 15 min was PSE $\geq$ BAC > CPC > EDTA > PCO.

Furrer et al. [121] tested currently used nonionic (Tween 20, Pluronic F127, Brij 35, and Mirj 51), anionic (sodium lauryl sulfate, sodium cholate), and cationic (benzalkonium chloride and cetrimide) surfactants for ocular irritation, in both rabbits and mice, using confocal laser scanning ophthalmoscopy, in which corneal lesions subsequent to instillation of surfactants are specifically marked by fluorescein and assessed by digital image processing. The test revealed the following irritation rankings: cationic > anionic > nonionic surfactants.

Rojanasakul et al. [88] found that most of these enhancers (in particular, EDTA, digitonin, and sodium deoxycholate) increased the corneal permeability, and they may lead to severe cellular membrane damage as indicated by laser scanning confocal microscopy and electrophysiological techniques.

25.8 CONCLUSION

Although penetration enhancers promise superior therapeutic efficacy, this approach should be introduced for clinical use only after considering the balance of risks and benefits. Furthermore, information on the different mechanisms of drug penetration, ocular metabolism, side effects, and the influence of ocular diseases on the specific drug absorption-enhancement techniques is required before the use of any enhancer.

REFERENCES

1. Raviola, G. 1977. The structural basis of blood ocular barriers. *Exp Eye Res* 25:27.
2. Stjernschantz, J., and M. Astin. 1993. Anatomy and physiology of the eye. Physiological aspects of ocular drug delivery. In *Biopharmaceutics of Ocular Drug Delivery*, ed. P. Edman, chap. 1. Boca Raton, FL: CRC Press.
3. Schoenwald, R.D. 1990. Ocular drug delivery. Pharmacokinetic considerations. *Clin Pharmacokinet* 18:155.
4. Aiache, J.M., et al. 1997. The formulation of drug for ocular administration. *J Biomater Appl* 11:329.
5. Mishima, S. 1965. Some physiological aspects of the precorneal tear film. *Arch Ophthalmol* 73:233.
6. Nicolaides, N., et al. 1981. Meibomian gland studies: Comparison of steer and human lipids. *Invest Ophthalmol Vis Sci* 20:522.
7. Gandhi, R.B., and J.R. Robinson. 1991. Permselective characteristics of rabbit buccal mucosa. *Pharm Res* 8:1199.
8. Mariott, C., and N.P. Gregory. 1990. Mucus physiology and pathology. In *Bioadhesive Drug Delivery Systems*, eds. Lenearts, V. and R. Gurney, chap 1. Boca Raton, FL: CRC Press.
9. Johnson, P.M., and K.D. Rainsford. 1972. The physical properties of mucus: Preliminary observations on the sedimentation behaviour of porcine gastric mucosa. *Biochim Biophys Acta* 286:72.
10. Milder, B. 1987. The lacrimal apparatus. In: *Alder's Physiology of the Eye, Clinical Application*, eds. Moses, R.A. and W.M.J. Hart, 15. St. Louis: The C.V. Mosby Company.
11. Lamberts, D.W. 1987. Basic science aspects: physiology of the tear film. In *The Cornea: Scientific Foundations and Clinical Practice*, eds. Smoli, G. and R.A. Thoft, 16. Boston/Toronto: Little, Brown and Company.
12. Doane, M.G., A.D. Jense, and C.H. Dohlman. 1978. Penetration routes of topically applied medications. *Am J Ophthalmol* 85:383.
13. Watsky, M.A., M.M. Jablonski, and H.F. Edelhauser. 1988. Comparison of conjunctival and corneal surface areas in rabbit and human. *Curr Eye Res* 7:483.
14. Ahmed, I., et al. 1987. Physiochemical determinants of drug diffusion across the conjunctiva, sclera and cornea. *J Pharm Sci* 76:583.
15. Wang, W., et al. 1991. Lipophilicity influence on conjunctival drug penetration in pigmented rabbit: A comparison with corneal penetration. *Curr Eye Res* 10:571.
16. Reinsten, D.Z., et al. 1994. Epithelial and corneal thickness measurements by high frequency ultrasound digital signal processing. *Ophthalmology* 101:140.
17. Klyce, S.D., and C.E. Crosson. 1985. Transport processes across the rabbit corneal epithelium: A review. *Curr Eye Res* 4:323.
18. Barry, B.W. 1983. Percutaneous absorption. In *Dermatological Formulations*, ed. Barry, B.W., 127. Marcel Dekker.
19. Junginger, H.E., and J.C. Verhoef. 1998. Macromolecules as safe penetration enhancers for hydrophilic drugs—a fiction? *Pharm Sci Technol Today* 1:370.
20. Gumbiner, B. 1987. Structure, biochemistry and assembly of epithelial tight junctions. *Am J Physiol* 253:749.
21. Anderson, J.M., and C.M. Van Itallie. 1995. Tight junctions and the molecular basis for regulation of paracellular permeability. *Am J Physiol* 269:467.

22. Shasby, D.M., M. Winter, and S. Shasby. 1988. Oxidants and conductance of cultured epithelial cell monolayers: Inositol phospholipid hydrolysis. *Am J Physiol* 255:781.
23. Mullin, J.M., and T.G. O'Brien. 1986. Effects of tumor promoters on LLC-PK1 renal epithelial tight junctions and transepithelial fluxes. *Am J Physiol* 251:597.
24. Rojanasakul, Y., and J.R. Robinson. 1991. The cytoskeleton of cornea and its role in tight junction permeability. *Int J Pharm* 68:135.
25. Muranishi, N., et al. 1981. Effect of fatty acids and monoglycerides on permeability of lipid bilayers. *Chem Phys Lipids* 28:269.
26. Tengamnuay, P., and A.K. Mitra. 1990. Bile-salt fatty acid mixed micelles as nasal absorption promoters of peptides. I. Effects of ionic strength, adjuvant composition and lipid structure on the nasal absorption of [D-Arg2]kyotorphin. *Pharm Res* 7:127.
27. Lichtenberg, D., R.J. Robson, and E.A. Dennis. 1983. Solubilization of phospholipids by detergents. Structural and kinetic aspects. *Biochim Biophys Acta* 737:285.
28. Nakanishi, K., et al. 1984. Mechanism of enhancement of rectal permeability of drugs by nonsteroidal anti-inflammatory drugs. *Chem Pharm Bull* 32:3187.
29. Martinez-Palomo, A., et al. 1980. Experimental modulation of occluding junctions in a cultured transporting epithelium. *J Cell Biol* 87:736.
30. Pitelka, D.R., B.N. Taggart, and S.T. Hamamotto. 1983. Effects of extracellular calcium depletion on membrane topography and occluding junctions of mammary epithelial cells in culture. *J Cell Biol* 96:613.
31. Saettone, M.F., et al. 1996. Evaluation of ocular permeation enhancers: *In vitro* effects on corneal transport of four β-blockers, and *in vitro/in vivo* toxic activity. *Int J Pharm* 142:103.
32. Hochman, J., and P.J. Artursson. 1994. Mechanisms of absorption enhancement and tight junction regulation. *J Control Release* 29:253.
33. Citi, S. 1992. Protein kinase inhibitors prevent junction dissociation induced by low extracellular calcium in MDCK epithelial cells. *J Cell Biol* 117:169.
34. Volberg, T., et al. 1986. Changes in membrane microfilament interaction in intercellular adherens junctions upon removal of extracellular Ca^{2+} ions. *J Cell Biol* 102:1832.
35. Volk, T., and B. Geiger. 1986. A-CAM: A 135-KD receptor of intercellular adherens junctions. II. Antibody mediated modulation of junction formation. *J Cell Biol* 103:1451.
36. Grass, G.M., and J.R. Robinson. 1988. Mechanisms of corneal drug penetration I: *In vivo* and *in vitro* kinetics. *J Pharm Sci* 77:3.
37. Ashton, P., S.K. Podder, and V.H.L. Lee. 1991. Formulation influence on conjunctival penetration of four β-blockers in pigmented rabbit: A comparison with corneal penetration. *Pharm Res* 8:1166.
38. Ashton, P., W. Wang, and V.H.L. Lee. 1991. Location of penetration and metabolic barriers to levobunolol in the corneal epithelium of the pigmented rabbit. *J Pharmacol Exp Ther* 259:719.
39. Sasaki, H., et al. 1995. Different effects of absorption promoters on corneal and conjunctival penetration of ophthalmic β-blockers. *J Pharm Res* 12:1146.
40. Kaur, I.P., M. Singh, and M. Kanwar. 2000. Formulation and evaluation of ophthalmic preparations of acetazolamide. *Int J Pharm* 199:119.
41. van Hoogdalem, E.J., et al. 1989. Intestinal drug absorption enhancement: An overview. *Pharmac Ther* 44:407.
42. Hosoya, K., and V.H.L. Lee. 1997. Cidofovir transport in the pigmented rabbit conjunctiva. *Curr Eye Res* 16:693.
43. Hochman, J.H., J.A. Fix, and E.L. LeCluyre. 1994. *In vitro* and *in vivo* analysis of the mechanism of absorption enhancement by palmitoylcarnitine. *J Pharmacol Exp Ther* 269:813.
44. Anderberg, E.K., and P. Artursson. 1993. Epithelial transport of drugs in cell culture VIII: Effects of sodium dodecyl sulfate on cell membrane and tight junction permeability in human intestinal epithelial (Caco-2) cells. *J Pharm Sci* 82:392.
45. Grzeskowiak, E. 1998. Biopharmaceutical availability of sulphadicramide from ocular ointments *in vitro*. *Eur J Pharm Sci* 6:247.

46. Harmia, T., P. Speiser, and J. Kreuter. 1987. Nanoparticles as drug carriers in ophthalmology. *Pharm Acta Helv* 62:322.
47. Morgan, R.V. 1995. Delivery of systemic regular insulin via the ocular route in cats. *J Ocul Pharmacol Ther* 11:565.
48. O'Hagan, D.T., and L. Illum. 1990. Absorption of peptides and proteins from the respiratory tract and the potential for development of locally administered vaccine. *Crit Rev Ther Drug Carrier Syst* 7:35.
49. Sakai, K., et al. 1986. Contribution of calcium ion sequestration by polyoxyethylated nonionic surfactants to the enhanced colonic absorption of *p*-aminobenzoic acid. *J Pharm Sci* 75:387.
50. Kaur, I.P., et al. 2004. Development of topically effective formulations of acetazolamide using HP-β-CD-polymer co-complexes. *Curr Drug Deliv* 1:65.
51. Aggarwal, D., A. Garg, and I.P. Kaur. 2004. Development of a topical niosomal preparation of acetazolamide: Preparation and evaluation. *J Pharm Pharmacol* 56:1509.
52. Morimoto, K., T. Nakai, and K. Morisaka. 1987. Evaluaton of permeability enhancement of hydrophilic compounds and macromolecular compounds by bile salts through rabbit corneas *in vitro*. *J Pharm Pharmacol* 39:124.
53. Burstein, N.L. 1984. Preservative alteration of corneal permeability in humans and rabbits. *Invest Ophthalmol Vis Sci* 25:1453.
54. Lee, V.H., and J.R. Robinson. 1986. Topical ocular delivery: Recent developments and future challenges. *J Ocul Pharmacol* 2:67.
55. Camber, O., and P. Edman. 1987. Influence of some preservatives on the corneal permeability of pilocarpine and dexamethasone, *in vitro*. *Int J Pharm* 39:229.
56. Green, K. 1993. The effects of preservatives on corneal permeability of drugs. In *Biopharmaceutics of Ocular Drug Delivery*, ed. Edman, P., 43. Boca Raton, FL: CRC Press.
57. Higaki, K., M. Takeuchi, and M. Nakano. 1996. Estimation and enhancement of *in vitro* corneal transport of S-1033, a novel antiglaucoma medication. *Int J Pharm* 132:165.
58. Podder, S.K., K.C. Moy, and V.H. Lee. 1992. Improving the safety of topically applied timolol in the pigmented rabbit through manipulation of formulation composition. *Exp Eye Res* 54:747.
59. Smolen, V.F., et al. 1973. Biphasic availability of ophthalmic carbachol. I. Mechanisms of cationic polymer and surfactant promoted miotic activity. *J Pharm Sci* 62:958.
60. Burton, G.D., and R.M. Hill. 1981. Aerobic responses of the cornea to ophthalmic preservatives, measured *in vivo*. *Invest Ophthalmol Vis Sci* 21:842.
61. Pfister, R.R., and N. Burstein. 1976. The effects of ophthalmic drugs, vehicles, and preservatives on corneal epithelium: A scanning electron microscope study. *Invest Ophthalmol* 15:246.
62. Green, K., and A. Tonjum. 1971. Influence of various agents on corneal permeability. *Am J Ophthalmol* 72:897.
63. Green, K., and A.M. Tonjum. 1975. The effect of benzalkonium chloride on the electropotential of the rabbit cornea. *Acta Ophthalmol* 53:348.
64. Mikkelson, T.J., S.S. Chrai, and J.R. Robinson. 1973. Competitive inhibition of drug–protein interaction in eye fluids and tissues. *J Pharm Sci* 62:1942.
65. Godbey, R.E., K. Green, and D.S. Hull. 1989. Influence of cetylpyridinium chloride on corneal permeability to penicillin. *J Pharm Sci* 78:815.
66. Pillion, D.J., et al. 1996. Structure–function relationship among Quillaja saponins serving as excipients for nasal and ocular delivery of insulin. *J Pharm Sci* 85:518.
67. Chiou, G.C., and C.Y. Chuang. 1989. Improvement of systemic absorption of insulin through eyes with absorption enhancers. *J Pharm Sci* 78:815.
68. Fiscum, G. 1985. Intracellular levels and distribution of Ca^{2+} in digitonin-permeabilized cells. *Cell Calcium* 6:25.
69. Wolosin, J.M. 1988. Regeneration of resistance and ion transport in rabbit corneal epithelium after induced surface cell exfoliation. *J Membr Biol* 104:45.
70. Liaw, J., and J.R. Robinson. 1992. The effect of polyethylene glycol molecular weight on corneal transport and related influence of penetration enhancers. *Int J Pharm* 88:125.

71. Kato, A., and S. Iwata. 1988. *In vitro* study on corneal permeability to bunazosin. *J Pharmacobiodyn* 11:115.
72. Muranishi, S. 1990. Absorption enhancers. *Crit Rev Ther Drug Carr Sys* 7:1.
73. Kato, A., and S. Iwata. 1988. Studies on improved corneal permeability to bunazosin. *J Pharmacobiodyn* 11:330.
74. Newton, C., B.M. Gebhardt, and H.E. Kaufman. 1988. Topically applied cyclosporine in azone prolongs corneal allograft survival. *Invest Ophthalmol Vis Sci* 29:208.
75. Tang-Liu, D.D., et al. 1994. Effects of four penetration enhancers on corneal permeability of drugs *in vitro*. *J Pharm Sci* 83:85.
76. Artursson, et al. 1994. Effect of chitosan on the permeability of monolayers of intestinal epithelial cells (Caco-2). *Pharm Res* 11:1358.
77. Illum, L., N.F. Farraj, and S.S. Davis. 1994. Chitosan as a novel nasal drug delivery system for peptide drugs. *Pharm Res* 11:1186.
78. Dodane, V., M.A. Khan, and J.R. Merwin. 1999. Effect of chitosan on epithelial permeability and structure. *Int J Pharm* 182:21.
79. Di Colo, G., et al. 2004. Effect of chitosan and of *N*-carboxymethylchitosan on intraocular penetration of topically applied ofloxacin. *Int J Pharm* 273:37.
80. Burgalassi, S., et al. 2000. Effect of xyloglucan (tamarind seed polysaccharide) on conjunctiva cell adhesion to laminin and on corneal epithelium wound healing. *Eur J Ophthalmol* 10:71.
81. Ghelardi, E., et al. 2000. Effect of a novel mucoadhesive polysaccharide obtained from tamarind seeds on the intraocular penetration of gentamycin and ofloxacin in rabbits. *J Antimicrob Chemother* 46:831.
82. Ghelardi, E., et al. 2004. A mucoadhesive polymer extracted from tamarind seed improves the intraocular penetration and efficacy of rufloxacin in topical treatment of experimental bacterial keratitis. *Antimicrob Agents Chemother* 48:3396.
83. Bijl, P., et al. 2001. Effects of three penetration enhancers on transcorneal permeation of cyclosporine. *Cornea* 20:505.
84. Flanagan, M.D., and S. Lin. 1980. Cytochalasins block actin filament elongation by binding to high affinity sites associated with F-actin. *J Biol Chem* 255:835.
85. Brown, S.S., and J.A. Spudich. 1981. Mechanism of action of cytochalasin: Evidence that it binds to actin filament ends. *J Cell Biol* 88:487.
86. Craig, S.W., and J.V. Pardo. 1979. α-Actinin localization in the junctional complex of intestinal epithelial cells. *J Cell Biol* 80:203.
87. Meza, I., et al. 1980. Occluding junctions and cytoskeletal components in a cultured transporting epithelium. *J Cell Biol* 87:746.
88. Rojanasakul, Y., J. Liaw, and J.R. Robinson. 1990. Mechanisms of action of some penetration enhancers in the cornea: Laser scanning confocal microscopic and electrophysiology studies. *Int J Pharm* 66:131.
89. Mitra, A.K.1983. PhD diss., University of Kansas, Lawrence, KS.
90. Pitelka, D.R., B.N. Taggart, and S.T. Hamamoto. 1983. Effects of extracellular calcium depletion on membrane topography and occluding junctions of mammary epithelial cells in culture. *J Cell Biol* 96:613.
91. Lehr, C.M., et al. 1992. Effects of the mucoadhesive polymer polycarbophil on the intestinal absorption of a peptide drug in the rat. *J Pharm Pharmacol* 44:402.
92. Borchard, G., et al. 1996. The potential of mucoadhesive polymers in enhancing intestinal peptide drug absorption. III: Effects of chitosan glutamate and carbomer on epithelial tight junctions *in vitro*. *J Control Release* 39:131.
93. Lehr, C.M., Y.H. Lee, and V.H. Lee. 1994. Improved ocular penetration of gentamycin by mucoadhesive polymer polycarbophil in the pigmented rabbit. *Invest Ophthalmol Vis Sci* 35:2809.
94. Hornof, M.D., and A.B. Schnurch. 2002. *In vitro* evaluation of the permeation enhancing effect of polycarbophil–cysteine conjugates on the cornea of rabbits. *J Pharm Sci* 91:2588.
95. Freedman, K.A., J.W. Klein, and C.E. Crosson. 1993. Beta-cyclodextrins enhance bioavailability of pilocarpine. *Curr Eye Res* 12:641.

96. Reer, O., T.K. Block, and B.W. Muller. 1994. *In vitro* corneal permeability of diclofenac sodium in formulations containing cyclodextrins compared to the commercial product Voltaren Ophtha. *J Pharm Sci* 83:1345.
97. Jarvinen, K., et al. 1994. The effect of modified beta-cyclodextrin, SBE4-beta-CD, on the aqueous stability and ocular absorption of pilocarpine. *Curr Eye Res* 13:897.
98. Loftssona, T., and T. Jarvinen. 1999. Cyclodextrins in ophthalmic drug delivery. *Adv Drug Deliv Rev* 36:59.
99. Tirucherai, G.S., and A.K. Mitra. 2003. Effect of hydroxypropyl beta cyclodextrin complexation on aqueous solubility, stability, and corneal permeation of acyl ester prodrugs of gancyclovir. *AAPS Pharm Sci Tech* 4:E45.
100. Aktas, Y. 2003. Influence of hydroxypropyl beta-cyclodextrin on the corneal permeation of pilocarpine. *Drug Dev Ind Pharm* 29:223.
101. Desai, S.D., and Blanchard, J. 1994. Ocular drug formulation and delivery, in *Encyclopedia of Pharmaceutical Technology*, vol. 3, eds. Swarbrick, J. and J.C. Boylan, 43. New York: Marcel Dekker.
102. Hayakawa, E., et al. 1992. Conjunctival penetration of insulin and peptide drugs in the albino rabbit. *Pharm Res* 9:769.
103. Pillion, D.J., et al. 1994. Efficacy of insulin eyedrops. *J Ocul Pharmacol* 10:461.
104. Yamamoto, A., et al. 1989. The ocular route for systemic insulin delivery in the albino rabbit. *J Pharmacol Exp Ther* 249:249.
105. Grass, G.M., R.W. Wood, and J.R. Robinson. 1985. Effects of calcium chelating agents on corneal permeability. *Invest Ophthalmol Vis Sci* 26:110.
106. Marsh, R.J., and D.M. Maurice. 1971. The influence of non-ionic detergents and other surfactants on human corneal permeability. *Exp Eye Res* 11:43.
107. Sasaki, H. 1994. Effect of preservatives on systemic delivery of insulin by ocular instillation in rabbits. *J Pharm Pharmacol* 46:871.
108. Grzeskowiak, E. 1998. Technology and biopharmaceutical availability of solid ocular inserts containing sulfadicramide and some promoters. *Acta Pol Pharm* 55:205.
109. Van Santvliet, L., and A. Ludwig. 1998. The influence of penetration enhancers on the volume instilled of eye drops. *Eur J Pharm Biopharm* 45:189.
110. Green, K., J.M. Chapman, and J.L. Cheeks, 1987. Ocular penetration of pilocarpine in rabbits. *J Toxicol* 2:89.
111. Grass, G.M., and J.R. Robinson. 1984. Relationship of chemical structure to corneal penetration and influence of low viscosity solution on ocular bioavailability. *J Pharm Sci* 73:1021.
112. Ismail, I.M., et al. 1992. Comparison of azone and hexamethylene lauramide in toxicologic effects and penetration enhancement of cimetidine in rabbit eyes. *Pharm Res* 9:817.
113. Merkus, F.W.H. M., et al. 1993. Absorption enhancers in nasal drug delivery: Efficacy and safety. *J Control Release* 24:201.
114. Draize, J.H., G. Woodard, and H.O. Calvery. 1944. Methods for the study of irritation and toxicity of substances applied topically to the skin and mucous membranes. *J Pharmacol Exp Ther* 82:377.
115. North-Root, H., et al. 1982. Evaluation of an *in vitro* cell toxicity test using rabbit corneal cells to predict the eye irritation potential of surfactants. *Toxicol Lett* 14:207.
116. Yang, W., and D. Acosta. 1994. Cytotoxicity potential of surfactant mixtures evaluated by primary cultures of rabbit corneal epithelial cells. *Toxicol Lett* 70:309.
117. Kruszewski, F.H., T.L. Walker, and L.C. DiPasquale. 1997. Evaluation of a human corneal epithelial cell line as an *in vitro* model for assessing ocular irritation. *Fundam Appl Toxicol* 36:130.
118. Parnigotto, P.P., et al. 1998. Bovine corneal stroma and epithelium reconstructed *in vitro*: Characterization and response to surfactants. *Eye* 12:304.
119. Saarinen-Savolainen, P., et al. 1998. Evaluation of cytotoxicity of various ophthalmic drugs, eye drop excipients and cyclodextrins in an immortalized human corneal epithelial cell line. *Pharm Res* 15:1275.

120. Burgalassi, S., et al. 2001. Cytotoxicity of potential ocular permeation enhancers evaluated on rabbit and human corneal epithelial cell lines. *Toxicol Lett* 122:1.
121. Furrer, P., et al. 2000. Application of *in vivo* confocal microscopy to the objective evaluation of ocular irritation induced by surfactants. *Int J Pharm* 207:89.
122. Monti, D., et al. 2002. Increased corneal hydration induced by potential ocular penetration enhancers: Assessment by differential scanning calorimetry (DSC) and by desiccation. *Int J Pharm* 232:139.
123. Rojanasakul, Y., J. Liaw, and J.R. Robinson. 1990. Mechanisms of action of some penetration enhancers in the cornea: Laser scanning confocal microscopic and electrophysiology studies. *Int J Pharm* 66:131.
124. Chetoni, P., et al. 2003. Ocular toxicity of some corneal penetration enhancers evaluated by electrophysiology measurements on isolated rabbit corneas. *Toxicol In Vitro* 17:197.

26 Iontophoresis for Ocular Drug Delivery

Esther Eljarrat-Binstock, Joseph Frucht-Pery, and Abraham J. Domb

CONTENTS

26.1 INTRODUCTION

Iontophoresis is a noninvasive technique in which a small electric current is applied to enhance ionized drug penetration into tissue. The drug is applied with an electrode carrying the same charge as the drug, and the ground electrode, which is of the opposite charge, is placed elsewhere on the body to complete the circuit. The drug serves as a conductor of the current through the tissue [1]. Figure 26.1 shows a diagram of ocular iontophoresis of a positively charged drug in a rabbit.

The ease of application, the minimization of systemic side effects, and the increased drug penetration directly into the target region resulted in extensive clinical use of iontophoresis mainly in the transdermal field. This technique has been utilized for administration of local anesthetics [2–5], sweat chloride testing in cystic fibrosis patients by transcutaneous delivery of pilocarpine [6,7], administration of vidarabine to patients with herpes orolabialis [8], fluoride administration to patients with hypersensitive dentin [9,10], and gentamicin delivery for the management of burned ears [11].

Ocular iontophoresis was first investigated in 1908 by the German investigator Wirtz [12], who passed an electric current through electrolyte-saturated cotton sponges placed over the globe for the treatment of corneal ulcers, keratitis, and episcleritis. By the end of the century, iontophoresis was extensively investigated for delivering ophthalmic drugs, including dyes, antibacterial, antiviral, antifungal, steroids, antimetabolites, and even genes. Ocular iontophoresis seemed to be the answer to the low bioavailability of drugs after topical administration

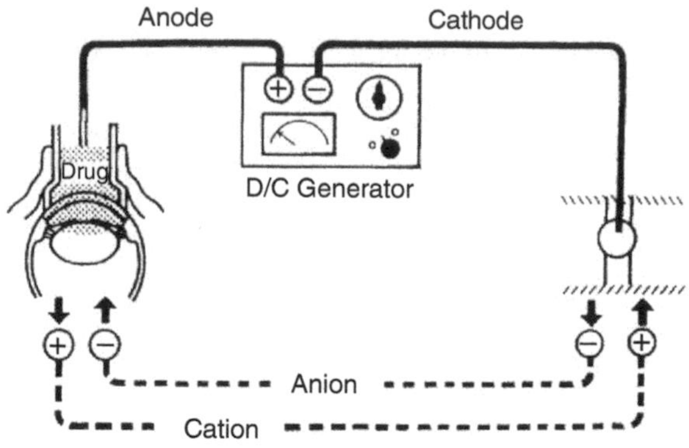

FIGURE 26.1 Diagram of ocular iontophoresis of a positively charged drug. Platinum electrode connected to the anode is placed in contact with the drug solution. The cathode is connected usually to the ear of the rabbit to complete the circuit. The positively charged anode drives the positively charged drug molecules from the solution into the eye. (Adapted from Myles, M.E. et al., *Ophthalmic Drug Delivery Systems*, 2nd ed. Mitra, A.K., Ed., Marcel Dekker, New York, 2003, p. 365. With permission.)

and to the potential serious complications after intraocular injections used for the treatment of many eye disorders.

Despite its widespread use and study during the first 60 years of the twentieth century, iontophoresis was never fully adopted as standard procedure. The lack of carefully controlled trials and the paucity of toxicity data were among the reasons that precluded its acceptance as an alternative for drug delivery. However, the last decade years have witnessed the development and optimization of the technology of ocular iontophoresis for fast and safe delivery of high drug concentrations in a specific ocular site [13].

26.2 LIMITATIONS IN OCULAR DRUG DELIVERY

Drug absorption through ocular tissues is dependent to a large degree upon a drug's physiochemical properties, such as octanol–water partition coefficient, molecular weight, solubility, and ionized state. In addition, the eye is well protected by a series of complex defense mechanisms, which may provide subtherapeutic drug levels at the intended site of action [14]. Systemically administered drugs have poor access to the eye because of the blood–ocular barrier, which physiologically separates the eye from the rest of the body by epithelial and endothelial components whose tight junctions limit transport from blood vessels to the eye. This barrier is comprised of two systems: (a) blood–aqueous barrier, which prevents drugs from entering the aqueous humor and (b) blood–retinal barrier, which prevents drugs from entering into the extravascular space of the retina and into the vitreous body. Subconjunctival and intravitreal injections of drugs are also applied clinically generating elevated intraocular concentrations with minimal systemic effects. However, these methods are still painful, inconvenient, and associated with severe complications, such as perforation of the globe and scarring of the conjunctiva [15,16].

The most common drug delivery method for treating ocular disorders is topical administration, due to its convenience and safety. However, the anterior segment of the eye also has various protective mechanisms for maintaining visual functions. After instillation of an ophthalmic drug, most of it is rapidly eliminated from the precorneal area due to drainage by the nasolacrimal duct and dilution by the tear turnover (approximately 1 μL/min) [17,18]. In addition, there is a finite limit to the size of the dose that can be applied and tolerated by the cul-de-sac (usually 7–10 μL) and the contact time of the drug with the absorptive surfaces of the eye. It has been determined that as much as 90% of the 50 μL dose administered as eyedrops is cleared within 2 min, and only 1%–5% of the administered dose permeates to the eye [19,20]. Moreover, the cornea itself is a highly selective barrier with five different layers to exclude compounds from the eye. The cornea consists of a hydrophilic stromal layer sandwiched between a very lipophilic epithelial layer and much less lipophilic endothelial layer. Drugs that are both lipid and water-soluble pass through the cornea readily. The drug finally absorbed may exit the eye through the canal of Schlemm or via absorption through the ciliary body into the episcleral space. Enzymatic metabolism may account for further loss, which can occur in the precorneal space or in the cornea [18]. Clearly, the physiological barriers to topical corneal absorption are formidable. Therefore, high administration frequency and high doses of drug are required resulting with fluctuations in ocular drug concentrations and local or systemic side effects. Either way, amounts of drug absorbed into the posterior segment of the eye will only be a minute fraction of the amounts achieved in the anterior segment due to the lens–iris diaphragm.

Although the corneal pathway is the primary route of intraocular entry for most drugs, penetration through the conjunctiva and sclera can also be important especially for drug absorption from adhesive gels and polymer inserts. A schematic diagram of ocular absorption is presented in Figure 26.2. The surface area of the conjunctival membrane is about 17 times greater than the surface area of the cornea in humans. The conjunctiva consists of squamous epithelium with tight junctions that are the main barriers for drug penetration but the intracellular spaces are wider than in the corneal epithelium. Thus, the conjunctiva is more permeable than the cornea especially for hydrophilic and large molecules, and so the sclera resistance is lower than the multilayer cornea [21,22]. After entering the conjunctival tissue, most drugs are rapidly removed by systemic uptake by the vessels embedded in that tissue, before diffusion to the intraocular tissues. This vascular clearance may be very important for

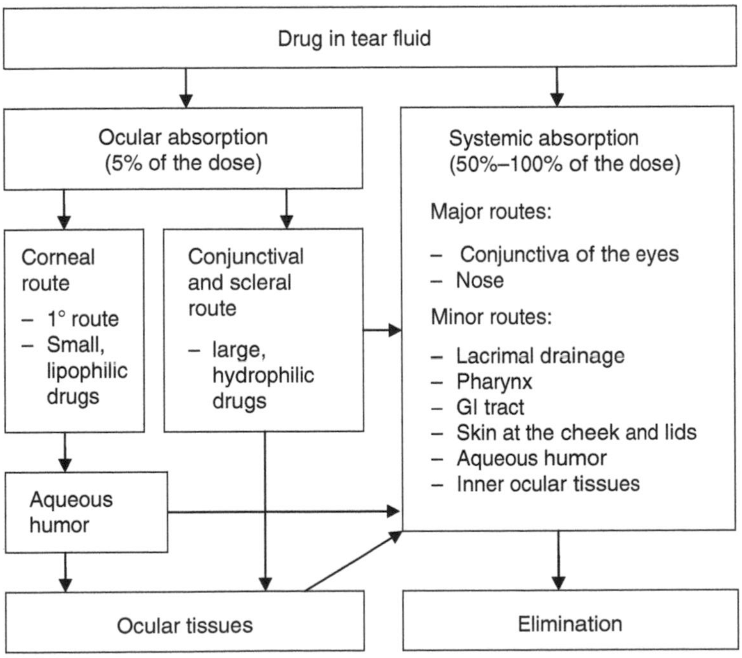

FIGURE 26.2 Schematic diagram of ocular distribution. (From Worakul, N. and Robinson, J.R., *Eur. J. Pharm. Biopharm.*, 44, 71, 1997. With permission.)

in vivo absorptions of drugs through the conjunctiva [23]. Coadministration of vasoconstrictors decreases the systemic absorption of topically applied drug [24].

In order to enhance drug permeability to the intraocular segments and improve ocular bioavailability of drugs, several approaches have been taken: ointments, collagen shields, liposome carriers, prodrugs, penetration enhancers, mucoadhesive gels, polymeric inserts, etc. Iontophoresis is another well-known method for enhancement of ionized drug penetration through tissue using a direct electrical current.

26.3 IONTOPHORESIS TRANSPORT MECHANISM

An iontophoretic device comprises a power source and two electrodes. The ionized drug is placed in the electrode compartment bearing the same charge, and the ground electrode is placed at a distal site on the body [25].

The basic electrical principle that oppositely charged ions attract and same charged ions repel is the central tenet of iontophoresis. The ionized substances are driven into the tissue by electrorepulsion either at the anode (for positively charged drug) or at the cathode (for negatively charged drug) [26]. This ionic–electric field interaction, also called the "Nernst–Planck effect," is the largest contributor to flux enhancement for small ions, but not the only one. Gangarosa and coworkers [27,28] demonstrated enhanced delivery of neutral species by application of an electric field. This brought the modification of the classical Nernst–Planck flux equation to include motion of the solvent called the "electroosmotic flow" [29]. Electroosmotic flow is the bulk fluid flow that occurs when a voltage difference is imposed across a charge membrane. Since the human membranes are negatively charged above pH 4, the electroosmotic flow occurs from anode to cathode, as the flow of the cationic counterions. For large monovalent ions, the electroosmotic flow is the dominant flow mechanism. The third mechanism that enhances drug penetration is the "damage effect" of the electric current that increases tissue permeability [30].

A few simple physics' laws can help to delineate the main parameters in iontophoretic transport. The first is Ohm's law:

$$V = IR$$

where V is the electromotive force in volts, I is the current in milliamperes, and R is the resistance in ohms. With iontophoresis, at a constant current supply, any change in the resistance of the tissue results in a change in the voltage.

The second important law is Coulomb's law:

$$Q = IT$$

where Q is the quantity of electricity, I is the current in milliamperes, and T is the time in minutes. The total current dosage in an iontophoretic treatment is expressed in milliamperes · minutes.

Finally, Faraday's law for determining the amount of drug transported by an electric current:

$$D = \frac{IT}{F\ IZI}$$

where D is the drug delivered in gram-equivalent, I is the current in milliamperes, T is the time in minutes, IZI is the valence of the drug, and F is Faraday's constant, which is the electrical charge carried by 1 gram-equivalent of a substance [13]. Hughes and Maurice [31,32] have expanded this law with factors governing the penetration of drugs through a tissue using iontophoresis, and formulated an equation to quantitate the amount of drug, M_d, which penetrates the epithelium after transcorneal iontophoresis:

$$M_d = \frac{IP_d C_d T}{F(P_d C_d + P_i C_i)}$$

where F is Faraday's constant. The current density, I, and the duration of iontophoresis, T, are the principal parameters determining penetration, if the concentration of the drug, C_d, is much higher than the concentration of the competing ions, C_i. The permeability of the tissue to the drug, P_d, and to the competing ions, P_i, is another variable governing penetration of the drug.

26.4 THE IONTOPHORETIC DEVICE

The basic design of the iontophoretic devices is a direct current power source and two electrodes. Platinum is the commonly used material for the electrodes, since it is nontoxic, has high oxidation voltage, slow degradation, and low ion release (Figure 26.1).

There are two approaches for retaining drug in the iontophoretic device. The more common approach is to fill an eyecup with the drug solution whereas a metal electrode extended from the current supply submerges into the solution. The eyecup with an internal diameter of 5–10 mm is placed over the eye and the drug solution is continuously infused into the cup during the iontophoretic treatment. The eyecup has two ports: one delivers the drug solution and the other holds the metal electrode and aspirates air bubbles that can disrupt the current supply, thus creating a slightly negative pressure that maintains the applicator in place. The ground electrode is attached usually to the ear of the animal, as close as possible to the former electrode, to obtain minimal resistance. Different eyecup shapes and sizes exist (from 0.4 mm^2 to 0.8 cm^2 contact surface areas) [33–37], including an annular-shaped silicone probe for transscleral iontophoresis (called Eyegate, Optis, France) with a 0.5 cm^2 contact area and a 13 mm opening to avoid contact with the cornea, used by Behar-Cohen et al. [38], Hayden et al. [39], and Voigt et al. [40]. The different iontophoretic devices are presented in Figure 26.3 through Figure 26.6.

The second approach uses a drug-saturated gel as the delivery probe. Jones and Maurice [41] first used this method to deliver fluorescein into the anterior chamber of the eye with a fluorescein-saturated agar gel. The gel was placed in a plastic tube and was partly extruded from the tube to make a direct contact with the eye, using a contact area of 0.2 cm^2. Later

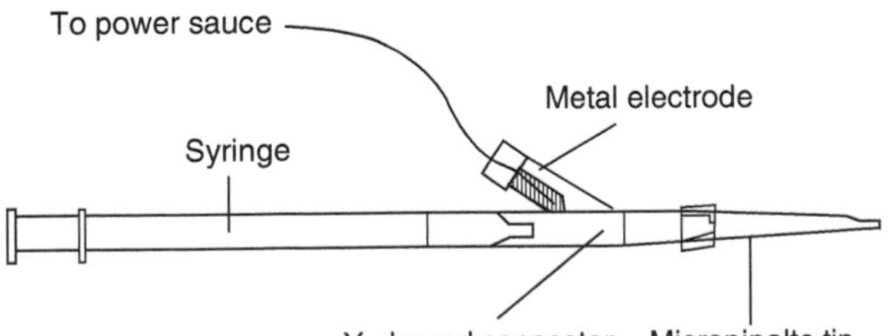

FIGURE 26.3 Diagram of the iontophoretic probe used by Sarraf et al. (Adapted from Sarraf, D. et al., *Am. J. Ophthalmol.*, 115, 748, 1993. With permission.)

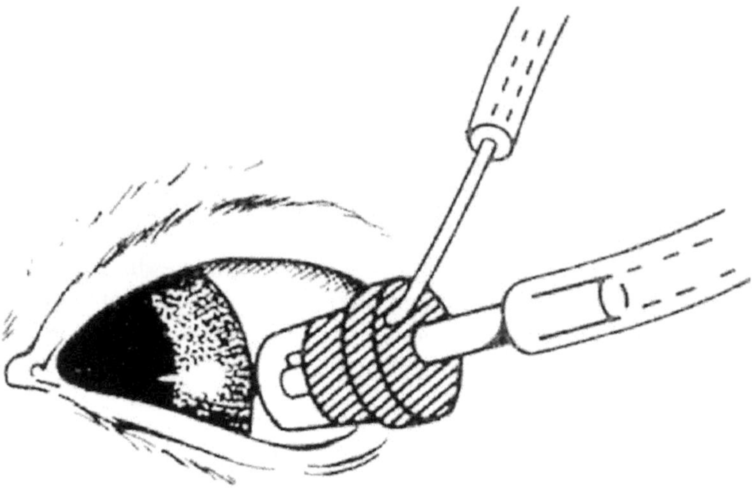

FIGURE 26.4 Diagram of the iontophoretic probe used by Barza. (Adapted from Barza, M., *Ophthalmology*, 93, 133, 1986. With permission.)

Grossman et al. [42] and Frucht-Pery et al. [43,44] used the same concept with gentamicin-saturated agar for transcorneal and transscleral iontophoresis in rabbits, using a contact surface area of 0.07 cm^2. However, the use of agar was abandoned since the agar material was fragile and left remains of agar on the eye surface.

In the past few years, several publications on drug-loaded hydrogels for ocular iontophoresis revealed novel iontophoretic applicators using the drug-saturated gel approach. Hydrogels are a three-dimensional network of hydrophilic polymers able to retain a large quantity

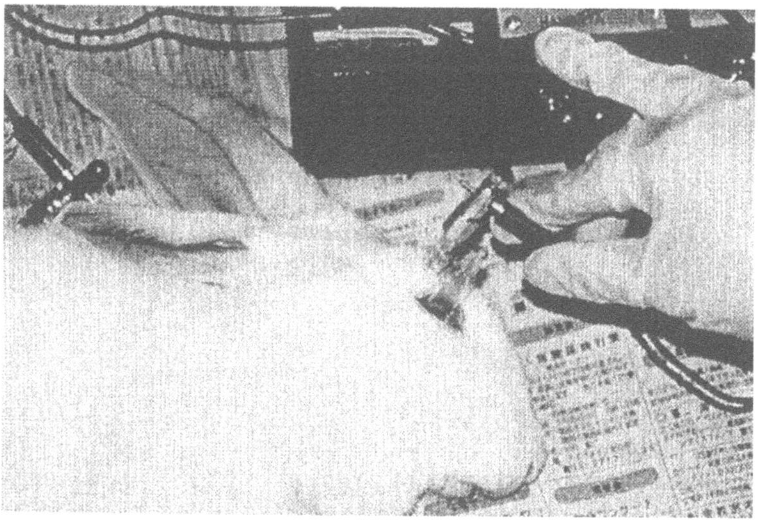

FIGURE 26.5 Ocular iontophoretic probe used by Asahara et al. (Adapted from Asahara, T. et al., *Jpn. J. Ophthalmol.*, 45, 31, 2001. With permission.)

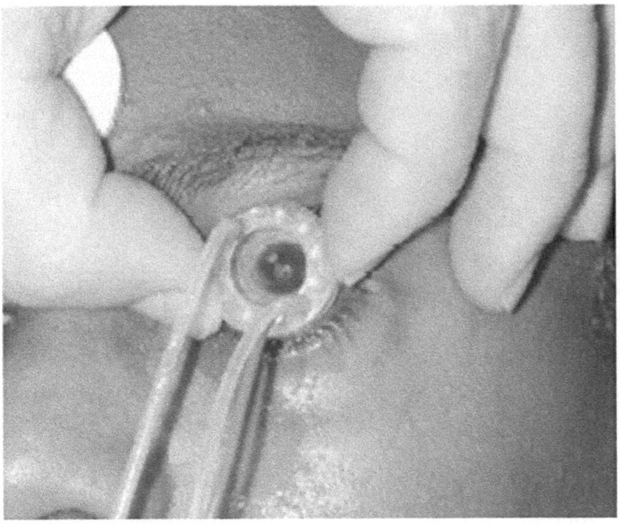

FIGURE 26.6 Annular-shaped silicone probe for transscleral iontophoresis (Eyegate, Optis) applied on the eye. Used by Behar-Cohen et al. [38], Hayden et al. [39], and Voigt et al. [40]. (Adapted from Behar-Cohen, F.F. et al., *Exp. Eye. Res.*, 74, 51, 2002. With permission.)

of water within their structure without dissolving. Due to their superior chemical and physical properties, hydrogels have received much attention for preparing biomedical materials, such as contact lenses, wound dressing, and drug delivery systems [45]. The monomer composition of a copolymer can be manipulated to influence the permeation and diffusion characteristics of the hydrogel, and through this manipulation, hydrogels can be synthesized to accommodate a variety of drug loaded into the matrix [46]. Hydrogels have been used for transdermal iontophoretic delivery of various drugs [47–50]. Hydrogels, rather than solutions, facilitate drug handling, minimize skin hydration, and allow for control of drug-release rate. In the ophthalmic field, hydrogels have been used for contact lenses with the cross-linked hydroxyethyl methacrylate (HEMA) hydrogel, the one most frequently used gel [45].

Fischer et al. [51], Parkinson et al. [52,53], and Vollmer et al. [54] used a custom manufactured OcuPhor hydrogel (Iomed Inc., Salt Lake City, UT, USA), composed of polyacetal sponge, for transscleral iontophoresis. The drug applicator is a small silicone shell that contains a patented silver–silver chloride ink conductive element, a hydrogel pad to absorb the drug formulation, and a small, flexible wire to connect the conductive element to the dose controller. At the time of administration, the dry hydrogel matrix is hydrated with the drug solution and placed against the sclera in the lower cul-de-sac of the rabbit eye. The treatment surface area of this applicator is 0.54 cm^2. The return electrode can be positioned anywhere on the body to complete the electrical circuit (Figure 26.7). A very similar applicator called Visulex (Aciont Inc., Salt Lake City, UT, USA) was developed for ophthalmic applications, and reported by Hastings et al. [55]. Visulex has been formulated with a unique selective membrane, which increases drug transport by excluding the transport of nondrug ions, thus making drug ions the primary carrier of electrical current through scleral tissue.

A polyacrylic–porous hydrogel saturated with gentamicin or dexamethasone solutions for transcorneal and transscleral iontophoresis has been reported [56–59]. A battery-operated portable device that applies a variable electrical current up to 1 mA, for preset periods of time is used (Figure 26.8). Hydrogels are prepared by radical polymerization of an aqueous

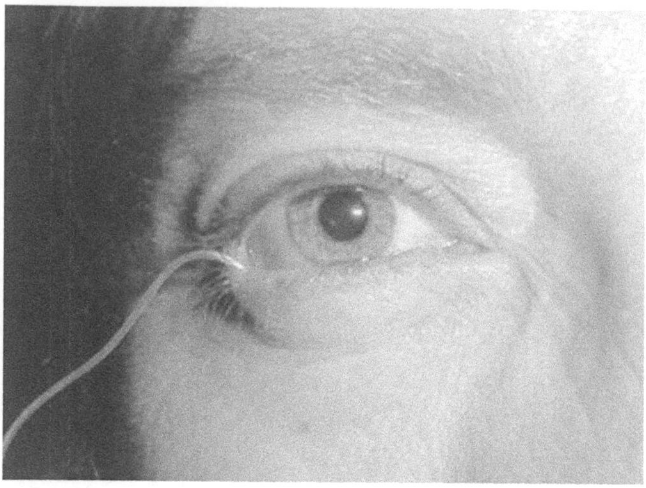

FIGURE 26.7 Ocular iontophoretic hydrogel applicator (OcuPhor) inserted into the inferior cul-de-sac of a human eye. Used by Fischer et al. [51], Parkinson et al. [52,53], and Vollmer et al. [54]. (Adapted from Fischer, G.A., Parkinson, T.M., and Szlek, M.A., *Drug Deliv. Tech.*, 2, 50, 2002. With permission.)

solution of HEMA and ethylene glycol dimethacrylate (EGDMA) divided into round wells. The hydrogels (5 mm height, 5 mm diameter) are dehydrated by lyophilization to form spongy–porous cylinders with a 10 to 20 μm diameter pore size, as observed using scanning electron microscope (Figure 26.9). The dry hydrogel is immersed in the drug solution before use and inserted into a well at the tip of the electrode of iontophoretic device. The hydrogel is placed onto the cornea or sclera and the complementary electrode is attached to the ear of the rabbit. The treatment surface area of this applicator is 0.2 cm^2. The use of a hydrogel applicator seems to be very convenient. It also encounters less current interruptions and is less harmful to the eye surface, compared with the eyecup approach.

When examining a new iontophoretic device, various parameters should be taken in mind. The ease to operate and apply on the human eye, patient convenience, surface contact area of

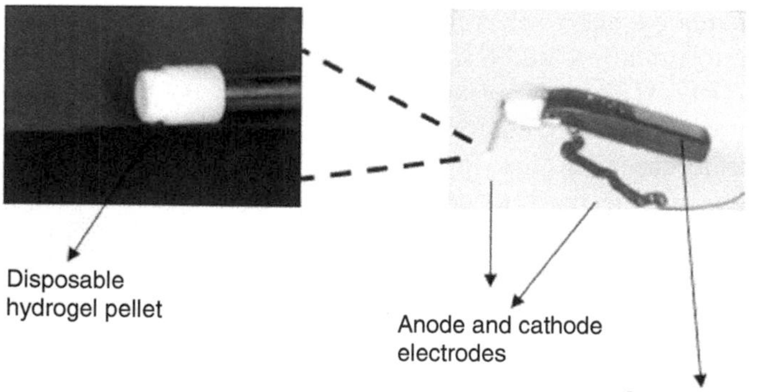

FIGURE 26.8 Ocular iontophoretic system using drug-loaded disposable hydrogels. Used by Eljarrat-Binstock et al. [57,58] and Frucht-Pery et al. [59]. (Adapted from Eljarrat-Binstock, E. et al., *Invest. Ophthalmol. Vis. Sci.*, 45, 2543, 2004. With permission.)

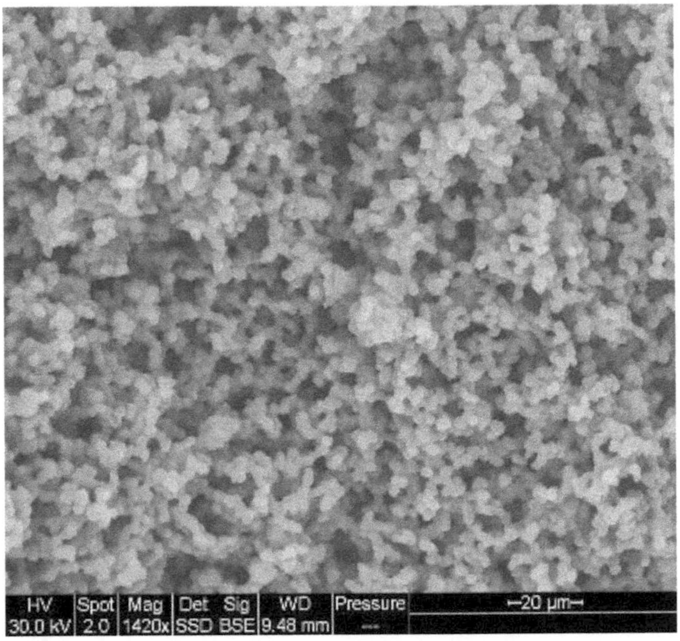

FIGURE 26.9 Representative scanning electron microscopy of HEMA hydrogel cross-linked with 2% EGDMA (×1420).

the drug with the application site which influences the current intensity applied and toxicity, the ability to maintain a constant current without disruptions, and the convenience of the drug reservoir to the physician.

26.5 *IN VITRO* MODELS

Ocular iontophoresis has been very poorly evaluated with *in vitro* models, in comparison to the work that has been done in the transdermal iontophoretic field. This general lack of a strong theoretical foundation for the practice of ocular iontophoresis has led to many useless animal experiments with no beneficial results due to drug or current unsuitability. Most of the preliminary experiments for optimization of the iontophoretic system were performed on animals.

One *in vitro* model used for the ocular drug penetration evaluation is the agarose gel. The agarose was previously used as a model for transdermal iontophoresis evaluation. Brouneus et al. [60] investigated the diffusion properties of some local anesthetics from the drug solution to a 1% agarose gel cast in syringes with tops cutoff, with relation to their molecular weights. The diffusion coefficients of the solutes were inversely proportional to their molecular weight. Anderson et al. [61] used the same method for evaluating dexamethasone phosphate penetration from a hydrogel patch to a cylindrical agar column, following the application of current of 4 mA for 10 min or 0.1 mA for 400 min. Two-millimeter-thick slices were cut parallel to the electrode surface immediately or 400 min after iontophoresis, in which the depth of penetration and the diffusion role were investigated. The active transport process of iontophoresis with dexamethasone phosphate appears to deliver drug ions immediately to a depth of 2 mm in the gel, while after 400 min the drug was delivered to a depth of 12 mm. The data results suggest that the drug is essentially "dropped off" as soon as it encounters the aqueous saline environment, resulting with drug depot formation. Deeper penetration of the drug apparently occurs not from the iontophoretic current, but from passive diffusion. Eljarrat-Binstock et al. [57] evaluated the usefulness of their hydrogel-based ocular iontophoretic device by using the agarose gel as a model resembling the eye surface. Iontophoresis was applied onto the solid-agar gels at different conditions using dexamethasone, hydrocortisone, and gentamicin salt solutions soaked in a hydrogel probe. The duration of iontophoresis, current density, and current direction applied was found to be correlated with the drug concentration achieved in the agar gel.

Kamath and Gangarosa [62] evaluated the mobility of seven positive and negative drugs (acyclovir, lidocaine, minoxidil, methylprednisolone, dexamethasone, adenosine arabinoside monophosphate, and metronidazole) suitable for iontophoresis in different pH values. Using a paper electrophoresis applying 10 mA current for 1.5 h, the mobility values (μ) of each drug were determined by measuring the spot migration. The variation in pH alters the maximum ionization of the drug and therefore the optimal mobility properties change. This information is essential when evaluating the suitability of drugs and the optimal conditions for iontophoresis.

Perfusion cell model is another well-known approach for drug penetration evaluation, yet limited in its use for ocular penetration after iontophoresis. Camber et al. [63] investigated the permeability and absorption rate of prostaglandin $F_{2\alpha}$ ($PGF_{2\alpha}$) and its methyl and benzyl esters through isolated pig cornea. They used a perfusion cell consisting of two compartments separated by the fresh cornea. The perfusion cell has two ports for introducing the iontophoretic electrodes on each corneal side. The drug solution was added to the donor compartment and a buffer solution to the receptor compartment, under sink conditions. Samples were taken every 40 min during a period of 4 h from the endothelial side. This study showed that the uptake to the cornea is dependent on the lipophilicity of the substance. The benzyl ester with the greatest octanol–water partition coefficient has the highest uptake whereas the parent prostaglandin has a low permeability rate across the cornea. TLC and GC-MS analyses showed that probably due to the high esterase activity in the corneal epithelium, the $PGF_{2\alpha}$ esters are hydrolyzed and only $PGF_{2\alpha}$ reaches the endothelial side of the perfusion apparatus.

Monti et al. [64] used the same perfusion cell for investigating the effect of a continuous and pulsed current of variable intensity and duration, on the permeation of two β-blocking agents across rabbit cornea. It was found that the two main factors governing drug penetration are the current density and overall treatment time, irrespective of the type of treatment (single or repeated) and current (direct or alternating). Iontophoresis effect on drug permeation was greater to the hydrophilic drug, accompanied by a significant increase in corneal hydration, indicating damage to the corneal texture.

Hastings et al. [55] used this same *in vitro* technology to assess the enhancement delivery of dexamethasone using the Visulex iontophoretic system. The Visulex applicator and a freshly excised rabbit sclera were positioned between two halves of a side-by-side diffusion cell with the conjunctival side of the sclera facing the applicator (Figure 26.10). The donor drug solution (1 mg of dexamethasone phosphate) was present in the applicator, and diluted vitreous humor was modeled in the receptor cell. One milliampere direct current was applied for 60 min, and samples were collected during different treatment periods. It was demonstrated that the Visulex system produced a twofold increase in the amount of dexamethasone phosphate delivered after 60 min, compared with a standard iontophoretic administration (without the Visulex applicator).

26.6 ANIMAL STUDIES

Iontophoresis of various classes of drugs including antibiotics, antivirals, antifungals, antimetabolics, and steroids has been investigated over the years applying transscleral or transcorneal iontophoresis. Table 26.1 and Table 26.2 summarize the results of these studies, with an emphasis on the current density applied, iontophoretic duration, and the sampling time after iontophoresis needed for achieving the peak drug concentration level in each segment.

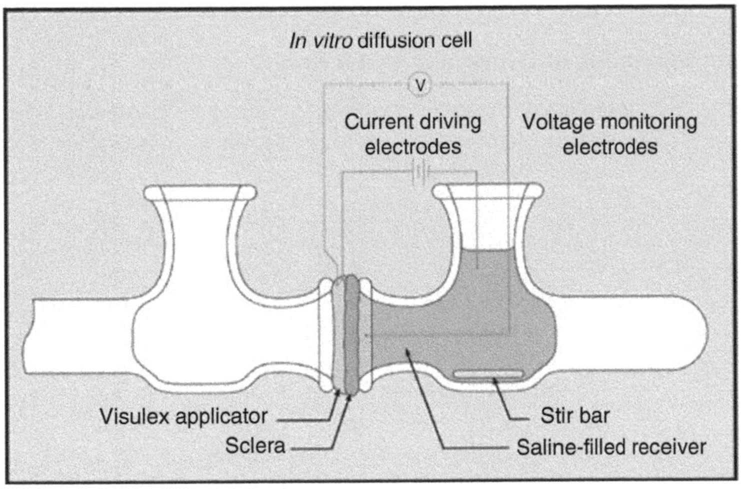

FIGURE 26.10 A side-by-side diffusion cell for *in vitro* experiments with the Visulex applicator is illustrated. Electric current carries the drug from the applicator through the sclera, and into the saline-filled half-cell representing the vitreous. (Adapted from Hastings, M.S. et al., *Drug Deliv. Tech.*, 4, 53, 2004. With permission.)

Current density was calculated from the current applied and the surface contact area of the iontophoretic probe, if mentioned in this chapter. It is clear from this summary that the two major parameters influencing drug penetration are the current density and iontophoretic duration. Ocular iontophoresis can be delivered by two approaches: transcorneal and transscleral iontophoresis.

26.6.1 Transcorneal Iontophoresis

Transcorneal iontophoresis delivers a high concentration of drug to the anterior segment of the eye (cornea, aqueous humor, ciliary body, iris, and lens), with the potential of treating anterior segment diseases, such as keratitis, glaucoma, dry eyes, corneal ulcers, and ocular inflammations (Table 26.1). Normally, after transcorneal iontophoresis drug cannot reach the posterior segment of the eye due to the lens barrier, unless aphakic rabbits are used as published by Fishman et al. [65].

Transcorneal iontophoresis of antibiotics for the treatment of corneal ulcers offers a potentially effective method of management. Numerous studies have demonstrated successful penetration of antibiotics into the anterior chamber after iontophoretic treatment, compared with topically eyedrop instillation of the drug [42,58,59,66,67]. The efficacy of the treatment was investigated in rabbit eyes after intrastromal injection of *Pseudomonas*. Gentamicin, tobramycin, and ciprofloxacin iontophoresis resulted in significantly fewer bacterial colonies in the cornea compared with frequent eyedrops instillation [68–71].

Iontophoresis of vancomycin [66], a complex glycopolypeptide antibiotic, resulted in poor corneal penetration compared with the other antibiotics, due to its high molecular weight (1448 Da) that highly influences the effectiveness of the iontophoretic drug delivery.

Steroids are poorly investigated for the transcorneal route despite the fact that dexamethasone seems to have a much higher corneal penetration than the positively charged antibiotics after corneal iontophoresis. Eljarrat-Binstock et al. [56] used dexamethasone phosphate-loaded hydrogels for the iontophoretic delivery, applying only 5.1 mA/cm^2 current for 1 and 4 min. Enormous amounts of dexamethasone (3077.5 and 1363.7 $\mu g/g$) were detected

TABLE 26.1
Transcorneal Iontophoresis of Drugs into Rabbit Eyes

Drug	Current (mA)	Density (mA/cm^2)	Duration (min)	Site of Sampling	Peak Level (μg/mL)	Sampling Time (h)	Ref.
Antibiotics							
Gentamicin	1	0.5	1	Aqueous	8.0	2	31
Gentamicin[a]	0.75	0.95	10	Cornea	72.0	0.5	65
				Aqueous	77.8	0.5	
				Vitreous	10.4	16	
Gentamicin	0.2	2.82	10	Cornea	376.1	2	42
				Aqueous	54.8	2	
				Vitreous	0.5	0.5	
Gentamicin	1.0	5.1	1	Cornea	363.1	0	58
				Aqueous	29.4	2	
Gentamicin	0.1	0.51	1	Cornea	9.5	5 min	59
	0.3	1.5	1	Cornea	34.0	5 min	
	0.6	3.1	1	Cornea	88.6	5 min	
Tobramycin	0.8	0.84	10	Aqueous	312.8	1	67
	0.8	0.84	5	Aqueous	145.3	1	
Vancomycin	0.5	7.1	5	Aqueous	20.2	2	66
				Cornea	12.4	0.5	
Antiviral drugs							
Vidarabine	0.5	?	4	Aqueous	0.45	1	84
				Cornea	15.6	20 min	
Antifungal drugs							
Ketoconazole	1.5	21.2	15	Aqueous	1.4	1	85
Steroids							
Dexamethasone	1.0	5.1	4	Cornea	3077.5	0	56
	1.0	5.1	1	Aqueous	19.9	0	56
				Cornea	1363.7	0	
				Aqueous	19.7	0	
Genes							
Oligonucleotide (S-ODNs)	1.5	?	5	Cornea	Detected	0	33
			5	Aqueous	Detected	0	
			10	Vitreous	Detected	0	
			20	Retina	Detected	0	
Antisense oligonucleotide	0.3	1.1	5	Cornea	Detected	5 min	72

Source: From Eljarrat-Binstock, E. and Domb, A.J., J. *Control. Release*, 110, 479, 2006. With permission.
[a]Aphakic rabbits.
?—Not mentioned in the article.

at the cornea immediately after the treatment, achieving a 30- to 60-fold higher concentration in the cornea than after the common treatment of frequent drop instillation (every 5 min for 1 h).

Pharmacokinetic studies to determine the distribution profile and elimination rate of the drug after transcorneal iontophoresis were conducted mostly with gentamicin. Fishman et al. [65] reported a peak concentration at the cornea and aqueous humor 30 min after the iontophoretic treatment (72.0 and 77.8 μg/mL, respectively), and 16 h after treatment at the vitreous (70.4 μg/mL), applying 0.95 mA/cm^2 current for 10 min on aphakic rabbits. Using a higher current density (2.8 mA/cm^2) for 10 min, Grossman et al. [42] reported higher gentamicin concentrations in the cornea (376.1 μg/mL) and aqueous humor (54.8 μg/mL), where peak concentrations were detected 2 h after treatment. A recent publication of Eljarrat-Binstock et al. [58] demonstrated a first-order kinetic distribution of gentamicin after 1 min of transcorneal iontophoresis (5.1 mA/cm^2) with a high peak concentration in the cornea (363.1 μg/mL) and aqueous humor (29.4 μg/mL) immediately and 2 h after treatment, respectively. Gentamicin's half-life ($t_{1/2}$) in the anterior chamber and clearance to the posterior segment of the eye (CL) were found to be 2.07 h and 1.73 μL/min, respectively. Grossman et al. and Eljarrat-Binstock et al. reported 8 h of therapeutic levels of gentamicin in the cornea after the transcorneal iontophoretic treatment.

Recently, new studies on gene induction into the animal eye by transcorneal iontophoresis as part of the antisense therapy have been reported [33,72]. Antisense migration to different ocular tissues and their stability in the eye were investigated. Asahara et al. [33] reported an oligonucleotide (S-ODNs) penetration to the anterior chamber of rabbit eyes after 5 min iontophoresis and to the vitreous and retina after 10 and 20 min iontophoresis, respectively. However, the transduction of bigger plasmids (4.7 kb) into the retina could not be definitively confirmed. After 20 min treatment with 1.5 mA, the big plasmids passed through the cornea and accumulated in the angle along the flow of the aqueous humor. Berdugo et al. [72] used rat eyes for evaluation of an oligonucleotide penetration to the cornea applying current density of 1.1 mA/cm^2 for 5 min. Five minutes after treatment, the fluorescence was apparent only in the superficial epithelial layers of the cornea. However, 90 min to 24 h after iontophoresis, the fluorescence intensity of the oligonucleotide was apparent on all corneal layers.

Despite these relatively successful results, transcorneal iontophoresis has never been studied in randomized, clinically controlled studies. The reason could be that the common topical drops treatment is relatively successful in treating most anterior ocular disorders with very few complications even though it requires frequent treatment.

26.6.2 Transscleral Iontophoresis

In aphakic animals, the lens–iris diaphragm limits the penetration of a topically applied drug to the posterior tissues of the eye, such as vitreous and retina. Transscleral iontophoresis overcomes this barrier and delivers drugs directly into the vitreous and retina through the choroid. The iontophoretic device is placed on the conjunctiva, over the pars plana area to avoid current damage to the retina. Figure 26.11 demonstrates the drug distribution to posterior segments of the eye after transscleral iontophoresis.

This iontophoretic route was investigated for various drugs challenging high vitreal drug penetration. Transscleral iontophoresis is a good potential alternative for multiple intravitreal injections or systemic therapy used for posterior ocular disorders, such as endophthalmitis, uveitis, retinitis, optic nerve atrophy, pediatric retinoblastoma, and age-related macular degeneration (AMD). The search for an alternative drug delivery system derives the serious intraocular complications occurring after recurrent intravitreal injections including retinal detachment, vitreous hemorrhage, endophthalmitis, and cataract. At the same time, systemic administration is often associated with side effects [73].

A number of antibiotics, including gentamicin, cephazolin, ticarcillin, amikacin, and vancomycin have been successfully delivered into the vitreous of rabbit eyes (Table 26.2). Barza et al. [74] further investigated the efficacy of transscleral iontophoresis of gentamicin for the treatment of *Pseudomonas* endophthalmitis in rabbits. They found that two sessions of iontophoresis in addition to an intravitreal injection of gentamicin resulted in a significantly lower number of bacterial colonies in the vitreous than by the injection alone.

Transscleral iontophoresis of steroids (dexamethasone and methylprednisolone) can be an alternative treatment for many ocular inflammations. Lam et al. [36] demonstrated high

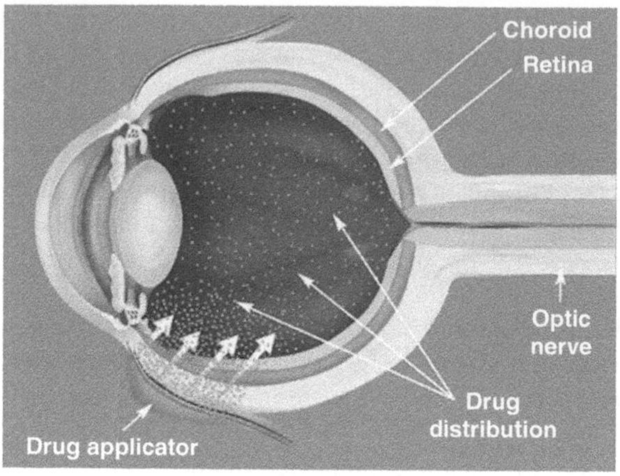

FIGURE 26.11 Illustration of drug distribution to posterior segments of the eye after transscleral iontophoresis. (Adapted from Hastings, M.S. et al., *Drug Deliv. Tech.*, 4, 53, 2004. With permission.)

dexamethasone penetration to the vitreous, 140 μg/mL immediately after the iontophoretic treatment. However, a very high current density was used (over 400 mA/cm^2) for 25 min. Eljarrat-Binstock et al. [56] used lower current density (5.1 mA/cm^2) for only 2 min at two close places in the sclera for delivering dexamethasone phosphate from a saturated hydrogel. Four hours after the transscleral iontophoresis, 341.4 μg/g of dexamethasone were measured in the retina and 166.8 μg/g in the sclera. Lower concentrations were found in the aqueous humor and vitreous (31.8 and 19.2 μg/g, respectively). The efficacy of dexamethasone iontophoresis was studied on rat and rabbit models for endotoxin-induced uveitis (EIU) by Behar-Cohen et al. [34] and Hastings et al. [55], respectively. Behar-Cohen et al. used a 6-mm diameter of eyecup covering the cornea and sclera of the rat whereas Hastings et al. used a saturated hydrogel applicator placed in the superior cul-de-sac. The applied electrical current was 0.4 mA (1.2 mA/cm^2) for 4 min by Behar's group, and 4 mA for 20 min by Hastings' group. Both results demonstrated that the iontophoretic treatment inhibited anterior and posterior signs of intraocular inflammation as effectively as systemic administration of dexamethasone, with a significant improvement over the control groups.

The direct influence of the current density and the iontophoretic time period on the drug levels achieved in posterior eye segments was demonstrated after transscleral iontophoresis of amikacin [54], gentamicin [37], and methylprednisolone [38] (Table 26.2).

Detailed pharmacokinetic studies were performed on transscleral iontophoresis of various drugs [38–40,42,75–78]. Each drug resulted in different patterns of distribution in the vitreous. Carboplatin distribution in the vitreous after iontophoretic delivery demonstrated heightened levels in a controlled manner from 1 to 6 h after treatment [39]. Foscarnet iontophoresis demonstrated a very low elimination rate, thus therapeutic levels in the vitreous were maintained for up to 60 h [78]. Methylprednisolone obtained a relatively low peak concentration 2 h after treatment [38], and gentamicin showed a peak concentration 16 h after the transscleral iontophoresis [42].

Foscarnet and ganciclovir are two antiviral drugs, which are investigated as candidates for iontophoretic administration for local treatment of cytomegalovirus (CMV) retinitis, which is most common in AIDS patient. These two drugs were successfully delivered into the vitreous, where high therapeutic levels were maintained for up to 24 h in the case of ganciclovir [76] and for 60 h after foscarnet delivery [78].

TABLE 26.2
Transscleral Iontophoresis of Drugs into Rabbit Eyes

Drug	Current (mA)	Density (mA/cm^2)	Duration (min)	Site of Sampling	Peak Level (μg/mL)	Sampling Time (h)	Ref.
Antibiotics							
Gentamicin	1.0	5.1	4	Aqueous	10.4	4	57
				Vitreous	3.6	4	
				Retina	110.8	4	
				Sclera	56.3	4	
Gentamicin	2.0	28.2	10	Aqueous	23.2	2	42
				Vitreous	53.4	16	
Gentamicin	2.0	44.2	12	Vitreous	10–12.0	1	93
Gentamicin	2.0	254.6	2	Vitreous	55.0	3	74
Gentamicin	2.0	254.6	5	Vitreous	91.0	3–3.5	37
	2.0	254.6	10	Vitreous	207.0		
	0.8	101.9	10	Vitreous	48.0		
	0.5	63.7	10	Vitreous	27.0		
	0.1	12.7	10	Vitreous	<2		
Gentamicin[a]	1.5	765.3	10	Vitreous	28.0	24	87
Amikacin	2.0	3.7	20	Aqueous	5.3	0.5	54
				Vitreous	1.0	0.5	
				Sclera (NT)	8.3	0.5	
				Sclera (T)	201.0	0.5	
				Retina	41.3	0.5	
	3.0	5.5	20	Aqueous	22.9	0.5	54
				Vitreous	3.9	0.5	
				Sclera (NT)	21.8	0.5	
				Sclera (T)	385.0	0.5	
				Retina	80.5	0.5	
	4.0	7.4	20	Aqueous	39.7	0.5	54
				Vitreous	5.4	0.5	
				Sclera (NT)	40.8	0.5	
				Sclera (T)	343.0	0.5	
				Retina	92.3	0.5	
Cephazolin	2.0	254.6	10	Vitreous	119.0	3	37
Ticarcillin	2.0	254.6	10	Vitreous	94.0	3	37
Vancomycin	3.5	49.5	10	Vitreous	13.4	2	66
Ciprofloxacin	5.0	70.7	15	Vitreous	0.1	3	94
Antiviral drugs							
Ganciclovir	1.0	259.8	15	Vitreous and retina	74.0	2	76
Foscarnet	1.0	526.3	10	Vitreous	200, 203 μM	4	78,95
Antimetabolic drugs							
5-Fluorouracil	0.5	1.3	0.5	Conjunctiva	480.0	0	96
				Sclera	165.0	0	
Carboplatin	2.5	5.0	20	Vitreous	1955.0	6	39
				Retina	45.3	1	
				Choroid	483.2	1	
				Optic nerve	37.7	2	
Antifungal drugs							
Ketoconazole	4.0–6.0	56.6–84.9	15	Vitreous	0.1	1	85

(*Continued*)

TABLE 26.2 (Continued)
Transscleral Iontophoresis of Drugs into Rabbit Eyes

Drug	Current (mA)	Density (mA/cm^2)	Duration (min)	Site of Sampling	Peak Level (μg/mL)	Sampling Time (h)	Ref.
Steroids							
Dexamethasone	1.6	415.7	25	Vitreous	140.0	0	36
Dexamethasone	1.0	5.1	4	Aqueous	31.8	4	56
				Vitreous	19.2	4	
				Sclera	166.8	4	
				Retina	341.4	4	
Methylprednisolone	0.4	0.8	4	Aqueous	16.1	2	38
				Vitreous	0.7[b]	2	38
	1.0	2.0	4	Aqueous	32.9	2	38
				Vitreous	0.9[b]	2	
	2.0	4.0	4	Aqueous	46.0	2	
				Vitreous	1.4[b]	2	
				Cornea	500[b]	2	
				Iris	130[b]	2	
				Sclera	40[b]	2	
				Choroid	18[b]	2	
				Retina	85[b]	2	
	2.0	4.0	10	Aqueous	48[b]	1	
				Vitreous	2.5[b]	1	
				Cornea	525[b]	1	
				Iris	200[b]	1	
				Choroid	452.4	6	
				Retina	446.2	18	38
Generic drugs							
Diclofenac	4.0	7.4	20	Vitreous	0.11		51
				Sclera (T)	52.0		
				Sclera (NT)	2.5		
				Retina and choroid	9.2		
Aspirin	2.5	5.0	10	Anterior uvea	1600[b]	1	77
				Retina	450[b]	1	
				Choroid	1250[b]	1	
Aspirin	2.5	5.0	10	Conjunctiva	452.3	0.5	40
				Anterior sclera	515.1	0.5	
				Anterior uvea	1614.2	0.5	
				Aqueous humor	495.9	0.5	
				Posterior sclera	208.8	0.5	
				Retina	443.0	0.5	
				Choroid	1276.0	0.5	
				Vitreous	9.1	0.5	

Source: From Eljarrat-Binstock, E. and Domb, A.J., *J. Control. Release*, 110, 479, 2006. With permission.
[a]Monkey.
[b]Estimated values from graphs.
T—Treated sclera (where the iontophoretic applicator was placed).
NT—Untreated sclera.

26.7 CLINICAL STUDIES

Several investigators conducted clinical studies using transscleral iontophoresis of the anti-inflammatory corticosteroid, methylprednisolone hemisuccinate (SoluMedrol). Chauvaud et al. [79] presented initial findings of phase II clinical trial for transscleral iontophoresis of the above drug using the coulomb-controlled annular applicator (Figure 26.6). The transscleral iontophoresis was safe, well tolerated, and easily applied for the treatment of severe ocular inflammation, thereby reducing the systemic side effects of corticotherapy. The same iontophoretic system and drug were used to assess the efficacy of the treatment on three subjects with severe acute corneal graft rejection. The patients were treated with SoluMedrol (methylprednisolone) iontophoresis (1.5 mA, 3 min) once a day for 3 days, under topical anesthesia, as supplement to topical dexamethasone drops. The treatment was tolerable, no side effects were observed, and visual acuity improved rapidly after the second treatment [80]. Behar-Cohen et al. [81] and Halhal et al. [82] presented similar results in a study with 17 patients with acute corneal graft rejection. The annular shape iontophoresis applicator (Eyegate, Optis, Paris, France), connected to the negative output, was placed on the eye. The positive return skin patch was placed on the patient's forehead. Iontophoretic treatment of methylprednisolone, using 1.5 mA (3 mA/cm^2) for 4 min, was performed once daily for 3 consecutive days with topical analgesia. After treatment, 15 of the 17 treated eyes (88%) demonstrated complete reversal of the rejection processes with no significant side effects. Tolerance of the course of iontophoresis treatment was good. In some instances, a mild or minimal pain sensation was reported, but this was easily tolerated. A transitory conjunctival redness due to vessel dilation was observed in most of the cases, resolving within 24 h.

Evaluation of ocular tolerance using transscleral iontophoresis on 27 healthy volunteers was reported by Parkinson et al. [52]. A novel OcuPhor hydrogel drug delivery applicator (Figure 26.7) was filled with balanced salt solution and placed against the sclera in the inferior cul-de-sac, previously treated with topical anesthesia. A hydrogel dispersive electrode was placed on the subject's skin on the right side of the neck or right shoulder, to complete the electrical circuit. Different current intensities were used (0, 0.1, 0.5, 1.0, 2.0, 3.0, or 4.0 mA) at negative and positive polarities for 20 or 40 min. The treatment surface area of this applicator is 0.54 cm^2, thus maximal current density applied is 7.4 mA/cm^2. The applicator and iontophoretic procedure were well tolerated and no clinically significant changes in symptomatology or in ophthalmic assessments were seen following exposure to 0–3 mA for 20 min or 1.5 mA for 40 min. Symptoms during dosing consisted primarily of tingling, burning, and aching sensations, which spontaneously resolved after dosing. Only when 4 mA current was applied, half of the subjects described intolerable discomfort and a severe burning sensation, which resolved after 22 h without any clinically significant effects on ophthalmic assessments.

An innovative clinical trial of iodide iontophoresis as a treatment for dry eye syndrome was reported by Horwath-Winter et al. [83]. This clinical study investigated the effectiveness of iodide iontophoresis treatment in moderate to severe dry eye patients, using the coulomb-controlled annular applicator. Sixteen patients were treated with iodide iontophoresis and 12 patients with iodide application without current for 10 days. A stronger positive influence was seen after iontophoretic application of iodide, as observed in a distinct improvement in breakup time, fluorescein, and rose bengal staining, and in a longer duration of this effect compared with the noncurrent group. A reduction in subjective symptoms and frequency of artificial tear substitute application could be observed in both groups.

26.8 TOXICITY

Iontophoresis is a local noninvasive method of administration associated with minimal discomfort for the patient. However, this procedure can produce complications, including epithelial edema, decrease in endothelial cells, inflammatory infiltration, and burns. Such damage is dependent upon the site of application, the current density, and the iontophoretic duration [1].

Transcorneal and transscleral iontophoresis, at the same conditions, can produce different toxicity levels due to the significant differences between cornea tissue and sclera. Any damage to the corneal surface immediately affects the vision and comfort of the patient, which is less pronounced when applied to the sclera. The clarity of the cornea is essential for interaction with light whereas the sclera is not relevant for light interaction. The cornea is an avascular and highly innervated tissue, unlike the sclera, and thus very sensitive to pain and hypoxia. On one hand, transcorneal iontophoresis endangers the front "window" of the eye, but on the other hand transscleral iontophoresis threatens the retina underneath the application site, which is essential for visual image formation.

This section summarizes the literature concerning the iontophoretic toxicity to the eye, when applied on the cornea or on the sclera.

Transcorneal iontophoresis was assessed for potential toxicity to the cornea by Hill et al. [84], demonstrating minimal surface pitting of the corneas immediately after 0.5 mA current application for 4 min. Hughes and Maurice [31] delivered current densities of 20–25 mA/cm^2 for 1 and 5 min through a 4-mm diameter polyacrylamide gel discs lying over the cornea of rabbits. They noted fluorescein staining of the epithelium and slight stromal swelling of the cornea immediately after treatment, which returned to normal after 24 h. No serious endothelial injury had occurred, as confirmed by scanning electron microscopy immediately after treatment and at 24 and 48 h afterwards. Rootman et al. [67] reported minimal surface epithelial abnormalities using light and electron microscopy immediately after 0.8 mA/cm^2 of transcorneal iontophoresis for 5 min, and focal areas of epithelial edema after 10 min of iontophoresis. The stroma and endothelium remained normal. Choi and Lee [66] found 8.8% and 5.4% decrease in endothelial cell count 4 days after transcorneal iontophoresis of vancomycin and saline solution, respectively, using a current density of 7.1 mA/cm^2 for 5 min. Grossman and Lee [85] noted persistent corneal opacities after transcorneal iontophoresis of ketoconazole using high current density of 21 mA/cm^2 for 15 min. Grossman et al. [42] assessed corneal toxicity 3 days after gentamicin iontophoresis, applying a current density of 2.82 mA/cm^2 for 10 min. No increase was found in corneal thickness, but a decrease in corneal cell count occurred only after saline iontophoresis and not after gentamicin iontophoresis. Eljarrat-Binstock et al. [57] demonstrated epithelial defects and stromal edema 5 min after applying 2.5–5.1 mA/cm^2 current for 1–2 min. When using the lower current density for 1 min these findings disappeared after 8 h whereas after applying a higher current density mild stromal edema was still observed at 8 h time point.

Transscleral iontophoresis toxicity was evaluated using higher current densities, where choroid and retinal damage were observed. Maurice [86] noted the destruction of the retinal layers and engorgement of the choroid, following transscleral iontophoresis of fluorescein using a 127–254 mA/cm^2 current density for 2–5 min. Barza [37] investigated the tolerance of gentamicin and cefazolin iontophoresis, 3 h after a high current density application (254.6, 127.0, and 101.9 mA/cm^2) for 5 or 10 min. Animals experienced no evident discomfort corresponding to all current densities applied with maximal voltage of 150 V. However, histopathological examination showed hemorrhagic necrosis, edema, and infiltration by polymorphonuclear cells in the retina, choroid, and ciliary body using 254.6 and 127.0 mA/cm^2 for 5 and 10 min. Another publication of this group [87], using an extremely high current density (765.3 mA/cm^2) for 10 min, demonstrated retinal burns but normal electroretinograms in monkeys several weeks after multiple sessions of gentamicin transscleral iontophoresis. Lam et al. [88] studied this issue using transscleral iontophoresis of saline applying a current density of 531 mA/cm^2. It was found that the size and severity of chorioretinal lesions in rabbits, 5 days after treatment, increased with the duration of application (2–25 min). No retinal lesions were observed by light microscopic examination, after 1 min of iontophoretic treatment. Iontophoretic treatment for 5 and 15 min resulted in necrotic retinal pigment epithelium (RPE), proliferation of RPE cells and macrophages activity, and loss of retinal nuclear layers that within 14 days resulted in eventual glial membrane. Behar-Cohen et al. [34] reported no histological damage or sign of thermal damage to the cornea and sclera after applying 1.2 mA/cm^2 current for 4 min, using a 6-mm diameter of eyecup covering the cornea and sclera of a rat. Also, when using 4 mA/cm^2 for 10 min no visible lesions in the area of application and no abnormalities or histological findings were observed 24 h after the iontophoretic treatment using the annular transscleral device [38]. Yet, in the same publication, Behar-Cohen's group reported a safe procedure with current intensity lower than 50 mA/cm^2 for transscleral iontophoresis [38]. Voigt et al. [40] and Kralinger et al. [77] evaluated the safety of single and repetitive transscleral iontophoresis of aspirin using a current density of only 5.0 mA/cm^2 for 10 min. Eight hours after a single iontophoretic treatment the anterior segment, vitreous cavity, and fundus were examined using slit lamp biomicroscopy and indirect ophthalmoscopy. No retinal detachment or other intraocular complications were observed, except for a slight conjunctival injection that disappeared after 8 h. Histologically, no signs of inflammation or tissue damage were found in the anterior or posterior segments of the eye [40]. Also, after serial iontophoretic treatment in 24 h intervals for a week, no evidence of inflammation or toxicity was observed. The retinal epithelium, photoreceptors, and nuclear layers were not altered, and the electroretinographic analysis showed a typical response [77].

Despite all these findings, very little attention has been given in patents and patent applications to the toxicity that the iontophoresis process might have using high current densities. Moreover, no current limitation supply or distinction between the iontophoretic application sites on the eye has been made.

The implications of high current density applied were emphasized in patent application WO 91/12049. It describes an iontophoretic system for focal transscleral destruction and perforation of living human tissue for the purpose of immediately decreasing eye pressure. A current of about 3.0–4.0 mA for about 30 s to 5 min is applied in dozens of locations on the sclera. The surface area that this apparatus is applied for is in the range of 0.2–2 mm, preferably 0.3–0.6 mm in diameter, which translates to 1000–5600 mA/cm^2 of current density [89].

US patent 4,564,016 describes an iontophoretic device using a solution chamber where current flow is adjusted to a desired value by adjusting a potentiometer for a flow passage of a diameter of 0.25–0.5 mm, which is equivalent to a range in current density of 500–2000 mA/cm^2 if only 1 mA current is applied [90]. This is an enormous amount of current that may electrify the patient or at least cause significant damage to the eye. A more restrictive patent, WO 99/40967, describes an invention wherein drugs are delivered to the eye using solid hydrogel discs in the size of 3 mm in diameter applied onto the cornea with a current of up to 1 mA, which is translated to about 14 mA/cm^2 [91]. A recent patent application for a safe iontophoresis describes a device comprising an arrangement that prevents operation of the device at a current density that is higher than a predetermined value depending on the tissue to be treated. The device includes a contacting member capable of transmitting a signal indicative of the contact surface area through which the current must pass [92]. This may be a good solution for preventing sudden current densities and voltage elevations during the iontophoretic treatment, due to poor contact of the drug container with the eye surface.

26.9 CONCLUSIONS

A better understanding of tissue interactions within the eye during electric current application, along with better designs of devices and probes adapted to the site of application have yielded efficient intraocular penetration of drugs and oligonucleotides using ocular iontophoresis. It is clear now that the ocular iontophoresis has clinical potential importance as a local delivery system for many drugs. It is only a matter of time until iontophoresis is routinely used in the ophthalmic field.

REFERENCES

1. Baeyens, V., et al. 1997. Ocular drug delivery in veterinary medicine. *Adv Drug Deliv Rev* 28:335.
2. Bridger, V.M., et al. 1982. A device for iontophoretic anesthesia of the tympanic membrane. *J Med Eng Technol* 6:62.
3. Zempsky, W.T., et al. 2004. Evaluation of a low-dose lidocaine iontophoresis system for topical anesthesia in adults and children: A randomized, controlled trial. *Clin Ther* 26:1110.
4. Moppett, I.K., K. Szypula, and P.M. Yeoman. 2004. Comparison of EMLA and lidocaine iontophoresis for cannulation analgesia. *Eur J Anaesthesiol* 21:210.
5. DeCou, J.M., et al. 1999. Iontophoresis: A needle-free, electrical system of local anesthesia delivery for pediatric surgical office procedures. *J Pediatr Surg* 34:946.
6. Gibson, L.E., and R.E. Cooke. 1959. A test for concentration of electrolytes in cystic fibrosis of the pancreas utilizing pilocarpine by iontophoresis. *Pediatrics* 23:545.
7. Carter, E.P., et al. 1984. Improved sweat test method for the diagnosis of cystic-fibrosis. *Arch Dis Child* 59:919.
8. Gangarosa, L.P., et al. 1986. Iontophoresis of vidarabine monophosphate for herpes-orolabialis. *J Infect Dis* 154:930.
9. Gangarosa, L., and N. Park. 1978. Practical considerations in iontophoresis of fluoride for desensitizing dentine. *J Prosthet Dent* 19:73.
10. Krauser, J.T. 1986. Hypersensitive teeth. 2. Treatment. *J Prosthet Dent* 56:307.
11. Rigano, W., et al. 1992. Antibiotic iontophoresis in the management of burned ears. *J Burn Care Rehabil* 13:407.
12. Wirtz, R. 1908. Die ionentherapie in der augenheilkunde. *Klin Monatsbl Augenheilkd* 46:543.
13. Myles, M.E., et al. 2003. Ocular iontophoresis. In *Ophthalmic drug delivery systems*, 2nd ed., ed. A.K. Mitra, 365. New York: Marcel Dekker.
14. Chastain, J.E. 2003. General considerations in ocular drug delivery. In *Ophthalmic drug delivery systems*, ed. A.K. Mitra, 59. New York: Marcel Dekker.
15. Raviola, G. 1977. The structural basis of the blood–ocular barriers. *Exp Eye Res* 25:27.
16. Sasaki, H., et al. 1999. Enhancement of ocular drug penetration. *Crit Rev Ther Drug Carrier Syst* 16:85.
17. Schoenwald, R.D. 1990. Ocular drug delivery—pharmacokinetic considerations. *Clin Pharmacokinet* 18:225.
18. Lee, V.H.L., and J.R. Robinson. 1986. Topical ocular drug delivery: Recent developments and future challenges. *J Ocul Pharmacol* 2:67.
19. Davies, N.M. 2000. Biopharmaceutical considerations in topical ocular drug delivery. *Clin Exp Pharm Phys* 27:558.
20. Ellis, P.P. 1985. Basic considerations. In *Ocular therapeutics and pharmacology*, 7th ed., ed. P.P. Ellis, 3. St. Louis, MO: The C.V. Mosby Company.
21. Hamalainen, K.M., et al. 1997. Characterization of paracellular and aqueous penetration routes in cornea, conjunctiva, and sclera. *Invest Ophthalmol Vis Sci* 38:627.
22. Worakul, N., and J.R. Robinson. 1997. Ocular pharmacokinetics/pharmacodynamics. *Eur J Pharm Biopharm* 44:71.
23. Sasaki, H., et al. 1997. *In-situ* ocular absorption of ophthalmic beta-blockers through ocular membranes in albino rabbits. *J Pharm Pharmacol* 49:140.

24. Luo, A.M., H. Sasaki, and V.H.L. Lee. 1991. Ocular drug-interactions involving topically applied timolol in thc pigmented rabbit. *Curr Eye Res* 10:231.
25. Kalia, Y.N., et al. 2004. Iontophoretic drug delivery. *Adv Drug Deliv Rev* 56:619.
26. Guy, R.H., et al. 2000. Iontophoresis: Electrorepulsion and electroosmosis. *J Control Release* 64:129.
27. Burnette, R., and D. Marrero. 1986. Comparison between the iontophoretic and passive transport of thyrotropin releasing hormone across excised nude mouse skin. *J Pharm Sci* 75:738.
28. Gangarosa, L., et al. 1980. Increased penetration of non-electrolytes into mouse skin during iontophoretic water transport (iontohydrokinesis). *J Pharmacol Exp Ther* 212:377.
29. Kasting, G.B. 1992. Theoretical models for iontophoretic delivery. *Adv Drug Deliv Rev* 9:177.
30. Pikal, M.J. 2001. The role of electroosmotic flow in transdermal iontophoresis. *Adv Drug Deliv Rev* 46:281.
31. Hughes, L., and D.M. Maurice. 1984. A fresh look at iontophoresis. *Arch Ophthalmol* 102:1825.
32. Maurice, D.M. 1983. Micropharmaceutics of the eye. *Ocul Inflamm Ther* 1:97.
33. Asahara, T., et al. 2001. Induction of gene into the rabbit eye by iontophoresis: Preliminary report. *Jpn J Ophthalmol* 45:31.
34. Behar-Cohen, F.F., et al. 1997. Iontophoresis of dexamethasone in the treatment of endotoxin-induced-uveitis in rats. *Exp Eye Res* 65:533.
35. Sarraf, D., and D.A. Lee. 1994. The role of iontophoresis in ocular drug-delivery. *J Ocul Pharmacol* 10:69.
36. Lam, T.T., et al. 1989. Transscleral iontophoresis of dexamethasone. *Arch Ophthalmol* 107:1368.
37. Barza, M. 1986. Transscleral iontophoresis of cefazolin, cicarcillin, and gentamicin in the rabbit. *Ophthalmology* 93:133.
38. Behar-Cohen, F.F., et al. 2002. Transscleral coulomb-controlled iontophoresis of methylprednisolone into the rabbit eye: Influence of duration of treatment, current intensity and drug concentration on ocular tissue and fluid levels. *Exp Eye Res* 74:51.
39. Hayden, B.C., et al. 2004. Pharmacokinetics of systemic versus focal carboplatin chemotherapy in the rabbit eye: Possible implication in the treatment of retinoblastoma. *Invest Ophthalmol Vis Sci* 45:3644.
40. Voigt, M., et al. 2002. Ocular aspirin distribution: A comparison of intravenous, topical, and coulomb-controlled iontophoresis administration. *Invest Ophthalmol Vis Sci* 43:3299.
41. Jones, R.F., and D.M. Maurice. 1966. New methods of measuring the rate of aqueous flow in man with fluorescein. *Exp Eye Res* 5:208.
42. Grossman, R., D.F. Chu, and D.A. Lee. 1990. Regional ocular gentamicin levels after transcorneal and transscleral iontophoresis. *Invest Ophthalmol Vis Sci* 31:909.
43. Frucht-Pery, J., et al. 1999. The distribution of gentamicin in the rabbit cornea following iontophoresis to the central cornea. *J Ocul Pharmacol Ther* 15:251.
44. Frucht-Pery, J., et al. 1996. Efficacy of iontophoresis in the rat cornea. *Graefes Arch Clin Exp Ophthalmol* 234:765.
45. Kishida, A., and Y. Ikada. 2002. Hydrogels for biomedical and pharmaceutical applications. In *Polymeric biomaterials*, 2nd ed., ed. S. Dumitriu, 133. New York: Marcel Dekker.
46. Kim, S.W., Y.H. Bae, and T. Okano. 1992. Hydrogels: Swelling, drug loading and release. *Pharm Res* 9:283.
47. Banga, A.K., and Y.W. Chien. 1993. Hydrogel-based iontotherapeutic delivery devices for transdermal delivery of peptides/protein drugs. *Pharm Res* 10:697.
48. Alvarez-Figueroa, M.J., and J. Blanco-Mendez. 2001. Transdermal delivery of methotrexate: Iontophoretic delivery from hydrogels and passive delivery from microemulsions. *Int J Pharm* 215:57.
49. Fang, J.Y., et al. 1999. Evaluation of transdermal iontophoresis of enoxacin from polymer formulations: *In vitro* skin permeation and *in vivo* microdialysis using Wistar rat as an animal model. *Int J Pharm* 180:137.
50. Fang, J.Y., et al. 2002. The effects of iontophoresis and electroporation on transdermal delivery of buprenorphine from solutions and hydrogels. *J Pharm Pharmacol* 54:1329.
51. Fischer, G.A., T.M. Parkinson, and M.A. Szlek. 2002. OcuPhor—the future of ocular drug delivery. *Drug Deliv Tech* 2:50.

52. Parkinson, T.M., et al. 2003. Tolerance of ocular iontophoresis in healthy volunteers. *J Ocul Pharmacol Ther* 19:145.
53. Parkinson, T.M., et al. 2000. The effects of *in vivo* iontophoresis on rabbit eye structure and retinal function. *Invest Ophthalmol Vis Sci* 41:S772.
54. Vollmer, D.L., et al. 2002. *In vivo* transscleral iontophores is of amikacin to rabbit eyes. *J Ocul Pharmacol Ther* 18:549.
55. Hastings, M.S., et al. 2004. Visulex: Advancing iontophoresis for effective noninvasive back-to-the-eye therapeutics. *Drug Deliv Tech* 4:53.
56. Eljarrat-Binstock, E., et al. 2005. Transcorneal and transscleral iontophoresis of dexamethasone phosphate in rabbits using drug loaded hydrogel. *J Control Release* 106:386.
57. Eljarrat-Binstock, E., et al. 2004. Hydrogel probe for iontophoresis drug delivery to the eye. *J Biomater Sci Polym Ed* 15:397.
58. Eljarrat-Binstock, E., et al. 2004. Delivery of gentamicin to the rabbit eye by drug-loaded hydrogel iontophoresis. *Invest Ophthalmol Vis Sci* 45:2543.
59. Frucht-Pery, J., et al. 2004. Iontophoresis-gentamicin delivery into the rabbit cornea, using a hydrogel delivery probe. *Exp Eye Res* 78:745.
60. Brouneus, F., et al. 2001. Diffusive transport properties of some local anesthetics applicable for iontophoretic formulation of the drugs. *Int J Pharm* 218:57.
61. Anderson, C.R., et al. 2003. Effects of iontophoresis current magnitude and duration on dexamethasone deposition and localized drug retention. *Phys Ther* 83:161.
62. Kamath, S.S., and L.P. Gangarosa. 1995. Electrophoretic evaluation of the mobility of drugs suitable for iontophoresis. *Methods Find Exp Clin Pharmacol* 17:227.
63. Camber, O., P. Edman, and L.I. Olsson. 1986. Permeability of prostaglandin F_2 and prostaglandin F_2 esters across cornea *in vitro*. *Int J Pharm* 29:259.
64. Monti, D., et al. 2003. Effect of iontophoresis on transcorneal permeation '*in vitro*' of two beta-blocking agents, and on corneal hydration. *Int J Pharm* 250:423.
65. Fishman, P.H., et al. 1984. Iontophoresis of gentamicin into aphasic rabbit eyes. *Invest Ophthalmol Vis Sci* 25:343.
66. Choi, T.B., and D.A. Lee. 1988. Transscleral and transcorneal iontophoresis of vancomycin in rabbit eyes. *J Ocul Pharmacol* 4:153.
67. Rootman, D.S., et al. 1988. Pharmacokinetics and safety of transcorneal iontophoresis of tobramycin in the rabbit. *Invest Ophthalmol Vis Sci* 29:1397.
68. Frucht-Pery, J., et al. 2006. Iontophoretic treatment of experimental *Pseudomonas* keratitis in rabbit eye using gentamicin-loaded hydrogel. *Cornea. In Press.*
69. Hobden, J.A., et al. 1989. Tobramycin iontophoresis into corneas infected with drug-resistant *Pseudomonas aeruginosa*. *Curr Eye Res* 8:1163.
70. Hobden, J.A., et al. 1990. Ciprofloxacin iontophoresis for aminoglycoside-resistant pseudomonal keratitis. *Invest Ophthalmol Vis Sci* 31:1940.
71. Rootman, D.S., et al. 1988. Iontophoresis of tobramycin for the treatment of experimental *Pseudomonas* keratitis in the rabbit. *Arch Ophthalmol* 106:262.
72. Berdugo, M., et al. 2003. Delivery of antisense oligonucleotide to the cornea by iontophoresis. *Antisense Nucleic Acid Drug Develop* 13:107.
73. Geroski, D.H., and H.F. Edelhauser. 2001. Transscleral drug delivery for posterior segment disease. *Adv Drug Deliv Rev* 52:37.
74. Barza, M., C. Peckman, and J. Baum. 1987. Transscleral iontophoresis as an adjunctive treatment for experimental endophthalmitis. *Arch Ophthalmol* 105:1418.
75. Chapon, P., et al. 1999. Intraocular tissues pharmacokinetics of ganciclovir transscleral coulomb controlled iontophoresis in rabbits. *Invest Ophthalmol Vis Sci* 40:S189.
76. Lam, T.T., et al. 1994. Intravitreal delivery of ganciclovir in rabbits by transscleral iontophoresis. *J Ocul Pharmacol* 10:571.
77. Kralinger, M.T., et al. 2003. Ocular delivery of acetylsalicylic acid by repetitive coulomb-controlled iontophoresis. *Ophthalmic Res* 35:102.
78. Sarraf, D., et al. 1993. Transscleral iontophoresis of foscarnet. *Am J Ophthalmol* 115:748.

79. Chauvaud, D., et al. 2000. Transscleral iontophoresis of cortcicosteoids: Phase II clinical trial. *Invest Ophthalmol Vis Sci* 41:S79.
80. Halhal, M., et al. 2003. Corneal graft rejection and corticoid iontophoresis: 3 case reports. *J Fr Ophthamol* 26:391.
81. Behar-Cohen, F.F., et al. 2002. Reversal of corneal graft rejection by iontophoresis of methylprednisolone. *Invest Ophthalmol Vis Sci* 43:U504.
82. Halhal, M., et al. 2004. Iontophoresis: From the lab to the bed side. *Exp Eye Res* 78:751.
83. Horwath-Winter J., et al. 2005. Iodide iontophoresis as a treatment for dry eye syndrome. *Brit J Ophthalmol* 89:40.
84. Hill, J.M., et al. 1978. Iontophoresis of vidarabine monophosphate into rabbit eyes. *Invest Ophthalmol Vis Sci* 17:473.
85. Grossman, R., and D.A. Lee. 1989. Transscleral and transcorneal iontophoresis of ketoconazole in the rabbit eye. *Ophthalmology* 96:724.
86. Maurice, D.M. 1986. Iontophoresis of fluorescein into the posterior segment of the rabbit eye. *Ophthalmology* 93:128.
87. Barza, M., C. Peckman, and J. Baum. 1987. Transscleral iontophoresis of gentamicin in monkeys. *Invest Ophthalmol Vis Sci* 28:1033.
88. Lam, T.T., J. Fu, and M.O.M. Tso. 1991. A histopathologic study of retinal lesions inflicted by transscleral iontophoresis. *Graefes Arch Clin Exp Ophthalmol* 229:389.
89. Dobrogowski, M.J., and M.A. Latina. 1991. Focal destruction of eye tissue by electroablation. World Patent WO 91/12049.
90. Maurice, D.M., and D. Brooks. 1986. Apparatus for introducing ionized drugs into the posterior segment of the eye and method. US Patent 4,564,016.
91. Domb, A., and J. Frucht-Pery. 1999. A device for iontophoretic administration of drugs. World Patent WO 99/40967.
92. Domb, A. 2006. Safe device for iontophoretic delivery of drugs. Pending Patent application.
93. Burstein, N.L., I.H. Leopold, and D.B. Bernacchi. 1985. Trans-scleral iontophoresis of gentamicin. *J Ocul Pharmacol* 1:363.
94. Yoshizumi, M.O., et al. 1991. Experimental transscleral iontophoresis of ciprofloxacin. *J Ocul Pharmacol* 7:163.
95. Yoshizumi, M.O., et al. 1996. Ocular iontophoretic supplementation of intravenous foscarnet therapy. *Am J Ophthalmol* 122:86.
96. Kondo, M., and M. Araie. 1989. Iontophoresis of 5-fluorouracil into the conjunctiva and sclera. *Invest Ophthalmol Vis Sci* 30:583.
97. Eljarrat-Binstock, E., and A.J. Domb. 2006. Iontophoresis: A non-invasive ocular drug delivery. *J Control Release* 110:479.

Part VIII

Drug Delivery to the Central Nervous System

27 Structure and Function of the Blood–Brain Barrier

David J. Begley

CONTENTS

27.1 INTRODUCTION

All organisms with a well-developed central nervous system (CNS) have a blood–brain barrier (BBB). In all mammals the BBB is created by the endothelial cells forming the capillaries of the brain and spinal cord microvasculature. The combined surface area of these microvessels constitutes by far the largest surface area for blood–brain exchange. This surface area, depending on the anatomical region, is between 150 and 200 cm^2/g of tissue giving a total area for exchange in the brain of 12–18 m^2 for the average human adult [1].

In addition, the epithelial cells of the choroid plexus facing the cerebrospinal fluid (CSF) constitute the blood–cerebrospinal fluid barrier (BCSFB). The BCSFB is also a significant area for exchange between the blood and the CSF. In rats the total calculated surface area of the choroid plexus is about 33% of that of the BBB [2]. In humans, based on the relative mass of the choroid plexus in comparison with the brain, the relative surface area of the choroid plexus may be in the region of 10% of that of the BBB. The CSF is secreted across the choroid plexus epithelial cells into the brain ventricular system [3]; the remainder of the brain extracellular fluid (ECF) and the interstitial fluid (ISF) are secreted at the capillaries of the BBB themselves [4]. The ratio of fluid production from these sites is 40%:60%, respectively [5].

The secretion of CSF and ECF is essentially driven by an osmotic gradient created by the Na^+/K^+-ATPase, expressed in the abluminal membrane of the BBB endothelium and the apical membrane of the choroid plexus epithelium, which produces water movement into brain ECF and produces volume secretion.

Finally, the avascular arachnoid membrane, underlying the dura and completely enclosing the CNS, completes the seal between the ECF of the CNS and that of the rest of the body.

Although the cells of the arachnoid membrane have tight junctions (TJ) between them, due to its avascular nature and relatively small surface area, the arachnoid membrane does not represent a significant potential interface for exchange between the blood and the CNS (Figure 27.1) [6].

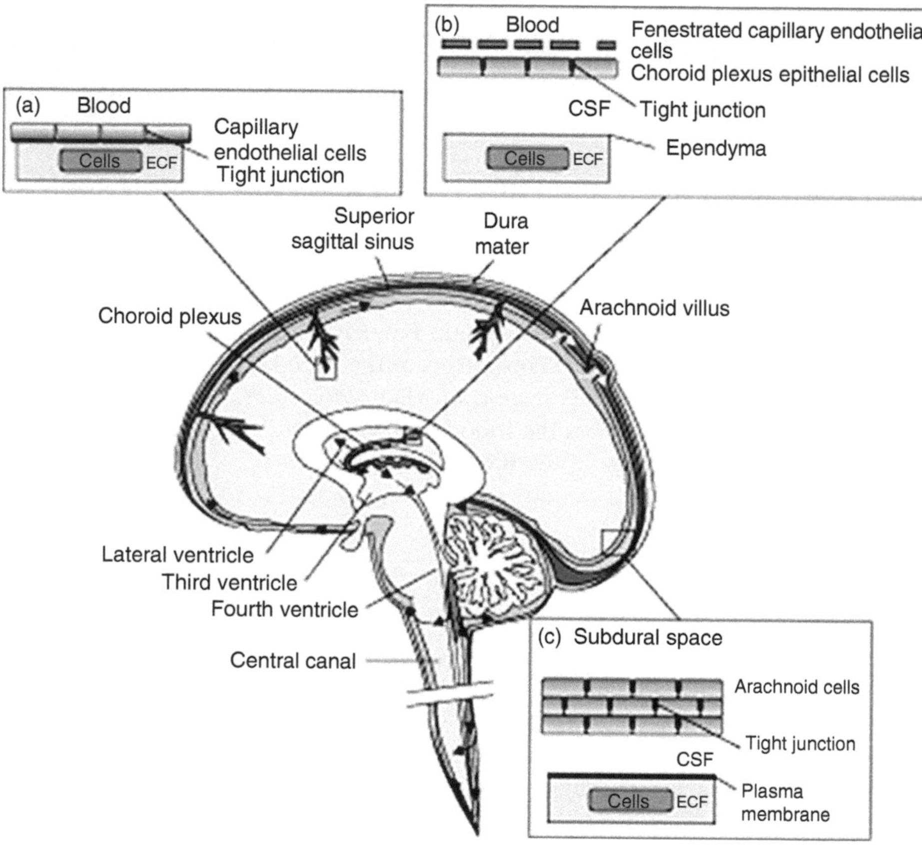

FIGURE 27.1 Barriers of the brain. There are three principal barrier sites between blood and brain. (a) The BBB proper, which is created at the level of the cerebral capillary endothelial cells by tight junction formation. It is by far the largest surface area for exchange and in the adult human is between 12 and 18 m^2 in surface area. No brain cell is farther than about 25 μm from a capillary, so once the BBB is crossed diffusion distances to neurons and glial cell bodies for solutes and drugs are short. Targeting a drug across the BBB is therefore the favored route for global delivery of drugs to all brain cells. (b) The blood–CSF barrier (BCSFB) lies at both the choroid plexi in the lateral and fourth ventricles of the brain where tight junctions are formed between the epithelial cells at the CSF-facing surface (apical surface) of the plexi. In the human, because of the microvillus nature of the apical surface of the choroid plexi, the total surface area of the BCSFB is approximately 10% of that of the BBB. Some drugs and solutes enter the brain principally across the choroid plexi into CSF, others enter via both the BBB and BCSFB. (c) The arachnoid barrier. The brain is enveloped by the arachnoid membrane lying under the dura. The arachnoid membrane is avascular but lies close to the superior sagittal sinus and is separated from it by the dura. The cells of the arachnoid membrane form multiple layers and have tight junctions between them that form an effective seal. Arachnoid villi project into the sagittal sinus through the dura and a significant amount of CSF drains into the sinus through these valve-like villi which only allow CSF movement out of the brain to blood. Transport across the arachnoid membrane is not an important route for the entry of solutes into brain. (Adapted from Kandel, E.R., Schwartz, J.H., and Jessel, T.M., *Principles of neural science*, 4th ed., McGraw-Hill, New York, 2000, 1294. With permission.)

A BBB is required because the CNS needs to maintain an extremely stable internal fluid environment surrounding the neurons. This stability of the internal environment of the brain is an absolute requirement for reliable synaptic communication between nerve cells. The quantal chemical communication that takes place at synapses, and the highly complex spatial- and temporal summation and integration of signals that continually occur between neurons cannot take place efficiently if brain ECF were allowed to fluctuate in its composition as markedly as does the somatic ECF. The concentration of potassium in plasma, for instance, is approximately 5.5 mM, and in the CSF, and probably also in the brain ISF this is maintained at 2.9 mM. This potassium concentration for CSF is precisely maintained in spite of significant changes that might occur in plasma potassium concentrations [7]. In addition, blood plasma contains high levels of the neuroexcitatory amino acids, glutamate, and glycine, which fluctuate significantly after the ingestion of food. If these excitatory neurotransmitters are released into the brain ISF in an uncontrolled manner, as for example they may be from neuronal sources during ischemic stroke, considerable and permanent neurotoxic and neuroexcitatory damage can occur to nervous structures.

The BBB also prevents many macromolecules from entering the brain. The total protein content of CSF when compared to plasma is very low and markedly different from that of plasma (Table 27.1). Plasma proteins such as albumin, prothrombin, and plasminogen are

TABLE 27.1
Typical Plasma and Cerebrospinal Fluid Concentrations for Some Selected Solutes

Solute	Units	Plasma	CSF	Ratio
Na^+	mM	140	141	~1
K^+	mM	4.6	2.9	0.63
Ca^{2+}	mM	5.0	2.5	0.5
Mg^{2+}	mM	1.7	2.4	1.4
Cl^-	mM	101	124	1.23
HCO_3^-	mM	23	21	0.91
Osmolarity	mOsmol	305.2	298.5	~1
pH		7.4	7.3	
Glucose	mM	5.0	3.0	0.6
Total amino acid	μM	2890	890	0.31
Leucine	μM	109	10.1–14.9	0.10–0.14
Arginine	μM	80	14.2–21.6	0.18–0.27
Glycine	μM	249	4.7–8.5	0.012–0.034
Alanine	μM	330	23.2–32.7	0.07–0.1
Serine	μM	149	23.5–37.8	0.16–0.25
Glutamic acid	μM	83	1.79–14.7	0.02–0.18
Taurine	μM	78	5.3–6.8	0.07–0.09
Total protein	mg/mL	70	0.433	0.006
Albumin	mg/mL	42	0.192	0.005
Immunoglobulin G (IgG)	mg/mL	9.87	0.012	0.001
Transferrin	mg/mL	2.6	0.014	0.005
Plasminogen	mg/mL	0.7	0.000025	0.00004
Fibrinogen	mg/mL	325	0.00275	0.000008
α 2-macroglobulin	mg/mL	3	0.0046	0.0015
Cystatin-C	mg/mL	0.001	0.004	4.0

Note: Compiled from various sources, values mostly human.

irritant and damaging to nervous tissue [8–10]. Factor Xa that is present in the brain, which converts prothrombin to thrombin, and the thrombin receptor PAR_1 are widely expressed in the CNS. Similarly tissue plasminogen activator is also present in central nervous tissues and converts plasminogen to plasmin. Thrombin and plasmin if present in brain ISF can initiate cascades resulting in seizures, cell death and glial activation, and glial cell division and scarring [9]. Thus, leakage of these large molecular weight serum proteins into the brain from a damaged BBB can have serious pathological consequences. It is interesting to note that the serine protease inhibitor cystatin-C is synthesized locally within the CNS and one of the few proteins to have a higher concentration in CSF than in plasma (Table 27.1). This high concentration of cystatin-C is probably a protective measure against microleaks in the BBB, which continually and spontaneously occur and allow plasma components to slowly ooze into the brain.

Also in many other senses, the BBB functions as a protective barrier that shields the CNS from neurotoxic substances that circulate in the blood. These neurotoxins may be endogenous metabolites or proteins or a multitude of xenobiotics that are ingested in the diet or otherwise acquired from the environment. A number of energy-dependent efflux ATP-binding cassette transporters (ABC transporters) actively pump many of these naturally occurring and ingested neurotoxins out of the brain. The adult CNS does not have a significant regenerative capacity if damaged and fully differentiated neurons cannot divide and replace themselves under normal circumstances. From birth, throughout life, there is a continuous steady rate of neuronal cell death in the healthy human brain. Any acceleration in the natural rate of cell death resulting from an increased ingress of neutoxins into the brain would become prematurely debilitating.

The formation of the BBB has the consequence of sealing off the brain from many essential water-soluble nutrients and metabolites that are required by the nervous tissue. Specific transport systems therefore are expressed in the BBB to ensure an adequate supply of these substances and discussed later. The formation of the BBB and the differentiation of the endothelium into a barrier tissue are largely induced by a close association with the end-feet of glial cells. This close association appears to promote the formation of tight junctions and the development of polarity in the endothelial cells arising from the differential expression of transporters in the luminal and abluminal membranes. The BBB is also closely associated with pericytes, microglia, and nerve endings. The arrangement of the cell association forming the BBB is shown in Figure 27.2.

Thus the BBBs together have a dual function, they firstly provide the especially stable fluid environment that is necessary for complex neural function, and secondly, they protect the CNS from chemical insult and damage.

27.2 BLOOD–BRAIN BARRIER TIGHT JUNCTIONS

The BBBs to macromolecules and most polar solutes are created by the formation of tight junctions between the cerebral endothelial cells, the choroid plexus epithelial cells, and the cells of the arachnoid membrane. These tight junctions are a key feature of the BBB and effectively abolish any aqueous paracellular diffusional pathways between the endothelial cells from the blood plasma, or somatic ECF, to the brain ECF. This removal of the paracellular pathway efficiently prevents the free diffusion of polar solutes from blood to brain [11].

The junctional complexes between endothelial cells consist of adherens junctions (AJ) where cadherin proteins span the extracellular space and are linked into the cell cytoplasm by the scaffolding proteins alpha, beta, and gamma catenin. These adherens junctions are

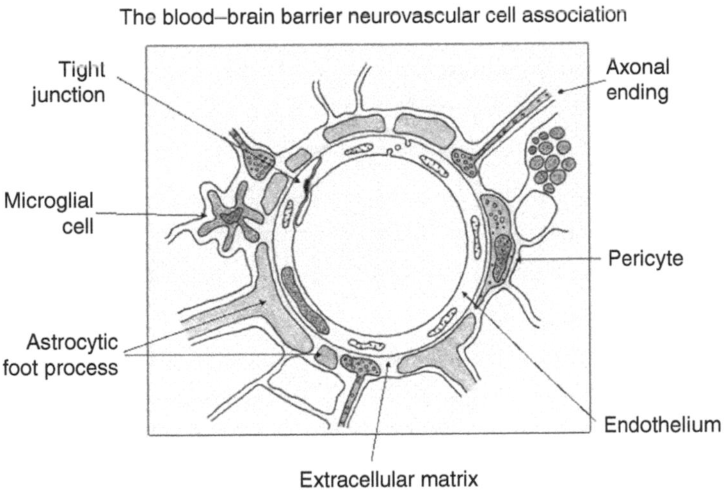

FIGURE 27.2 The neurovascular unit (cell association) forming the BBB. The cerebral endothelial cells form tight junctions at their margins which completely seal the aqueous paracellular diffusional pathway between the cells. Pericytes are distributed discontinuously along the length of the cerebral capillaries and partially surround the endothelium. Both the cerebral endothelial cells and the pericytes are enclosed by, and contribute to, the local extracellular matrix which forms a distinct perivascualar matrix, different in its composition from the extracellular matrix of the brain parenchyma. Foot processes from astrocytes form a complex network completely surrounding the capillaries and it is this close cell association that is thought to maintain barrier properties. Axonal projections from neurons are also applied closely to the capillary endothelial cells and contain vasoactive neurotransmitters and peptides. These axonal endings may play a part in modulating BBB permeability on a short-term basis by release of vasoactive peptides and other agents. Microglia (perivascular macrophages) are the resident immunocompetent cells of the brain and derived from circulating monocytes and macrophages. The movement of solutes across the BBB is either passive, driven by a concentration gradient from plasma to brain, with more lipid-soluble substances entering most easily, or may be facilitated by passive or active transporters in the endothelial cell membranes. Efflux transporters in the endothelium may limit the CNS penetration of a wide variety of solutes. (From Begley, D.J., *Pharmacol. Ther.*, 104, 29, 2004.)

present between cells in all tissues and hold the cells together giving the tissue structural support. They do not appear to participate directly in the formation of occlusive tight junctions between the cells that block the paracellular pathway. The tight junctions proper consist of a further complex of proteins spanning the paracellular space and of junction-associated molecules (JAM) and of the proteins occludin and claudin [12] (see Figure 27.3). The occludins and claudins are linked to cingulin and a number of cytoplasmic scaffolding and regulatory proteins ZO1, ZO2, ZO3. There are some 20 known isoforms of claudin (claudins 1–20) [13]. It has been shown [14] that in experimental allergic encephalomyelitis (EAE) and glioblastoma multiforme (GBM) there is a selective loss of claudin-3, from the tight junctions, but not claudin-5 or occludin, and this disappearance is associated with a loss of BBB integrity together with some functional barrier loss. In the case of EAE, changes in BBB permeability are associated with inflammatory events at the BBB. Also, genetically altered mice that have claudin-5 knocked out have a severely compromised and leaky BBB and these animals die shortly after birth [15], although their death is probably not solely related to the BBB defects. Therefore it appears clear that disappearance of either claudin-3 or claudin-5 from the tight junctional complexes can result in a compromised BBB. It appears to be a unique feature of the tight junctions of the BBB that they are extremely sensitive to

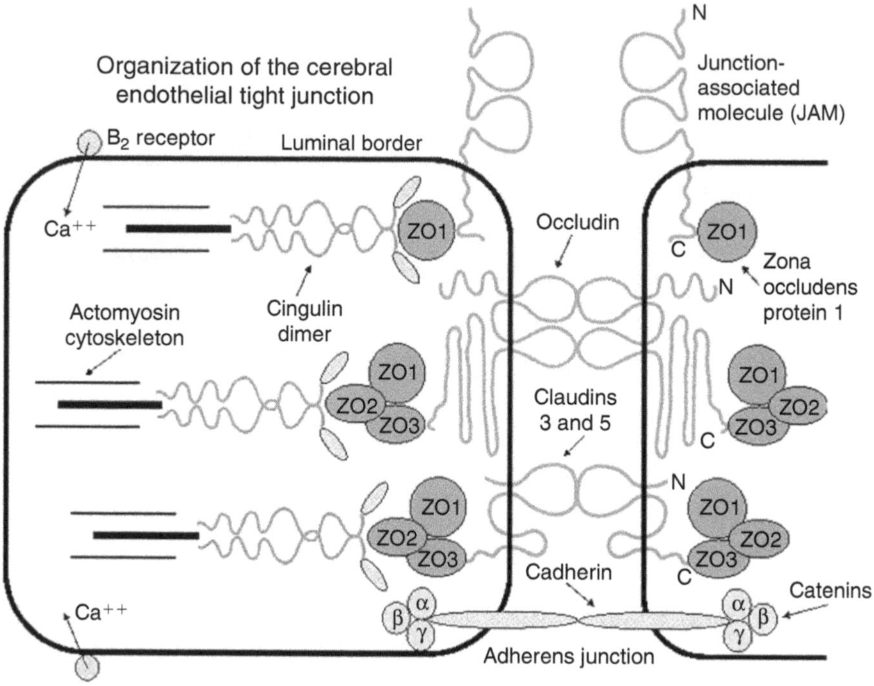

FIGURE 27.3 Structure of BBB tight junctions. The tight junctional complexes comprise the junction-associated molecules (JAM, which are members of the immunoglobulin superfamily), occludin, claudins 3 and 5, and the cadherin molecules. The cadherins provide structural integrity and attachment between the cells but do not appear to contribute to barrier function. The barrier to diffusion and the high electrical resistance of the BBB appear to be largely due to the properties of claudins 3 and 5 (see text). The claudins, as do the occludins, appear to associate and bind to each other in the paracellular cleft. Whether claudins 3 and 5 always associate with each other (homodimerization) or whether 3 to 5 pairings are possible (heterodimerization) is not known. A different ratio of the claudin mix may subtly alter tight junctional properties and their tightness. The claudins and occludin are linked to the scaffolding proteins ZO1, ZO2, and ZO3. These scaffolding proteins are in turn linked via cingulin dimers to the actin/myosin cytoskeletal system within the cell. Activation of the actin/myosin system may be initiated by a rise in free intracellular calcium, for example arising from ligand binding to the B_2 bradykinin receptor, and the actin/myosin system may withdraw the claudins and occludin from the paracellular space thus directly modifying the tight junctional properties. The JAM molecules are members of the immunoglobulin superfamily and thought to act as cell-adhesion molecules for leukocytes.

locally produced CNS circulating factors, which can on a minute-to-minute basis modulate the properties and the function of the paracellular pathways [14]. Clearly, the barrier function of the tight junctions is not solely related to the expression and presence of claudins and occludin in the paracellular cleft, but also to their organization and interaction [16].

It is the presence of these tight junctions that occludes the aqueous paracellular diffusional pathway between the endothelial cells, and blocks the free diffusion of macromolecules, polar solutes, and ions from blood plasma into the ECF of the brain. It is this impediment to the movement of ions that results in the high *in vivo* electrical resistance of the BBB, of approximately 1800 Ω cm^2 [17]. This high electrical resistance or low conductance of the potential paracellular pathway emphasizes the extreme effectiveness of the tight junctions in occluding this pathway by effectively reducing the movement of ions. The radius of a sodium ion (Na^+) is 0.95 Å but it may be effectively larger when fully hydrated. The *in vivo* BBB for an extracellular marker such as sucrose is extremely effective, with a measured *in vivo* permeability of 7.2 mL/s/g $\times$ 10^6, for a molecule of molecular weight 342 Da that has a molecular radius of 4.7 Å [18]. Ionic lanthanum (ionic radius 1.16 Å), when introduced into the cerebral capillary lumen, can be shown by electron microscopy to penetrate the paracellular cleft as far as the tight junctional complexes and then its movement is arrested [19]. The effectiveness of the tight junctions appears to be regulated via the intracellular regulating scaffolding proteins ZO1, ZO2, and ZO3, which link both the junctional molecules claudin and occludin via cingulin to intracellular actin and the cytoskeleton [12,20]. Alterations in both intracellular and extracellular calcium concentration can modulate the tight junction assembly [21,22] and alter the electrical resistance across the cell layer and the effectiveness of the tight junctions as a barrier. The nerve endings that terminate adjacent to the extracellular matrix surrounding the endothelial cells of the BBB contain vesicles that in turn contain peptides and other vasoactive agents that can modify the tight junction assembly and barrier permeability [23].

Induction of many BBB properties, including formation of the tight junctions, and the polarized expression of transporters in the luminal and abluminal membranes of the cerebral endothelial cells, is believed to result from a close association between astrocytes and the endothelial cells [24]. Tight junction formation between cerebral endothelial cells in culture can be induced by the use of astrocyte conditioned media *in vitro*, and therefore some of the BBB inducing factors are thought to be soluble in nature [25]. It is probable that these factors are multiple and complex, involving both a two-way exchange of signaling molecules between glia and endothelia and a cell–cell contact [26,27].

27.3 TRANSPORT ACROSS THE BLOOD–BRAIN BARRIER

27.3.1 Passive Partitioning into Brain

A wide range of lipid-soluble molecules can diffuse through the BBB and enter the brain passively. There is a general correlation between the lipid solubility of a compound, usually determined as the log D octanol or buffer partition coefficient at pH 7.4, and the rate at which most solutes will enter the CNS. Factors that restrict the entry of compounds into the CNS are a high polar surface area (PSA) greater than 80 $Å^2$ and a tendency to form more than six hydrogen bonds, a factor that greatly increases the free energy requirements of moving from an aqueous phase into the lipid of the cell membrane. The presence of a number of rotatable bonds in the molecule and a molecular weight in excess of 450 Da also appear to restrict BBB permeability. A high affinity of binding to plasma proteins with a low off-rate can also significantly reduce CNS penetration. However, these molecular and physicochemical factors are not always an absolute indication for CNS penetration and activity; there are many examples of effective CNS active drugs in clinical use, which do not comply with these general rules for BBB penetration [28]. Bases, which carry a positive charge, have an advantage over acids when penetration of the BBB is considered; it is probably the cationic nature of these molecules and an interaction with the negatively charged phospholipid head groups of the outer leaflet of the cell membrane that provide the compounds with an advantage.

The movement of the blood gases oxygen and carbon dioxide across the BBB is diffusive and the dissolved gases move down their concentration gradients. Oxygen supply to and carbon dioxide removal from the brain are thus blood-flow dependent when it comes to delivery and removal; as long as cerebral blood flow remains within physiological limits, gas transport is adequate. Due to the negative charge and the presence of the tight junctions, the BBB has an extremely low permeability to bicarbonate.

27.3.2 Solute Carriers in the Blood–Brain Barrier

The barrier to paracellular diffusion potentially isolates the brain from many essential polar nutrients such as glucose and amino acids that are required for metabolism; and, therefore, the BBB endothelium must contain a number of specific solute carriers (transporters) to supply the CNS with its requirements for these substances. The formation of tight junctions essentially confers on the BBB the properties of a continuous cell membrane, both in terms of the diffusional characteristics imposed by the lipid bilayer, and the directionality and properties of the specific transport proteins, and solute carriers (SLC) that are present in the cell membrane. Examples of BBB solute carriers (SLC transporters) are listed in Table 27.2.

Most polar molecules cannot diffuse through the membrane of the cell freely and thus all cells express a large number of SLCs in the cell membrane [29]. The endothelial cells forming the BBB express a large number of these transport proteins in their cell membranes, which can transport a wide variety of solutes and nutrients into and out of the brain. These are summarized in Table 27.1. Some of these transport proteins are polarized in their expression and inserted into either the luminal or the abluminal membrane only, and others are inserted into both membranes of the endothelial cells [1,11,26,30–32]. The orientation of these transporters may therefore result in preferential transport of substrates into or across the endothelial cell, and the direction of the transport may be from blood to brain or brain to blood. (see Figure 27.4 and Table 27.2 and Table 27.3). A further function of the tight junctions in the lateral cell membranes may be to act as a fence in the membrane and segregate transport proteins and lipid rafts, to either the luminal or the abluminal membrane domain, and to prevent their free movement from one side of the endothelium to the other, thus strictly preserving the polarity of the barrier.

27.3.3 ATP-Binding Cassette Transporters in the Blood–Brain Barrier

When comparing brain penetrance with lipid solubility (lipophilicity), a very large number of solutes and drugs have a much lower CNS rate of uptake than might be suggested by their log D value. Members of the ABC transporter actively efflux these substances, and many of their metabolites, from the brain and the capillary endothelium forming the BBB [33]. The strategy of increasing the lipid solubility of a drug to make it more brain penetrant may sometimes be counter-productive as it may also increase the likelihood of the molecule becoming a substrate for ABC efflux transporters [33]. The ATP-binding cassette of transporters in humans is a superfamily of proteins containing 48 members, which on the basis of structural homology are grouped into 7 sub-families (Table 27.2) [34]. In the BBB the ABC transporters of greatest significance for efflux transport are P-glycoprotein (Pgp, Multidrug Resistance Protein, ABCB-1), the multidrug resistance-associated proteins (MRPs, ABCC-1, 2, 3, 4, 5), and Breast Cancer Related Protein (BCRP, ABCG-2) [33]. These are summarized in Table 27.3. The ABC transporters ABCA1 and ABCG1 transport cholesterol and are also expressed in the brain and the BBB. ABCA2 is also expressed in the CNS and has been reported to be associated with drug resistance [34], although its functional significance has not been explored. The major role of the ABC transporters Pgp, MRPs , and BCRP in the BBB is to function as active efflux pumps consuming ATP and transporting a diverse range of lipid-soluble chemical compounds out of the CNS and brain capillary endothelium. In this role, they are removing from the brain potentially neurotoxic endogenous or xenobiotic molecules and carrying out a vital neuroprotective and detoxifying function. In addition many drugs are substrates for these ABC efflux transporters and their brain penetration is significantly reduced [33] by their activity. P-glycoprotein and BCRP are expressed in the luminal

TABLE 27.2
Examples of Transporters Present in the Blood–Brain Barrier

Transporter (SLC)	Abbreviation and/or Transporter Subtype	Equivalent Human Gene Name	BBB Location	Orientation*	Example of Endogenous Substrates/Mechanism
Glucose	$GLUT_1$	SLC2A1	Luminal Abluminal	Blood to brain	Glucose (facilitative, bidirectional)
Sodium-dependent glucose transporter	$SGLT_1$	SLC5A1	Abluminal	Brain to endothelium	Glucose
Sodium myoinisitol cotransporter	HMIT SMIT	SLC5A3 SLC2A13	Luminal	Blood to endothelium	Myoinisitol (sodium dependent)
Aminoacid	L_1/LNNA	SLC7A5	Luminal Abluminal	Blood to brain	Glutamate, histidine, isoleucine, leucine, methionine, phenylalanine, tryptophan, threonine, tyrosine, valine (facilitative, bidirectional)
Amino acid	y+	SLC7A6	Luminal Abluminal	Blood to brain	Arginine, lysine, ornithine (facilitative, bidirectional)
Amino acid	A	SLC6A5/8	Abluminal	Brain to endothelium	Alaninine, glutamate, glycine, proline (sodium dependent)
Amino acid	ASC	SLC1A4/5	Abluminal	Brain to endothelium	Alanine, leucine (sodium dependent)
Amino acid	$B^{o,+}$	SLC7A9	Abluminal	Brain to endothelium	Alanine, serine, cysteine (sodium dependent)
Amino acid	X^{-}_{AG}	SLC1A1/2/3/6/7	Luminal Abluminal	Blood to endothelium	Glutamate, aspartate (sodium dependent)
Amino acid	β	SLC6A6	Luminal Abluminal	Brain to endothelium	Taurine, β-alanine (sodium dependent)
Nucleosides, nucleotides, and nucleobases	ENT_1 ENT_2	SLC29A1 SLC29A2	Luminal	Blood to endothelium	Nucleosides, nucleotides, nucleobases (facilitative, equilibrative)

(Continued)

TABLE 27.2 (Continued)
Examples of Transporters Present in the Blood–Brain Barrier

Transporter (SLC)	Abbreviation and/or Transporter Subtype	Equivalent Human Gene Name	BBB Location	Orientation[a]	Example of Endogenous Substrates/Mechanism
Nucleosides, nucleotides, and nucleobases	CNT_1 CNT_2 CNT_3	SLC28A1 SLC28A2 SLC28A3	Abluminal	Endothelium to brain	Nucleosides, nucleotides, nucleobases (sodium-dependent exchange)
Monocarboxylic acids	MCT_1	SLC16A1	Luminal Abluminal	Blood to brain	Ketone bodies
Monocarboxylic acids	MCT_2	SLC16A7	Abluminal	Brain to endothelium	Lactate (proton exchanger)
Organic anion transporters	OAT_3	SLC22A8	Luminal	Blood to brain	Dicarboxylate exchange with α-ketoglutarate, bicarbonate, Cl^-
Organic anion-transporting polypeptide	OATP-B	SLC02B1	Luminal Abluminal	Blood to endothelium Endothelium to brain	Organic anion/bicarbonate exchangers
Organic cation transporters	OCT	SLC22A2/3	Luminal	Blood to endothelium	Organic cation/proton exchange
Novel organic cation transporter	$OCTN_1$ $OCTN_2$	SLC22A4 SLC22A5	Luminal Abluminal	Blood to endothelium Endothelium to brain	Organic cation/proton exchange

[a]In order to transport a substrate across the blood–brain barrier from blood to brain the transporter must be expressed in both cell membranes and be bidirectional. Alternatively one transporter may carry substrate into the **BBB** endothelial cells and another out of the cells. If a transporter is inserted into one membrane only it will transport out of the endothelium or accumulate substrate within the endothelium. Exchangers are driven by the concentration gradient of substrate and exchange ion/molecule and can reverse if the concentration gradient is reversed.

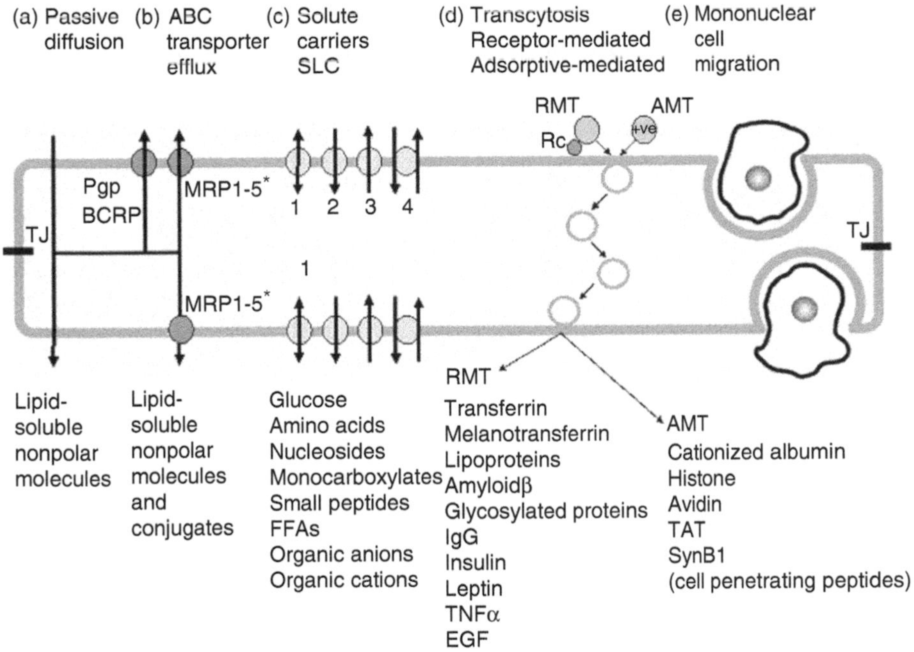

FIGURE 27.4 Routes of transport across the BBB. (a) Solutes may passively diffuse through the cell membrane and cross the endothelium. A greater lipid solubility and several other physicochemical factors favor this process (see text). (b) Active efflux carriers (ABC transporters) may intercept some of these passively penetrating solutes and pump them out of the endothelial cell either as they diffuse through the cell membrane itself or from the cytoplasm. Pgp and BCRP are strategically placed in the luminal membrane of the BBB endothelium. *MRPs 1–5 are apparently inserted into both luminal and abluminal membranes. There would appear to be some species differences in both the polarity and the isoforms of MRPs expressed at the BBB [33]. (c) Carrier-mediated influx via solute carriers (SLCs) may be passive or primarily or secondarily active and can transport many essential polar molecules such as glucose, amino acids, and nucleosides into the CNS. The solute carriers may be bidirectional the direction of transport being dictated by the substrate concentration gradient (1), unidirectional either into or out of the cell (2/3), or involve an exchange of one substrate for another or driven by an ion gradient (4). In this case the direction of transport is also reversible depending on concentration gradients. (d) RMT requires receptor activation and can transport a variety of macromolecules such as peptides and proteins across the cerebral endothelium (transcytosis). AMT appears to be induced in a nonspecific manner by negatively charged macromolecules and can also transport across the endothelium. Both RMT and AMT appear to be vesicular-based systems which carry their macromolecule content across the endothelial cells. (e) Leukocytes cross the BBB by a process of dipedesis through the endothelial cells. They penetrate close to the tight junctional regions, but not through them. The junction-associated molecules (JAM) and the cell surface protein CD99 are thought to interact with the leukocytes to initiate dipedesis. Tight junction modulation can also occur either endogenously within the cell association or be induced pharmacologically, in which molecular rearrangement of the proteins of the tight junctions is induced, resulting in a whole or partial opening of the paracellular aqueous diffusional pathway.

membrane of the BBB and clearly their function is to transport substrate from endothelium to blood. Some of the MRP isoforms appear to be expressed in the luminal or the abluminal membranes and sometimes in both membranes. Because they favor water-soluble conjugates as substrates, a bidirectional export from the endothelium may be acceptable as conjugation

TABLE 27.3
Human ABC Transporter Superfamily and Relevance for BBB and BSCFB

Sub-Family	BBB Significant Members	Alias	Substrates	Expression
ABCA1–12	ABCA2		MDR surfactants	Brain and lung
ABCB1–11	ABCB1	P-glycoprotein Pgp Rodent mdr1a/mdr1b	MDR (amphiphilic /lipid soluble)	BBB luminal/choroid plexus apical membrane
ABCC1–12	ABCC1	MRP1	MDR/OA/conjugates	BBB luminal and abluminal/choroid plexus
	ABCC2	MRP2/MOAT	MDR/OA/conjugates	
	ABCC3	MRP3	MDR/OA/conjugates	
	ABCC4	MRP4	Nucleosides	
	ABCC5	MRP5	Nucleosides	
ABCD1–4				Peroxisomes
ABCE1				
ABCF1–3				Ubiquitous
ABCG1+2; 4+5; 8	ABCG2	BCRP/MXR		BBB luminal

Notes: MDR-multidrug resistance; OA—Organic acids

by drug transforming enzymes has rendered them less cytotoxic. There may also be some species variation in the expression of the MRP isoforms [33].

27.3.4 Blood–Brain Barrier Transport of Macromolecules

Transcytosis of macromolecules across the BBB via endocytic mechanisms provides the only significant route by which large molecular weight solutes such as proteins and peptides can enter the CNS intact. Although the majority of large blood-borne molecules are physically prevented from entering the brain by the presence of the BBB and tight junction formation, specific and some nonspecific transcytotic mechanisms exist to transport a variety of large molecules and complexes across the BBB. A summary of a number of known transcytotic mechanisms is presented in Table 27.4 [35–50] (see also Figure 27.2). These vesicular mechanisms involve either receptor-mediated transcytosis (RMT) or absorptive-mediated transcytosis (AMT). In RMT, the binding of macromolecular ligands to specific receptors on the cell surface triggers an endocytic event; the receptors and their bound ligand cluster together, and a caveolus is formed, which pinches off into a vesicle, and both ligand and receptors are internalized into the endothelial cell and routed across the cytoplasm to be exocytosed at the opposite pole of the cell. Dissociation of the ligand and receptor presumably occurs during cellular transit or the exocytic event. In AMT, an excess positive charge on the molecule, which renders it highly cationic, appears, in this case, to induce endocytosis and subsequent transcytosis. In both cases, to achieve transcytosis of an intact protein or peptide, the lysosomal compartment within the cell needs to be avoided by routing the primary sorting endosome and its contents away from this degradative compartment. This routing away from the lysosome does not seem to occur in many peripheral endothelia and it may be a specialized phenomenon of the BBB where the intact transcytosis of a significant number of macromolecules becomes a necessity [1]. In many cells and tissues, an avoidance of the intracellular

TABLE 27.4
Examples of Transcytosis/Transport of Large Molecules and Complexes across the BBB

Transport System	Abbreviation (Receptor)	Example Ligands	Type	BBB Direction	Refs.
Transferrin	Tf_R	Fe-transferrin	RMT	Blood to brain	[35]
Melanotransferrin	MTf_R	Melanotransferrin (p97)	RMT	Blood to brain	[36]
Lactoferrin	Lf_R	Lactoferrin	RMT	Blood to CSF	[37]
Apolipoprotein E receptor 2	ApoER2	Lipoproteins and molecules bound to ApoE	RMT	Blood to brain	[38]
LDL-receptor related protein 1 and 2	LRP1	Lipoproteins, amyloid-β, lactoferrin, α 2-macroglobulin, melanotransferrin (p97), ApoE	RMT	Bidirectional	[38,39]
	LRP2				
Receptor for advanced glycosylation end-products	RAGE	Glycosylated proteins Amyloid-β, S-100, amphoterin	RMT	Blood to brain	[40,41]
Immunoglobulin G	Fcγ-R	IgG	RMT	Blood to brain	[42]
Insulin	–	Insulin	RMT	Blood to brain	[43]
P-glycoprotein	Pgp	Amyloid-β	RMT	Brain to blood	[44]
Leptin	–	Leptin	RMT	Blood to brain	[45]
Tumour necrosis factor	–	TNFα	RMT	Blood to brain	[46]
Epidermal growth factor		EGF	RMT	Blood to brain	[47]
Heparin-binding epidermal growth factor-like growth factor (diptheria toxin receptor)	HB-EGF (DTR)	Diphtheria toxin and CRM197 (protein)	RMT	Blood to brain	[39]
Leukaemia inhibitory factor	LIFRa (gp190)	LIF	RMT	Blood to brain/spinal cord	[48]
Cationized proteins	+	Cationized albumin	AMT	Blood to brain	[49]
Cell penetrating peptides	+	SynB5/pAnt-(43–58)	AMT	Blood to brain	[50]

Many of the receptors involved in RMT are poorly defined and are multifunctional and multiligand in nature. Thus some ligands may be transported by more than one system and some receptors may with time turn out to be one-and–the same. (–receptor uncharacterized; + not receptor-mediated, nonspecific) lysosomal compartment does not occur, and the contents of the primary endosome are routed to the acidic lysosome where enzymic degradation takes place [51–53].

Most electron microscopic studies of BBB endothelial cells suggest the presence of relatively few observable endocytic vesicles in the cytoplasm of these cells compared with other endothelia. For example, the BBB contains only a fifth to a sixth of the endocytic profiles seen in muscle capillary endothelia [54], although they may increase to comparable levels with inflammation of the BBB [55]. However, when a comparison is made of the ability of capillary endothelia in a variety of different tissues to trancytose protein, there is a very poor correlation between the protein permeability of a microvessel and the number of observable endocytic profiles [54]. Brain capillary endothelia are very thin cells, the luminal and abluminal membranes only being separated by some 500 nm or less (5000 Å), and caveoli are 50–80 nm in diameter and thus the events of transcytosis may be difficult to capture within the cell using conventional electron microscopical techniques.

Smaller peptides may be transported across the BBB either by nonspecific fluid-phase endocytosis or RMT mechanisms. It may also be possible for them to use a peptide-specific transporter protein, directly inserted into the cell membrane in a similar manner to the solute transporters, which flips them across the membrane [43,56]. In some cases the receptor that transduces the signal at the cell membrane may also act as the transporter for the peptide and is co-opted to initiate RMT, or another transport system; in other cases the membrane transporter for a peptide may be quite distinct in structure from the receptor that transduces signals at the cell membrane from a signaling peptide [45].

27.3.5 Cell Movement across the Blood–Brain Barrier

Mononuclear leukocytes, monocytes, and macrophages are continually penetrating the BBB and taking up residence in the CNS. The microglia, derived from monocytes, are also able to return to the general circulation via the CSF drainage, and thus there is a slow continuous turnover and exchange of the microglial cell population with immunologically competent cells from the peripheral immune system. The CNS microglia together with the perivascular macrophages are the immunologically competent cells of the CNS and become activated in inflammation and other pathological states. Circulating mononuclear cells and neutrophils are attracted to sites of BBB inflammation, penetrate the barrier, and form cuffs in the perivascular space around small vessels. In the normal BBB, mononuclear cells appear to penetrate by a process of diapedesis directly through the cytoplasm of the endothelial cells and not by a paracellular route involving rearrangement and opening of the tight junctional complexes as has been previously suggested [57,58]. This mechanism enables the mononuclear cells to cross the BBB without temporarily disrupting it as tight junctions are not affected. During diapedesis the leukocyte enters the endothelial cell with the luminal membrane closing over it before creating an opening in the abluminal membrane, so a fluid-filled channel through the cell is never created [58]. In inflammatory pathological states that involve the BBB, the tight junctions between endothelial cells may be relaxed by cytokines and other agents and mononuclear cells may then enter by both transcellular and paracellular routes [59,60].

27.4 DEVELOPMENT AND ONTOLOGY OF THE BLOOD–BRAIN BARRIER

The BBB develops during fetal life and is well formed by birth, especially to proteins and macromolecules [61–68]. In the mouse, the BBB begins to form between E11 and E17, by which time identifiable tight junctions are present. The presence of the tight junction will be highly restrictive to any transendothelial movement of both polar- and macromolecules. In mammals that are born in a relatively immature state, such as the rat and mouse, many of the characteristic BBB transport mechanisms may still continue to mature and only become fully expressed and functional in the peri- or postnatal period.

In the mouse an RT-PCR signal for the ABC transporter mdr2 is present in brain tissue by E13 and all three mdr isoforms, mdr1a, mdr1b, and mdr2 show strong signals by E18 [69]. In the rodent, the mdr1a and mdr1b isoforms are the principal efflux and drug transporting isoforms. In the human, there is only a single MDR P-glycoprotein gene product MDR1 that effluxes and transports drugs [33]. MDR1 in the human and mdr1a in the rodent are both expressed in the luminal membrane of the BBB endothelial cells and mdr1b is expressed by glia and elsewhere in the CNS of rodents [33]. Expression of P-glycoprotein (mdr1a) in the luminal endothelial cell membranes of the rat BBB can be detected by immunoblotting at P7 and reaches a plateau of expression by P28 [70].

The high electrical resistance, characteristic of the BBB, is exhibited in the BBB of rats by E21 [71] indicating that functional tight junctions are formed prenatally and this low conductivity demonstrates that they already form an effective barrier to the movement of ions. It has been demonstrated [65] that the BBB permeability to ^{86}Rb is significant at E21with an influx rate constant of 42.5 ± 4.3 μL/g/min but within 2 d postnatally this has declined to 12.2 ± 0.6 μL/g/min and then followed by a further slow decline to 7.0 ± 0.3 μL/g/min by 50 d postnatally. Tight junction formation seems to be a very early feature of BBB development and a barrier to the free movement of proteins and macromolecules is formed at the primary stages of brain development [67]. Tight junctions are formed as blood vessels invade the brain at E10 in the mouse and E11 in the rat. As gliogenesis in the rat does not begin until E17 and is still occurring postnatally, tight junction formation per se would appear to be initiated primarily by neural signals rather than glial signals in the first instance [67].

Preston et al. [66] have shown that BBB permeability to the nonmetabolizable, but slowly BBB penetrant, tracer mannitol (182 Da), is between 0.19 and 0.22 μL/g/min in the brain of rats of 1 week of age and this permeability is identical to that of adult rats. The vascular space occupied by the tracer mannitol (the initial volume of distribution V_i) falls from 1.23 mL, at 1 week of age, to 0.75 mL per 100 g brain in the adult rat [66], indicating either a larger vascular volume, resulting from a greater capillary density or capillary diameter in the neonatal rat, or to a significant degree of internalization of the mannitol by the endothelium, possibly by fluid-phase endocytosis into the cerebral capillary endothelial cells in the newborn, compared to the adult.

There is nothing to suggest that the human BBB is not at least equally well formed at birth than it is in the rat. Occludin and claudin-5 expression is detected in the capillary endothelium of the brain of the 14 week human fetus, and it has the same pericellular distribution as seen in the adult [72]. Pioneering studies by Grontoft [73] in stillborn human fetuses from approximately 12 week gestation and perinatal deaths have demonstrated a postmortem BBB to trypan blue present from at least the start of the second trimester, which is comparable to that of the adult human.

REFERENCES

1. Nag, S., and D.J. Begley. 2005. Blood–brain barrier, exchange of metabolites and gases. In *Pathology and genetics. Cerebrovascular diseases*, ed. H. Kalimo, 22. Basel: ISN Neuropath Press.
2. Keep, R., and H.C. Jones. 1990. A morphometric study on the development of the fetal lateral ventricle choroid plexus, choroid plexus capillaries and ventricular ependyma in the rat. *Brain Res Dev Brain Res* 56:47.

3. Speake, T., et al. 2001. Mechanism of CSF secretion by the choroid plexus. *Microsc Res Tech* 52:49.
4. Cserr, H., et al. 1981. Efflux of radiolabeled polyethylene glycol and albumin from rat brain. *Am J Physiol* 240:F319.
5. Milhorat, T.H., et al. 1971. Cerebrospinal fluid production by the choroid plexus and brain. *Science* 173:330.
6. Kandel, E.R., J.H. Schwartz, and T.M. Jessel. 2000. *Principles of neural science*, 4th ed., 1294. New York: McGraw-Hill.
7. Bradbury, M.W.B., et al. 1963. The distribution of potassium, sodium, chloride and urea between lumbar cerebrospinal fluid and blood serum in human subjects. *Clin Sci* 25:97.
8. Nadal, A., et al. 1995. Plasma albumin is a potent trigger of calcium signals and DNA synthesis in astrocytes. *Proc Natl Acad Sci USA* 92:1426.
9. Gingrich, M.B., and S.F. Traynelis. 2000. Serine proteases and brain damage—Is there a link? *Trends Neurosci* 23:399.
10. Gingrich, M.B., et al. 2000. Potentiation of NMDA receptor function by the serine protese thrombin. *J Neurosci* 20:4582.
11. Begley, D.J., and M.W. Brightman. 2003. Structural and functional aspects of the blood–brain barrier. *Prog Drug Res* 61:39.
12. Wolburg, H., and A. Lippoldt. 2002. Tight junctions of the blood–brain barrier: Development, composition and regulation. *Vasc Pharmacol* 38:323.
13. Mitic, L.L., C.M. van Itallie, and J.M. Anderson. 2000. Molecular physiology and pathophysiology of tight junctions. I. Tight junction structure and function: Lessons from mutant animals and proteins. *Am J Physiol* 297:G250.
14. Wolburg, H., et al. 2003. Localisation of claudin-3 in tight junctions of the blood–brain barrier is selectively lost during experimental autoimmune encephalomyelitis and human glioblastoma multiforme. *Acta Neuropathol* 105:586.
15. Nitta, T., et al. 2003. Size-selective loosening of the blood–brain barrier in claudin-5 deficient mice. *J Cell Biol* 161:653.
16. Hamm, S., et al. 2004. Astrocyte mediated modulation of blood–brain barrier permeability does not correlate with loss of tight junction proteins from the cellular contacts. *Cell Tissue Res* 315:157.
17. Butt, A.M., H.C. Jones, and N.J. Abbott. 1990. Electrical resistance across the blood–brain barrier in anaethetised rats: A developmental study. *J Physiol* 429:47.
18. Smith, Q.R., et al. 1988. Kinetics and distribution volumes for tracers of different sizes in the brain plasma space. *Brain Res* 462:1.
19. Brightman, M.W., and T.S. Reese. 1969. Junctions between intimately apposed cell membranes in the vertebrate brain. *J Cell Biol* 40:648.
20. Bauer, H.C., A. Traweger, and H. Bauer. 2004. Proteins of the tight junction in the blood brain barrier. In *Blood–spinal cord and brain barriers in health and disease*, eds. H.S. Sharma and J. Westman, 1. San Diego: Elsevier.
21. Balda, M.S., et al. 1991. Assembly and sealing of tight junctions: Possible participation of G-proteins, phospholipase C, protein kinase C and calmodulin. *J Membr Biol* 122:193.
22. Abbott, N.J. 1998. Role of intracellular calcium in regulation of brain endothelial permeability. In *Introduction to the blood–brain barrier: Methodology and biology*, ed. W.H. Pardridge, 345. UK: Cambridge University Press.
23. Rennels, M.L., T.F. Gregory, and K. Fugimoto. 1983. Innervation of capillaries by local neurones in the cat hypothalamus: A light microcopic study. *J Cereb Blood Flow Metab* 3:535.
24. Rubin, L.L., et al. 1991. Differentiation of brain endothelial cells in cell culture. *Ann N Y Acad Sci* 633:420.
25. Neuhaus, J., W. Risau, and H. Wolburg. 1991. Induction of blood–brain barrier characteristics in bovine brain endothelial cells by rat astroglial cells in transfilter coculture. *Ann N Y Acad Sci* 633:578.
26. Abbott, N.J. 2002. Astrocyte–endothelial cell interactions and blood–brain barrier permeability. *J Anat* 200:629.
27. Begley, D.J. 2004. Delivery of therapeutic agents to the central nervous system: The problems and the possibilities. *Pharmacol Ther* 104:29.

28. Bodor, N., and P. Buchwald. 2003. Brain-targeted drug delivery: Experiences to date. *Am J Drug Target* 1:13.
29. Zhang, E.Y., et al. 2001. Structural biology and function of solute transporters: Implications for identifying and designing substrates. *Drug Metab Rev* 34:709.
30. Betz, A.L., J.A. Firth, and G.W. Goldstein. 1980. Polarity of the blood–brain barrier: Distribution of enzymes between the luminal and abluminal membranes of brain capillary endothelial cells. *Brain Res* 192:17.
31. Begley, D.J. 1996. The blood–brain barrier: Principles for targeting peptides and drugs to the central nervous system. *J Pharm Pharmacol* 48:136.
32. Mertsch, K., and J. Maas. 2002. Blood–brain barrier penetration and drug development from an industrial point of view. *Curr Med Chem—Central Nervous System Agents* 2:187.
33. Begley, D.J. 2004. ABC transporters and the blood–brain barrier. *Curr Pharm Des* 10:1295.
34. Dean, M., A. Rzhetsky, and R. Allimets. 2001. The human ATP-binding cassette (ABC) transporter superfamily. *Genome Res* 11:1156.
35. Visser, C.C., et al. 2004. Characterisation of the transferrin receptor on brain capillary endothelial cells. *Pharm Res* 21:761.
36. Demeule, M., et al. 2002. High transcytosis of melanotransferrin (P97) across the blood–brain barrier. *J Neurochem* 83:924.
37. Talkuder, J.R., T. Takeuchi, and E. Harada. 2003. Receptor-mediated transport of lactoferrin into the cerebrospinal fluid via plasma in young calves. *J Vet Med Sci* 65:957.
38. Hertz, J., and P. Marchang. 2003. Coaxing the LDL receptor family into the fold. *Cell* 112:289
39. Gaillard, P., C.C. Visser, and A.G. de Boer. 2005. Targeted delivery across the blood–brain barrier. *Expert Opin Drug Deliv* 2:299.
40. Stern, D., et al. 2002. Receptor for advanced glycation endproducts: A multiligand receptor magnifying cell stress in diverse pathologic settings. *Adv Drug Del Rev* 54:1615.
41. Deane, R., Z. Wu, and B.V. Zlokovic. 2004. RAGE (Yin) versus LRP (Yang) balance regulates Alzheimer amyloid β-peptide clearance through transport across the blood–brain barrier. *Stroke* 35:2628.
42. Zlokovic, B.V., et al. 1990. A saturable mechanism for transport of immunoglobulin G across the blood–brain barrier of the guinea pig. *Exp Neurol* 107:263.
43. Banks, W.A., 2004. The source of cerebral insulin. *Eur J Pharmacol* 490:5.
44. Lam, F.C., et al. 2001. β-amyloid efflux mediated by P-glycoprotein. *J Neurochem* 76:1121.
45. Banks, W.A., et al. 2002. Leptin transport across the blood–brain barrier of the Koletsky rat is not mediated by a product of the leptin receptor gene. *Brain Res* 950:130.
46. Pan, W., and A. Kastin. 1999. Entry of EGF into brain is rapid and saturable. *Peptides* 20:1091.
47. Pan, W., and A. Kastin. 2002. TNFα transport across the blood–brain barrier is abolished in receptor knockout mice. *Exp Neurol* 174:193.
48. Pan, W., A.J. Kastin, and J.M. Brennan. 2000. Saturable entry of leukaemia inhibitory factor from blood to the central nervous system. *J Neuroimmunol* 106:172.
49. Pardridge, W.M., et al. 1990. Evaluation of cationised albumin as a potential blood–brain barrier drug transport vector. *Exp Neurol* 255:893.
50. Drin, G., et al. 2003. Studies on the internalisation mechanism of cationic cell-penetrating peptides. *J Biol Chem* 278:31192.
51. Broadwell, R.D., B.J. Balin, and M. Salcman. 1988. Transcytotic pathway for blood-borne protein through the blood–brain barrier. *Proc Natl Acad Sci USA* 85:632.
52. Mellman, I. 1996. Endocytosis and molecular sorting. *Annu Rev Cell Dev Biol* 12:575.
53. Mukherjee, S., R.N. Ghosh, and F.R. Mayfield. 1997. Endocytosis. *Phys Rev* 77:759.
54. Stewart, P.A. 2000. Endothelial vesicles in the blood–brain barrier: Are they related to permeability? *Cell Mol Neurobiol* 20:149.
55. Claudio, L., et al. 1990. Increased vesicular transport and decreased mitochondrial content in blood–brain barrier endothelial cells during experimental autoimmune encephalomyelitis. *Am J Pathol* 135:1157.
56. Kastin, A., and W. Pan. 2003. Peptide transport across the blood–brain barrier. *Prog Drug Res* 61:79.

57. Engelhardt, B., and H. Wolburg. 2004. Transendothelial migration of leukocytes: Through the front door or around the side of the house? *Eur J Immunol* 34:2955.
58. Wolburg, H., K. Wolburg-Bucholz, and B. Engelhardt. 2005. Diapedesis of mononuclear cells across cerebral venules during experimental autoimmune encephalomyelitis leaves the tight junctions intact. *Acta Neuropathol* 109:181.
59. Anthony, D.C., et al. 1997. Age-related effects of interleukin-β on polmorphonuclear neutrophil-dependent increases in blood–brain barrier permeability in rats. *Brain* 120:435.
60. Bolton, S.J., D.C. Anthony, and V.H. Perry. 1998. Loss of tight junction proteins occluding and zonula-occludens-1 from cerebral vascular endothelium during neutrophil-induced blood–brain barrier breakdown *in vivo*. *Neuroscience* 86:1245.
61. Olsson, Y., et al. 1968. Blood–brain barrier to albumin in embryonic, new born and adult rats. *Acta Neuropathol* 10:117.
62. Tauc, M., X. Vignon, and C. Bouchaud. 1984. Evidence for the effectiveness of the blood–CSF barrier in the fetal rat choroid plexus. A free-fracture and peroxidase diffusion study. *Tiss Cell* 16:65.
63. Saunders, N.R. 1992. Development of the blood–brain barrier to macromolecules. In *Barriers and fluids of the eye and brain*, ed. M.B. Segal, 128. London: Macmillan Press.
64. Moos, T., and K. Mølgård. 1993. Cerebrovascular permeability to azo dyes and plasma proteins in rodents of different ages. *Neuropathol Appl Neurobiol* 19:120.
65. Keep, R.F., et al. 1995. Developmental changes in blood–brain barrier potassium permeability in the rat: Relation to brain growth. *J Physiol* 488:439.
66. Preston, J.E., H. Al-Saraff, and M.B. Segal. 1995. Permeability of the developing blood–brain barrier to ^{14}C-mannitol using the rat *in situ* brain perfusion technique. *Brain Res Dev Brain Res* 87:69.
67. Saunders, N.R., G.W. Knott, and K.M. Dziegielewska. 2000. Barriers in the immature brain. *Cell Mol Neurobiol* 20:29.
68. Ballabh, P., A. Braun, and M. Nedergaard. 2004. The blood–brain barrier: An overview: Structure, regulation, and clinical implications. *Neurobiol Dis* 16:1.
69. Scheingold, M., et al. 2001. Multidrug resistance gene expression during the murine ontogeny. *Mech Ageing Devel* 122:255.
70. Matsuoka, Y., et al. 1999. Developmental expression of P-glycoprotein (multidrug resistance gene product) in rat brain. *J Neurobiol* 39:383.
71. Butt, A.M. 1995. Effect of inflammatory agents on electrical resistance across the blood–brain barrier in pial microvessels of anaesthetised rats. *Brain Res* 696:145.
72. Virgintino, D., et al. 2004. Immunolocalisation of tight junction proteins in the adult and developing human brain. *Histochem Cell Biol* 122:51.
73. Grontoft, O. 1964. Intracranial haemorrhage and blood–brain barrier problems in the newborn. *Acta Pathol Microbiol Scand* Suppl. C:1.

28 Strategies to Overcome the Blood–Brain Barrier

Elena V. Batrakova and Alexander V. Kabanov

CONTENTS

28.1 INTRODUCTION

The brain is the most mysterious and sacred organ in the body. Hidden behind the blood–brain barrier (BBB) and consequently inaccessible to the majority of substances circulated in the blood, it controls and regulates almost all critical processes within the organism.

The BBB plays a key role in maintaining brain function by allowing selective access to essential nutrients and signaling molecules from the vascular compartment, and restricting the entry of xenobiotics. The protective function of the BBB, however, becomes a major obstacle to the treatment of many devastating diseases of the central nervous system (CNS) such as brain tumors, HIV encephalopathy, cerebrovascular diseases (stroke), and neurodegenerative disorders (Parkinson's and Alzheimer's diseases) by restricting drug delivery to the CNS. Disorders such as depression, severe pain, fungal infections, obesity, and lysosomal diseases require drug transport across the BBB as well. Efficient delivery of imaging agents across the BBB is also necessary for accurate diagnosis of neuropathology, monitoring of disease progression, localization for surgical intervention, and introduction of therapeutic agents.

Brain microvessel endothelial cells (BMVEC) are important structural and functional components of the BBB, having tight extracellular junctions, relatively low pinocytic activity [1], and efflux protein systems (such as P-glycoprotein [Pgp] and multidrug resistance proteins [MRPs] [2]), that rapidly transport many small molecule compounds back into the bloodstream. There is an additional enzymatic barrier for drugs with low stability that is maintained by the high density of mitochondria supplying the BMVEC with ATP necessary for enzymatic reactions. The activities of many enzymes participating in the metabolism of endogenous compounds, such as γ-glutamyl transpeptidase, alkaline phosphatase, and aromatic acid decarboxylase, are also elevated in cerebral microvessels [3]. Although small lipophilic molecules ($MW < 400$ kDa) can cross the BBB in pharmacologically significant amounts, effective concentrations of lipid-insoluble drugs (polar molecules and small ions) as well as high-molecular-weight compounds (peptides, proteins, and DNA) cannot be delivered to the CNS within the limits of clinical toxicity. Thus, there remains a need to develop delivery technologies that are capable of transporting drugs across the BBB.

Strategies to overcome these obstacles and enhance CNS drug delivery across the BBB can be divided into two principal groups: enhancing drug influx and restricting drug efflux from BMVEC.

28.2 ENHANCING DRUG INFLUX IN THE BLOOD–BRAIN BARRIER

Possible approaches for enhancing drug influx in the BBB include (i) modification of a drug chemical structure, (ii) drug solubilization in nano- or microcontainers, and (iii) transient increase of BBB permeability.

28.2.1 Drug Modification

28.2.1.1 Lipophilic Drug Modification

Since drug lipophilicity strongly correlates with its cerebrovascular permeability [4], one of the possible strategies is to modify the drug with a lipophilic moiety. The best application for this approach was shown for proteins that were modified with fatty acid residues [5–8]. Specifically, this approach involves point modification of lysine or N-terminal amino groups with one or two fatty acid residues per protein molecule. As a result, the protein molecule remains water soluble but also acquires lipophilic anchors that can target even very hydrophilic proteins to cell surfaces [5]. To obtain low and controlled degrees of lipid modification, reverse micelles of a surfactant—sodium *bis*-(2-ethylhexyl)sulfosuccinate (Aerosol OT)—in octane can be used [5]. Proteins solubilized in such a colloidal system (less than 1% H_2O w/w) become entrapped in the water phase of the reverse micelle interior and covered with a monolayer of hydrated surfactant molecules (Figure 28.1) [9]. The water volume of the reverse micelle interior can be altered by changing the ratio $[H_2O]/[\text{Aerosol OT}]$ to adjust to the size of the protein. The lipophilic reagent localized in the organic solvent incorporates into the surfactant layer of the micelle and comes into contact with the modified group of the protein. Over a dozen water-soluble polypeptides (enzymes, antibodies, toxins, cytokines) have been modified using this method [5,6,10–17]. Overall, modification of water-soluble polypeptides results in enhanced cell surface binding and increased rates of internalization into many cell types.

The relevance of this technology to CNS delivery emerged from the studies of Chekhonin et al. [11]. It was demonstrated that stearate modification of the Fab fragments of antibodies against both gliofibrillar acid protein (GFAP) and brain-specific α 2-glycoprotein increased the accumulation of modified Fab fragments in the rat brain. Subsequent studies using bovine BBMEC as an *in vitro* model of BBB [16] indicated that stearoylation of ribonuclease

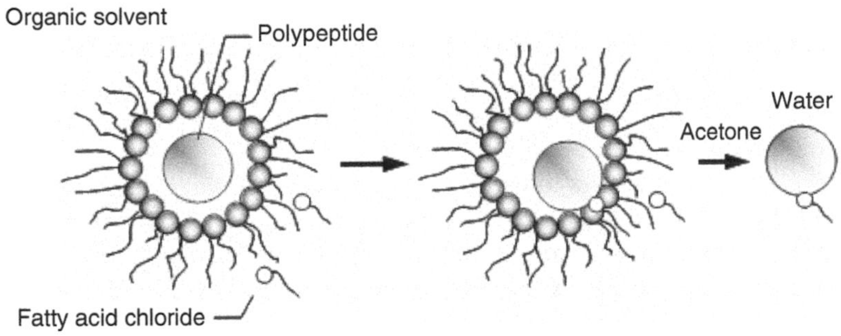

FIGURE 28.1 Chemical modification of the protein with a water-insoluble reagent in the reverse micelles of Aerosol OT in octane. The protein molecule is entrapped in the reverse micelle surrounded by a cover of hydrated surfactant molecules. The water-insoluble reagent is located in the bulk organic phase and can be incorporated into the micelle surface layer coming into contact with the reactive group in the protein. After completion of the reaction the reverse micelle system is disintegrated and the protein is precipitated by cold acetone.

A (approximately 13.6 kDa) increases passage across the BBB by almost ninefold. Of the three fatty acid derivatives analyzed (myristic, palmitic, and stearic), the stearic modification had the greatest effect. The most likely mechanism for the entry of fatty acylated polypeptides to the brain is adsorptive endocytosis.

In spite of all the benefits, this approach has several limitations. First, lipophilic modification of the drug is likely to decrease specific pharmacological activity and increase binding to the plasma proteins, thereby promoting drug clearance from the bloodstream. Second, lipophilic modifications increase side effects due to the accumulation of the modified drug in other organs such as the liver and kidneys. Finally, along with increasing influx to the brain, lipophilic modification can increase drug efflux out of the brain through the BBB or blood–cerebrospinal fluid (CSF) barrier.

28.2.1.2 Prodrugs and Chemical Delivery Systems

The use of prodrugs has been proposed as an effective approach to trapping CNS drugs in the brain. Prodrugs are pharmacologically active compounds that are chemically modified to be inactive until they are converted inside the target tissue by a single activating step. These modifications improve a deficient property (i.e., membrane permeability, stability, or water solubility) of a CNS drug, so that water-soluble drugs attached to lipid-soluble carriers by cleavable bonds may be carried across the BBB and released into the brain. Ideally, the released and processed prodrug is then unable to be transported across the BBB and back into the vascular compartment. Thus, esterification or amidation of hydroxy-, amino-, and carboxylic acid containing drugs enhances their lipid solubility and, as a result, transport to the brain [18]. For example, morphine has a relatively low brain uptake. Replacing one of the hydroxyl groups of morphine with a methyl group increases the lipid solubility and significantly increases the brain uptake. The addition of another acidic group to the molecule to form heroin increases brain uptake to very significant levels. Heroin in the brain is rapidly metabolized to 6-acetyl-morphine and then to morphine, which interacts with the opioid receptor. Morphine, being much more polar than heroin, becomes effectively locked into the brain as it cannot diffuse back across the BBB. This lock-in principle is a major feature of the prodrug approach to CNS delivery. Following transport to the brain, an active compound is released by hydrolysis in the CNS.

A drug can also be linked to a delivery vector, which may render the drug acceptable to a BBB transport system. This linker can then be broken once the prodrug has penetrated the brain [19]. This extension of the prodrug principle has been termed as a chemical delivery system. An example of an anionic chemical delivery system has been developed and applied to deliver testosterone to the CNS [20]. The delivery of a targeted compound was achieved through the use of a specific vector moiety, which is an (acyloxy)alkyl-phosphonate-type functional group. This prodrug readily penetrates the BBB by passive transport due to its increased lipophilicity and enters the brain. Hydrolytic cleavage by esterases provides a negatively charged, hydrophilic intermediate phosphona compound, which is locked in the CNS where it can provide sustained site-specific release of the drug. In particular, *in vitro* and *in vivo* studies in rats demonstrated that methyl-pivaloyloxymethyl-17-testosterylphosphonate is metabolized mainly to testosterone-phosphonate that appears in the brain 5–10 min after intravenous administration. Maximum concentration of the decomposition product was achieved 30 min following administration, and did not decrease significantly during the study. Authors suggested that this design principle can also work for other CNS compounds. Recently one more example of brain-specific chemical delivery system based on 1,4-dihydroisoquinoline targetor moiety was proposed by Bodor et al. [21].

Problems associated with prodrugs are related to the complexity of synthesis, high cost, and the possibility of reactive metabolites. Poor selectivity also impedes extensive use of CNS prodrug systems.

28.2.1.3 Vector-Mediated Drug Delivery

An attractive strategy to improve CNS drug delivery is to link a nontransportable drug with a vector to the BBB. These moieties can work as molecular Trojan horses to transport across the BBB attached proteins, DNA molecules, and drug micro- and nanocarriers facilitating their penetration through the BBB. The choice of a vector moiety and a type of a linker is crucial for the success of this method of drug delivery.

It was demonstrated that some natural peptides can effectively pass the BBB [22,23]. Insulin and transferrin were among the first polypeptides for which binding to the brain microvessels and transcytosis through the BBB was suggested [24–26]. Vectors to these receptors have been considered to enhance the passage of compounds to the brain. In particular, conjugation of insulin with anticancer drug, methotrexate, resulted in receptor-mediated endocytosis of the conjugate to human hepatoma BEL7402 cells [27]. Insulin fragments were also suggested as a peptide carrier for delivery of a model peptide, horseradish peroxidase (HRP), across the BBB [28]. However, the short serum half-life and hypoglycemic effect of insulin are likely to limit the suitability of insulin as a carrier for CNS drug delivery.

Transferrin (Tf) was also suggested as a vector for the drug transport to the brain [29–31]. Conjugation of transferrin with a point mutant of diphtheria toxin resulted in a considerable increase in brain tumor responses by reduction in tumor volume [29]. However, transferrin itself is limited as a vector for CNS delivery as its receptors are almost saturated under physiologic conditions [32]. To this end, utilizing an antibody for these receptors as a brain-specific vector is the most promising strategy. Particularly, the monoclonal antibody to the transferrin receptor, OX26 that binds to the receptor's extracellular domain, was shown to pass across the BBB [33]. Conjugation of potential therapeutic polypeptides, basic fibroblast growth factor and brain-derived neurotrophic factor, with OX26 resulted in their increased entry to the brain and improved neuroprotective effects [30]. In particular, a conjugate of OX26 with fibroblast growth factor protects cortical cell cultures against hypoxia or reoxygenation insult in a dose-dependent manner *in vitro* indicating that the biological activity of this polypeptide was retained following conjugation [30]. *In vivo* studies demonstrate that a single intravenous injection of this conjugate produces an 80% reduction in stroke volume in the brain of rats with a significant improvement of neurological deficit. By contrast, native fibroblast growth factor must be given directly into the brain to perform its therapeutical effect because it does not enter the brain following intravenous administration. Overall, the monoclonal antibody to the transferrin OX26 appears to be an efficient vector candidate in CNS drug delivery.

Short naturally derived peptides, D-penetrain and pegelin (SynB1) were also suggested for CNS vector-mediated drug delivery [34]. The ability of an antineoplastic agent, doxorubicin (Dox), to cross the BBB was studied by an *in situ* rat brain perfusion technique and intravenous injections in mice. The vectorization of Dox with both peptides led to a 20-fold increase in the brain accumulation compared with free Dox [34]. It is noteworthy that vectorization also resulted in a significant decrease of drug concentration in the heart suggesting reduced cardiotoxicity, the main side effect of Dox. Switching to carrier-mediated transport and escaping one of the most therapeutically relevant drug efflux transporters, Pgp, was suggested as a possible mechanism for the enhanced CNS drug transport. This and other drug delivery systems aimed at overcoming drug efflux transporters in the BBB will be described in more detail later. Furthermore, some homing peptides exhibiting specific targeting to the brain can be selected from phage display libraries [35,36]. The BBB binding single-domain antibodies (sdAbs) were selected from a nonimmunized llama sdAb phage display library. The average molecular mass of these peptides was 14 kDa, approximately 10-fold smaller than IgG molecules. It was demonstrated that selected phages (FC5 and FC44) were able to transmigrate across the human *in vitro* BBB model about 20 times better than control sdAb (NC11). FC5 and FC44 phages also showed a high ability to target the brain *in vivo*. Thus, the brain uptake of phage-displayed FC5 and FC44 was 4.5% and 2.9% of injected dose/gram tissue, respectively, in contrast to negligible brain uptakes of wild-type phage (<0.1%). In general, utilizing the specific peptides for targeting of macromolecules and their receptor-mediated transcytosis across BBB could be a successful strategy for improving drug delivery to the brain.

One of the most elegant approaches to improve CNS drug transport across the BBB utilizes specific transport systems supplying the brain with nutrients and essential compounds, such as tryptophan, choline, adenosine, etc. Numerous evidences support high neurological requirements of these compounds in the brain, although it is unable to synthesize them de novo in sufficient amounts [37]. As a result, multiple receptors with a high affinity and selectivity for these molecules are expressed on cerebral endothelial cells to ensure efficient transport of these nutrients from plasma to brain. Conjugation of a drug molecule with these low-molecular-weight ligands does not cause significant alterations in the drug structure that can affect the pharmacological activity of the drug. For example, choline carrier has been suggested as a brain-specific vector for CNS therapeutics [38]. Thus, nanoparticles vectorized with choline accumulated four times more efficiently in the brain than nonvectorized nanoparticles [39]. Further, adenosine A_1 receptor agonists can also be used for improved transport of CNS drugs [40]. Other transporters such as glucose GLUT1 [41,42], large neutral amino acid transporter LAT1 [43,44], monocarboxylic acid transporter MCT1 [45], and cationic amino acid transporter CAT1 [46] were suggested for vector-mediated drug delivery to the brain.

Along with increased cerebrovascular permeability, targeted drug delivery has several limitations that make this approach less available for therapeutic applications. The main setback is an insufficient specificity of the vector moiety to the brain endothelial cells; most receptors used for brain delivery are more or less present in endothelial cells of other organs. As a result, the drug is not targeted specifically to the brain and has decreased efficiency and increased side effects in other organs. In addition, drug conjugation with a vector moiety usually decreases its biological activity due to steric hindrance by the vector group or changes in the chemical structure. Nevertheless, vector-mediated drug delivery is a high potential strategy for CNS therapeutics.

28.2.1.4 Drug Delivery through Adsorptive-Mediated Transcytosis

Since the luminal side of the plasma membrane of brain endothelial cells is negatively charged, the presence of a cation in the drug can considerably increase its interactions with these membranes and initiate adsorptive-mediated transcytosis. This mechanism is actually utilized by the HIV virus that penetrates across the BBB with high efficiency. The virus particles have a glycoprotein (gp120) coat that is capable of inducing adsorptive endocytosis and greatly potentiate the uptake by and passage across the brain endothelial cells [47]. Such an approach also proved to be realistic and practically relevant for targeted drug delivery to the brain. Thus, coupling of cationized albumin with nontransportable peptide resulted in a considerable increase of drug transport across the BBB [48]. Instead of cationized albumin, a positively charged molecule of protamine can also be used for enhancing drug transport across the BBB. For example, conjugation of HRP [49] or native albumin [50] with protamine resulted in their successful delivery to the brain though adsorptive-mediated uptake by BMVEC. Moreover, when nonconjugated protamine sulfate was administered along with plasma albumin into mice, a considerable increase of the transport of albumin across the BBB was recorded by Vorborodt et al. [51]. The authors reported that as early as 10 min after infusion of protamine sulfate solution, the adsorption of blood plasma albumin to the endothelial luminal surface increased by about three times. Simultaneously, the immunolabeling of the endothelial profiles and the subendothelial space was significantly increased. These results suggest that coadministration of protamine with albumin leads to enhanced adsorption of albumin or albumin–protamine complexes to the luminal plasmalemma, followed by transendothelial vesicular transport. Another example of high affinity receptor binding to the brain endothelial cells was demonstrated for cationized macromolecules (antibodies) [50].

The luminal side of BMVEC is also rich in galactosylated glycoconjugates that bind lectins. Therefore, coupling of nontransportable drugs with lectins (such as wheat germ agglutinin [WGA]) can enhance drug transport to the brain via adsorptive-mediated endocytosis. It was reported that the WGA–HRP conjugate was absorbed on the endothelial cells and had entered the brain about 10 times more rapidly than native HRP [52] undergoing adsorptive transcytosis through cerebral endothelia.

The main disadvantage of this approach is the low specificity of modified drugs to the BBB along with decreased biological activity of a drug upon chemical modification or due to the steric hindrance. This problem could be avoided when an appropriate linker between a drug and a modulating group is used. Biodegradable linkers such as avidin–biotin [53] or disulfide linkers [54] that can be cleaved to release an active drug molecule in the brain are advantageous for noncleavable (amide) linkers. Thus, a disulfide biodegradable linker that could later be cleaved in the neuropil by brain disulfide reductases was successfully used *in vivo* [54]. Moreover, a vector moiety and a drug can be separated by an extended poly(ethylene glycol) (PEG) linker (longer than 200 atoms) to eliminate steric interruptions [55].

28.2.2 Drug Solubilization in Nano- or Microcontainers

The need for a protective drug carrier for CNS drugs is highlighted by the fact that many of them have a low hydrolytic stability and are subject to degradation by blood proteins or enzymes encountered in the BBB. Further, a drug carrier can be targeted by brain-specific vector moieties using the receptor-mediated transport. A single unit of a given drug carrier can incorporate many drug molecules, resulting in high payloads per one targeting moiety. By increasing the payload of the carrier, one might improve the efficacy of the delivery while maintaining a relatively low level of involvement of a number of targeted moieties and receptors. Three main types of vehicles are used for transport of neuropharmaceuticals across the BBB: liposomes, micelles, and nanoparticles (or nanogels).

28.2.2.1 Liposomes

The use of liposomes for drug delivery across the brain capillaries has been extensively reported [56–67]. In general, encapsulation of a drug into liposomes prolongs the time of drug circulation in the bloodstream, reduces adverse side effects, and enhances therapeutic effects of CNS agents. Conventional liposomes containing hydrocortisone were demonstrated to penetrate the BBB in experimental autoimmune encephalomyelitis [58]. Further, liposomes prepared using lecithin, cholesterol, and *p*-aminophenyl-α-mannoside were efficiently transported across the BBB in mice via mannose receptor-mediated transcytosis [56]. An interesting approach was utilized for thermosensitive liposomes loaded with an antineoplastic agent, Adriamycin, for treatment of malignant gliomas [66]. These carriers released their contents when the tumor core was heated to 40°C by the brain heating system. Elevated accumulation of the drug in the brain of the heated animals resulted in significantly longer overall survival times compared to the nonheated animals. This method was suggested as a promising approach for malignant gliomas chemotherapy. Unfortunately, conventional liposomes are normally cleared rapidly from circulation by the reticuloendothelial system. Extended circulation of these carriers can be accomplished with small-sized liposomes (<100 nm) composed of neutral, saturated phospholipids, and cholesterol. It was demonstrated that water-soluble polymers such as PEG attached to the surface of long-circulating liposomes (or stealth liposomes) reduce adhesion of opsonic plasma proteins, which induce recognition and rapid removal of liposomes from circulation by the mononuclear phagocyte system in the liver and the spleen [57,68,69]. Using this approach, PEG-coated long-circulating liposomes were shown to remain in circulation with a half-life of as long as 50 h in humans [70]. Specifically, commercial pegylated liposomal-encapsulated Dox, Doxil (ALZA Co., Mountain View, CA), has already been approved for use in the treatment of recurrent ovarian cancer and AIDS-related Kaposi's sarcoma [71]. Long-circulating PEG-liposomes were also used for the delivery of high doses of glucocorticosteroids to the CNS to treat multiple sclerosis [64]. Furthermore, the PEG-coated lipid liposomes encapsulating prednisolone have provided selective targeting to the inflamed CNS in rats (up to 4.5-fold higher accumulation compared to healthy animals). The mechanism of preferential accumulation of these lyposomes in the brain is not fully understood. It appears to be crucial that liposomes exhibit a long-circulating behavior, and relate to either a change of the pharmacokinetics of the encapsulated drug or an indirect effect via monocytes and macrophages in the liver, spleen, or blood.

Attachment of immunoreactive moieties to PEG-modified liposomes can target them to the BBB. Thus, efficient delivery of PEG-liposomes conjugated with transferrin to the postischemic cerebral endothelium was achieved in rats [65]. The expression of transferrin receptors in the cerebral endothelium was reported to increase with a peak at the first day after induced transient middle cerebral occlusion, which makes them a promising tool for drug delivery to the brain after a stroke. Pegylated immunoliposomes have also been successfully employed to target and nonpermanently transfect β-galactosidase and luciferase to the brain tissue [59]. The gene was encapsulated within the liposome that was coated with PEG. Packaging of the gene in the interior of the liposome can prevent the degradation of the therapeutic gene by the ubiquitous endonucleases *in vivo*. These liposomes were directed to transferrin receptor-rich tissues, such as brain, liver, and spleen by monoclonal antibody OX26 linked to about 2% of the PEG strands. When the β-galactosidase gene was also coupled with a brain-specific glialfibrillary acidic protein promoter, its expression was achieved predominantly in the brain [72]. Similar immunoliposome constructs using OX26 transferrin receptor monoclonal antibodies have also been used to deliver doxorubicin [57], digoxin [73], biotinylated oligonucleotides (ODNs) [62], and neurotrophin peptides [74] to the brain. OX26-conjugated liposomes were selectively distributed to BMVEC leaving choroids plexus epithelium, neurons, and glia unlabeled [67]. Another example of brain targeting vector has been reported by Coloma et al. [75]. A chimeric monoclonal antibody to the human insulin receptor, 83-14 MAb, was replaced by human antibody sequence. The intravenously administered antibody showed robust transport to the brain in rhesus monkey. This vector was also used for targeting liposomes with β-galactosidase or luciferase gene to the brain [63]. As a result, the level of luciferase gene expression in the monkey brain was 50-fold higher as compared to the rat. Overall, liposomes hold a significant promise as an efficient tool for CNS drug delivery.

28.2.2.2 Polymeric Micelles

While liposomes can be successfully used for the transport of water-soluble or hydrophilic drugs encapsulated in the inner water volume, polymeric micelles are used in general for carrying water-insoluble or hydrophobic drugs [76,77]. For example, amphiphilic block copolymers are perfectly suited for producing micellar containers as they can self-assemble in water solutions above the critical micelle concentration (CMC) forming drug delivery complexes of nanoscale size. These micro- and nanocontainers can incorporate hydrophobic and amphiphilic drugs and serve as carriers for drug delivery (micellar nanocontainers). The core-shell architecture of polymeric micelles is essential for their utility as novel functional materials for these applications. The core is a water-incompatible compartment that is segregated from the aqueous exterior by the hydrophilic chains of the corona, thereby forming, within the core, a cargo hold for the incorporation of various therapeutic or diagnostic reagents. The hydrophilic shell contributes greatly to the pharmaceutical behavior of block copolymer formulations by maintaining the micelles in a dispersed state, as well as by decreasing undesirable drug interactions with cells and proteins through steric-stabilization effects. Therefore, polymeric micelles can solubilize water-insoluble drugs and shelter them from metabolic degradation in a physiological environment. Polymeric micelles have been evaluated in multiple pharmaceutical applications as drug and gene delivery systems [78–82], as well as carriers for various diagnostic imaging agents [83].

One of the early studies of targeted drug delivery to the brain used Pluronic block copolymer (BASF Co., Mount Olive, New Jersey) micelles as carriers for CNS drugs [78,84]. These micelles were conjugated with either polyclonal antibodies against brain α_2-glycoprotein (Ab-α_2-gp) or insulin targeting to the receptors at the lumenal side of BMVEC. Both the antibody-conjugated and insulin-conjugated micelles were shown to effectively deliver a drug incorporated into the micelles to the brain tissue *in vivo* [84]. Studies of animal behavior reactions estimating precisely the pharmacological activity of dopaminergic compounds, such as mobility and grooming, were performed. It was demonstrated that incorporation of a neuroleptic, haloperidol, into the Pluronic micelles that were vectorized with insulin resulted in a 25-fold stronger effect as compared with the free drug. Vectorization of the drug-loaded micelles with Ab-α_2-gp led to a much more pronounced (up to 500-fold) neuroleptic effect. Subsequent studies demonstrated that the vectorized micelles undergo receptor-mediated transport in BMVEC [85].

Polymeric micelles formed by Pluronics, PEG–phospholipid conjugates, PEG-b-polyesters, or PEG-b-poly-L-amino acids were proposed for drug delivery of poorly water-soluble compounds, such as amphotericin B, propofol, paclitaxel, and photosensitizers [77,86,87]. It was also emphasized that using polymeric micelles can significantly increase the drug transport into the brain.

28.2.2.3 Nanoparticles

In contrast to polymeric micelles, nanoparticles usually have a cross-linked core and do not disassemble upon dilution in the bloodstream. The use of nanoparticles as vehicles for drug and gene delivery has been an area of intensive research and development for over a decade [88–97]. The surface of such carriers is also often modified by PEG brush (PEGylation) to increase the stability of nanoparticles in dispersion and extend circulation time of nanoparticles in the body [90–92]. To allow for efficient transcytosis across the BMVEC, the carrier particles have to be small with a size not exceeding approximately 100–200 nm. Thus, poly (butyl)cyanoacrylate (PBCA) nanoparticles have been successfully used to deliver a wide range of drugs that were either incorporated into the particle structure or absorbed onto the surface to the CNS [92,97,98]. The particles were then coated with Tween 80, which appears to promote the binding of apolipoprotein E to the surface of the particle. This final product was transported across the cerebral endothelial cells by endocytosis via the low-density lipoprotein (LDL) receptor. Analgetics, dalargin and loperamide; antineoplastic agent, Dox; the NMDA receptor antagonist MRZ 2/576; and peptides, dalargin and kytorphin have been successfully delivered to the CNS in these constructs [98]. Bypassing of the P-glycoprotein efflux system is also suggested as a mechanism for enhanced transport of CNS drugs incorporated into the nanoparticles. It was demonstrated that the nanoparticles penetrated into the brain to a high extent and localized in the ependymal cells of the choroids plexuses, and in the epithelial cells of pia mater and ventricles, and to a lower extent in the capillary endothelial cells of the BBB. Coating of the nanoparticles with polysorbate 80-coated led to elevated transport across the BBB due to, in part, increased permeabilization induced by the surfactant. Therefore, the possibility of a general toxic effect can impede the therapeutic applications of these carriers [99].

A new family of carrier systems, Nanogel was recently developed for targeted delivery of drugs and biomacromolecules to the brain [93,96]. Nanogel represents a nanoscale size polymer network of cross-linked ionic polyethyleneimine (PEI) and nonionic PEG chains (PEG-*cl*-PEI). It forms swollen cross-linked networks dispersed in solution and collapses upon binding of a macromolecular drug through electrostatic interaction of this drug with PEI chain resulting in decreased volume and size of the particles. Because of the effect of PEG chains, the collapsed Nanogel forms a stable dispersion with the particles size of approximately 80 nm. Nanogel can absorb spontaneously, through ionic interactions, a broad set of biomacromolecules, including negatively charged ODNs. A key advantage of the Nanogel is that it displays efficient loading of macromolecules (40%–60% by weight), resulting in high payloads not achieved with conventional carrier systems. The ODNs incorporated into the Nanogel were protected against degradation by nucleases. However, upon delivery within a target cell the ODNs were released and they exhibited specific activity against their molecular targets, as demonstrated using several cell models [93]. The study using bovine BMVEC monolayers, as an *in vitro* model, demonstrated that following incorporation in the Nanogel particles, the transport of ODNs across BBB was significantly increased compared to the free ODNs transport [93]. A recent study tested the Nanogel system for the receptor-mediated delivery of ODNs across BMVEC monolayers [96]. Specifically, to target the receptors displayed at the BMVEC, the surface of the Nanogel particles was modified by either transferrin or insulin using avidin–biotin coupling chemistry. Both peptides were shown to increase transcellular permeability of the Nanogel and enhance delivery of ODNs across BMVEC monolayers (Figure 28.2). Recent *in vivo* studies demonstrated that incorporation of phosphorothionate ODN in the Nanogel particles resulted in brain increases by over 15-fold, whereas the liver and spleen decreased by twofold compared to the free ODN [94,96]. Overall, Nanogel was shown to be a promising system for delivery of ODN to the brain.

Transport of carboxylated polystyrene nanospheres (20 nm) across the BBB was studied *in vivo* following cerebral ischemia and reperfusion [100]. A microdialysis probe was

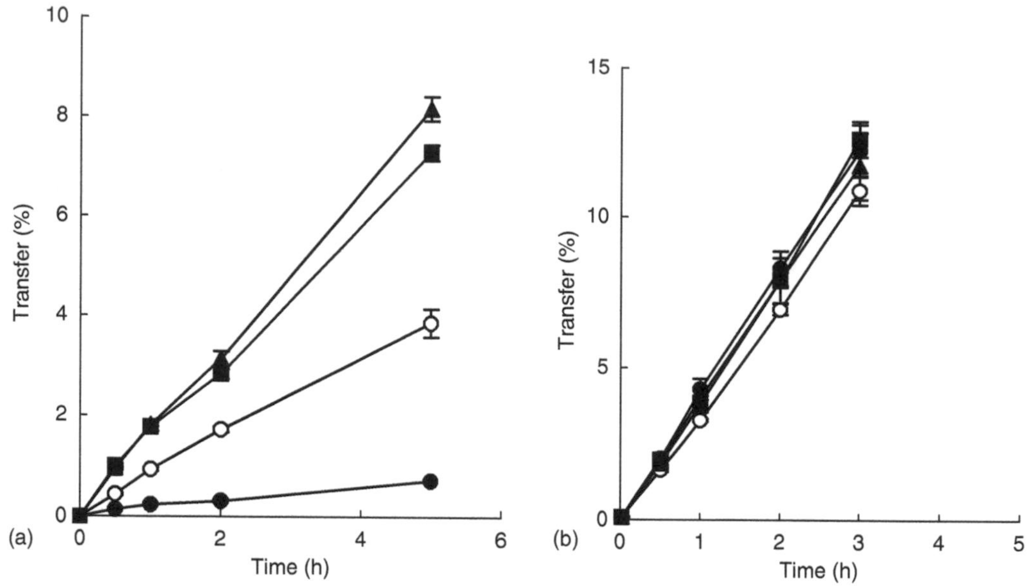

FIGURE 28.2 (a) Transport of free ODN and ODN formulated with Nanogel across BBMEC monolayers. (b) Transport of paracellular marker, ^{3}H-mannitol, in the presence of the ODN and its formulations with Nanogel. The symbols correspond to free ODN (black circles); Nanogel–ODN complex (open circles); transferrin-vectorized Nanogel–ODN complex ODN (squares); insulin-vectorized Nanogel–ODN complex (triangles). All Nanogel–ODN complexes are prepared at $N/P = 8$. Data are means $\pm$ SEM ($n = 3$). (From Vinogradov, S., Batrakova, E., and Kabanov, A., *Colloids Surf. B Biointerfaces*, 16, 291, 1999. With permission.)

implanted in the cerebral cortex of an anesthetized rat injected with fluorescent nanospheres. It was demonstrated that cerebral ischemia and reperfusion induced transient increase in fluorescence intensity indicating accumulation of nanocarriers in the brain. Atomic force microscopy (AFM) images confirmed presence of nanospheres in the microdialysates collected from extracellular fluids of rat brain after cerebral ischemia. Noteworthy, the intravenously injected nanospheres remained in the vasculature under normal conditions. The increase of nanosphere extravasation was suggested to be due to cerebral ischemia-induced stress that leads to the opening of the tight junctions between endothelial cells [101]. Therefore, polystyrene nanospheres have potential clinical applications for CNS delivery of drugs and imaging agents during ischemia and stroke.

28.2.2.4 Cell-Mediated Central Nervous System Drug Delivery

An extreme case of the vehicle-mediated CNS drug delivery represents the use of specific cells or carriers that can incorporate micro- and nanocontainers (usually liposomes) loaded with drugs and act as perfect Trojan horses migrating across the BBB and carrying drugs to the site of action [102–105]. It is documented that many neurological diseases, such as Alzheimer's disease, Parkinson's disease, Prion disease, meningitis, encephalitis, and AIDS related dementia, have in common an inflammatory component [106]. The process of inflammation is characterized by the extensive recruitment of the leukocytes (neutrophils and monocytes). These cells have a unique property of migrating toward the inflammation site via the processes known as diapedesis and chemataxis [107]. Their combat arsenal consists of endocytosing the foreign particle, producing toxic compounds, and liberating substances stored in intracellular vesicles via exocytosis. Therefore these cells can be used for cell-mediated CNS drug delivery when loaded with a required drug and administered into the bloodstream.

It has been shown that cells capable of phagocytosis, such as macrophages or monocytes or neutrophils, can endocytose liposomes [102,104]. To accelerate transport of monocytes and neutrophils loaded with drug-containing liposomes to the brain, these carriers were additionally loaded with magnetic particles [104]. Magnetic liposomes demonstrated about a 10-fold increase in brain levels compared with nonmagnetic controlled carriers when local magnetic field was applied. It is noteworthy that both cell types showed preferential uptake of liposomes containing negatively charged lipids, such as phosphatidylserine [103], than liposomes containing only neutral lipids, such as phosphatidylcholine. Similar results were shown for liposomes consisting of neutral lipids modified by poly(acrylic acid) (PAA). PAA-conjugated liposomes were rapidly endocytosed by the macrophages and digested in acidic compartments [103]. Recognition of PAA by the macrophage surface receptors, including the scavenger receptor, seems to be crucial for the enhanced uptake of PAA-modified liposomes. Subsequently, part of their aqueous component was exocytosed. Thus, phosphatidylserine-containing negatively charged liposomes was shown to increase therapeutic activity of an encapsulated antifungal agent, chloroquine, against *Cryptococcus neoformans* infection in the mouse brain [105]. The chloroquine-loaded liposomes accumulated inside macrophage phagolysosomes and resulted in a remarkable reduction in fungal load in the brain even in low doses compared to free drug in high doses thus increasing antifungal activity of macrophages.

28.2.3 Disrupting of the Blood–Brain Barrier

One of the most invasive strategies for CNS drug delivery involves transient disruption of the BBB along with the drug systemic administration. This procedure is associated with a high risk of adverse effects and appropriate only in extreme cases of rapidly growing high grade gliomas [108]. A variety of techniques that transiently disrupt the BBB have been investigated, including osmotic opening of the BBB tight junctions using hypertonic solutions of mannitol, arabinose, lactamide, saline, urea, and several radiographic contrast agents. Because it is approved for administration to patients, intracarotid injection of mannitol has become the choice in both preclinical and clinical studies [109]. It initiates endothelial cell shrinkage and opening of BBB tight junctions for a period of a few hours thereby permitting the delivery of antineoplastic agents to the brain [110,111]. Thus, the median survival for patients with primary CNS lymphoma was increased from 17.8 to 44.5 months when osmotic disruption of the BBB was employed [110]. It is emphasized that patient toxicity was manageable in this intensive therapeutic regimen. Further, when intensive consolidation therapy with carboplatin and etoposide was given in conjunction with mannitol-induced osmotic BBB disruption, complete response was received in most of the patients with disseminated CNS germinoma [112]. One reason for the unfavorable toxic or therapeutic ratio often observed with hyperosmotic BBB disruption is that this methodology results in only a 25% increase in the permeability of the tumor microvasculature, in contrast to a 10-fold increase in the permeability of normal brain endothelium. Nevertheless, the osmotic disruption of the BBB has shown efficacy in increasing the CNS drug delivery and a potency to improve clinical outcome in certain neoplastic diseases.

Recently, a new and potentially safer technique of disruption of the BBB, called a biochemical disruption, has been developed. In contrast to osmotic disruption, this method is based on the observation that some substances can selectively open only brain tumor capillaries leaving normal brain capillaries unaffected. The substances used for this procedure are vasoactive leukotrienes, such as leukotriene C4 [113], vasoactive amines such as bradykinin, histamine, and the synthetic bradykinin analog RMP-7, Cereport [114], or insulin [115]. The mechanism of leukotriene C4 effect was shown to be related to the abundance of g-glutamyl transpeptidase (g-GTP) in normal capillaries and its decreased amount in tumors.

This results in the reduction of the enzymatic barrier in tumor endothelial cells, and in the elevated effect of leukotriene C4. In the case of Cereport, the opening of the BBB is mediated specifically through bradykinin B_2 receptors. Using an electron dense marker, lanthanum, the electron microscopy studies demonstrated that an intravenous injection of Cereport resulted in the opening of the tight junctions between the BMVEC and increasing the paracellular transport of lanthanum to the brain [116]. It is noteworthy that effects of the drug occur rapidly, within minutes of initiation of the infusion, and restoration of the barrier begins almost immediately upon termination of infusion and is essentially complete within 2–5 min [117]. Thus, Cereport increases the transport of a wide variety of antineoplastics and CNS diagnostic agents, including carboplatin [118], loperamide [119], cyclosporine A [120], etc. The enhanced uptake in the tumor and, especially, the area of the brain immediately around the tumor boundary is important since this is the area that tumor cells infiltrate, often escaping detection, surgery, and therapeutic levels of antineoplastic agents [121].

Another approach for opening the BBB for targeted drug delivery to the brain using focused ultrasound exposures was developed recently [122,123]. It was shown that sonication of the brain, applied in the presence of an ultrasound contrast agent injected intravenously, caused an increased number of vesicles and vacuoles fenestration and channel formation, and reversible opening of the BBB tight junctions in rabbits. The electron microscopy findings demonstrated HRP passage through vessel walls via both transendothelial and paraendothelial routes. Overall, adjunct therapy to pharmacological treatments of CNS disorders proves to be efficacious although it is associated with risk factors including passage of plasma proteins, the altered glucose uptake, the expression of heat shock proteins, microembolism, and abnormal neuronal function [124].

28.3 RESTRICTING DRUG EFFLUX IN THE BLOOD–BRAIN BARRIER

28.3.1 Role of Drug Efflux Transport Systems in the Blood–Brain Barrier

There are a number of efflux mechanisms within the CNS that influence drug concentrations in the brain. Some of them are passive whereas others are active. Recently much attention has been focused on the so-called multidrug efflux transporters: Pgp and MRP, breast cancer resistance protein (BCRP), and the multispecific organic anion transporter (MOAT) that belong to the ATP-binding cassette (ABC) family [1,2]. Cerebral capillary endothelium expresses a number of efflux transport proteins, which actively remove a broad range of drug molecules before they cross into the brain parenchyma. Pgp is the most thoroughly investigated brain efflux transport protein with broad affinity for dissimilar lipophilic and amphiphilic substrates [125]. As a consequence, the therapeutic value of many promising drugs, such as protease inhibitors for HIV-1 encephalitis (ritonavir, nalfinavir, and indinavir) [126], anti-inflammatory drugs (prednesolone, dexamethasone, and indomethacin) for treatment of microglial inflammation during idiopathic Parkinson's disease and Alzheimer's disease [125,127], neuroleptic agents (amitriptyline and haloperidol) [128], analgesic drugs (morphine, beta-endorphin, and asimadoline) [129], as well as antifungal agents (itraconazole and ketoconazole) [130], is diminished, and cerebral diseases have proven to be most refractory to therapeutic interventions. Moreover, delivery of many anticancer agents (doxorubicin, vinblastine, taxol, etc.) to the brain for the treatment of brain tumors [131], and drugs for the treatment of epilepsy (carbamazepine, phenobarbital, phenytoin, and lamotrigine) [132] are also restricted by the Pgp drug efflux transporter.

An emerging strategy for enhanced BBB penetration of drugs is coadministration of competitive or noncompetitive inhibitors of the efflux transporter together with the desired CNS drug. First generation low-molecular-weight Pgp inhibitors (cyclosporine A, verapamil,

PSC833, etc.) are substrates of the drug efflux transporter, which compete for the active site with the therapeutic agent [133]. Second generation inhibitors (LY335979, XR9576, and GF120918) are noncompetitive inhibitors, which allosterically bind to Pgp, inactivating it and increasing drug transport to the brain [134]. Despite their high efficiency in cell culture models, the small therapeutic range of these inhibitors, high *in vivo* toxicity, and fast clearance are the main obstacles for their therapeutic application.

28.3.2 Pluronic Block Copolymers

Recently, a new class of inhibitors (nonionic polymer surfactants) was identified as promising agents for drug formulations. These compounds are two- or three-block copolymers arranged in a linear ABA or AB structure. The A block is a hydrophilic poly(ethylene oxide) chain. The B block can be a hydrophobic lipid (in copolymers BRIJs, MYRJs, Tritons, Tweens, and Chremophor) or a poly(propylene oxide) chain (in copolymers Pluronics [BASF Corp., N.J., USA] and CRL-1606). Pluronic block copolymers with various numbers of hydrophilic EO (*n*) and hydrophobic PO (*m*) units are characterized by distinct hydrophilic–lipophilic balance (HLB). Due to their amphiphilic character these copolymers display surfactant properties including ability to interact with hydrophobic surfaces and biological membranes. In aqueous solutions with concentrations above the CMC, these copolymers self-assemble into micelles.

Studies in multidrug resistant (MDR) cancer cells, polarized intestinal epithelial cells, Caco-2, and polarized BMVEC monolayers provided compelling evidence that selected Pluronic block copolymers can inhibit drug efflux transport systems [85,135–139]. Specifically, in primary cultured BMVEC monolayers, used as an *in vitro* model of BBB, the inhibition of drug efflux systems, Pgp, and MRP was associated with an increased accumulation and permeability of a broad spectrum of drugs in the BBB, including low molecular drugs [85,140] and peptides [141]. These effects were most apparent at concentrations below the CMC [85,135]. It was suggested that unimers, i.e., single block copolymer molecules, are responsible for the inhibition of Pgp and MRPs efflux transport system (CMC for Pluronic P85 is 0.03%wt). Incorporation of the probe into the micelles formed at high concentrations of the block copolymer decreases its availability to the cells and reduces the transport of this probe in BMVEC.

Recent findings suggest that effects of Pluronic on drug efflux transport proteins involve interactions of the block copolymers with the cell membranes [137,142]. It was demonstrated that a fine balance is needed between hydrophilic (EO) and lipophilic (PO) components in the Pluronic molecule to enable inhibition of the drug efflux systems [139]. Overall, the most efficacious block copolymers are those with intermediate lengths of the PO block and a relatively hydrophobic structure ($HLB < 20$), such as Pluronic P85 or L61 [143]. The hydrophobic PO chains of Pluronic immerse into the membrane hydrophobic areas, resulting in alterations of the membrane structure and decreases in its microviscosity (membrane fluidization). Pluronic at relatively low concentrations (e.g., 0.01%) inhibits the Pgp ATPase activity, possibly due to conformational changes in the transport protein induced by the immersed copolymer chains in the Pgp-expressing membranes [137]. In particular, Pluronic P85 displayed the effects characteristic of a mixed type enzyme inhibitor—decreasing the maximal reaction rate V_{max}, and increasing the Michaelis constant K_m, for ATP as well as Pgp-specific substrates such as vinblastine [144]. The magnitude of these effects for vinblastine was as high as over 200-fold V_{max}/K_m change (interestingly, MRP1 ATPase activity was affected less, which could explain somewhat smaller effects of Pluronic on this transporter). In contrast, at the high concentrations (e.g., 1%), binding of Pluronic to the membrane actually resulted in restoration of Pgp ATPase activity. This could be due to the segregation of the block copolymer molecules in the 2D clusters in the membrane, which diminishes its interactions with the transport proteins.

Various drug resistance mechanisms, including drug transport and detoxification systems, require consumption of energy to sustain their function in the barrier cells. Because of this fact, mechanistic studies have focused on the effects of Pluronic block copolymers on metabolism and energy conservation in BMVEC [145]. The basis for such studies was the earlier reports that Pluronic block copolymers can affect mitochondria function and energy conservation in the cells [146]. Recent studies have demonstrated that exposure to Pluronic P85 induced significant decrease in ATP levels in BMVEC monolayers [137]. The observed energy depletion was due to inhibition of the cellular metabolism rather than a loss of ATP in the environment. The study by Rapoport et al. [147] suggested that Pluronic P85 can be transported into the cells and decrease the activity of electron transport chains in the mitochondria. Remarkably, the ATP depletion induced by Pluronic appears to be tightly linked to the specific cell genotype, because this effect is observed selectively in the cells that overexpress Pgp (as well as MRPs) [145,148–150]. The explanation for this relationship still needs to be found, although it was speculated that inhibition of ATP production in high energy-consuming cells, such as cells overexpressing Pgp, results in the rapid exhaustion of intracellular ATP, i.e., ATP depletion [145]. Overall, the energy depletion (decreasing ATP pool available for drug transport proteins) and membrane interactions (inhibiting of ATPase activity of drug transport proteins) are critical factors collectively contributing to a potent inhibition of the drug efflux systems by Pluronic [145].

The effect of Pluronic P85 on drug transport into the brain was evaluated in animal experiments [138]. Brain delivery of a Pgp substrate, digoxin, administered intravenously in the wild-type mice expressing functional Pgp, was greatly enhanced in the presence of Pluronic P85. It was found that the digoxin brain/plasma ratios in the Pluronic treated animals were practically the same as those in the knockout mice, an animal model that is deficient in both mdr1a and mdr1b isoforms of Pgp (Figure 28.3). This suggests that coadministration of Pluronic with the drug in mice resulted in inhibition of Pgp in the BBB of the wild-type animals [138] (Figure 28.4).

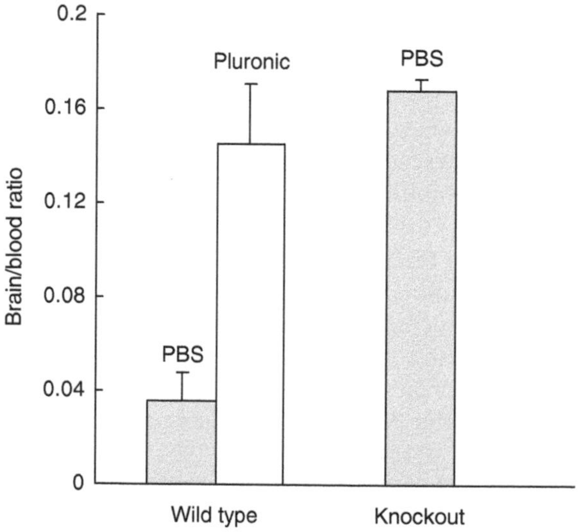

FIGURE 28.3 Effect of Pluronic P85 on the digoxin brain/plasma ratio in the wild-type mice when compared to the mdr 1a/b (−/−) knockout mouse at 5 h postdose. A tracer dose of [^{3}H]-digoxin (4 μCi; 7.8 μg/kg) in PBS control or 1% Pluronic P85 solution (100 μL) was administered intravenously into wild-type control group or Pgp knockout mice. Four hours following injection animals were sacrificed and the amount of digoxin in the blood in the brain was assayed. (From Batrakova, E., et al. *J. Pharmacol. Exp. Ther.*, 296, 551, 2001. With permission.)

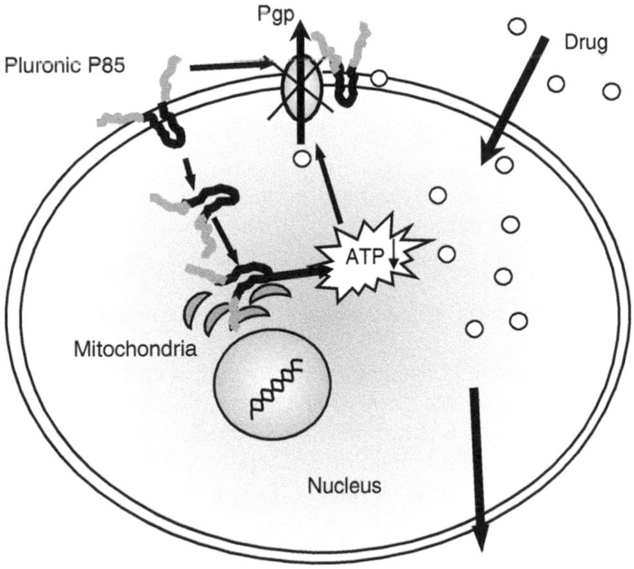

FIGURE 28.4 Schematic illustration twofold effects of Pluronic block copolymer on Pgp drug efflux system. Inhibition of Pgp activity is due to a combination of two cellular pathways: depletion of ATP levels and membrane fluidization in the cells.

One possible concern in these studies is that by virtue of inhibiting the ATP in BMVEC, the copolymer may display toxic effects on the BBB. However, the ATP depletion was found to be transient; following removal of the block copolymers from BMVEC monolayers, the initial ATP levels were restored [145]. Although there were significant decreases in cellular ATP following Pluronic treatment, even during peak depletion of ATP by Pluronic there was no evidence of loss of barrier functions of BBB as demonstrated using ^{3}H-mannitol as a permeability marker both *in vitro* and *in vivo* [85,138]. Moreover, Pluronic does not affect the glucose transporter, GLUT1, and only slightly inhibits the lactate transporter, MCT1, the two transporters playing an important role in the brain metabolism [150]. Pluronic also does not inhibit amino acid transporters, LAT1, CAT1, and SAT1, in the BBB (unpublished data). A histochemical examination of the tissue sections obtained from animals treated with Pluronic revealed no pathological changes in the BBB. Importantly, no cerebral toxicity of any kind has been observed in the human phase I studies of SP1049C, a Pluronic-based formulation of doxorubicin to treat MDR tumors [151]. After completion of phase I clinical trials, SP1049C is now undergoing several phase II clinical trials. It is possible that this formulation, evaluated in human trials, can be adopted for use with CNS drugs to enhance drug delivery to the brain.

28.4 CONCLUSION

Tremendous efforts in the last several decades have resulted in numerous inventions of CNS drug delivery systems creating a feeling of unlimited potential. Most of them have a significant potential for the development of new technologies. The wide variety of strategies reflects the inherent difficulty in transport of therapeutic and imaging agents across the BBB. In fact, the effective combination of several approaches, such as encapsulation of drugs into nanoparticles conjugated with vector moieties or using micelles of Pluronic block copolymers along with Pluronic unimers that will inhibit drug efflux transporters in the BMVEC, may give the most promising therapeutic outcomes.

ACKNOWLEDGMENT

This work was supported by National Institutes of Health grant NS36229.

REFERENCES

1. Pardridge, W. 1998. *Introduction to the blood–brain barrier, methodology, biology and pathology*, 486. New York: University Press.
2. Tamai, I., and A. Tsuji. 2000. Transporter-mediated permeation of drugs across the blood–brain barrier. *J Pharm Sci* 89:1371.
3. Minn, A., et al. 1991. Drug metabolizing enzymes in the brain and cerebral microvessels. *Brain Res Brain Res Rev* 16:65.
4. Levin, V.A. 1980. Relationship of octanol/water partition coefficient and molecular weight to rat brain capillary permeability. *J Med Chem* 23:682.
5. Kabanov, A., A. Levashov, and K. Martinek. 1987. Transformation of water-soluble enzymes into membrane active form by chemical modification. *Ann N Y Acad Sci* 501:63.
6. Kabanov, A., A. Levashov, and V. Alakhov. 1989. Lipid modification of proteins and their membrane transport. *Protein Eng* 3:39.
7. Hashimoto, M., et al. 1989. Synthesis of palmitoyl derivatives of insulin and their biological activities. *Pharm Res* 6:171.
8. Colsky, A., and J. Peacock. 1991. Palmitate-derivatized antibodies can specifically "arm" macrophage effector cells for ADCC. *J Leukoc Biol* 49:1.
9. Kabanov, A.V., and E.V. Batrakova. 2004. New technologies for drug delivery across the blood–brain barrier. *Curr Pharm Des* 10:1355.
10. Alakhov, V., et al. 1990. Increasing cytostatic effects of ricin A chain and *Staphylococcus aureus* enterotoxin A through *in vitro* hydrophobization with fatty acid residues. *Biotechnol Appl Biochem* 12:94.
11. Chekhonin, V., et al. 1991. Fatty acid acylated Fab-fragments of antibodies to neurospecific proteins as carriers for neuroleptic targeted delivery in brain. *FEBS Lett* 287:149.
12. Kabanov, A.V., et al. 1992. Pluronic micelles as a tool for low-molecular compound vector delivery into a cell: Effect of *Staphylococcus aureus* enterotoxin B on cell loading with micelle incorporated fluorescent dye. *Biochem Int* 26:1035.
13. Slepnev, V., et al. 1995. Fatty acid acylated peroxidase as a model for the study of interactions of hydrophobically-modified proteins with mammalian cells. *Bioconjug Chem* 6:608.
14. Melik-Nubarov, N.S., et al. 1993. Interaction of hydrophobized antiviral antibodies with influenza virus infected MDCK cells. *Biochem Mol Biol Int* 29:939.
15. Ekrami, H., A. Kennedy, and W. Shen. 1995. Water-soluble fatty acid derivatives as acylating agents for reversible lipidization of polypeptides. *FEBS Lett* 371:283.
16. Chopineau, J., et al. 1998. Monoacylation of ribonuclease A enables its transport across an *in vitro* model of the blood–brain barrier. *J Control Release* 56:231.
17. Kozlova, N., et al. 1999. Catalytic properties and conformation of hydrophobized alpha-chymotrypsin incorporated into a bilayer lipid membrane. *FEBS Lett* 461:141.
18. Oldendorf, W.H. 1971. Brain uptake of radiolabeled amino acids, amines, and hexoses after arterial injection. *Am J Physiol* 221:1629.
19. Bodor, N., and P. Buchwald. 2003. Brain-targeted drug delivery: Experiences to date. *Am J Drug Targ* 1:13.
20. Somogyi, G., P. Buchwald, and N. Bodor. 2002. Targeted drug delivery to the central nervous system via phosphonate derivatives (anionic delivery system for testosterone). *Pharmazie* 57:135.
21. Bodor, N., et al. 2002. *In vitro* and *in vivo* evaluations of dihydroquinoline- and dihydroisoquinoline-based targetor moieties for brain-specific chemical delivery systems. *J Drug Target* 10:63.
22. Pardridge, W. 2002. Targeting neurotherapeutic agents through the blood–brain barrier. *Arch Neurol* 59:35.
23. Banks, W., and C. Lebel. 2002. Strategies for the delivery of leptin to the CNS. *J Drug Target* 10:297.

24. Miller, D., B. Keller, and R. Borchardt. 1994. Identification and distribution of insulin receptors on cultured bovine brain microvessel endothelial cells: Possible function in insulin processing in the blood brain barrier. *J Cell Physiol* 161:333.
25. Shin, S.U., et al. 1995. Transferrin–antibody fusion proteins are effective in brain targeting. *Proc Natl Acad Sci USA* 92:2820.
26. Banks, W.A., J.B. Jaspan, and A.J. Kastin. 1997. Selective, physiological transport of insulin across the blood–brain barrier: Novel demonstration by species-specific radioimmunoassays. *Peptides* 18:1257.
27. Ou, X.H., et al. 2004. Receptor binding characteristics and cytotoxicity of insulin–methotrexate. *World J Gastroenterol* 10:2430.
28. Fukuta, M., et al. 1994. Insulin fragments as a carrier for peptide delivery across the blood–brain barrier. *Pharm Res* 11:1681.
29. Laske, D.W., R.J. Youle, and E.H. Oldfield. 1997. Tumor regression with regional distribution of the targeted toxin TF-CRM107 in patients with malignant brain tumors. *Nat Med* 3:1362.
30. Song, B.W., et al. 2002. Enhanced neuroprotective effects of basic fibroblast growth factor in regional brain ischemia after conjugation to a blood–brain barrier delivery vector. *J Pharmacol Exp Ther* 301:605.
31. Qian, Z.M., et al. 2002. Targeted drug delivery via the transferrin receptor-mediated endocytosis pathway. *Pharmacol Rev* 54:561.
32. Seligman, P.A. 1983. Structure and function of the transferrin receptor. *Prog Hematol* 13:131.
33. Friden, P.M., et al. 1991. Anti-transferrin receptor antibody and antibody-drug conjugates cross the blood–brain barrier. *Proc Natl Acad Sci USA* 88:4771.
34. Rousselle, C., et al. 2001. Enhanced delivery of doxorubicin into the brain via a peptide-vector-mediated strategy: Saturation kinetics and specificity. *J Pharmacol Exp Ther* 296:124.
35. Pasqualini, R., W. Arap, and D.M. McDonald. 2002. Probing the structural and molecular diversity of tumor vasculature. *Trends Mol Med* 8:563.
36. Muruganandam, A., et al. 2002. Selection of phage-displayed llama single-domain antibodies that transmigrate across human blood–brain barrier endothelium. *FASEB J* 16:240.
37. Pardridge, W.M. 1999. Vector-mediated drug delivery to the brain. *Adv Drug Deliv Rev* 36:299.
38. Allen, D.D., and P.R. Lockman. 2003. The blood–brain barrier choline transporter as a brain drug delivery vector. *Life Sci* 73:1609.
39. Lockman, P.R., et al. 2002. Nanoparticle technology for drug delivery across the blood–brain barrier. *Drug Dev Ind Pharm* 28:1.
40. Schaddelee, M.P., et al. 2003. Functional role of adenosine receptor subtypes in the regulation of blood–brain barrier permeability: Possible implications for the design of synthetic adenosine derivatives. *Eur J Pharm Sci* 19:13.
41. Agus, D.B., et al. 1997. Vitamin C crosses the blood–brain barrier in the oxidized form through the glucose transporters. *J Clin Invest* 100:2842.
42. Halmos, T., et al. 1997. Synthesis of *O*-methylsulfonyl derivatives of D-glucose as potential alkylating agents for targeted drug delivery to the brain. Evaluation of their interaction with the human erythrocyte GLUT1 hexose transporter. *Carbohydr Res* 299:15.
43. Cornford, E.M., et al. 1992. Melphalan penetration of the blood–brain barrier via the neutral amino acid transporter in tumor-bearing brain. *Cancer Res* 52:138.
44. Takada, Y., et al. 1992. Rapid high-affinity transport of a chemotherapeutic amino acid across the blood–brain barrier. *Cancer Res* 52:2191.
45. Terasaki, T., and A. Tsuji. 1994. Drug delivery to the brain utilizing blood–brain barrier transport systems. *J Control Release* 29:163.
46. Xiang, J.J., et al. 2003. IONP-PLL: A novel non-viral vector for efficient gene delivery. *J Gene Med* 5:803.
47. Banks, W.A., V. Akerstrom, and A.J. Kastin. 1998. Adsorptive endocytosis mediates the passage of HIV-1 across the blood–brain barrier: Evidence for a post-internalization coreceptor. *J Cell Sci* 111 (Pt 4):533.
48. Kang, Y.S., and W.M. Pardridge. 1994. Brain delivery of biotin bound to a conjugate of neutral avidin and cationized human albumin. *Pharm Res* 11:1257.

49. Nagy, Z., H. Peters, and I. Huttner. 1981. Endothelial surface charge: Blood–brain barrier opening to horseradish peroxidase induced by the polycation protamin sulfate. *Acta Neuropathol Suppl (Berl)* 7:7.
50. Pardridge, W.M., et al. 1993. Protamine-mediated transport of albumin into brain and other organs of the rat. Binding and endocytosis of protamine–albumin complex by microvascular endothelium. *J Clin Invest* 92:2224.
51. Vorbrodt, A.W., et al. 1995. Immunocytochemical studies of protamine-induced blood–brain barrier opening to endogenous albumin. *Acta Neuropathol (Berl)* 89:491.
52. Banks, W.A., and R.D. Broadwell. 1994. Blood to brain and brain to blood passage of native horseradish peroxidase, wheat germ agglutinin, and albumin: Pharmacokinetic and morphological assessments. *J Neurochem* 62:2404.
53. Zhang, Y., and W.M. Pardridge. 2001. Conjugation of brain-derived neurotrophic factor to a blood–brain barrier drug targeting system enables neuroprotection in regional brain ischemia following intravenous injection of the neurotrophin. *Brain Res* 889:49.
54. Pardridge, W.M., D. Triguero, and J.L. Buciak. 1990. Beta-endorphin chimeric peptides: Transport through the blood–brain barrier *in vivo* and cleavage of disulfide linkage by brain. *Endocrinology* 126:977.
55. Deguchi, Y., A. Kurihara, and W.M. Pardridge. 1999. Retention of biologic activity of human epidermal growth factor following conjugation to a blood–brain barrier drug delivery vector via an extended poly(ethylene glycol) linker. *Bioconjug Chem* 10:32.
56. Umezawa, F., and Y. Eto. 1988. Liposome targeting to mouse brain: Mannose as a recognition marker. *Biochem Biophys Res Commun* 153:1038.
57. Huwyler, J., D. Wu, and W.M. Pardridge. 1996. Brain drug delivery of small molecules using immunoliposomes. *Proc Natl Acad Sci USA* 93:14164.
58. Rousseau, V., et al. 1999. Early detection of liposome brain localization in rat experimental allergic encephalomyelitis. *Exp Brain Res* 125:255.
59. Shi, N., et al. 2001. Brain-specific expression of an exogenous gene after i.v. administration. *Proc Natl Acad Sci USA* 98:12754.
60. Mora, M., et al. 2002. Design and characterization of liposomes containing long-chain N-acylPEs for brain delivery: Penetration of liposomes incorporating GM1 into the rat brain. *Pharm Res* 19:1430.
61. Thole, M., et al. 2002. Uptake of cationized albumin coupled liposomes by cultured porcine brain microvessel endothelial cells and intact brain capillaries. *J Drug Target* 10:337.
62. Wu, D., et al. 2002. Pharmacokinetics and brain uptake of biotinylated basic fibroblast growth factor conjugated to a blood–brain barrier drug delivery system. *J Drug Target* 10:239.
63. Zhang, Y., F. Schlachetzki, and W.M. Pardridge. 2003. Global non-viral gene transfer to the primate brain following intravenous administration. *Mol Ther* 7:11.
64. Schmidt, J., et al. 2003. Drug targeting by long-circulating liposomal glucocorticosteroids increases therapeutic efficacy in a model of multiple sclerosis. *Brain* 126:1895.
65. Omori, N., et al. 2003. Targeting of post-ischemic cerebral endothelium in rat by liposomes bearing polyethylene glycol-coupled transferrin. *Neurol Res* 25:275.
66. Aoki, H., et al. 2004. Therapeutic efficacy of targeting chemotherapy using local hyperthermia and thermosensitive liposome: Evaluation of drug distribution in a rat glioma model. *Int J Hyperthermia* 20:595.
67. Gosk, S., et al. 2004. Targeting anti-transferrin receptor antibody (OX26) and OX26-conjugated liposomes to brain capillary endothelial cells using *in situ* perfusion. *J Cereb Blood Flow Metab* 24:1193.
68. Kozubek, A. et al. 2000. Liposomal drug delivery, a novel approach: PLARosomes. *Acta Biochim Pol* 47:639.
69. Voinea, M., and M. Simionescu. 2002. Designing of "intelligent" liposomes for efficient delivery of drugs. *J Cell Mol Med* 6:465.
70. Gabizon, A., et al. 1994. Prolonged circulation time and enhanced accumulation in malignant exudates of doxorubicin encapsulated in polyethylene-glycol coated liposomes. *Cancer Res* 54:987.

71. Gabizon, A., H. Shmeeda, and Y. Barenholz. 2003. Pharmacokinetics of pegylated liposomal doxorubicin: Review of animal and human studies. *Clin Pharmacokinet* 42:419.
72. Shi, N., and W.M. Pardridge. 2000. Noninvasive gene targeting to the brain. *Proc Natl Acad Sci USA* 97:7567.
73. Huwyler, J., et al. 2002. By-passing of P-glycoprotein using immunoliposomes. *J Drug Target* 10:73.
74. Wu, D., R. Boado, and W. Pardridge. 1996. Pharmacokinetics and blood–brain barrier transport of [^{3}H]-biotinylated phosphorothioate oligodeoxynucleotide conjugated to a vector-mediated drug delivery system. *J Pharmacol Exp Ther* 276:206.
75. Coloma, M.J., et al. 2000. Transport across the primate blood–brain barrier of a genetically engineered chimeric monoclonal antibody to the human insulin receptor. *Pharm Res* 17:266.
76. Torchilin, V.P. 2000. Drug targeting. *Eur J Pharm Sci* 11 Suppl 2:S81.
77. Kwon, G.S., 2003. Polymeric micelles for delivery of poorly water-soluble compounds. *Crit Rev Ther Drug Carrier Syst* 20:357.
78. Kabanov, A.V., et al. 1989. The neuroleptic activity of haloperidol increases after its solubilization in surfactant micelles. Micelles as microcontainers for drug targeting. *FEBS Lett* 258:343.
79. Kabanov, A., E. Batrakova, and V. Alakhov. 2002. Pluronic block copolymers as novel polymer therapeutics for drug and gene delivery. *J Control Release* 82:189.
80. Jones, A.T., M. Gumbleton, and R. Duncan. 2003. Understanding endocytic pathways and intracellular trafficking: A prerequisite for effective design of advanced drug delivery systems. *Adv Drug Deliv Rev* 55:1353.
81. Torchilin, V.P., et al. 2003. Immunomicelles: Targeted pharmaceutical carriers for poorly soluble drugs. *Proc Natl Acad Sci USA* 100:6039.
82. Allen, T.M., and P.R. Cullis. 2004. Drug delivery systems: Entering the mainstream. *Science* 303:1818.
83. Kataoka, K., A. Harada, and Y. Nagasaki. 2001. Block copolymer micelles for drug delivery: Design, characterization and biological significance. *Adv Drug Deliv Rev* 47:113.
84. Kabanov, A., et al. 1992. A new class of drug carriers: Micelles of poly(oxyethylene)-poly(oxupropylene) block copolymers as microcontainers for drug targeting from blood in brain. *J Control Release* 22:141.
85. Batrakova, E., et al. 1998. Effects of pluronic P85 unimers and micelles on drug permeability in polarized BBMEC and Caco-2 cells. *Pharm Res* 15:1525.
86. Kwon, G., and T. Okano. 1999. Soluble self-assembled block copolymers for drug delivery. *Pharm Res* 16:597.
87. Lavasanifar, A., J. Samuel, and G.S. Kwon. 2002. Poly(ethylene oxide)-block-poly(L-amino acid) micelles for drug delivery. *Adv Drug Deliv Rev* 54:169.
88. Moghimi, S., L. Illum, and S. Davis. 1990. Physiopathological and physicochemical considerations in targeting of colloids and drug carriers to the bone marrow. *Crit Rev Ther Drug Carrier Syst* 7:187.
89. Gref, R., et al. 1994. Biodegradable long-circulating polymeric nanospheres. *Science* 263:1600.
90. Peracchia, M., et al. 1998. Pegylated nanoparticles from a novel methoxypolyethylene glycol cyanoacrylate–hexadecyl cyanoacrylate amphiphilic copolymer. *Pharm Res* 15:550.
91. Torchilin, V. 1998. Polymer-coated long-circulating microparticulate pharmaceuticals. *J Microencapsul* 15:1.
92. Calvo, P., et al. 2001. Long-circulating PEGylated polycyanoacrylate nanoparticles as new drug carrier for brain delivery. *Pharm Res* 18:1157.
93. Vinogradov, S., E. Batrakova, and A. Kabanov. 1999. Poly(ethylene glycol)–polyethyleneimine NanoGel particles: Novel drug delivery systems for antisense oligonucleotides. *Colloids Surf B Biointerfaces* 16:291.
94. Lemieux, P., et al. 2000. Block and graft copolymers and NanoGel copolymer networks for DNA delivery into cell. *J Drug Target* 8:91.
95. Gupta, A.K., et al. 2003. Receptor-mediated targeting of magnetic nanoparticles using insulin as a surface ligand to prevent endocytosis. *IEEE Trans Nanobioscience* 2:255.

96. Vinogradov, S.V., E.V. Batrakova, and A.V. Kabanov. 2004. Nanogels for oligonucleotide delivery to the brain. *Bioconjug Chem* 15:50.
97. Kreuter, J. 2004. Influence of the surface properties on nanoparticle-mediated transport of drugs to the brain. *J Nanosci Nanotechnol* 4:484.
98. Kreuter, J., et al. 2003. Direct evidence that polysorbate-80-coated poly(butylcyanoacrylate) nanoparticles deliver drugs to the CNS via specific mechanisms requiring prior binding of drug to the nanoparticles. *Pharm Res* 20:409.
99. Olivier, J.C., et al. 1999. Indirect evidence that drug brain targeting using polysorbate 80-coated polybutylcyanoacrylate nanoparticles is related to toxicity. *Pharm Res* 16:1836.
100. Yang, C., et al. 2004. Nanoparticle-based *in vivo* investigation on blood–brain barrier permeability following ischemia and reperfusion. *Anal Chem* 76:4465.
101. Kreuter, J. 2001. Nanoparticulate systems for brain delivery of drugs. *Adv Drug Deliv Rev* 47:65.
102. Daleke, D.L., K. Hong, and D. Papahadjopoulos. 1990. Endocytosis of liposomes by macrophages: Binding, acidification and leakage of liposomes monitored by a new fluorescence assay. *Biochim Biophys Acta* 1024:352.
103. Fujiwara, M., J.D. Baldeschwieler, and R.H. Grubbs. 1996. Receptor-mediated endocytosis of poly(acrylic acid)-conjugated liposomes by macrophages. *Biochim Biophys Acta* 1278:59.
104. Jain, S., et al. 2003. RGD-anchored magnetic liposomes for monocytes/neutrophils-mediated brain targeting. *Int J Pharm* 261:43.
105. Khan, M.A., et al. 2005. Enhanced anticryptococcal activity of chloroquine in phosphatidylserine-containing liposomes in a murine model. *J Antimicrob Chemother* 55:223.
106. Perry, V.H., et al. 1995. Inflammation in the nervous system. *Curr Opin Neurobiol* 5:636.
107. Kuby, J. 1994. *Immunology*. New York:W.H. Freeman.
108. Fortin, D. 2003. Altering the properties of the blood–brain barrier: Disruption and permeabilization. *Prog Drug Res* 61:125.
109. Kroll, R.A., and E.A. Neuwelt. 1998. Outwitting the blood–brain barrier for therapeutic purposes: Osmotic opening and other means. *Neurosurgery* 42:1083.
110. Neuwelt, E.A., et al. 1991. Primary CNS lymphoma treated with osmotic blood–brain barrier disruption: Prolonged survival and preservation of cognitive function. *J Clin Oncol* 9:1580.
111. Williams, P.C., et al. 1995. Toxicity and efficacy of carboplatin and etoposide in conjunction with disruption of the blood–brain tumor barrier in the treatment of intracranial neoplasms. *Neurosurgery* 37:17.
112. Neuwelt, E.A., et al. 1994. Therapeutic dilemma of disseminated CNS germinoma and the potential of increased platinum-based chemotherapy delivery with osmotic blood–brain barrier disruption. *Pediatr Neurosurg* 21:16.
113. Chio, C.C., T. Baba, and K.L. Black. 1992. Selective blood–tumor barrier disruption by leukotrienes. *J Neurosurg* 77:407.
114. Borlongan, C.V., and D.F. Emerich. 2003. Facilitation of drug entry into the CNS via transient permeation of blood–brain barrier: Laboratory and preliminary clinical evidence from bradykinin receptor agonist, Cereport. *Brain Res Bull* 60:297.
115. Witt, K.A., et al. 2000. Insulin enhancement of opioid peptide transport across the blood–brain barrier and assessment of analgesic effect. *J Pharmacol Exp Ther* 295:972.
116. Sanovich, E., et al. 1995. Pathway across blood–brain barrier opened by the bradykinin agonist, RMP-7. *Brain Res* 705:125.
117. Fike, J.R., et al. 1998. Cerebrovascular effects of the bradykinin analog RMP-7 in normal and irradiated dog brain. *J Neurooncol* 37:199.
118. Elliott, P.J., et al. 1996. Unlocking the blood–brain barrier: A role for RMP-7 in brain tumor therapy. *Exp Neurol* 141:214.
119. Emerich, D.F., et al. 1998. Central analgesic actions of loperamide following transient permeation of the blood brain barrier with Cereport (RMP-7). *Brain Res* 801:259.
120. Borlongan, C.V., et al. 2002. Bradykinin receptor agonist facilitates low-dose cyclosporine-A protection against 6-hydroxydopamine neurotoxicity. *Brain Res* 956:211.
121. St Croix, B., et al. 2000. Genes expressed in human tumor endothelium. *Science* 289:1197.

122. Sheikov, N., et al. 2004. Cellular mechanisms of the blood–brain barrier opening induced by ultrasound in presence of microbubbles. *Ultrasound Med Biol* 30:979.
123. Hynynen, K., et al. 2005. Local and reversible blood–brain barrier disruption by noninvasive focused ultrasound at frequencies suitable for trans-skull sonications. *Neuroimage* 24:12.
124. Miller, G. 2002. Breaking down barriers. *Science* 297:1116.
125. Tsuji, A. and I. Tamai. 1997. Blood–brain barrier function of P-glycoprotein. *Adv Drug Deliv Rev* 25:287.
126. Kim, R.B., et al. 1998. The drug transporter P-glycoprotein limits oral absorption and brain entry of HIV-1 protease inhibitors. *J Clin Invest* 101:289.
127. Perloff, M.D., L.L. von Moltke, and D.J. Greenblatt. 2004. Ritonavir and dexamethasone induce expression of CYP3A and P-glycoprotein in rats. *Xenobiotica* 34:133.
128. Uhr, M., et al. 2000. Penetration of amitriptyline, but not of fluoxetine, into brain is enhanced in mice with blood–brain barrier deficiency due to mdr1a P-glycoprotein gene disruption. *Neuropsychopharmacology* 22:380.
129. Moriki, Y., et al. 2004. Involvement of P-glycoprotein in blood–brain barrier transport of pentazocine in rats using brain uptake index method. *Biol Pharm Bull* 27:932.
130. Miyama, T., et al. 1998. P-glycoprotein-mediated transport of itraconazole across the blood–brain barrier. *Antimicrob Agents Chemother* 42:1738.
131. Tsuji, A. 1998. P-glycoprotein-mediated efflux transport of anticancer drugs at the blood–brain barrier. *Ther Drug Monit* 20:588.
132. Potschka, H., M. Fedrowitz, and W. Loscher. 2002. P-Glycoprotein-mediated efflux of phenobarbital, lamotrigine, and felbamate at the blood–brain barrier: Evidence from microdialysis experiments in rats. *Neurosci Lett* 327:173.
133. Kemper, E.M., et al. 2003. Increased penetration of paclitaxel into the brain by inhibition of P-glycoprotein. *Clin Cancer Res* 9:2849.
134. Kemper, E.M., et al. 2004. The influence of the P-glycoprotein inhibitor zosuquidar trihydrochloride (LY335979) on the brain penetration of paclitaxel in mice. *Cancer Chemother Pharmacol* 53:173.
135. Miller, D., et al. 1997. Interactions of Pluronic block copolymers with brain microvessel endothelial cells: Evidence of two potential pathways for drug absorption. *Bioconjug Chem* 8:649.
136. Batrakova, E., et al. 1999. Fundamental relationships between the composition of pluronic block copolymers and their hypersensitization effect in MDR cancer cells. *Pharm Res* 16:1373.
137. Batrakova, E., et al. 2001. Mechanism of pluronic effect on P-glycoprotein efflux system in blood–brain barrier: Contributions of energy depletion and membrane fluidization. *J Pharmacol Exp Ther* 299:483.
138. Batrakova, E., et al. 2001. Pluronic P85 enhances the delivery of digoxin to the brain: *In vitro* and *in vivo* studies. *J Pharmacol Exp Ther* 296:551.
139. Batrakova, E., et al. 2003. Optimal structure requirements for pluronic block copolymers in modifying P-glycoprotein drug efflux transporter activity in bovine brain microvessel endothelial cells. *J Pharmacol Exp Ther* 304:845.
140. Miller, D., E. Batrakova, and A. Kabanov. 1999. Inhibition of multidrug resistance-associated protein (MRP) functional activity with pluronic block copolymers. *Pharm Res* 16:396.
141. Witt, K.A., et al. 2002. Pluronic p85 block copolymer enhances opioid peptide analgesia. *J Pharmacol Exp Ther* 303:760.
142. Batrakova, E.V., et al. 2003. Sensitization of cells overexpressing multidrug-resistant proteins by pluronic P85. *Pharm Res* 20:1581.
143. Batrakova, E., et al. 1999. Pluronic P85 increases permeability of a broad spectrum of drugs in polarized BBMEC and Caco-2 cell monolayers. *Pharm Res* 16:1366.
144. Batrakova, E.V., et al. 2004. Effect of Pluronic P85 on ATPase activity of drug efflux transporters. *Pharm Res* 21:2226.
145. Batrakova, E.V., et al. 2001. Mechanism of sensitization of MDR cancer cells by Pluronic block copolymers: Selective energy depletion. *Br J Cancer* 85:1987.
146. Kirillova, G., et al. 1993. The influence of pluronics and their conjugates with proteins on the rate of oxygen consumption by liver mitochondria and thymus lymphocytes. *Biotechnol Appl Biochem* 18 (Pt 3):329.

147. Rapoport, N., A. Marin, and A. Timoshin. 2000. Effect of a polymeric surfactant on electron transport in HL-60 cells. *Arch Biochem Biophys* 384:1000.
148. Kabanov, A.V., E.V. Batrakova, and V.Y. Alakhov. 2003. An essential relationship between ATP depletion and chemosensitizing activity of Pluronic block copolymers. *J Control Release* 91:75.
149. Kabanov, A.V., E.V. Batrakova, and D.W. Miller. 2003. Pluronic block copolymers as modulators of drug efflux transporter activity in the blood–brain barrier. *Adv Drug Deliv Rev* 55:151.
150. Batrakova, E., et al. 2004. Effects of Pluronic P85 on GLUT1 and MCT1 transporters in the blood brain barrier. *Pharm Res* 21:1993.
151. Danson, S., et al. 2004. Phase I dose escalation and pharmacokinetic study of pluronic polymer-bound doxorubicin (SP1049C) in patients with advanced cancer. *Br J Cancer* 90:2085.

Index

D

O

Q

R

Milton Keynes UK
Ingram Content Group UK Ltd.
UKHW052030071024
449327UK00027B/2502

9 780367 389826